Heavy-Duty Wheeled Vehicles

Other SAE books of interest:

Heavy-Duty Wheeled Vehicles: Design, Theory, Calculations

By Boris N. Belousov and Sergei D. Popov

SAE INTERNATIONAL

Warrendale, Pennsylvania, USA

400 Commonwealth Drive
Warrendale, PA 15096-0001 USA

E-mail: CustomerService@sae.org
Phone: 877-606-7323 (inside USA and Canada)
724-776-4970 (outside USA)
Fax: 724-776-0790

ISBN 978-0-7680-7723-0
SAE Order Number R-419
DOI 10.4271/R-419

Library of Congress Cataloging-in-Publication Data

Belousov, Boris N.
Heavy-duty wheeled vehicles : design, theory, calculations / Boris N. Belousov, Sergei D. Popov.
pages cm
ISBN 978-0-7680-7723-0
1. Trucks--Design and construction. 2. Four-wheel drive vehicles--Design and construction. I. Popov, Sergei D. II. Title.
TL230.B382 2013
629.224--dc23
2013021860

Translated from Russian to English by Leonid Schukin.

To purchase bulk quantities, please contact:
SAE Customer Service
E-mail: CustomerService@sae.org
Phone: 877-606-7323 (*inside USA and Canada*)
724-776-4970 (*outside USA*)
Fax: 724-776-0790

Visit the SAE Bookstore at
BOOKS.SAE.ORG

Dedication

In memory of
Professors P.V. Aksyonov, N.F. Bocharov, G.A. Smirnov, and Yu.V. Pirkovsky

Table of Contents

PART II OVERVIEW OF HDWVS

PART III FUNDAMENTALS OF HDWV THEORY AND DESIGN

Preface

This book narrates the history of heavy-duty wheeled vehicles (HDVWs) and presents the basics of the design, theory, and calculation of their individual units and systems; as well as the peculiarities of common design solutions used.

HDWVs are all-wheel-drive vehicles that have a load capacity of 25 tons or more and three or more axles. Such vehicles are used for transporting heavy, indivisible, bulky cargos and as transport platforms for various technical equipment.

HDWV designs have their own development requirements, which are associated with high loads, huge dimensions, and specific operation conditions. Naturally, the approaches to designing such vehicles differ from those used in designing other multipurpose all-wheel-drive automobiles. Designers and manufacturers of HDWVs use a custom approach, such as designing a chassis for a particular purpose requested by the customer. The main elements of the chassis, namely, the undercarriage, the carrying system, the transmission type and composition, the engine, and resilient elements of the wheel suspension are flexible and vary. Other design variables include the number of axles and which are the driving and driven ones, and what size and type of tire are needed. Due to the individual character of the design process, design solutions are often different than those used to design passenger cars. Therefore, every positive and negative aspect of designing an HDWV deserves a thorough study for future use in designing new vehicles.

An enormous contribution into the theory and practice of designing HDWVs was made by Professors N.F. Bocharov and G.A. Smirnov of the N.E. Bauman MSTU and Professors P.V. Aksyonov and Yu.V. Pirkovsky, who closely cooperated with MSTU. This book presents certain elements of updated theory, calculations, and related general design solutions. The concept of general design solutions, introduced by Professor Aksyonov, comprises of the number and location of axles, steering and transmission system layout, and general vehicle layout principles.

The authors hope that this book will be helpful for the engineers, technicians, and scientists involved in the motor industry; for those engaged in designing and building wheeled transportation vehicles; and for students and trainees at universities where specialists of respective branches of industry are trained.

The authors will be grateful for any critical remarks and wishes expressed by the readers.

Acknowledgments

The authors gratefully acknowledge the following reviewers:

G.I. Gladov

S.B. Shukhman

Introduction

Heavy-duty wheeled vehicles (HDWVs) are crucial for the economy and defense of any country. The *Dictionary of Motor Transport* by Yu.I. Egorov and A.N. Narbut defines this group of vehicles as those that can transport cargos of 25 tons or more. The Russian fleet of HDWVs includes quarry dumpers made by BelAZ and MoAZ; heavy-duty road trains for transporting long-sized cargoes weighing 50 tons or more and consisting of tow cars made by the Open Joint-Stock Company (OJSC), "Rusich," KZKT, or MZKT and semi-trailers by ChMZAP or FGUP "Motor KB"; multi-axle special wheeled chassis (SWCs) made by OJSC "BAZ" and MZKT; and truck-mounted cranes built on foreign-made multi-axle chassis. Abroad, HDWVs include quarry dumpers, multi-support wheeled transportation vehicles, and multi-axle chassis vehicles that have truck-mounted cranes.

One way to boost the efficiency of a truck is to increase its load capacity. That is why multi-axle and multi-support vehicles with 4–12 axles, more than 20 supports, over 100 wheels, and a load capacity of 500 tons or more have appeared since the middle of the twentieth century. Multi-support vehicles can transport prefabricated modules of production shops, factories and plants, dockage facilities, and other structures of huge dimension and weight.

Historically, vehicle designs have advanced by increasing the number of axles. To improve the adaptability of wheeled vehicles to poor roads, and because of legislatively established limits on load per axle, axles totaling three or more were successively added.

Powerful hydraulic excavators placed increased demands on quarry transport. To fully benefit from the greater capabilities of such excavators, which are able to dig into all types of rocks, new models of super-heavy-duty quarry dumpers were developed that can work with such excavators. The HDWVs class is essentially different from the class of general transportation automobiles. Like other complex machines, they have their own theory, which sets the general rules of design and direction for further development and improvement.

HDWV theory stems from the general automobile theory developed by the Academician E.A. Chudakov, which is based on the works by N.R. Brilling, V.I. Grinevetsky, N.E. Zhukovsky, G.V. Zimelyov, B.S. Falkevich, and others. It considers the general motion principles of mainly a two-axle vehicle with one driving axle.

The general design theory of multi-axle HDWVs is based on the works of the following Soviet scientists and engineers: Ya.S. Ageikin, A.S. Antonov, D.A. Antonov, P.V. Aksyonov, N.F. Bocharov, M.S. Vysotsky, A.I. Grishkevich, V.P. Efimov, K.S. Kolesnikov, N.I. Korotonoshko, L.G. Krasnevsky, V.I. Medvedkov, Yu.V. Pirkovsky, V.F. Platonov, R.V. Rotenberg, V.M. Semyonov, G.A. Smirnov, Yu.A. Stepanov, B.L. Shaposhnik, V.E. Chvyalyov, Ya.S. Farobin, N.N. Yatsenko, and others. Currently, Russian and Belarusian scientists and engineers are in a leading position when it comes to addressing theoretical and practical problems of future and promising HDWVs.

The fact that a HDWV has many systems and mechanisms that use different energy types (mechanic, hydraulic, and electric), leads to an essential increase in weight, a higher complexity of design, and problems in operation, maintenance, repair, etc.

The need to design multi-axle HDWVs for the military, for transporting various weapons and combat vehicles, has caused another qualitative breakthrough in HDWV design and has given birth to radically new technical solutions.

A special contribution to HDWV designs was made by the Belarusian school of machine-building. Belarus is home to motor industry giants such as BelAZ, MAZ, MoAZ, and MZKT. The achievements of Russian motor factories—the KZKT, ZIL, BAZ, and Motor KB—also deserve mention. Specialists at these enterprises have designed and made—in cooperation with respective scientists—a number of modern transportation vehicles that have no equal abroad.

The improvement of multi-axle vehicles and a broad introduction of road trains took place in parallel. Road trains are most efficient on primary highways. The theory and practice of manufacturing highway road trains are nicely presented in scientific and technical literature, for example, in the works by M.S. Vysotsky. At the same time, there are practically no general publications on designing multi-axle vehicles, thus this book deals mainly with the theory and practice of designing multi-axle HDWVs. Heavy-duty wheeled road trains are described only in the first part of the book, where the design peculiarities of multi-axle vehicles intended for towing trailers and semi-trailers with a total weight of 60 tons or more are presented.

Figures I.1 and I.2 present several models of multi-axle HDWVs and heavy-duty road trains that have a total weight of 90 tons or more; Tables I.1, I.2, and I.3 present some of their specifications.

The book consists of three parts. The first part sets out the design history and general methodology, including the basics of CAD (computer-aided designing), the evaluation of HDWVs areas and conditions for use, general requirements for operational and consumer properties, and a brief analysis of structural components (units, assemblies, and systems) of existing batch-manufactured models and prototypes.

The second part presents the general information necessary for designing HDWVs.

The third part reviews the theory basics and calculation of non-traditional technical solutions, as well as individual and special requirements for transportation vehicle components.

HDWV manufacturers prefer an individual approach, which assumes making a product for a particular mission, demanded by a particular customer; thus, the bulk of the reference data provided may be of practical interest for designers of future models.

Fig. I.1. General views of some HDWVs: a—A multi-support Kommeto transportation platform; b—7904 model chassis made by MZKT (Belarus); c—A quarry dumper made by BelAZ (Belarus); d—A pipe carrier by Foremost (Canada); e—Tansportation vehicle, MAZ-2000 "Perestroika" (Belarus)

Fig. I.2. Heavy-duty wheeled road trains: a—A road train with a KZKT-7427 tow car and a ChMZAP trailer; b—An experimental multi-support road train; c—A road train with a KZKT-74281 tow car and a semi-trailer; d—A "Supercargo" road train made by Steinwinter

Table I.1 Main specifications of HDWVs

Parameter	MAZ-543 (Belarus)	KZKT-80051 (Russia)	7906 (Belarus)	7907 (Belarus)
Wheel arrangement	8×8	8×8	16×16	24×24
Full weight, t	40.0	50.0	220.0	220.0
Engine power, kW	386.0	321.5	11,043.0	882.0
Maximum speed, km/h	65.0	60.0	40.0	40.0
Overall dimensions, mm:				
length	11,300	12,150	27,900	28,500
width	3070	3070	4800	4700
height	2970	3245	4500	4500
wheelspan	2375	2200	3820	3504
Road clearance, mm	450	500	500	500
Tires	1500×600–635	18.00–24	1980×750–737	1660×670–687
Roadability formula	12–00	12–00	1234–0000	1234–0000–9101112
Minimum turning radius, m	13.5	18.0	30.0	27.5
Transmission type	HMT	HMT	HMT (two)	Electromechanical

Note. HMT stands for hydromechanical transmission.

Table I.2 Main specifications of heavy-duty wheeled road trains

Parameter	KZKT-74281-9990 (Russia)	KZKT-7427 (Russia)	MAZ-73132-9990 (Belarus)	T 6066/05 "Mall" (Belgium)
Wheel arrangement	8×8	8×8	8×8	6×6
Full weight, t:				
road train	94.0	180.0	100.0	150.0
tow car	24.0	55.0	45.15	60.0
Engine power, kW	478	478	386	318
Maximum speed, km/h	70.0	45	60	25
Overall dimensions, mm:				
length	10,208	10,208	18,015	10,350
width	3070	3070	3070	3350
height	3100	3100	2920	4070
wheelspan	2800	2800	2375	2800
Road clearance, mm	450	450	400	400
Tires	18.00–24	18.00–24	1500×600–635	14.00–24
Roadability formula	12–00	12–00	12–00	1–00
Minimum turning radius, m	18.0	18.0	13.5	10.68
Transmission type	HMT	HMT	HMT	HMT

Table I.3 Main specifications of quarry dumpers

Parameter	MAZ-525	BelAZ-540	MAZ-530	BelAZ-548
Wheel arrangement	4×2	4×3	6×4	4×2
Load capacity, t	25	27	40	40
Kerb weight, t	24	21	38	28
Tare weight ratio	0.96	0.78	0.95	0.70
Wheelbase, mm	4780	3550	4900	4200
Overall length, mm	8220	7180	10,515	8160
Minimum turning radius by front wheelspan, m	12.0	8.4	14.0	9.5
Nominal engine power, kW	221	265	331	368
Specific power, kW	4.4	5.5	4.2	5.4
Transmission type	Mechanical	HMT	HMT	HMT
Suspension type:				
front	Leaf (spring)	Air-hydraulic	Leaf (spring)	Air-hydraulic
rear	No suspension	No suspension	No suspension	No suspension
	17.00–32	18.00–25	18.00–32	21.00–33
Maximum speed, km/h	30	53	40	55

Symbols used

P, R—active force and reaction

F—internal forces in suspension and tires arising from deformation of resilient elements and displacement of damping members

m—mass

G—gravity

M—torque (moment)

J—inertia moment

N—power

D—dynamic factor

U—transmission ratio

η—efficiency factor

S—travel

t—time

V, v—travel speed

$\dot{V}$—longitudinal acceleration

g—gravity acceleration

w_i—angular velocities and accelerations

n_p, n_t, w_p, w_t—rotation speeds and angular velocities of the pump and turbine wheels of the hydraulic torque converter (transformer)

z—change in amplitude of vertical oscillations

γ, φ—change in amplitude of lateral- and longitudinal-angular oscillations

ψ_0—relative oscillation decay coefficient (aperiodicity factor)

C, c—rigidity (stiffness)

Δ, $\dot{\Delta}$—deflection value and deflection speed of the suspension

K_t—tire damping factor

δ—angle of wheel drift

$\Delta\delta_i$—lateral tire deformation

h_{st}, h_d—static and dynamic travels (strokes) of the suspension

$S_H(\ell_B)$– irregularity (wave) length

$k(\lambda)_i$—spectral density of road micro-profile heights

λ_{Ti}—travel frequency of the border of zones of medium and short irregularities

D_i—factor of spectral densities

$F_r(n)$—integral density of radius probabilities

$F_\theta(n)$—integral density of turning angle probabilities

$F_{Np}(n)$—integrated density of turning direction probabilities

$S(\lambda_i)$—spectral density depending on travel frequency

$\rho(\lambda_i)$—correlation coefficient of micro-profile parallel cross-sections

$\psi(i)$—phase of harmonic realization components

M_z and M_ψ—expectations of realization of a half-sum and half-difference of rut micro-profiles

r—wheel radius

R—radii: of turning, passing ability, lateral, etc.

α, β, δ, θ, γ—turning angles (of steered wheels, torsion bar spring tension of the suspension, etc), lateral skid, ascend, descend, and roll

L, B, ℓ—linear dimensions

h_c—height of center of mass

h_r—rut depth

x_p—distance from the front (rear) axle of the HDWV to the steering system pole in the longitudinal direction

l_d, h_d—longitudinal and vertical coordinates of driver's workplace

K_o—current curvature of path of motion

ξ—vertical linear coordinate of wheel displacement

F—cross-section area of special wheeled chassis (SWCs)

$C_{\Gamma i}$—specific ground resistivity of indentation (ground settlement factor)

f—coefficient of rolling resistance of wheel axle

φ—wheel adhesion factor

S_k—wheel sliding factor

K_b—blocking factor of the differential

Note. All formulas use SI units of measure, that is, forces are measured in newtons (N); pressures in pascals (Pa); rigidity (stiffness) in N/m; linear dimensions in meters (m); and speed (velocity) in m/s and rad/s (s^{-1}). Exceptions are some individual examples, empirical expressions and curves taken from 1970s literature, and—in certain cases—those used for the sake of preserving narration continuity.

Enterprises and institutions

IKTP—Institute for Complex Transport Problems

NAMI—Central Scientific-Research Automobile and Motor Institute

MADI—Moscow Automobile and Road Institute

KZKT—Kurgan Works of Wheeled Tow Cars

MZKT—Minsk Works of Wheeled Tow Cars

MAZ—Minsk Automobile Works

BelAZ—Belarusian Automobile Works

YaMZ—Yaroslavl Engine Plant

KamAZ—Kama Automobile Works

ZIL—I.A. Likhachov Automobile Works

UralAZ—Urals Automobile Works

BAZ—Bryansk Automobile Works

MAMI—Moscow Auto-Mechanical Institute

VABTV—Military Academy of Armored Forces

MIE&A—Moscow Institute of Electronics and Automation

ChZDM—Chelyabinsk Factory of Road Machinery

ChMZAP—Chelyabinsk Mechanical Plant of Automobile Trailers

MoAZ—Mogilyov Automobile Works

Part I

Evolution of heavy-duty wheeled vehicles

Chapter 1
History of HDWVs

1.1 General design-layout solutions

The progress of the armament and military equipment (A&ME) industry has always been a strong impetus for the rapid development of accompanying technologies, and in particular the automobile industry. This is why the most unique heavy-duty wheeled vehicles (HDWVs) designs have been and are being designed for militaries to meet their specific operation requirements.

Both specialized and general motor companies are engaged in HDWV manufacture. Several companies also offer a broad range of vehicle units and assemblies to adapt to various customer requirements. Typical designs, layouts, and a brief review of HDWV's main units and systems are given below.

Starting from the late 1960s to early 1970s, leading NATO countries adopted the concept of mobile A&ME, with priority given to vehicles associated with air and naval deployment. As to the ground, NATO took the direction of stationary super-protected and super-deep launching pads for strategic missiles and mobile operative-tactical vehicles. Although the U.S. attempted to create vehicle launching pads, for example, for the MX series of intercontinental ballistic missiles (ICBMs); these programs were not realized in practice.

An example of using multi-axle vehicles as launching pads for ICBMs is the U.S. Midgetman missile program.

This program included a mobile basing for protected transporter-erector launchers (TELs) and stationary super-protected silo launchers. For the mobile protected TELs, it was assumed to deploy them in U.S. territory and have them capable to withstand, in case of attack, the pressure of a 1.8–2.1 kPa shock-wave front.

The TEL, made by Martin Marietta, is a road train consisting of a wheeled trailer, also made by Marietta, and a caterpillar tow car (Fig. 1.1). The

Fig. 1.1. Road train comprising of a wheeled trailer and a caterpillar tow car for Midgetman TELs

tow car, hydraulic system, and suspension were made by Caterpillar Tractor. The dimensions were as follows: a total length of 27.7 m, a width of 3.65 m, a tow car height of 2.9 m, a tow car weight of 32.2 t, and a total weight (TEL and missile (13.6 t)) of 47.4 t. It was assumed that once at the combat position, the TEL would lower to the ground and its side sections would tilt off and form a streamline shaped body, which was 7.3 m wide with a steel skirt adjoining the support surface.

The tow car was equipped with a four cycle diesel engine, model 3412 V-12, made by Caterpillar Tractor, with a power of 559 kW (750 hp) used to drive the tow car chassis and trailer. The tow car-TEL coupling was quick-release, which allowed the tow car to move quickly to the hideout after bringing the TEL into combat position.

The Boeing Company, unlike Marietta, preferred to use a wheel chassis for both the tow car and the trailer mounted TEL (Fig. 1.2), which allowed for transporting the TEL at a speed of about 100 km/h. The tow car had four axles (eight wheels), while the TEL was mounted on a three-axle trailer (six wheels, all driving). However, in the finished experimental prototype, only the tow car wheels were steered. The steel bandaged tires were 1.37 m high and 0.61 m wide. The design also envisioned a centralized system for adjusting the air pressure in the tires while in motion. The tow car was equipped with a 559 kW (750 hp) diesel engine and the TEL trailer with the same engine of 410 kW (550 hp). The total length of the TEL and tow car was 28 m, the width was 3.65 m, the height was 2.8 m, the weight of the tow car was 20.4 t, and the total weight of the TEL and missile was 74.8 t.

When in combat position, this TEL also lowers to the ground: initially to about 40 cm above the support surface and then makes a controlled fall to the ground. The lower edges of the TEL are sharp and partially submerge into the soil under the weight. Then, the skirt unfolds, covers the edges, and ensures compaction. Ventilation openings allow air flow, which prevents the missile launcher (ML) from tipping-over from the blast wave. The lateral cross-section of the ground ML is a triangle with a rounded peak.

The variety of A&ME vehicles also assumes a broad use of special wheeled trailers and semi-trailers, ground handling bogies, and special "wheeled carts" where the missile container was used as the bearing element.

An example is an active road train with electric rear wheel drive of the articulated tow car and the semi-trailer (Fig. 1.3), designed in the U.S. in the 1970s by the WABCO Company. The weight of the vehicle was about 200 t and it was equipped with a 987 kW (1325 hp) diesel engine.

Fig. 1.2. Road train comprising of a wheeled trailer and tow car for Midgetman TELs

Fig. 1.3. Transporter for MX series ICBMs

Information about the design and manufacture of commercial multi-axle chassis also deserves attention. The FAUN GmbH Company (Germany) is manufacturing batches of three-, four-, and five-axle chassis for truck cranes with a load capacity of up to 125 t. Single-frame six-, seven-, and eight-axle chassis are also supplied in small batches for truck cranes with load capacities of up to 400 t (Fig. 1.4). General design specifications of crane chassis are presented in Table 1.1 [1].

Almost all FAUN chassis use V6 or V8 diesel engines made by DEUTZ AG, with air cooling and a power of 127 to 313 kW (170–420 hp). The seven- and eight-axle chassis use diesel engines made by MTI Diesel Inc. with water cooling and a power of 335 to 395 kW (450–530 hp).

Truck crane chassis are provided with all-wheel or rear-wheel driving. All-wheel drive has two-stage ordinary or planetary gears, with most of the axles steered. For example, the eight-axle chassis has seven steered axles. The steering system also has a hydraulic booster, which is powered by a special pump driven by the chassis wheels when the chassis is being towed.

All HDWV chassis frames are made from special steels. The frames are welded and the

Fig. 1.4 FAUN (Germany) six-axle chassis

Table 1.1 General design solutions of crane chassis made by FAUN

Group	Chassis layout (bogie)	Formula			
		Wheels	Axles	Steering	Axle drive
I		10×6	2-3	12-300	10045
II	or	12×6	3-3	123-000	000456
		12×6	2-4	12-3006	000456
III		14×8	4-3	1234-007	1204500
IV		16×8	5-3	12340-678	00305670

frame girders are replaced by a single box-shaped frame, which has higher bend and twist resistance than a girder-based frame [2].

The MAN Group (Germany) makes a whole range of heavy-duty vehicles. The main specifications of their multi-axle chassis are presented in Table 1.2.

MAN is the major European supplier of special-purpose army tactical trucks used by European NATO member-states. The company offers two lines of military vehicles manufactured in parallel, modernized batch-produced trucks and special-purpose vehicles.

The SX2000 series of heavy-duty tactical trucks includes 4×4, 6×6, 8×8, and 10×10 chassis with a full weight of 22–50 tons and a load capacity of 12–27 tons. They are equipped with serial MAN-made in-line or V-shaped diesel engines with 6–12 cylinders and a power of 194 to 745 kW (260–1000 hp); multi-stage mechanical or automatic gearboxes; lever-spring or hydro-pneumatic wheel suspensions; and a centralized system of pumping the tires, sized at 14.00R20,

Table 1.2 Main specifications of chassis made by MAN Group (Germany)

Specification	Chassis model				
	30.321 VFAK	35.365 VFAK	40.321 DFAK	40.365 DFAK	48.365 VFAK
Wheel formula	8×8	8×8	6×6	6×6	8×8
Full weight, t	35.0	35.0	40.0	40.0	48.0
Load capacity, t	24.7	24.3	30.0	28.6	37.0
Dimensions, mm:					
length	8235	8235	7635	7635	8260
height	3100	3180	3264	3264	3160
Wheel span:					
front wheels	1990	1990	1990	1990	1990
rear wheels	1800	1800	1800	1804	1804
Wheelbase	1500+ +3100+ +1350	1500+ +3100+ +1350	3825+ +1400	3825+ +1400	1500+ +3125+ +1400
Engine power, kW	235	268	268	268	268
Maximum speed, km/h	116	121	116	116	116

16.00R20, and 24.00R22.5. These vehicles are able to conquer slopes up to 60 percent and can wade up to 1.5 m deep. Special models are equipped with fully armored cabins, all steered wheels, and automatic tire pumping. Low-frame, airborne, or special narrow models are also available upon request. The SX2000 series is used to transport special equipment and heavy armaments. Model 45.1000FFAEG (with a 10×10 wheel formula), is the heaviest model and is 13.2 m long. It is intended for carrying heavy howitzers, bridge-building and crane equipment, fire-fighting appliances, and reusable freight containers weighing up to 27 t.

The Alvis Car and Engineering Company Ltd. (Great Britain) manufactures heavy-duty full-drive chassis and tow cars with 6×6 and 8×8 wheel formulas. The Unipower (8×8) is the core of the product line, and is a bogie-type chassis with two front steered axles. It can be equipped with either the Cummins M-380E or Perkins 410 Tx in-line six-cylinder diesel engine with turbo-supercharge. The *Cummins* M-380E has a power of 279.4 kW (375 hp) and an active volume of 11 liters. The Perkins 410Tx has a power of 300.2 kW (403 hp) and an active volume of 12.2 liters. The automatic gearbox is a ZF Automat-6HP with six forward gears and one reverse gear; the latter being a hydrodynamic brake-decelerator with hydraulic pumps for driving two power take-off shafts. There is a one-stage planetary torque divider and the wheel suspensions are built on parabolic springs. The wheels use 475/80R20 large cross-section tires and there is an automatic system for changing the air pressure. The all-metal cabin has five seats, an autonomous heating system, and an air conditioner. The chassis has a load capacity of 24 t, a total weight of 39 t, and a maximum speed of 91 km/h. The full weight of the road train with the chassis-based tow car is 110 t and has a maximum speed of 80 km/h. The company also makes a modernized version with a 301.7 kW (405 hp) Cummins engine and a 12×12 model for transporting bridge sections.

The M series was used as the basis for designing the MP family of articulated tow cars, intended to work as a road train with a full weight of 110–117 t for quick transportation of tanks under ambient temperatures from –31 to +55°C and able to cross water as deep as 1.2 m. The maximum speed with a full load is 85 km/h. The MH-6660 (6×6) tow car is equipped with a 447.0 kW (600 hp) in-line six-cylinder Cummins KTA-19 diesel engine with an active volume of 19 liters. The MH-8875 (8×8) model uses a 558.8 kW (750 hp) Cummins QSK-19 diesel engine with an active volume of 19 l. Both tow cars have ZF automatic five-change gearboxes united with two-stage torque dividers. Inter-axle and inter-wheel differentials equipped with pneumatically driven blocking devices are also provided. The tow car is equipped with two independent capstane engines with a pulling force of 250 kN each.

Iveco S.P.A. (Italy) manufactures a range of models that are subdivided into serially made civil trucks, adapted for use in the army, and tactical and special-purpose trucks. The Euro Transporter civil truck series are notable for their reinforced frames and spring suspensions. For load capacities over 16 t, they manufacture 6×4 and 6×6 models weighing 26–30 t with a 16–24 t load capacity and 8×4 and 8×8 models weighing 32–40 t with a 22–29 t load capacity. The Euro Transporters three-axle tow cars are equipped with 224–313 kW (300–420 hp) engines. It should be noted that all Euro Transporter models are road automobiles and cannot be conventionally regarded as off-road vehicles.

The Oshkosh Corporation (USA) specializes in manufacturing heavy-duty transportation chassis for the military that are used for missions performed in close vicinity of combat operation zones. The types of vehicles include several full-drive multi-axle vehicles intended for quick deployment of heavy armored equipment, military hardware, and for carrying special armaments and missiles. Oshkosh also makes the M-911 (6×6) articulated tow car, which operates as part of a road train that can weigh up to 102.5 t. The M-911's are considered Heavy Equipment Transporters (HETs) and are standardized for the U.S. Army. The vehicle has a 335 kW (450 hp) V8 two-stroke engine made by Detroit Diesel and an active volume of 12.1 liters. It also has an automatic five-change gearbox made by Allison Transmission. The weight of the tow car is 18 t, its dimensions are 9373×2896×3404 mm, and the maximum speed is 72 km/h. In 1990, the company made a new articulated tow car, model M-1070 (8×8), designed for the quick transportation of the main American tank, the M-1A1, by means of a road train with a total weight of up to 105 t. The tow car was included into the HET System. It has single wheels that are suspended on semi-elliptic springs and is all-wheel drive. It has the same automatic gearbox and diesel engine as the M-911, but is equipped with a 372.5 kW (500 hp) turbo charge. The cabin is semi-armored and holds six persons, including the crew of the transported tank. The tow car weighs 18.6 t, is 9195 mm long, and 3735 mm high.

Renault (France) has two chassis series, one civil and one military. The civil chassis is the 385.40 (8×4) model that is used for repair and evacuation equipment. It has a 287 kW (385 hp) in-line six-cylinder diesel engine with a turbo charger, an active volume of 11.1 l, and a mechanical 16-change gearbox. In late 1998, they started manufacturing the 400.32 (8×8) army chassis with a 292 kW (392 hp) engine, a working load of 20–26 t, and a full weight of 32–38 t.

For more than 40 years, the MOL Company (Belgium) has been designing and manufacturing various specialized vehicles, including tow cars, trucks for drilling complexes, articulated tractors, crane chassis, self-propelled platforms, etc. The tow cars for large-sized trailers (weighing 50–70 t) and ballast prime movers are an essential part of the production program. These are typically full-drive, three- and four-axle, and have the cabin mounted behind and above the engine. Articulated tow cars used for road trains weigh up to 150 t and those used with ballast prime movers weigh up to 250 t. The road vehicles have double wheels in the rear bogie with 14.00×24 tires and off-road vehicles have single wheels with 26.5R25 or 29.5×25 tires. The main specifications of certain MOL tow cars are presented in Table 1.3.

MOL Company equips its vehicles with different engines and transmissions based on customer requirements. The road tow cars, the articulated T 5264/05 and the ballast TC 250 models (Fig. 1.5), have Cummins, in-line, six-cylinder petrol engines with turbo charge. The articulated tow cars models intended for heavy operations, the T 6066/05 and the TB 800, have 12-cylinder, V-shaped, diesel engines made by KXD and MWM Diesel respectively. The TG 250 and TB 800 models can also be equipped with Daimler-Benz or Detroit Diesel 71, 72, and 92 series engines combined with an Allison automatic transmission.

The TG 250 and T 6066/05 models have a Clark automatic transmission with a lock-up torque converter, a 2.536 transformation ratio, and an eight-change gearbox. The T 2564/05 tow car has a two-disk dry clutch and a 15-change Fuller gearbox; however, an automatic gearbox can also be installed. The TB 800 tow car has a Renk AG automatic transmission, a seven-change gearbox, a lock-up torque converter, a hydraulic decelerator, and a 2.5 transformation ratio.

The driving axles are manufactured by Kessler and are equipped with planetary wheel-hub drives. The rear bogie front axles of all tow car models have a blocked differential. The TB 800 model has blockings of inter-wheel differentials on all axles.

The front suspensions are built on semi-elliptical leaf springs with double-action hydraulic shock absorbers, while the rear ones are balance type. Only the TB model has hydro-pneumatic suspension on all wheels.

The operational braking system is double-loop pneumatic—it operates on all wheels (using drum brakes)—while the parking brake uses spring brakes based on the rear bogie wheels.

Table 1.3 Main specifications of tow cars made by the MOL Company				
Specification	T 5264/05	T 60066/05	NG 250	TB 800
Wheel formula	6×4	6×6	8×8	6×6
Weight, kg:				
unloaded tow car	16,400	23,700	28,480	–
full tow car	52,000	60,000	60,000	66,000
full road train	150,000	150,000	250,000	154,000
Dimensions, mm:				
length	8500	103,350	8190	–
width	2996	3350	3000	–
height	3200	4070	3190	–
Wheel base, mm	4225+1550	4550+1900	1450+3040+1510	–
Engine:	Cummins	KXD	Cummins	MWM
model	NTS400	BF12L413	KT450	TBD234
active volume, cm^3	14000	19144	14000	21600
power, kW (hp)	298 (405)	318 (432)	335 (456)	600 (815)
torque, Nm	1559	1470	1831	2950
Gearbox transmission ratio ranges	10.93–0.89	4.43–0.59	4.07–0.54	–
Axle transmission ratio	10.77	15.19	14.4	11.915
Maximum speed of the tow car (road train), km/h	60 (25)	72 (25)	60 (20)	80 (30)
Minimum turning radius, m	10.68	–	15	15

Fig. 1.5. MOL Company TC 250 tow car

The ladder-type frame is made of heavily alloyed steel. A Jost Machinery articulated hooking device, with openings for the 88.9 mm pivot stud, is intended for towing a ballast axle with a platform or a platform with drilling equipment. Cross-members are bolted to frame rails and the T 5264/05 and TG 250 models have frames with reinforced front bumpers (for push-pull freight transportation). There is room behind the cabin for two mechanical capstane engines, each with a 30 t pulling force, and for electric and pneumatic sockets for trailers and semi-trailers (in the push-pull models the sockets are also available on the front surface of the frame).

The conventional (bonneted) tow cars are equipped with a two-seat, all-metal cabin on a three-point suspension, consisting of rubber elements and shock absorbers. The seats for the driver and passenger are set on springs and are provided with additional hydraulic shock absorbers. The bonnetless (minivan layout) version has a seven-seat, three-door (the third in the right rear section) tipping cabin. The cabin is tipped by a hydraulic lift. The TC 250 model has three front seats on hydraulic suspensions, while the rear one, designed as a bench, can be used as a bed. All cabins have thorough noise and vibration insulation, and are equipped with a protective sunlight shield, an air conditioner, a tank with drinking water, a radio, and a heater.

The Scania Group (Sweden) makes its special-purpose vehicles by strengthening its standard chassis with more powerful engines, multistage transmissions, all-wheel drives or additional lift-up axles, and armored cabins. In the heavy-duty class, the company manufactures the P124CB 8×6 HZ chassis weighing up to 40.5 t with an additional pneumatically suspended axle. Scania's most powerful model is the T144GB 6×4 NZ articulated tow car, equipped with a 394.9 kW (530 hp) V8 diesel engine with an active volume of 14.2 l, a mechanical 14-change gearbox with a built-in hydrodynamic decelerator, two capstane engines with a pulling force of 196.2 kN each, and a six-seat cabin. It is designed to work in a road train weighing up to 120 t and, in particular, for transporting tanks on a five-axle trailer made by the Loor Company.

The heavy-duty production program of Tatra a.s. (Czech Republic) includes two models of special-purpose chassis. Model T816G3LIWA (8×8) is designed for carrying special equipment. The latest version weighs 32 t and is equipped with a 405.3 kW (544 hp) DEUTZ 1015C V8 diesel engine with liquid cooling, turbo charge, an active volume of 15.9 l, and an automatic Twin Disc gearbox. The T816-6VWN9T-43.610 (8×8) articulated tow car is equipped with a 2100 rpm, 618.4 kW (830 hp), MTU V12 diesel engine with turbo charge; an automatic Twin Disc gearbox; a double cabin for 6–8 persons; and two capstane engines each with a 245.3 kN pulling force. It is designed for operating a road train weighing up to 120 t, has a cruising range of 1000 km, and a maximum speed of 85 km/h.

Articulated vehicle design has been known since 1913; however, it didn't become popular for heavy-duty vehicles until the mid-1970s and major developments, especially for articulated dumpers, occurred by the end of the 20th century. At present, over 30 percent of dumpers with a load capacity of 18–40 t are articulated. They differ from non-articulated vehicles by their "breaking" (hinged or pin-joined) frame and from road trains by their jointing individual links. The articulated joint is non-separable, of small length, and has controlled degrees of freedom in the horizontal and vertical planes.

These basic design distinctions allows for using the same low-pressure, big tires on all axles; decreasing the frame rigidity and weight; and achieving better adaptability to the terrain. These design advantages also allows for an increase in the load capacity, the passing ability, and the maneuverability, which makes for a smoother motion, a reduction in the dynamic loads on the undercarriage, and a decrease in the metal consumption per unit [3].

Due to their design layouts, applications, and load capacities (from 0.5 to 180 t), articulated vehicles are classified under the same principle as automobiles [1]. Classification by joint connection type is split into two categories—saddle (bolster-type) and tow-type vehicles. In saddle-type, individual links are connected by a united cargo platform, which is connected to the links by saddle-type joints that are used for road trains with semi-trailers.

Tow-type articulated vehicles can have one, two, or three degrees of freedom in the joint.

The first group includes models made by Foremost Industries (Canada), which has built and tested—in permafrost and desert conditions—a family of batch-manufactured articulated vehicles with 4×4, 6×6, and 8×8 wheel formulas (Fig. 1.6) and load capacities of 9.0 to 45.5 t intended for operation in extreme road and various climate conditions. All models are comprised of two links and have low-pressure tires (sort of pneumatic rollers), which can have a different number of tire cord layers (depending on axial load). The turning of the vehicle is "powered" by folding its links with the help of hydraulic cylinders. The maximum turning angle of links relative to each other is 38–45° on each side.

Fig. 1.6. Articulated vehicle Delta of Foremost Industries

The Lauster (Germany) articulated vehicle (Fig. 1.7) is a two-link tow car with a combined wheel-pneumatic-roller mover and an electric

Fig. 1.7. Lauster (Germany) tow car

transmission. The link joint has two degrees of freedom. In the horizontal plane, the link folding angle is ±43° and the lateral-vertical is ±30°. The vehicle turns by folding its links in the horizontal plane. The lateral-vertical allows the wheels to follow the profile of the road surface. The power unit, comprising of a 480 kW (653 hp) diesel engine and a synchronous generator with oil cooling, is mounted in the rear of the chassis. Two pneumatic rollers with strong, herringbone-shaped tire grousers under the body ensure high traction efficiency on adverse road conditions. Every roller houses a traction electric motor (TEM), a gearbox (multiple-reduction gear unit), and a differential. The TEMs are asynchronous, four-pole, oil-cooled, electric motors, which have a maximum rotation frequency of 10,000 min^{-1} and a nominal power of 150 kW (204 hp).

The roller axles are connected to the wheel axles by casings, which act as pendulum arms. Hydraulic cylinders are mounted between the pendulums and the vehicle body. The hydraulic cylinder cavities are connected with a pneumatic chamber, which acts as a resilient suspension element. The hydraulic cylinders allow for variation in the pendulum levers inclination angle, which adjusts the gap between the supporting surface and the rollers from the maximum value when on improved roads, to the required value when on roads with adverse conditions. In most cases, the traction force can be increased by lowering the rollers until they slightly touch the road surface.

Depending on the supporting surface condition, the electric drive in combination with regulated suspension ensures vehicle movement in several modes: on the wheels only, on pneumatic rollers only, on the wheels and one roller, and on all wheels and two rollers. The front roller can also be used as a capstane engine (a winch) for dragging out stuck vehicles or for self-winching. Switching from one mode to another takes a couple of seconds.

The main technical parameters of articulated vehicles made by different companies are presented in Table 1.4.

Table 1.4 Specifications of articulated vehicles

Parameter	D-250 (England)	"Commander-3" (Canada)	WD-3 (Holland)	D-550 (England)	"Magnum-4" (Canada)
Function	Dumper	Chassis	Dumper	Dumper	Transporter
Wheel formula	6×6	6×6	6×6	6×6	8×8
Load capacity, t	25.0	27.2	30.0	55.0	63.5
Kerb weight, t	18.6	29.5	18.0	37.0	37.2
Dimensions, mm:					
length	9820	13,500	9150	11,330	16,260
width	2062	3500	3500	3650	3990
height	3100	3100	3150	4370	4150
Minimum turning radius, m	7.5	9.5	10.5	10.8	17.2
Engine power, kW	174	258	189	333	342
Maximum speed, km/h	50	32	45	48	37
Tires	26.5×25	66×43.00-25	23.5×25	33.25×29	66×43.00-25
Degrees of freedom in joint	2	1	2	2	1

or branches of big corporations are also not able to make specialized, non-traditional components and units for such vehicles and must outsource to specialized companies, which explains the conservative nature of most technical solutions. However, multi-support transporters, articulated wheeled vehicles, and crane chassis are excluded [4].

A search for new general designs of heavy-duty vehicles is ongoing, including efforts to design non-traditional technical solutions for their components, parts, systems, and units.

1.1.1 Progress in designing vehicle components

Companies involved in supplying heavy-duty wheeled vehicles are facing an essentially limited demand. In terms of the motor industry, the scale of purchases excludes mass or medium-batch production. As a result, there are no big manufacturers of heavy-duty and super-heavy-duty wheeled vehicles. Certain small companies

One of the main problems in designing HDWVs is to ensure the required traction and speed properties. This includes optimization of the power unit and transmission, selection of systems for power distribution among the wheels, and transmissions for every load capacity class depending on the functions and conditions of operation. This is especially crucial for multi-axle wheeled vehicles and special wheel chassis (SWCs).

In most cases, it is not enough to increase the power of the engine system. That is why leading motor companies are limiting the engine power and instead try to increase the number of stages in gearboxes and automate the gearing. Intensive research and development (R&D) is underway towards creating an automated motion control system (AMCS) of the vehicle as a whole. The greatest attention is given to designing a system for rational power (torque) delivery from the engine to each wheel based on its specific rolling (working) conditions.

There is no doubt that essential improvements in traction and speed properties can be reached, but only if design solutions ensure the required (optimal) combination of the torque delivered to the wheels and the reaction of the driving wheels. The answer to this challenge is in the layout of the vehicle undercarriage; the transmission design and component specifications, which distributes the torque among the driving wheels; vehicle characteristics; and principles of the control system and its implementation.

Using the transmissions (hydromechanical, traction electric, and hydrostatic) and steering systems of heavy-duty vehicles made in Russia and abroad as examples, we will consider the evolution of the main units and system of a HDWV.

1.2 Hydromechanical transmissions (HMT)

The most notable manufacturer of components for HDWVs is Allison Transmission, which was founded in 1915 and bought soon after by General Motors. For more than 30 years, the company mainly designed and manufactured aviation engines; however, in 1944, Allison designed a HMT that was later modernized into the CD 850 model used in M460 tanks and other self-propelled vehicles. In 1950, the company's automatic HMTs were first used in urban buses, and, in the 1960s, in medium load capacity trucks (HMTs of the MT series). In the early 1970s, Allison released its second generation of HMTs, the AT, MT, and HT series, and recently the third generation WT series [18, 59 and 60].

Since 1948, Allison has made over 1.5 million HMTs. Currently, they are used in 52–1678 kW (70–2250 hp) engines.

The company uses many different designations for its transmissions, which are divided into brand names and industrial HMTs. The brand names are explained in the following list:

- CT stands for HMTs with hydraulic torque converters (HTCs);
- CLT stands for HMTs with HTCs and friction couplings;
- CLBT stands for HMTs with HTCs, friction couplings, and hydraulic brake-decelerators;
- DP stands for a version of CLBT with parallel power flows in the mechanical transmission;
- TT stands for HMTs with two-turbine HTCs;
- TRT is the same as TT, but with the same number of direct and reverse gears;
- CRT is the same as TRT, but with a usual HTC; and
- AT, MT, and HT are definitions of HMTs intended for small, medium, and high power.

The designations for industrial HTCs are as follows: TC stands for ordinary HTCs, while VTC stands for HTCs with rotating reactor blades.

Letters are usually followed by three digits: an index of the transmitted power or torque, the number of direct gears, and a modification number. A fourth digit may be present in front which defines a dimension of the HTC.

Functionally, HMTs are divided into three groups: trucks and buses (AT, MT, HT, and V series), traction vehicles (CLBT and DP series), and cycle-action machines.

Let us consider the operation principle and main parameters of these HMTs.

The trucks and buses series (Table 1.5), like all other Allison series, have standardized HTCs, friction and control system units, casings, and planetary mechanisms. For example, two front planetary mechanisms with four friction units (Fig. 1.8a) are used in all series and modifications to obtain three top gears (by engaging F1T2, F1T1, and F1F2) and the rear gear (by engaging F2T2); only the tooth ratios vary, i.e., transmission ratios or additional planetary mechanisms with friction units are introduced.

For example, in a four-change HMT (Fig. 1.8b), an additional lower (first) gear is achieved by a third planetary mechanism with a brake T3 (F1 and T3 are engaged). The reverse gear is engaged by F2 and T3.

The five-change HMTs differ from the four-change by the presence of an extra planetary mechanism with brake T4 (located in an additional casing, connected to the main one). There are two versions: the first (Fig. 1.8c), defined by the index CR (MT654CR or HT754CR), has gears with a dense (1.25–1.33) geometric series (in MT654CR it is also close to the harmonic series) of transmission ratios, except the first gear (1.8–1.9), which is below the general range of 3.7–4.2. This version has total automatic gearing and is recommended for tow cars of inter-city road trains. The second version

Table 1.5 Main parameters of HMT series for trucks and buses

HMT model	Maximum power, kW/ maximum torque, Nm	Maximum rotation frequency, min^{-1}	Dry weight, kg	Length, mm	HTC model (conversion ratios)	Transmission ratios of direct and reverse gears
AT 545	175/603	2400–4000	125	691	TC230(2.6), TC270(2.0), TC290(1.7)	3.45–2.25–1.41–1/5.02
MT 643	186/867	2200–4000	231	773	TC350(3.0), TC360(2.9), TC380(1.8)	3.58–2.09–1.39–1/5.67
MT 653DR	186/867	2200–4000	273	920	TC350(3.0), TC360(2.9), TC378(2.3)	8.05–3.58–2.09–1.39–1/5.67
MT647, MT648	186/1058	2000–3000	255	773	TC430(3.6), TC470(3.0), TC495(2.2)	3.58–2.09–1.39–1/5.67
MT654CR	224/1288	1800–3000	310	920	TC430(3.6), TC495(2.2), TC497(2.7)	4.17–2.21–1.66–1.27–1/5.67
HT740, HT741	332/1871	1900–2400	381	950	TC470(2.7), TC487(1.5), TC497(2.7)	3.69–2.02–1.38–1/6.04
HT747, HT748	272/1491	1900–2400	381	950	TC488(1.7), TC495(2.4), TC486(1.8) TC497(2.7), TC499(1.9), TC487(1.9)	3.69–2.02–1.38–1/6.04
HT754CR, HT755CR	332/1871	1800–2600	426	1041	TC488(1.7), TC495(2.4), TC496(1.8) TC497(2.7), TC470(2.7), TC487(1.5)	3.69–2.0–1.58–1.25– 1.0/9.65 or 2.96
HT750DR, HT755DR	332/1871	1900–2500	426	1041	TC488(1.7), TC495(2.4), TC487(1.5)	1.39–1.0/4.72 or 17.22

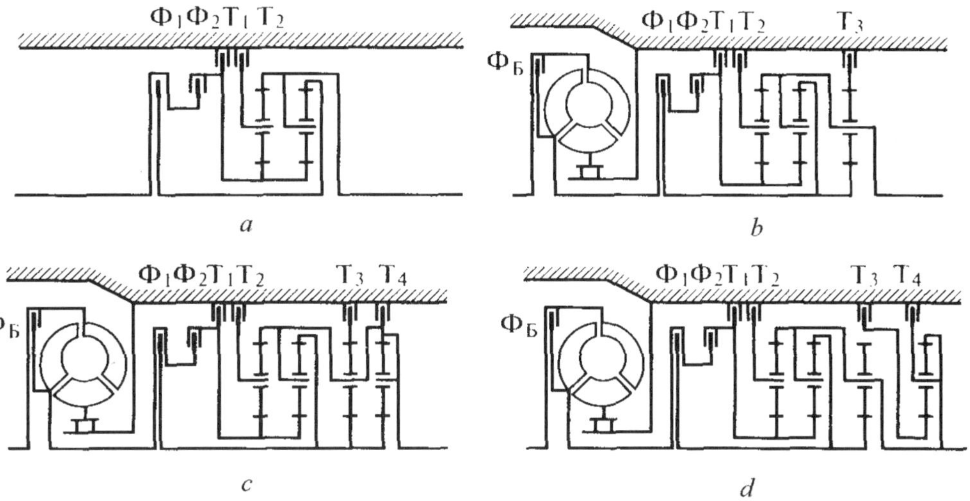

Fig. 1.8. Engagement diagrams of planetary mechanism friction units of Allison AT, MT, HT, and V HMTs

(Fig. 1.8d), designated by DR (MT653DR or HT50DR), has a number of transmission ratios that are close (like in four-change HMTs) to the harmonic series, it starts in first gear with 2.3–2.5, and its range is ~8. The gearing, starting from the second gear, is automatic. These HMTs are intended for buses and trucks.

Upon customer's demand, the HT700, CLBT, and CP series are provided—in front of the mechanical reducing gear, in the additional casing-spacer—with a hydraulic two-chamber brake-decelerator with its rotor fixed on the HTC turbine shaft. The MT600 and HT700 series, with a two-chamber two-change decelerator with its rotor fixed on the driving shaft of the transmission (in which case the letter B is added to the HMT model, MTB600 or HTB700) is also available on demand. The first stage of the decelerator—a two-chamber hydraulic brake—is efficient at medium and high speeds, while the second one—a friction disc brake operating in oil—decelerates the vehicle at low speeds to a complete stop.

The traction HMTs (Table 1.6) are designed to transfer a power of 332–678 kW (445–2250 hp). The 3000 series, manufactured in the early 1970s (except for CZT3461), and the 4000, from 1977, were replaced by the 9600 and 9800 series. These have five, six, or eight direct gears and one reverse gear, except for the 9800, which lacks the reverse gear. They have a friction coupling to block the HTC, some models have a hydraulic decelerator, and they are equipped with automatic or manual gearing based on customer's needs. The manual gearing system comprises of a switch and control valve connected by an electric cable. It is air-tight, powered by a voltage of 12 or 24 V, and requires no maintenance.

The company also makes the 5000 and 6000 series of traction HMTs, used in oilfield installations. Here, manual hydraulic control is used since electronic control is not permitted.

Automatic transmissions have logical (microprocessor-based) and hydraulic executive parts. The microprocessor has a programmable read-only memory (ROM) that stores particular conditions for using the HMT. The system elements are under permanent control of the built-in diagnostic unit, thus, should a malfunction be detected, its location is fixed.

Table 1.6 Main parameters of traction HMTs

HMT model	Maximum power, kW/ maximum torque, Nm	Maximum rotation frequency, min^{-1}	Dry weight, kg	Length, mm	HTC model (conversion ratios)	Transmission ratios of direct and reverse gears
CL(B)T754 CL(B)755	332/1764	2500	444	1387	TC470(3.0) TC490(2.6)	5.18–3.19–2.02–1.38–1.0/9.93 or 4.72
CL(B)T5961	373/2136	2500	954	1380	TC570(3.0) VTC(1.8–2.4)	4.0–2.68–2.01–1.35–1.0–0.67/5.12
DP8962	727/4474	2300	1678	1836	TC840(2.3) TC1070(1.7)	4.24–2.32–1.69–1.31–1.0–0.73/5.75
CLBT9681	933/5017	2100	1678	1836	TC1060(1.8) TC1070(1.7) TC1080(1.6) TC1080(1.6)	4.24–3.05–2.32–1.67–1.0–0.67/5.75 or 4.13

Prior to production, the automatic gearing system underwent tests equivalent to 140,000 operation hours, which confirmed the increased efficiency and fuel economy of the vehicle, as well as the high reliability of the HMT as a whole.

The basic unit of the Allison traction HMTs comprises of two planetary mechanisms, a friction F2 and brakes T2 and T3, or F1, T1, and T2 (Fig. 1.9a).

The company continues improving its traction HMTs, with the DP series, intended for engines with a power of up to 718 kW (975 hp), as an example of an updated HMT. These HMTs (Figs. 1.9b and c) have an additional planetary mechanism with a brake in front, and the links are connected a bit differently. This modernization resulted in a split power flow in planetary mechanisms at all gears, except for the first and reverse gears, which decreases the

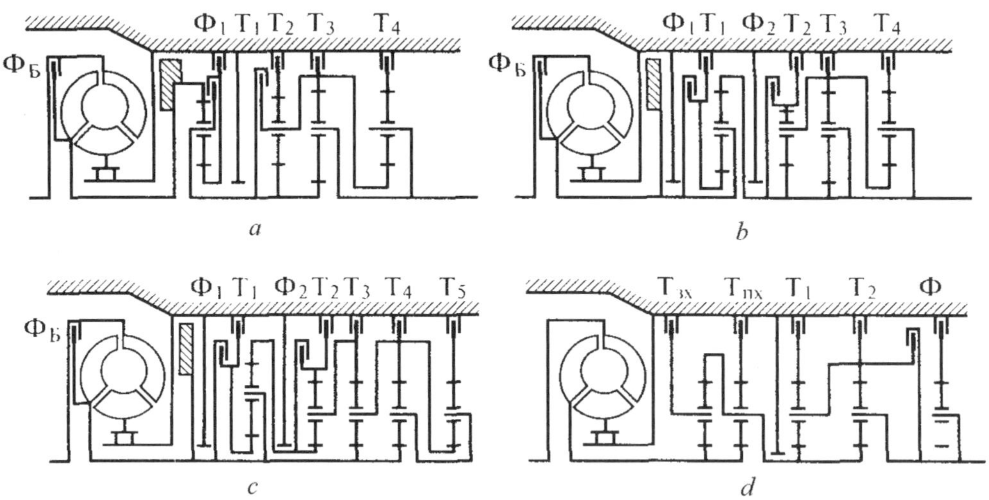

Fig. 1.9. Friction engagement diagrams in Allison traction HMTs

sprocket load. However, when the CLBT 9600 series, designed for up to 920 kW(1250 hp), was produced in the late 1970s, the company reverted to their earlier solution without the branching power flow as shown in Fig. 1.9d. In particular, this was done in the CLBT9681 model, which has six direct and one reverse gear. According to Allison, this HMT is best for HDWVs.

1.3 Electric transmissions (ETs)

Based on theoretical and experimental studies, more and more specialists are switching from mechanical transmissions to the individual delivery of power to each wheel. This is because the possibilities for further improving mechanical links in multi-wheel movers have been exhausted.

Two drive variants, electric and hydrostatic, are considered and both have been used in some special transport vehicles.

However, HDWV designers are exhibiting a permanent interest in using electric and electromechanical drives as transmissions. This is because there are a number of essential advantages, including the fact that they are continuous, easily automated, and demand no great physical efforts from the driver. Also we can obtain practically all required characteristics, such as regulation flexibility, through respective adjustment. Finally, and perhaps most important, an ET may be as easily designed, due to simple links between the units, as a multi-drive and remote transmission.

A transmission is considered electric if the power flow conversion coming from the engine (an internal-combustion engine (ICE), gas-turbine engine (GTE), and others) is achieved by an electric torque transformer. The simplest transformer consists of two main electric machines, a generator of electric power, and an electric motor.

If a transmission is equipped, in addition to the ET, with stepped (friction-tooth) gear reducers or other mechanical distribution mechanisms, the transmission is considered electromechanical.

The motor industry has always tried to use electric transmissions. Multiple prototypes and vehicles with ETs were manufactured both in Russia and abroad, mainly in quarry dumpers. Designers are especially interested in using ETs in heavy-duty multi-drive vehicles, active road trains, and dumpers.

However, the positive features are not able to completely compensate for the essential drawbacks of ETs, such as their huge dimensions and weight. The first ETs had specific weights of 11–15 kg/kW (15–20 kg/hp), while modern ETs have 4–6 kg/kW (6–8 kg/hp) and mechanical gearboxes have 1.5–2.5 kg/kW (2–3 kg/hp). In spite of this, ETs are often used in road trains with active trailers, in multi-axle machines, and quarry dumpers. Installation of electric machines into driving wheel hubs (motor-in-wheel) provides huge layout advantages compared to other transmission types.

LeTourneau Technologies, Inc. (USA) designed the first transport vehicles with electrically driven wheels back in the 1930s. They were multi-link (up to 13 links), over 100 m long, and had a load capacity of up to 180 t. Diesel or GTEs were used as primary power sources and drove DC or AC generators. The maximum power of an energy unit comprising of two GTEs was 3456 kW (4700 hp). The traction generators and AC traction electric motors (TEMs) were powered by a 300 V three-phase current and a 120 Hz frequency. An AC TEM is an induction motor with a short-circuited rotor and one AC phase is used for controlling the transmission.

Figure 1.10 presents a cross-section of a motor-in-wheel made by LeTourneau, where we see a two-line ordinary reducing gear (2), built-in to a fixed bracket of the electric motor (1) (the epicyclic wheel of the reducer is attached to the wheel hub), and an electromagnetic parking brake (3) (the operational braking is achieved by switching the motor into generator mode).

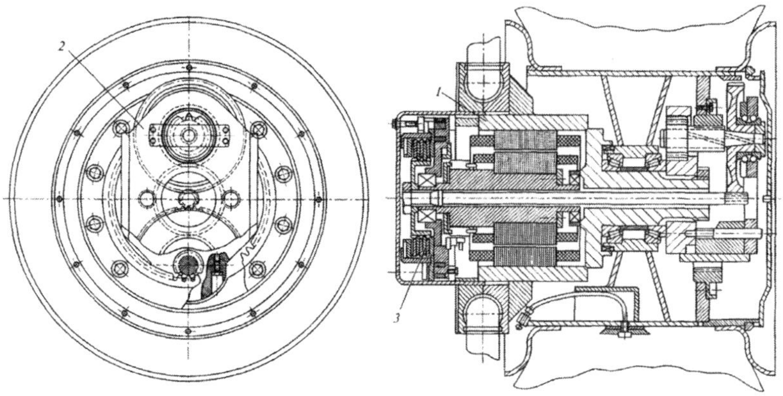

Fig. 1.10. Electric motor-in-wheel made by LeTourneau

To decrease the sprocket modules and reduce radial loads, the reducing gear has a branched power flow and two parallel blocks of intermediary wheels.

LeTourneau makes several motors-in-wheel sizes and builds various vehicles based thereon, including multi-link active road trains.

The General Electric Company motor-in-wheel (Fig. 1.11) is of different design. Here, the motor body is the wheel axle, while the reducing gear casing is located outside the wheel, and only the second-line driving sprocket is located in cantilever fashion and intermeshes with the wheel hub crown. Arrows show the air flow that cools the motor. The motor-in-wheel reducing gear is located outside the wheel rim; it has a small axial dimension and does not reduce road clearance.

The reducing gear body has a flange (8) for attaching to the frame, which can be lifted much higher than the wheel axle. In a large-diameter reducing gear, we can use a higher transmission ratio and a small high-speed electric motor. The

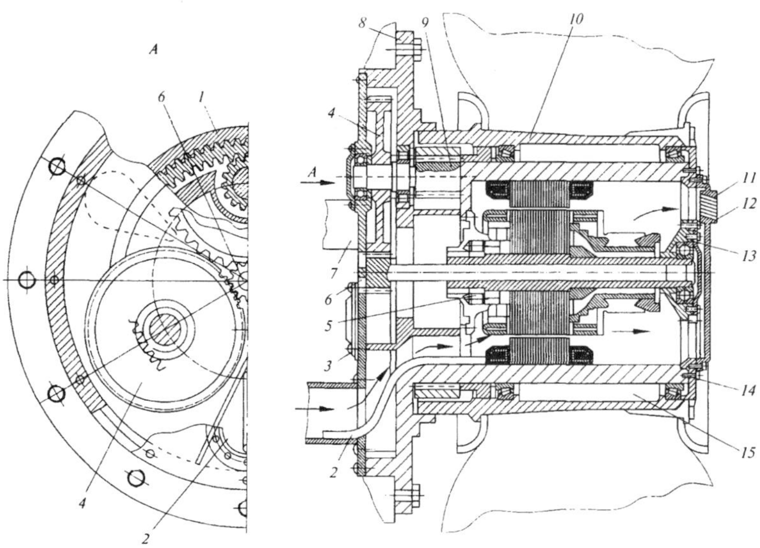

Fig. 1.11. Motor-in-wheel made by General Electric

electric motor is inside the rim (5) and its body (10) is the wheel axle, and carries the large-diameter bearings (14). Air cooling is supplied through the pipes (2 and 7) and is exhausted through the gills (11), which are provided with a protective cap (12). The reducing gear cavity is separated by partitions (3). The crown (1), connected to the wheel rim by a gear clutch, is linked to the motor by sprockets (6, 4, and 9). The electric motor shaft rests on bearings (13 and 5). The cavity (15) is filled with the cooling liquid. This design does not assume a brake; therefore, braking is achieved by the electric motor when the vehicle is moving. However, other General Electric models' have disc brakes on the commutator side.

The large number of patents on electric motors-in-wheel (EMsiW) shows an increased interest by designers for using electric transmissions in HDWVs. Figures 1.12 and 1.13 show some of the patented EMsiW [6].

The Berliet Company (France) manufactures vehicles of high cross-country ability, comprising of the TX-40 model single-axle tow car and a single-axle road scraper-trailer with a 40 t load capacity. Both axles are equipped with EMsiW provided with one- or two-speed reducing gears. When using an EMsiW with a one- speed reducing gear, the maximum torque is 12 times higher than the minimum and 36 times higher in a two-speed reducing gear. In the first case, the maximum speed is 40 km/h and the maximum conquered slope is 20 percent; in the second case, 70 km/h and 40 percent accordingly. The power source is a 316 kW (430 hp) Cummins T8 diesel engine combined with a Brown Boveri (Switzerland) made generator. The

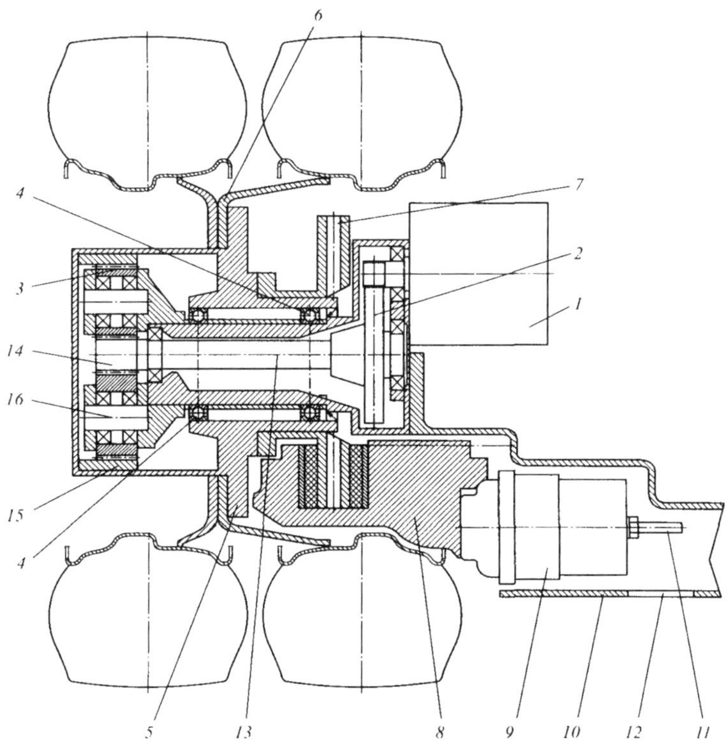

Fig. 1.12. Electric motor-in-wheel according to patent DE 19709577C2: 1—Drive electric motor; 2—Sprocket; 3—Satellite; 4—Ball bearing; 5—Wheel hub; 6—Wheel rim; 7—Combined brake (pneumatic and hydraulic); 8—Floating clamp; 9—Hydraulic drive; 10—Casing; 11—Pipeline; 12—Mounting hole; 13—Shaft of sprocket 2; 14—Shaft-sprocket; 15—Crown sprocket (epicycle); 16—Satellite axle

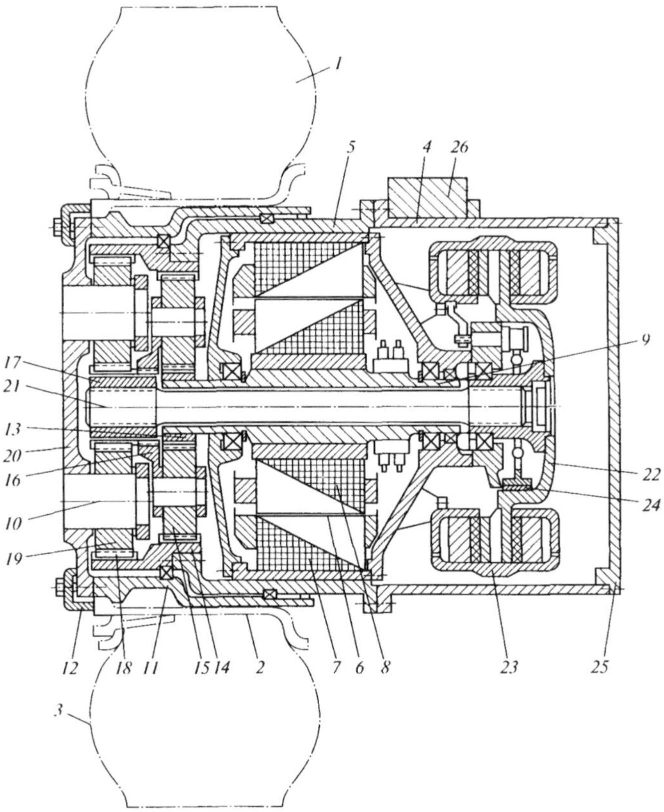

Fig. 1.13. Electric motor-in-wheel according to patent 2109372: 1—Mechanism in assembly; 2—Wheel rim; 3—Tire; 4—Brake mechanism casing; 5—Electric motor casing; 6—Air gap; 7—Stator winding; 8—Rotor Winding; 9—Hollow shaft; 10—Power distributing mechanism; 11—Planetary mechanism as a unit; 12—Second stage of the planetary mechanism; 13—Sun pinion; 14—Second-stage epicycle; 15—First-stage satellite; 16—Second-stage epicycle; 17—Sun pinion; 18—Second-stage satellite; 19—Wheel hub pinion carrier; 20—Planetary mechanism lid; 21—Electric motor shaft; 22—Brake disc; 23—Drum spacer; 25—Lid of the casing 4; 26—Vehicle frame

engine and generator are mounted on a removable additional frame. The generator is six-pole, with independent excitation, and a nominal power of 220 kW (300 hp). The TEMs of the EMsiW are four-pole, with independent excitation, and a nominal and maximum rotation frequency of 1800 and 3600 rpm accordingly.

Siemens AG (Germany) has designed a turbo-electric drive for vehicles with a load capacity of 20 t, a 300 kW single-shaft GTE, and a unipolar four-pole generator with air cooling. Jointly with Zander Fabrik (Germany), they designed an EMsiW comprising of a DC TEM, a two- or three-change reducing gear, a working disc, and a drum parking brake. The TEM with parallel excitation has a nominal power of 132 kW (180 hp) at 2550 rpm. The short-term starting torque, when starting from stop, is 2.5 times higher than the nominal torque. This jointly designed vehicle of high cross-country ability uses a combined drive where some wheels are driven mechanically and others electrically. The mechanical drive is used when the vehicle is on a normal road, while the electric drive engages in off-road conditions to improve passing ability. An example is the M-437 (USA) road train with a 13.6 t load capacity that comprises of a single-axle tow car and a trailer connected by a ball coupling. The drive to the tow car wheels is mechanical from a 200 kW (268 hp) diesel engine. An AC generator with a 518 V nominal voltage and an up to 200 Hz frequency is mounted on the same shaft as the diesel engine, which powers the trailer. The wheels have low-pressure tires with a 1.83 m diameter and the brakes are multi-disc with a pneumatic drive.

Oshkosh, in their latest generation of tactical vehicles, attempted to benefit from the advantages of ETs. These vehicles are radically different from their predecessors; only the layout is the same: the power unit is behind the cabin, axles are mounted in pairs, and wheel suspensions are independent of road clearance adjustment.

In these vehicles, the wheel traction drive is diesel-electric. A 335 kW (450 hp) six-cylinder turbo diesel engine is connected with a 335 kW AC generator; this power unit is feeding four electric motors installed on each axle (Fig. 1.14).

A set of batteries, intended to accumulate power, is mounted between the cabin and the power unit. These batteries help the diesel-generator at acceleration and when moving uphill and accumulate power during deceleration and braking.

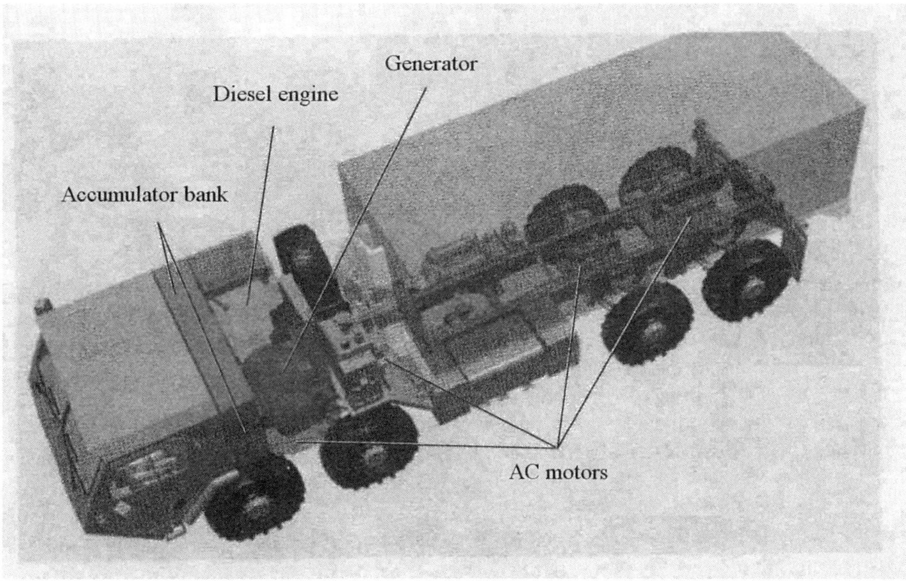

Fig. 1.14. Layout of Oshkosh's third-generation vehicle

This layout is used in a family of various-purpose vehicles.

The above examples show that there is no single answer to the most convenient way to use electric machines. However, to decrease weight and reduce the dimensions of ET elements, AC machines are recommended because of their high efficiency factor and their ability to be easily sealed and cooled. At low rotation frequencies (below 1000 rpm), synchronous generators, which are widespread in traction electric drives, can be used. At higher rotation frequencies, unipolar or asynchronous generators with capacitor excitation may be used.

Originally in EMsiW design, DC TEMs with serial or parallel excitation were used, which have attractive weight and dimension parameters. The next step was to use asynchronous TEMs with a short-circuited rotor, which also have good weight and dimension parameters. However, to regulate them, a complicated semiconductor control system, which still fails to ensure the required response rate and individual control of the wheel drive, was needed. In many countries, R&D works are underway to create EMsiW to use with other TEM types, for example, synchronous electric motors, which work as AC converter-fed motors, and motors on permanent magnets made with rare-earth metals [6].

The advantages of traction electric drives have predetermined how it can be used in various vehicles, including those used with heavy one-piece cargos ("monocargoes") in off-road conditions. More important directions, in our opinion, is designing new electric machines with broad regulation ranges and automated control systems based on microprocessors that ensures rational control and power delivery to the wheels based on road conditions.

1.4 Hydrostatic transmissions (HSTs)

1.4.1 General provisions

The HST has a number of positive features, which make it possible to use in future HDWVs [3]:

- Simple design, stepless, and infinitely smooth rotation speed and torque variation of hydraulic machines and displacement speed and power force of hydraulic cylinders;
- Fast acceleration, braking, and reversing due to very small inertia moments of hydraulic units;
- Possibility to completely stop hydraulic machines without using friction brakes;
- High power density and low specific weight of the pumps: 0.5–1.5 kg/kW (0.7–2 kg/hp) and hydraulic motors: 0.2–0.5 kg/kW (0.3–0.7 kg/hp), thus, a possibility to develop high forces at small dimensions;
- Able to obtain high transmission ratios without using reducing gears;
- High operation stability of certain hydraulic machines at low rpms, their ability to overcome high starting resistance of driving mechanisms, and the ability to operate at "creeping" speeds;
- Ease of automation and the ability to follow preset algorithms;
- High precision and quick response to command signals and the possibility to amplify these signals;
- Small friction and wear of certain hydraulic units, as they work immersed in oil;
- Small amount of nonferrous metals needed;
- Broad room for hydraulic unit standardization and parts thereof; and
- A sophisticated and nicely studied toolkit available for designing hydraulic drives and a huge accumulated experience in building various hydraulic machines.

It is well-known that the hydrostatic drive is superior to the mechanical drive in many of its features. It is also better than the electromechanical drive, at least in its manufacture adaptability, lower consumption of nonferrous metals, higher starting torques, and possibility of achieving a "creeping" speed.

Hydrostatic transmissions exhibit their advantages in 8×8, 10×10, 12×12, etc., vehicles for which the mechanical and hydromechanical transmissions are too complex, have low efficiency factor, poor adaptability to automation, and expedient power distribution. Given these conditions, an HST built under the modular principle is not inferior to a mechanical one in weight, cost, and efficiency factor; it also provides a number of additional advantages.

In 1897, the first patent using a hydrostatic transformer instead of toothed gearing appeared in German patent literature. Experiments were then conducted aimed at using them in transport vehicles.

The first automobile show in Berlin (1897) demonstrated an automobile with the Pittler hydrostatic transmission. In 1909, tests were conducted on a vehicle with a HST designed by Lenz. During WWI, one of the British tanks had a hydraulic transmission designed by Jenny, which they later tried to put it into a car. However, all early attempts to use HSTs in vehicles faced a number of technological problems and drawbacks such as big dimensions and weight, the need to seal up and filter the fluid, a low efficiency factor, restrictions of the temperature operation mode, and the absence of rational algorithms for the automated control system (ACS). At the same time, hydraulic drives were broadly used in machine tools, aviation, and railway transport, where HSTs operated under long-term stationary regimes. Among the first to use HST theory in practical locomotive designs were the experts from the N. E. Bauman Moscow State Technical University (MSTU) Loco Lab. Today, many companies, for example Alstom (France), make locomotive HSTs with 643 kW (862 hp) diesel engines. By using a HST instead of an electromechanical drive, Alstrom was able to reduce the undercarriage weight by 7 t and the non-spring-loaded weight by 2 t. The company has also built a locomotive HST with a 6400 kW (8576 hp) diesel engine. The fluid pressure in such hydraulic units reaches 60 MPa.

By the late 1950s to early 1960s, the world motor industry began using HSTs in the additional wheels of a road train trailer. Studies and experiments were also conducted

Deere & Company (USA), in the 1960s, was one of the first companies to mass produce vehicles with hydrostatic drives of additional wheels. In 1967, the Soviet Union made the ZIL 137-137B road train with hydrostatic drive of the semi-trailer wheels.

In the 1970s, many companies began to produce hydrostatic drive vehicles, for example Levi (Canada), Robert Bosch (Germany), and Zizu (Finland). The latter's HST is used by a number of companies in Italy, Holland, and Finland to drive the trailers of log and pipe carriers, container trucks, and other vehicles.

According to Zizu, a log carrier with a hydrostatic instead of traditional transmission generates a 25 percent higher traction force; conquers a 1.5 times steeper slope; and pays back faster, despite the high cost and technical maintenance, because it can work more days out of the year due to its resistance to weather conditions. Zizu also notes that using a hydrostatic transmission allows for reducing the minimum turning radius if necessary.

The ZIL 137-137b road train showed a 200 percent increase in traction force on wet roads, wet sandy surfaces, and snow; the ability to climb 1.5 times steeper slopes; and the ability to cross boggy areas at depths of 0.3–0.4 m. It could also easily climb and turn on sandy surfaces.

The currently existing hydrostatic transmissions for additional driving wheels can be divided into two groups:

- HSTs with asynchronous drive, where the main pump is driven by the primary gearbox shaft, i.e., the pump shaft rotation frequency is independent of the transmission used;
- HSTs with synchronous drive, where the main pump is driven by the secondary gearbox shaft, i.e., the pump shaft rotation frequency varies in synchrony with the rotation frequency of the main driving wheels.

In order to kinematically match the rotation frequencies of the main and additional driving wheels, the HSTs with asynchronous drive need to use a regulated pump or a regulated pump and hydraulic motors. This makes the transmission more expensive since the cost of regulated axial-piston machines are 50–60 percent higher than non-regulated ones. Also, the cost increases because of the complicated control system needed.

In many cases, HSTs with synchronous drives use non-regulated hydraulic units, such as the ones used by Zizu and ZIL in the driving wheels of their semi-trailers.

The main advantages of synchronous drive HSTs are smaller costs and the possibility to not use a complex control system. This is because the pump rotation frequency changes a certain number of times when switching gears; therefore, we could expect automatic synchronism in the rotation of the main and additional driving wheels. However, this is not always the case since any change in the pump shaft rotation frequency strongly affects its efficiency factor. As a result, we still need a control system to compensate for the above dependencies of the hydrostatic drive parameters on its operating conditions. In some designs, the control system is also used for obtaining technological speeds or ensuring the braking mode.

In spite of the above distinctions, both types of hydrostatic transmissions are very similar.

Due to the specific working processes of hydrostatic systems, HSTs are only used in a narrow range of vehicle speeds. Thus, the HST speed range of the Zizu log carrier is 10 percent of the total vehicle speed range. In the ZIL 137-137B road train, the HST works only when the lowest gear in the transfer gearbox is engaged, or when the first and second gears in the transmission gearbox is engaged (speeds up to 7 km/h). The design of the transmission provides for the mechanical or hydraulic blocking of the hydrostatic transmission control system, which excludes its use at higher operating and transportation speeds.

The most widespread use of HSTs are with motors-in-wheel, which are built on low-speed, high-torque, radial-piston, hydraulic motors (made by Zizu or Poclain) without additional mechanical reducing gears; or on high-speed, axial-piston, hydraulic motors (made by Sundstrand) with a mechanical reducing gear. In this case, the motor-in-wheel has a mechanism intended to disconnect the wheel from the hydraulic motor when the vehicle moves at high speed.

Currently, the most interesting version is a hydrostatic drive of semi-trailer wheels, where the transmission ratio is varied by changing the volume of an auxiliary regulated pump. This pump is driven by the primary gearbox shaft, while the main non-regulated pumps are driven by the secondary shaft. This design meets most of the requirements for HSTs of additional driving wheels, simplifies the ACS in managing the operation modes, and automates the blocking and unblocking of the HST inter-wheel links without degrading their characteristics.

In the late 1950s to early 1960s, theoretical and experimental works were conducted aimed at designing HSTS for vehicles as a whole.

In 1955–56, a Moscow-based research institute made a three-wheel vehicle prototype with an HST. Two regulated axial-piston pumps driven by a 15 kW (20 hp) engine fed working fluid to two non-regulated piston hydraulic motors that rotated the rear wheels. The prototype was tested in parallel with a similar model with a mechanical transmission to compare their traction-dynamic parameters and it was established that they were higher in the HST prototype.

Starting in 1960, the M.I. Kalinin Factory in Moscow began manufacturing batches of the D-456 model small-sized tow car intended for loading and earthwork. Its transmission included a non-regulated, screw-type pump and stepwise, regulated, screw-type, six-section motors. The pump was connected to the engine by a step-up gearbox, while the hydraulic motors were mounted in the rear axle and connected to the wheels by a step-down gearbox. The hydrostatic transmission comprised of two manually regulated piston pumps, driven by a 12.0 kW (16 hp) diesel engine. Each pump was connected by a pipeline to a non-regulated piston hydraulic motor. The motors made use of onboard gear reducers, which drove two driving wheels on each side of the tow car. Hydraulic drives were provided for driving interchangeable implements. The tow car weighed 1640 kg and had a maximum traction force of 14.0 kN. The aim was to assess the option of using HSTs based on screw-type hydraulic units in motor vehicles. It was established that a screw-type pump can work at up to 12,000 rpm and the screw-type hydraulic motors can work up to 6000 rpm. However, the efficiency factor proved to be low.

In 1961, the U.S. made an experimental 4×4 amphibian vehicle, weighing 540 kg, with a fiberglass body. Its maximum speed was 80 km/h and it had floating speed of 10 km/h in which the water propeller was hydraulically driven. It had an HST with a pressure of 2.1 kPA (210 kgf/cm^2) in which a manually regulated piston pump was directly connected to an air-cooled 30 kW (40 hp) Kohler K-622 carburetor engine. Non-regulated, piston hydraulic motors were mounted in the wheels and the front wheel motors could be disengaged.

In 1962, a transportation amphibian with an HST and a maximum speed of 56 km/h was made intended to support oil prospecting works. The HST had a pressure of 4.2 kPa (420 kgf/cm^2) and the units were used to drive the propellers and capstane engine. The HST contained an automatically regulated axial-piston pump directly linked to a 335.3 kW (450 hp) engine. Each wheel contained two piston, non-regulated, hydraulic motors of different working volumes and two two-change gear reducers. The motors were serially connected in a crosswise manner.

In 1963, the American company Gar Wood made an aerodrome tow car with an HST and manufactured a batch of 237 vehicles. The HST design comprised of an automatically regulated piston pump with a power of 89.4 kW (120 hp) and four piston hydraulic motors-in-wheel.

In 1964, the Moscow Auto-Mechanical Institute (MAMI) conducted experiments with an automatic HST built on axial-piston hydraulic units. The transmission was installed into a Moskvich-410 passenger car. The aim was to master the automation system and rational regulation. It was established that under rational automatic regulation, the fuel efficiency of the HST was similar to a mechanical transmission on good roads and even better on adverse road conditions. Another Moscow-based research organization designed, made, and tested an experimental HST based on screw-type hydraulic units that were installed into a ZIL-150 truck.

At the same time, U.S.-based Niagara Frontier Transit System made a 78-seat city bus weighing 13,900 kg. It used a 141.6 kW (190 hp) diesel engine that drove an axial-piston pump and two hydraulic motors. A branch of the Regal-Beloit Corporation (USA) built a log carrier with an HST that had a pressure of 2.1 kPA (210 kgf.cm^2). It used a regulated twin vane pump directly connected to the engine. The regulated hydraulic motors drove the axles via gear reducers and each motor was powered by a separate pump section. The HST was used to drive not only the wheels, but also the capstane engine, the bulldozer, attached implements, other equipment, and appliances. The maximum speed of the vehicle was 26 km/h and the total weight was 850 kg.

In 1964, in Canada, a two-link, air-cushion vehicle with a motor-in-wheel HST was built. The U.S. Armored Vehicle Center, ATAC, also made an experimental three-axle cargo amphibian with an HST. The HST design comprised of a servo regulated pump directly connected to a carburetor engine. The servo regulated motors installed on the wheels were provided with disconnection devices. The vehicle had hydraulic braking, wheel and axle blocking, and water-jet propellers.

During the 1960s and later, many companies designed and built HSTs for wheeled and caterpillar tractors, forklifts, and road-building machinery. One of the first vehicles equipped with HSTs were wheeled tractors made by Fordson Major Ferguson, the Varnz-90 Porsche caterpillar, the Hydromatic motor truck, the Hidrostal road-building machines, and others. Work towards improving HST designs were conducted by Vickers, Denison, Kellot, Gar Wood, Bullard & Co., Lockheed, Marmon Hier (U.S.); Lucas, Hamilton, Doughty, Hitlok (England); Porsche, Hydromechanik, Gansmag, Hydromatik, Demag, Pittler-Toma (Germany); and others.

HSTs made abroad for tractors are built mainly under the integral-aggregate design. Occasionally, a two-change reducing gear is placed behind the hydraulic motor. The pressure in such HSTs is 30–40 MPa.

Comparative tests conducted by International Harvester (USA) on two universal tractors—one with an HST and one with a mechanical transmission—established that an HST increases a tractor's performance at hay pick-up and briquetting by 28.9 percent, at loading-unloading by 24.9 percent, at forage storage by 25.4 percent, and at tillage by 7.4 percent.

However, in spite of its advantages, HSTs have not been broadly used in tractors until recently. Only 2–3 percent of manufactured tractors have HSTs. This is due to the high cost of the hydraulic drive; an HST tractor is 20 percent more expensive than a traditional one. HSTs are more widespread on small-sized vehicles used as horticultural and garden tractors (40–50 percent in the U.S.), grass-mowers, tow cars, crop collectors, snow and garbage removers, etc. These vehicles weigh 170–600 kg and have an engine power of 6–15 kW.

According to foreign experts, a promising idea is to use HSTs in urban buses and special-purpose municipal vehicles. By using hydraulic accumulators and energy recuperation systems, HSTs allow for reducing a bus's fuel consumption by 30 percent. The combination of a control and energy recuperation system results in a 40–50 percent decrease in the fuel consumption of a urban travel cycle. However, using a special ACS for the HST is mandatory.

Many experts currently believe that using HSTs in multi-axle vehicles and heavy-duty dumpers is both economically and technically feasible. The Alstom Company (France) makes HSTs for dumpers with a load capacity of 120 t (full weight, 308 t). The dumper has a 985 kW (1320 hp) engine, can move at 50 km/h, and overcome slopes of up to 25 percent.

The advantages of HSTs are especially obvious when used in multi-axle vehicles with 8×8 and up wheel formulas, for which mechanical or hydromechanical transmissions are very complex in design, expensive, and have a low efficiency factor because they cannot ensure a rational power distribution among the wheeled movers. The former USSR took advantage of this and built a self-propelled 25-ton crane with an HST on an 8×8 automobile chassis.

The first Soviet-made prototype of a 12×12 chassis with an HST was the E547AG model, which weighed 3250 kg less than a similar model with a hydromechanical transmission. The full weight of the HST units and working fluid was 2697 kg.

HSTs are currently widely used in self-propelled platforms (multi-support vehicles) with load capacities of up to 1000 t intended to transport huge monocargoes such as nuclear reactors, ship bodies, spacecrafts, furnace units, drilling rigs, etc. Super heavy-duty vehicles with HSTs are made by Sheyerle and Kamag (Germany), Nicolas (France), and others. Generally, their bogies are modular-based and specific platforms are assembled from individual modules based on the load capacity needed to transport a particular cargo. For example, the multi-support platform Otomag made by Nicolas for a load capacity of 500 tons has 192 wheels, with all or almost all of them driven. Usually, every pair of double-tire wheels has a common independent suspension and a driving hydraulic motor installed between the wheels. The HST is comprised of several pumps and the system for controlling the HST channels is microprocessor-based. HST designs for multi-support HDWVs will be considered in more detail later.

This brief review and analysis shows that it is favorable to use HSTs in the driving wheels of tractors, road-construction machines, urban buses, heavy-duty multi-axle and multi-support vehicles, and semi-trailers.

In our opinion, the most promising uses of HSTs are in small-sized and multi-support HDWVs. These vehicles benefit from such HST advantages like the relatively low weight and dimensions, the possibility to build according to a modular concept, the stepless (continuous) regulation, the ability to recuperate energy, and the implementation of optimum control algorithms. These factors can result in lower fuel consumption, improved traction, and dynamic parameters in special wheel chassis (SWCs).

1.4.2 Design principles of hydrostatic transmissions

Currently, wheeled vehicles use different types of HSTs. They differ in the layout of hydraulic units (mutual positions of pumps and motor and how they are connected to the engine and wheels) and the types of elements used (single or composite pumps and motors).

As a rule, the motors used are high-speed and high-torque with varying designs (e.g., axial-piston, radial-piston, impeller, gear-type, and others) and regulation methods (e.g., stepwise, stepless, or mixed). The pumps, motors, or all hydraulic units can be regulated with equal and non-equal regulation ranges.

For multi-drive vehicles, transmissions with a broad regulation range (12–36) and an ability rating (dynamic factor) above 0.7 are needed. The regulation ranges should be efficient so that the HST efficiency factor never gets below the preset value (about 0.7).

In order to reduce the working volume and weight of hydraulic units and increase the HST efficiency factor, the general regulation range must be broadened by increasing the range of the hydraulic motors. Let us understand the range as the ratio of the maximum to the minimum rpm values (i.e., speed range) or torque values (i.e., power range) of motors.

In this context, it is not reasonable to recommend non-regulated hydraulic motors or stepless regulated motors with narrow regulation ranges for use in multi-drive SWCs.

Regulated hydraulic units not only ensure high traction and dynamic characteristics of HDWVs, but also allows for disconnecting the engine at zero gearing for starting and idling; continuously varying the vehicle speed; using the reverse gearing; engine braking; partial parking braking; implementing any preset motion control algorithm; and meeting all HST ACS requirements.

Non-regulated and stepwise-regulated pumps and motors do not allow for disconnecting the engine; therefore, they fail to ensure a smooth start and the "creeping" motion. The pump is not reversible, which means the vehicle has no reverse gearing and the motor cannot decelerate for park braking. It is also impossible to implement multiple control algorithms. Since the HST needs to operate steplessly, one or all of the hydraulic units must be regulated steplessly or in a mixed way.

To ensure the preset control algorithms, mixed regulation is better than stepless.

In general, only two HST designs for multi-drive machines are expedient—a mixed regulation of the pump and a stepwise-mixed regulation of the hydraulic motors.

To ensure the work of the vehicle (engine start and idling, run-down, braking, etc.), HSTs use different control methods and auxiliary equipment units.

However, an orderly theory for building an automobile HST based on the main principles of vehicle theory was missing.

1.4.3 Designing principles of HST automated control and user experience

All current HST automated control systems (ACSs) are intended to control the gearing ratio of the hydraulic transmission. Most of the algorithms and structures of HST ACSs were borrowed from the control theory of hydraulic servo systems, which are broadly used in robotics, machine tools, etc. However, these algorithms are notable for their single-line (single-channel) character and are a poor fit with the highly dynamic and spacious parameters of HSTs. Despite intensive studies, there is currently an absence of clarity in many aspects of the theoretical bases for controlling a multi-channel HST in different operation modes and environments. For example, S.B. Shukhman and V.I. Solovyov have offered the following HST control principles depending on the input signal.

I. ACSs of the permanent power (N) class. The control is affected based on the following deviations:

a) the angular rate of the engine shaft;
b) the torque of the engine load;
c) the angular rate of the vehicle's wheels; and
d) the pressure in the HST pump unit.

II. ACSs of the permanent speed (V) class, which ensures the constant speed of the vehicle.

Class N systems ensure engine operation under the constant preset (optimum) mode for the regulatory characteristics of all positions of the system adjustment pedal. Systems of this class try to correct the moment of resistance to the load by varying the HST transmission ratio. Special literature is available that offers about 50 systems of this class, which differ in the methods of correcting the resistance moment and the load evaluation depending on the vehicle's intended specific operation. An advantage of these systems is the possibility to automatically adjust the HST to the dominant engine operation mode under certain conditions. However, they are not able to work with diesel engines; they need complex equipment to monitor parameters and control the HST actuating devices. In order to use these principles in control systems, the optimum parameters of the multi-factorial space must be known; however, this problem has no single-valued theoretical solution today.

One of the first experimental Class N ACSs was created by V.V. Kalinnikov and V.A. Petrov in the early 1960s for the Moskvich-410 passenger car with an experimental HST. A similar system was designed in the mid-1980s by Hydromatik (Germany), which found broad application in many vehicles.

The same control principles were implemented in the HST ACS of the Utpak-2006 tractor, made jointly by Deutz and Bosch (Germany). However, because of its high price and low reliability, it was replaced by an ACS made by Linde AG, where only the actuating devices were updated and not the control algorithms.

Automated control systems, which are regulated based on deviations in the pumping unit pressure, are used in Russian- and foreign-made excavators with HSTs and in a number of other machines and process equipment with hydraulic systems. Their advantages are in simplicity and operation reliability. However, the application area of these ACSs is restricted by the use of non-regulated hydraulic motors. Respectively, their use as control systems of multi-channel HSTs is not expedient.

The ACSs in Class V maintain engine operation under the regulatory characteristic preset modes and ensure a permanent travel speed. These systems are found in tractors and vehicles that work at a constant and, in most cases, "creeping" speed. Such systems were experimentally designed for the NTS-55 tractor model made by Lipetsk Tractor Works. At the Kuban Agricultural Institute (Krasnodar Territory, Russia), L.I. Grom-Maznichevskiy designed a similar system for the DT-14B, NTS-50, MTZ-50B, and MTZ-80B tractor models; V.A. Petrov did the same for the experimental K-700 tractor equipped with a YaMZ-238NB engine.

The main drawback of Class V ACSs is in their complexity, i.e., the mandatory presence of Class N systems, which decreases reliability, dynamics, and other properties.

If we compare different HST ACSs based on their functionality, we see that a major advantage of HSTs—the possibility to control individual power distribution among wheel movers depending on operation condition and modes—is not embodied in existing ACSs. This is because there is no single methodology tool for defining the optimum characteristics needed to control a separate transmission power drive channel and control the channel interaction when they work asymmetrically.

The above principles for controlling HSTs also disregard the global motor industry trend of aggregating the control systems into a single modular communication network. Therefore, these systems fail to account for the information and power exchange of other control systems (of the engine, suspension, steering system, etc.).

Thus, it is necessary to formulate the theoretical grounds for controlling the HST. This methodological toolset should define the minimum optimum criteria and take the following into account: the individual addressable ACS reaction to an asymmetrical parameter deviation in all channels of a multi-channel HST; the channel interaction rules; the quasi-stationary engine operation principles; and the HST ACS interaction parameters with all other ACSs and the general automated motion control system (AMCS) of the vehicle.

Therefore, future HDWV hydrostatic transmissions and their control systems should make use of both the extensive design experience and use of such systems by the world motor industry and the achieved level of controlling servo drives. This experience shows that vehicles using HSTs possess many advantages, which become especially important in heavy-duty vehicles.

Multi-support vehicle chassis built under the modular principle may supply the closest prototype of future multi-axle HST chassis. HSTs for small-sized vehicles also deserve design attention.

The most promising designs are those with a mixed control over the HST pump and motor sections. These designs ensure the required traction and dynamic properties of SWCs and HSTs and implement various ACSs. There is no doubt that modern HST design is characterized by boosting the positive HST properties through creating a rational ACS and developing its functional algorithms.

The knowledge gained by using HST ACSs in experimental and serial models shows that the available control principles allow us to meet a number of requirements for future ACSs and ensure reliable operation with acceptable efficiency and traction dynamics parameters. To fully implement these requirements, the control principles should complement the wheel rolling modes asymmetry correction, account for the controlled autonomous channel interactions, and the ACS should operate within the integral AMCS of the vehicle.

1.4.4 Composition of HSTs and their automation systems. Hydraulic machines

In general, a hydrostatic transmission comprises of the following main elements: a pump, a motor, auxiliary equipment, and elements of the automated control system.

The pump and motor are based on hydraulic units, which are typically rotary hydraulic units. The main types of hydraulic units used in hydraulic drives will be considered in detail later. Based on the requirements for HSTs and ACSs and the need to choose a rational mixed-regulated hydraulic unit design to ensure regulation and reversibility, minimum weight and dimensions and a high efficiency factor (0.85–0.95) are desirable. Axial-piston and radial piston machines currently meet the above requirements and are used throughout the world to build HSTs.

Axial-piston machines with an inclined block, including those with two non-power double cross-joint couplings, also known as a cardan, are the most widespread in HSTs. The cardan connects the input shaft of the hydraulic units with a cylinder block.

The cardan is considered non-powered because it does not participate in the transmission of power from the shaft to the fluid; instead it synchronizes the shaft rotation with the cylinder block. However, in transient modes when the cylinder block experiences high accelerations, the cardan is loaded and may fail. There are also hydraulic units with a power cardan, but they are less popular because of large weight and high inertia moments.

A drawback of axial-piston pumps with a double cardan joint is the rather complicated system of fluid delivery. Also, when adjusting the feed rate, the cradle rotates along an arc, thus increasing the space (dimensions) occupied by the hydraulic unit. This design is widely used in Type II hydraulic units (Table 1.7). Type II D (D stands for remote control) axial-piston hydraulic units and Type II M (M stands for local manual control) hydraulic motors are the main hydraulic drive elements in various automated, semi-automated, manual, or remote control systems. They have modest volumetric and mechanical losses and a rigid kinematical link of the shaft with the piston mechanism, which ensures perfect rotation stability at low flow rates and speeds. A drawback is the relatively low maximum pressure at 16 MPa. This restriction is caused mainly by the load capacity of the rod joint link with the piston.

Type II hydraulic machines are expedient in small-power hydrostatic drives where there are no tough requirements for the weight and dimensions.

The Praga Company offers a popular range of Type II M hydraulic motors (Table 1.8). These non-regulated, reversible motors are intended for mobile and stationary objects.

Table 1.7 Specifications of Type II M hydraulic units

Parameter	Classes of hydraulic units			
	5	10	20	50
Hydraulic motor flow rate, cm^3/rev.	71	142	251	790
Maximum pump shaft rotation frequency, min^{-1}	1440	1440	1440	980
Maximum pump yield and flow rate of hydraulic motor (theoretical), l/min	102	204	361	774
Pressure in decreasing space (of the pump), MPa:				
nominal	-	-	-	-
maximum permissible	10	10	10	10
Torque developed on the shaft, Nm:				
at nominal pressure	105	210	370	1170
at maximum pressure	158	315	556	1760
Maximum weight, kg	31.5	54.5	83	210
Dimensions, mm	421×130×200	527×176×246	615×200×305	825×276×425

Table 1.8 Specifications of the Praga AM series hydraulic motors						
Parameter	Motor classes					
	AM12	AM16	AM28	AM56	AM105	AM125
Pressure, MPa:						
nominal	25	25	25	25	25	25
maximum	35	35	35	35	35	35
Hydraulic motor flow rate, cm^3/rev.	12.5	16	28.5	56	105	125
Inertia moment, $kg \times m^2 \times 10^{-3}$	0.456	0.688	1.80	5.54	15.8	21.1
Rotation frequency, min^{-1}:						
nominal	1920	1920	1920	1500	1500	1500
maximum	6000	6000	4800	3600	3000	3000
Torque, Nm:						
nominal	47	60	108	209	397	473
maximum	65	85	149	293	550	662
Weight, kg	6.5	10.1	12.5	23.5	42.5	54

The main parameters of Type II D hydraulic units are summarized in Table 1.9.

These hydraulic units have a working pressure of 21 MPa, a maximum working pressure of 35 MPa, and the maximum efficiency factor is 0.9–0.92.

Many foreign-made products, such as the axial-piston hydraulic pumps with an inclined disc made by Denison Hydraulics (USA) (Table 1.10), are similar to the Type II D units and can be used for making hydrostatic transmissions.

Radial hydraulic units are the best fit for the planetary series and allow for addressing the problem of weight and dimension parameters, even if applied in transmissions of caterpillar vehicles. Currently, the Russian industry has no convertible, reversible, and regulated hydraulic units that could be used in HST vehicle designs.

Table 1.9 Main characteristics of hydraulic units with inclined disc				
Standard size	Working volume, cm^3	Maximum working rotation frequency, min^{-1}	Limit rotation frequency, min^{-1}	Distance between shaft supports, mm
20	33.3	3800	4800	68
21	51.6	3500	4400	80
22	69.8	3200	4000	88
23	89.0	2900	3660	95
24	118.7	2700	3400	104
25	165.8	2400	3000	117
26	227.3	2100	2650	130
27	333.7	1900	2400	147

Table 1.10 Specifications of pumps made by Denison Hydraulics			
Parameter	Models of hydraulic pumps		
	P6W	P7W	P8W
Hydraulic pump flow rate, cm^3/rev.	98	119	131
Working pressure, MPa	34.5	34.5	31.0
Maximum pressure, MPa	41.4	41.4	31.0
Maximum rotation frequency, min^{-1}	3000	3000	2100
Working rotation frequency, min^{-1}	2050	1950	1800
Inertia moment of rotating parts, kg×m^2	0.027	0.027	0.027
Weight, kg	81	81	81

The most interesting designs from other countries are the hydraulic units with pistons that have a spherical working surface. They attract consumers by their simple design and relatively low price, and are used in hydromechanical transmissions for high cross-country ability vehicles and in transmissions of caterpillar vehicles, particularly the M2 caterpillar machine (USA).

Currently, two types of hydraulic units with spherical pistons are used: one with a piston block and one with a cylinder block.

In hydraulic units of the first type, the piston block is a rotor. The pistons are shaped as spherical belts attached to radial rotor protrusions. The cylinders rest on a yoke, and the cylinders connections to the yoke have hydrostatic unloading. The fluid is delivered to the channels through radial openings in the rotor protrusions.

In hydraulic units of the second type, pistons are shaped as spheres (balls) and placed in the cylinder block. The pistons roll over the yoke, where the radial groove is similar to those in roller bearings.

Modern hydraulic unit design is characterized by an active development of these types, including for vehicles notable for stringent limiting parameters (a specific weight of 0.14–0.18 kg/kW, a specific volume of 22000–29000 kW/m^3, and a service life of 6000–8000 hours) and built-in control systems.

Other types of hydraulic units (vane, screw, and gear types), are not typically used as main HST units. This is due to their drawbacks, namely, increased volumetric and mechanical losses, low efficiency factors, poor rotation stability at low rpms, pulsation, low values of transmitted pressures and torques, and large dimensions.

1.4.5 Working fluids (WFs)

In hydrostatic transmissions, WFs have two main missions: first, as the working medium of the HST, i.e., they are its element; and, second, ensuring the operability and reliability of all hydraulic drive units by acting as a lubricating and cooling agent. The recent appearance of a broad assortment of working fluids has led to the widespread use and success of hydraulic drives.

Among Russian-made working fluids, the most appropriate for self-propelled vehicles is the AMG-10 thickened mineral oil. It is a solution of a polymer additive (dope) in a light and low-viscous oil base. The light oil fractions have a low-angle viscosity-temperature characteristic and a low solidification point; however, their viscosity at high temperatures is insufficient. The AGM oil contains additives, which ensure anticorrosion protection, nice lubricating properties, and prevents deterioration of rubber parts. The only obstacle to the broad use of such oils may be their high cost.

For hydrostatic transmissions intended for construction and road-building vehicles, VMGZ and MG-30 WFs are recommended. The former is better for negative temperatures, while the latter is better for positive temperatures.

VMGZ is used as an all-season WF in extreme northern areas and as a winter option for regions with moderate climate. It ensures normal operation of hydraulic units in temperatures ranging from –40 to +65°C, and short-term operation at temperatures of –50 to +70°C. At temperatures below –40°C, the working fluid must be warmed. VMGZ is a crude oil-based, thickened fluid and contains additives that ensure its antioxidant, anticorrosion, and antifoaming properties.

MG-30 is used season-wide for southern regions and in summer in areas with moderate climate. It ensures continuous operation of hydraulic units within a temperature range of –50 to +70°C and a short-term operation within a range of –10 to +70°C. MG-30, like VMGZ, contains various additives to improve its properties.

The service life of VMGZ and MG-30 is 3500–4000 hours, which is 2–3 times longer than other non-standard oils. AU and IZOA oil grades can be used to replace these working fluids.

Recent improvements in oil fraction processing technologies have resulted in new hydrocarbon fluids with exceptionally useful properties. These oils, such as M10 7415, have better properties than synthetic oils. However, these oils are rather expensive. Also, some hydrostatic transmissions are known to use diesel fuel as the working fluid. This can be done for short-term operation in certain forced exclusive conditions.

1.4.6 Auxiliary systems of hydrostatic transmissions

An important precondition for an efficient HST is the presence and rational layout of its auxiliary systems, which maximizes its efficiency factor. This equipment includes an oil storage tank; a refilling circuit; control system power elements; valves; flow rate regulators; a system for circulating (pumping or flushing) the transmission; a system to drain, collect, and bring oil leaks back into the tank; a radiator for cooling the oil; filters; etc. These should be thoroughly designed as any defect leads to an essential drop in the efficiency factor and an increase in the dimensions and weight of the HST and vehicle.

The charging (refilling) circuit is one of the major auxiliary systems. It serves to ensure the normal operation of the looped hydraulic piping system by maintaining the required pressure and compensating for leaks. In most designs, the charging circuit works as follows: the charging pump sucks the WF from the tank through the filter located at the pump entrance, and feeds it through the charging valve to the hydraulic transmission pump input. In addition to the filter, a diverter valve is usually provided, which operates if the filter clogs and the pressure increases beyond the preset value. In charging systems, an impeller, roller-vane, or gear-type pump is used. It is expedient to use pumps with mixed control and reversing, which are able to achieve high pressures.

In hydrostatic machines, control system power elements make it possible to vary the active volume and, accordingly, the HST gearing ratio. This is achieved by changing the inclination angle of the control body with the help of the control mechanism that uses hydraulic and electric amplifiers.

Hydrostatic transmissions also make use of return (back-pressure), relief, and safety valves. At a pressure of 40–50 MPa, the use of direct-acting valves is unreasonable since the increasing force requires too large a spring. In order to do without a spring, indirect-acting valves (reverse and safety ones) with an additional needle valve to control the main valve are used. If a separate pump is used to feed the hydraulic boosters that manage the hydraulic units, then a relief valve with a variable discharge control pressure to decrease the driving power can be used. The function of the relief or safety valve is to drain the working fluid into the tank whenever the pressure exceeds the preset value.

When the hydrostatic transmission comprises of several hydraulic motors, a flow rate regulator is used to drive individual wheels. In this case, the regulator acts a differential blocking mechanism and is responsible for the differential link among the wheels.

The oil circulating (flushing) system works to eliminate mechanical particles and keep air away from the HST. It typically comprises of a circulating valve (tap), filters, diverter valves, and pipelines. The flushing is done periodically as a preventive measure, every time a semi-trailer or trailer is connected to the vehicle, after dismantling-installation works, in case of a lengthy idling, etc.

Thus, the development of modern hydraulic units, and the systems for controlling them, is creating preconditions for building rational HSTs and implementing new control algorithms.

The engine unit of perspective future HST vehicles may comprise of high-speed engines with reducing gears or high-torque ones without reducing gears. The requirements for good controllability, reversibility, and convertibility should be preserved.

1.5 Steering systems

To ensure the high passage ability of HDWVs, apart from the transmission (traction drive) type and mover used, the steering system is of paramount importance [5].

In spite of the fact that a design with all-wheel steering was used for the first time in 1901 in the steam-driven Cotta automobile, the idea was very ahead of its time. From time to time, some interest in it arose, but was not stimulated by the market or supported by available technologies. Today, however, the situation is much different.

In August 1985, Nissan (Japan) was the first to mass manufacture all-wheel-steered vehicles by using the electrohydraulic high capacity actively controlled steering (HICAS) system in its Skyline passenger cars. The company is currently busy designing its second-generation Super-*HICAS* system.

Apart from Nissan, all-wheel-steered passenger cars are currently manufactured by Mitsubishi, Mazda, and Honda (all from Japan). They are actively steered by changing the turning angle of the rear wheels. This change depends on either the travel speed (Nissan, Mitsubishi, and Mazda) or the turning angle of the steering wheel (Honda).

Steering systems regulated by travel speed are turning the rear wheels in a counter-phase manner against the front wheels at a speed of up to 30–50 km/h and in an in-phase manner at higher speeds by an angle of 0.5–8.0°. Steering systems based on the turning angle of the steering wheel are turning the rear wheels counter-phase against the front wheels at up to 250° of the steering wheel turning angle and in the in-phase manner at angles above 250°.

Trucks with all-wheel steering are being made as "next year vehicles" (Leyland, Goldhofer, and Nissan).

Goldhofer believes these steering systems are most promising in the crane chassis and trucks operating in Western Europe on its highly loaded roads. Chassis with all-wheel steering are manufactured by Kommeto, Sheyerle, Kamag, and Kato and are intended for transporting huge single-piece cargoes and for mounting cranes.

Studies conducted by scientists from the University of California indicate that when all wheels are turned (the rear wheels to the same side as the front), the phase response of the vehicle to the control input is faster; while the speed of turning the steering wheel may be not so rapid when quickly changing lanes. Tire resistance to breakaway is also higher. When rear wheels are steered in the counter-phase manner to the front wheels when moving at low speeds, the minimum turning radius is reduced, which simplifies maneuvering in a confined space and decreases the resistance-to-turning torque.

Bases on development trends, all-wheel steering systems can be conditionally divided into two groups.

Group 1: Steering systems with an engaged rear axle drive. In the main transportation modes, these vehicles are controlled traditionally, i.e., by turning the front wheels. The rear wheels engage automatically, or by a driver maneuvering in a confined space at low speeds. The drive system ensures the rear wheels are turned to the same angle as the front wheels and usually have a reverse mechanism. Thus, the steering control has two operation modes—transportation and maneuvering. The latter halves the minimum turning radius and—once the reverse mechanism is engaged—allows for the so-called "crabbing" mode, when all wheels are turned to the same side and angle (Fig. 1.15). The media has reported on such steering systems designed by Volkswagen.

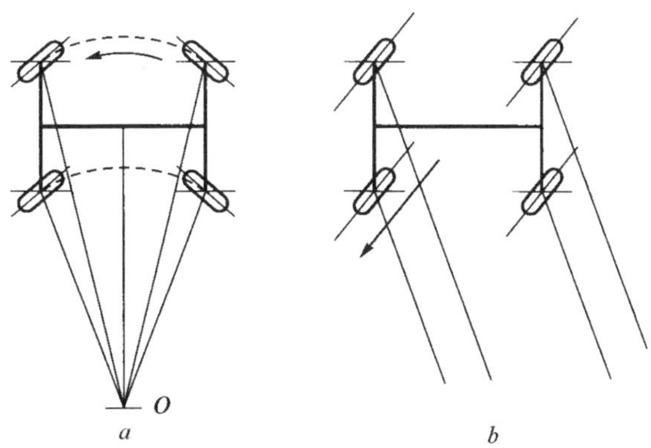

Fig. 1.15. Additional functionality when rear wheels are steered: a—Turning under the minimum radius relative to the turning center O; b—"Crabbing" motion (shown by an arrow)

Group 2: Steering systems with the drive continuously affecting all wheels. This group may be subdivided into two subgroups: steering systems with a "rigid" and a "flexible" turning geometry.

In a "rigid" turning geometry, the drive of certain wheels (often the front) remains traditionally mechanical, while other wheels are controlled under a certain law. Such systems are used, for example, in multilink vehicles. The front link has traditional steering, while the following ones are steered individually by receiving a signal from the front steered wheels. The steering drive can be mechanical, hydraulic, or electric and with or without a follower.

A "rigid" turning geometry can be applied in all-wheel drive vehicles. In this case, the front wheels are steered traditionally, while the rear ones are steered under a definite preset law. The onboard computer calculates a definite dependence between the transmission ratios of the rear and front axles and accounts for different factors.

This type of steering systems makes it possible to improve vehicle maneuverability and preserve high-speed motion stability. An example is the works of Furujawa Osami and Sano Seiti from Honda. A typical example of such steering systems is presented in Fig. 1.16.

Certain functional restrictions can be regarded as drawbacks of such systems. The "rigid" turning geometry is characterized by a single fixed operation algorithm of the control system, which defines a single-valued functional dependence of each wheel's turning angle to the steering wheel turning angle. At different steering wheel turning angles, the correlations between the front and rear wheels' turning angles may be quite different. For example, at small steering wheel turning angles in a Honda Prelude and Mazda-626, the rear wheels are turning in the in-phase manner, i.e., in the same direction as the front ones (Fig. 1.17, curve 2), but by a smaller angle, which ensures good motion

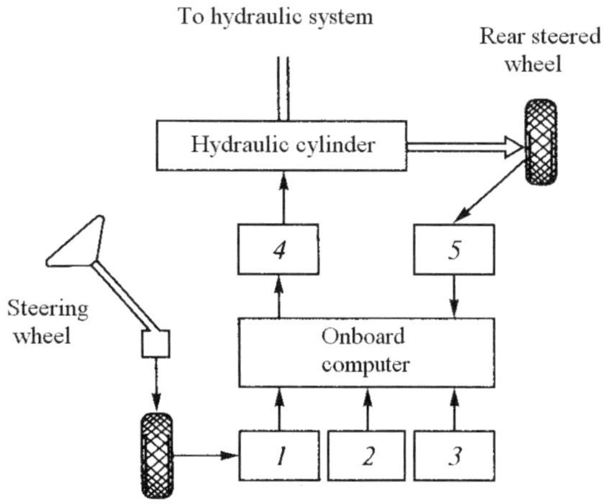

Fig. 1.16. Steering system made by Honda: 1—Steering wheel turning angle sensor; 2, 3—Correcting sensors; 4—Sliding valve (spool); 5—Feedback sensor

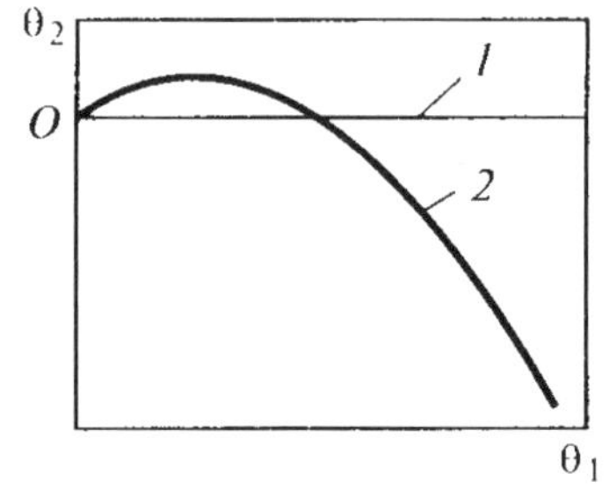

Fig. 1.17. Turning angle of rear wheels 01 versus the turning angle of front wheels 02 in the Honda steering system

stability. At higher steering wheel turning angles, i.e., at maneuvering, the rear wheels start turning in the counter-phase manner to the front wheels and decrease the turning radius (straight line 1 in Fig. 1.17 is the abscissa).

The "flexible" geometry is characterized by the control system operation algorithm, which allows for multiple turning options: by means of the front or rear wheels only, by the "crabbing" motion, a turn relative to the base center, etc. To ensure this variety of steering modes, we need an autonomous power drive of all the controlled wheels, and an onboard computer that calculates the forces sent to each drive depending on the required turn kinematics. Such steering systems have been implemented in special vehicles only, for example, in especially heavy-duty self-propelled modules made by the Italian Cometto Company, which will be considered in detail later.

Apart from the aforementioned systems, steering systems which incorporate a hydraulic mechanism of forced jack-knifing of articulated wheeled vehicles (AWVs) are also broadly used. Such systems allow for using big tires (even pneumatic rollers) and achieve high parameters of maneuverability and turnability.

Chapter 2
Multi-support vehicles

2.1 Overview

Multi-support vehicles intended for transporting large-sized indivisible loads (monocargoes) are widespread in the construction industry [19].

Such transporters should meet a number of special requirements: ability to use the existing road network while carrying heavy oversized cargoes, ability to take and fix special cargo items, and ability to load and unload items.

Using these requirements, France has designed a special transporter with a 40 t load capacity for transporting bridge sections and other oversized structural elements. This four-axle vehicle with a diesel engine and hydrostatic transmission is made by Poclain. The load on each axle is 20 t. The Sheyerle Company (Germany) has also built a family of heavy-duty special transporters (Fig. 2.1).

The general design specifications for special heavy-duty multi-support transporters are presented in Tables 2.1 and 2.2 [1] and in Chapter 9.

Sheyerle has the following product line:

- Low-frame trailers with a up to 400 t load capacity;
- Low-frame semi-trailers with a up to 120 t load capacity;
- Bogies with individual (independent) control with a up to 70 t load capacity;
- Wheeled platforms with a up to 200 t load capacity;
- Special vehicles with a up to 400 t load capacity; and
- Transporters with built in hydrostatic drives and driving and driveless axles capable of carrying 30–40 t loads.

Vehicle modifications are obtained by adding axles and modular bodies (Fig. 2.2) both length- and crosswise. Thus, these vehicles can have two, three, or four rows of bogies and each row can contain one to ten bogies.

Fig. 2.1. Multi-support heavy-duty MT 120.8.2 transporter made by Sheyerle

Table 2.1 Main specifications of multi-support vehicles made by Sheyerle (Germany)

Parameter	Vehicle model			
	MT 120.8.2	MT 180.12.2	MT 300.20.4	MT 400.26.4
Load capacity, t	113	180	270	370
Dead weight, t	47	60	130	150
Number of supports:				
driving	6	10	16	22
driven	2	2	4	4
Number of wheels	32	48	80	104
Tires	10.00-20	10.00-20	10.00-20	10.00-20
Overall dimensions, m:				
length	18.3	22.48	26.06	26.06
width	5.0	6.0	9.4	10.0
Road clearance, mm	200	200	250	250
Turning radius, m:				
external	10.25	13.6	17.1	17.5
internal	2.25	3.0	4.15	3.5
Maximum speed, km/h:				
with full load	8	8	8	8
without load	20	20	20	20
Power unit:				
engine type	Diesel	Diesel	Diesel	Diesel
Power, kW (hp)	212 (285)	253 (340)	212×2 (285×2)	253×2 (340×2)
Transmission	HST	HST	HST	HST

Table 2.2 General design solutions of multi-support vehicles [1]

Group	Conventional chassis type	Number of supports		
		Total	Driving supports	Driven (controlled) supports
1	Single-link	8-26	2-4	2-26
2	Articulated trailer type	24	6	12
3	Articulated saddle type	48	12	48

The starting module is the K-50 model, a two-axle, 3000×3275 mm, welded frame bogie weighing 7.5 t with a 40 t load capacity and eight wheels. The wheels—all steered—have a hydraulic suspension. These basic bogies allow for combining a vehicle of different load capacities.

Sheyerle's single-link, heavy-duty vehicle model has 20 axles; four of which are driving. The

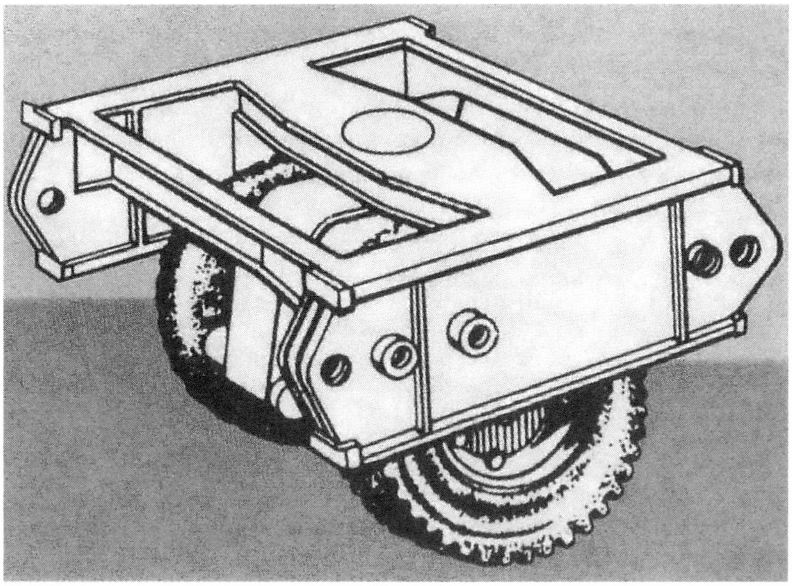

Fig. 2.2. Modular bogie with a 25 t load capacity

engines are mounted in the middle of the frame, which allows for decreasing the loading height. The wheel drive is hydrostatic and all wheels are steered. The vehicle has two cabins—one at the front and rear. A HST is used while the hydraulic motors are built into the support axles.

The French company Nicolas has designed a heavy-duty vehicle named Automac Routier (see Fig. I.2b in the Introduction) which consists of three sections: the front and rear are support platforms which carry the middle section. Each platform is a trailer and has 96 wheels located in 12 rows on 24 swinging axles.

The front six rows (48 wheels) are driven by 24 hydrostatic motors designed by the company. The number of driving wheels can be 16 to 48 and is selected by the driver depending on road conditions. An electronic device helps to synchronize the front and rear chassis sections.

Each platform has a diesel engine located behind the driver's cabin that is connected to the hydraulic system pump, which has its hydraulic motors built into the support axles. The bridge-type carrying system is telescopic; its width can vary from 2.6 to 4.3 m and its height from 3.53 to 5.07 m.

The side members of the carrying system are made from doped steel. The cargo platform length is 14 m, the maximum length of a cargo item is 10 m, and the loading height is 1.08 ± 0.27 m. The engine power is 2×477 kW (2×640 hp), the specific power is 1.27 kW/t (1.7 hp/t), the maximum travel speed is 18 km/h, and the vehicle can climb a slope of 5%. The tires are 8.25–15-RT8. A transporter with a 500 t load is 58 m long and weighs 760 t.

Kamag Company (Germany) also makes multi-support vehicles intended for transporting huge ship sections. These vehicles have lift-up platforms driven by hydraulic cylinders which simplifies loading and unloading.

The minimum platform height above the road surface is 1520 mm with a maximum of 2130 m; therefore, the lift height is 610 mm.

Kamag transporter vehicles have 2 to 24 axles and each axle has four wheels. The maximum load capacities vary, as necessary, from 68 to 443 tons depending on the number of axles. A transporter with 24 axles is 24 m long and 6.1 m wide. All axles are steered ensuring high maneuverability in confined areas. The steering system is provided with hydraulic boosters, the wheels can turn at angles of up to 70°, and there is a hydraulic suspension.

The transporters are equipped with Deutz or Cummins 127 to 253 kW (from 170 to 340 hp) diesel engines.

Recently, Kamag has made two new transporter models with 450 and 500 t load capacities, also intended for shipbuilding yards. The latter has a 9×20 m sized platform with a maximum height of 2.7 m. The transporter has 28 axles with four wheels on each axle; eight of them steered. The 323 kW (440 hp) engine combined with a hydrostatic transmission ensures climbing slopes of 6% and a speed of up to 6 km/h.

Although foreign companies address design issues and the layout of individual units and systems differently, they all adhere to the unified approach to the vehicle as a whole caused by particular transportation needs.

Thus, Nicolas, Cometto (see Fig. I.1a), and Kamag make use of the principle of a monoblock multi-support vehicle. Sheyerle, however, prefers vehicles assembled from individual bogies (as shown in Fig. 2.2) and installs traction hydraulic motors uniformly in each module. The control module—together with the driver-operator cabin—is attached to the front of the first module and the power unit to the rear of the last module; thus, achieving a high standardization level of the whole range of manufactured designs.

Cometto is using a finite number of active and passive (driveless) modules (Figs 2.3 and 2.4) when building modular transporters with an up to 1880 t load capacity. A detailed analysis of the layouts used by the four companies is presented in Chapter 9.

In order to decrease the weight and ensure the required strength, all multi-support vehicle frames are lattice-shaped (Fig. 2.5) and comprise of longitudinal, power welded, box-sectioned

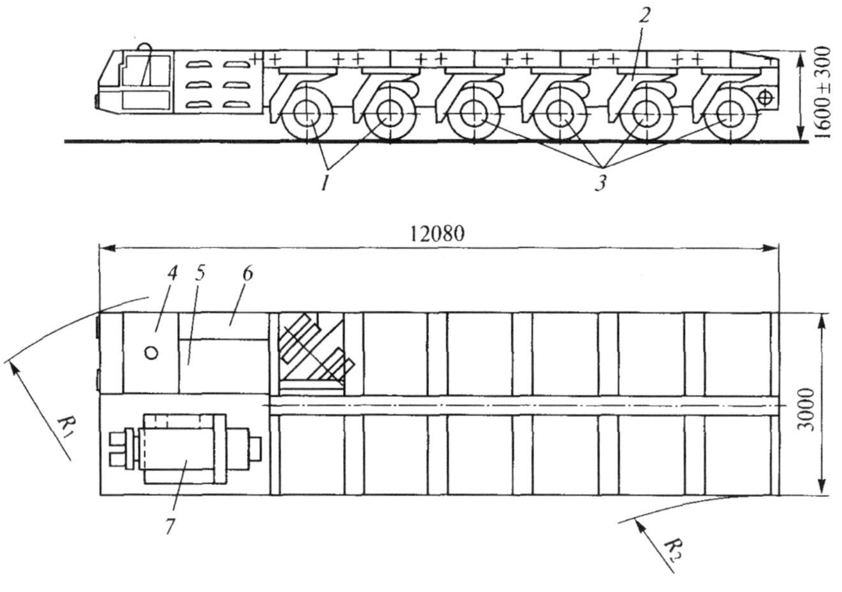

Fig. 2.3. Active module A made by Cometto: 1, 3—Driving and driven wheels respectively; 2—Suspension; 4—Driver's cabin; 5, 6—Fuel and oil tanks respectively; 7—Power unit; R_1 = 10,200 mm—Minimum external dimension turning radius; R_2 = 4200 mm—Minimum internal dimension turning radius

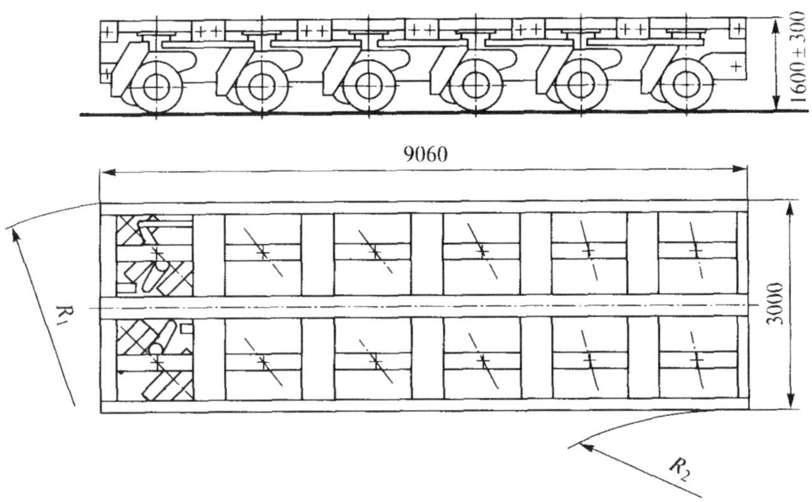

Fig. 2.4. Passive module D made by Cometto: R_1 = 14,400 mm—Minimum external dimension turning radius; R_2 = 7600 mm—Minimum internal dimension turning radius

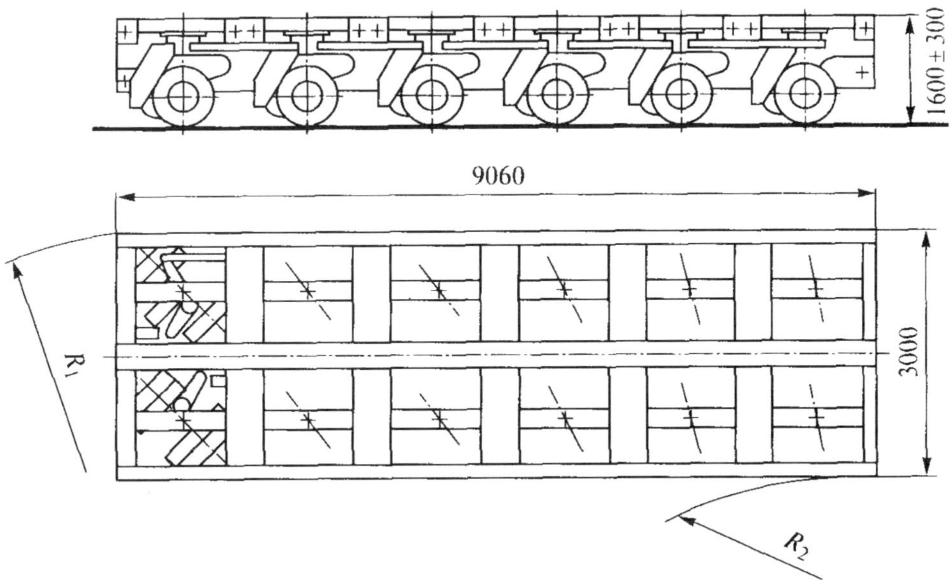

Fig. 2.5. Frame of the vehicle made by Nicolas: 1—Trailed module; 2—Connection cross-beam; 3—Towing beam (for trailers); 4—Frame

beams with lateral links of various profiles welded to them and bearing wheel suspensions.

The frame elements are made of rolled steel sheets that have a yield point of 353 MPa and a tensile strength of 608 MPa. The applied special steel chemical composition (given in percentages) is: C = 0.19, Si = 0.65, Mn = 0.92, P = 0.013, S = 0.005, Cr = 0.9, and Mo = 0.43.

is responsible for lifting and leveling the cargo platform and the vehicle's braking system.

The main diagram of a multi-support HST is presented in Fig. 2.6. Engine power is delivered directly to the hydraulic pump shaft H, which generates—in pressure line A—a high pressure of about 350–400 kg/cm^2. Two hydraulic motors—M1 (left-hand) and M2 (right-hand)—

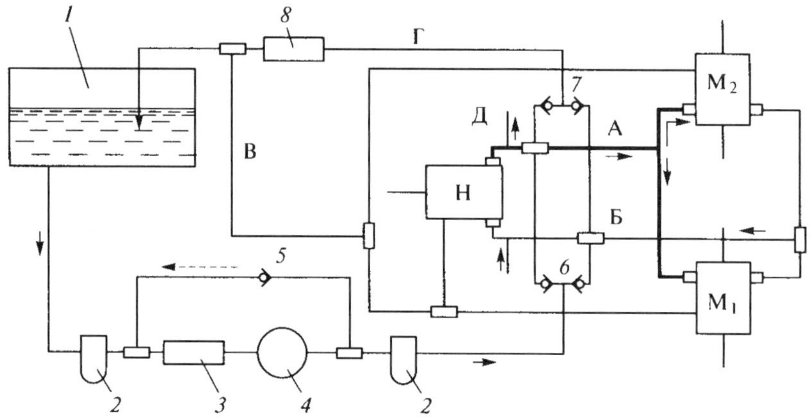

Fig. 2.6. Basic diagram of the HST: A—High-pressure line; B—Low-pressure line; C—Cross-over line; D—Drain line; E—Fluid line for the control system and platform lift; M1, M2—Hydraulic motors of the left- and right-hand of the vehicle supports respectively; H—Hydraulic pump; 1—Hydraulic tank of the HST; 2—Filter; 3,
8—Working fluid coolers; 4—Top-up pump; 5, 7—Safety valves; 6—Reverse valves; ← - Fluid workflow; < - - - - - Fluid drainage

The central beams (most loaded) are made of special steels with a breaking strength of 784–931 MPa, a yield point of 686 MPa, and an impact viscosity at ambient temperature (45°C) reaching 4.15 kg×m/s.

2.2 Transmission

Only a hydrostatic or electric transmission can be used in multi-support vehicles because of their ability to bring power to any point of the vehicle. A hydrostatic transmission (HST) ensures the autonomous drive of the driving support wheels and operates the control system which

are connected to pipe A. It is assumed that part of the fluid is taken away—through pipeline E—to drive the control system and lift the loading platform. Both ends of the hydraulic motor shaft are connected to the transporter's driving wheels. The working fluid, used in the hydraulic motors and other systems, is fed—through the low-pressure line B—to tank 1. Leaks that may happen through the glands are collected by pipe C and sent to the tank.

To compensate for leaks and to cool down the oil, the HST has a top-up (refilling) and cooling system comprising of a top-up pump (4), one or two filters (2), and working fluid coolers (3 and 8). To protect the top-up pump from overload, a safety valve (5) is provided. Hydrostatic machines are reversible: they can work both as pumps and as motors. For example, if a vehicle is decelerated (braked), its hydraulic motors can be made to work as pumps, and its pump to work as hydraulic motors. In this case, the high-pressure line (A) becomes the low-pressure line, while the low-pressure line (B) turns into the high-pressure line. The top-up system should be connected to the high-pressure line (A). The line switchover is made automatically with the help of reverse valves (6).

To protect the transmission from overload, safety valves (7) are connected to pressure pipelines (A and B); they operate when the pressure reaches the maximum value preset for the system. When the valve (7) is open, the working fluid is drained into the tank through the drainage line (D). Since at braking the working fluid should take heat away from the hydraulic motors, an additional cooler (8) is provided in the drainage line (D).

The basic diagram is simple; however, the real system is much more complex since it has automation elements that control fluid delivery and power transformation. Another variant of a multi-support transporter HST is presented in Fig. 2.7.

2.3 Supports

The supports have a hydraulic suspension, which is used to move the platform up and down by ±350 mm at loading and unloading. A special system is also provided for leveling the platform.

Design variants of a controlled wheel support are presented in Fig. 2.8. The base of the support is a shaft carter and a profiled, C-shaped pivot lever. The upper section of the lever has a slewing journal (a sling) connected to the cargo platform. The lower section of the lever is connected—by means of a joint—to the wheel axle carter. The lever's upper section also has a second link that is connected to the wheel axle carter by a hydropneumatic element, which acts as a resilient suspension element and as the hydraulic lifting device.

The most widespread suspension design is the so-called three-point design, where wheel suspensions are divided into three loops and their geometrical center forms the hydraulic point. The lines connecting these centers form a stability triangle.

Connecting suspensions into a four-point loop is only allowed for short periods of time when moving on poorly paved roads. This ensures increased motion stability with a load. By mounting cut-off valves on each hydraulic cylinder of the suspension, it is possible to disconnect one cylinder from the rest of the loop if support module repairs are needed.

Cometto favors suspensions designed in a parallelogram shape (Fig. 2.8a) and applies them in nearly all of its vehicles by varying the dimensions.

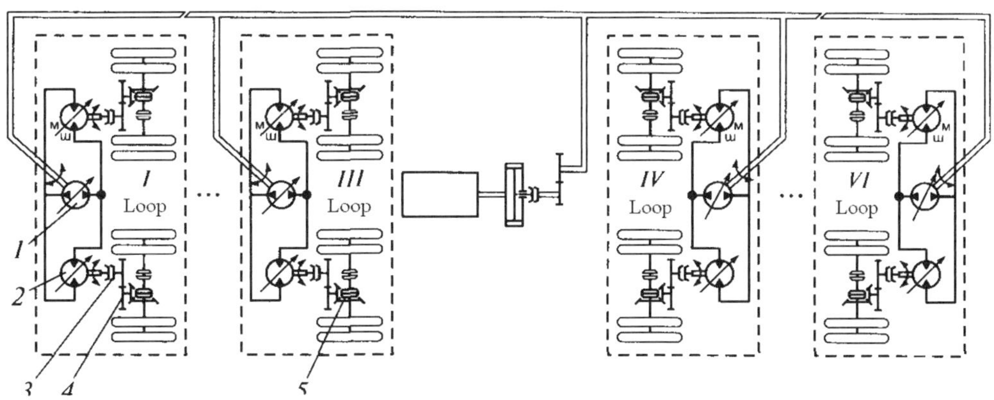

Fig. 2.7. Variant of an HST in a multi-support transporter: 1—Pump; 2—Hydraulic motor; 3—Disconnection coupling; 4—Reduction gear; 5—Differential

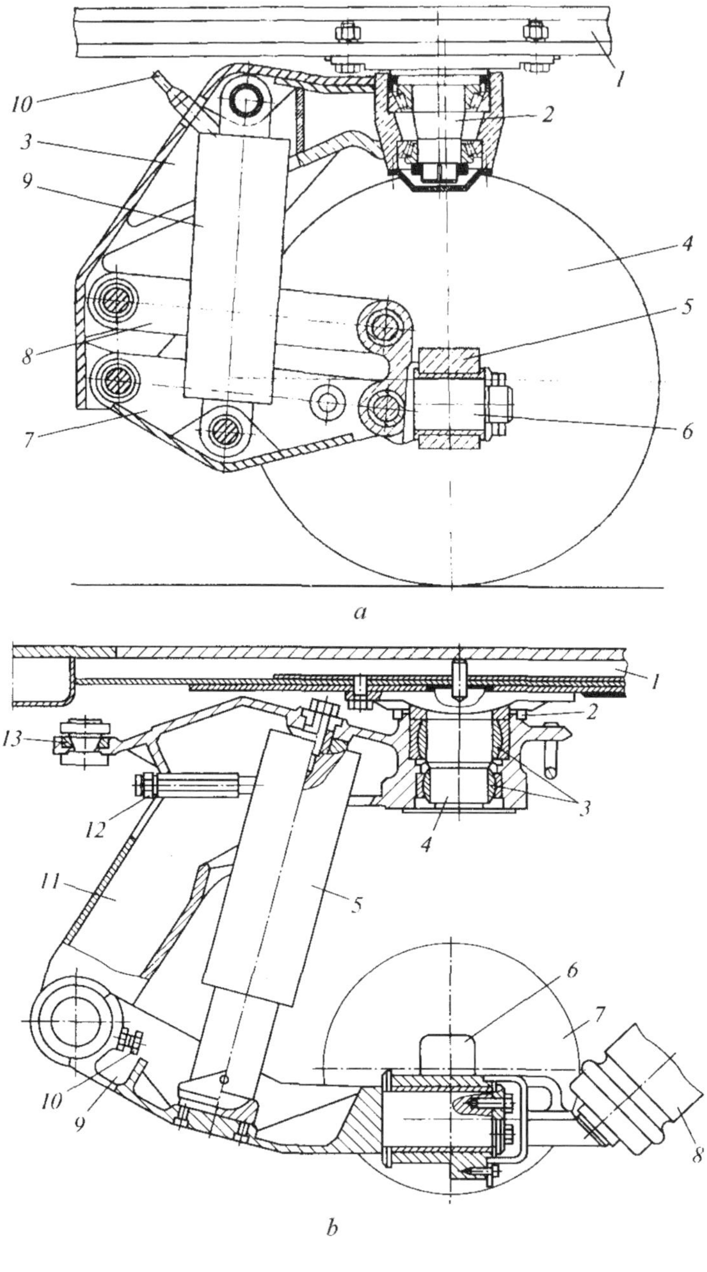

Fig. 2.8a–b. Supports of multi-axle transporters:
a—Structural layout of the Cometto: 1—Vehicle frame; 2—Suspension rotation axis; 3—Support; 4—Wheel with brakes; 5—Balancing lever; 6—Balancing lever axis pin; 7, 8—Lower and upper levers respectively; 9—Hydraulic cylinder; 10—Union.
b—Structural layout of the Nicolas: 1—Vehicle frame; 2—Thrust bearing; 3—Spherical bearing; 4—Suspension rotation axis; 5—Hydraulic cylinder; 6—Balancing lever; 7—Wheel with brakes; 8—Brake cylinder; 9—Lever; 10—Blocking bolt; 11—Support; 12—Union with reverse valve; 13—Joint to control wheel turning.

Fig. 2.8a shows that the parallelogram is formed by the lower and upper levers (7 and 8, respectively) fixed pivotally on one side of the support (3). On the other side, the balancing lever's (5) axis pin (6) is attached to the levers. This ensures—when moving on very poor roads—the horizontal position of the balancing lever axis pin, resulting in a load reduction in the suspension elements. A power hydraulic cylinder (9) is installed between the support (3) and the lower lever (7) of the suspension, which absorbs the load pressing on it from the vehicle. The load can equal 11 t at a speed of 30 km/h and the pressure in the hydraulic cylinder chamber reaches 19 MPa. The maximum

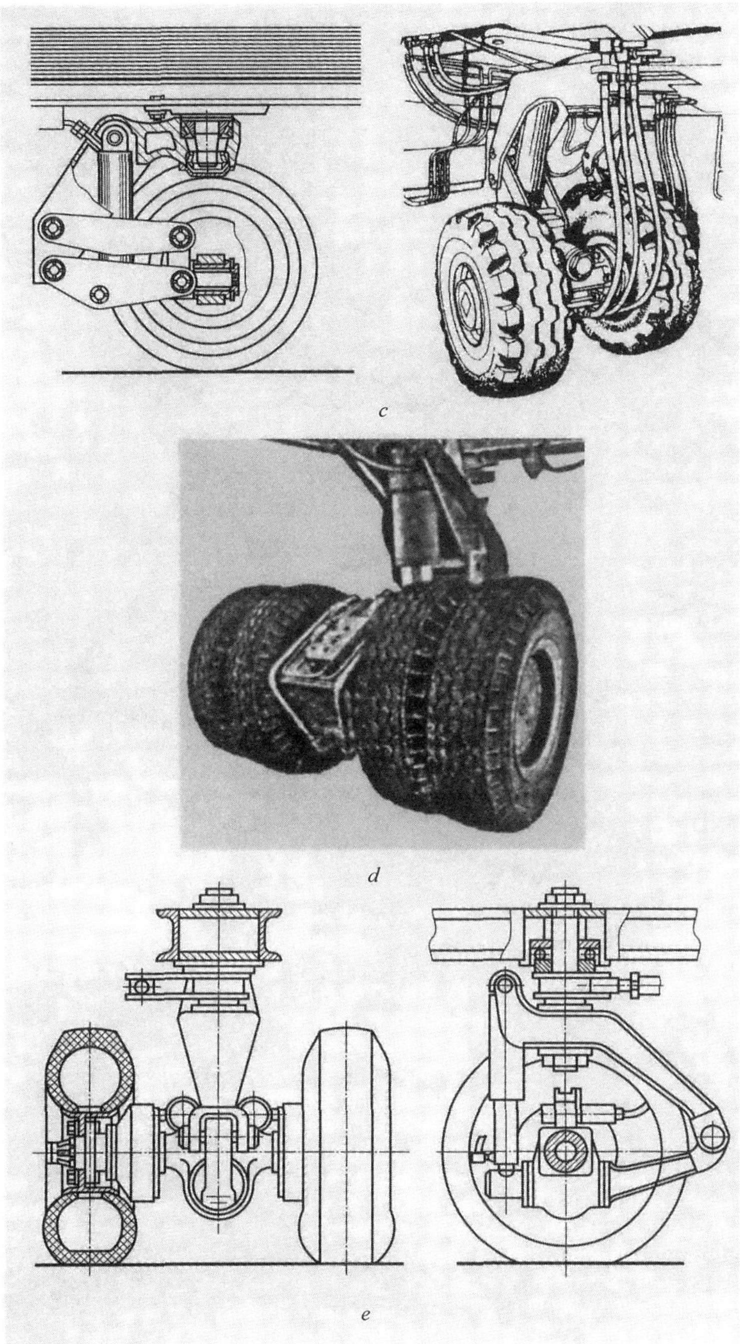

Fig. 2.8c–e. c—Layout and general view of the Cometto variant. d and e—Layouts and general views of the Nicolas variant

stroke of the suspension is 600 mm. For normal travel conditions, the suspension's hydraulic cylinder should be set in its middle position; the suspension range is ±300 mm. A deviation from this range can be admitted for a short time when moving on a flat surface.

The wheel suspension support rests on the vehicle frame by means of a bearing unit, which includes either a roller taper bearing (as used by Cometto and Sheyerle, Fig. 2.8a) or a combination of spherical and ball thrust bearings (as used by Nicolas, Fig. 2.8b). A bearing with a large-diameter support ring can be used to extend the bearing's service life when the vehicle works on poor road conditions. Goldhofer uses this technique in its vehicles intended for adverse work conditions.

Additional layouts and general views of vehicle supports used by Cometto and Nicolas are presented in Fig. 2.8c, d, and e.

Vehicles with small loading heights use the lateral parallelogram suspension with travel equal to ±180 mm, which ensures the wheels rotate around the pivot stud's vertical axis. The maximum load delivered by this suspension on the road surface is 59 kN at a speed of 20 km/h. The working pressure in suspension cylinders is 19 MPa.

In multi-support vehicles, the load on one wheel support can reach 150 kN at a speed of 5 km/h. In spite of the fact that the hydraulic system controls fluid pressure through pressure gauges in each of the three supporting points, Cometto recommends to use—when operating its vehicles—a nomogram to define the load limit on the suspension and road surface. When using a nomogram (Fig. 2.9), the suspension load can be defined if the fluid pressure (given by the pressure gauge) is known. Figure 2.9 shows the static pressure values under the three possible positions of the wheel suspension lever, where the nominal limit values can reach 22.5 MPa at the middle position and 19.0–25.5 MPa when the platform is raised and lowered. When driving on an irregular road surface, or when raising and lowering the platform, the pressure can increase up to 28 MPa for a short period of time.

Using a nomogram for practical calculations is illustrated in the following examples.

Example 1. The vehicle is loaded at the suspension's middle position and the pressure

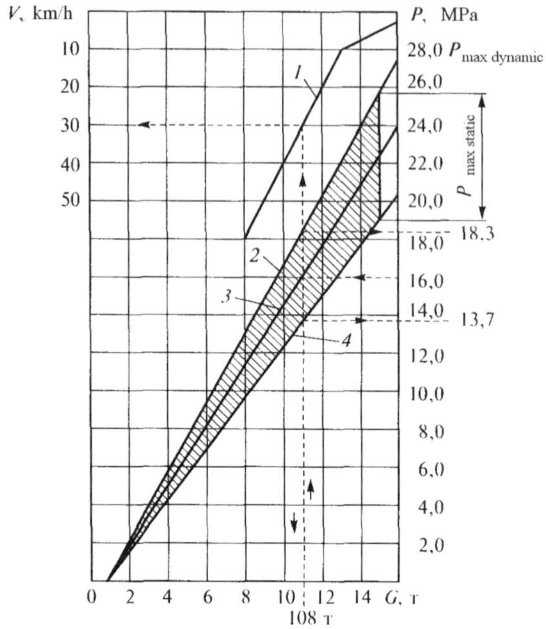

Fig. 2.9. Nomogram applied to Cometto vehicles to define the speed and pressure in the hydraulic system and the axle load: 1—Broken line for defining the vehicle speed and pressure in the wheel drive system; 2, 3, and 4—Straight lines to define the suspension pressure and the axle load under the middle, lower, and upper positions of the suspension respectively

gauge shows 16.0 MPa. In order to define the suspension load and possible travel speed, we draw a horizontal line through the P = 16.0 MPa ordinate and find where it crosses with straight line 2. From here, we draw a vertical line and find a load of G = 10.8 t. If we draw a vertical line from this point to where it crosses with broken line 1, we find a maximum travel speed of V = 30 km/h.

Example 2. Under special loading conditions, the wheel suspension load is C = 10.8 t. To check the accuracy of this value, we draw a vertical line crossing the abscissa of G = 10.9 t until it crosses straight line 4 (here, we find a pressure of 13.7 MPa) and continue until it crosses straight line 3 (here, we find a pressure of 18.3 MPa). Consequently, the static pressure, indicated by the pressure gauge, is within the limits of 13.7 to 18.3 MPa during loading. In travel conditions, the pressure gauge will show a static pressure of P = 16 MPa for the wheel suspension. Thus, overloading the vehicle is prevented.

This calculation process is easily automated.

Tires and disc wheels are important elements of the wheel suspension. All types of passive rolling stock used in and with HDWVs (trailers and semi-trailers) use mainly 7.50×15 or 8.25×15 tires with 5.00×15 disc wheels. Multi-support vehicles use 7.50×20 tires with 5.00×20 disc wheels. The number of tires on one support is generally four, thus, there are eight tires per one axle.

The wheel suspension's most loaded parts are made of special steels with the following chemical compositions (given in percentages): C = 0.25, Mn = 1.11, Si= 1.35, P = 0.024, S = 0.015, Ni = 0.24, Cr = 0.63, and Mo = 0.34.

2.4 Steering system

Multi-support vehicles with more than two axles face the problems of restricting lateral tire skidding, when moving along a curvilinear trajectory, and maintaining the required maneuverability. These problems are resolved by designing a mandatory turning of all or some wheels around the vehicle's respective united instant turning center. This turning can be executed with the help of a purely mechanical drive, or by means of hydromechanical or electronic-hydraulic system.

There are two types of mechanical drives used for controlling the wheels of the passive modules with articulation linkage.

Fig. 2.10 presents a design with the side location of longitudinal steering rods (4) fixed with the help of spherical joints to the steering levers (3) and mounted on the upper section of the wheel suspension. At turning, the control pulse is transmitted by the front rod to the lever of the front wheel suspension (5) and further—through rods (4) and levers (3)—to the remaining wheels.

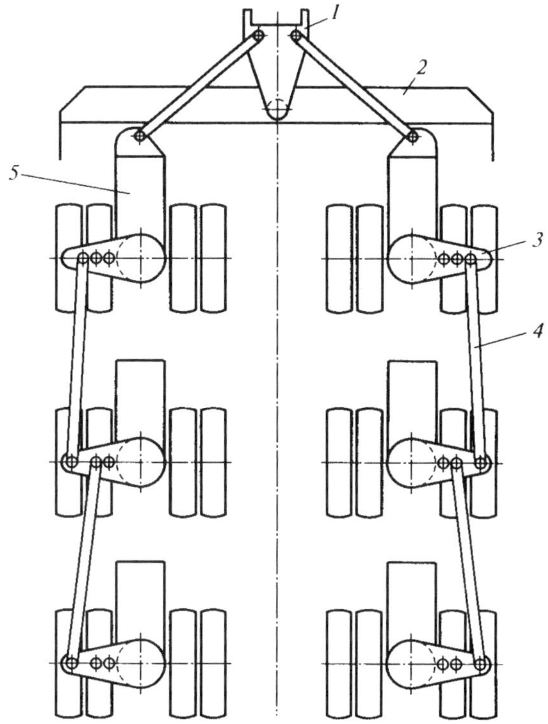

Fig. 2.10. Mechanical wheel steering system with side location of control rods: 1—Carrier; 2—Trailer frame; 3—Control levers; 4—Longitudinal rods; 5—Lever of the front wheel suspension

Fig. 2.11 presents a design with a central location of the steering longitudinal rods (3). The links (5) are mounted in the module's middle section, usually in a special niche of the frame's longitudinal power beam. In both types of mechanical drives used to turn the wheels of a trailer-type module, the steering levers have a number of coordinated holes into which the longitudinal (3) (see Fig. 2.10) or lateral (4) (see Fig. 2.11) steering rods are installed, which ensures all wheels are turned under the preset law.

The use of side location control rods allows for reducing the number of joints and rods; simplifying maintenance, installation, and dismantling of the modules when they are being re-docked; and using an ordinary box-type design, without windows and niches, of the central trailer longitudinal power beam.

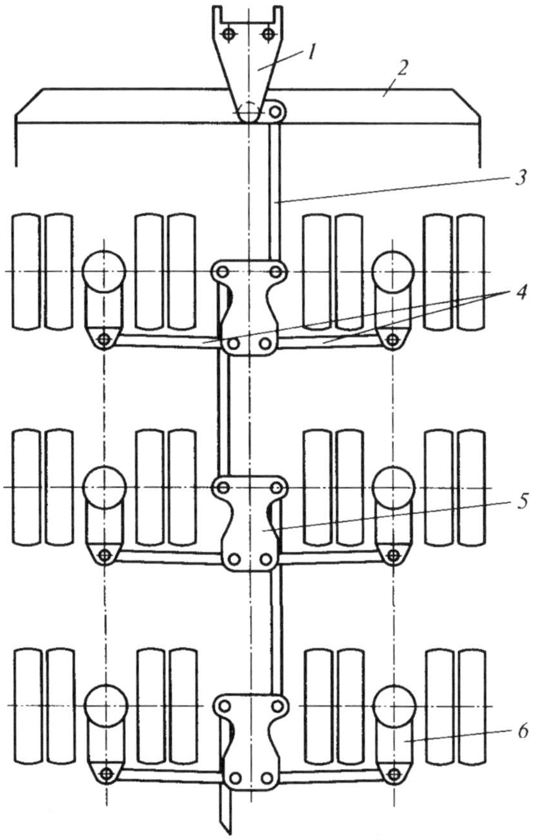

Fig. 2.11. Mechanical wheel control system with central location of control rods: 1—Carrier; 2—Frame; 3 and 4—Longitudinal and lateral control rods respectively; 5—Link; 6—Suspension with wheels

However, the loading height of the platform in Fig. 2.10 is 150 mm higher than the one in Fig. 2.11. That is why, in spite of a number of advantages, the design of Fig. 2.10 does not allow for building low-frame modules.

Hydromechanical systems are now broadly used in vehicles to control the curvilinear motion.

Fig. 2.12 shows a hydraulic system for steering the wheels of a passive module, which consists of two interlinked modular bogies. The system includes four double-action cylinders (5) interconnected by pipelines (6). The rods of the hydraulic cylinders are attached to the control levers (4); one of which is connected—via the rod (3)—to the carrier (2) with the towing beam (1).

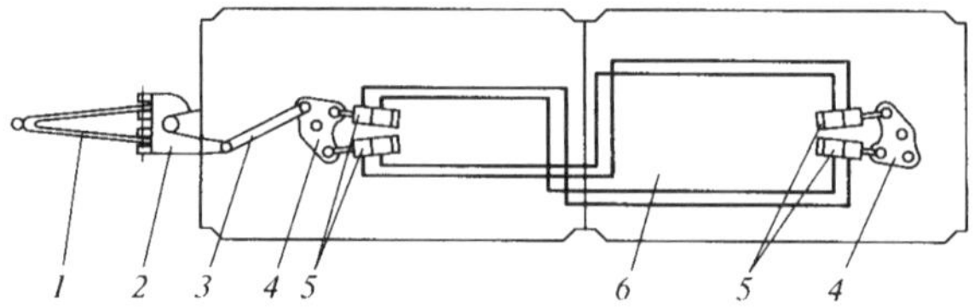

Fig. 2.12. Hydraulic steering system for the wheels of a passive module: 1—Towing beam; 2—Carrier; 3—Rod; 4—Control lever; 5—Hydraulic cylinder; 6—Pipelines

The hydraulic system for controlling the wheels also includes a manual pump, high pressure valves, and tanks with the working fluid, which are also the vessels of the wheel suspension's hydraulic system. This latter hydraulic system comprises its own high pressure valves that connect the hydraulics with the power accumulators.

All of these systems are simple in design, and, while in operation, require only external visual examination for detection of working fluid leaks.

Hydromechanical systems in their classical form cannot be used in designs of active modules, since each wheel support should be turned independently. That is why electronic-hydraulic control systems, based on onboard microcomputers, are broadly used to control the turning of the wheels. These systems ensure, with the help of sensors installed on the wheel supports' turning shafts and steering columns, a coordinated execution of the vehicle's motion geometry in accordance with the prescribed law.

A multi-support vehicle made by Cometto is representative of the hydromechanical drive used by most foreign companies to control the turning of the wheel supports. Fig. 2.13 presents the design of the mechanism used to control active module A's turning of the wheel supports. The frame (1) bears an articulation linkage, comprising of the two-arm lever (4) that rotates around the axis (5) and is connected by the rod (3) to the lever (2) that turns the wheel support.

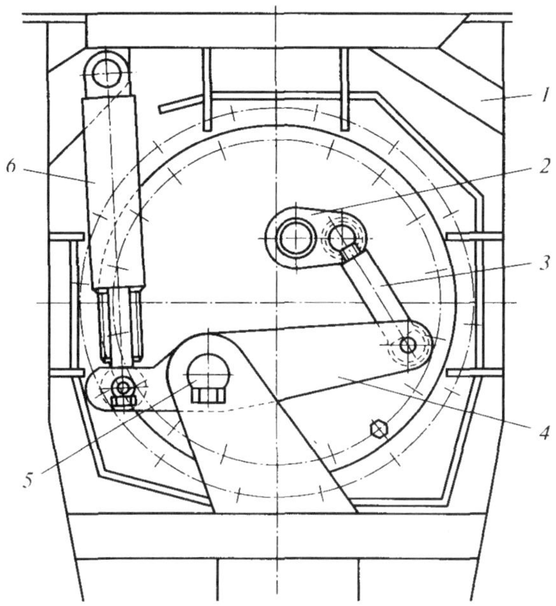

Fig. 2.13. Mechanism to control wheel turning in Cometto's multi-support active module A: 1—Frame; 2—Lever; 3—Rod; 4—Two-arm lever; 5—Lever rotation axis 4; 6—Power hydraulic cylinder

The lever's (2) rotation axis bears a sensor of the wheel turn angle's actual values. The hydraulic drive is powered by the hydraulic cylinder (6), which is controlled by a microcomputer with the help of a servo valve.

Fig. 2.14 presents the mechanism used to control passive module D's turning of the wheel supports. Since module D can be used independently, its design is somewhat different from module A. This is because it has an elongated lever (4), used for turning the wheel support (3), that is provided with a number of openings intended for the mechanical connection of the longitudinal connection rods, same as in Fig. 2.10.

The universal character of the mechanism used to control the turning of the wheel supports in module D is ensured by the use of the lever (6), which rotates on the axis (2) and is connected by the rod (5) to the turning lever (4) of the wheel support. When module D is used together with module A, the control mechanism for turning

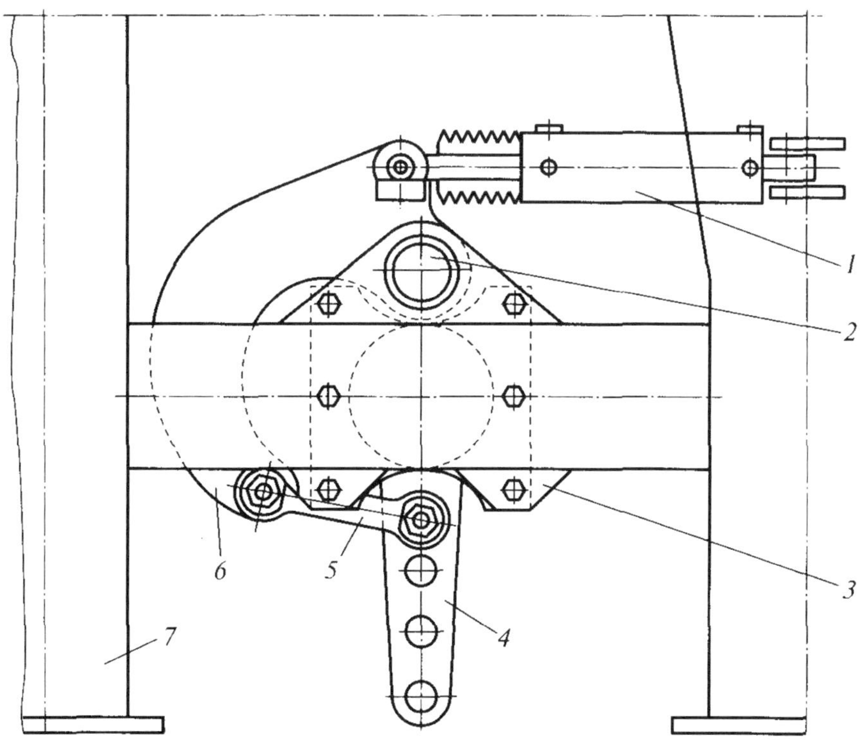

Fig. 2.14. Mechanism to control the drive for turning the wheels Cometto's passive module D: 1—Power hydraulic cylinder; 2—Pivot stud; 3—Wheel support; 4 and 6—Turning lever; 5—Rod; 7—Frame

the wheel support is driven by the hydraulic cylinder (1), mounted on the frame (7), under the same principle as in module A. The sensor of the turning angle's actual values is installed on the rotation axis of the wheel support.

Nicolas Company uses different mechanisms to control the rotation of the wheel suspension supports in its multi-support vehicles. They use a hydraulic motor, the exit shaft of which bears a sprocket that rotates another sprocket mounted on the movable section of the wheel suspension support.

Fig. 2.15 presents a flow chart showing the various drive types used to control the turning of wheel supports. Fig. 2.15II shows the design used in multi-support vehicles made by Kamag. Its main distinction from the one used by Nicolas in Fig. 2.15I is in the type of turning drive mechanism used (positions 10, 11, and 12).

In both flow-charts, the electronic-hydraulic systems are identical and based on microprocessor control. This allows for a coordinated turning of wheel suspensions under the preset programs for the normal, parallel (crabbing), and sinusoidal (Indian file) motion both of an individual module and the vehicle as a whole (Fig. 2.16).

The following parameters are the initial data needed for the calculations made by the microprocessor:

- The required turning angle (set by the steering wheel sensor);
- The number of modules in the vehicle;
- The lateral distance between the individual wheel lines of the modules; and
- The total length of the vehicle.

Each vehicle support has its own regulation circuit, which allows for measuring—with the help of the feedback sensor (12, see Fig. 2.15)—the current value of the turning angle of the support; comparing—with the help of the proportional servo valve (9)—the difference between the preset and current positions of the support; and controlling the operation of the hydraulic turning cylinders or motors.

The electronic-hydraulic control system for turning the wheel support operates in the following way. The regulated axial piston pump, driven by the diesel engine, delivers the working fluid, under the pressure corresponding to the turning angle of the steering wheel (1, see Fig. 2.15), to the servo valves (9). During the straight motion, the valves are closed. When the steering wheel (1) is rotated to either side, the built-in generator (17) supplies voltage to the inductive offset sensor (13), which affects the proportional electric magnets (14) that offset the control spools of the servo valve's (9) pass valve. The displacement value is controlled by the proportional control unit (15) installed on each support. The microprocessor (5) calculates the current turning angles of the vehicle's individual wheel supports based on the preset function and depending on the steering system programs and downloaded trigonometric function.

If the driver turns the steering wheel faster than the supports of the wheel suspensions (11) can turn, a mismatch arises. The mismatch is perceived by the microprocessor (5), which with help from the motion mode analyzer (6) and proportional deceleration unit (8), generates an additional coordinated effort on the braking device (3) in the steering shaft which is opposite of the steering wheel's rotation. Thus, the driver receives the feedback.

The vehicle modules, which follow the internal trajectory, are the setting devices for the microprocessor. The value of the remaining wheels' turning angle is calculated from the turning angle of the steering wheel. The quick response of the computer (0.1s) allows for defining the turning angle of the wheel suspension supports for each module and the vehicle as a whole. The turning angle values for all wheels are sent—through the control circuit—to the hydraulic proportional valves, which feeds the corresponding working fluid volume

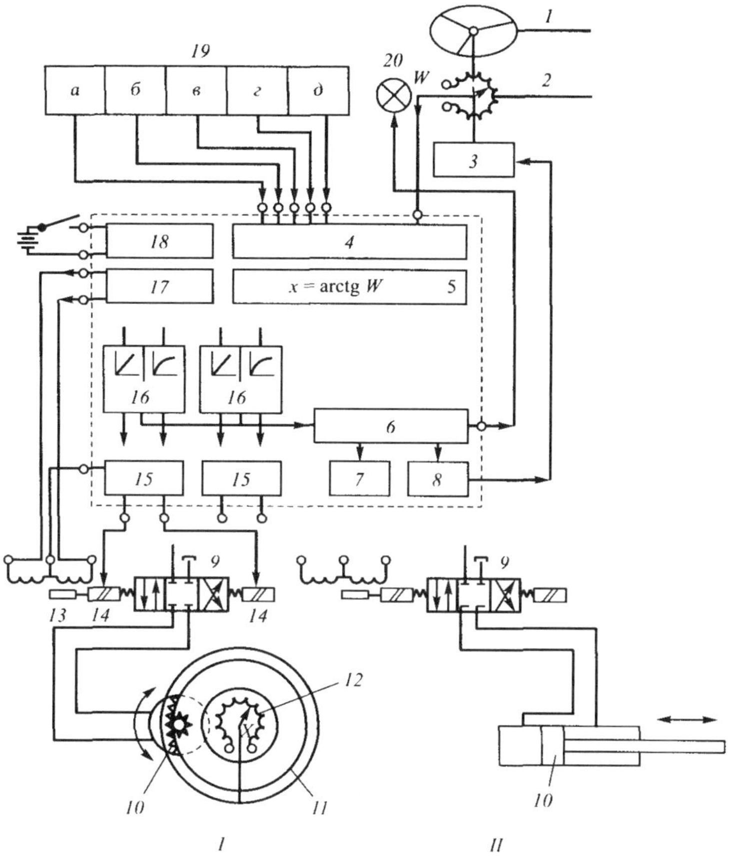

Fig. 2.15. Flow-chart of electronic-hydraulic control system of wheel turning in multi-support vehicle with different drives: I—Hydraulic motor; II—Hydraulic cylinder; 1—Steering wheel; 2—Setting potentiometer; 3—Steering shaft's braking device; 4—Unit to search logic mistakes; 5—Microprocessor; 6—Motion mode analyzer; 7—Emergency stop unit; 8—Unit of proportional deceleration of the steering shaft; 9—Proportional servo valve; 10—Executive drive of the turning mechanism; 11—Support of the wheel suspension; 12—Potentiometer of the turning angle's actual values; 13—Displacement (offset) inductive sensor; 14—Proportional electric magnets; 15—Units to control spool displacement proportionality; 16—Indicators of current and preset turning angle values; 17—AC generator; 18—Power unit; 19—Selector of motion programs (a—Normal motion; b—Crabbing motion, α = 90°; c and d—Normal and lateral diagonal motion; e—Indian file motion); 20—Excess of preset parameters signal lamp

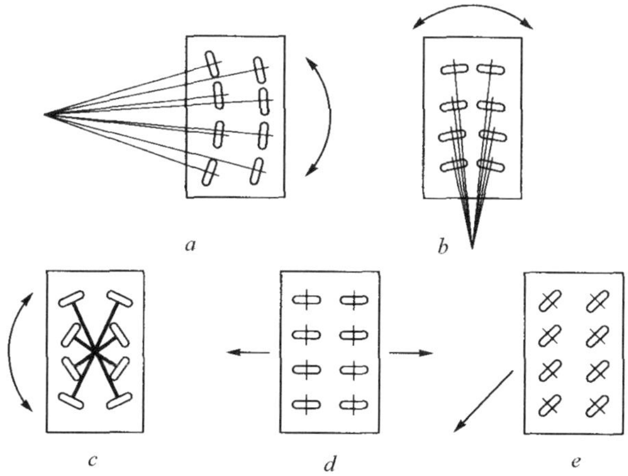

Fig. 2.16. Turning diagrams implemented by the steering control of the Cometto multi-support vehicle: a—Classical turning; b—Lateral (side) turning; c—Central turning; d and e—Crabbing motion

control valve consists of four functional parts: a servo valve with control spools (9), two proportional electric magnets (14), and an inductive sensor (13, see Fig. 2.15).

In principle, the controlling servo valve is designed as a hydraulic valve with electromagnetic control; however, it uses proportional electric magnets. The current acting in the magnet windings displaces the spool to the right or left and compresses the springs until the spring compression and magnetic forces becomes equal. Proportional electric magnets are not completely precise; therefore, the spool displacement should be controlled through an inductive sensor. The proportional electric magnets voltage is controlled by the electronics as a function of the spool displacement. As a result, there is a tough dependence between the spool displacement and the working fluid volume passing through the valve.

In case of malfunctions, the control valve can be switched from automatic to emergency manual control.

The electronic control system is fed from the power unit (18, see Fig. 2.15), which converts the non-stable voltage of the 12 V (Kamag) or 24 V (Nicolas) battery into a strictly defined voltage of several ratings. For example, the internal voltage of the Kamag unit is ±15 V, the potentiometers—of both actual and preset values—are powered by ±10 V, and the microprocessor and its periphery are powered by ±5 V.

The above brief design analysis of multi-support transport vehicles shows that the road pavement can be saved from destruction by a uniform distribution of supports. The vehicle should fit well into the overall road dimensions and curvature by using all steered wheels and ensuring a turning angle of up to 90°.

The ease and comfort of placing, loading, and unloading cargo is achieved by using an oversized cargo platform and a variable loading height achieved by the support hydraulic system.

However, multi-support vehicles have an essential drawback: high loads on wheel supports and the pressure in the wheels' contact spots make it impossible to use these vehicles on road and land surfaces with low carrying capacities. However, the demand for HDWVs with high passing abilities on bridges and off-road conditions is very high. To solve this problem, it is possible to partially unload the mover using an air cushion.

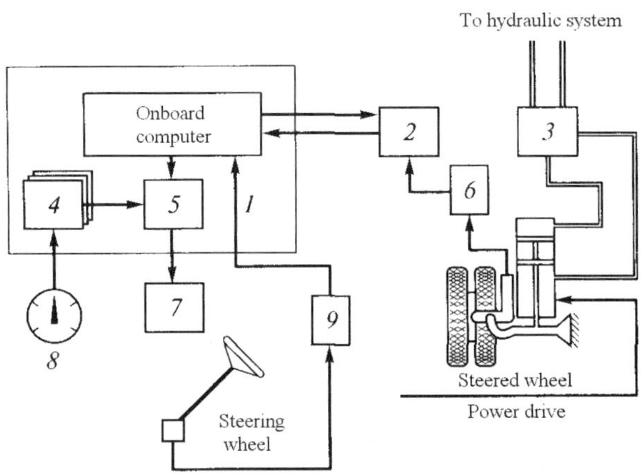

Fig. 2.17. Steering system used in Cometto vehicles: 1—Onboard computing complex; 2—Feedback Amplifier; 3—Controlling spool; 4—Software package; 5—Automatic steering wheel loader setting; 6—Feedback sensor; 7—Automatic steering wheel loader; 8—Turning diagram selector; 9—Steering wheel turning angle sensor

Chapter 3
Heavy- and super-heavy-duty dumpers

3.1 Overview

The open method of extracting fossil minerals has led to the global progress of the mining industry. This progress was also facilitated, beginning in the 1960s, by the development of the extraction and transportation industries which provided mining operations with efficient machinery, excavators, and special heavy-duty quarry dumpers.

The first attempt of open field development in the USSR dates back to the early 1950s to the Zhirnov Quarry, which was used to extract building materials for the Volga-Don Canal. The quarry used MAZ-525 dumpers with a 25 t load capacity made by Minsk Automobile Works (MAZ) in small batches up to 1958. In 1959, Belarusian Automobile Works (BelAZ) began an assembly-flow production of dumpers. Thus, the USSR launched a new highly promising area of motor-building—heavy-duty dumpers for the extracting industries and large construction sites. However, MAZ-525 dumpers did not account for the specificities of transport operations in quarries. They were traditionally designed (e.g., mounting the cabin behind the engine) and similar to small load capacity dumpers which resulted in a number of drawbacks: high weight, large turning radius (see Table I.3), low speeds, and non-rational characteristics of the suspension.

The designers' lack of awareness about the specificities of mining enterprises, geological conditions of quarries, and parameters of extracting equipment resulted in the fear of exceeding the axial load adopted for the MAZ-525. Thus, the next dumper was designed with a 40 t load capacity, three axles, and a 6×4 wheel formula. It was only after several years and extensive research that designers returned to the two-axle model.

By the mid-1960s, the design principles of a new family of quarry dumpers were clearly and expressly formulated.

The basic models, launched in the mid-1960s, were the BelAZ-540 with a 27 t load capacity and the BelAZ-548 with a 40 t load capacity. They radically changed the efficiency of transport operations in quarries; in particular, they doubled performance (as calculated per 1 t of the vehicle's load capacity). The high level of standardization contributed to a fast start in mass production, in spite of the use of a number of unique units, such as the hydromechanical transmission, pneumohydraulic suspensions of the front and rear axles, high-pressure hydraulic drives, etc., and promoted their use at hundreds of mass mining and construction operations in the USSR and 40 foreign countries [9, 10, and 11].

Table I.3 presents a number of specifications of the BelAZ-540, BelAZ-548, and the earlier MAZ-525 and MAZ-530 dumper models, while Fig. 3.1 shows a later modification, the BelAZ-7547.

The family is also comprised of BelAZ-7510 coal-carriers (with a 40 t load capacity), a single-axle tow car for operation with a 15–18 m^3 capacity scraper, aerodrome tow cars, and many others.

Fig. 3.1. BelAZ-7547 Dumper with a load capacity of 45 tons

All models had the "cabin next to the engine" layout, which became traditional for quarry dumpers. The hydromechanical transmission (HMT), used instead of a mechanical one, allowed for increasing the average travel speed at climbing by 15–20%, almost halving the dynamic loads in the transmission elements, and made steering easier on the poor roads inside quarries. The introduction of a pneumohydraulic suspension made it possible to achieve higher travel speeds in all operation modes and to create necessary driver comfort.

BelAZ later switched to manufacturing the BelAZ-7540 dumper with a 30 t load capacity and the BelAZ-7548 dumper with a 40–45 t load capacity; each were equipped with a 308 and 477 kW, respectively, YaMZ-840 engine (made in Yaroslavl, Russia) and a five-step hydraulic transmission.

New BelAZ dumpers with a 30–55 t load capacity and specialized vehicles (loading machines, slag cars, aerodrome tow cars, etc.) are equipped with HMTs made by BelAZ. The HMTs were made by the axle diagram which is known to have a number of advantages over the planetary diagram when used in short-base vehicles. These advantages include the ability to create a series of transmission ratios and correlations; an ability to reduce the HMT length and the distance between the input and output flanges since friction clutches can be mounted on parallel shafts; the reduction of power losses in open friction clutches; the reduction of power flow break time and dynamic loads during gear change by reducing the moment of inertia of rotating masses; and a possibility to standardize the friction couplings.

The first BelAZ-made HMT (3+1 type) was developed in the early 1960s. It had a particularly simple design, easy maintainability, and was cheap to manufacture. It was available with different modifications of input and output power takeoff gearboxes and was mounted in irrigation machines, heavy-duty vehicles, slag cars, aerodrome tow cars, etc.

The production of the second HMT (5+2 type) was launched in the 1980s and replaced the three-speed transmissions on quarry dumpers. This resulted in reducing the fuel consumption by 10–15% per 1 t of transported load and a 20–30% increase in the dumper's overall performance.

This was achieved due to the rational choice of transmission gear ratios, an increase in the number of gears, and introduction of automatic hydraulic torque converter (transformer) (HTC) blocking.

The new HMT has a transmission ratios power range equal to 6.14, whereas the three-speed was 3.5. It is equipped with a complex, four-wheel, non-transparent HTC (Fig. 3.2) with a centripetal type turbine. The HTC blocking and unblocking is performed automatically at all gears depending on the turbine shaft's rotation frequency. The HTC pump wheel (1), made of an aluminum alloy casting, is mounted on the ball radial-thrust bearing (2), which fixes it against axial displacements. It is supported by the roller bearing (3) on the left side. The pump wheel is bolted to the jacket (4) which is rotated by the engine.

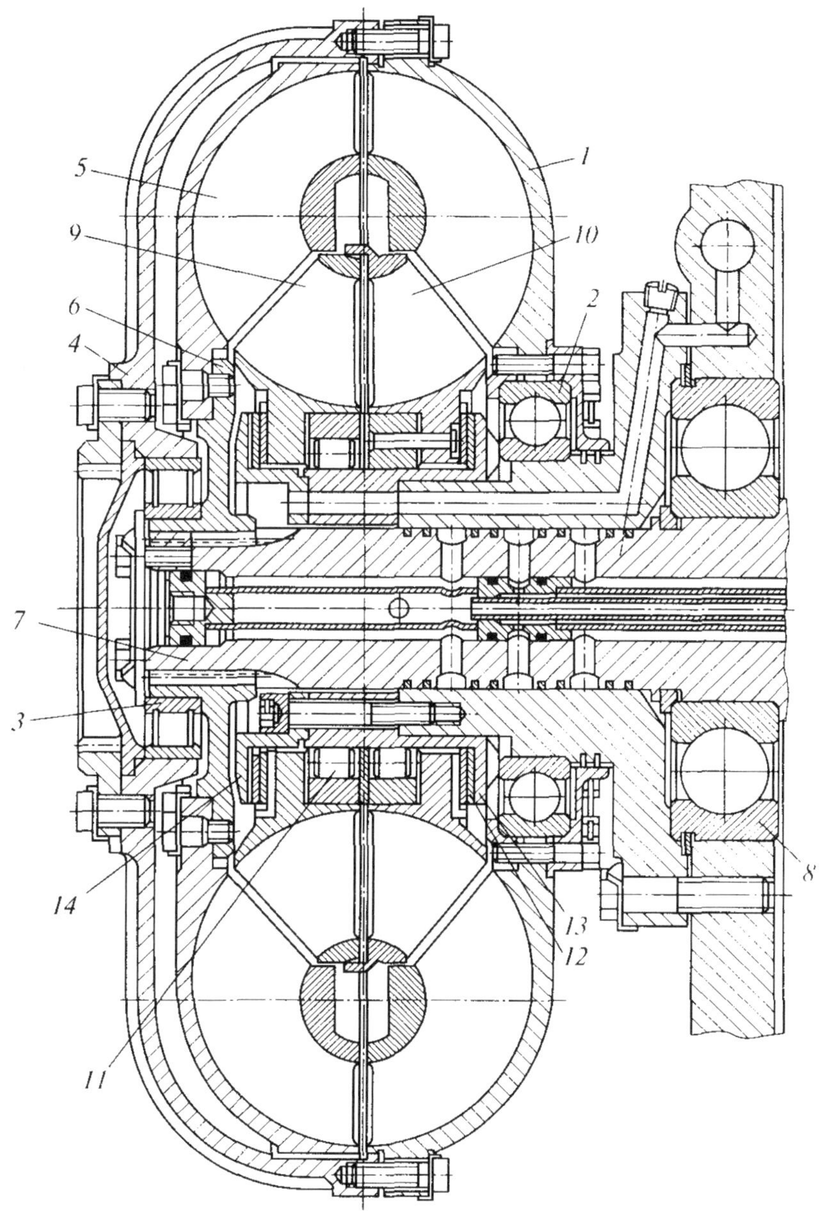

Fig. 3.2. Standardized HTC LG-470: 1—Pump wheel; 2, 8—Ball radial-thrust bearings; 3—Roller bearing; 4—Jacket; 5, 7—Turbine wheels; 6—Turbine wheel hub; 9, 10—Reactor wheels; 11, 13—Freewheel clutches; 12—Bronze floating washers; 14—Thrust clutch

The HTC turbine wheel (5), cast from an aluminum alloy, is bolted to the turbine wheel hub (6). The hub is rigidly mounted by splines on the turbine wheel (7) shaft. The shaft rests on the ball radial-thrust bearing (8), which fixes it against axial forces, and on the roller bearing (3).

The reactor wheels (9 and 10), also aluminum alloy castings, are mounted on the roller freewheel clutches (11). Axial forces are transmitted—through bronze floating washers (12)—to the freewheel clutch (13) hub (on the right) and the thrust clutch (14) (on the left). To reduce wear, the area of the floating washers' bearing surfaces was significantly increased compared to the earlier design. There is also a HTC version with no torque conversion (in the hydraulic clutch mode), when the turbine and pump wheels are blocked.

In total, three HTC modifications are currently made for 260–320, 370–450, and 510–600 kW engines where the crankshaft's nominal rotation rate is 2100 rpm.

The HMT mechanical gearbox has three shafts (primary, range, and reverse) with electrohydraulic gear switching and two friction clutches. All gear pinions are straight-toothed and in constant mesh. One pinion of each pair is rigidly geared to the shaft, while the other is mounted on a bearing on a parallel shaft. The reverse shaft has a rigid kinematical connection to the primary shaft through the constant-mesh pinions; the range and output shafts have no rigid kinematical connection to the primary and reverse shafts.

HMTs designed to work with 500 kW (680 hp) engines have each shaft mounted on two bearings, while the shafts for 515–600 kW (700–815 hp) engines have an additional intermediate support.

Gear switching is electrohydraulic, both manual and automatic, and is carried out under load.

The friction clutches have a fast and reliable engagement and a guaranteed disengagement at high rotation frequencies. This is ensured by the piston's stepwise design, the smooth engagement mechanism, and by using a special device to compensate for the fluid pressure's centrifugal force which affects the piston and prevents its return to the original position.

A standardized friction clutch with 14 friction pairs is used in the three-speed and five-speed HMT for a transmitted power of up to 500 kW (680 hp). The company also manufactures a 6+1 HMT, designed for 515–600 kW (700–815 hp) engines, with 18 friction pairs, better (down to 10 μm) oil filtration, the presence of intermediate shaft supports, and reinforced bearings and pinions.

The 6+1 design uses a hydromechanical blade type brake-retarder. It is mounted on the driving shaft of the gearbox and acts as an auxiliary braking system intended to maintain a constant speed when moving downhill. The retarder was also modernized: the version mounted on dumpers with a 45–55 t load capacity has about a15% higher deceleration efficiency and has become the main braking system.

It is also possible to use a built-in, multi-disc, oil-cooled, friction brake with hydraulic drive instead of the blade-type brake retarder.

BelAZ also makes several other 6+2 type HMT models with a forward-directed power output. However, these are used in MoAZ-7405 underground trains and not dumpers. There is also an HMT model with two-sided power output used on JSC Amkodor loaders with an up to 10 t load capacity and on the 43DM model autograder.

All new quarry dumpers with a 45–55 t load capacity and some modifications with 30 and 42 t load capacities are equipped with an automated HMT electronic control system that increases performance efficiency and ensures the optimal modes of switching gears. This frees the driver from manual switching and protects the HMT from erroneous control actions. For example, it prevents the engine from starting when in gear, switching out of first gear when the brake-retarder and working braking system are engaged, an attempt to engage any gear except for first and reverse in a cold HMT, and allows a change in motion direction only when the HMT's output shaft's rotation rate exceeds 80 rpm. The electronic control system also engages the brake-retarder when the engine's crankshaft's rotation rate exceeds the permitted level (for example, at downhill motion) and also controls the electromagnet circuits.

Possible malfunctions or erroneous driver actions are displayed on the information panel mounted in the cabin.

In the late 1970s, dumpers with 27 and 40 t load capacities in the largest quarries were replaced with more powerful models. This was caused by a rapid increase in transportation volumes in quarries, growth of enterprises' processing capacity, significant changes in the technical conditions of mining, deepening of the quarries to 250–350 m, and an increase in the average transportation distance to up to 3–4 km. In the 1980s, the old models were replaced (in medium and large quarries) by dumpers of especially high load capacities (75–180 t). These dumpers, designed by BelAZ, include three basic models with a 4×2 wheel formula and load capacities of 75, 110, and 180 t. All are equipped with traction electric drives and motors-in-wheel. The BelAZ-549 dumper with a 75 t load capacity and the BelAZ-7519 (Fig. 3.3 shows the BelAZ-7548 modification) with a 110 t load capacity were launched in 1977 and 1981, respectively. The first batch of BelAZ-7525 (Fig. 3.4) with a 180 t load capacity was manufactured in 1983. In 1993, BelAZ released the BelAZ-7550 dumper with a 280 t load capacity. It was designed with the motors-in-wheel of its predecessor, the BelAZ-75214 with a 180 t load capacity; however, it contained a number of entirely new design solutions. For example, it has a 4×4 wheel formula instead of the classic 4×2; an articulated frame; a laterally mounted power plant; the suspension's pneumohydraulic cylinders are connected in pairs by the lateral pneumatic balancing device; and the vertical articulated joint of the front and rear semi-frames is based on two spherical joints, one of which is height-regulated. The use of two driving axles with eight tires has allowed for creating a load capacity of 280 t, while a powerful (2355 kW or 3200 hp) diesel engine provides high traction-dynamic power.

In 1995, BelAZ expanded its production line to include the BelAZ-7555 dumper with a 55 t load capacity (Fig. 3.5) and the BelAZ-75303 dumper with a 200 t load capacity and an AC-DC transmission (Fig. 3.6).

Fig. 3.3. BelAZ-7548 dumper with a 42 t load capacity

Fig. 3.4. BelAZ-75215 dumper with a 180 t load capacity

Fig. 3.5. BelAZ-7555 dumper with a 55 t load capacity

Fig. 3.6. BelAZ-75303 dumper with a 200 t load capacity

The main specifications of very high load capacity dumpers are presented in Tables 3.1 and 3.2.

3.2 Transmission

The BelAZ-549, BelAZ-7519, and BelAZ-7521 quarry dumpers are equipped with six- and eight-cylinder DM21A model diesel engines (773 and 957 kW, respectively). Dumpers with 75 and 110 t load capacities are equipped with 810–880 kW engines manufactured in the Czech Republic (Pielstick 8RACh-185). The BelAZ-7521 model is equipped with a 9-26DG locomotive diesel-generator unit with a 1700 kW 124H26/26 engine. The traction generators are 600–1350 kW and the traction electric motors are 560 kW.

The engine and traction generator are mounted with rubber shock absorbers on a common subframe and form a single power plant (diesel-generator). The shock absorbers are inclined when mounted to reduce engine vibrations and increase reliability. The engine is equipped with a starting PZhD-600 preheater and a fuel heater which are connected to the cooling system.

The transmission is electrical and DC operated. It consists of a GPA-600 traction generator, motors-in-wheel, and regulation and control equipment (housed in a heat-insulated cabinet mounted on the upper platform of the hood). The transmission allows for automatically varying the traction

Table 3.1 Main specifications of heavy-duty dumpers

Parameter	BelAZ-549	BelAZ-7519	BelAZ-7521
Wheel formula	4×2	4×2	4×2
Load capacity, t	75	110	180 (170)
Road weight, t	66	85	145 (155)
Wheelbase, mm	4450	5300	6650 (6700)
Overall dimensions, mm:			
length	10,250	11,250	13,500 (13,815)
width	5360	6100	7600 (7800)
height	4750	5000	6930 (6450)
Minimum turning radius of the front wheel, m	11.5	12.0	15.0
Nominal power, kW	773	957	1700 (1700)
Specific power, kW/t	5.5	5.0	5.2
Maximum speed, km/h	50	50	50
Electrical transmission type	DC	DC	With AC generator
Suspension	Pneumohydraulic	Pneumohydraulic	Pneumohydraulic
Tires	27.00–49.00	33.00–51.00	40.00–57.00

Note. The data for the modification with a 9-26DG diesel-generator unit are given in parenthesis.

Table 3.2 Dynamics of heavy-duty dumper specifications

Parameter	BelAZ model						
	7522	7523	7526	7527	7909	75221	75501
Basic model	BelAZ-540	BelAZ-548	BelAZ-7510	BelAZ-7525	BelAZ-549	—	—
Start of manufacture	1965	1967	1972	1972	1976	1982	1993
Load capacity, t	30	42	30	42	75	180	280
Road weight, t	21.5	29.5	21.4	29.3	67.0	145	200
Body capacity, m^3	15.0	21.0	19.0	27.3	35.0	70.0	110
Overall dimensions, mm:							
length	7250	8120	7435	8395	10,250	13,500	15,250
width	3480	3787	3480	3787	5300	6300	6900
height	3480	3910	3620	4135	4750	7700	6750
Wheelbase, mm	3550	4200	3550	4200	4450	6650	6650
Wheel span, mm:							
front wheels	2800	2800	2800	2800	4100	6000	6900
rear wheels	2400	2540	2400	2540	3730	5200	6900
Engine power, hp	360	500	360	500	1050	2300	3150
Transmission type	Mechanical				Electrical		
Maximum speed, km/h	55	55	55	55	60	60	60
Fuel consumption, l/100 km	100	138	100	138	235	420	600

force value on the driving wheels depending on road conditions and is fully compliant with all quarry dumper requirements. A peculiarity of electric transmissions is the stepless regulation of the traction electric motors' excitation when they work in the electrodynamic braking mode. The traction and braking characteristics of the BelAZ-7519 transmission are presented in Fig. 3.7. The figure shows the electrodynamic braking (P_B) dependencies at both full and unloaded conditions. The curve *D* shows the dynamic factor versus the travel speed.

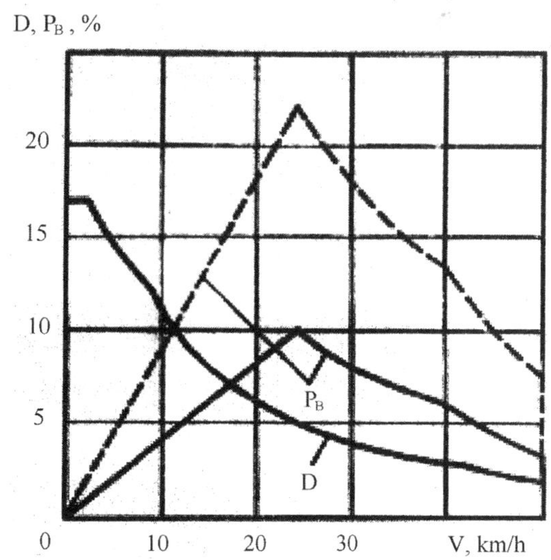

Fig. 3.7. Traction and braking characteristics of the BelAZ-7519 transmission: - - -—Empty vehicle; -----—Loaded vehicle

The electric motor-in-wheel (Author's Certificate of the USSR No. 185216, B60P1/02) comprises of a 360 kW traction DK-722 electric motor, a two-line differential gearbox with a total ratio of 31.63, parking and wheel braking mechanisms, and wheel hubs. The design of the gearbox as a part of the motor-in-wheel is shown in Fig. 3.8.

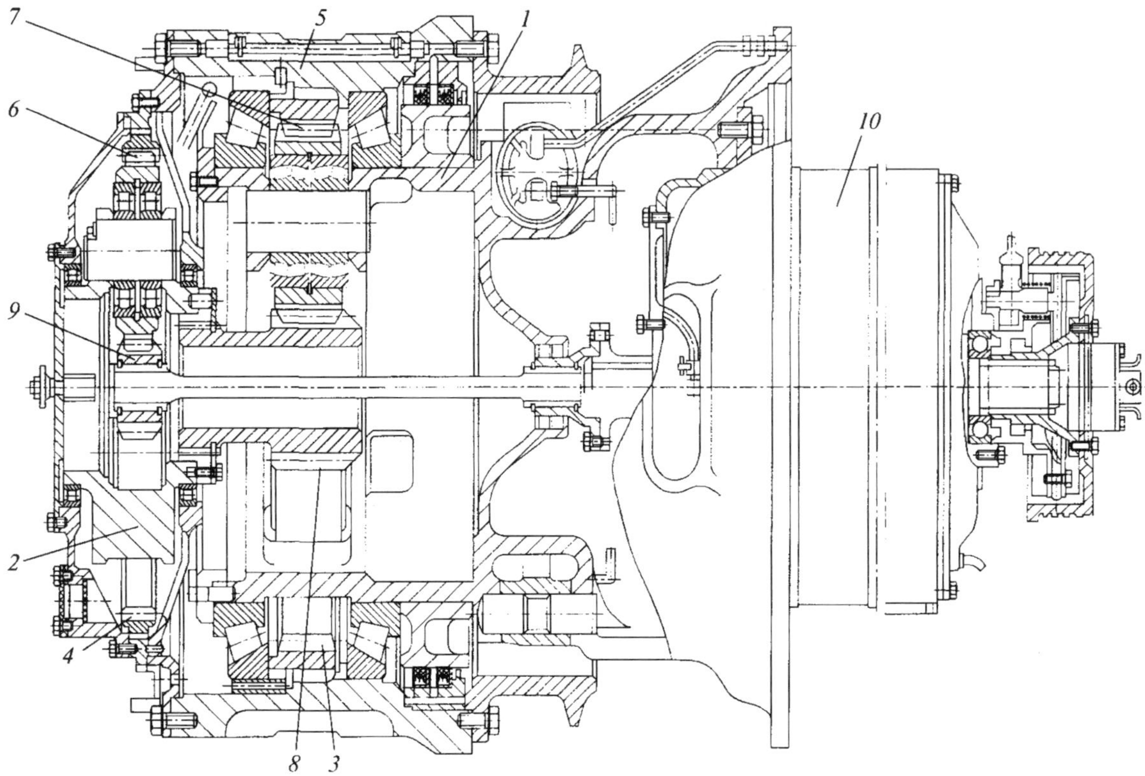

Fig. 3.8. BelAZ-7519 electric motor-in-wheel: 1, 2—Carrier of the first and second steps respectively; 3, 4—Crown of the first and second steps respectively; 5—Gearbox body of the motor-in-wheel; 6—Satellite of the high-speed step; 7—Satellite of the low-speed step; 8, 9—Sun pinion of the first and second steps respectively; 10—Electric motor

The torque to the driving (sun) pinion of the gearbox's first line is transmitted from the traction electric motor through the torsion shaft. The carrier, placed in the lid on ball bearings, has three satellites of the first line mounted on axles; each rests on two cylindrical roller bearings. The crown pinion is connected to the lid by a splined bushing. Thus, torque from the first line's sun pinion is transmitted through the satellite gears and crown pinion to the rear wheel hub. The module of the first-line pinions is 7 mm and the transmission ratio is 5.83.

The internal flange spline opening of the gearbox's first-line carrier bears the gearbox's second line sun pinion (9). It meshes three second line satellite gears mounted on axles inside the body (1). These satellite gears are meshing the crown pinion, which is connected to the wheel hub (6) through a splined bushing. The module of the second line pinions is 10 mm and the transmission ratio is 3.77.

The wheel hub (6) rests on two tapered roller bearings and the brake drum (2) is attached to its internal side. The gearbox body is mounted on the rear axle beam.

The fixed connections of the gearbox parts are sealed with rubber gaskets, the torsion shaft flange is fixed with two sleeve gaskets, and the wheel hub is fixed with two oversized sleeve gaskets (working diameter of 920 mm). To prevent dirt from penetrating the sleeve gaskets (5), the hub cap has a dirt-protection seal (4). To extend the service life of the hub seal, the external surface of its stopper ring bears a stuck and screwed heat-treated under-sleeve tape.

The first-line pinions and bearings are lubricated by spraying oil and oil catchers that capture the lubricant from the peripheral oil ring formed when the hub is rotated. The second-line pinions and bearings are lubricated by spraying oil when the gearbox elements are rotated.

The differential unit is one of the main components of a mechanical transmission. It redistributes the angular velocities of the wheels depending on their rolling or turning radii so that the half-sum is always equal to the angular velocity of the cardan shaft. Heavy-duty wheeled vehicles with electric transmissions are made without the cardan shaft and mechanical differential since the driving wheels are rotated by the electric motors built into the wheels (motors-in-wheel). The motors of motors-in-wheel are connected in parallel to the common generator. Angular velocity redistributions at a curvilinear motion were a problem in first generation BelAZ dumpers with electric transmissions. This problem was solved by using the electric differential design (ED), which is explained as follows: When moving along a straight section, the wheels' angular velocities differ slightly. However, at turning, the ratio of angular velocities of the left- and right-hand wheels is:

$$\frac{\omega_1}{\omega_2} = \frac{R}{R-A},$$

where: ω_1 and ω_2 are the angular velocities of the internal and external wheels respectively, R is the turning radius, and A is the distance between the axles.

This ratio is valid when the slipping between the wheel and the ground can be neglected, and the wheels have the same rolling radius. For the BelAZ-549 dumper, the minimum turning radius along the front external wheel axis is 10 m and the distance between axles is 3.9 m. Thus, the ratio of angular velocities ω_1/ω_2 will be 1.64.

Fig. 3.9 presents the relative angular velocity versus the relative load moment for the DK-717 electric motor mounted in the BelAZ dumpers' motors-in-wheel.

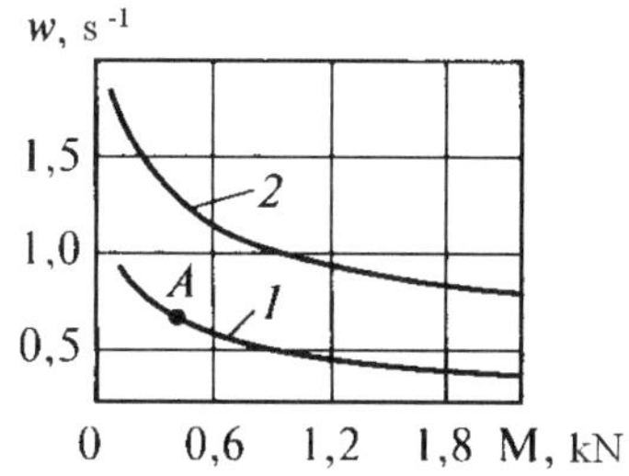

Fig. 3.9. Angular velocity ω versus load moment M of the DK-717 electric motor: 1—Motor armature voltage 300 V; 2—Motor armature voltage 550 V; point A is the engine temperature (155°C), the resistance of the armature circuit is 0.1 Ohm, and the load of the external wheel is 0.4 of the nominal moment

When braking the internal wheel, which is on the smaller turning radius, to the speed at which the above ratio is observed, the electric motor torque will increase 4.5-fold compared to that of the external wheel. In this electric drive design, the internal wheel's excess electric motor torque is hampering turning and causing premature wear of tire treads and elements of the kinematical chain.

In a well-known design of a two-motor electric drive, the armature circuits of the electric motors are not connected to the common shaft, but are instead serially connected and linked to the common generator. This means that the generator current can be freely redistributed between the electric motors depending on their load functions. Accordingly, the angular velocities of the electric motor shafts will change. This is known as the electric differential (ED) design, since its mechanical characteristics are equivalent to those with a mechanical differential. The ED design was implemented beginning with the BelAZ-549 dumper.

A similar design is also used in foreign-made quarry dumpers. The flow-chart of the ED Statex, used in dumpers made by Euclid (USA) with a 170 t load capacity, is presented in Fig. 3.10.

The AC-DC electric drive is built with a synchronous generator (SG), which feeds the motors-in-wheel of the independent excitation through a three-phase rectifier bridge (RB). The excitation circuits of traction electric motors are connected to the other SG winding through a controlled rectifier U_{rec}.

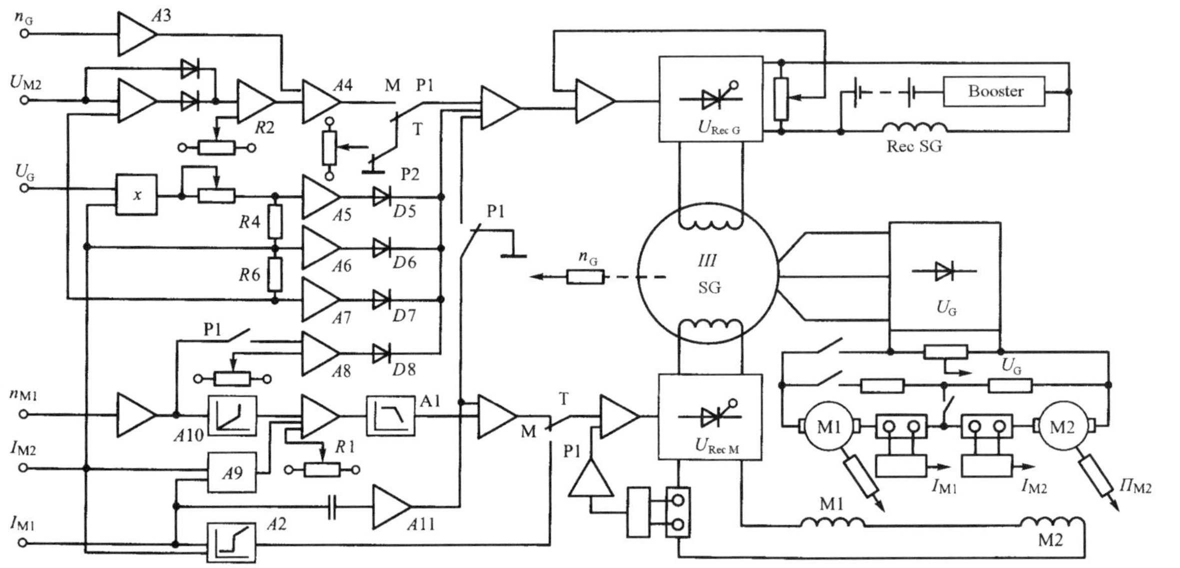

Fig. 3.10. Flow-chart of the ED Statex system of traction engines made by Euclid

The SG poles are connected during start-up by a boosting device to the onboard circuit's accumulator battery. A subsequent switchover of the synchronous generator's excitation winding in the current function to SG winding III occurs through the controlled rectifier of the generator's excitation U_{erec}. The U_{rec} circuit uses the current negative feedback, while the U_{erecd} circuit uses the excitation voltage U_{recsg} feedback. In the motion mode, the P1 relay is on. The feedbacks are as follows: the armature current I_{M2} through the A6 amplifier; the generator voltage U_g through the A7 amplifier by the rotation rate of the electric motor ω_{M1} through the A8 amplifier; and the load powers through A5 form a signal on the A10 amplifier. The diodes (D5–D8) perform a logical operation (OR).

In the beginning of the motion mode, the current in the armatures is 2500 A, the generator's voltage is relatively low, the power is low, and the signal at the A10 amplifier is controlled by the A6 amplifier. When the travel speed increases, the bottom corner of the dependence between the voltage and generator current is reached; as a result, the power limiter with the A5 amplifier begins working. At a voltage of 1600 V, the A7 amplifier is triggered and limits the voltage on the SG; the restriction of the generator power by the voltage is ensured through the R6 resistor and by the armature current through R4. Depending on the speed, the generator's power is limited by the A8 amplifier, which is tuned to a speed 3.2 km/h below the triggering point of the protection relay from the maximum allowed speed. The reference load signal, in the function of the diesel rotation rate, is formed by the A4 amplifier. The A3 amplifier forms a reference signal curve with three sections of different slope. The negative signal of the voltage difference, ΔU, between U_g and U_{M2} (the voltage difference of electric motors armatures) is summed up at A4 with the positive signal A3 and at a significant value of ΔU; the generator voltage is limited to 800 V. The reference engine excitation signal is generated by the A2 block. In the braking mode (T), the R2 relay is enabled. The amplitude of the A1 block's reference braking signal is defined by the position of the R1 brake pedal. In high current areas of the first or second traction motors (M1 or M2), selected by the logic operation (OR) of the A9 block, a current cutoff is triggered under the sharp-falling characteristic. The cutoff starting point displaces along the axis of currents, depending on the velocity of the electric motor.

The A1 amplifier is fully locked when the total signal of the vehicle speed and engine armature current exceeds the signal level of the R1 pedal, which is supplied with a time delay. In both the motion and braking modes, the armature current is differentiated and the rate of change in the load current is controlled through A11.

Operational experience has confirmed the high efficiency of electric drives connected under the ED circuit.

BelAZ is constantly working to improve the traction electric drive of its heavy-duty dumpers. The aim is to improve their traction-dynamic properties and increase reliability and efficiency. The results are utilized in several models, with the AC-DC drive (Fig. 3.11) of dumpers with a 130 t load capacity considered to be the most successful.

The power circuit is based on the traction generator (G1) which is driven by a diesel engine (not shown in the flow-chart). Both generator stator windings are connected to a rectifier (UZ1 and UZ2, respectively). The output rectifier circuits in the traction mode are connected in

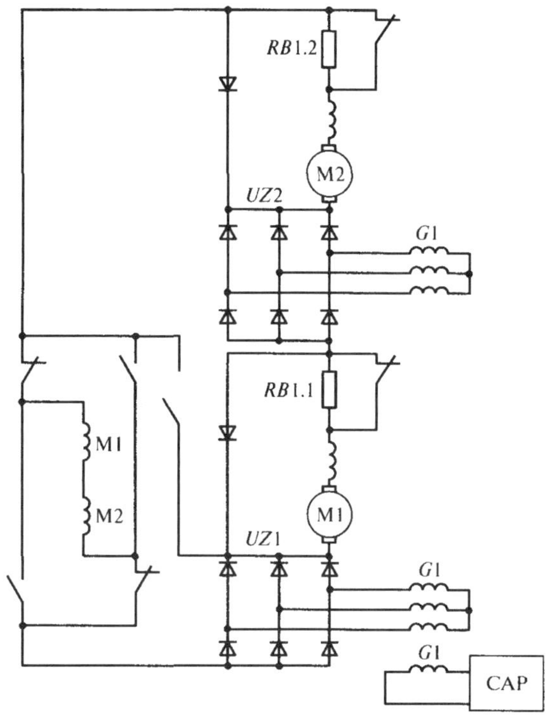

Fig. 3.11. Flow-chart of the AC-DC electric drive of BelAZ dumpers with a 130 t load capacity

series with the armature and excitation windings of the traction electric motors (Ml and M2).

The generator excitation current is controlled by the ACS. In the standard electrodynamic braking mode, it connects the output circuit of the rectifier (UZ1) to the excitation windings of the traction electric motors.

The traction electric drive of a dumper with a 130 t load capacity has a number of principal features that distinguish it from the electric drives used, for example, in dumpers with a 110 and 120 t load capacities.

First, a brushless single-support AC generator of enhanced specific capacity is used as a power source. It is proportionate to the GPA-600 generator and is adapted for coupling with diesel engines with a nominal rotation rate of 1500 and 1900 rpm.

Second, it has no rotating excitation device; the excitation system is static. The power circuit is designed according to the ED circuit, which ensures a forced equality of their currents and moments because of the serial engagement of the traction electric motors. This improves the steering at turning and expands the service life of the traction electric motors.

Third, the control system ensures a smooth, non-contact, automatic regulation of the traction motor's magnetic flow within the maximum voltage limit zone; as a result, the diesel engine's free power is fully engaged at all possible speeds.

Fourth, the effective electric braking range is significantly expanded: the braking force is preserved at low speeds and even at zero speed. This reduces the number of times the mechanical brakes work in the field, thus, expanding their service life.

Fifth, the control system allows for controlling the external characteristic of the diesel-generator by loading it on its own braking resistors. This means that the system can be tuned without using an external loading rheostat, and, since an electronic control system is used, it can be adjusted during manufacturing. This reduces the time and fuel consumption used for adjustments. It is also possible to maintain the constant input power (instead of the output) of the traction generator, which avoids overloading the diesel engine, especially when the dumper moves on roads with a heavy longitudinal profile.

The considered traction electric drive is intended for two dumper modifications, which differ by the transmission ratio, u_p, of the motors-in-wheel gearbox (31.63 and 38.1). The limit dynamic characteristics for the traction (a and b) and braking (c and d) modes are shown in Fig. 3.12.

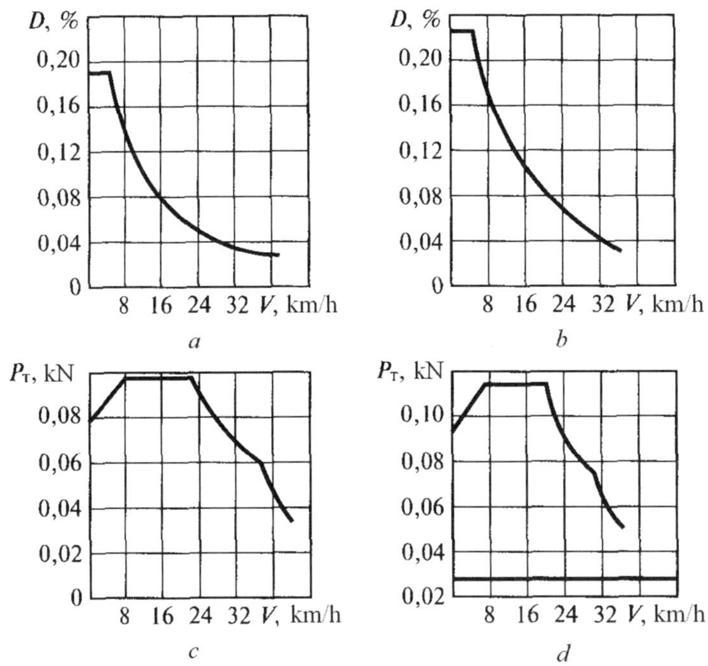

Fig. 3.12. Limit dynamic characteristics for the traction (a and b) and braking (c and d) modes of the AC-DC electric drive of the BelAZ dumper: a—Transmission ratio of the gearbox, u_p = 31.63; b—u_p = 38.1; c—u_p = 31.63; d—u_p = 38.1; D—Dynamic factor; P_m—The braking force

By using the traction mode characteristic, we can define the steady travel speed on a road with the preset longitudinal slope *i* and the specific resistance to rolling *f* as the speed, corresponding to the dynamic factor:

$$D = i + f;$$
$$f = 0.012 + 0.0003\ \mathrm{V}.$$

The dynamic characteristic of the braking mode allows for defining the steady speed at braking on a road with the longitudinal slope *i* and specific resistance to rolling *f* as the speed, corresponding to the specific braking force *b*, that is, the ratio between the aggregated force on the driving wheels and the dumper's weight ($b = i - f$).

3.3 Undercarriage

Let us consider examples of the functional use of the main systems, units, and components of the undercarriage of BelAZ dumpers.

The dumpers' frame is welded from 10HSND grade sheet steel with a yield point of 4000 MPa. It consists of two box-like sections of variable height linked with four cross bars. The thickness of the frame rail walls is 14 mm and the upper and lower shelves are 25 mm.

The suspensions are pneumohydraulic with the front independent and the rear dependent. The front wheels and the rear axle are mounted on four pneumohydraulic cylinders which act as pneumatic springs and hydraulic shock absorbers.

The suspension cylinders respond to vertical forces only. Longitudinal and lateral loads and torques are transmitted in the front by a pivot axle mechanism and in the rear by a central lever and a lateral bar.

The front suspension's pivot axle has a diameter of 260 mm and is rigidly fixed to the swivel cam; when moving with the wheel along its axis, it rotates in its bracket which is fixed to the frame. Metal-fluorine plastic bushings, inserted into bores in the bracket, serve as slide bearings and guides for the pivot axle. The bushings are sealed with rubber sleeve gaskets, which are protected from damage by polyurethane wipers. The upper end of the pivot axle is connected by a hinge to the suspension cylinder.

The rear driving axle suspension (Fig. 3.13) comprises of two pneumohydraulic cylinders (2), the central lever, and the lateral bar (1). The cylinders are placed behind the axle casing and are hinged to the axle and frame by spherical joints with a 150 mm diameter. The central lever of the box-like cross section is welded in the rear to the axle casing and in the front through an eyelet (a joint bearing and a pin) connected to the frame's cross bar. The hinged unit is covered by a protective jacket and the spherical bearing is additionally sealed. One end of the lateral bar (1), placed above the axle, is connected through a hinge bearing and pin to the axle support and the other end is connected to the frame support. The pin is tightened by means of a slotted conical plug. Each hinge unit is closed with two rubber dust caps and the bearings are sealed with additional seals. The rear suspension ensures a vertical displacement of the axle relative to the frame by the full travel of the cylinder (290 mm) and a lateral distortion of the axle by 8° in both directions.

The suspension cylinder (Fig. 3.14) is a piston-type pneumatic spring in combination with a hydraulic shock absorber. Gaseous nitrogen serves as the elastic working element and MGP-10 is used as the WF. The front and rear cylinders are structurally identical; only the dimensions are different. The front cylinder's rod and piston have diameters of 220 and 250 mm, respectively; the rear cylinder's rod and piston are 250 and 280 mm, respectively. The full travel of the rod is 320 mm.

The suspension cylinder (Author's Certificate of the USSR No. 572110, 5V60R1/02) consists of the main cylinder pipe (24) and the rod (6) with the piston and shock absorber divider welded to it. The lower part of the rod fixes the lower cover (15), which clamps the jacket (26) to the butt end

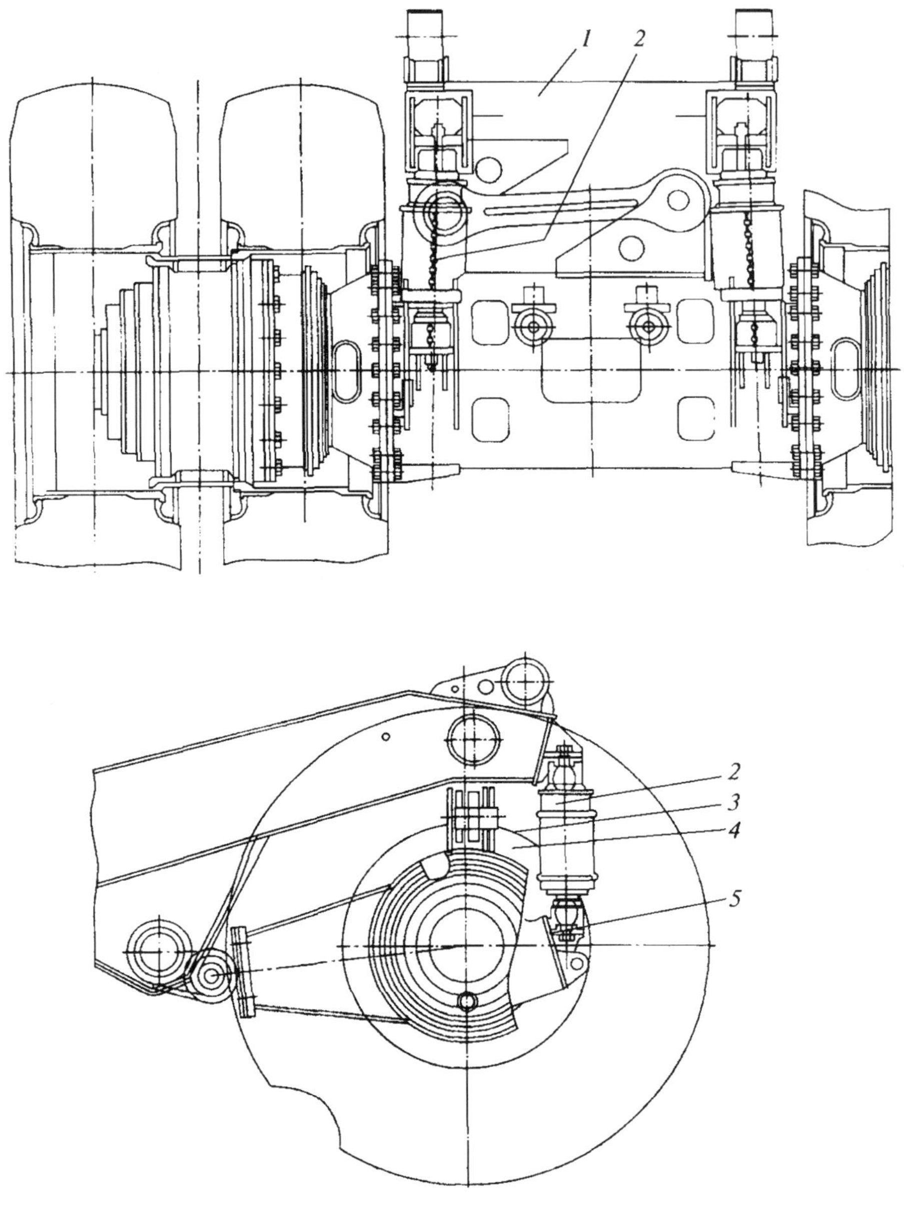

Fig. 3.13. Suspension of the driving axle: 1—Lateral bar; 2—Hydraulic cylinder of the suspension; 3—Cover; 4—Joint; 5—Ball support of the hydraulic cylinder

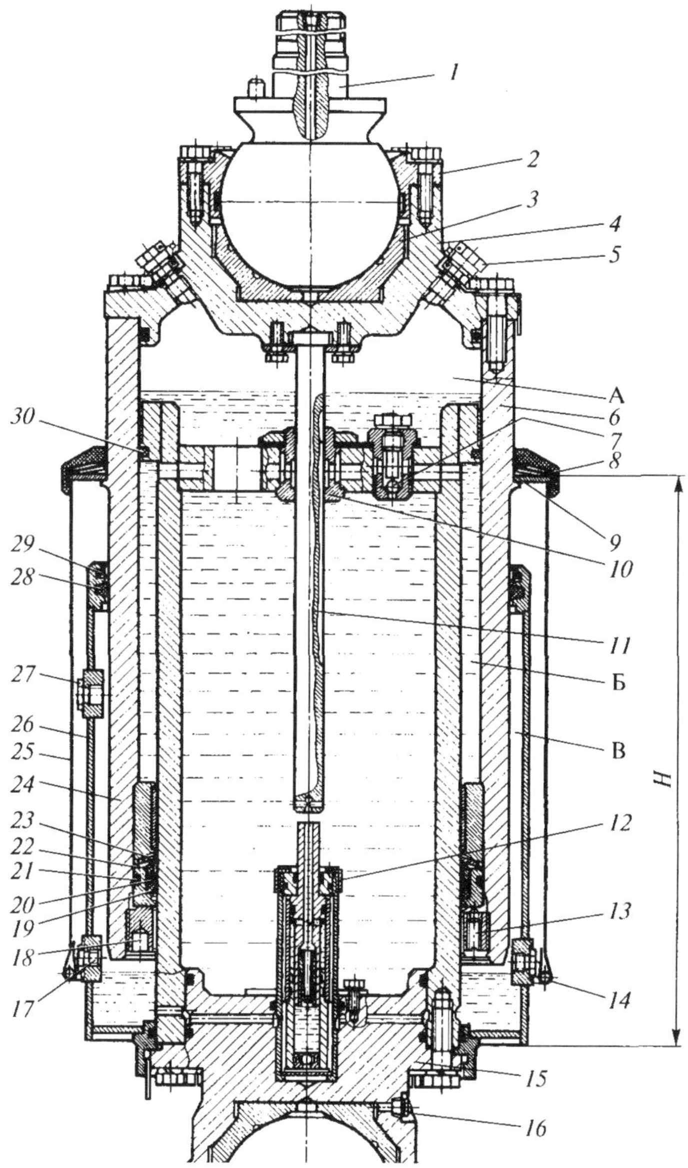

Fig. 3.14. Suspension cylinder: A, B, and C—Cavities; 1—Spherical support; 2—Cover; 3—Thrust block; 4—Upper cover; 5—Filling valve; 6—Rod; 7—Compression valve; 8—Gasket; 9, 14, and 28—Rings; 10—Retreat valve body; 11—Rod; 12—Pump; 13—Stopper screw; 15—Lower cover; 16 and 27—Safety valves; 17—Plug; 18—Pressure nut; 19—Rod packing; 20—Spacing ring; 21—Gasket ring; 22—Regulation spacers; 23—Spring; 24—Main cylinder pipe; 25—Protective jacket; 26—Jacket; 29—Scraper; 30—Washer

of the rod; the upper part of the main cylinder fixes the upper cover (4). The upper and lower covers bear thrust blocks and spherical joints with a diameter of 120 (the front cylinder) and 150 mm (the rear cylinder).

The shock absorber divider bears two compression valves (7) and the retreat valve body (10). The rod (11), with two longitudinal tapered slots that act as the retreat valve throttle, moves within the retreat valve body. The rod is also intended for driving the plunger pump (12).

The jacket (26) forms a cavity (C) that is filled by WF to the level of the check plugs (17) to compensate for leaks in other cavities (A and B) and for lubricating the rod seal. The air-tightness of the movable connection of the main cylinder (24) and the jacket (26) is ensured by the scraper (29) and ring (28). To ensure that the main cylinder's (24) moveable connection and the jacket (26) are airtight, a scraper (29) and ring (28) are used. The protective jacket (25), fixed between the ring (9) and gasket (8), protects the main cylinder's external surface from dust and dirt. Rubber O-rings ensure all fixed connections are airtight. Fluoroplastic packing (19), the working edges of which are unclamped by the spacing ring (20), ensures that the connection between the rod and main cylinder is airtight. The tightness of the packing can be adjusted by a set of regulating spacers (22) and a spring (23); new packing is typically set with a tightness (negative allowance) of 2.0–2.3 mm. Cavity B is made airtight by using fluoroplastic washers (30). The jacket (26) has a safety valve (27) which protects it from overloads due to increased pressure in cavity C caused by possible oil leaks and changes in volume. When in working conditions, the filling valves (5) fill the cylinder with nitrogen.

The suspension cylinder operates as follows: At the compression stroke (vehicle runs over an obstacle), the rod moves up in the cylinder and compresses the gas in cavity A. Due to the pressure increase above the piston, the compression stroke is elastically limited. At the retreat stroke (after passing the obstacle), the load on the rod decreases and it moves down.

For reducing the oscillations created by the vehicle's movement, the cylinder has a hydraulic shock absorber consisting of two compression valves (7), the body of the retreat valve (10), and the rod (11) with two longitudinal grooves of variable cross-section. The compression valves allow the WF to flow from cavity A to the ring cavity B at the compression stroke. At the retreat stroke, they are closed; the fluid comes from cavity B to cavity A through holes in the retreat valve's body and the rod's longitudinal variable cross-section grooves, which change their flow passage area as the cylinder expands. This creates the necessary flow resistance of the working fluid from cavity B to cavity A and impedes the full unclamping of the cylinder. The shock absorbers rods of the front and rear cylinders differ in their configuration and the depth of the variable cross-section groove.

The plunger pump (12), mounted in the lower cover (15) bore, serves to maintain the constant level of working fluid in cavity A. It starts working when there are significant vehicle oscillations and decreased levels of the working fluid in cavity A. At normal levels, the rod (11) does not touch the pump plunger. When the oil level decreases due to vehicle oscillations, the rod (6) moves the pump plunger down until the valve ball blocks the WF from exiting the pump cavity. The pressure from the oil forces the reverse valve to open; this allows the WF to flow through channels in the pump plunger and rod to cavity A. The cylinder decompresses due to the WF pressure and causes the pump plunger to shift up; thus, pressure below atmospheric one occurs under it and oil from cavity C comes under the plunger.

3.4 Steering system

The steering system (StS) of BelAZ dumpers consists of mechanical and hydraulic drives. The mechanical drive is comprised of a planetary gearbox with a 12.25 ratio, a steering wheel, a steering wheel column, angular gearboxes, and a cardan shaft, which, when turned, results in the action of the hydraulic distributor spool. This directs the working fluid to the turning cylinders, which are mounted on each front wheel, run simultaneously, and contain their own independent oil circuits.

The operation principle of the StS is as follows: The pumps (19) (Fig. 3.15) deliver the working fluid from the tank (17) to the hydraulic spool distributors (16 and 21) and the StS distributor (8). Part of the oil fills in the pipelines and cavities of the turning cylinders (3) and the rest is drained into the tank through filters (14).

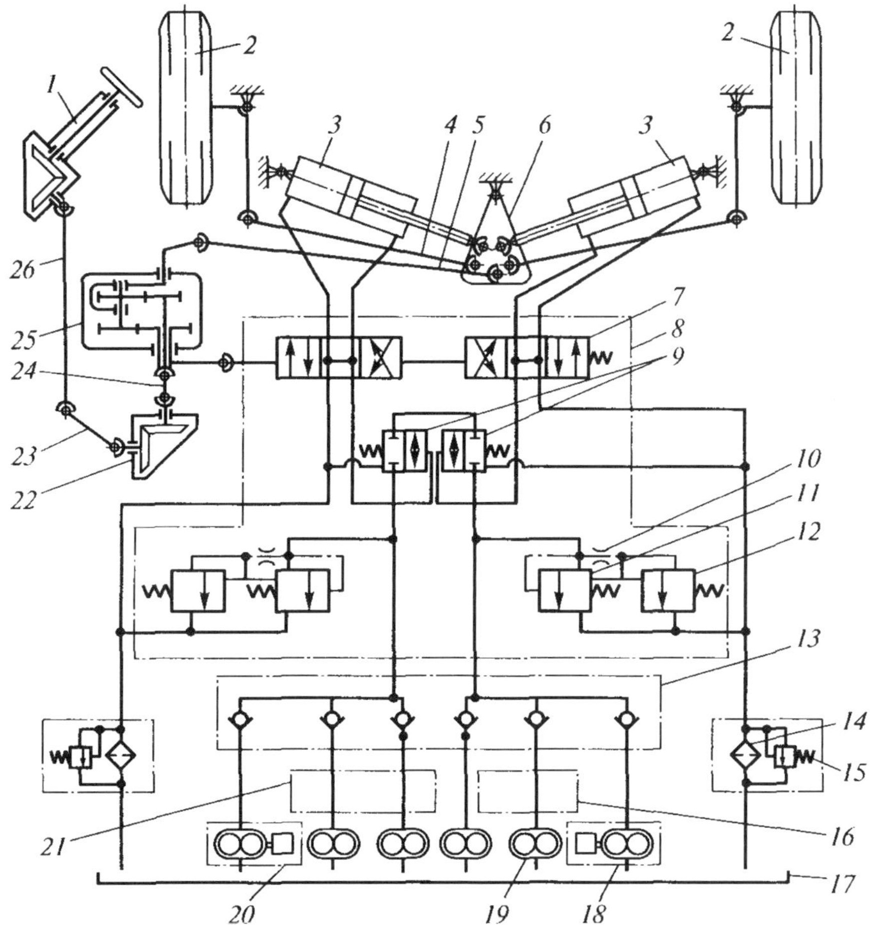

Fig. 3.15. Kinematical and hydraulic diagrams of the BelAZ StS: 1—Steering column (with angular gearbox); 2—Steered (controlled) wheels; 3—Hydraulic cylinders; 4—Steering linkage rods; 5—Follow-up linkage; 6—Central lever; 7—Spool; 8—Hydraulic distributor; 9—Leveling valves; 10—Throttles; 11—Relief valves; 12—Safety valves; 13—Block of reverse valves; 14—Filters; 15—Diverter valves; 16 and 21—Hydraulic distributors of the dumping mechanism; 17—Oil tank; 18, 20—Units of the back-up drive; 19—Pumps; 22—Angular gearbox; 23, 24, and 26—Cardan shafts; 25—Planetary gearbox

The cavities of the turning cylinders simultaneously communicate—through the gaps and channels in the distributor (8) spool pair—with the pressure and drain hydraulic lines; the pressure created by the throttles, the resistance of the hydraulic lines, and the gaps in the distributor spool pair cause the cavities to fill with oil.

When the steering wheel is turned right, the planetary gearbox's (25) output shaft lever shifts the distributor spool (7) left through the steering linkage rod. At the same time, the right cavities of the cylinders remain connected to the pressure hydraulic lines coming from the pumps (19) and the left cavities are connected to the drain pipeline. The piston of the right cylinder extends and the piston of the left cylinder retracts; the central lever (6) rotates in a clockwise direction and, through the steering linkage rods (4), turns the steered wheels of the vehicle right.

When the steering wheel is turned left, the distributor spool (7) shifts right and oil from the pumps is delivered to the left cavities of the turning cylinders. At the same time, the right cavities are connected to the drain hydraulic line. The left cylinder is decompressed and the right cylinder is compressed. The central lever (6) rotates in a counter-clockwise direction and, through the steering linkage rod (4), turns the steered wheels left.

When the steering wheel stops turning, the hydraulic distributor spool remains in the end position and oil flows to the cylinders causing the wheels and central lever (6) to continue to turn. The planetary gearbox carrier, connected to this lever with a follow-up linkage, also continues turning. Since the steering wheel is motionless, a slight movement of the carrier causes the distributor spool to shift to the neutral position and connect the discharge pipeline to the drain pipeline; thus, the steered wheels stop turning.

If pressure in one of the circuits exceeds 10 MPa, the safety valve (12) opens and the fluid drains through it and the spring cavity of the relief valve (11) into the tank. When the pressure in the pressure hydraulic line decreases, the safety valve (12) closes and the draining stops. The pressure hydraulic line and drain line disconnect when the WF no longer flows through the throttle hole (10), the pressure in the cavities of the relief valve (11) equalizes, and the spring of the relief valve's spool returns it to its original position.

Leveling valves (9) are designed for equalization in the circuits and for disconnecting a sealed-off circuit from a serviceable one. In working pumps, the spools of the equalization valves connect to the pressure hydraulic lines when the WF pressure exceeds their spring force. In this case, pressure in both circuits becomes equalized. When a circuit becomes depressurized, the WF pressure decreases causing the spool of the equalization valve to shift and cut off the channel; thus, disconnecting the pressure pipelines and the damaged circuit.

Vehicles are also equipped with a duplicate StS drive generally used for towing the vehicle in case of engine damage, damage to the main pumps of the StS, or damage to elements of other systems outside of the undercarriage. The electric motors of the duplicate drive's pumps are connected to accumulator batteries and oil flows from them to the circuits of the StS.

3.5 Brake system

The latest generation BelAZ dumpers are equipped with four braking systems: working, parking, reserve, and auxiliary.

The working system comprises of drum and plate type braking gears on the front and rear wheels and the hydraulic drive (Fig. 3.16). Braking gears have an automatic adjustment of the brake pad and brake drum gap; the two-circuit hydraulic drive has separate braking of the front and rear wheels. Its principle is based on using the power of the fluid pressure that accumulates in special devices called pneumo-hydro-accumulators (pressure tanks).

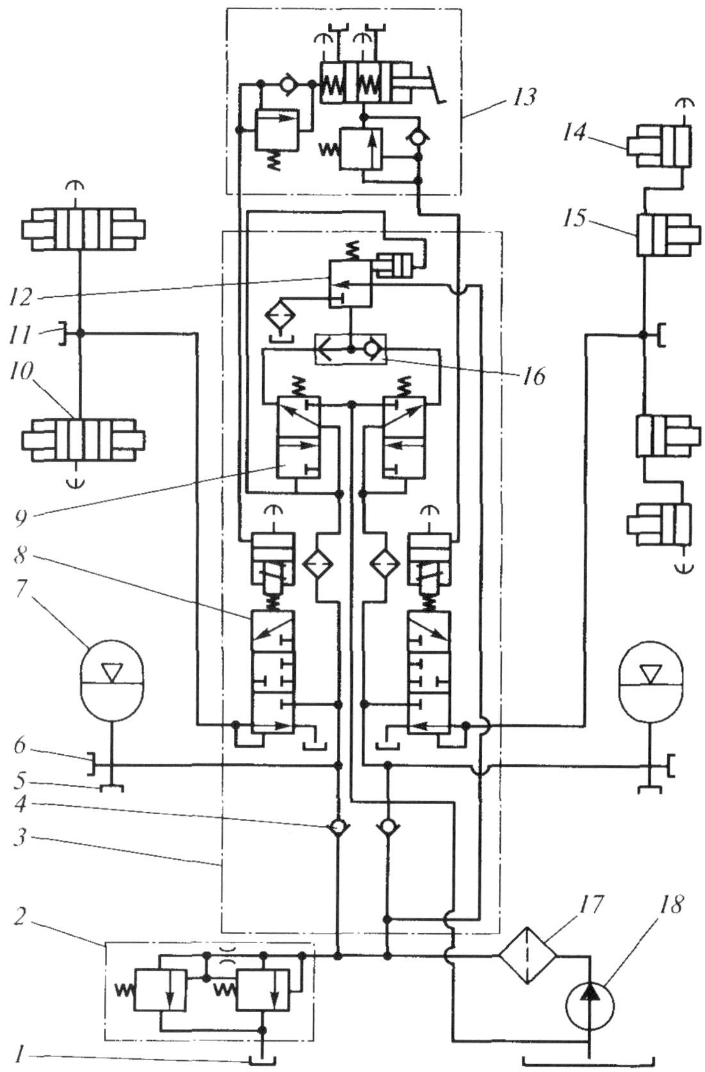

Fig. 3.16. Hydraulic diagram of the working braking system: 1—Oil tank; 2—Safety valve; 3—Control unit; 4—Reverse valves; 5, 6, and 11—Connections of respective pressure sensors in pneumatic accumulators, pressure gauge, and braking signal; 7—Pneumatic accumulators; 8—Braking valves; 9—Pressure regulators; 10, 14, and 15—Cylinders of wheel braking mechanisms; 12—Pump disconnection spool; 13—Control cylinder; 16—Dividing valve; 17—Filter; 18—Pump

The hydraulic circuit of the working braking system operates as follows.

The pump (18) uses the filter (17) to deliver the oil from the tank (1) to the control unit (3) and safety valve (2). At the absence of pressure or when low pressure in the pneumo-hydro-accumulators (7) occurs, the spool (12) blocks the drain to the tank and working fluid is delivered to the pneumo-hydro-accumulators through the reverse valves (4). Once the preset pressure (11.0–11.5 MPa) defined by the pressure regulators (9) is reached, the spool (12) connects the pressure line of the pump to the drain.

The reverse valves (4) and dividing valve (16) ensure that the hydraulic drive is divided onto two independent circuits.

If the pressure exceeds the preset limits (because of faults in the regulatory system), the working fluid is drained into the tank through the safety valve (2).

When the pedal is pressed, the braking fluid arrives from the main braking cylinder (13) to the cylinders driving the sections of the braking valves (8). These sections ensure supply of the working fluid to the wheel cylinders (10, 14, and 15) from the pneumo-hydro-accumulators. When the pedal is released, the working fluid from the wheel cylinders is discharged into the hydraulic tank.

Branches 5 and 6 are designed for connecting the pressure gauge sensors and the emergency pressure drop relay; branch 11 is intended for the "brake light" switches.

The parking braking system consists of two braking gears, a braking drive with pneumatic control, and monitoring devices. The constantly closed circuit braking gears are placed on shafts of traction electric motors. The parking braking system has an additional increase in braking torque and is used as an emergency braking system.

The auxiliary braking system is used for maintaining a constant speed when moving downhill. It represents a traction electric drive, which operates in the electrodynamic braking mode by switching the traction electric motors to the generator mode. In this mode, braking resistors are partially cooled by a special fan and serve as an external load.

3.6 Cabin

BelAZ is constantly working to improve the design of the cabin and its equipment, namely, the seat, controls, air conditioning, and noise-insulation systems. As a result, three cabins with a high level of unit standardization and improved components have been created.

The first is a single-seat cabin designed for dumpers with a 30–55 t load capacity. Its internal width is 1240 mm and it has a volume of 3 m^3. The second is a two-seater, with an internal width of 1486 mm and a volume of 3.6 m^3; it is installed on dumpers with a 65–90 t load capacity. The third, a two-seater with the same internal dimensions as the second, is intended for dumpers with a 120–220 t load capacity.

The single-seat cabins have an additional small-sized seat for a trainee and the two-seaters are equipped with a standard folding seat.

These cabins are externally distinct from the first models because their windscreens are tilted vertically back 15° (instead of forward) which increases their internal volume by 15%.

The cabin bodies are a box-type cross section and are made of high-strength steel with an 800 MPa (8000 kgf/cm^2) yield point. Cabins are mounted to the frame by four rubber shock absorbers. The welded bodies, their strong attachment to the frame, and the protective visor of the platform reliably protect the driver in case of tipover (the ROPS safety system under the ISO 3471 and SAE J1040C Standards) and at loading (FOPS under the ISO 3449 Standard). Large windows and spherical mirrors ensure good visibility both in front and behind the vehicle. By request, the internal surface of the front glass can also be covered with an opaque sun-protective film.

A spacious workplace with comfortably located controls, a height- and tilt-adjustable steering column, a pneumatically sprung seat, soft upholstery covering the side walls and ceiling, and reliable noise and vibration isolation create comfortable conditions for the driver-operator. The microclimate in the cabin is maintained in the winter by the heater, powered from the engine's cooling system, and in summer by two electric ventilators of the heater. The outside air sucked by the ventilators passes through a cardboard filter that removes the dust. The cabin also has room for installing a roof-mounted KL 2 C/4 model air conditioner made by Konvekta (Germany) with a cooling performance of 4800 kcal/h.

3.7 Dumpers made by MoAZ

The Mogilyov Automobile Works (MoAZ) specializes in the manufacture of earthwork-transporting road trains towed by single-axle tow cars, self-unloading road trains intended for operation in mines, and full-drive (4×4) dumpers (Fig. 3.17–3.19).

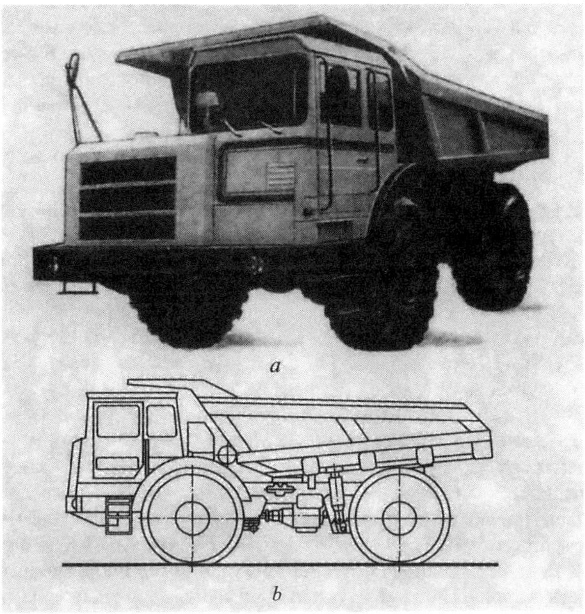

Fig. 3.17. MoAZ-7505 dumper: a—Outside appearance; b—Layout

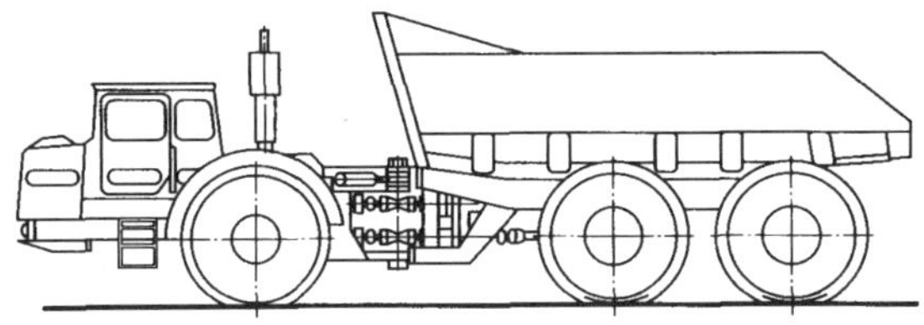

Fig. 3.18. Layout of the MoAZ-7503 dumper

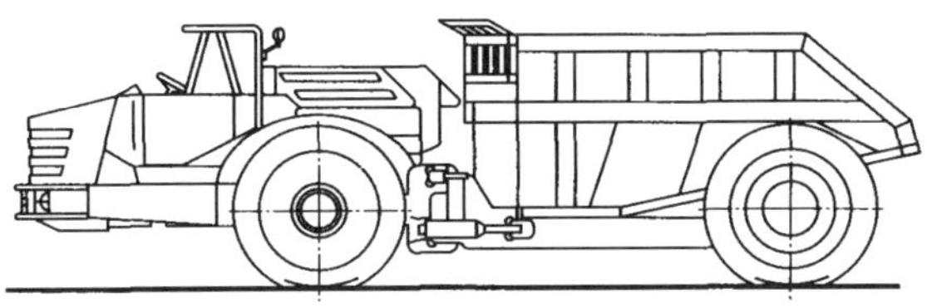

Fig. 3.19. Layout of the MoAZ-7405-9586 dumper-road train

The MoAZ dumpers are used in applications where the non-full-wheel-drive BelAZ quarry dumpers cannot be used efficiently: major construction sites with huge volumes of earthwork, transporting the overburden at open extraction of minerals, and in other situations with adverse conditions.

These dumpers are capable of doing the above work due to their design peculiarities (see Table 3.3), namely: powerful engines, short wheelbases, all drive axles, wide-profile low-pressure tires, large road clearances, etc.

The MoAZ dumpers are equipped with hydromechanical transmissions and pneumohydraulic suspensions of all axles, which make them notable for high dynamic qualities and soft driving. A comfortable cabin (a single-seat with an additional seat for a trainee); a friendly location of controls; small efforts required to turn the wheels, drive the pedals, and control the levers; good ventilation and heating; perfect thermal and noise insulation; and nice visibility both forward and backward all guarantee comfort and safe work for the driver.

The MoAZ-7405-9586 dumping road train (see Fig. 3.19) is intended to work in mines, tunnels, and other confined areas. Its design feature is the presence of parallel controls, which allows for using it as a shuttle (moving forward and backward without turning around).

This road train is equipped with a 140 kW (190 hp) YaMZ-238KM2 diesel engine, which allows for a maximum travel speed of 40 km/h. It has a 22 t load capacity, weighs 41.5 t, can conquer (at full load) an angle of 9°, and has a 7.5 m minimum turning radius.

The general trends discussed above in dumper design are applicable for the whole global motor industry and reflect the most advanced direction. The formation and technical appearance of foreign-made dumpers are considered in more detail in Chapter 9.

Table 3.3 Specifications of MoAZ dumpers

Parameter	MoAZ-7503	MoAZ-7505	MoAZ-7506
The engine:			
model	YaMZ-8401.10-01	YaMZ-238B or (customized) Cummins TA-10	YaMZ-8424.10
power, kW (hp)	441 (600)	220 (299) or 246 (335)	312.5 (425)
Weight of the dumper, kg:			
full	78,000	42,600	57,400
transported load	33,000	19,600	25,400
Wheel formula	4×4	4×4	6×4
Body:			
volume, m^3	30	15.5	20.5
loading height, m	2.9	3.04	3.65
dumping mechanism	Hydraulic, two-cylinder		
Transmission	Hydromechanical		
Wheels	Rimless		
Tire dimensions, inch	26.5–25		
Maximum speed, km/h	50	56	50
Conquered slope, degrees	12	7	11

Chapter 4
Evolution of the multi-axle chassis

4.1 History and evolution of the multi-axle chassis for monocargoes

Initial intensive studies related to designing vehicles intended for transporting heavy one-piece items (monocargoes) date back to the 1960–1970s. This was when the main ideas and basic design solutions that are still used today to design and manufacture heavy-duty special wheeled chassis (SWCs) appeared.

In the late 1960s, the USSR initiated studies aimed at creating large-size wheeled vehicles for transporting monocargoes weighing 60–100 t or more. The aim was to develop and approve the general designs and layouts of wheeled vehicles, their units and components, and define their specifications.

A thorough analysis was made of all means of transporting monocargoes both in the USSR and abroad. Conditions such as the bearing ability of roads and soils and the characteristics of roads, road structures, and bridges were defined and influenced the vehicle's layout.

The potentials were examined for making wheeled vehicles for the following cargo weights: 60, 79–84, 109–114, 136–141, and 164–169 tons.

During the initial stage of the studies, a conclusion was made to design a particular single-link chassis for each of the above load groups.

The following principles were adopted as starting points for designing the vehicle:

- The increase in load capacity should be achieved by increasing the number of axles;
- The load should be uniformly distributed among the axles;
- The following should be used to reduce the profile loads on the undercarriage: placing the axles according to the bogie-type layout, the use of balancing lever-type (pivotal) links in wheel suspensions, and using two- and three-link chassis with various types of coupling hitches (clutches); and
- The vehicle should move along prepared routes consisting of available roads and duly investigated localities (terrains).

It also became clear that road trains are not suitable for off-road transportation of monocargoes because of their high weight, large dimensions, and poor maneuverability; therefore, they were excluded from the design process.

Prior to these studies, a number of restrictions on the main specifications of the future vehicle were formulated.

Some of them, interesting from the viewpoint of future vehicles, are listed below:

Overall dimensions	
maximum height with the load	4500 mm
maximum width	3400 mm
length	minimum possible
Wheel span	2600 mm
Maximum load per axle	125–150 kN
Distribution of axles along the base	bogie-type
Transmission	electric, hydrostatic, or hydromechanical
Suspension (with partial or complete hydraulic balancing of the resilient elements of the bogie of one board)	independent hydropneumatic
Overall wheel travel	up to 500 mm
Frame	girder-type, combined (backbone, pipe-type with a subframe), or step frame of variable cross-section
Angles of approach and departure	at least 20°
Angle of lateral static stability (without the yield of tires and springs)	27–30°
Maximum turning angle of steered (controlled) wheels	38°
Number of seats in the cabin	at least 2

Note. Single-link chassis were considered only for loads of 60–84 t.

As a result of the studies and calculations made in analogy with the then existing heavy-duty vehicles, the specifications were obtained for an eight-axle vehicle with a 60 t load capacity (Fig. 4.1) and for a 12-axle vehicle (Fig. 4.2) with a 84 t load capacity, which are presented in Table 4.1. For comparison, the table gives some specifications of the currently manufactured 79221 (16×16) chassis model [8].

Table 4.1's data is interesting in the sense that, until now, none of the hypothesized main parameters of the chassis have been achieved in current chassis of the same load capacity.

It was also considered a necessity to use an electrical or hydrostatic transmission, a gas-turbine, or a combined power unit to achieve the applicable travel speed.

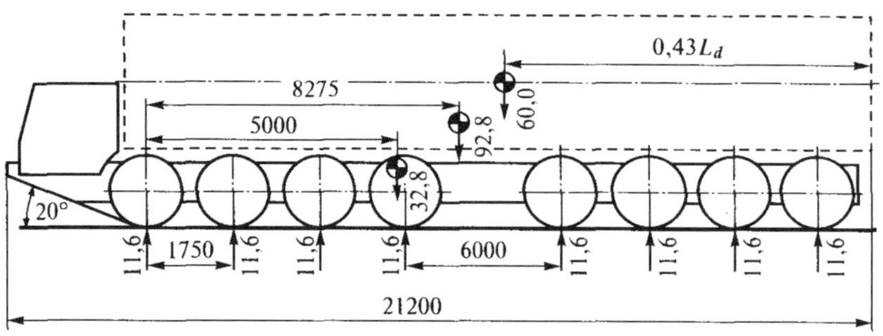

Fig. 4.1. Design layout of the 16×16 chassis with a 60 t load capacity: L_d—Length of mounted equipment (cargo)

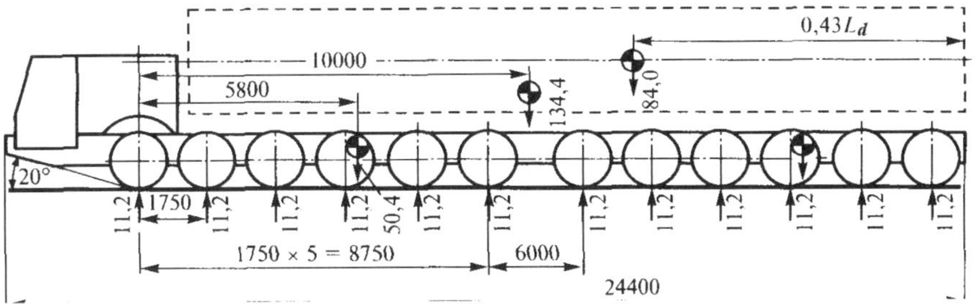

Fig. 4.2. Design layout of the 24×24 chassis with a 84 t load capacity: L_d—Length of mounted equipment (cargo)

Table 4.1 Main specifications of SWCs

Parameter	Single-link eight-axle chassis with a 60 t load capacity	Chassis 79221	Single-link 12-axle chassis with a 84 t load capacity
Full weight, t	93.0	120.0	135.0
Weight of transported cargo, t	60.0	-	84.0
Load per axle, t	11.6	15.0	11.2
Road weight, t	33.0	-	50.0
Wheel formula	16×16	16×16	24×24
Overall dimensions, mm:			
length	21,200	-	24,400
width	3350	3350	3350
height (with the cargo)	4000	-	4000
Wheelbase, mm	16,500	-	20,000
Wheel span, mm	2600	2700	2600
Turning radius of external wheel, m	14.5	17.0	19.25
Dimension corridor, m	4.1	-	6.1
Engine:			
type	combined (diesel+GTE)	diesel	combined (diesel+GTE)
power, kW (hp)	894 (1200)	633 (850)	1267 (1700)
specific power, kW/t (hp/t)	9.61 (13.0)	5.28 (6.67)	9.39 (12.65)
Maximum travel speed, km/h	60	40	60
Average speed, km/h:			
on roads with hard surface	40–45	25–30	40–45
on profiled or ground roads	25–30	20-25	25–30
Fuel distance, km	250	500	250
Parameters of passing ability:			
ascend, degrees	15	10	15
angle of approach, degrees	20	-	20
angle of departure, degrees	at least 75	-	at least 75
Depth of conquered wade, m	1.2	1.1	1.2
Road clearance, mm	500	475	500

Within the framework of the scientific and research studies, a possibility was studied—both theoretically and experimentally—of achieving the preset technical parameters. In particular, a six-axle prototype was built on the basis of the Tatra-813 to study the combine operation of different power unit types with one transmission. Upon mounting two two-shaft gas-turbine engines (GTEs), namely, TS-20 model aircraft turbine starters, the prototype received the name model 103G.

A brief characterization of the 103G chassis is presented in Table 4.2.

The authors of the above studies' arguments for using combined power units in heavy-duty vehicles are summarized below.

According to the technical requirements for the chassis of the near future, the power units should be within 1120–2235 kW (1500–3000 hp). Although the diesel engine was the most popular and well-studied main power unit element, the dimensional volume allotted today requires that only a 374–484 kW (500–650 hp) diesel engine be used.

The desire to obtain more power from these engines by using turbo charging leads to shorter lifetimes, decreased reliability, and more complex design.

Gas-turbine engines (GTEs) are an alternative to diesel that can ensure a power of 745–1200 kW (1000–1500 hp) at acceptable overall dimensions. However, powerful unitary GTEs can only be used in heavy vehicles economically if the engine's continuous operation is under a load of

Table 4.2 Brief specifications of the 103G chassis

Parameter	Value
Wheel formula	12×12
Axle formula	3-3
Road weight of the chassis, t	13.86
Load capacity, t	22.0
Power unit:	
main diesel engine model 5D-20:	
nominal power, kW (hp)	224 (300)
maximum torque, kg×m	96
Auxiliary gas-turbine engine:	
type	two-shaft
model	TS-20
nominal power, kW (hp)	2×60 (2×80)
maximum angular velocity of the turbo-supercharger, rpm	50,500
maximum angular velocity of the drive shaft, rpm	3000
weight, kg	37.5
Overall dimensions, mm:	
length	888
maximum diameter	236
Tires, mm	1200×500-508
Maximum speed, km/h	70

0.6–1.0 N_{max}. This mode is possible when vehicles operate long-term on good roads with a close to maximum speed. However, vehicles that carry special equipment must work in different road conditions. The studies showed that in such conditions, the GTE will be loaded to 0.3–0.6 N_{max} during half of its service life. The specific fuel consumption is 186–251 g/(kW×h) (250–350 g/(h.p.×h)) and the operation time at the maximum power mode is only 5–6%.

The experimental studies conducted on the 103G chassis confirmed the theoretical possibility of making a combined power unit for heavy vehicles. The obtained results led to the following conclusions.

A combined diesel-gas-turbine unit (CDGTU) allows for reaching a power of about 894–1341 kW (1200–1800 hp) without any detriment to the mounted equipment's layout options.

In the CDGTU, the diesel engine is the main and constantly operating unit. When moving on surfaces with a high overall resistance to motion, a supportive GTE without a heat exchanger is added. The authors of the studies referred to foreign sources and asserted that the CDGTU's fuel efficiency factor is somewhat better than that of a single diesel engine of comparable power.

If a power of about 1490–2235 kW (2000–3000 hp) is needed, it is expedient to replace the main diesel engine with a GTE with a heat exchanger. The high loading of the main GTE and the short-term operation of the supportive engine (10–25% at most) allows for obtaining adequate fuel efficiency practically throughout the whole power range.

The best power unit for current and future heavy and super-heavy load capacity chassis is still being debated since the diesel engine remains the most popular (but limited) choice, while available GTEs suffer from low efficiency factors.

In our opinion, using a combined power unit should not be ruled out when designing future heavy-duty SWCs.

In the late 1960s to early 1970s, the studies of electric and hydrostatic transmissions were rapidly developing because of their many advantages over mechanical transmissions, such as: simplicity and ease of layout; the absence of transmission elements with rigid kinematical interlinks (allowing for individual control of all elements); the smaller dimensions and ability to standardize the used aggregates; and improved traction-dynamic characteristics, passing ability, and maneuverability.

It is interesting for practical purposes to compare the weights of different transmissions used previously (Table 4.3). It follows from the table that the electric and hydrostatic transmissions were comparable with each other and that both were better than the mechanical and hydromechanical ones.

Table 4.3 Weights of different transmissions used in heavy-duty chassis

Transmission type	Approximate weight of the transmission used in chassis with the following wheel formulas, t:			
	16×16	20×20	24×24	32×32
Mechanical or hydromechanical	15.0	20.6	25.0	34.0
DC electric	12.6	16.0	20.0	25.0
AC electric	10.0	13.0	15.0	20.0
Hydrostatic with regulated motors-in-wheel	11.5	14.4	17.3	23.0
Hydrostatic with non-regulated motors-in-wheel	8.2	10.3	12.4	16.5

An even greater difference in weight parameters favoring the electric and hydrostatic transmissions can be expected given today's technologies, its promise, and the quality of currently used materials.

The above studies, however, did not solve the problem of how to effectively use the advantages of these transmissions. To do so, deep studies on the interaction of wheeled movers with the support surface in various motion modes and conditions must be conducted; based on the findings, algorithms for the control system on how to distribute the power in the most rational way must be developed. The experience of designing and using electric transmissions has confirmed that most of the advantages of these transmissions remain unutilized since the relevant studies have not been conducted.

The analytical (kinematical) studies of the chassis turning ability showed that single-link chassis designed for loads of 79–84 t can move satisfactorily on Category 4 and 5 roads only if all wheels are steered. However, it was noted that making all wheels steered worsened motion stability at straight sections, at turning, and made the control system more complicated. Further studies showed that this can be rectified by using a turning delay for the rear wheels against the front wheels or by using other control laws. Simplification of the drive design can be achieved—practically at any control algorithm—by using steering drives, which operate under other physical principles (for example, an electrohydraulic or electromechanical drive with wheel turning controlled with the help of software).

In order to choose and justify a proper springing system, a broad range of studies were conducted to examine the influence of the different systems' parameters on soft driving and the vibration load on the crew and undercarriage. Important conclusions were made about the positive influence that increasing the suspension stroke to 500 mm had on soft driving and the vibration load on the crew. The necessity to use hydraulically balanced wheel links of one board in order to reduce the profile load on the undercarriage was also confirmed. The studies yielded recommendations about the rational parameters of springing systems for chassis of different load capacity classes that were used by the industry to design future generations of chassis.

Thus, the scientific and practical studies conducted in the USSR in the late 1960s to early 1970s on vehicles intended to transport monocargoes allowed designers to do the following [1]:

- Formulate the technical appearance of the vehicle as a single-link multi-axle chassis designed for a specific type of equipment (see Figs. 4.1 and 4.2);
- Formulate and justify the main technical requirements which became the basis for the industrial manufacturing of a whole generation of SWCs; and
- Accumulate an essential scientific experience for later theoretical and experimental studies aimed at improving the main parameters of the chassis.

4.2 Hovercrafts

In spite of the fact that the technical appearance of transportation complexes was completely formed by the mid-1970s, large-scale research studies continued aimed at creating vehicles capable of transporting cargoes and complexes in off-road conditions. These included air-born vehicles, known as hovercrafts, intended to partially unload the vehicle when traveling across marshland and to ensure the observance of road legislation on the maximum permissible axle load when traveling on general use roads (Categories 1–4).

According to technical requirements, hovercrafts should have a load capacity of 5, 20, or 60 t and be able to transport long cargoes with the help of a generated air cushion intended to unload the mover and increase the passing ability on soft ground conditions.

In those days, this was new and complicated for the following reasons: First, the need to create an air-cushion under vehicles with a long supporting base (20–30 m) excluded any possibility of using a caterpillar mover. Second, specific requirements for creating a hovercraft precluded the ability of using the well-known design methods for SWCs. Third, designers had no experience designing and manufacturing such vehicles.

However, through the cooperation of a number of research and scientific institutions and enterprises, an attempt was undertaken to develop an articulated multi-wheel hovercraft prototype with a 5 t load capacity. Since practically all of the prototype's components were unique, both then and now, we present the main design solutions and test results below. Such a detailed description may have a positive effect on designs of certain future chassis units.

4.2.1 General layout

The hovercraft prototype was a two-link, 14×14 vehicle with all steered wheels, equipped with the components for generating the air cushion, and intended for transporting cargoes over soft grounds, on public (general-purpose) roads, and off-road terrains. The general layout is presented in Fig. 4.3 and general views in Figs. 4.4 and 4.5.

The body's links were connected to each other by means of a joint, which allowed for a ±17.5° jackknifing angle in the longitudinal-vertical plane. To increase the contact area with the ground, the hovercraft used 14 driving wheels with OI-28 sized 100×600 arch tires model. The individual hydropneumatic suspension

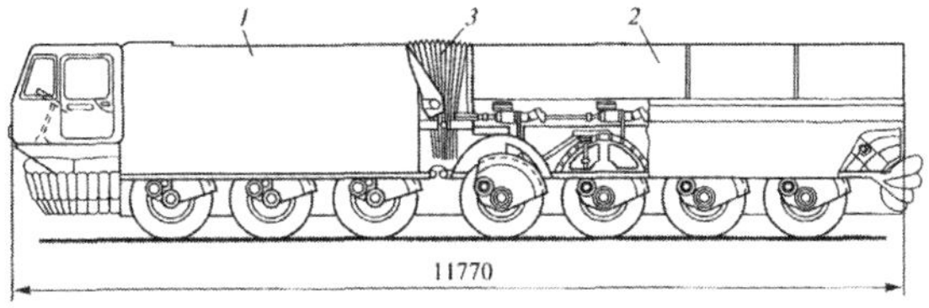

Fig. 4.3. Design layout of a 14×14 prototype: 1—first link; 2—second link; 3—sealing between links

Fig. 4.4. Overall view of a 14×14 hovercraft prototype

Fig. 4.5. Overall view of a 14×14 prototype without the air cushion

was installed in the turning sections with a maximum turning angle of ±83° and ensured the "crabbing" motion.

The front link housed a four-seat cabin with the corresponding controls, a power compartment with a four-cycle 5D20 model diesel engine, a fan-engine compartment with a GTE-350 model gas-turbine engine, and two fans for creating the air cushion.

The bottom part of the front link housed six turning sections, equipped with balancing levers, pneumatic springs, and wheels with arch tires. Each wheel was mechanically driven through a distribution reduction gearbox, an intermediate gear quadrant, and a wheel gearbox. Wheels were turned within their turning sections by means of an electric drive.

The rear link was a cargo platform made of aluminum alloys; its lower part housed eight turning sections, same as the front link. This link contained the transmission units and aggregates, the differential unit, and the hydraulic systems that changed the road clearance. The back section of the rear link contained water-jet propellers for ensuring travel when afloat.

The air cushion was formed in the flexible fencing chamber mounted along the perimeter of the prototype. The front and side sections of the fence were completely under the vehicle's body, while its rear section partially protruded outside the body. In order to provide a satisfactory lateral static stability, the air cushion was divided into two parts by a balloon type longitudinal elastic curtain. Each fan sucked in air from the atmosphere and injected it into the corresponding air cushion cavity through holes in the vehicle's bottom.

4.2.2 Power unit

The power unit included a standard 179 kW (240 hp) 5D20 model diesel engine, an integral gearbox, and supporting systems.

4.2.3 Transmission

The transmission was chosen with the following considerations in mind. Analysis showed that the classical mechanical transmission with a high efficiency factor was not acceptable because of the large number of driving wheels and the necessity to ensure differential links. Simplification of this transmission leads to power circulation, wear of tires, and a decrease in the efficiency factor.

The designers concluded, after analyzing different drives, that the hydraulic drive used in the transmission would practically eliminate the dependence of the transmission's complexity from the number of driving wheels; could easily ensure the controlled differential links and protect against overloads; and the elements could be nicely accommodated inside the body, while the power units could be directly mounted in the driving wheels. However, in those days there was no hydrostatic transmission with acceptable weight, reliability, and efficiency factor characteristics; thus, they had to develop a transmission which could combine the positive features of the mechanical and hydraulic drives.

As a result, the prototype used a specially designed combined transmission with individual mechanical drive of each wheel. The hydrostatic drive (in the second power flow) ensured a blocked differential link of all driving wheels and protected each wheel drive against overloads.

The prototype's transmission is presented in Fig. 4.6.

The transmission's operation principle is as follows: When the vehicle moves, engine power is delivered through the main friction unit (1), the gearbox (2), and the angular reduction gearboxes (3) to the longitudinal shafts (4) and, through a pair of sprockets and satellite gears (7) of sum-up gearboxes (5), to the epicycle gears (6) of the planetary mechanisms. Power is driven to the driving wheels by the carrier (8) of each sum-up gearbox through the conical pair (not shown in Fig. 4.6). The traction forces developed on the wheels and caused by the resistance to motion create reactive moments on the sun sprockets of the planetary mechanisms (shafts of hydraulic machines) and their respective reactive pressures in hydraulic machines.

If the wheels rotate at the same rpm (rectilinear motion), the planetary mechanism's carriers are also rotating at the same rpm causing the shafts of the hydraulic machines to stop. Since the hydraulic machines' (9) chambers and collectors are interconnected (the differential unit is unblocked), the reactive pressure, which is the same for all the hydraulic machines, will set in the power manifold. This causes equal traction forces on the wheels and the presence of the hydraulic link ensures the simultaneous appearance of the traction force in the wheels' contact point with the support surface.

If the wheels rotate at different rpms (turning, skidding, or conquering obstacles), the power delivered to the sum-up gearboxes of the lagging wheels is divided by the planetary mechanisms into two flows. Part of the power is delivered to the lagging wheels in accordance with their actual rpm; the rest is delivered to the hydraulic machines and is transformed into hydraulic power. At the same time, WF from the hydraulic

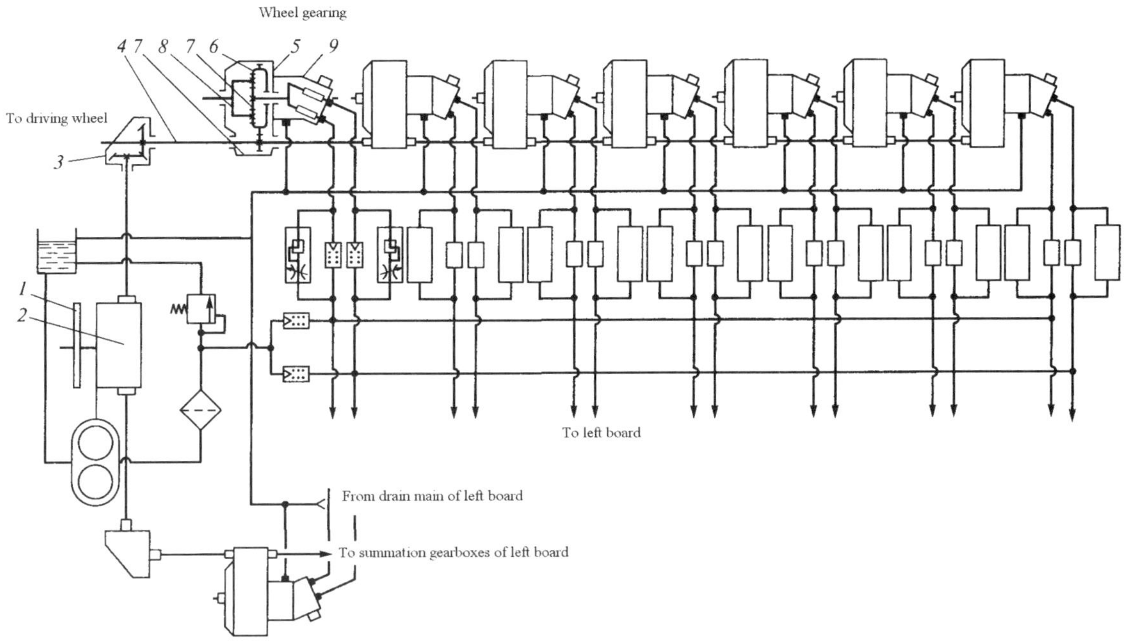

Fig. 4.6. Diagram of the 14×14 prototype transmission: 1—Internal-combustion engine (ICE); 2—Step-up gearbox; 3—Angular gearbox; 4—Mechanical transmission; 5—Wheel planetary gearbox; 6—Epicycle; 7—Sun pinion and satellite gears; 8—Carrier; 9—Hydraulic motor

machines of the lagging wheels, which operate in the pumping mode, is squeezed through the flow regulators into the unified manifold. It is delivered further by the reverse valves into the hydraulic machines of the leading wheels, operating in the motor mode, which ensures the required rpm increments needed by the leading wheels to make a turn.

A system to control the blocking provides a possibility to block the differential unit both jointly and separately (by links). The forced complete blocking of the differential unit is carried out by closing the throttles that regulate the working fluid's flow rate. In this case, power does not pass through the hydraulic differential unit because the fluid's exit is blocked. Therefore, all wheels, irrespective of the external resistance, rotate at the same angular rate (rpm).

Some results of the factory and field tests of the transmission are rather interesting. Analysis showed that when in the "Unblocked" position, the hydraulic differential unit ensures the different rpms of individual wheels when turning and reliable operation in all modes.

The conducted tests also revealed an increase in the free traction force when blocking the hydromechanical differential in conditions of worn-down and wet ground roads by 17%, in conditions of virgin snow and boggy terrain by 25%, and in conditions of marshland (with the use of tire vacuuming) by 10%.

The idea of designing such a transmission is very interesting and requires a deeper study and theoretical justification. While designing the above prototype, the authors failed to solve the problems of oil purification and making the hydraulic section completely airtight.

4.2.4 Undercarriage

The prototype had 14 driving, steered wheels with OI-28 model arc tires. The tires' inner cavities were equipped with star-shaped metal structures (bandages), which, when vacuumed, forced the tires to take its shape and allowed for improving the passing ability over marshlands. Each wheel was installed on a U-shaped balancing lever. The transfer of the vertical load from the wheels to the body was achieved through a pneumohydraulic spring. The full stroke of the spring was 300 mm and its dynamic stroke was 250 mm. The variable road clearance was within the limits of 490–740 mm and regulated by the road clearance system. The idea of using vacuumed wheels when designing new vehicles deserves close attention. At first glance, it ensures quite a few advantages when moving over terrains with soft grounds. However, its broad application requires a profound scientific study and justification.

4.2.5 Turning control mechanism

The prototype made use of a principally new steering system (StS), both in programming the wheel turning and in the executive drives. All 14 wheels were steered and ensured an 83° maximum turning angle. The control system also allowed for making a turn by eight methods: relative to each of the seven axles and the plane-parallel ("crabbing") motion (Fig. 4.7).

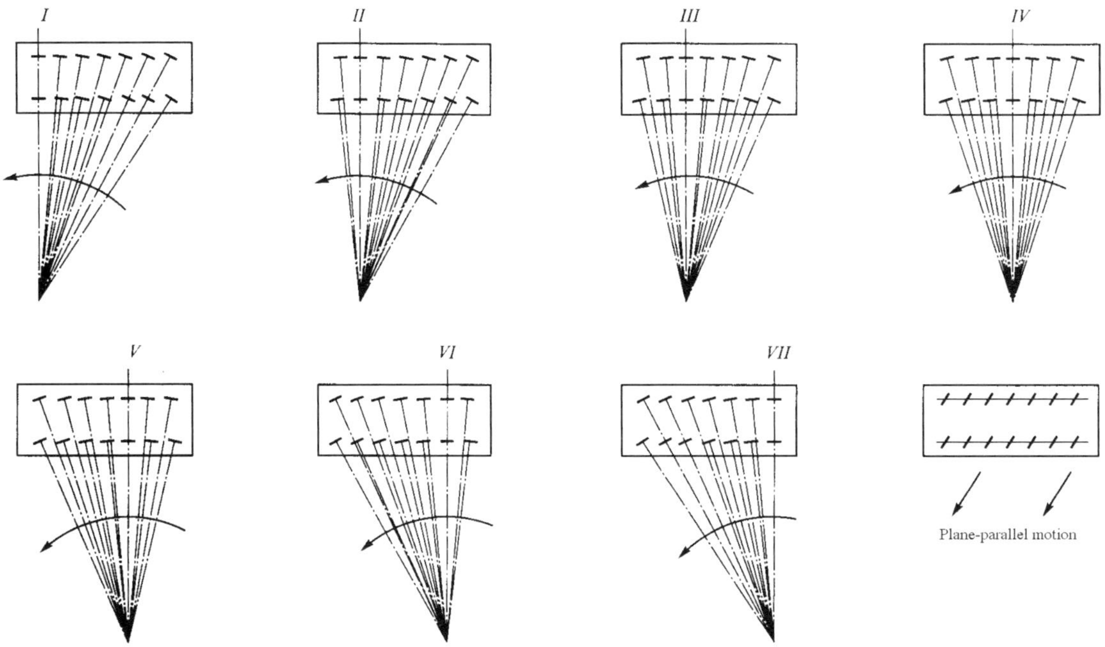

Fig. 4.7. Kinematic diagrams of turning methods relative to axles I–VII

Each wheel's turning control channel (Fig. 4.8) was an electric servo drive and contained the following elements: a selsyn transmitter (1), a commutation block (2), a phase-sensitive detector (4), a vibration amplifier (6), a transistor block (7), a contactor block (8), a block of resistors (9), a mode switching block (3), an electric motor of the hydraulic pump (10), a tacho-generator of the drive (11), a selsyn receiver (5), and a wheel gearbox (12).

The (1) rotor receives the turning angle from the programming device, while the servo system ensures the wheel connected to the (5) rotor turns by this angle. The performance speed automatically changes depending on the mismatch angle (the difference between the turning angles of the (1) and (5) rotors).

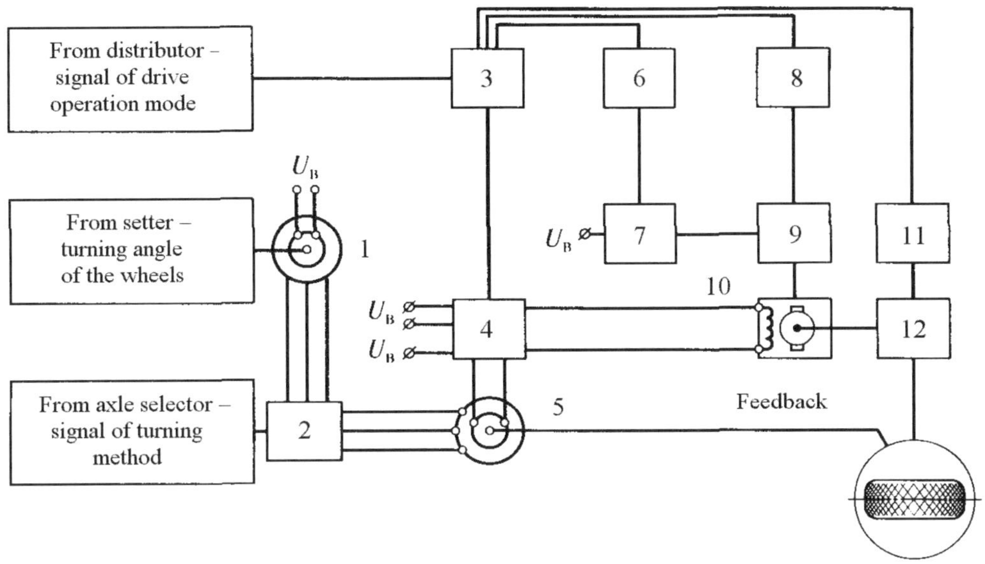

Fig. 4.8. Flow-chart of the servo drive of the 14×14 prototype

The kinematics of the wheel turning mechanism is shown in Fig. 4.9.

The StS's operation was tested both at the factory and in the field. The results of the factory tests on the lifted prototype were as follows:

- It took 14–18 seconds to bring the wheels to a coordinated position after two sharp turns of the steering wheel.
- The command execution's mismatch was 1.5–2.0 seconds;
- The average precision of bringing the wheels into the initial position (when returning from the left or right) was +0.5° (individual deviations were up to 1.5–2.0°); and
- The average precision of bringing the wheels into the position required under the kinematics law was 1.5° (individual deviations were up to 4°).

The field tests also included experimental studies aimed at assessing the controllability of the prototype. At the maximum turning angles of the steered wheels (83°) and different turning methods, the external dimensional (clearance)

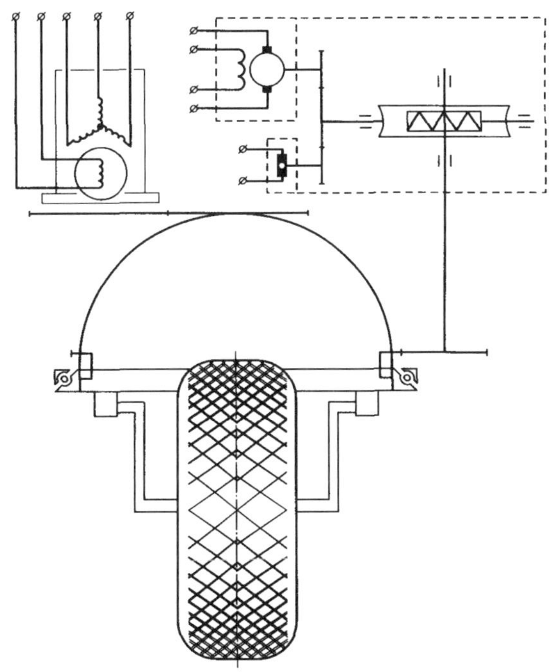

Fig. 4.9. Kinematical diagram of the wheel turning mechanism of the 14×14 prototype

turning radius was 8.0–12.9 m, the internal radius was 0.25–2.00 m, and the minimum turning radius was 6.15 m.

The maximum deviation from the rectilinear motion depending on different speeds and turning methods was as follows:

At a speed of 10 km/h and turning around the fourth axle: +1.45 m;

- At a speed of 20 km/h: +0.65 m;
- At a speed of 10 km/h and disengaging the turning wheels of the third and the seventh axles: ±0.82 m. With disengaging the wheels of the fourth and seventh axles: ±0.5 m; and
- At a speed of 10 km/h and transfer of the turning axle to the sixth axle: ±0.45 m.

The turning parameters obtained in the experimental studies can be considered positive, except for the turning speed of the steered wheels and the execution precision of the prescribed signal.

Since a great number of failures were detected on the prototype, it could not complete the full scope of testing. The defects stemmed from trying to implement several novelties in one prototype without a detailed theoretical and experimental study of the individual units and systems and without simulating their interactions. Thus, a good idea was compromised.

The fact that the customers' needs were covered by simpler vehicles greatly contributed to the termination of the work on the above prototype; there were also no vacant industrial facilities capable of embodying the new technical ideas implemented in the prototype. All of the above factors predetermined the fate of this advanced direction in designing vehicles. Although the process of designing hovercrafts continued, the fact that some original self-propelled and trailed complexes with load capacities of 10, 20, and 60 t were manufactured and tested in the Tyumen Region, and that unique theoretical and experimental materials were accumulated, the models failed to reach any stage of mass production due to the above reasons.

4.3 General design and layout solutions

In parallel with designing vehicles for monocargoes, an intensive study was carried out on the general design and layout solutions of SWC for special equipment. These would receive official status (see Figs. 4.1 and 4.2) and later become traditional.

In this context, an experimental chassis, model ZIL-135E with an electric transmission, which was designed and made at the B.A. Grachov Special Design Bureau in the late 1960s, deserves mentioning. It was the first, deepest, and most complete attempt to design a SWC with an electric transmission (ET).

For this chassis, special traction 22 kW DT-22 DC motors and 120 kW GET-120 DC generators driven by two ZIL-375 gasoline engines were designed. The transmission was controlled by varying the excitation of the generator. For that time, the elements of the ET had satisfactory weight, dimensional, and power characteristics. For example, the 7907 model chassis with a specific power of 0.167 kW/kg designed later (in the 1980s) made use of the promising types of electric machines (30 kW traction AC electronic motors), while the DT-22 motors used on the ZIL-135E chassis had a specific power of 0.244 kW/kg.

The comparative tests of the ZIL-135E prototype and the ZIL-135LN chassis (with an HMT) showed a decrease in fuel consumption by 14–20% on all road types, an increase in the average travel speed by 10–20%, and an improvement of certain passing ability indicators (for example, the maximum ascent angle). It is safe to assume that by using the respective theoretical studies directed at the complete use of all the potentials of the ET and by using today's elements, even more considerable advantages over traditional transmissions can be achieved.

In those days, the poor fate of this technical solution was predetermined by the "unwillingness" of industrialists to deal with a new electrotechnical product planned for production in small batches.

The experimental and theoretical studies conducted during the initial development stage yielded the basic design, layout solutions, and specifications for SWCs which met, for the time being, the demands of the equipment to be installed. Thus, at that stage, a stereotype of the vehicle was formed, which used the following elements (see Figs. 4.1 and 4.2):

- A diesel tank (or automobile) engine;
- A hydromechanical transmission with all-wheel-drive;
- An independent lever-torsion or hydropneumatic wheel suspension; and
- A steering system with a hydromechanical drive and steered wheels of the front or both the front and rear axles.

These technical solutions can be considered traditional since several generations of SWCs were created based on them and since the demand for greater load capacities was satisfied by adding more axles. Other original designs of SWCs created on the basis of non-traditional technical solutions remained at the prototype stage due to a number of reasons, such as:

- The absence of deep theoretical studies and a breakaway from the available production facilities;
- The intention to combine different units and systems in one prototype without any prior study and the difficulty of bringing them to perfection; and
- The poor quality, technical and technological levels, and, therefore, low reliability—or even absence of—the elements used to make the original elements of the chassis.

We should also note that current and future SWC designs are based on many solutions used in those days for building prototypes. Many ideas put forward then can now be implemented.

4.4 Designs of special wheeled chassis in the 1980s

In the 1980s, mobile equipment was designed by increasing the weight and overall dimensions. In those days, wheeled chassis were needed to transport special equipment weighing 150–220 t. Thus, designers attempted to utilize the seemingly simplest and most logical idea of scaling up the chassis' load capacity by increasing its mass and number of axles.

Initially, based on experimental studies, the possibility of driving a chassis with an up to 600 kN axle load was evaluated and the characteristics of the support surfaces (roads and off-road localities) were defined to select the travel routes. Later, designers created the 12×12 7904 full-scale prototype with a 220 t load capacity based on traditional technical solutions (see Fig. I.1b). In order to deliver engine power to the vehicle, a hydromechanical transmission with two hydromechanical gearboxes were used. Test results showed that the 7904 had a low efficiency when using traditional solutions (in particular, of the transmission) for a chassis weighing a total of 360 t. Therefore, two types of chassis for transporting equipment weighing 220 t were designed: the 16×16 7906 model (Fig. 4.10) based on traditional technical solutions and the 24×24 7907 model (Fig. 4.11) of original design (articulated layout, GTE, etc.). Comparative tests were used to select the best solution.

Fig. 4.10. Overall view of the 7906 chassis

Fig. 4.11. Overall view of the 7907 chassis

The 7904, 7906, and 7907 chassis models are interesting because MZKT used both traditional and original design and manufacturing solutions in certain components.

The 7904 and 7906 models were examples of the classical approach to designing the main components.

Their engine-transmission units (ETUs) (Fig. 4.12) did not fundamentally differ from each other; they included a diesel engine, a matching gearbox (1), two hydromechanical gearboxes (4), a sum-up gearbox (6), distribution gearboxes on the front (5) and rear (7) axles, six axle gearboxes (3) for the 7904 chassis and eight for the 7906, and wheel gearboxes (2) equal to the number of wheels.

The 7904 chassis had a specific power of 3.75 hp/t (2.75 kW/t) while the 7906 had a specific power of 6.8 hp/t (5.0 kW/t). This ensured a maximum speed in the 7904 of 25.2 km/h and 30.0 km/h in the 7906.

Due to the absence of an HMT able to deliver a high power with acceptable weight and dimensions, both models used a block of hydromechanical transmissions comprised of two similar HMTs working parallel to the power flow. That is why matching and sum-up gearboxes were added to the transmission. Also, these models used two transmission gearboxes, one for the front and rear bogies, instead of the traditional single transfer gearbox (or power splitting case). In general, the ETUs of the 7904 and 7906 models were complex and bulky, with huge weight and dimensions, and low specific power making them unfit for practical use.

Within a search of new technical solutions of SWC ETUs, prototypes of the 547E (12×12), 7907 (24×24), and 7923 (16×16) chassis models with electric transmissions were successively created.

Comparison tests indicate that SWCs with ETs are superior to those with HMTs in a number of parameters, including the passing ability and traction-speed properties; layout flexibility since the ET elements can be placed in any suitable area of the chassis; suitability for maintenance and making repairs due to its modular design; and survivability due to the possibility to continue travel even when several traction engines fail.

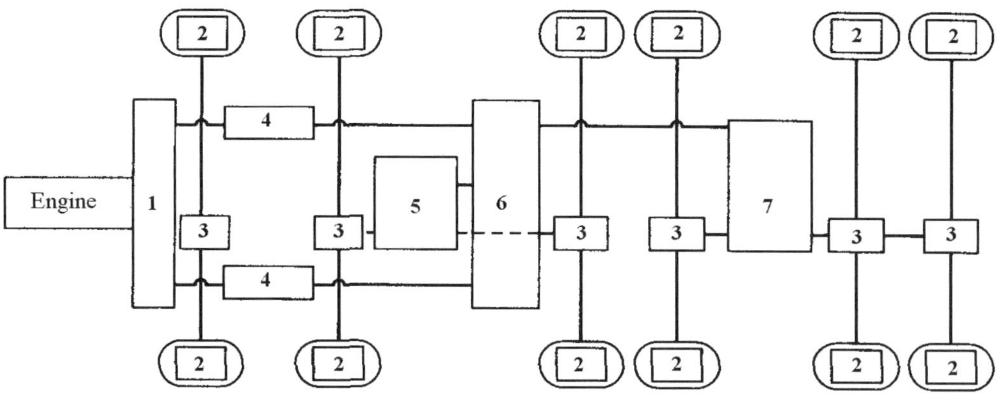

Fig. 4.12. General kinematics diagram of the transmission used in the 7904 and 7906 chassis

However, the creation of these transmissions mainly used adaptations of existing designs of industrial electric machines, which did not take into account the nature and operation modes of the chassis. Thus, the design analysis and test results revealed a number of essential drawbacks, especially in the 7907 model.

First, the transmission was made with intermediary elements, namely, mechanical gearboxes (6 in Fig. 4.13), which transfers the torque from the traction engine to the wheel gearboxes (5). This complicates the transmission, makes it heavier, and reduces its efficiency factor. A more advanced technical solution was used in the 547E and 7923 models, where the traction motors were built into the wheels creating a module consisting of a motor and a wheel (Fig. 4.14).

Second, the primary gas-turbine engine power on the 7907 was 882 kW and the generator's power was 625 kW, which led to incomplete use of engine power.

The main drawbacks of the electric transmissions used in the prototypes were as follows:

- An imperfect design of the generator and traction electric motors, which limits the maximum speed and the free traction force on the hook;
- An insufficient reliability of ET elements because of imperfect design and absence of requirements for them; and
- An imperfection of the automated control and regulation systems, which resulted in the inability to benefit from the main advantages of electric transmissions—

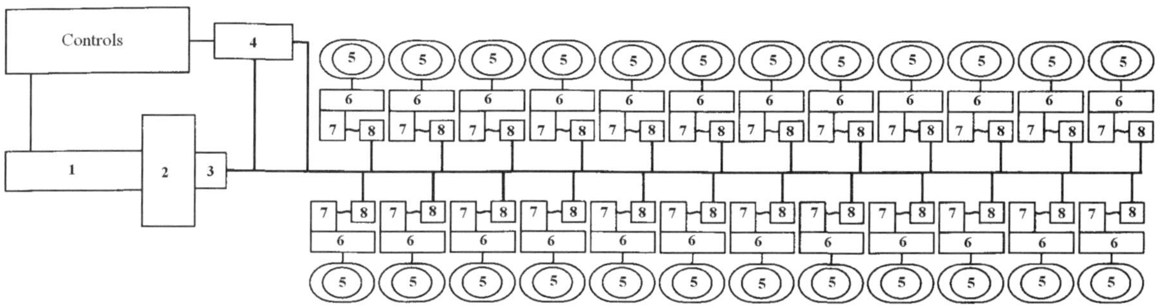

Fig. 4.13. Basic diagram of the electromechanical transmission of the 7907 chassis: 1—Gas-turbine engine; 2—Step-down gearbox; 3—Generator; 4—Automated control system; 5—Wheel gearbox; 6—Mechanical gearbox; 7—Traction engine; 8—Converter;

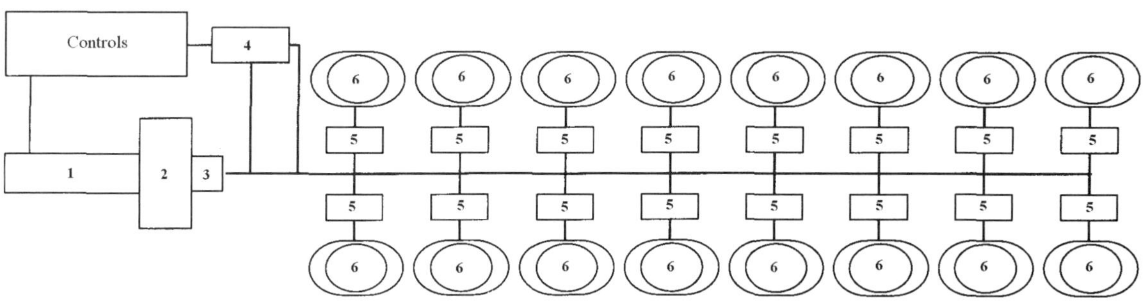

Fig. 4.14. Basic diagram of the electromechanical transmission: 1—Gas-turbine engine; 2—Step-down gearbox; 3—Generator; 4—Automated control system; 5—Converter; 6—Electric motor-in-wheel

individual control of wheel drives depending on road conditions and maximum load on the engine.

The analysis of these and other materials allows us to conclude that there is an insufficient theoretical base for designing ETs for SWCs due to their design and usage peculiarities, while the traditional approach to designing transmissions for the load capacities required of SWCs (like in the 7904 model) convinces yet again that it has no future. The weight parameters of various types of ETUs in SWC prototypes are presented in Table 4.4.

The approach to the development of the steering systems for the considered chassis was not new. Although the 7904 used steered wheels of two front and two rear axles, the drive used to turn the wheels, which was comprised of conventional levers and rods and stretched along a length of more than 20 m (Fig. 4.15), was a complicated and bulky structure. The steering system of the 7907 with steered wheels of four front and four rear axles was designed in a similar manner. Model 7906, which was more than 26 m long, had steered wheels of the four front axles only. It gave a 34.0 m minimum turning radius and created a high resistance to turning.

Table 4.4 Absolute and relative weight parameters of ETUs used in SWCs

Model, types of components	Power unit power, kW (hp)	Chassis	Weight, kg	Specific weight, kg/kW (kg/hp)	Specific weight (ETU weight/ chassis road weight), %	Overall volume, m^3	Power per volume, kW/m^3 (hp/m^3)
Diesel engine:							
KamAZ-740.10	309 (420)	BAZ-69501	1278	4.15 (3.05)	7.1	-	-
YaMZ-8401	478 (650)	KZKT-74281	1870	3.91 (2.88)	7.4	-	-
UTD-25 for tanks	294 (400)	BAZ-6944	800	2.73 (2.0)	4.7	-	-
A-38 for tanks	478 (650)	MAZ-547B	1100	2.3 (1.69)	4.0	-	-
V-58 for tanks	522 (710)	MAZ-7917	1300	2.49 (1.83)	4.06	-	-
M-351 for ships	1100 (1500)	MAZ-7906	7220	6.56 (4.82)	10.3	5.94	185 (445)
Gas-turbine engine:							
GTD-1000A	736 (1000)	MAZ-547Э	960	1.3 (0.96)	3.4	-	-
GTD-1100TMF	880 (1200)	MAZ-7907	1550	1.76 (1.3)	2.22	2.7	326 (445)
Transmission:							
mechanical	309 (420)	BAZ-69501	3680	11.9 (8.86)	20.4	-	-
HMT	386 (252)	MAZ-543	6230	16.14 (11.87)	30.6	-	-
HMT	478 (650)	MAZ-7911	6210	12.99 (9.57)	29.0	-	-
HMT	478 (650)	MAZ-547B	8760	13.5 (9.95)	31.8	-	-
HMT (two)	1100 (1500)	MAZ-7906	17600	14.0 (11.8)	25.0	12.0	91.8 (125)
Electric:							
DC	736 (1000)	MAZ-547Э	8060	9.95 (8.06)	29.3	-	-

Table 4.4 Absolute and relative weight parameters of ETUs used in SWCs

Valve-type:							
with wheel gearboxes	880 (1200)	MAZ-7907	20050	22.7 (16.7)	27.8	16.0	55 (74.7)
with motors-in-wheel	880 (1200)	MAZ-7923	14650	16.7 (12.2)	21.5	11.3	77.8 (116)
ETU:							
Diesel+HMT	1100 (1500)	MAZ-7906	24820	22.6 (16.6)	35.5	17.94	61.4 (83.5)
GTE+ET	880 (1200)	MAZ-7907	21600	24.9 (18.2)	30.9	18.0	49.0 (66.5)

An example of addressing the above problems was a prototype of the 79085 (8×8) model chassis with all-wheel steering created at the N.E. Bauman MSTU in the 1990s (Fig. 4.16). It was able to make turns and move relative to any axle. Its wheels were turned by the electrohydraulic servo drive (EHSD). The StS drive used in the 7904 chassis is presented in Fig. 4.17 and can be compared in its complexity with the standard drive (see Fig. 4.15).

In our opinion, the best steering system available today was designed by the Division of Wheeled Vehicles of the N.E. Bauman MSTU with a "flexible" turning geometry. It was realized in a full-scale prototype, model E79085, weighing 56 t (see Figs. 4.16 and 4.17b). As a result of scientific studies, an algorithm was designed for the onboard computer based on the most advanced achievements in the motion theory of wheeled vehicles.

The algorithm assumes the possibility of obtaining any required dependence of the transmission ratio from the turning angle of the steering wheel, which is practically impossible in mechanical steering systems. An example of this characteristic is presented in Fig. 15.12.

Fig. 4.16. Overall view of the 8×8 chassis with an all-wheel StS

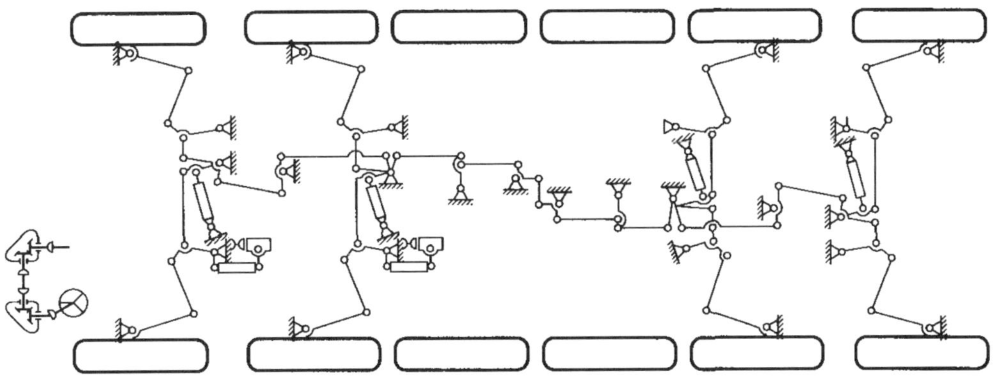

Fig. 4.15. Diagram of the 7904 chassis StS drive

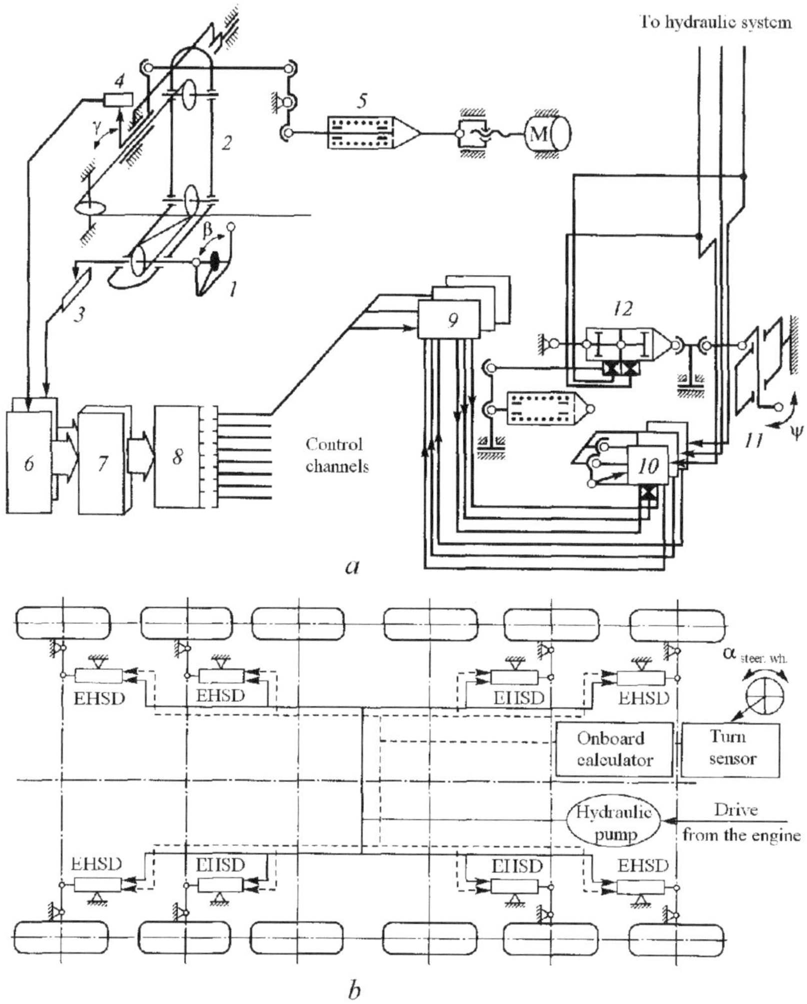

Fig. 4.17. StS with an EHSD: a—Overall StS diagram of the object E79085; 1—Steering wheel; 2—Steering column; 3—Course sensor; 4—Turning pole position sensor; 5—Steering wheel loading mechanism; 6—Analog-to-digital converter (ADC); 7—Onboard computer; 8—Digital-to-analog converters (DACs); 9—Electronic control units; 10—Three-channel steering unit; 11—Turning cam; 12—Hydraulic cylinder; b—Layout of StS with EHSD on a six-axle chassis

When in the zone corresponding to small turning angles of the steering wheel, i.e., close to rectilinear motion and typical for high travel speeds, the vehicle has low sensitivity to steering (high route stability). At larger turning angles, the maneuvering properties sharply increase.

The main task of the onboard computer algorithm and its calculation program is to coordinate the turning angles of each wheel in accordance with the selected method of steering. Since the position setter of the turning pole allows for continuously changing the place of the pole (not in the discrete manner, like in Cometto), the algorithm becomes universal and able to realize practically any steering method. Figure 4.18 presents a calculation example of a steered wheel's turning angle, where i is the serial number of the wheel; ψ is the output

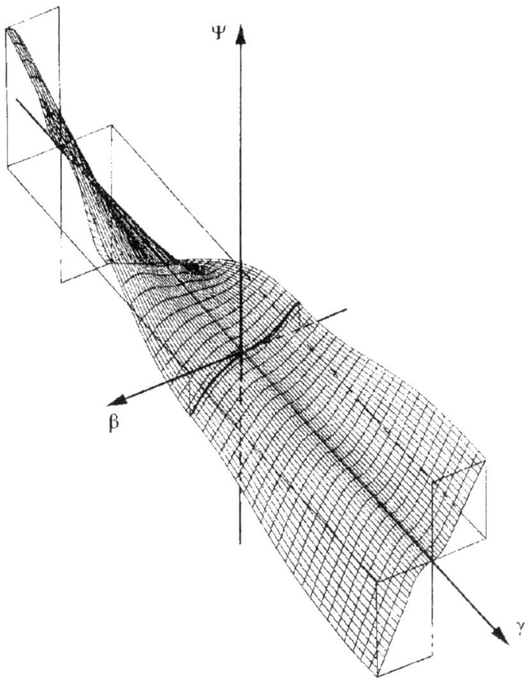

Fig. 4.18. Graphic illustration of the operation algorithm of the StS onboard computer in the 79085 chassis: β—Turning of the steering wheel; γ—Angle corresponding to angle β; ψ—Turning angle of the wheel

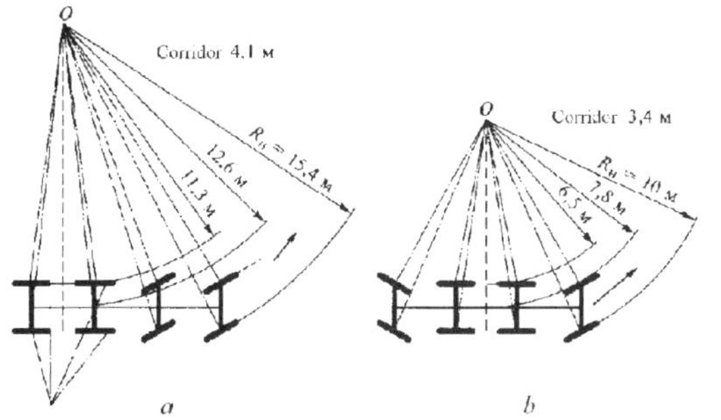

Fig. 4.19. Comparison of the maneuvering properties of vehicles with different turning methods: a—MAZ-7908 prototype; b—E79085 model

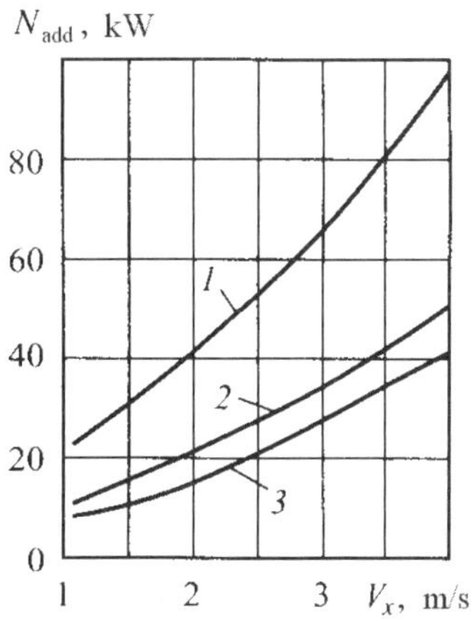

Fig. 4.20. Comparison of the power consumption N_{add} versus the travel speed V_x with different steering models: 1— MAZ-7908 prototype; 2— E79085 model; 3—Rectilinear motion

signal proportional to the turning angle of the ith wheel; β is the input signal proportional to the turning angle of the steering wheel; and γ is the input signal proportional to the position of the turning pole.

The structure of the algorithm is such that the turning angle of each wheel is calculated independently. This is applicable to vehicles with any number of axles and wheels. The implementation of the StS main diagram (see Fig. 4.17a) and the operation algorithm of the onboard computer (Fig. 4.18) in the E79085 model has provided it with the following advantages compared to the prototype:

a. Decreased turning radius and overall dimensional corridor (Fig. 4.19); and
b. Decreased power consumption of the engine when making small-radius turns (Fig. 4.20).

Since all wheels of the model are turned in accordance with the required kinematics, there is no tire breakaway and no slippage in the contact spot of the tire tread with the road. On the other hand, the prototype, which has—like all other automobiles—several axles with non-turning wheels, consumes additional power for tire deformation and friction in the contact point. The experimental dependence, presented in Fig. 4.20, was obtained at a turning radius of 25.6 m—identical for both the prototype and the E79085 model. The power required for turning the prototype (curve 1) is double the power needed for turning the E79085 model (curve 2), which is almost equal to the power needed for a rectilinear motion.

The E79085 model also has the following additional advantages compared to its prototype:

1. A more rational maneuvering of the vehicle on the route.

   This advantage is most relevant for multi-axle long-base vehicles, especially when reversing during parking. The traditional steering system used in the prototype caused blocking of the chassis between adjacently parked vehicles; while the new system has an ability to change the position of the pole, thus changing the position of the turning center, which creates broad opportunities for maneuvering. The model did not need as much space for maneuvering when leaving the yard. The area sufficient for the E79085 model to make a right-angle turn and subsequently leave the yard was not enough for the prototype. If the prototype tried to move within the dimensions of the E79085, it was forced to begin turning earlier, then, when approaching the gate, it failed to make a 90° turn and could not fit into the gate opening.

2. Decreased the effort spent by the driver when steering the vehicle.

   Bypassing an obstacle is an example of this advantage in action. The prototype's driver has to turn the steering wheel to the right and move around the center of turning, located in one point, and then quickly turn the steering wheel in the opposite direction in order to return to the course by turning around another turning center. The E79085 model is able to bypass an obstacle without problems by using the "crabbing" motion.

The most advanced steering systems are equipped with a built-in electronic system controlled by an onboard computer. They are able to cope with a number of required turning schemes and allow for obtaining a nonlinear characteristic of the transmission ratio. This feature allows for reducing the vehicle's sensitivity to steering along straight sections, along sections with a large turning radius (which is close to a straight section), and increases the sensitivity at maneuvering. These control algorithms increase course stability at high travel speeds in combination with good maneuverability.

In our opinion, the electrohydraulic steering drive, especially its autonomous version, can be considered—both today and in the foreseeable future—as the most promising due to its multiple advantages.

When summing up the progress in creating SWCs starting from early 1980s, we should note that none of chassis prototypes designed both with traditional and non-traditional technical solutions were approved for batch or mass production. This fact can be explained only in the retrospect of the political and economic events that occurred in the USSR and worldwide during that period.

The main results of the evolution of SWCs can be summed up in the following way:

Based on the huge amount of theoretical and experimental studies performed by many organizations in the USSR (first of all, by MZKT), the country managed to design and launch the production of vehicles, chassis, and tow cars with 4 to 12 driving axles and load capacities of 20 to 220 t. All were full-drive models equipped with powerful engines (from 177 to 1104 kW, or from 240 to 1500 hp); hydromechanical transmissions; two-step distribution boxes and axle drives with planetary wheel transmissions; articulated axles with independent torsion or pneumohydraulic suspensions; wide-profile, low-pressure, heavy-duty tires with regulated air pressure; and differential drives of the driving wheels, which comprised of self-blocking and blocked inter-wheel and inter-axle differentials. However, when creating these HDWVs, designers followed stereotypes of the overall layout, based on particular earlier technical solutions (see Figs. 4.1 and 4.2).

Any further improvements in SWC designs based on traditional technical solutions (obtaining a new quality at a higher level) seems impossible since the chassis operated today and planned for production in the near future have reached their limit of design excellence and have practically identical specifications.

The same concept is underlying the family of chassis and tow cars manufactured by KZKT, which makes use of many technical solutions initially suggested by MZKT.

Specifications of the main models of the KZKT family are presented in Table 4.5.

Today, the design of SWCs results from the theoretical and experimental studies and the technical and technological achievements accumulated and reached by many branches of the defense industry. This creates a sufficient foundation for creating a new generation of chassis that will meet the most stringent requirements of today and the future.

Table 4.5 Basic specifications of the KZKT family

Parameter	KZKT-8005	KZKT-8014
Wheel formula	8×8	10×8
Load capacity, t	22.5	40
Full weight, t	50	66
Maximum climb, degrees	18.0	21.5
Engine:		
type	Diesel	Diesel
model	TMZ-8424.10	TMZ-8424.10.08
power, kW (hp)	312.5 (425)	349 (460)
Tire dimensions, inches	18.00-24	18.00-24
Maximum speed, km/h	60	60

Part II

Overview of HDWVs

Chapter 5
HDWV general design methods

5.1 General provisions

The material in Chapters 1–4 shows that in spite of the accumulated experience, the best way to create new HDWVs has not yet been determined.

The need to design new wheeled chassis arises from the following:

- Existing vehicles quickly become obsolete; and
- New applications appear which cannot be coped with by the existing chassis models for technical or economic reasons.

Designing a new chassis means that a set of technical documents to organize the proper manufacturing and operations of the vehicle must be created. These technical documents can be viewed as a model of the chassis presented in a graphic, numerical, or other form. Modern computer-based engineering has opened up radically new opportunities to automate the design process, where the degree of automation is defined by the formalization degree of the project assignment (a rather strict formulation of the problem and the availability of its solution algorithm).

At the design stage, initial data from the following are known:

- The assumed areas and conditions of using the vehicle in its main function;
- The current and potential future requirements for the operational and consumer properties of the vehicles within a broad range of their applications in the economy;
- The efficiency of using the vehicle for its main purpose, as well as the choice of the technical criterion (criteria) and method (methods) of assessing the efficiency; and
- The design solutions of possible prototypes and the basic trends of the world motor industry as regards the respective load capacity class.

Today, algorithms to directly create complex technical objects are not known; therefore, they are designed through multiple iteration and repeated analysis of various project options. The creation of a new wheeled vehicle is no exception and generally consists of the following basic steps (Fig. 5.1).

1. The design aims and objectives are defined.
2. The main idea (project, concept) of the future chassis is formulated and the solution variants (options) are assumed. This is based on the ability to synthesize new ideas and principles and achieve a stepwise improvement of the object characteristics under design. This means that at the initial design stage, the role of the project manager and his or her experience, intuition, knowledge, and qualifications are of paramount importance. For example, the success of SWCs created in the USSR and Russia, which in many aspects exceeded their foreign-made analogs, was possible

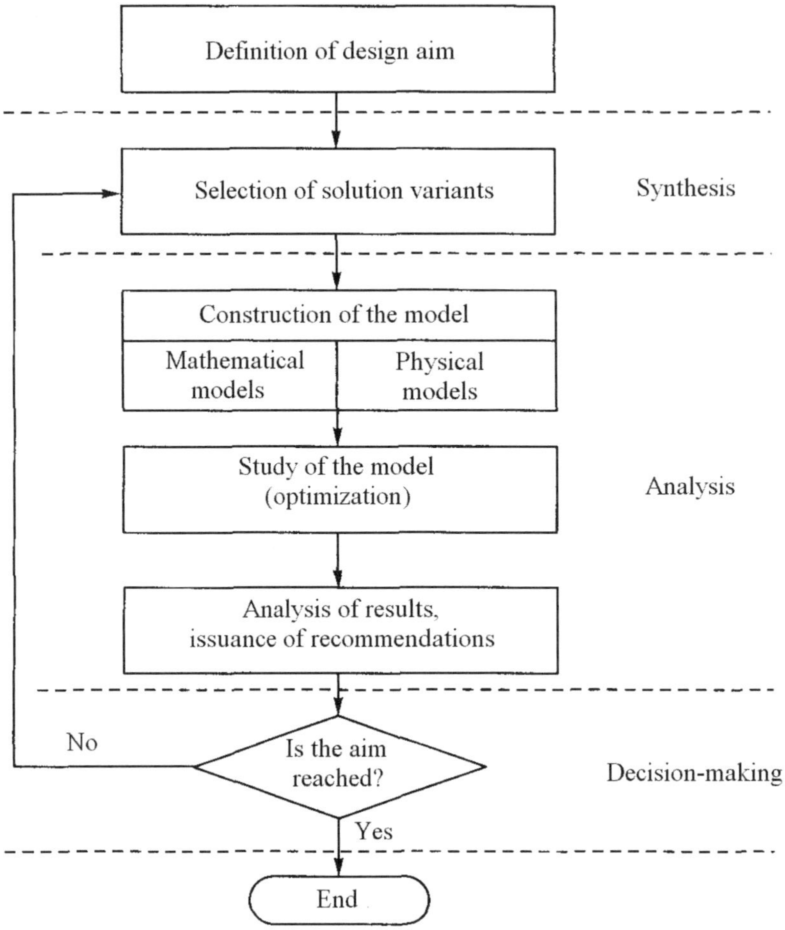

Fig. 5.1. Algorithm of the process of designing multi-axle wheeled chassis

thanks to the efforts of outstanding designers like B.L. Shaposhnik and V.A. Grachov, who acted as project managers and successfully used the results achieved by the scientific team headed by P.V. Aksyonov.

However, the general progress of engineering and the history of multi-axle SWCs, in particular, show that very few projects are absolutely new and even those are using technical solutions that are based on previous experience.

3. The engineering analysis is carried out during which the following is performed:
   - The definition of the structure and parameters of the chassis; and
   - A check of the project idea's compliance with physical laws and other restrictions imposed by the specific requirements of HDWVs.

This stage uses both physical (material) and mathematical (abstract) models that reflect—to an approximate degree—real vehicles in their interaction with the environment and operating conditions. These models yield necessary information about the complex system, which is commonly known as the "driver-vehicle-environment."

The physical models most commonly used in the engineering analysis of wheeled vehicles and their main units are:

- Chassis prototypes (full-scale, dimensional-layout, external appearance, etc.);
- Blow-off models both of the chassis and the vehicle as a whole;
- Models of the heavy-duty vehicle; and
- Various testing complexes.

Physical modeling (simulation) requires significant amounts of material and time; however, it gives the best and most reliable understanding of the processes under study and is the only way to define new regularities. When chassis with new concepts (non-traditional technical solutions) are designed, physical models become even more important.

Mathematical (analytical) modeling allows for conducting studies and obtaining results in a general form suitable for analyses. However, creating these models requires significant simplification and schematization of the objects under study, which reduces the reliability of the results. Numerical modeling has broader possibilities and has become one of the main methods of studying complex objects and processes because it yields high precision and reliable results.

4. Choosing the final unique solution among all possible project options. This choice requires compromise and is ambiguous since every option has pros and cons. This stage entails a multi-step process of choice and analysis, with each progressive step narrowing the search field by limiting both the number of options and the scope of factors taken into account.

The second design stage—generating potential solutions—is the least formalized. Only if a solution is based on known elements can the search for a rational combination be formalized.

The number of mathematical and physical models in many ways defines the chances of formalizing the engineering analysis. If we design a principally new chassis, the role and scope of the physical modeling increases.

Formalizing the decision-making process is possible if the design aim is described using indicators of the project parameters. Yet simply choosing quality indicators of the project parameters does not formalize the process. The task is then to find a combination of parameters in which the indicators reach their extreme value. The problem is that a global extreme that can be used in designing complicated objects, such as wheeled chassis which are members of even more complex vehicles (Fig. 5.1), does not seem to exist even if an undeniable, universal, and unambiguous quality indicator is found. However, the history of these objects (see Chapters 1–4) show that several nearly equivalent local extremes exist, but that the choice among them is far from obvious. Finally, it should be noted that the quality indicators of the project solution are not constant—they tend to vary with time, meaning, for example, that the chassis that was once recognized as ideal may no longer be the best choice. Therefore, the role of the project manager at this stage is highly important since a wrong decision results—in the best case—in early termination of the project. Any scientifically justified methodology on how to create the vehicle's general appearance and how to automate analysis of project options should help the project manager in decision-making by narrowing—as much as possible—the scope of rational (competitive) project options.

The systemic approach is the best way to design HDWVs and SWCs, in particular, and assumes the following:

- Studying the system under design as a united object, which performs definite functions in particular conditions (Fig. 5.2);
- A possibility to split complex systems into relatively independent subsystems (for example, levels 1–5 in Fig. 5.2); and
- A possibility to further split subsystems into lower-level subsystems (these are the elements, which can't be used to solve problems but whose properties affect the

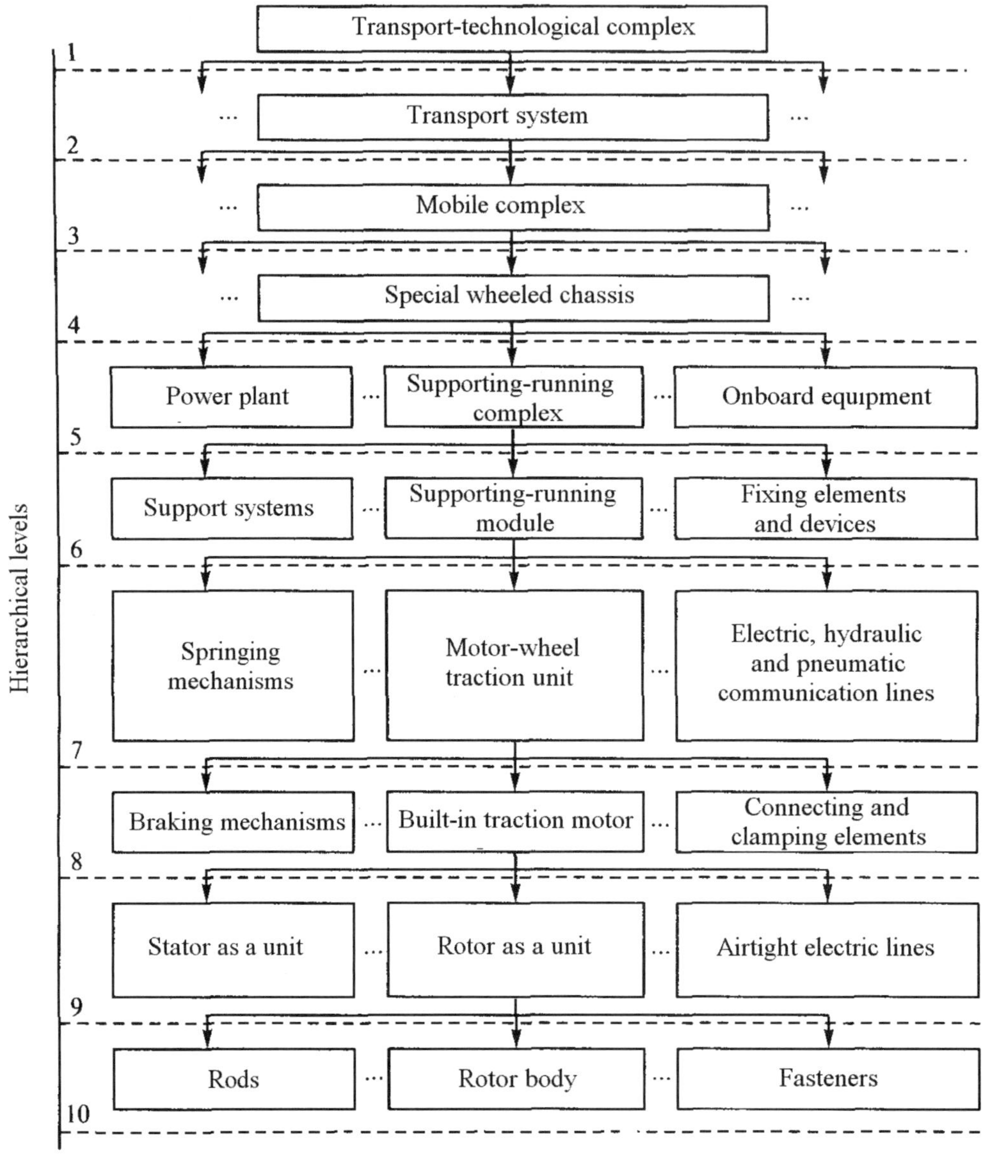

Fig. 5.2. Hierarchical structure of a multi-support wheeled chassis model with electric drive of the wheels (a fragment)

properties of other subsystems and the system as a whole (levels 6–10 in Fig. 5.2).

Thus, creating acceptable mathematical models, which set up the quantitative relationships among the system's parameters and characteristics, is the most important task in automating SWC design. As we move down the hierarchy, the parameters of the upper level become characteristics of the lower level. For example, the chassis' load capacity is a parameter when studying the vehicle as a whole. It also becomes a characteristic when forming the chassis' general appearance, which is an element of the vehicle.

The mathematical model is generated as follows:

1. A list of parameters which can significantly affect the outcomes of modeling is defined (the number of parameters should provide sufficient information required for decision-making at the current stage of project development);
2. The interconnections between the parameters and between the parameters and characteristics are defined (the major challenge is to identify the most important interrelationships, which depends on the design stage, the task at hand, the designers' experience, and the completeness of the a priori knowledge about the chassis under design);
3. Parameters that are not associated with other parameters and characteristics are excluded from the formalized description of the model (they are left for consideration at subsequent design stages);
4. The problem solving algorithm is developed;
5. The required software is created, which includes debugging and optimizing software complexes, test problem solutions, etc.;
6. The necessary calculations are made;
7. The calculation results are analyzed; and
8. The materials are prepared for decision-making by the project manager.

When designing a chassis, the following basic provisions of system optimization should be used as guidelines:

- A system should be optimized as a whole, united object with a predefined functionality (a system consisting of optimal subsystems is not always optimal itself);
- A system should be optimized under a unique, predefined, and quantitative criterion, which reflects the design aim; and
- The obtained optimum system is always relative, since optimization is carried out with certain quantitative restrictions imposed on the optimized parameters.

In this context, the most important task of HDWV design methodology is the synthesis of interrelated criteria, which should allow for the following:

- Objectively evaluating the design results at each hierarchical level; and
- Finding parameters for each element of the subsystem, which ensures high efficiency of the entire system.

5.2 Hierarchical structure of SWC as a complex technical system

5.2.1 Common approach to describing the SWC structure

A SWC is a complex technical system with a highly developed hierarchical structure (see Fig. 5.2, levels 4–10).

It is well known that when creating final products, there is no need to build the entire system hierarchy—it is enough to study the systems and subsystems located above and below the required level. When creating a new SWC model, it is necessary to consider at least the transportation system within which the SWC is designed to work which has its own hierarchy of complex interactions.

A chassis, when viewed as the initial system, can be split into subsystems, which can subsequently be divided into a number of elements, aggregates, and units (Fig. 5.2).

The hierarchical levels are interconnected by two types of relationships:

1. Structural relationships, which put its elements, blocks, aggregates, and interrelations into an order (solid lines in Fig. 5.2); and

2. Functional relationships, which characterize the functions for which the given subsystem is designed (dotted lines in Fig. 5.2).

Each hierarchical level corresponds to its own list of questions whose answers can be used for making design decisions about that level.

At the highest hierarchical levels (1 and 2), it is necessary to define the types, technical and economic characteristics, and number of chassis used; the operation areas and conditions; the most probable traffic routes; and the permanent and temporal garages for SWCs and equipment thereof, future maintenance methods, etc. The answers to these questions are compiled in a document called the Terms of Reference (TOR) (Fig. 5.3). The following three levels (3–5) are dedicated to forming the technical appearance of the chassis, its layout diagram, the overall and basic design-layout dimensions, and the structure and composition of its main aggregates and systems. The results of which are documented as the Technical Proposal, which describes the SWC's general appearance, and the Preliminary Design, which confirms the general design feasibility.

The subsequent levels are dedicated to solving the design and engineering problems of all units and parts of the chassis' aggregates and systems. This is the Detailed Engineering stage and the results are documented in the Technical Documentation intended for manufacturing and operating the chassis.

Thus, when the stages are properly organized, design solutions are made step by step through the levels of the hierarchy (see Fig. 5.2), ensuring a more detailed chassis design.

The structure of a typical design bureau specializing in designing new vehicles is to some extent reproducing the hierarchical structure of the object under design (see Fig. 5.2, levels 4–10). Traditionally, design bureaus comprise of subdivisions in charge of formulating the general appearance of the vehicle (divisions of general views, overall layout, etc.), which are formally or actually above the other structural subdivisions. Design bureaus in charge of accompanying serial manufacture usually have a more fuzzy structure without a clear hierarchy of subdivisions.

The appearance of a chassis, its structure, layout, geometry, and dimensions can be described by a finite number of symbols called chassis parameters. These are, for example, the coefficient of the total resistance to motion in particular conditions, the specific traction force (dynamic factor), the load capacity coefficient, and aerodynamic coefficients. The chassis parameters are designated by both symbols and numerals, e.g., the layout of the support-running complex (SRC) is described with the help of a system of specially developed signs/codes.

The chassis' properties can be described by another set of symbols called the Chassis Characteristics (for example, the travel speed, fuel distance, service life, etc.). Most often, these properties are described by systems of numerals. In general, characteristics are functions of parameters and other characteristics.

5.2.2 Nature of the links at the Terms of Reference (TOR), Technical Proposal, and Preliminary Design steps

Direct links (solid lines in Fig. 5.3) deliver the results of work conducted at higher hierarchical levels to the lower levels and contain the initial data for organizing work at these levels. This information is also the basis for formulating the criteria and restrictions for solving problems on this level (the so-called discipline conditions).

For example, direct links between the "Development of the Terms of Reference" and the "Development of the Technical Proposal" levels are delivering variables that characterize the future chassis properties to be regulated. Direct links between the "Development of the Technical Proposal" and "Preliminary Design" levels transmit the general description of the

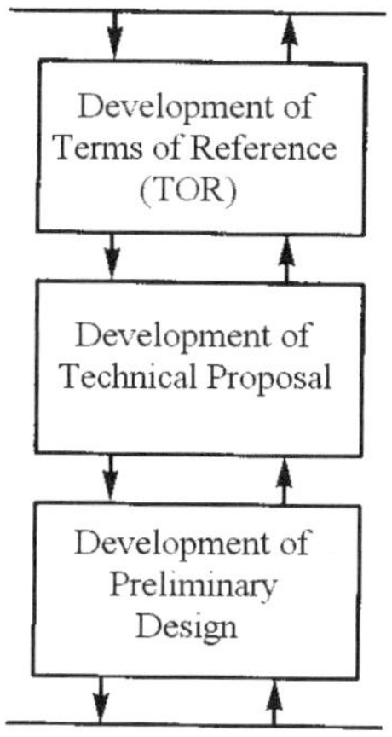

Fig. 5.3. Structural links of SWC design stages

future chassis, which confirms the full or partial fulfillment of the Terms of Reference. This description includes the general design of the chassis, its general layout, areas and volumes of cargo platforms or boxes, mass and centering characteristics, full weight report, aerodynamic characteristics with and without load, stability and maneuverability characteristics, results of preliminary strength calculations, performance characteristics, common knowledge about construction, chassis systems and aggregates, power unit, onboard and special equipment, etc. It is also formalized in accordance to the system of design and technical documentation exchange that is being used.

The inverse (return) links for the "Development of the Terms of Reference" level transfers the forecasted technical-economic characteristics of the new chassis to the designers of the entire vehicle and is expressed through generalized efficiency indicators (i.e., through the load capacity coefficient, specific fuel consumption, average travel speed along typical routes, etc.).

The inverse links running to the "Development of the Technical Proposal" level deliver information about the design and layout characteristics of different chassis diagrams; characteristics of engines, transmissions, equipment, and systems; properties of structural materials approved for application, etc.

As a rule, the chassis is designed in a top-down manner, where the work at each level should confirm the characteristics applied and formulated at a higher level. At the same time, the forecast nature of information requires running iteration cycles among different levels. Therefore, if certain links are ignored (for subjective reasons, the impossibility of taking certain factors into account, or a lack of reliable information available at this design stage), there is a danger of making incorrect or unjustified decisions.

For example, in the 1970s, the designers of the ZIL-135LM wheeled chassis model, intended for mounting and transporting the "Luna" (Moon) tactical missile complex, were required to ensure high chassis maneuverability, as well as high off-road passage ability (a tactical requirement) and strength (rigidity) of the carrying system (to ensure the functionality of the target equipment). The width of the chassis was also strictly limited by the dimensions of the cargo hatch of the An-12 transportation aircraft (the requirement for ensuring operative mobility).

In order to meet the requirements of a higher hierarchical level, the designers chose to use an SRC with the middle axles brought closer, oversized tires, and a rigid and wide girder frame. To achieve the acceptable maneuverability of the chassis, a steering system (StS) with steered front and rear wheels, with the wheels of the front and rear axles turning synchronously, was used.

Thus, the desired maneuverability was ensured; however, the critical speed (the highest speed possible without losing stability) was 22–23 km/h, whereas the chassis' maximum speed defined by the tactical and technical assignment was 60 km/h, while the maximum possible speed of the energy unit was 80–85 km/h. As a result, the chassis was capable of high speeds, but these speeds greatly exceeded safe operating conditions. In the 1970s, the travel properties of vehicles with rear steered wheels had not been studied and the theory of non-stationary lateral skidding was only just being developed; therefore, the designers were unaware of the stable motion idiosyncrasies that occur when a vehicle has steered front and rear wheels. Thus, a wrong decision was made for objective reasons (absence of necessary information at the moment of decision-making) and the manufacture of the ZIL-135LM was later transferred to Bryansk Automobile Works (BAZ).

The idea of using steered rear wheels became popular due to its obvious design and layout advantages—perfect maneuverability under limited turning angles and small wheel niches. This is especially beneficial when designing floatable chassis (amphibias). By that time (mid-1970s), the stability theory of multi-axle wheeled vehicles, including those with rear steered wheels, was sufficiently mature and the theory of non-linear and non-stationary lateral tire skidding was in place. In particular, the stability loss of vehicles with rear steered wheels had been studied and recommendations for calculating and designing StSs for them drafted. However, the BAZ-made three-axle wheeled chassis was equipped with a steering system that ensured synchronous turning of the front and rear steered wheels and they are still manufactured with this system in place. This is an example of a subjective decision-making mistake.

The "Luna" missile complex design also provides an instructive story related to designing a chassis with two engines. The mobility requirements of the "Luna" complex could only be met by achieving a specific power of 11.2–12.6 kW/t (16–18 hp/t) while the full weight of the carried equipment was 20–22 t. This meant that the chassis had to be equipped with at least a 224 kW (320 hp) engine with a 268–290 kW (360–390 hp) one preferred. However, the Soviet industry could not offer any batch-manufactured automobile engines of this power, while a diesel tank engine was unacceptable because of its low service life. A subjective factor was the designers' intent to use domestic production facilities for building the chassis (this meant using in-line 77 kW (110 hp) petrol ZIL-120 engines and the 105–126 kW (150–180 hp) ZIL-130 engine family).

Therefore, it was decided to use a two-engine power unit with a respective transmission. This solution was initially implemented with ZIL-120 engines and later used ZIL-375Ya4 engines. The designers tried to maximize the chassis' passing ability which was a notable feature and reflected the work of ZIL's Special Design Bureau. Hence, the chassis' transmission was made in the block board scheme which reflected the viewpoints of many experts in the field, including the top managers of ZIL's Special Design Bureau. As a result, the engines worked separately for each transmission board and were linked only through the driving wheels' interaction with the support surface.

The designers of the "Luna" complex had no scientific justification on the work of multi-engine power units in systems with elastic wheels and could not predict the consequences of their decision. Later it was established that these systems have an intense power circulation among the engines caused by a kinematical mismatch of their operation modes due to wheel elasticity. In this case, it was intensified by the technical and regulation distinctions of the engines.

For the chassis using the ZIL-135 engine, this situation was worsened by a mismatch in the engines' operation when the chassis turned and the wheels of different boards rolled along different radii. The carburetor system of fuel supply did not allow for regulating the engines at partial and full loads, while the idle run systems were essentially autonomous. The carburetors had a notable scatter of regulatory characteristics; therefore, attempts to ensure synchronous engine operation by regulating their idle run were completely useless.
The prototypes of the ZIL-135 chassis used hydromechanical gearboxes (HMGBs) which somewhat remedied the situation.

BAZ, where the chassis' batch production was transferred, failed to cope with the complex transmission and replaced the HMGB with a mass-manufactured mechanical gearbox. When operated, the BAZ-135LM chassis consumed huge amounts of expensive fuel (100 l/100 km of AI-93 or B-70 petrol) and had a decreased service life of the power unit. Specific problems in the steering also appeared, for example, switching gears was extremely difficult especially when driving on soft grounds. Also, if the front and rear wheels were left hanging (such as when overcoming ledges and rollers, beginning a steep descent, or finishing a steep ascent), the chassis could start to spontaneously turn, independent of the driver, due to the power circulation among the engines. This was especially dangerous when the complex was loaded into a transport airplane or was leaving a narrow box.

Similar reasons forced designers of the first Soviet batch-manufactured, floating armored personnel carrier to apply a two-engine power unit based on the 63 kW (90 hp) GAZ-40 petrol engine. For certain reasons, mostly due to production, the engines were grouped along axles instead of sides (boards). When moving on the ground, the left-hand engine drove the first and third axles while the right-hand engine drove the second and forth axles. The inter-axle sections of these "bogies" had a blocked connection. Later, it was experimentally established that this variant is somewhat better than the detached drive of the boards from independent engines.

The described technical solutions are difficult to qualify as mistakes due to the lack of information available; these solutions were the only way to achieve the assigned tasks within the set terms without organizing an auxiliary production of special engines

In the 1980s, the operation of wheeled vehicles with two engines vividly revealed deficiencies. By that time, the peculiarities of multi-engine power transmission units operating with elastic wheels had been studied properly by N.E. Bauman MSTU and a range of other organizations.

The theory of automated engine control was also developing and allowed for theoretically describing and analyzing the work of several internal combustion engines (ICEs) of different power in blocked loops. Design experience in creating ship multi-engine power units, where engines of different power (and even of different types) worked for one propeller shaft was already available. Designers also knew how to build multi-engine caterpillar (Sherman tank, armored personnel carrier (APC) M-559) and wheeled (Twister, etc.) vehicles. Bryansk Automobile Works had built prototypes of the BAZ-135 wheeled chassis with a YaMZ-238 diesel engine and had designed and put into batch production the BAZ-135MB wheeled chassis with a YaMZ series diesel engine. The powerful YaMZ-240 engine already existed and was put into large-scale production. Prototypes of a 12-cylinder engine were made by Kama Automobile Works (KamAZ) and its future production program included 10-cylinder charged and non-charged modifications of the KamAZ-740 engines.

In general, Soviet engine-builders were able to equip the BAZ wheeled chassis with automobile engines of the required power. Corresponding proposals regarding the overall layout of the engine section for a four-axle BAZ chassis using several different engines were also developed, in particular by MSTU.

However, a so-called "discretionary decision" was made and the four-axle BAZ chassis were equipped with two KamAZ-740 engines and batch-manufactured mechanical KamAZ gearboxes. Since the basic transmission diagram of these chassis reproduced the transmission of the BAZ-135 chassis and was built by the board principle, the engines also worked for each of its boards. As a result, all the earlier drawbacks were still present in the new BAZ series chassis.

These problems were further aggravated by the use of diesel engines equipped with all-mode regulators of the high-pressure fuel pumps (HPFP) instead of petrol engines with relatively good adaptability (especially at working on partial characteristics). The HPFPs maintained a constant shaft rotation speed with an accuracy of at least 5–7% at all positions of the accelerator pedal. For this reason, when the chassis was making a turn, the diesel regulator of the forward board reduced the fuel supply and decreased the engine power, sometimes down to a full stop.

From a technical viewpoint, this project is one of the best examples of a wrong decision made for subjective reasons. However, it should be taken into account that the problem is a specific engineering decision and that multi-engine power units for special and fighting wheeled chassis are not as simple as they are usually considered to be.

To reduce the risk of making wrong decisions for subjective reasons, it is necessary to expand the amount of scientific-technical knowledge accumulated on the perspective development directions of off-road vehicles. This will make it easier to test and master new technical solutions and reduce the risk of mistakes made due to unjustified decisions

5.2.3 SWC models

The appearance of numerical (digital) methods of analysis and decision-making allows for the creation of better models that adequately reflect the objects or processes under study.

Experience has shown that making models more complicated does not always lead to better quality design decisions; in other words, the additional design expenses are not justified by higher quality practical designs. A reasonable compromise between accuracy, time, and calculation costs is required. At present, it is not possible to create a universal model of a special wheeled chassis that is a suitable solution for all types of design problems. At present, the use of private (particular) models that allow for thoroughly and deeply analyzing certain particular aspects and properties of processes or objects (see Fig. 5.2) is still very topical. This means that each design phase (Fig. 5.3) uses its own models, which differ in the level of detailing, i.e., in the number of factors to be considered and the links among them. These models are also "hierarchical."

At each hierarchical level, a chassis model is a set of sub-models, which reflect different sides of the vehicle's structure, operation, and subsystems (Fig. 5.2).

The top hierarchical levels (levels 1–3 in Fig. 5.2) use mainly functional models (describing the functionality) of the entire SWC with all the attached or mounted process equipment (the so-called transportation-technological complexes). In these cases, the chassis models should have the maximum degree of generality without losing information about the specific properties of the particular SWC. However, these models should be sensitive to changes in the main tactical and technical parameters and properties of the chassis. This allows the designer of the equipment to be mounted on the chassis, who works at this level, to evaluate the efficiency of the potential tactical and technical assignment solutions. When generating these models, it is important to use compact and adequate descriptions of the chassis and vehicle's operational (use) conditions.

At the "Development of the Technical Proposal" (Fig. 5.3) step, a chassis model with a much higher level of detailing and completeness is needed since it is intended for obtaining reliable information about the possibility of achieving the set goals, the ability to implement the offered design and engineering solutions, and for evaluating the comparative efficiency of these solutions.

This SWC model is formed by obtaining the following calculations: weight, dynamic and aerodynamic properties, strength, stability and controllability, etc. These are combined into separate blocks, which are sub-models (particular models) of the general chassis model.

The early stages of designing a SWC use the following basic models (Fig. 5.4).

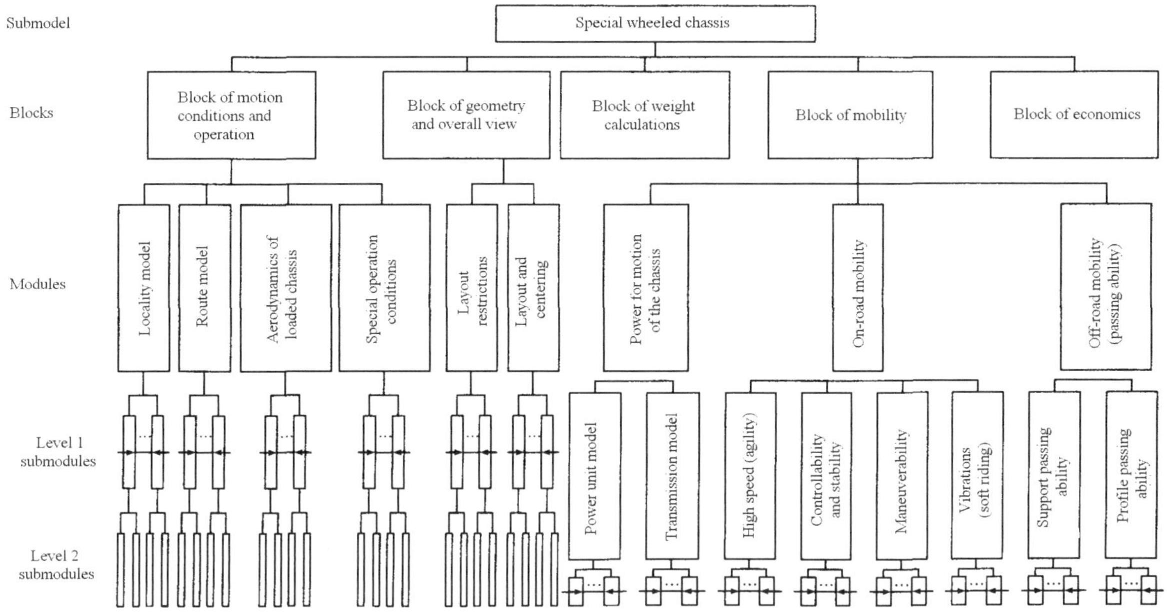

Fig. 5.4. Structural diagram of the "Special wheeled chassis" model (top hierarchical levels)

1. **The geometric model** forms a visual image of the chassis under design and is usually documented in general view drawings and the so-called sheer drawings of the chassis' overall layout. The model describes the interrelationships of the chassis parameters and its shape and dimension characteristics. It helps to define the contours, areas, volumes, and cross-sections of the chassis and its aggregates. The model is used for the weight, aerodynamic, and strength calculations and for displaying the design outputs. At later stages, it is used to develop the technological documentation, tooling, and programs for numerically-controlled (NC) machine tools. At present, there are several powerful software tools that allow for generating perfect geometric models and using them at all stages of the automated development and pre-production process within the framework of the so-called "paperless technology" (complexes of AutoCAD (computer-aided design), Compass, etc.).
2. **The weight model** provides information about the weight of the chassis and its components and the results are documented in the "weighted summary report," with the level of detailing defined by the project stage. The model describes the interrelationships of the chassis geometry, loads, power circuit peculiarities, target load value, equipment and hardware composition and location, as well as the weight of the chassis and its components. The model also reflects, in some hidden form, the technological level of the production facilities where the designed chassis will be manufactured.
3. **The aerodynamic model** provides information about the displacements and load of the static chassis or of the chassis as it moves with/without load under the action of wind or other factors of an aerodynamic nature. For multi-axle wheeled chassis, which have large dimensions, relatively low travel speeds, and extended requirements for operating conditions, the aerodynamic model becomes more important than for regular vehicles. It should provide the designer

not only with information necessary for calculating the fuel efficiency and related efficiency indices, but also with the data required for calculating the controllability and stability of the chassis, which is blown by squally winds from different sides and at different angles. However, the uniqueness of design, the large dimensions, and the tight development schedule often hamper using aerodynamic tubes and other similar devices on the chassis or its models. This requires developing specific combined models, which allow for conducting theoretical estimates at early design stages and phases without high material costs.

4. **The chassis mobility model** provides information about the SWC's ability to fulfill the transportation mission with the required efficiency indices in the predefined locality or when moving along the predefined route. Unlike regular vehicles, SWCs often move in groups along the same route. They have heavy axial loads which are intensely destructive to the support surface, and the damage scales up under the snowballing effect after each successive vehicle.

   In this context, SWC mobility models should provide the designer and customer with data on how many chassis and what modes are able to overcome the predefined route (locality) in the predefined environmental and weather conditions. The loads from chassis with large dimensions and huge weights (for example, the 7904 chassis model, Fig. I.1*b*) affect the support surface much stronger than regular vehicles; thus, their mobility models should take into account the deep structural properties of the localities to be overcome.

5. **The chassis maneuverability model** provides information about the SWC's ability to move in constrained road conditions or limited areas. Particular features of multi-axle wheeled chassis maneuverability models are concerned with their dimensions and weight. In most cases, the geometrical dimensions of the chassis significantly complicate its motion on the regular road network. This necessitates using special algorithms for turning the steered wheels. The turning diagram can also be changed directly by the driver in the course of the turn, which is atypical for regular vehicles.

   Multi-axle wheeled chassis are often designed as multi-link road trains. In this case, fitting very long and wide multi-link trains (for example, Fig. 1.2*b*) into a winding road's limited width contour becomes a problem (ordinary road train designers rarely face this problem).

   Finally, multi-axle chassis require a particular accuracy when performing maneuvers. This is related to the following circumstances.

   First, the driver of an oversized vehicle has difficulties assessing the motion of different points of the body by using just his or her own perceptions; therefore, it is advisable to use special observation systems for monitoring critical points. For a number of reasons, the risk of making a mistake in defining the positions of certain vehicle section points with respect to surrounding objects and buildings is increased. Maneuvering an oversized heavy-duty chassis almost always takes place when surrounding objects are much closer to it than they would be to a regular vehicle in the same conditions (this difference is even more notable in relative values).

   Second, because a loaded SWC is so heavy, the damage caused to surrounding objects and buildings due to a minor mistake by the driver (typically unnoticed by him or her) can be extensive.

   Finally, it is typical for multi-axle chassis to maneuver in such constrained and dangerous conditions that a driver's small mistake may cause catastrophic consequences. For example, a maneuvering mistake when loading the chassis onto railway platforms can cause the chassis to fall onto the railway track, resulting in damage to both the chassis and surrounding objects.

6. **The controllability and stability model** provides information about the moving chassis' ability to follow control commands in the broad sense, as well as to keep its travel parameters within acceptable limits when encountered with random external factors. Traditionally, this group also includes descriptions of special cases like lateral tripping of the chassis with/without load and a number of other situations. At present, it has been established that SWC travel in typical situations is significantly different from that of regular automobiles, and that this applies to all SWC models.

   Multi-axle chassis also face special requirements for control precision. For example, the width of a regular vehicle, even when traveling at high speeds, is always smaller than the traffic lane. However, the widths of practically all SWCs exceed 3.0 m and their maximum speed is around 40–45 km/h and is increasing. Thus, when traveling on narrower roads, the risk of touching an oncoming vehicle is high and the results can be rather serious.

7. **The model of the chassis' power-transmission system** provides information about the efficiency of transforming the fuel's internal energy into the SWC's plane-parallel motion energy in the predefined direction. The components of the power-transmission systems are built based on various physical principles. Unlike regular automobiles, energy losses in multi-axle chassis transmissions are extremely high and should be taken into account in the early design stages.

   To increase mobility, automated transmission control systems that ensure, in particular, the individual control of each wheel's operating modes must be created. The majority of modern transmissions require common control with the SWC's power unit. This has led to the appearance of united automated onboard control systems; an adequate SWC model is required to design this control system.

8. **The model of the chassis' oscillations** provides information about the motion of the SWC's components when it moves in predefined modes on uneven roads, as well as the influence of the oscillations on the crew, load, and target equipment in these conditions. The number of axles, the huge dimensions, and the influence of external factors essentially affect these models. In a number of cases, the yielding ability of the carrying system, especially in frame-based chassis, more strongly affect the crew and cargo than in ordinary vehicles.

   A case particular to SWCs, and practically never seen in oscillation models of regular vehicles, is the oscillations of a fixed chassis with equipment under the action of a short-term (pulse) load.

   A modern automated onboard control system includes a control subsystem of the wheel's suspensions, which should be designed together with an adequate oscillations model and the synthesizing principles of an automated control system.

9. **The technical and economic excellence model** has its characteristics defined by the specific tasks to be solved by the transportation-technological complex as a whole, by the batch or mass production scales, and by the comparison of the chassis' economic effect with the cost of making the chassis.

These models are units (blocks) of the general chassis model.

Each of these blocks (blocks 1–9) consists of a number of calculation modules. How the blocks are split into modules is defined by tasks of the considered stage.

The following pages describe step by step the modern generation process of the basic initial data for the main units of a SWC at different design stages.

5.3 Process of forming the appearance of SWCs

5.3.1 General approaches to forming the SWC appearance

The SWC's appearance (general layout) is developed in the Technical Proposal. This stage plays a critical role in the design process and occupies an intermediate position between the Terms of Reference and the Preliminary Design. In the process of forming the SWC's appearance, more than 70% of the total project decisions are made and these decisions define the chances of creating the chassis within the predefined time and budget limits.

Implementing multi-parameter optimization is crucially important because it allows for finding the best combination of parameters; as a rule, optimizing the chassis' parameters is translated into a one-criterion problem, which significantly simplifies its solution. The generalized indicators, which characterize the chassis' technical and economic excellence, are usually taken as criteria, while the choice of a particular criterion as unique is defined by the specific formulation of the problem.

The practice of creating new wheeled chassis is based on two approaches to addressing the preset tasks (the same as when designing the majority of complicated transportation objects).

The first approach can be conventionally treated as designing based on the Technical Assignment.

In this case, the composition and structure of the Technical Proposal are defined by the following objectives:

1. Initially forming the chassis appearance to meet the Technical Assignment requirements;
2. Optimizing the design parameters under the selected criterion and accounting for the restrictions of the Technical Assignment requirements and existing norms and regulations;
3. Studying the changes in the chassis' technical and economic characteristics, which should occur after applying the newest technical solutions and estimating the significance of these changes; and
4. Calculating the chassis' technical and economic characteristics with the selected parameters.

The first objective is to form the project's admissible existence area, i.e., where the initial parameter's acceptable solution is located.

The second objective's aim is to create the optimum design.

The third objective assumes a parametric analysis aimed at assessing the project's sensitivity to changing alternatives (different design solutions), parameters, and restrictions.

The fourth objective assumes clarifying the chassis' characteristics and documenting the design results.

When using this approach, formulating the project alternative (project hypothesis) is the most important precondition for obtaining the best result. The choice (selection) of productive (viable) hypotheses is based on the multiple analyses of alternative variants and is the most difficult, interesting, and insufficiently explored design area.

In practice, the most frequently used is the second approach to problem-solving, which can be treated as designing based on a prototype.

This design approach is used for making SWC modifications in an existing prototype (basic chassis). Therefore, the majority of design solutions and parameters remain unchanged (fixed). For example, Chapter 4 shows that all of the serial MAZ (MZKT) multi-axle wheeled chassis, the KZKT heavy wheeled tow cars, and some other products were designed and built according to this approach.

One way to use this approach is to equip the chassis with a more advanced engine or with an engine of another type. For example, replacing a diesel tank engine with an automobile one or switching from a diesel power unit to a gas-turbine unit. In this case, it is necessary to define the amount and nature of the minimum required modifications of the prototype, related to the redesign, the change in weight, as well as the technical and economic characteristic of the modified vehicle.

Another example is an update (upgrade) of the project, which assumes replacing the materials in some aggregates, installing a new generation of the control system, replacing equipment, etc.

A separate group within the prototype-based approach is the creation of design modifications for different chassis applications (e.g., at different loads and operation conditions) with the maximum utilization of the prototype's aggregates, systems, and compartments (the modular design principle). An example of this approach can be found in the development of the MZKT Volat family of multi-axle wheeled tow cars and chassis, which are based on the standardized units and aggregates of the MZKT-547V wheeled chassis model, as well as some units and aggregates of Minsk Automobile Works (MAZ) and some foreign companies. A similar example is the Rusich-KZKT family of chassis and tow cars.

The solution algorithms of the above approaches can be quite different since each has its own initial data, required parameters, restrictions, and criteria. The degree and information sources of the tasks at hand are also different.

However, in almost all of the algorithms, a system of invariants (modules) and their combinations can be identified for solving various problems. These modules are used to compile the basis of the generalized chassis model, which ensures the realization of random design strategies under the control of the monitoring systems (Fig. 5.5).

The algorithmic flow-chart in Fig. 5.5 admits both a partial and a complete automation of the process. So far, the automotive industry is not using automated design systems built according to these algorithms; however, adjacent branches are. For example, the aviation industry has and uses software complexes that ensure the automated synthesis of aircraft appearance based on algorithms of a similar structure. The aviation industry has mastered the process rather deeply, for example, they already have educational and research software packages like the CAD MAI (Moscow Aviation Institute).

At the initial design stage, the information about the created chassis is basically limited to the scope of knowledge about its necessary characteristics (obtained from the Technical-Tactical Assignment, also referred to as the Request for Proposal (RfP)), the required consumer properties, the accumulated design experience from similar chassis, and the results of the latest studies and technical solutions.

This circumstance requires the use of empirical and semi-empirical calculation methods based on simplified theoretical formulas and approximations that reflect the most important parametric links. Most of these formulas are derived from processing and generalizing the statistical and experimental materials. When using these formulas and theoretical assumptions, it is very important not to go beyond the restrictions and assumptions that have been adopted in the course of development. When staying within these limits, the calculation results are often very accurate and reliable. However, if this requirement is neglected, major mistakes or unreliable conclusions are possible.

The algebraic theory of lateral skidding developed by D.A. Antonov in the 1980s is an example of a semi-empirical theory that can be used to calculate a SWC's controllability and stability parameters at early design stages. Its advantage is that important characteristics of oversized tires can be evaluated based on their overall dimensions, which allows for making important

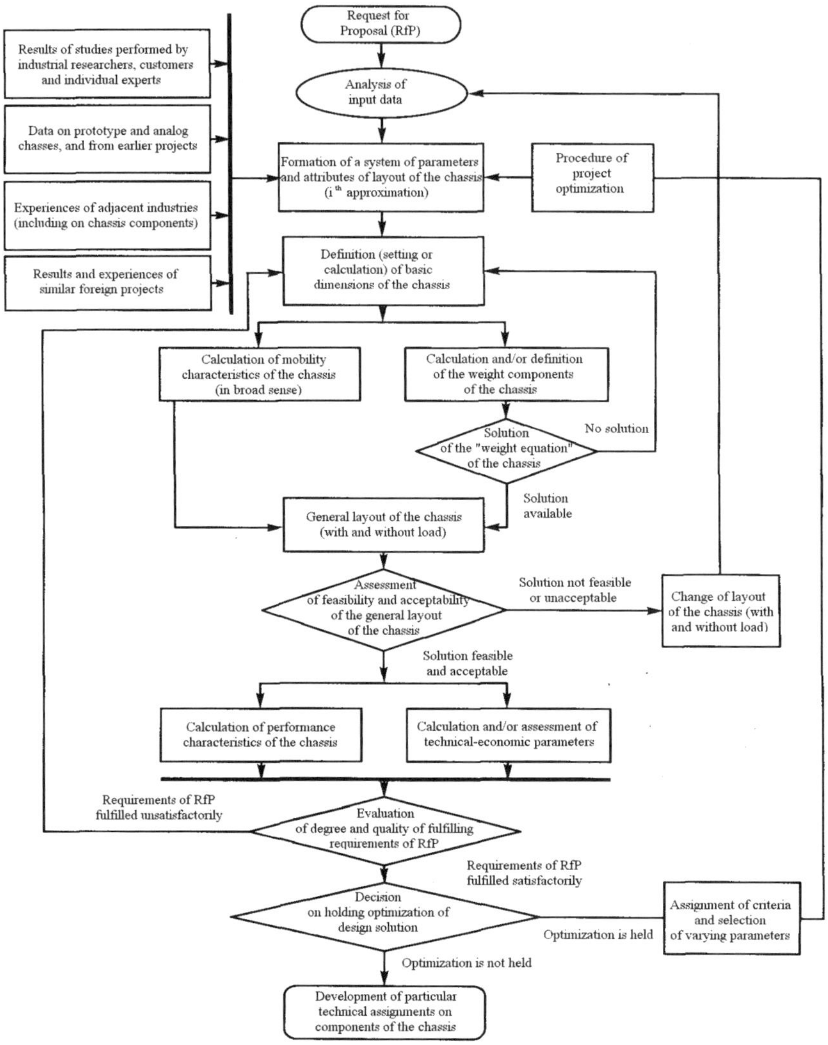

Fig. 5.5. Flow-chart of the formation process of a special wheeled chassis

vehicle design calculations without testing a real tire or even having such a tire at all. The accuracy of such calculations fit nicely within the technological scatter of real tire characteristics.

In particular, MSTU has used this theory for calculating the turning ability of a chassis with wide-section, 1600 mm diameter tires; the data was subsequently verified through experiments that showed the data from the theory was quite satisfactory. This allowed for calculating the basic characteristics of super-oversized tires with a 3100 mm diameter (creating an experimental vehicle with these parameters was impossible for technical and economical reasons). However, certain parameters of these tires were later defined experimentally. Experiments were also conducted aimed at assessing the turning ability of dumpers equipped with these tires. The experimental results coincided with the ones calculated based on the algebraic theory of lateral skidding.

However, one should remember that these formulas were obtained from the generalization of vast experimental data accumulated in the course of test bed and road trials of tires of a certain class (mainly diagonal toroidal and wide-section tires of high passing ability vehicles of similar design). These tires may be markedly different from the tested ones in design and size, but can still be used to characterize other numeric parameters within the formulas of the same structure; however, a qualitative change in their characteristics is also possible. Specialists at MSTU faced this effect in particular when testing super-oversized radial tires for quarry dumpers.

5.3.2 Formation and coding of a SWC's appearance when designing from the Terms of Reference

The generalized task of forming the SWC's appearance, which is designed based on the TOR, is formulated as finding the vector of the parameters characterizing the layout, structure, and dimensions of the chassis that meet the design requirements and ensure efficiently achieving the extremum of the target function.

The vector of the searched parameters is a set of independent variables that expressly define all other parameters and characteristics, as well as the value of the target function. The scope of the parameters describing the structure, architecture, and dimensions is rather large at the Technical Proposal (or the Feasibility Study) stage; processing of such a scope is far from trivial and is especially far from its complete solution in regards to HDWVs.

The appearance of the chassis is characterized, first of all, by its schematic attributes (general design and layout solutions), which define the following:

- The principal design scheme of the chassis (a single-link chassis, a road train, etc.);
- The design and layout characteristics of the support-running complex (SRC) (number of axles, their distribution along the base, number and placement of steered wheels, etc.);
- The fundamental characteristics of the carrying system (frame or body-based, etc.);
- The appearance of the power unit (type and number of engines, location of air inlets and engines radiators, etc.); and
- Other specific characteristics of the chassis' appearance.

The design attributes are discrete, and, accordingly, are subject to coding by a system of symbols (codes); for example, certain numerals, groups of numerals, or more complicated mathematic objects (vectors, tensors, etc.). The number of codes describing the chassis under design is fixed for each calculation variant.

The split and formalization of SWC design attributes are rather delicate, poorly developed, and controversial issues. In particular, there are evident difficulties in creating a strict border between design attributes that define the appearance of the chassis as a whole, and the design attributes that define the appearance of its components and, accordingly, are subject to analysis and application at the lower hierarchical levels of the design process.

As far as we know, the problem of developing a system of attributes for multi-axle SWCs that oriented toward formalization and further design automation has never been considered with sufficient completeness and detailing. The general classification issues of multi-axle vehicles were first considered by P.V. Aksyonov [1] within an applied technical discipline, which he started to develop under the title "The theory of general design solutions of multi-axle automobiles." His proposed classification and coding system for attributes were borrowed here for building a formal model of the general appearance of a SWC. In general, the below considerations should be viewed as a starting point, which needs further development and refinement.

In our opinion, the main design attributes of a multi-axle wheeled chassis are as follows:

- The end use (purpose or destination);
- The number of links;

- The SRC category;
- The SRC class;
- The longitudinal layout diagram of the SRC (layout diagram of the vehicle);
- The method of controlling the maneuver en route, i.e., the way of making a turn divided into two parts: the formula to control flexibility and the formula to control the turning of the wheels; and
- The carrying system class.

The **End Use** attribute as a classification method is based on the former motor industry standard that defined automobile classification groups by their end use or purpose (trucks, dumpers, tow cars, etc.).

This book considers only Group 9 of the above standard, namely, "Special Automobiles," which can be more accurately defined as "Special-Purpose Vehicles." When describing the SWC's appearance, this group is further divided into three "Functional Purpose" sub-attributes (second level attributes, see Fig. 5.2): transportation vehicles (engaged only in transporting cargo and not operating or using the cargo), transportation-technological vehicles (engaged in transporting the cargo and using the cargo's components in the course of transportation), and technological vehicles (engaged in using built-in equipment for their operations).

Transportation-technological vehicles are engaged mainly in transportation; this is a priority when forming the vehicle's appearance. These vehicles usually preserve distinct universal features and form various types of families. They can be manufactured and used in isolation from the technological (target) equipment. These are, for example, transportation-charging vehicles, wheeled chassis for different second generation special equipment (the "Luna" vehicle), self-propelled drilling rigs, etc.

Technological vehicles are usually functionally and structurally integrated with the target equipment and are used only in combination; without the equipment they have no independent functional value. These include, for example, certain engineering vehicles and specialized geological complexes.

The **Number of Links** (Fig. 5.6) attribute reflects another fundamental vehicle classification, into single- and multi-link vehicles, which fundamentally defines the way their general appearance is formed. Under the proposal of Aksyonov, vehicles are further classified based on the **Type of Link Connection** sub-attribute and are divided into saddle-type and trailer-type linkages (joints).

When there are a large number of links, the probability of combining trailer and saddle joints increases dramatically since a saddle cargo platform can be split and cover several links at a time (but not the entire vehicle). In this case, the extent to which the cargo platform is split should be defined in advance with the help of the **Platform Fragmentation** sub-attribute. Also, one should not forget that some links may have their own SRCs.

The proposed classification structure is convenient since it ensures a simple coding and decoding with the help of numerical sets; it is also easily developed, added, and refined. This coding is integrated into most widespread software products.

The classification nodes that contain no information are coded as 0 and those containing information are coded as 1…n. The description of an attribute's full structure assumes indication of its own code (the first digit in the code of the sub-attribute) and of the codes of all its constituent sub-attributes. If necessary, the depth of indicating sub-attributes can be restricted from below as presented in Fig. 5.6.

For example, the code structure for the MAZ-547 wheeled chassis (Fig. 5.7) is {2, 1, 0, 0, 1}, which corresponds to the following: a wheeled chassis, single-link, with a united non-split platform. The

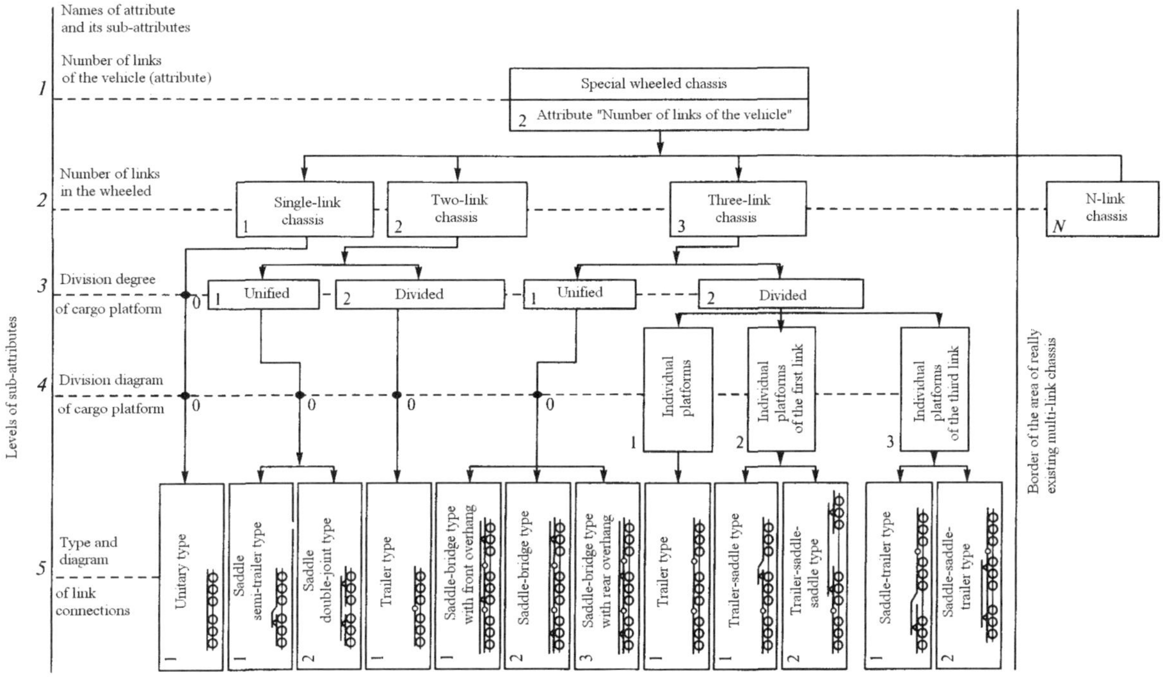

Fig. 5.6. Structure of the "Number of Links" classification attribute

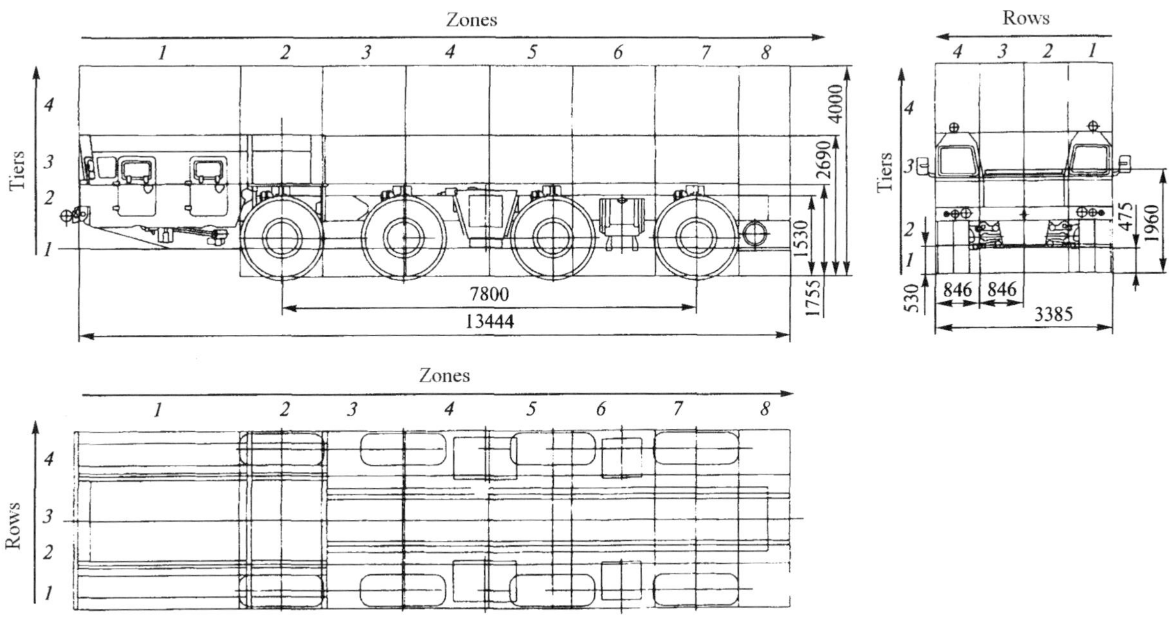

Fig. 5.7. Basic diagram of forming the "Vehicle Design Layout" attribute code

code structure of the American Twister wheeled vehicle looks different: {2, 2, 2, 0, 1}, which is a wheeled chassis, with two-links, and individual loading platforms, of trailer type.

The classification scheme offered by one of the authors of this book is generalized, in principle, for a random number of links; however, at the same time, the possible number of sub-attributes in the lower levels is rapidly increasing. This generalization is almost formal in character since the vast majority of existing wheeled chassis have no more than three links; all of the known super-multi-link vehicles were built completely under trailer type, or, in much rarer cases, under semi-trailer type. Thus, the structure's complication due to its expansion to the right (Fig. 5.8) is practically unjustified.

The latter remark concerns the conventional chassis designation in the blocks of the last structural level. This designation scheme was offered by Aksyonov and seems to be very suitable and demonstrative. The number and location of wheels in every link in this case does not matter (the left side of the figure corresponds to the front (nose) section of the chassis).

Under the same scheme, the **End Use** attribute (attribute code 1) is structured by using the codes of the first level's sub-attributes (**Functional Purpose**) as follows:

{1, 1}—transportation vehicle;
{1, 2}—transportation-technological vehicle; and
{1, 3}—technological vehicle.

Thus, the more complete description of the MAZ-543 chassis (its appearance code) is as follows: {{1, 2};{2, 1, 0, 0, 1}}: special wheeled transportation-technological vehicle, single-link, with a non-split cargo platform. The Twister wheeled vehicle is described as: {{1, 3};{2, 2, 2, 0, 1}}: a special wheeled technological (combat) vehicle, two-linked, with individual cargo platforms, with a trailer type linkage.

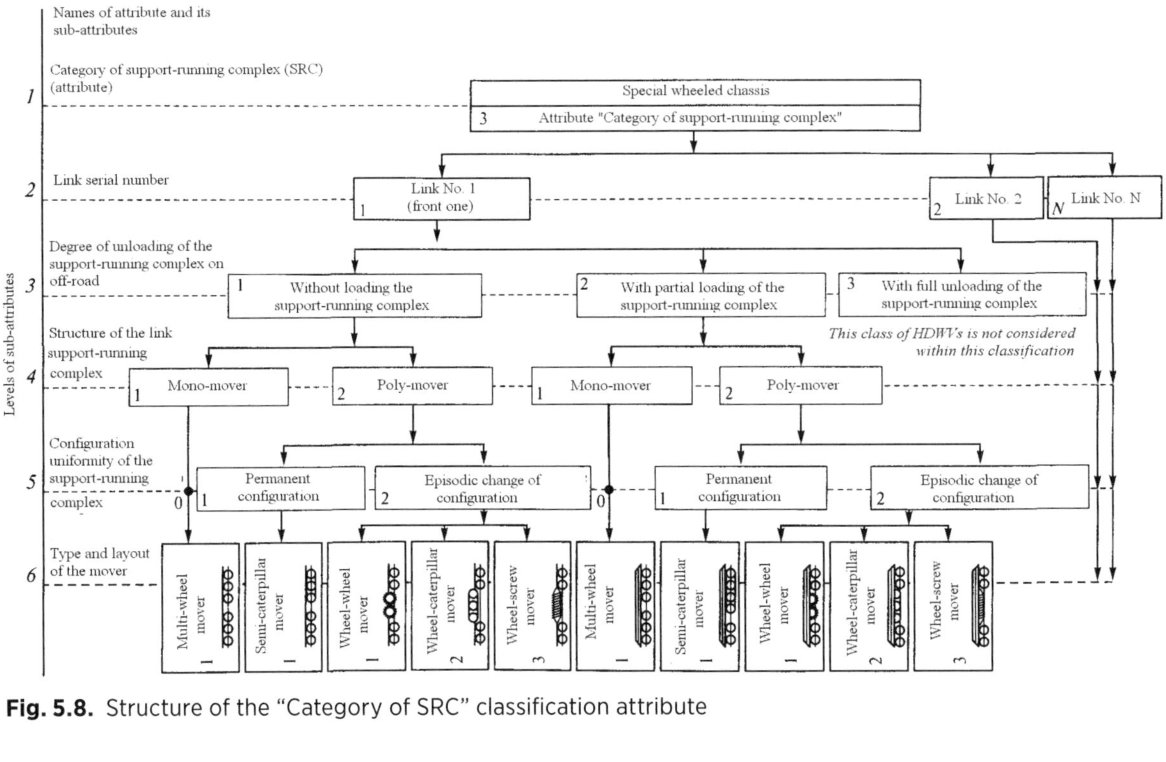

Fig. 5.8. Structure of the "Category of SRC" classification attribute

The **SRC Category** attribute adds details to the general appearance of the wheeled chassis by describing and codifying its SRC (see Fig. 5.8). In general, the development of wheeled vehicles advances in two different directions that "globally" define their general appearance—vehicles with SRCs of one type (a mono-mover) and vehicles with SRCs of different types (poly-mover) that can operate in parallel or one by one, and perform the support-running functions in full or in part.

Combinations such as a parallel use of wheeled and caterpillar movers, of wheeled and screw movers, a wheeled mover and an air cushion (a hovercraft), and a wheeled mover and skis are all known. Theoretically, it is possible to combine a wheeled mover with a mover on pneumatic rollers. Vehicles that changed movers based on changing road conditions were also implemented (Swedish Landsverk tanks, British Kristy wheeled-caterpillar tanks, Soviet BT series tanks, Soviet Object 911 caterpillar-wheeled combat vehicle, and others). This book does not consider these models and they are not included in the classification. Implementing these ideas in heavy-duty ground chassis designs is so technically complicated that any progress seems to be practically hopeless, with the only exception being when additional removable units are used for a short-term increase in passing ability (removable band, lamellar, or chain-caterpillar contours). It should also be noted that the experience of building vehicles with high passing ability able to mechanically switch from the wheeled to an alternative (preferably caterpillar) mover and vice versa proved to be completely negative.

There are models in which movers of various types operate in parallel in all road conditions (for example, the semi-caterpillar vehicles broadly used in Germany during World War II). Soviet and Russian literature assess this variant, as a rule, negatively. However, this assessment is not quite accurate since such combinations are reasonable while designing some types of special off-road transportation-technological complexes.

Since this book deals with designing wheeled vehicles, chassis with combined movers will not be considered further. In this context, the issues of placing auxiliary movers are not considered within this classification attribute's structure.

One should also keep in mind that, in principle, every link of a multi-link system may be equipped with its own SRC.

The structure of the **SRC Category** attribute is shown in Fig. 5.8. It accounts for all of the described limitations on the variants' application related to the characteristics of the considered objects. Also, to avoid an excessive diagram, the structure has been deployed for just the first link of the articulated vehicle since it is obvious how to continue. In this attribute structure, the code for the 103G vehicle (see Figs. 4.3 and 4.5) described in Chapter 4 is as follows: {{3, 2, 2, 1, 0, 1};{3, 2, 2, 1, 0, 1}}. This translates to: the first link having a SRC with a possibility of periodic partial unloading, the SRC is a mono-mover, and the mover type is multi-wheeled; the second link also has a SRC with a possibility of periodic partial unloading, the SRC is a mono-mover, and the mover type is multi-wheel.

The **SRC Class** attribute (Fig. 5.9) specifies the SRC's appearance by describing the distribution of the support-running elements within the horizontal projection of each vehicle link.

All varieties of wheeled vehicle SRCs can be divided into two types: those with clearly expressed linear wheel locations along the sides (boards) and the same expressed axial wheel groupings in the lateral direction (axial double-track layouts), and those with a more or less free distribution of wheel groups along the whole horizontal projection (multi-truck and multi-support layouts).

There is some uncertainty when applying this sort of classification: some multi-support transporters are equipped with post-type supporting modules with two wheels placed on the same axle and rather far apart. These

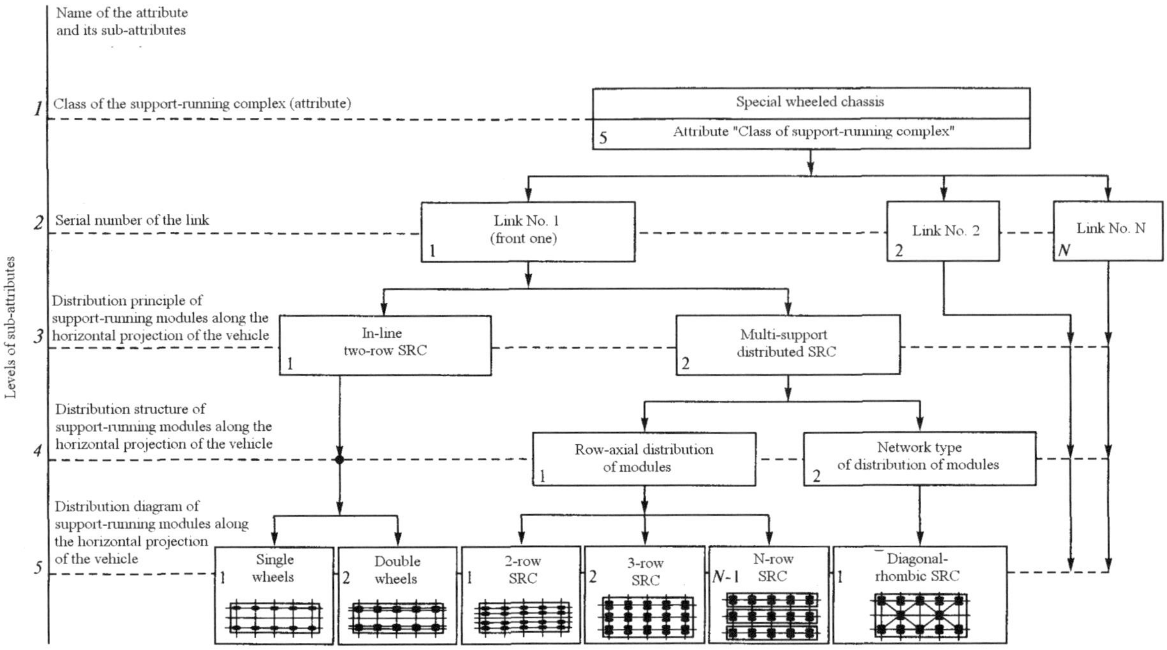

Fig. 5.9. Structure of the "SRC Class" classification attribute

modules are placed in two rows along the length of the vehicle, thus, this type of SRC may be referred to as double-track designs. However, in most cases, the width of these vehicles is limited. As a result, the distance between the modules is comparable to the distance between the wheels in one module; the wheels are also more or less uniformly spread within the vehicle's supporting surface. In this case, the SRC is classified as multi-support.

Each post (or pillar) of the support-running module (SRM) may have two or more wheels. Thus, MSTU has designed a SRM variant for an off-road vehicle with two symmetrical two-wheel bogies (size 12.00–18 toroidal tires with adjustable air pressure) located on longitudinal swinging levers symmetrically relative to the central post. On the other hand, an SRC with a linear location of wheel modules may have all double-wheel axles (for example, the American Giant four-axle tow car).

We prefer symmetrical wheel module designs to those with linear wheel module designs, where the distance between the internal sides of the tires may be considered as significantly larger than the total width of the tires mounted on the module (accounting for the gap between the tires on one module). Double (twin) tires are an example of a wheel design where the distance between the adjacent tire sides is comparable or less than the width of one tire.

Thus, according to this attribute, the code for the MAZ-543 chassis is {4, 1, 1, 0, 1} (a single-link vehicle with a linear-axle wheeled SRC with single wheels). The code for the Sheyerle 400.26.4 transporter is {4, 1, 2, 1, 1} (a single-link vehicle with a wheeled multi-support distributed SRC; the distribution of the supporting modules is row-axial and double-row).

The **Longitudinal Layout of the SRC** attribute considers the main diagram of the wheels' location and grouping along the vehicle. This attribute is also called the chassis' axle formula.

In most cases, the wheels of linear-axle SRCs are distributed along the vehicle's board in groups of closely located wheels. These groups are traditionally treated as bogies, reflecting the original meaning of the word as it applied to very old vehicles. Bogie-type designs still exist in some multi-purpose three- and four-axle vehicles. However, multi-axle wheeled chassis have no wheeled bogies as isolated aggregates or components; thus, in this sense, the term has a traditional character and means a group of closely located wheels.

In principle, the wheel projections of multi-support complexes onto the vertical longitudinal plane may overlap each other. However, if each wheel support row is considered separately, the same terminology and designations used for linear-axle layouts can be applied.

For conventional designation (symbolic coding) of the axle formula, the following system was suggested by Aksyonov: a figure defines the number of approximate wheels and a hyphen defines a gap between bogies. For example, the axle formula of a six-axle vehicle with approximate wheels of the second, third, fourth, fifth, and sixth axles (counting from front to rear) looks as follows: 1-2-3. This diagram is well compatible with the rules for coding the appearance applied during computer-aided design. For coordinating the systems, designate the gap between the bogies by a 0 instead of a hyphen and to present the axle formula as mono-code: {1, 0, 2, 0, 3}.

When describing multi-support systems, the code must be complemented with the row index. Also, the nose wheels in certain rows may be moved slightly backwards. If a perpendicular line is drawn through the most forward wheel and used as the starting line, then the above displacement backwards is also coded as 0. With this approach it is no longer reasonable to differentiate among linear-row and multi-support complexes. If the axle formula is given for one row (e.g., the first one), then a linear-axle SRC is meant since there is no bicycle type multi-axle wheel chassis in practice. In all the other cases, a multi-support complex is meant. This significantly simplifies the structural diagram of the attribute, since all intersecting interrelations are excluded.

For example: a SRC longitudinal diagram of the first link of a multi-link, multi-support vehicle (the fifth attribute), which has a three row module location diagram with approximate extreme axles of the four support external and four support internal rows and two approximate supports of the middle row, positioned close to the center of the platform. This design has the following code: {5, 1, {1, {2, 0, 2}, 2, {0, 2, 0}, 3, {2, 0, 2}}}. Its classification structure is presented in Fig. 5.10.

The **Control Method of Maneuver en Route** attribute, i.e., the turning method of the vehicle at curvilinear motion, defines the basic methods for generating the forces and moments (torque), which cause the chassis to change its travel trajectory.

When analyzing the appearance of ordinary automobiles, it is enough to describe the number and location of steered wheels since only one method is typically used—the steering wheel is rotated which forces the rolling planes of some wheels to deviate from the longitudinal axis of the vehicle. For a schematic description of the turning system in ordinary automobiles, Aksyonov suggested a system of symbolic coding called the control formula. In this formula, a digital code is awarded to a steered wheel; the code formally corresponds to the wheel's ordinal (counting from the front to the back of the vehicle). The 0 code is awarded to the non-steered wheel. The codes of steered and non-steered wheels are separated by hyphens for better visualization. For example, the control formula for a four-axle vehicle with front and rear steered wheels is as follows: 1-00-4.

When analyzing special off-road vehicles, the situation becomes more complicated because of the different methods of carrying out a turn (for example, the board turn). Vehicles with combined movers may have variants, while every type of mover has its own way of affecting

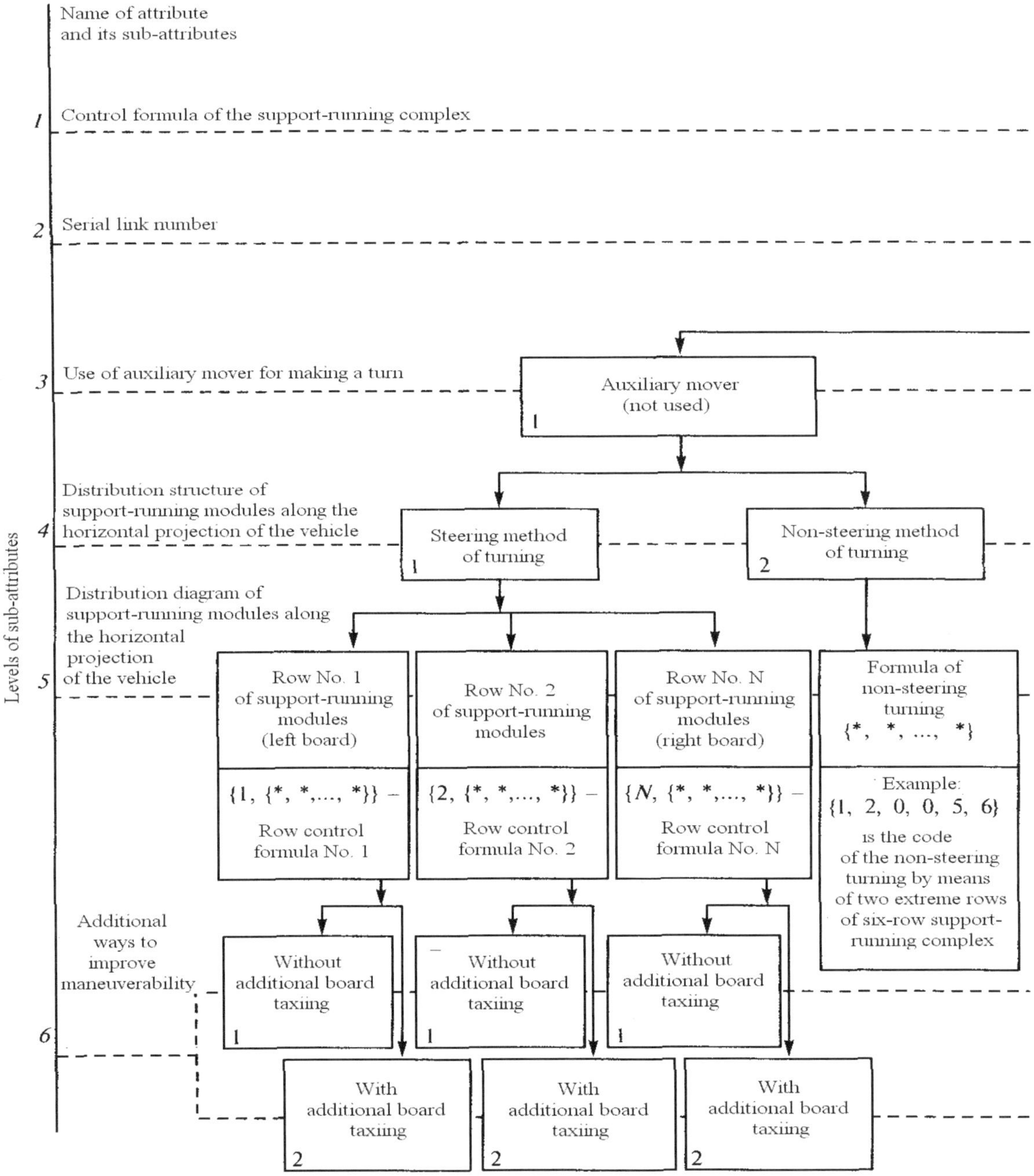

Fig. 5.10a. Structure of the "SRC Control Formula" classification attribute: a)—First part of the figure

the vehicle's travel trajectory. For example, a combined wheel-caterpillar vehicle may occasionally be steered by combining a steered turn (for the wheeled part of the SRC) with a board turn (executed by the caterpillar section).

Multi-link vehicles must also consider another turning principle—a purposeful change in the angle

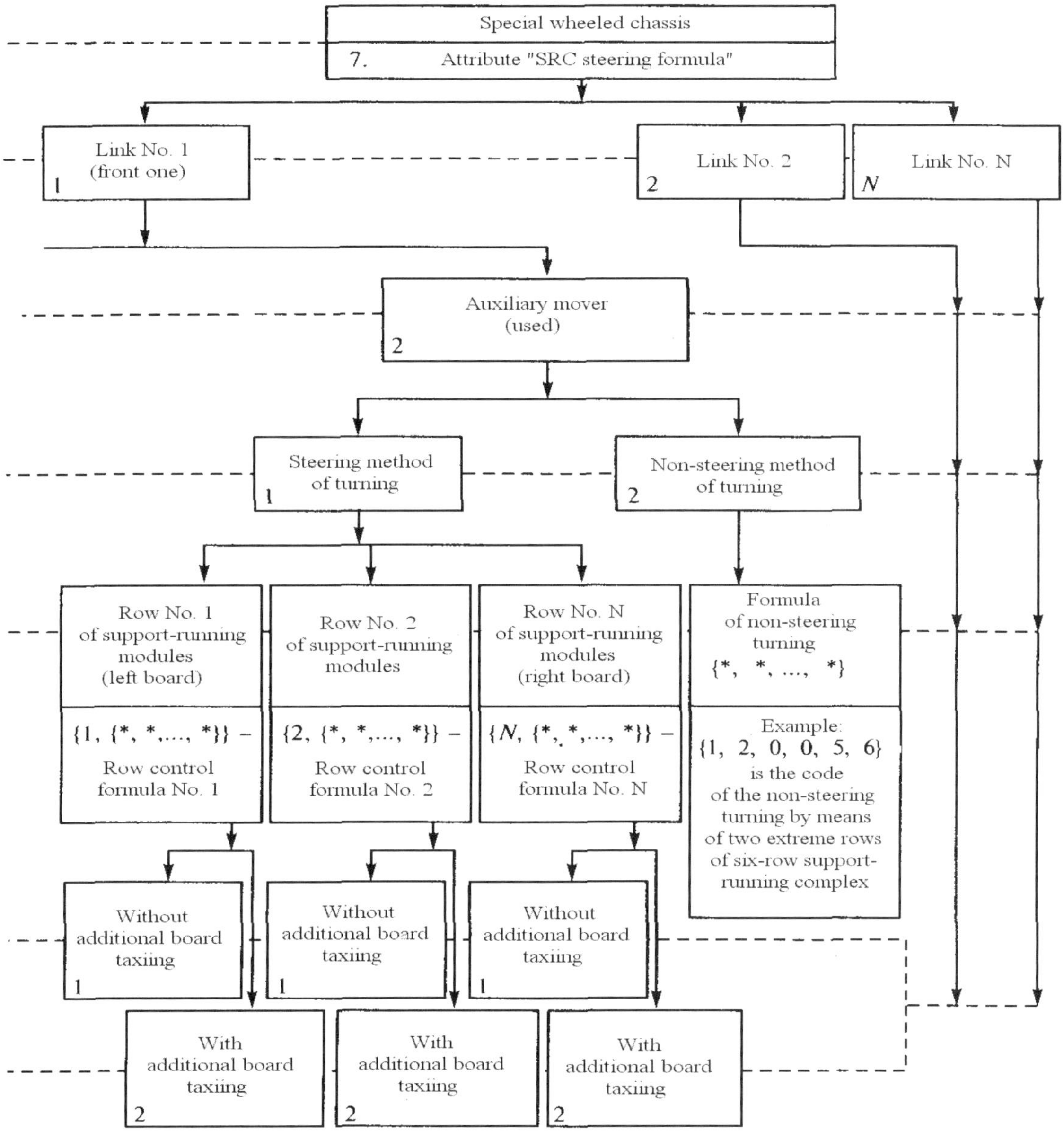

Fig. 5.10b. Structure of the "SRC Control Formula" classification attribute: b)—Second part of the figure

between the longitudinal axes of adjacent links. In this case, the following options are possible:

1. A special drive is completely and expressly defining the actual angle between the links (a forced jackknifing of the chassis);
2. A special system restricts either the changing range of the angle between the links or the jackknifing speed of the links, which take place due to the interaction of the adjacent links; and

3. Only the extreme permissible jackknifing angle is restricted by means of stops; the jackknifing is completely defined by the interaction of the links (a sort of a taxiing system).

Multi-link systems can also be steered by means of linked SRCs. In this case, steering is executed through the movers of all or some of the links and by a joint operation with the link jackknifing drives. For example, a two-link transporter with high passing ability designed by MSTU had completely steered wheels of the front link and a system to restrict the jackknifing angle and control the jackknifing speed. This system's operation algorithm took into account the motion modes, the chassis' actual current configuration in the plane view, and the actual turning angles of the front link's steered wheels.

A single-link vehicle is also moving in the vertical plane by copying to a certain degree the road profile. The driver is unable to radically influence this process except for controlling the roll and pitch in a limited way by controlling the wheel suspensions or caterpillar rollers (for example, keeping the cannon vertical in the Swedish S tank series).

A multi-link vehicle, however, allows for a forced change in the mutual angular and linear interrelationship of the links within a broad range in both the vertical and horizontal planes. Designers are widely using these opportunities, for example, the forced jackknifing in the vertical plane of the D-30 two-link caterpillar transporter made by Ishimbai Machine-Building Plant).

Thus, when characterizing the appearance of a multi-link vehicle, it is necessary to describe its angular flexibility in all three coordinate planes and its longitudinal compliance (the possibility of a controlled, including forced, change in the distance between the adjacent links); this procedure should be performed for each inter-link joint (hinge).

The can be addressed in several ways.

The first assumes forming an independent code for each degree of freedom (for each coordinate plane common for the whole vehicle). This option increases the code length, especially in multi-link vehicles with complicated joints. It also has a poor visualization, since each joint description is repeated many times and included into different parts of the rather complicated and long code.

A more rational way assumes a complete coding of each joint with a transition from one joint to another. However, such code will no longer characterize the maneuver en route. Practically, this code will characterize the ability of a multi-link vehicle to perform a spatial maneuver using all of its links. From a general viewpoint, the second approach seems more natural, since it corresponds with the design aims of multi-link vehicles with completely or partially controlled joints. In this case, the code's visualization and compactness are also essentially better.

However, when designing the code structure, one should also account for the possibility of using split joints. In these cases, two joints with typically the same kinematics are interconnected by a short intermediate link, whose length is smaller than normal links. It is unreasonable to consider this as a three-link design; moreover, the connecting link has no mover and no cargo platform.

Thus, it seems reasonable to split the **Control Method of Maneuver en Route** attribute into two independent parts: the flexibility control formula and the mover control formula. Another justification is the fact that when designing multi-link vehicles, the principles describing these two attributes are independently realized in the vehicle's appearance.

According to the suggested structure, the code of the flexibility control formula for the two-link transporter designed by MSTU is as follows: {6, {2, 2}, {1, 1, 1}, 3, 1} (a non-split by jackknifing in the horizontal plane joint between the first and the second links, equipped with a system for restricting the angle and speed of jackknifing; a joint, split by jackknifing in the vertical plane between the first and the second links; the first and the second links are equipped with a mechanism for forced displacements; a joint, non-split by jackknifing in the lateral plane, with a free relative link displacement; a forced system of moving links apart (stepping according to the "push-pull" principle)). A three-link transporter of the same type will have the following code: {{6, {2, 2}, {1, 1, 1}, 3, 1}, {6, {2, 2}, {1, 1, 1}, 3, 1}}.

The structure of the **Mover Control Formula** attribute is obvious for the suggested diagram (see Fig. 5.10). It accounts for the following characteristics. In accordance with the idea of this book, it is assumed that the main SRC of all links of the multi-link vehicle is wheel-based. In this case, steering is the main method to control turning. The non-steered turning method for vehicles with wheeled SRCs is not popular, but its use in HDWVs is quite probable. Some time ago, the Military Academy for Armored Troops (D.A. Antonov et al.) justified and experimentally verified the possibility of improving the maneuverability of wheeled vehicles with steered turning by adding additional steering (the so-called taxiing) under the board principle. At present, this method is implemented in the new BTR-90 model armored personnel carrier (APC).

However, the additional movers actually used in HDWVs only allow for two-row (by one set along each board) installation and the board turning method. The structure of the attribute accounts for this characteristic and this allows for eliminating all the knowingly unrealizable variants at an early stage.

Finally, this coding scheme uses direct recording of the control formulas' digital code, which allows for simplifying the system and reducing the length of the full code. In a multi-support wheeled module with board turning, a different, including incomplete, number of rows of support-running modules can be used. Thus, a special code was introduced for this sub-attribute and is composed in the same way as the code for the axle steering formula. The use of these codes is analogous.

The place and location of the driving axles (wheels) are often referred to as secondary SRC attributes, which are to be considered only at further design stages. Thus, Aksyonov proposes to consider these attributes as the reason for identifying the classification subgroups of vehicles. Among HDWVs, non-all-wheel drive is only used in vehicles intended for hard roads (quarry dumpers and multi-support platforms and then not always). Nevertheless, traditional descriptions of automobiles use this information in basic classification attributes by including the number of driving wheels in a vehicle's technical characteristics presented in catalogues and reference books.

The same tradition is used for describing the drive principle of the wheel formula, composed as the total number of wheels multiplied by the number of driving wheels, for example, 8×8. It is easy to see that this structure is not adequate for describing the drive diagram of a multi-support multi-link wheeled vehicle, which is needed for a clear presentation of its appearance. Information about the location of the driving supports within the contour of the SRC and about the characteristics of each link's SRC is needed. It also fails to account for, among other things, the possibility of supports with pair wheels that may have separate drives and there is no clear definition of the double wheels.

If we take into account that, at the initial design stage, the number of activated wheel supports, which is different from the number of driving tires, is needed and that real SRCs have a clearly expressed row scheme, then we can propose the following classification structure for the **Vehicle Wheel Formula** attribute (we preserve the traditional name of the attribute): {code of **Vehicle Wheel Formula** attribute } = {{number of the attribute Vehicle Wheel Formula, 1 (number of the first link), {1 (number of the first SRM row of the first link), {0, 0, … 4, 5, … N1}}, … {M (number of the last SRM row of the first link), {0, 0, … 4, 5, … NM}}, … number of the Vehicle Wheel Formula attribute, Q (number of the last row), {1 (number of the first SRM row of the last link), {0, 0, … 4, 5, … R1}, P (number of the last SRM row of the last link), {0, 0, … 4, 5, … RP}}}. The core of the structure is the modified wheel formula for the SRM row, which contains information about the number of supports in one row (the number of members in the list), the presence of non-activated supports (coded as "0") and their position along the row (the position of the "0" in the list), and the presence and position of activated supports within the row (the non-zero code, which is the serial number of the activated support).

Once this code is complete, this can be considered the end of describing the general appearance of the SRC of a multi-link multi-support HDWV and the compilation of its corresponding code.

Longitudinal SRC Design (Vehicle Design Layout). In general, the appearance of a wheeled transportation-technological vehicle defines the mutual location of its main elements. In most cases, these elements occupy rather large spaces with clearly expressed borders in the form of partitions, e.g., the bottom, the roof, etc. These spaces are usually defined by the common term "compartment" (habitation compartment, crew compartment, cargo compartment, etc.). To form the principal idea on the vehicle's appearance, it is necessary to describe the mutual location of the largest and functionally most important compartments: that of the crew (cabin), the cargo (or passengers), the power unit (engine compartment), and the transmission (transmission compartment).

Unlike aircrafts, wheeled chassis compartments are rather freely placed within the contour of the vehicle, occupying different positions along its length, width, and height. They also have different dimensions and shapes depending on the design of the main equipment (for example, the engine) or functionality. The shape of these compartments can be rather intricate depending on their main equipment's configuration. This significantly complicates the process of forming the chassis' description, which is subject to formalization by the adopted system of coding the attributes.

When the SRC's appearance is being formed, it is important to have a very general idea about the configuration of the complex. Therefore, replace the intricate shapes of the compartments with simple approximations in the form of parallelepipeds and describe their mutual location in the most general way.

To build a parallelepiped that embraces the contours of the whole complex, draw planes through its front and rear perpendiculars, along the boards through their extreme points, along the roof or upper point of the cargo, and include the supporting surface. Then, conventionally split this parallelepiped into tiers (vertically), zones (lengthwise), and rows (across the vehicle). Now, the generalized compartments can be easily attached to the obtained areas.

The split into tiers (Fig. 5.7) is obtained naturally, since all vehicles have three obvious horizontal planes: the road surface; the bottom plane (or the lower plane of the frame girder), i.e., the plane passing at the level of the wheel in its extreme upper position; and the plane passing horizontally across the vehicle's upper point with the maximum admissible cargo height. However, the height of multi-axle chassis with cargo in most cases is rather high, 3.8–4.6 m or more. With a road clearance of 500–600 mm, a tire diameter of 1600–1900 mm, and a wheel stroke of 250–500 mm, the height of the third tier is 2.5–2.9 m. Any sharp difference in the grid pitch is undesirable, especially when the obtained difference is comparable to the size of typical compartments for placing components. It seems reasonable to additionally split the upper tier into two approximately equal horizontal sub-tiers. Thus, four natural tiers are obtained, which can be numbered (1–4), for example, in the bottom-up manner with the number indicating the tier code.

For different reasons, most wheeled vehicles have limited width. At the same time, the lateral dimensions of elements like the cabin cannot be freely reduced since they are defined by the dimensions of the human body and requirements for habitability (the width of the wheeled mover is usually large enough). As a result, the frontal projection of the wheeled chassis has a clearly expressed split into vertical "rows" (see Fig. 5.7). In a classical chassis, at least three such rows are distinguishable and the wide middle row overlaps the vehicle's longitudinal plane of symmetry. Most wheeled chassis tend to have lateral symmetry; therefore, the frontal projection is split into four symmetrically vertical rows of approximately equal width and are assigned corresponding numbers (1–4), counting, say, from left to right. These are row codes.

The side projection of a classical wheeled chassis has the most complicated configuration. However, two obvious zones can be differentiated that almost always exist: the front (nose) and rear (back) overhangs, limited by the vertical line, tangent to the extreme point of the extreme wheels. Another natural division is the vertical line passing through the complex's center of mass (practically through the middle of the distance between the vehicle's extreme axles). This line divides the chassis into the front and rear parts. Such a split is made in most calculations of weight, mobility, aerodynamics, etc.

For large multi-axle chassis with long side projections, such division is too rough. Thus, for example, the front overhang of a typical multi-axle wheeled chassis may have a length of 1.5–3.0 m, and within this length a one- or a two-seat tandem cabin can be placed, but nothing more. The distance between the extreme points of the extreme wheels of a six-axle chassis may be 12–16 m and in chassis with more axles the distance may be even bigger. With a simple four-member split, the length of the second zone is 6–8 m, which is too much since it may house several compartments without overlapping each other and will not be reflected by the placement code.

There are several possible ways out of this situation. The simplest is to apply a smaller split. Then, it is reasonable to be guided by the length of the driver's or passenger's workplace as recommended by motor industry standards and normative documents in force. They state that the length of the workplace should be 1200–1400 mm, accordingly, and the length of the one-row cabin should be close to this range. Thus, it is reasonable to additionally split the second and the third zones into two (or better three) sub-zones.

The length of a single-link chassis is restricted by road traffic legislation. Experience shows that a body vehicle (using the worst layout) with a length of more than 30–35 m with a front overhang of 3.0 m and a rear overhang of 1.5–2.0 m is close to the maximum allowable dimensions. This means that the suggested scheme of splitting into zones ensures a pitch (frame spacing) of about 4.0–6.0 m (no more) in the middle section. Thus, divide each zone between the overhangs and the middle line into their sub-zones. By awarding them with numerals 1–8, starting from the front overhang and working toward the rear one, we can refer to each numeral as the zone (sub-zone) code.

Finally, the vehicle's overall volume is split into non-overlapping sub-volumes, each characterized by its tier, row, and zone codes (see Fig. 5.7). Any larger volume will be described by a succession of codes of its volumes-components. To preserve the generality of presentation, the volumes of one row and one zone are placed in the order of increased tier number; then the volumes of the next row in the same order; and then the volumes of the next zone in the same order (agreement on the order of coding). For example, the volume of where the left-hand cabin is placed in the MAZ-543 four-axle wheeled chassis (Fig. 5.7) has the code: {{2, 1, 1}, {3, 1, 1}}, i.e., it occupies approximately the volume of the second and third tiers of the left row of the front overhang.

Since only the fundamental characterization of the chassis' general appearance is needed, this precision in defining the cabin location is sufficient.

Thus, the order of using this coding system is as follows:

- Determining the united list of the placed components and awarding each with its identification code (a serial number in the list);
- Approximating the volume occupied by the component's compartment in the chassis (or link);
- Defining the code sequence of these volumes;
- Formatting the list of component codes for one link; and
- Formatting the list of attribute codes for all links.

We suggest the following list of components and their codes (the system allows for expanding this list downwards):

1. The power unit (engine compartment);
2. The transmission system (transmission compartment; in electric transmissions with motors-in-wheel, it is the compartment with transformers complete with auxiliary equipment);
3. The SRC;
4. The cabin; and
5. The cargo platform or compartment (the volume).

If the complex includes several split components with the same purpose, their main code is added with the serial code, for example: 42—the second cabin.

The structure of the location code for the compartment looks as follows: {compartment location code) = {compartment code, {{location code of the left lower front sub-volume}, {location code of the next upward left front sub-volume}, ... {location code of the upper right rear sub-volume}}.

The code structure of the **First Link Design Layout** sub-attribute is as follows: {sub-attribute code} = {Design Layout attribute code, link code (number), {{location code of compartment 11}, {location code of compartment 12}, ... {location code of compartment 21}, ... {location code of compartment 5N}}}.

The code structure of the **Vehicle Design Layout** attribute: {attribute code} = {{Design Layout of link No. 1 sub-attribute code}, ... {Design Layout of Link No. 2 sub-attribute code}}.

The **Carrying System Class** attribute is necessary since there are chassis with three fundamentally different versions of carrying systems (the digit represents the sub-attribute code):

- Frame-based (having a clearly expressed linear carrying construction, which unites all of the chassis' components and cargo) – 1;
- Platform-based (having a clearly expressed two-dimensional or "plane" carrying construction, which unites all of the chassis' components) – 2; and
- Body-based (having a clearly expressed spatial carrying construction, which unites all of the chassis' components and cargo) – 3.

The code structure of this attribute is simple and does not need special explanatory diagrams. At the same time, one should remember that it should still be compiled for each link (in the order they are linked together) according to the following order: {**Carrying System Class** attribute code} = {{**Carrying System Class of Link No. 1's** attribute code}, ... {**Carrying System Class of Link No.** N's attribute code}} = {{9 (number of the **Carrying System Class** attribute), 1 (number of link No. 1), **Carrying System Class** sub-attribute code for link No. 1}, ... {9 (number of the **Carrying System Class** attribute), N (number of link No. N), **Carrying System Class**'s sub-attribute code for the last link}}. For example, the two-link transporter designed at MSTU has the following code: {{9, 1, 3},{9, 2, 3}}—a wheeled two-link transporter, the first link's carrying system is body-based, and the second link's carrying system is also body-based.

Strictly speaking, the scientifically justified classifications should be built according to the principle of dichotomy, which excludes any chance of the attributes overlapping. In a complicated technical object like a wheeled chassis, the links between the components are numerous and varied, mainly due to the multipurpose character of the technical solutions and the units and aggregates that embody these solutions. As a result, all attempts to create a strict classification of wheeled chassis, undertaken more than once from different viewpoints, ended in failure and had to create blocks with indeterminate content (e.g., "Mixed Constructions" or "Combined Items"), which broke the integrity of the classification structures. However, absolute classification strictness is not needed for creating the algorithm for building the chassis' general appearance. It is sufficient that parasitic (from the viewpoint of the classification system) interlinks of the attributes have a secondary nature. In other words, if the interlink of some attributes have no radical effect on the character of the corresponding general design-and-layout solution, it is treated as weak and temporarily ignored.

Consequently, we can introduce a list of HDWV layout attributes: Attr (WV) = {$Attr_1$, $Attr_2$, … $Attr_N$} and a plurality of admissible HDWV design variants, which have an identical set of layout attributes X_0. In this case, the X_0 plurality defines the plurality of all admissible HDWV designs of the given configuration. Setting a particular plurality of attributes (of the full HDWV code) will define a specific implementation of the HDWV, named X.

Defining the X_0 plurality is achieved by using a system of layout restrictions—a system of equations, inequalities, and logical correlations that interconnect the elements of the list Attr(HDWV) and defines the possibility of their simultaneous implementation. The layout restrictions are based on the design experience from creating similar technical objects and also includes the **Discipline Requirements** downloaded from higher designing levels (for example, the overall restrictions on the dimensions based on the chassis' transportation via railway, air, etc.). With the help of these restrictions and requirements, incompatible attributes are selected and eliminated, which reduces the number of admissible layout variants. As applied to multi-axle wheeled chassis, the described process is poorly studied, which is a serious obstacle for automating design. Moreover, as far as we know, these issues have not been duly developed within the automotive industry as a whole.

Thus, these coding systems can be used as the basis for an automated design system for HDWVs when the Technical Proposal stage is being elaborated.

5.4 Structure of a HDWV mathematical model

The formalization of the design process demands development of a digitized model of the HDWV, in which the information detailing grows during the course of the project development.

The principles of building such models are well developed in the aviation industry. The model is built by presenting the HDWV as a design object in the form of a finite combination of blocks, or "constituent elements," and their formalized interactions. Each block is associated with a separate part of the design that performs certain functions and has the following:

- A definite "name," which allows for uniquely identifying it among other blocks; and
- A list of features, containing all of the necessary information about the object for analysis and synthesis of the project.

To describe the model, a tree-like oriented graph called the "design tree" is used. The design object as such (in this case, the HDWV) is the graph's root vertex. The major subsystems, blocks, or modules of the design are the first level vertexes. Their subsystems are the second level vertexes, and so on. An element is a "primary block" when it is not subject to further segmentation at the current design stage (level 10 in Fig. 5.2).

It was shown in point 5.3 that using a system of design restrictions allows for forming a plurality of admissible variants on the appearance of the HDWV. A formalized description of these restrictions demands a mathematical model of the HDWV, which consists of a mathematical description of its properties and their interrelations.

First it is necessary to form a list of properties for each model level, giving complete information about the object at this particular hierarchical level and design stage Thus, the HDWV model now contains two design tree graphs: one containing the structure of the properties and traditionally placed on the left-hand side of the figure and one of the objective structure placed on the right-hand side of the figure.

The structure of the properties tree should be split until it reaches the "elementary properties" level and every property can be can be set by a certain numerical value (a parameter or characteristic).

In the design tree, terminal vertexes (those which end) correspond to elementary, aggregate, structural parameters and are represented by even or real numbers.

These issues are very poorly developed in HDWV design theory, although a large number of multi-axle wheeled chassis parameters and characteristics are broadly used in the actual design practice and in relationships between customers and designers. Thus, it seems possible to form the required plurality of parameters by structuring the traditional technical characteristics of vehicles and their aggregates. A system of parameters in even the simplest version of the HDWV's appearance will have a very large number of components. The training literature typically only gives general outlines of particular graphs for the design object's models. It is possible to compile them in a separate diagram, which has notable methodological advantages.

Analyzing the parameters which characterize the HDWV model shows that many at the early design stages can only forecast or estimate their character. In such conditions, we have to address the task step by step, leaving the decision-making on a number of parameters for subsequent design stages or basing them on previous experience.

The main factors that define the restrictions placed on the project and are either explicitly defined in these factors or are restrictions based on design choice are as follows:

- The restrictions dictated by the conditions of the physical implementation of the chassis, which are the same for all vehicles within the limits of a certain concept (defined by the "equations of existence");
- The restrictions dictated by the prescriptive requirements, which are the same for all vehicles of a certain type or class; and
- The restrictions arising from the requirements of the Tactical and Technical Assignment to develop a particular chassis and, accordingly, the specific restrictions for this particular chassis.

At present, the requirements for the Terms of Reference (TOR) content is generally defined by the respective standards of the motor industry.

The decisions on the numerical values of the initial parameters should be made before the design is started.

5.5 HDWV appearance formation algorithm

The design process for a HDWV is realized within the algorithm presented in Figs. 5.1 and 5.5.

The initial data required for the algorithm to function properly are as follows:

- The numeric values defining the tactical-technical characteristics of the chassis (taken from the Tactical and Technical Assignment); and
- The attributes and parameters of design solutions (defined by the designer on the basis of his or her intuition, analysis of the chassis' requirements, and his or her experience accumulated while designing similar chassis of the same purpose and from studying the results of scientific researches).

The algorithm's functionality is ensured by the codification system of the indicators and parameters described above.

Without going into a detailed description of the algorithm (it has a general nature and is described in detail in literature on CAD), let us now consider the characteristics of the optimization process, which depends on the status and development level of the theory and available methods of calculating the class of multi-axle wheeled vehicles. The issues of compiling its weight specifications, their analysis, and the solution of the weight equation will be considered in further chapters.

It is well known that, in practice, making a design layout choice for further elaboration is very difficult. The reason is that there are a large number of "efficient" design layouts. At subsequent design levels, there is a snowballing effect that increases the number of layout options, all of which preserve the characteristics and aggregated parameters of the first level. Thus, there is a problem of rejecting the unfit design layouts when passing from one hierarchical level to the other.

The first method to reduce the number of "efficient" design layout options is the decomposition of partially efficient criteria.

As applied to HDWV design, there is a problem in forming a system of acceptable criteria for assessing the efficiency of the design solutions. These criteria should meet the requirements of a systemic and hierarchical structure. In the ideal case, it is desirable to have one generalized criterion that can assess the design outcome as a whole. Such criterion should be dimensionless, vary within the limits of 0–1, and be able to split into partial criteria according to the principle of dichotomy.

Many researchers were and are trying to find such criteria and there is plenty of literature available. In particular, N.M. Balin raised the issue of the feasibility of a single criterion able to assess to an approximate degree the design result which arises from the use of this criterion. He proposed to use the complex indicator of chassis purpose (CIP) for this criterion. There are special (partial) efficiency criteria, which were offered by Ya.S. Ageikin, P.V. Aksyonov, M.I. Ipatov (for example, the criterion of reduced costs), and by a number of other researchers. Unfortunately, it must be stated that, at present, the structured system of efficiency criteria for multi-axle wheeled chassis has not been created and requires further intensive development. More details on this issue will be discussed in further chapters.

An example of a successful efficiency criterion and its decomposition is the criterion of "tactical mobility" and its sub-criterion "off-road mobility," suggested by the Military Academy of Armored Troops (VABTV).

Taking a SWC's support passing ability as an example, consider the principal difference between a *parameter* and a *criterion*. In general, a *parameter* is the ability of a wheeled vehicle to move on soft ground, a specific parameter is the average pressure p_{av} in the wheel's contact point with the support surface. However, the knowledge of this parameter is not enough to evaluate the necessary level of the chassis' passing ability for operating in these particular conditions.

To answer this question, both individual criteria for the passing ability and systems of criteria were suggested (see details in Chapter 8). Here, only the reference on some of the variants intended to show the difference between the parameter that characterizes the chassis' passing ability and the particular efficiency criterion of "passing ability" which should be used when performing the design optimization is presented.

The simplest version of the passing ability criterion is the "free specific traction force" criterion, defined as the difference between the maximum possible adhesion of the driving wheels with the ground in these conditions and the total resistance to rolling of the mover in the same conditions:

$$Pa_1 = K_{ad}\varphi - f,$$

where: k_{ad} is the chassis' adhesion weight coefficient in the predefined configuration; φ is the adhesion coefficient; and f is the resistance coefficient to rolling.

Also, the dimensionless criterion of passing ability is used:

$$Pa_2 = \frac{M_\varphi - M_f}{M_\varphi},$$

where: M_f is the resistance moment to wheel rolling and M_φ is the moment of the wheel adhesion with the soil.

Many researchers have considered systems of indicators (criteria) for evaluating the passing ability (which in fact means a certain decomposition of the initial notion), for example:

$$Pa_{bc} = 1 - \frac{P_{ay}}{p_s};$$

$$Pa_s = 1 - \frac{f_g}{\varphi},$$

$$Pa_{pf} = 1 - \frac{f}{D_{max}},$$

$$Pa_h = 1 - \frac{h_g}{h_K},$$

where: Pa_{bc} is the passing ability indicator by the soil's bearing capacity; p_s is the soil's resistance to stamp indentation; Pa_s is the passing ability indicator by the wheel's adhesion with the soil; h_r is the depth of the stamp indentation in the soil; Pa_{pf} is the passing ability indicator by the engine's pulling force; Pa_h is the passing ability indicator by the wheel's deepening into the soil; and h_K is the depth of the rut in the soil after the vehicle passes.

There are more complex criteria capable of evaluating not only the ability of a vehicle to overcome various obstacles en route, but also the efficiency of using this vehicle on the particular route.

The existing multi-axle wheeled chassis efficiency criteria will be considered in more detail in the following chapters.

The second method to reduce the number of design options is an isolation of the conceptually oriented sub-plurality of the vehicle's design layouts.

This approach is based on the fact that is easier to reduce the dimensionality of efficient design layouts if the list of particular efficiency criteria is defined in advance and the area that embraces the weight coefficients of each criteria is indicated. The existing methodologies for general design cases are described in specialized literature. The same methodologies as applied to designing multi-axle wheeled chassis, including compiling a list of particular efficiency criteria, comparing them with the system of scientifically-based weight coefficients, and the practical methodology for isolating the conceptually oriented sub-pluralities of design layouts are still at a very early stage of development.

The third method assumes a joint analysis of the efficiency and cost of the transportation-technological complex as a whole and of the chassis, in particular. This analysis is carried out if there is a single-valued vector of the chassis' concepts (from the higher model levels) and if the minimum possible value of the efficiency level is known. The fundamentals of this approach have been developed by M.I. Ipatov and is recognized among specialists as the "evaluation methodology based on reduced costs."

The influence of this criterion on design decision-making can be very high. For example, MSTU was designing a special chassis weighing 20–25 t with a high passing ability. Earlier chassis' of this type typically used caterpillar variants. The use of the "evaluation methodology based on reduced costs" criterion has shown that a switch to a wheeled mover is able to increase the efficiency of using the vehicle in certain cases by 9–12 times, which justified a non-traditional, alternative (competing) design solution.

The fourth method is the "metrics method," which states: a plurality of chassis with different layout concepts can be characterized by different pluralities of particular efficiency criteria.

These pluralities are built in such manner as to admit certain "distances between the elements within each plurality" (a metric for this plurality) and the "distances between them," i.e., the "deviations of one plurality from another"; such deviation can be expressed by some numeral. It is assumed that a chassis of one concept is "no worse" than one of another if such deviation is sufficiently small. This allows for creating, within the space of the particular efficiency criteria, an area containing the variants of "the most efficient chassis," which has advantages over other chassis in all metric combinations and admissible domains of the particular efficiency criteria.

5.6 Characteristics of the "Formation of HDWV Appearance" project stage

The necessity to complete this project stage is connected to the fact that when designing new vehicles, the requirements for the external design do not always match the possibilities of the internal design—they are either too high and, therefore, not feasible, or too low and fail to benefit from all available opportunities. Thus, the external design requirements should correctly agree with the capabilities of the internal design.

For this purpose, it is necessary to do the following:

- Eliminate obviously bad options;
- Form a range of feasible options; and
- Identify among the feasible options the variants that achieve the goals set at the external design stage and are reflected in the Tactical and Technical Requirements for the chassis.

It follows from here that the aim of this stage is to form the optimum appearance of the HDWV, which is always a compromise of solving particular problems. The numerous examples in Chapters 1–4 have shown that, unlike mass-produced multi-purpose automobiles, the creation of HDWVs is almost always individual and directed towards a specific purpose, i.e., it is conceptual. In this aspect, the HDWV creation process, especially for special-purpose vehicles, is similar to the creation processes of complex transportation objects such as aircrafts and helicopters.

The literature shows that in aircraft design, the decisions responsible for the continuation of the project are made in the initial stages and in conditions of incomplete information, not only in regards to how to implement the Tactical-Technical Assignment, but also in regards to the proper justification of the Assignment as such.

Analyzing the reasons behind the uncertainty that appears in the external design stage for

HDWVs is beyond the goals and objectives of this book.

However, experience shows that uncertainty at the internal design stage has to do with the following factors:

- The delay in submitting reliable information about the characteristics of the chassis' components designed by co-contractors and the distortion of such information by them for advertising purposes or because of a wrong assessment of the technological capabilities and experience of an enterprise (an example is the design of the engine for the T-64 tank model); and
- The insufficient study of certain processes and physical laws, which is most clearly revealed when making new technical solutions (an example is an insufficient knowledge in the control processes of the power flows in the electric drive of wheels in a multi-support wheeled vehicle).

The attribute matrices of design layouts show that the number of basic variants in HDWVs of the same purpose is very high. At present, it is already clear that the process of choosing the proper design cannot be formalized; however, the preparation of the support variant of the vehicle's general appearance can be automated. It is suggested that the customer of a new transportation-technological complex set up an ordered sequence of particular efficiency criteria, which characterize the HDWV only as a vehicle.

In order to achieve their aims, both the customer and designer should possess the same model of both the transportation-technological complex and of the vehicle itself. The structure and principles of building such a model and the algorithm for using it for internal design were described in this chapter. Further chapters deal with the model of a heavy-duty wheeled vehicle's main structural blocks.

Chapter 6
Conditions and areas for using HDWVs

6.1 Techno-economic feasibility for using HDWVs for cargo hauling

The basic starting point for designing vehicles, HDWVs in particular, is the data about their areas and conditions of use. While a HDWV's area of use (its main purpose) is set by the client in the Terms of Reference (TOR), the road and climatic conditions are typically presented in the most general form. For example, the temperature range of the ambient air can be indicated as from +50 to –50°C and an ability to move on all roads and localities can be indicated. When making a solution choice and studying the model (see Fig. 5.1), the designer needs to know particular values (at least averaged and typical ones) of the possible road and climatic conditions faced by the HDWV. The available technical literature does not mention using summarized data or analyze its effect on the vehicle's design.

However, the designer needs to know the prototype's economic efficiency to make a feasible analysis and justification of the selected technical solution.

A deep techno-economic justification of developing HDWVS to meet the economic needs of the USSR was made in the mid-1970s by the Institute of Complex Transport Problems (ICTP) at the USSR "Gosplan" (State Planning Committee) and the Central Scientific and Research Automobile and Motor Institute (NAMI).

The calculations made by ICTP showed that increasing the average load capacity of the USSR's dumper fleet by switching to heavy-duty dumpers with a 4.7–7.0 t load capacity allowed for decreasing—for every 100 million ton-kilometers (t·km)—the annual transportation costs by two million U.S. dollars, cutting the demands for drivers and maintenance staff by 750 persons, and reducing fuel consumption by 6600 t.

According to foreign experts, a 20% increase in the vehicle's full weight (for example, from 45.5 to 55.4 t) cuts the prime cost of 1 t·km by 54% (from 1.39 to 0.9 U.S. cents per 1 t·km).

Every country has accepted the trend towards higher load capacity motor vehicles; however, the growth of the vehicle's full weight is restricted by the existing road legislation, which prescribes the maximum axial load and overall dimensions. Thus, the European Union (EU) allows vehicles and road trains with the following overall dimensions to travel across all EU countries: a maximum length of 18.5 m, a maximum width of 2.59 m, and a maximum height of 4.12 m. The maximum allowed loads are as follows: 7.5 to 10 t per axle (respectively, for single and double (twin) tires); 11.5 to 20 t for a two-axle bogie depending on the span between the axles; and up to 24 t for a three-axle semi-trailer bogie. Five countries (France, Spain, Greece, Belgium, and Luxembourg) have established a 13 t axial load limit.

It has been theoretically established and confirmed experimentally that increasing the permitted axial load from 10 to 13 t is economically justified, because while it requires increasing road construction costs by 2% on average, it decreases transportation costs and improves the overall quality of the transport infrastructure. Many scientists have proved that in order to have Russian-made vehicles with progressive parameters, it is necessary to admit a 13 t load per axle and up to 21–23 t on twinned driving axles.

Thus, according to existing road legislation, all HDWVs are considered off-road vehicles and are typically only allowed to pass once on general-purpose roads. This is why an increase in load capacity is often accompanied by an increase in the number of axles, both driving and driven. We see that Russia and other countries are building more and more multi-drive vehicles with four, five, six, eight, and more axles.

The need for further HDWV development is defined by the need to transport respective cargoes. The development of natural resources and the construction of unique energy facilities and industrial complexes in cold climate areas are resulting in an increase in the volume of cargo in need of transportation. The typical characteristics of cargoes transported by motor vehicles in cold climates are as follows:

- An increased volume in mass cargoes (food and industrial goods; building materials; equipment; petroleum, oil, and lubricants (POL); etc.) to be transported both long distances (up to 1500–2000 km) and locally;
- Essential volumes of transported cargoes with practically unlimited batch sizes (timber, coal, ores, overburdens, etc.) over small distances (3–5 km); and
- Transportation of especially heavy one-piece items ("monocagroes") weighing100–500 t to new development areas or construction sites (of the petroleum, gas, forest, and chemical industries; new power plants; and other facilities).

Changes in the fleet's load capacities are also indicative. The number of vehicles with a 3–5 t load capacity continuously goes down while there is a simultaneous increase in the number of vehicles with a full weight of up to 3 t and of heavy- and super-heavy-duty vehicles. This shows the intention to increase the average load capacity of the fleet, consolidate shipments, improve transportation efficiency, and reduce the number of drivers involved.

The practice of using heavy-duty dumpers with an increased load capacity shows a high economic efficiency. For example, using dumpers with a 108 t load capacity in the technological cycle of the mining works at the Tsentralny Mine of the Apatit Production Association (PA) has shown the ability to considerably improve the technical and economic indicators compared to 40-ton dumpers. There has been a 3.1 times increase in labor productivity, a 18% cost reduction of 1 t·km, and a threefold cut in the maintenance staff. The economic feasibility of using heavy-duty dumpers has been recognized worldwide.

The development of the leading branches of the world economy—electrical power engineering, chemical and petroleum industries, metallurgy, and power machine-building—relies on the build-up or re-equipment of the existing capacities, the construction of new plants, and the use of the equipment with a high degree of ex-works readiness. The introduction of the modular-complex construction method has also caused a necessity to ensure the delivery of finished factory-assembled units to construction sites. The operational reliability of factory-assembled industrial equipment and units is much higher compared to the equipment assembled directly at the construction site. The installation and start-up of the projects can also be reduced by 2–3 months or more. Finally, it predetermines the efficiency of manufacturing ever larger indivisible items, for example, transformers, nuclear reactors, synchronous condensers, ship hulls, kiln assembly units, boring machines, gasholders, and other articles weighing 40–60 to 200 t or more. The amount of freight traffic of similar items in a developing economy should be constantly on the rise, as estimated by domestic and foreign experts.

Prior to the collapse of the USSR, the transportation of oversized heavy cargoes in the RSFSR (the then Russian Federation—part of the USSR) was entrusted to a specialized enterprise Spetstyazhavtotrans, founded in 1978, which operated under the Ministry of Automotive Transports.

During its existence, the enterprise transported more than five million tons of unique cargoes, with the economic effect reaching 700 dollars per ton of transported cargo. Below are just a few examples of its transportations.

In 1982, the enterprise delivered equipment from the Uralkhimmash and Penzakhimmash plants to the Dimitrovgrad Chemical Combine in Bulgaria. Separate items of that equipment weighed more than 500 t and were over 30 m long. Examples of transporting oversized heavy cargoes are presented in Figs. 6.1 and 6.2.

The enterprise also had to work in severe conditions, for example, when delivering transformers weighing 100 t each from Termez to Kabul (Afghanistan). It also performed 15 hauls from Poland to Klaipeda, Sumy, Sochi, and Pskov; and more than 10 hauls with equipment for the USSR industry from Czechoslovakia. In 1987–1988 only, Spetstyazhavtotrans performed 86 hauls from Finland to the city of Zapolyarny, in the Murmansk Region, to bring equipment for the Pechenganikel Metallurgy Combine. A considerable economic effect was achieved. For example, the transportation of 60 ball mills in 1988 with overall dimensions of 4500×6000 mm and weighing 150 t from Czechoslovakia to the site of the Krivorozhsky Mining-and-Processing Combine of Oxide-Bearing Ores, had an overall haulage cost of 1.6 million rubles, with a savings of 905,000 rubles.

Fig. 6.1. Road train comprising of two trailers and two tow cars with an especially heavy load

Fig. 6.2. Road train comprising of a trailer and tow car with an especially heavy load

Analysis of the transportation needs of extra-heavy oversized cargoes allows for defining the cargo distributions by batch sizes (Table 6.1).

For transporting heavy bulky goods, various types of HDWVs are used, e.g., trailers, semi-trailers, and self-propelled ground rolling stock (see Chapters 1–4). While the theory of designing heavy-duty trailers and semi-trailers is well developed and there are advanced production facilities in Russia for implementing this theory, the theory of general design solutions for self-propelled HDWVs has been poorly developed and needs serious advancement and generalization.

Since the above analysis shows that using HDWVs for transporting cargoes is economically feasible and technically expedient, the development of HDWV theory based on new engineering solutions, non-traditional for the Russian automotive industry, must be made in order to organize the manufacture of HDWVs at Russian production facilities.

Table 6.1 Distribution of oversized cargoes by batch weight

Batch weight, t	Share in the total haulage volume,%
40-60	46.30
60-80	28.80
80-100	12.50
100-140	11.10
140-300	1.02
300-600 and more	0.28

6.2 Conditions for using HDWVs

The efficiency of using an automobile as a means of transport is affected by climatic factors, the landscape, and hydrological and road conditions of the place of use. Accounting for the possible consequences of these factors, i.e., characterizing the vehicle's usage conditions, is a major task both in designing, calculating, and testing the vehicles and in the choice of the most efficient vehicle for a specific mission.

Climatology distinguishes between the climate-forming factors (the radiation background, atmospheric circulation, and local physical and geographical characteristics) and climatic elements (the air temperature and relative humidity, precipitations, cloudiness, etc.). In technical literature and this book, however, the variety of climate-forming factors and climatic elements affecting the vehicle's performance is defined by the generalized term *climatic factors* [13].

According to GOST (Soviet Standard) 16350, zoning of the former USSR territory for industrial purposes was made according to the monthly average temperature in the coldest (January) and warmest (July) months of the year accounting for the relative air humidity measured at 1 p.m.

The following climatic factors most affect the vehicle dynamics: the air temperature and humidity, the wind speed, the atmospheric pressure, and precipitates, and all should be taken into account when designing a vehicle.

Variations in air temperature within the year (the annual curve) at any given location depend on a number of factors. The annual curve of the average air temperature for continental climate zones has an expressly periodic nature (Fig. 6.3). Only in zones of tropical low-continental climate is the annual temperature curve practically constant (Curve 2) [13].

The distribution of average air temperatures in winter in the former USSR shows that in January negative temperatures cover over 95% of the territory. A characteristic feature of the Russian Federation is that the lowest temperatures

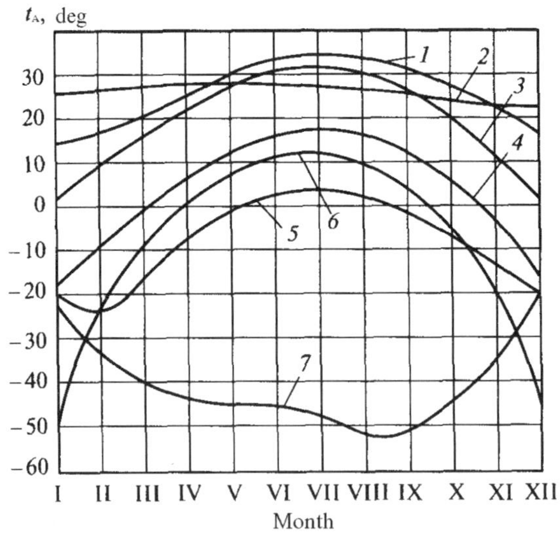

Fig. 6.3. Annual curve of air temperature in various climatic areas: 1—Aswan; 2—Singapore; 3—Ashkhabad; 4—Krasnoyarsk; 5—Uelen; 6—Verkhoyansk; 7—Antarctic, Station "Pionerskaya"

occur in the sub-arctic zone (the Verkhoyansk-Oimyakon region) instead of the arctic zone.

In summer, positive air temperatures cover all former USSR territory, while the highest temperatures are in the subtropical regions of Central Asia.

For a heavy-duty wheeled vehicle with a cargo, wind is a major climatic factor. The most important wind parameters are its speed and gustiness, which defines the specific wind pressure, the pressure exerted by the wind on the vehicle's windward side. The strong wind areas in the former USSR are in the majority of coastal areas of the Arctic and Pacific Oceans, while the weak wind areas are in Siberia and the central regions. Territory zoning for the former USSR according to wind speed and dynamic wind pressure was made according to GOST 1451 and allocated seven areas where the calculated values of non-operational wind speed ranged from 21 to 40 m/s.

Average monthly wind speeds in Russia range from 0.3 m/s (Oimyakon) to 8.4 m/s (Uelen). Monthly variations in average wind speed in some locations, borrowed from the *Reference Book of the USSR Climate*, are presented in Table 6.2, while Table 6.3 lists the probabilities of the maximum wind speed occurring.

Air pressure varies depending on the altitude above sea level. As presented in Table 6.4, the definite nominal air pressure corresponds to every altitude above sea level.

Table 6.5 presents generalized data on climatic factors in various climate zones that should be taken into account when creating HDWV design solutions.

A support surface's suitability for vehicle travel is defined by its degree of flatness, deformability, and friction properties. The efficiency of using HDWVs on dry non-deformable surfaces depends on the traction-speed properties, the suspension's quality, and the geometrical parameters of the passing ability; on deformable and slippery surfaces the efficiency depends on the parameters of the mover's interaction with the ground and the ground's mechanical properties.

The ground's friction properties define the mover's traction capabilities, which are estimated by the adhesion coefficient, φ_{adh}, of the mover with the ground, while the deformability of both the ground and a wheeled mover with an elastic tire defines the rolling resistance, which is estimated by the rolling resistance coefficient, f.

The key parameters describing the mechanical properties of the ground (soil) include the following:

- The ground's weight density γ;
- The deformation module E (measured in Pa);
- The internal friction angle φ_o;
- The internal ground adherence C_o;
- The rubber/ground friction coefficient φ_r;
- The thickness of the ground's soft layer Hr;

Table 6.2 Average monthly wind speed

Dwelling settlement	Average wind speed (m/s) by months of the year											
	I	II	III	IV	V	VI	VII	VIII	IX	X	XI	XII
Verkhoyansk	0.5	0.4	0.8	1.4	2.4	2.4	2.2	1.8	1.4	0.9	0.6	0.6
Novosibirsk	3.7	3.5	4.2	3.6	3.6	2.9	2.3	2.3	2.8	3.9	4.1	4.0
Kustanai	4.8	5.2	5.8	5.2	5.4	5.0	4.5	4.2	4.6	5.2	5.0	5.2
Fort Shevchenko	6.9	6.9	6.9	6.4	5.7	5.2	5.0	5.0	5.8	6.15	6.8	6.7
Norilsk	6.4	5.8	6.6	6.4	5.8	5.2	4.25	4.1	5.2	6.4	5.7	6.2
Uelen	6.3	5.8	5.4	5.7	5.0	5.4	7.1	6.4	7.1	8.2	8.4	6.3

Table 6.3 Maximum wind speed in different dwelling settlements

Settlement	Wind speed (m/s) possible once every			
	1 year	5 years	10 years	20 years
Oimyakon	13	16	17	18
Yakutsk	15	17	20	22.5
Krasnoyarsk	21	25	26	28
Novorossiysk	32	42.5	46	50
Nakhodka	36	45	49	53

Table 6.4 Effect of atmospheric pressure (altitude above sea level) on decreased engine power

Parameter	Value at the given altitude above sea level, m								
	0	500	1000	1500	2000	2500	3000	3500	4000
Average atmospheric pressure, kPa	102	95.7	89.7	84.5	79.5	74.6	70.3	65.8	61.8
Average relative air density	1.0	0.935	0.907	0.85	0.822	0.772	0.742	0.698	0.669
Average air temperature at 15˚C at the earth surface, ˚C	+15	+11.75	+8.5	+5.25	+2.0	-1.25	-4.65	-7.75	-11.0
Decrease of engine power,%:									
diesel engine without supercharging;	0	1.0–1.5	3.0–3.5	6.0–7.0	8.0–9.0	11.0–12.0	15.0–16.0	20.0–21.0	28.0–29.0
diesel engine with supercharging;	0	1.0	2.0	3.5	5.0	7.0	9.0	11.0	14.0
carburetor engine	00	6.0–7.0	16.0–17.0	25.0–26.0	33.0–34.0	40.0–41.0	46.0–47.0	52.0–53.0	56.0–57.0

Table 6.5 Parameters of ground surfaces for motion of HDWVs

Ground surfaces	Weight ground density, N/m^3	Deformation module, MPa	Thickness of ground soft layer, m	Angle of internal friction, degrees	Internal ground cohesion, MPa	Rubber/ground friction coefficient
Ground roads:						
spring	14,000	8.0	0.32	11.0	0.03	0.15
summer	15,000	15.0	0.60	12.5	0.045	0.60
autumn	14,000	12.0	0.29	12.0	0.04	0.20
Swollen ground road	13,000	1.0	0.25	12.0	0.005	0.20

Table 6.5 Parameters of ground surfaces for motion of HDWVs

Dry sand	15,000	2.0	0.50	32.0	0.001	0.30
Swollen tillage	11,000	0.5	0.40	5.0	0.006	0.20
Marshy meadow	12,000	1.5	2.00	3.0	0.02	0.20
Virgin snow	2500	0.1	0.60	16.0	0.003	0.10
Ice	-	-	-	-	-	0.15

- The ground parameter *B*, describing the influence of the loading time on deformation;
- The ground's shear modulus *E'*; and
- The ground's bearing capacity q_s.

The ground parameter *B* can be approximately defined through the internal friction angle from the following formula:

$$B \approx K/2\varphi_o,$$

where the coefficient $K \approx 1$ °/s, and φ_o is expressed in degrees.

The shear modulus *E'* can be roughly defined as:

$$E' \approx K_1 + K_2C_O,$$

where the coefficients $K_1 = 0.4$ MPa and $K_2 \approx 20$.

The ground's mechanical properties are defined by their grain-size classification (grading, the dimensions of solid particles), humidity, density, and sod matting. Soil grading is the most stable ground parameter and is usually the basis for classification. Humidity, density, and sod matting all vary over a broad range during the year.

Grounds with a high clay particle content are called cohesive grounds. Water, which is always present in grounds, renders considerable influence on their properties. The cohesion of cohesive grounds increases with an insignificant rise in humidity; once a certain humidity limit (which is individual for each ground) is exceeded, certain parameters change drastically; therefore, the condition of cohesive grounds is defined by their humidity (consistency), where the parameters are the yield point and the plasticity limit.

The yield point corresponds to the humidity at which the ground passes into fluidity. At this point, the ground's load bearing capacity and cohesion become insignificant. The plasticity limit characterizes the humidity at which the ground passes from a plastic to a solid state.

In dry conditions, cohesive grounds have fairly high mechanical properties. These grounds include light loam, loam, and clayey ground. At the humidity approximately equal to their plasticity limit, the bearing capacity average is $q_s = 0.29–0.7$ MPa, the internal ground adherence is $C_o = 0.035–0.1$ MPa, the internal friction angle is $\varphi_o = 21–17°$, and the deformation modulus is $E = 25–50$ MPa. Here, the first values in the range refer to light loams and the second to clayey grounds. At a humidity approximately equal to the yield point, these parameters have the following values: $\varphi_o = 10–0°$, $q_s = 0.07–0.09$ MPa, $C_o = 0.01–0.015$ MPa, and the deformation modulus $E = 3–4$ MPa. The most adverse conditions for movement are created when the ground's humidity is above its yield point, when $q_s = 0.03–0.01$ MPa, $C_o = 0.0–0.01$ MPa, $\varphi_o = 10–0°$, and $E = 0.3–1.5$ MPa [13].

This condition is characteristic of the spring and autumn water logging periods (the periods of road impassability and rains). The depth of the ground's waterlogged layer reaches 30–50 cm in spring, 20–30 cm in the autumn rainy season, and up to 10 cm in summer after rain. The heaviest periods for driving are the first weeks in spring after the snow melts. In other seasons, cohesive grounds have high plasticity.

The properties of sandy (non-cohesive) grounds are less dependent on humidity. When the humidity increases, their resistance to external loads first rises and then decreases once the yield point is reached. Compared to cohesive grounds, sands differ by their increased internal friction angle, φ_o = 32–43°, and by the almost complete absence of the internal adherence, C_o = 0.001–0.004 MPa. Their main deformation type is shear deformation. Their bearing capacity is also more dependent than cohesive grounds on the dimensions of the body (stamp) impressed into the ground and rises with its increase. The bearing capacities of sands increase considerably in the presence of an underlying solid layer.

Excessively waterlogged surface areas covered by a layer of peat at least 10 cm thick are called bogs and those with a peat layer less than 10 cm or absent are called marshy soils. The mechanical properties of these surfaces for driving can be compared to cohesive grounds in the spring during road impassability conditions. Bogs are subdivided into three types:

- Continuous bogs, where peat lies directly on the mineral bottom;
- Sapropelic bogs, where the peat crust rests on organic or semi-organic silt (sapropels); and
- Floating bogs, where the peat crust floats in water.

The bearing capacity of sapropelic bogs do not exceed q_s = 0.02–0.03 MPa at peat layer thickness of $H_r \geq 2.5$ m. At a peat layer thickness of $H_r \geq$ 3.5–4.0 m, the sapropel does not affect the bogs' bearing capacity. Floating bogs with a peat layer thickness of $H_r \leq$ 5–6 m are practically impassable for vehicles.

The main differences of peat from mineral grounds when impressing the stamp are the very high compaction, the absence of lateral displacements, and a noticeable effect of peat shearing strength along the stamp's (wheel) perimeter. Unlike mineral grounds, the peat's bearing capacity does not increase with an increase in the stamp's dimensions; on the contrary, it goes down.

Virgin snow also presents great difficulties for vehicle travel. The ability to move on snow is defined by the snow's mechanical properties and depth, which widely varies. The bearing capacity of loose snow, q_s, typically does not exceed ≤ 0.02–0.03 MPa and for a settled snow, $q_s \geq$ 0.10–0.15 MPa and up.

Ice-covered travel surfaces are characterized by low adhesion coefficient, φ_{adh}, values and, as non-deformable surfaces, by small traction resistance coefficient, f, values. If both values are close to each other, movement can become impossible. These surfaces are treated as almost impassable and are dangerous for motion because of the loss of traction and braking properties and a loss of stability.

The approximate values of the mechanical parameters of grounds most difficult for motion (according to the Scientific and Production Association (SPA) Gidrospetsgeologia) are presented in Table 6.5.

Thus, for areas of planned wheeled vehicle traffic (when choosing transportation routes), a ground map and the results of continual, seasonal monitoring of the changes in the ground parameters are needed.

A simplified diagram of the climatic factors and atmospheric precipitations that influence the "driver-automobile-road" system is presented in Fig. 6.4, where the climatic factors include the following:

- Solar radiation, air temperature, humidity, and atmospheric pressure variations during wind and thunderstorms (different air ionization);
- Air temperature during fogs, snowstorms, dust storms, mist, and wind;

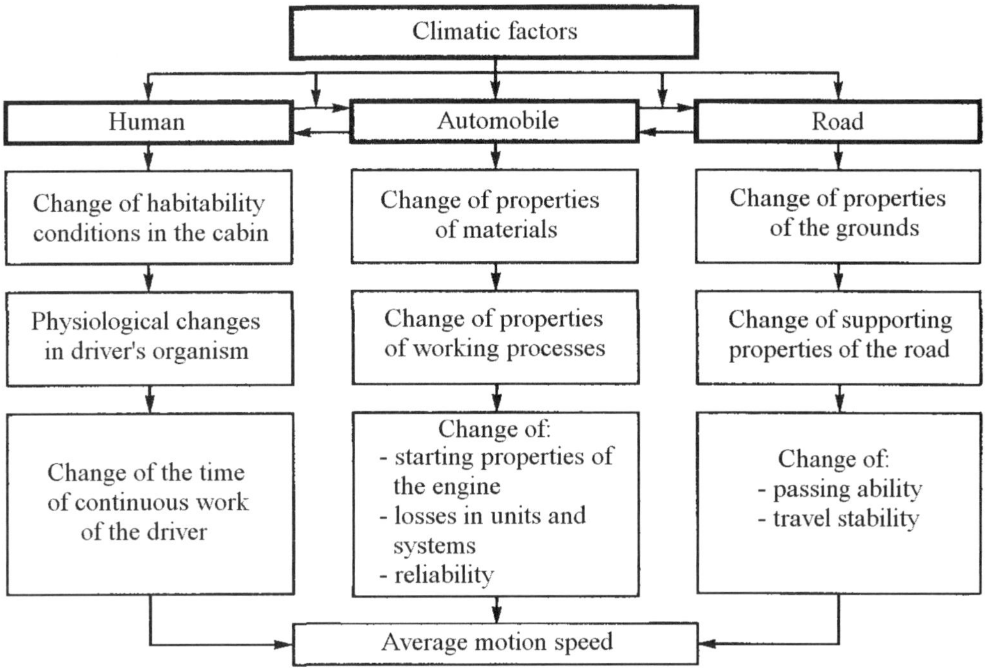

Fig. 6.4. Integrated diagram of the effects of climatic factors on the "human-automobile-road" system

- Solar radiation, low and high air temperatures, air humidity, and precipitates during wind, dust storms, and mist;
- The air humidity, precipitates, glaze ice, snow cover height, and the depth of soil freezing during snowstorms; and
- High and low temperatures, air humidity, precipitates, and wind during normal conditions.

As a result of these interactions, a complex of direct links and feedbacks arise in the "human (driver)-automobile-road" system: a decrease in the vehicle's steering quality; the influence of heat and gases (emitted by the vehicle), the dust (formed by the vehicle at work), and vibration and noise on the driver; and the destruction of the bearing (support) surface, an increased resistance to motion, and a decreased adhesion coefficient of the wheels with the bearing surface affect the road.

Thus, climatic factors affect the "human-automobile-road" system in a complex way by significantly influencing its functioning efficiency (the average travel speed is the integral indicator) and they must be taken into account when developing the general design solutions theory for HDWVs.

6.3 Basic requirements for traffic routes

The basic documents describing the locality where HDWV operation or testing is planned should contain the following:

- The traffic route's region and area;
- The traffic route's altitude above sea level (maximal and minimal);
- The locality's relief type;
- The locality's horizontal classification;
- The types of grounds (map of soils);

- The humidity condition of the grounds;
- The duration and depths of snow cover;
- The density of snow;
- The bearing capacity of ice on rivers;
- The density of the river network;
- How often rivers cross the traffic route;
- The density of lakes and the vegetation type; and
- The prevailing way of travel (along the roads or in the off-road mode).

A major factor in using HDWVs is the motor road. Here, the vehicle's operational properties are displayed in its interaction with the road surface's macroprofile, microprofile, and roughness. The macroprofile of solid-coated roads includes road irregularities of over 80–120 m long (ascends and descends), the microprofile includes road irregularities from 0.25–0.5 to 50-100 m long, and the roughness includes regularities less than 0.20–0.25 m long.

Motor roads can be subdivided by their features into the following basic groups:

- General purpose motor roads, which are permanently operated roads with an artificial ground bed and coating (highways);
- Planed (graded) roads;
- Trackway-formed roads;
- Plank (corduroy) roads;
- Slab roadways;
- Winter snow roads;
- Country roads; and
- Intra-quarry and dumping (stockpile) roads.

All general-purpose motor roads are divided into five technical categories depending on the expected average daily density of traffic in both directions in accordance with the SNiP 2.05.02 (Construction Norms and Rules).

Modern motor roads are complex engineering structures. When organizing the traffic and testing of HDWVs, the mathematical model of the vehicle's dynamics should include, apart from road conditions, the road as a description of a traffic route with the basic perturbation affection, which is needed for the theoretical justification of the chosen technical solutions. As demonstrated by numerous experimental studies, the dynamics of HDWVs are mostly affected by the macroprofile. The smallest is considered to be an irregularity of 0.5 m long, since a smaller roughness is not perceived by oversized tires.

A road, as a spatial curve, is characterized by the longitudinal profile, plan, and lateral profile. The longitudinal profile defines the steepness of separate road sections, the plan defines the road sinuosity (tortuosity), and the lateral profile defines the road's lateral steepness and carriageway width.

The basic element of a road's longitudinal profile is the slope. The longitudinal slope value is one of the major characteristics of a motor road's traffic qualities. The breaks in the longitudinal profile, formed at slope changes, also cause a number of travel speed restrictions in HDWVs. Therefore, the value of the vertical curve's radius, the type of transient vertical curve, and its length can be considered as the speed-limiting factors and should be included in the motion model of HDWVs.

The basic component of a road plan is the turning radius. The turning radius's value is also one of the major characteristics of a motor road's traffic qualities. At curves with small radii, the centrifugal force causes tire deformation in the lateral direction, which increases the power spent on the wheels' rolling movement, increases fuel consumption, and decreases driving speed. Since the curvature radius forms the resistive motion force at a turn, its value can be considered as a travel speed restrictor and should be included in the HDWV's movement model.

The core elements of the lateral road profile are the lateral slope and carriageway width. The road's carriageway width is defined by the width and number of traffic lanes. The width of the traffic lane is another major characteristic of the motor road's traffic qualities.

Thus, analyzing the links between the road elements and the dynamics of HDWVs allows for making the following conclusion. The key road elements that restrict the HDWV's potential motion capability and that should be included into the vehicle's dynamics model are the longitudinal slope and its length, the lateral slope, the carriageway width, and the roughness of the road pavement.

As demonstrated by experimental studies, the road's technical category cannot serve as a criterion for typifying motor roads. The longitudinal and lateral profile elements and the plans of motor roads negatively differ by 15–20% from the values regulated by the respective normative documents.

Thus, when scientists at MADI studied the highways of the 2nd technical category (Moscow–Minsk, Moscow–Yaroslavl, Moscow–Saint Petersburg, and Moscow–Simferopol), it was established that quite often the longitudinal slopes were 10–12%. It should also be noted that during the study of mountain motor roads, there were frequent ascends of 10–12% that were 600 to 1000 m long and also ascends of 20% less than 600 m in length.

The lateral profile slope of the studied roads varied from 0–4%, while regulatory documents stipulate it to be within the range of 1.5–2.0%. The radii of horizontal curves observed on studied roads also often fell outside the value range regulated by the normative documents.

The typical elements of roads of the second technical category are 2.2° ascents that are 495 m long; 2.3° descents that are 512 m long; a vertical curve connecting longitudinal slopes built along the circle with a radius of 3310 m; a horizontal curve built along the circle with a radius of 1170 m; and a 0–2.3° lateral profile slope.

Planed (graded) motor roads prevail in relatively dry steppe and forest-steppe areas and some extend for tens and even hundreds of kilometers. The width of graded roads is usually 6–10 m and slopes seldom exceed 8–10°. These roads can be used for traffic in dry weather and winter seasons.

Trackways and plank roads are made from logs placed longitudinally and laterally, respectively and are used for passage of single vehicles and light-weight road trains.

Slab roadways are made from logs, saw-timber, metal corrugated sheets, or concrete slabs. These roads are usually assembled from separate components (units) and are delivered to the site by motor transport. A slab road's load-carrying capacity and characteristics are defined by the vehicles with the largest dimensions and axial loads, which transport at least 50% of cargoes, or have a density that equals 25% of the general vehicle density that pass on these roads.

In severe northern climatic conditions, winter snow roads, i.e., seasonal motor roads, are widespread and used by multi-purpose vehicles and road trains.

Country roads are usually made by automobiles themselves during the seasons favorable for their travel. Top-soil (natural) country roads become impassable in certain periods (autumn and especially spring) and during rains. The slopes on these roads are defined by the specific locality but usually do not exceed 15–18°.

Intra-quarry and dumping roads do not differ much from mountain roads in their macroprofile, where the longitudinal profile varies from 5 to 20%. It should also be noted that during a review of quarries located in different climatic zones (in the Far North, Kazakhstan, Siberia, and Central Russia), quite often ascents of 10–12% were met with 600–1000 m lengths and ascents of 20% did not exceed 600 m in length. The results of studying the microprofile of quarry roads and their classification are presented in Tables 6.6 and 6.7.

The climatic conditions, insufficient road widths, unsatisfactory traffic-bearing surfaces (pavement), small turning radii, and steep slopes (17, 12, and 9%) create extreme conditions for using HDWVs, and must be considered when designing and testing prototypes [3].

Thus, the increase in the efficiency of using HDWVs should advance in two directions:

- Creating a general design-layout solution theory accounting for the maximal influence of external factors (conditions of use); and
- Creating a theory to justify the requirements for the HDWVs traffic routes and operation conditions for the maximal realization of the capabilities provided for in the vehicle design.

In general, the mathematical traffic route model, intended for use during studying and testing HDWVs, should include a typical route with the following elements:

- A straight (rectilinear) section of a road of the 2nd–4th technical category with a sufficient length for defining the maximal travel speed (3–5 km long);
- A straight section of a road that simulates the microprofile of quarry roads (for the characteristics for rocky ground, see Table 6.7) and a sinusoidal profile with a wave length equal to one to two lengths of the HDWV's wheelbase (3–5 km long);
- A straight section of a road simulating a top-soil road of the 5th technical category and reflecting the locality (for the characteristic as per the soil map, see Table 6.6) (3–5 km long);
- A curvilinear section of a road (without longitudinal slopes) with 15–60 m long

Table 6.6 Results of studying the microprofile of quarry roads

Surface		Place of measuring	Irregularity coefficient, mm²×s		Irregularity height, mm
Type	Quality		empirical	recommended	
Rubble	Good	Central Russia	2.58-29.15	2.0-30.0	1.4-5.5
Rubble	Good	Siberia	3.85-29.15	2.0-30.0	1.4-5.5
Sandy soil	Medium	Central Russia	41.95-178.7	30.0-450.0	5.5-21.2
Small gravel	Medium	Kazakhstan	38.81-165.9	30.0-450.0	5.5-21.2
Rubble	Bad	Siberia	333-2215	450.0-6750.0	21.2-82.2
Rocky ground	Very bad	Kazakhstan	7641-16950	6750 and more	87.4-130.2 and more

Table 6.7 Classification

Classification of quarry roads	Height of irregularities on vehicle's two wheelbases section, mm	Decrease in driving speed,%
Normal	1-5	10.0
With average passage ability	6-20	10.0
With heavy passage ability	21-85	20.0
With extra heavy passage ability	over 85	40.0

curves with alternating curve values times the minimal turning radius, and starting from this radius;

- A section of mountain road with ascents and slopes (descents) ranging from 8.7–30% and each being 600–1000 m long; and
- A section of the following macro-irregularity types: an "embankment" with an 8.7–30% ascent angle and an entrance length up to the vehicle's wheelbase length; an "embankment with uphill slope" with an 8.7–30% entrance angle and a 4.4–17.5% lateral slope; a 0.15–0.50 m high "road hump"; an "entrenchment" profile "ditch" with a width of up to half the wheelbase of a HDWV with four or more axles; and a locality section with the ground's bearing characteristics for the area of the HDWV's future use.

As a brief review of this chapter, it is possible to state that using the above materials with concrete values and recommendations enabled the experts at MSTU to solve several serious problems in designing HDWV prototypes, which were both highly praised by the customers and won recognition during demonstrations at specialized exhibitions.

Chapter 7
Basic requirements for HDWV consumer properties

7.1 General provisions

The efficiency of using HDWVs depends on the level of their performance and the quality of their consumer properties. Just like any other automobile, the following main requirements are imposed on HDWVs: simple design, reliability, survivability, acceptable overall dimensions and weight, efficiency, and ease of maintenance and repair.

The performance (operating) properties usually include the following [4]: the ability to conveniently and easily use and maintain, traction-speed properties, fuel efficiency, beneficial weight and dimensional parameters, maneuverability and motion stability, passing ability when moving on and off roads, passive and active safety, environmental safety, and reliability.

Heavy-duty wheeled vehicles, intended for installing and transporting special equipment, also have to meet certain special technical requirements (diagram in Fig. 7.1), which include the following:

- Mobility: defined as its ability to move from one point to another for a definite time (diagram in Fig. 7.2);

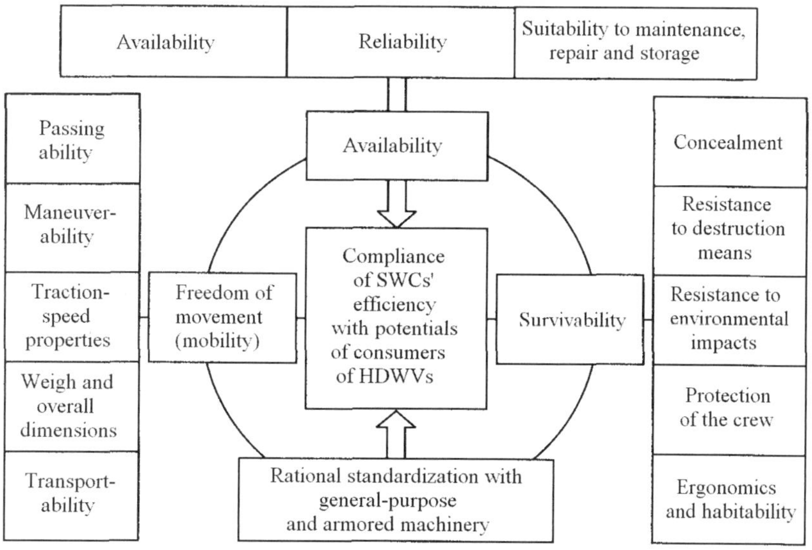

Fig. 7.1. Exemplary structure of special technical requirements for HDWVs

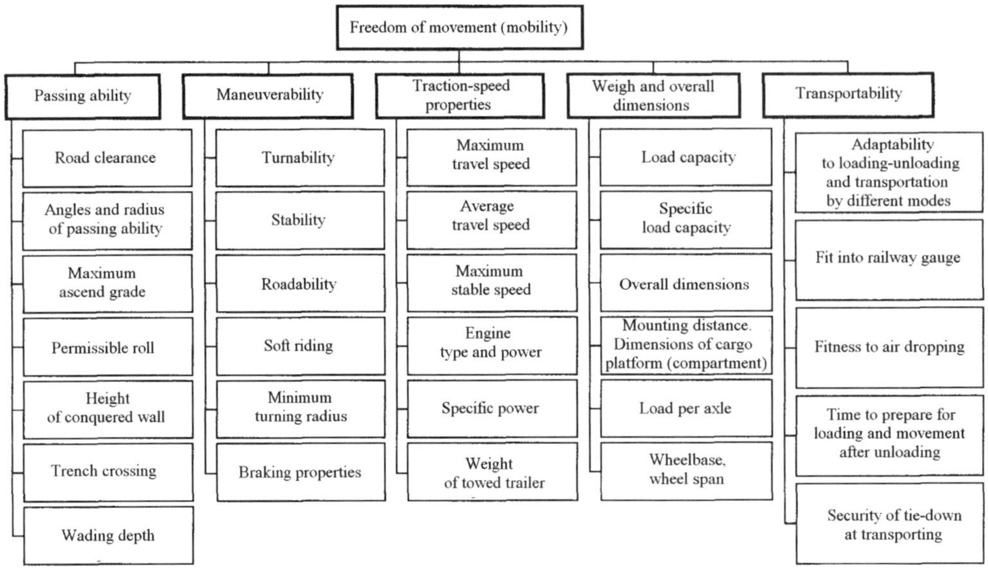

Fig. 7.2. Exemplary structure of mobility parameters of HDWVs

- Readiness to use (availability): defined by the time needed for the vehicle to get ready for motion (diagram in Fig. 7.3);
- Habitability: defined and ensured by the respective dimensions of the cabin, rational location of controls, illumination, and standard values of microclimatic mechanical-acoustic and chemical factors;
- Reliability: defined as the vehicle's ability to perform the preset functions while keeping the values of the established performance indicators within the limits that correspond to various operation modes and conditions, maintenance, repair, storage, and transportation (GOST 13377);
- Survivability: defined as the special motor equipment's ability to withstand attack by modern military weapons and preserve its working capacity to perform the intended mission (diagram in Fig. 7.1).

Most of the requirements for HDWV performance and consumer properties are defined according to international and national standards. However, the specific uses of these vehicles demand updating the values of individual indicators associated with the areas and conditions of these uses. Only the justifications and values of HDWV performance properties not prescribed by official documents are considered below.

7.2 Traction-speed properties and fuel efficiency

One of the main advantages of HDWVs, or any other automobile, compared to other means of ground transport, is their high operative and technical mobility. Naturally, the consumer requires not only fast and safe delivery of the cargo and equipment, but also the HDWV's ability to travel in various operating conditions with minimal transportation costs. This provision is characterized by both the vehicle's traction capabilities and by the amount of fuel consumed when transporting cargo and equipment.

Despite the numerous works and regulating documents on traction-speed properties, clear definitions fully describing the traction-speed properties of vehicles are seldom found in technical literature. For the modern vision of

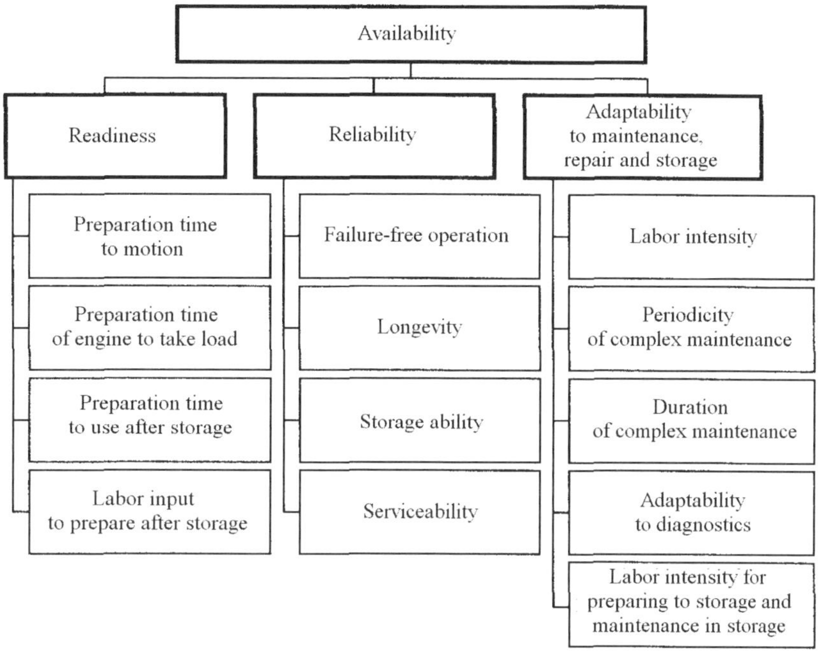

Fig. 7.3. Exemplary structure of readiness-to-use (availability) parameters of HDWVs

this issue, please see titles 3, 4, 15, and 16 in the Bibliography. However, as applied to HDWVs, these materials need supplementing since none of the suggested indicator values are regulated by normative documents.

Table 7.1 presents the traction-speed properties and fuel efficiency indicators and values suggested in the works of M.S. Vysotsky, L.Kh. Gileles, and S.G Khersonsky; and A.A. Tokarev [14, 30]. This data is supplemented with indicator values obtained from numerous tests on HDWVs conducted with participation of the authors.

The proposal to introduce hourly performance fuel consumption for estimating the fuel efficiency of HDWVs demands explanation.

The fuel efficiency of HDWVs should be defined according to existing methods. At the same time, one of the specific features of using special wheeled chassis (SWCs) is that the engine in any

Table 7.1 Classification of main measures, indicators, and values for HDWVs

Measure	Parameter	Value
F_k - traction force on the driving wheels, N	$F_{k\ max}$ - maximum traction force on the driving wheels, N	At least 0.6–0.7 of the full weight of the loaded vehicle
F_k/m_a - specific traction force on the driving wheels, N/kg	$F_{k\ spec\ max}$ - maximum specific traction force on the driving wheels, N/kg	6–7
	D_{max} - maximum dynamic factor	$D \geq 0.50$–0.6
N_e/m_a - the specific power, kW/t	$N_{spec\ max}$ - maximum specific power, kW/t	4.0–5.8
m_a/N_e - specific weight of HDWVs, t/kW	$G_{spec\ max}$ - maximum full specific weight, t/kW	0.031–0.038

Table 7.1 Classification of main measures, indicators, and values for HDWVs

i - road slope climbed by HDWVs, %	I_{max} - maximum slope climbed by HDWVs, %	≥7
V - travel speed, km/h	V_{max} - maximum speed, km/h	40–60
	V_{av} - average technical speed on roads, km/h:	
	with rigid pavement	20–30
	on ground roads	15–25
	V_{min} - minimum stable speed, km/h	4–6
Q_s - fuel consumption per 100 km, l	Q_{sk} - control fuel consumption, l/100 km	120–220
	Q_{sav} - average performance fuel consumption on typical route, l/100 km	280–350
	Q_{sye} - specific route fuel consumption, l/t 100 km	3.0–3.5
Q_h - hourly fuel flow rate, kg/h	Q_{hpc} - hourly performance fuel consumption at HDWV transport work (with account of expenses to maintain thermal regime and preparation of the engine), kg/h	100–200

season should be constantly ready to start-up and take the load. In this context, the calculations of the SWC's total fuel efficiency should account for both the fuel consumption necessary for movement and the fuel used to warm the diesel engine and keep it in a state of permanent readiness, including at negative air temperatures. A comparison of the fuel characteristics of two HDWVs, having the same full weight, but equipped with a gas-turbine engine (GTE) (chassis model 7907, 24×24) and a diesel engine (chassis 7906, 16×16), is presented in Fig. 7.4 and Table 7.2.

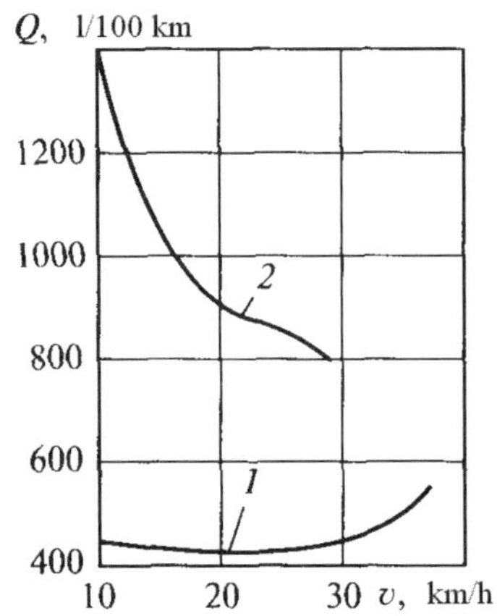

Fig. 7.4. Fuel characteristics of established motion of chassis 7906 (1) and 7907 (2)

In its efficiency (specific fuel consumption, g/kW), the GTE is slightly worse than the M-351 ship diesel engine (by 1.2 times). However, in on-site conditions (in a chassis), the control fuel consumption by the GTE is almost twice as high as the diesel engine.

According to Fig. 7.4, the 7906 chassis model has a considerably lower fuel consumption in the established travel mode than the 7907 chassis model in the same travel mode (by 2.17 times at V = 20 km/h). This indicates that the 7907 chassis is equipped with a transmission that does not allow for fully implementing the capabilities of the GTE.

When moving on a concrete-paved road, the operating fuel consumption is as follows:

- 901 l/100 km in the 7907 chassis; and

Table 7.2 Fuel efficiency of diesel engine and GTE in site conditions [18, 57]			
Chassis, wheel formula, transmission	Engine, power, kW (hp)	Specific fuel consumption, g/kW (g/hp)	Control fuel consumption, l/100 km
7906, 16×16, HMT	Diesel M-351, 1100 (1500)	239 (176)	446 at V = 30 km/h
7907, 24×24, ET	GTE-1100TF, 880 (1200)	288 (212)	860 at V = 25 km/h 800 at V = 29 km/h

- 610 l/100 km in the 7906 chassis, which is 1.47 times less than the 7907.

Hence, the difference between the operating fuel consumption values of the GTE chassis and the diesel engine chassis at unsteady travel speeds significantly decreases compared to the difference at established travel speeds. This is because of an increase in power consumption due to the higher rotation speed of the HMT elements used in the 7906 chassis, which has a high combined inertia moment, a low efficiency factor of the hydrotransformer at variable modes, and a higher rotational inertia coefficient of the mechanical transmission.

At idling, the GTE's fuel consumption in the 7907 chassis is 3.8 times higher than that of the diesel engine used in the 7906 chassis.

The average operating fuel consumption [18]:

- In the 7906 chassis at V_{av} = 17.6 km/h: 800 l/100 km; and
- In the 7907 chassis at V_{av} = 16.7 km/h: 1270 l/100 km, which is 1.4 times higher than the 7906 chassis.

Further, it requires 26–36 liters of fuel and 1.5–2.0 h to achieve a long-term warming by a heater of the diesel engine used in the 7906 chassis at low air temperatures. In these conditions, a GTE is ready to take the load in 8 minutes. The starting properties of a diesel engine and GTE are presented in Table 7.3. When including the fuel consumed for preparing the vehicle for travel (the period of start-up and maintaining the permanent readiness of the chassis' engine), the total annual

Table 7.3 Starting properties of the 7906 chassis with a diesel engine and the 7907 chassis with a GTE		
Parameter	7906	7907
1. Winter period		
1.1. Ambient air temperature, °C	From -26 to -36	From -21.5 to -22
1.2. Readiness time of the chassis to motion, min, including:	77-121	13-15
time to prepare and start the engine	47-60	1-2
time to warm-up the engine	25-55	7
run at low gear to warm up oil in transmission	5-6	5-6
2. Summer period		
2.1. Ambient air temperature, °C	16	12
2.2. Readiness time of the chassis to motion, min, including:	49	5-7
time to prepare and start the engine	24	2-3
time to warm-up the engine	25	3-4

operating fuel consumption of the 7907 chassis with a GTE can be 30% less than the 7906 chassis.

When using a GTE with warm accumulator batteries, the start-up does not require any preliminary heating; for this reason, the fuel consumed for warm up should not be considered while calculating the fuel efficiency of a SWC with a GTE.

As has been noted above, diesel engines should be sufficiently heated for start-up in a cold season, which is especially important for engines with short service lives.

For example, a B-38 model diesel engine can take a full load when the cooling fluid is at least 45° C. The operating experience of SWCs confirms that at an ambient air temperature of –40° C and below, the engine's cooling fluid should be warmed by tubular electric heaters (TEHs).

There is a linear dependence of the TEHs' operating time versus the ambient air temperature (Table 7.4).

Studies have shown that the dependence of the area of the diesel engines' cooling surface on their power is practically linear and is expressed by the formula $F = 0.01\ N_e$. If the $F = f(N_e)$ dependence, the TEHs' total power, and the capacity of the electric motor's pump used for fluid circulation in the engine's cooling system (for the B-38 engine it is 13.2 kW) is known, then the TEHs power required for heating the cooling fluid of any other engine can be calculated. Having defined the TEHs power and their work time depending on the ambient air temperature (see Table 7.4), the electric power can be found, and thus, the fuel consumption at a known efficiency factor of the electric generator and the auxiliary internal combustion engine to drive the generator.

The daily average fuel consumption by the auxiliary engine used for heating the cooling fluid can be presented as follows:

$$G_t^{day} = A\ N_{NEH}\ \tau \frac{1}{\eta_{gen}\ \eta_{eng}\ Q_n}$$

where: A = 860 kcal/kW and is the constant coefficient; N_{TEH} is the TEHs electric power, in kW; τ is the working time, in hours; η_{eng} is the auxiliary engine's efficiency factor; η_{gen} is the generator's efficiency factor; and Q_n = 10,300 kcal/kg and is the fuel calorific value.

The daily average fuel consumption for heating up to 1000 kW diesel engines in the climatic areas at 80° northern latitude and northward reaches 100 kg (4.2 kg/h).

Thus, the hourly operating fuel consumption is comprised of the hourly fuel consumed by the vehicle's sustained thermal heating (particularly in special purpose HDWVs), the pre-start-up preparation of the power unit and transmission, and the transportation of cargo.

Experience has also shown that increasing the engine power in order to increase N_{spec} slightly improves the SWC's traction and speed properties, but it also requires strengthening the transmission elements, frame, and suspension, i.e., an inevitable increase in the chassis' total weight. That is why we introduce the parameter of the full maximum specific weight, $G_{spec\ max}$.

7.3 Weight and dimensions

The weight and dimensions characterize, along with other HDWV parameters, the degree of design excellence for the given operating conditions and the rational utilization of

Table 7.4 Duration of TEH operation depending on the ambient air temperature [18]

Temperature, °C	40	30	20	10	0	-10	-20	-30	-40
TEH working time during 24 hours, h	0	4	7	10	13	16	18	21	24

material. The vehicle's full weight and load capacity are the main classification parameters.

By their load capacity, HDWVs form a standardized range from 25 to 200 t. These are multi-support vehicles composed from individual modules and can be further divided into two large groups. The first group includes vehicles with axle loads not exceeding the current legislative restrictions; therefore, they can be used on general-purpose roads. The second group includes vehicles that exceed the existing restrictions, so they work off-road or use general-purpose roads under special permits issued by corresponding authorities. Since legislative restrictions on both the weight and overall dimensions exist in all countries, the dimensional parameters of HDWVs are divided into the same two groups as the weight parameters.

It should be noted that the absolute values of the vehicles' dimensions and weights or masses are not sufficient for assessing their weight and dimensions parameters. An important characteristic of the HDWV's weight or mass is the specific load capacity coefficient, K_{avl}, i.e., the ratio of the working load to the vehicle's road weight. For modern heavy-duty vehicles, this ratio is 1.1–2.5. The same approach is used to assess the overall dimensions. In this case, the overall area utilization coefficient K_{ava} is applied, which is defined as the ratio of the cargo platform's area used to accommodate the working load, to the area formed by the vehicle's overall length and width. The distribution of the vehicle's full weight among the wheels is of great value for assessing the weight and dimension parameters. As a rule, in modern HDWVs, the weight distribution among the axles is uniform. The values of multi-support HDWVs' weight and dimensional parameters, calculated according to work [19], are presented in Tables 7.5–7.8.

Table 7.5 HDWVs used for transporting various installations

Manufacturer, model	Load capacity, t	K_{avl}	Platform length, mm	Platform width, mm	Loading height, m	Vehicle length, mm	K_{ava}	Axle load, N	Number of tires
Kamag, *U*1504	115	3.97	7000	3300	1.6±0.3	10,000	0.70	180	32
Sheyerle, *IMK*120.8.2	90	3.00	15,000	2750	1.4±0.3	15,000	1.00	300	32
Nicolas, *ASM* 19-4A	130	5.91	9300	3250	$1.5^{+0.6}$	10,850	0.86	190	16
Kamag, *U*1505	145	4.14	8500	3300	$1.6_{-0.3}$	11,500	0.88	180	40
Nicolas, *ASM* 19-5A	162	5.79	11,450	3250	$1.5^{+0.6}$	13,000	0.71	190	20
Cometto, *EIL*	120	4.00	7550	3200	1.6±0.3	10,600	0.77	150	40
Kamag, *U*1506	170	3.70	10,000	3300	$1.6_{-0.3}$	13,000	0.96	180	48
Nicolas, *ASM* 19-6A	194	5.71	13,100	3300	$1.5^{+0.6}$	14,650	0.89	190	24

Table 7.6 HDWVs used for long-distance hauls

Manufacturer, model	Load capacity, t	K_{avl}	Platform length, mm	Platform width, mm	Loading height, m	Vehicle length, mm	K_{ava}	Axle load, N	Number of tires
Cometto, *ETU*	110	3,93	8000	3200	1.35±0.3	10,800	0.76	115	48
Sheyerle, *M*.6	128	5,57	8400	2430	1.5±0.3	12,675	1.66	250	24
Nicolas, *ASM* 12-6A	119	4,76	10.350	2500	$1.2^{+0.65}$	12,350	0.88	240	24

Table 7.6 HDWVs used for long-distance hauls

Cometto, *ETL*1	150	4,41	11,000	3200	1.35±0.3	14,000	0.81	115	64
Nicolas, *ASM* 12-8A	158	4,65	14,050	2500	$1.21^{+0.65}$	16,050	0.91	240	32
Cometto, *ETL*2	190	4,75	14,000	3200	1.35±0.3	17,200	0.83	115	80
Cometto, *ETL*3	204	5,67	13,100	3000	1.15±0.3	16,300	0.80	120	80
Nicolas, *ASM* 19-10	283	4,96	19,100	3250	$1.5^{+0.65}$	21,100	0.91	340	40

Table 7.7 HDWVs used at shipbuilding yards

Manufacturer, model	Load capacity, t	K_{avl}	Platform length, mm	Platform width, mm	Loading height, M	Vehicle length, mm	K_{ava}	Axle load, N	Number of tires
Kamag, *U*1402	103	4.12	8000	5000	1.6±0.3	15,000	0.89	320	16
Sheyerle, *MT*120.8.2	113	2.40	15,000	5000	1.7±0.5	18,360	0.82	200	32
Nicolas, *ASM* 25-4A	158	3.76	12,400	8000	$1.52^{+0.6}$	12,400	1.00	250	32
Kamag, *U*1406	328	5.86	14,000	6100	1.6±0.3	20,000	0.70	320	48
Nicolas, *ASM* 25-6A	240	3.00	17,750	8000	$1.52^{+0.6}$	17,750	1.00	250	48
Kamag, 1408*S*	440	6.11	18,000	6100	1.6±0.3	24,000	0.75	320	64
Nicolas, *ASN* 25-8B	320	4.00	21,850	8000	$1.52^{+0.6}$	21,850	1.00	250	64
Kamag, 1412*S*	653	5.68	26,500	6100	1.6±0.3	32,000	0.83	320	96
Kamag, 1412	1316	5.96	26,500	10,000	1.6±0.3	32,000	0.83	320	192

Table 7.8 HDWVs used at metallurgy plants

Manufacturer, model	Load capacity, t	K_{avl}	Platform length, mm	Platform width, mm	Loading height, m	Vehicle length, mm	K_{ava}	Axle load, N	Number of tires
Kamag, *I* 1602	38	3.80	7000	2700	1.6±0.3	13,000	0.54	120	8
Sheyerle, *IMK* 48.2.2	38.5	4.50	4000	2500	1.25±0.3	5500	0.73	240	8
Cometto, *ETL* 3.1	56.0	3.50	6000	2700	1.5±0.30	8400	0.71	120	12
Kamag, *I* 1604	78.0	4.33	9000	2700	1.6±0.30	14,000	0.86	120	16
Sheyerle, *IMK* 120.10.2	95.5	3.90	8500	2700	1.25±0.30	10,050	0.85	240	20
Cometto, *ETL* 6	115	3.92	9000	2700	1.5±0.30	11,400	0.79	120	24
Kamag, *I* 1606	118	4.54	12,000	2700	1.6±0.30	17,000	0.70	120	24
Sheyerle, *IMK* 144.12.2	116	4.14	9300	2700	$1.5_{-0.2}$	10.850	0.86	240	24
Kamag, *I* 1607	136	4.25	14,000	2700	1.6±0.30	19,000	0.74	120	28
Cometto, *ETL* 7	137	4.42	10.000	2700	1.5±0.30	12,400	0.81	120	28
Cometto, *ETL* 8	158	4.65	11,000	2700	1.5±0.30	13,400	0.82	120	32

7.4 Maneuverability, ease of steering, and road passing ability

The use of multi-axle, multi-support vehicles with especially high load capacities, notable for their huge dimensions and axle loads, raises the problem of their ability to fit into the curves and turns of the existing road network and their maneuverability in localities, within construction sites, and in other conditions.

In this book, the term maneuverability characterizes the vehicle's ability to change its position within a confined area and in restricted passages [1, 2]. To assess maneuverability, the parameters of a steady turn measured at a low travel speed are used.

The maneuverability of HDWVs is characterized by the following indicators:

- The minimum turning radius along the axis (relative to the center of turning) of the front external wheel's trace;
- The external dimensional turning radius; and
- The width of the corridor taken by a vehicle or a road train when turning at the preset external overall radius.

The main assessment indicator of a HDWV's maneuverability is the width of the corridor. It is equal to the difference between the radii of two points of the vehicle—the most distant from and the closest to the turning center—and characterizes the width of the corridor required at sharp turns and the HDWV's ability to move within passages of a predefined shape and dimensions.

Documents of the EU prescribe the minimum external and internal overall turning radii, which should not exceed 12.5 and 5.3 m, respectively. Thus, it defines the permissible corridor width for road vehicles as 7.2 m.

So far, there is no generally accepted definition of "ease of steering" (in other words, roadability or controllability), because of the complexity of the steered (controlled) motion process. That is why this book interprets the ease of steering according to the following definition, borrowed from the respective Russian Branch Standard (OST): "The ease of steering (controllability) is the property of a vehicle driven by a driver to maintain—under the respective road and climate situation—the preset direction of travel, or to change it in response to the affection applied to the steering wheel. [31]."

The value limits for the ease of steering and stability of motor vehicles, recommended by Rule No. 79 of the Inland Transport Committee of the United Nations Economic Commission for Europe (UNECE) and OST 37.001.471, are presented in Table 7.9.

The passing ability is interpreted as the HDWV's ability to move on the roads and structures of the terrain depending on the support surface's carrying capacity (the support passing ability) and the geometry of the road conditions, structures, and available obstacles (the profile passing ability).

Table 7.9 Main value limits for ease of steering (controllability) and stability recommended by Rule No. 79 of the UNECE and OST 37.001.471

Test name	Force P_p	Value for motor vehicles of Category N_3
Force at the steering wheel, N	At the steering wheel with properly operating amplifier (booster), N	No more than 200
	At driving with a failed amplifier (booster), N	450
	"At rest" with properly operating amplifier (booster), N	No more than 250

For several reasons, the travel of a wheeled vehicle follows a complicated trajectory consisting of straight and curvilinear sections of various natures and orientation. The possible conditions for using HDWVs were considered in Chapter 6. The motor road characteristics of the Commonwealth of Independent States (CIS) are presented in Table 7.10, while those of automobile bridges are presented in Fig. 7.5 and Table 7.11. The bridge dimension is the contour outside which no bridge elements should protrude (see Fig. 7.5). The bridge dimensions are designated by the Russian letter Г and the numbers corresponding to the value are related to the width of the carriageway (width between the curbs). If there is a median between

Table 7.10 Motorway characteristics in CIS countries

Road category	Permissible speed, km/h	Number of traffic lanes	Lane width, m	Carriageway width, m	Shoulder width, m	Road bed width, m	Turning radius, m		
							Main	Off-road	In mountains
I	150	Up to 4.8	3.75	2×7.5	3.75	22.5	1000	600	250
II	120	2	3.75	7.5	3.75	15	600	400	125
III	100	2	3.5	7.0	2.5	12	400	250	100
IV	80	2	3.0	6.0	2.0	10	250	125	60
V	60	1	4.5	4.5	1.75	8	125	60	30

Note. According to GOST 21624, motorways are divided into 5 categories by their operation conditions.

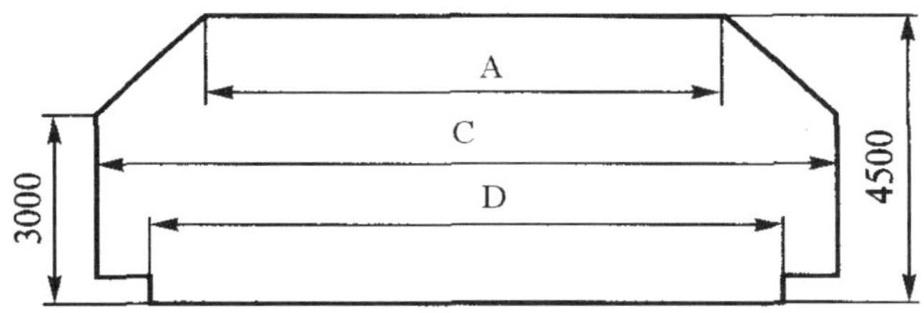

Fig. 7.5. Crosswise contour of bridges: A—bridge dimension at the upper overpass; C—overall width of the bridge; D—bridge dimension by the sidewalks

Table 7.11 Dimensions of automobile bridges [3]

Bridge dimension	Distance by overhang (Fig. 7.2), m		
	between curbs Г	between supports Г	along the top A
Г-21	21	21.5	20
Г-14	14	14.5	13
Г-10.5	10.5	11.0	9.5
Г-9	9	9.5	8
Г-8	8	8.5	7
Г-7	7	7.5	6
Г-6	6	6.5	5
Г-4.5	4.5	5.0	3.5

the lanes, it is designated by the letter C. For example, Г-9+C+9. Ten bridge dimensions are specified and the main ones are presented in Table 7.11. Dimension Г-6 is applied only to wooden bridges and dimension Г-4.5 is applied only in exceptional cases in mountain localities. The span height of all bridges is established as 4.5 m. In general, the bridges' bearing capacity is 40–60 t and occasionally 80–100 t [3].

Stringent requirements can be presented for the HDWVs' operative mobility, defined by the ability to transport them by rail, water, and air transport. These requirements strictly limit the vehicle's overall dimensions.

The operative mobility requirements are satisfied depending on the HDWV's main purpose by using various design methods (transportation of vehicles by modules, partial disassembly, etc.) or, for example, by the corresponding preparation of railroad transport-carrying facilities.

Ensuring the HDWV's required operative mobility is complicated for both the designers and their users. A mutually acceptable compromise has to be found in each particular case.

7.5 Safety

The safety of the vehicle is one of the major performance characteristics. The safety requirements stipulated by normative documents should cover HDWVs since safety is responsible for human life and health, and the security of vehicles, cargoes, and other material assets. Traditionally, specialists distinguish between the active, passive, and ecological (environmental) safety of the vehicle design.

The active safety is ensured by the properties and design qualities that help the driver to prevent accidents. These include the braking properties, audio and light warning systems, the efficiency of the headlights, etc.

The passive safety is ensured by the design properties that reduce to a minimum the traumatizing of the driver and passengers in case of an accident. They include the rigidity (strength) of the cabin design, reliability of door locks, etc.

The ecological safety is ensured by the vehicle properties which decrease its harmful influence on the traffic participants and the environment in the process of operation. It includes limiting the quantity of emitted harmful substances, possibility of soil destruction, etc.

Vehicle safety is ensured by the failure-free performance of all controls and mechanisms, and also by the vehicle's ergonomic parameters, which increase the driver's comfort and decrease fatigue.

A set of UNECE rules, which elaborates on the agreement of the UNECE's Inland Transport Committee on the introduction of uniform, generally recognized rules on the admissibility of vehicle equipment and accessories has been developed which concerns practically all aspects of a motor vehicle's active, passive, and ecological safety.

Official confirmations of a motor vehicle's conformity with current UNECE rules is a necessary precondition for export.

In general, the official confirmation contains the following five elements: the sign *E*; the serial number of the country that issued the official confirmation; the number of the UNECE Rule according to which the confirmation has been issued; the number of the latest series of approved amendments to the specified Rules; and the number of the official confirmation.

The countries-members of the EU have adopted the corresponding directives, which, like the UNECE Rules, regulate the admissible motor vehicle safety parameters. The indicator of the vehicle's conformity to the EU Directives is the sign *E*, which certifies its right to be sold in the common market countries [15].

The process of the conformity check to the UNECE Rules or to the EU Directives is called homologation.

The national standards of some countries on design safety are even tougher than the requirements of the UNECE Rules and EU Directives (Sweden, Japan, and the U.S.). This must be taken into account by the countries exporting vehicles.

Braking properties. The demands to increase HDWV safety in growing traffic speeds and density and the increasing weight of transported cargoes entails further improvements in the vehicle's brake systems.

The efficiency parameters of the service and emergency brake systems are the braking path and the steady braking deceleration, depending on the initial speed when braking begins, while the total braking force developed by the corresponding system is the indicator of the parking and secondary brake systems efficiency.

The brake system of a wheeled vehicle is a combination of devices intended for exercising the process (phase) of braking (according to a known braking diagram). In its turn, the brake systems combination is treated as the brake control.

In Russia, the established efficiency standards of automobile brake systems are stipulated by GOST 22895, GOST 4364, and GOST 23181, while OST 37.001.067 *Braking properties of automotive rolling stock. Test methods to define the brake system's efficiency* contain the description of the braking test conditions and the definition of the brake systems efficiency indicators.

The main international European document containing the requirements for brake systems is Rule No. 13 of the UNECE's Inland Transport Committee; in EU countries, Directives Nos. 71/320.75/524, 79/489, 74/132, 85/647, and 88/194 contain the requirements for brake systems, which are basically identical to the UNECE's Rule No. 13 requirements.

The HDWV's brake control should consist of four brake systems:

- The service brake system is intended for regulating the travel speed in any motion condition. The efficiency requirements for this system are established by normative documents or specified by the customer in the Technical Assignment or TOR;
- The emergency brake system stops the vehicle in case the service brake system fails. The emergency brake system can be a part (loop) of the service brake system and should ensure at least 30% of the service brake system's efficiency with a maximal deceleration $j_A = 2.8\ m/s^2$;
- The parking brake system is intended for holding the fully loaded vehicle on the maximum slope indicated in the vehicle specifications. For a vehicle in operation, as per the requirements of GOST 25478, the parking brake system should hold a fully weighted vehicle parked on a slope of at least 16%. For vehicles with the curb (unloaded) weight, the vehicle should be held by the parking brake system on road slopes of 31% or more, while the manual effort applied to the parking lever should not exceed 400 N; and
- The secondary (auxiliary) brake system serves for extended braking on long descents without using other brake systems. It should ensure the vehicle's descent with a speed of 30±2.0 km/h along a 7% slope 6 km long. The deceleration in this case should be $j_A = 0.6–2.0\ m/s^2$.

It is allowable for two or several brake systems to simultaneously use common elements (for example, brake mechanisms or energy sources). However, each HDWV should have at least two independent controls of different brake systems.

The elements of the brake control system are designed so that they are able to perform their functions irrespective of wear, aging, corrosion, vibration, radioactive land contamination, light, and electromagnetic irradiation, which can occur in the process of operation.

According to the requirements of GOST 22895, a number of brake control system elements should have a guaranteed strength for ensuring the safety of travel, i.e., their failure as a result of breakage should be completely excluded during the vehicle's entire service life. These elements include the brake pedal and its attachment, the brake valve, the main brake cylinder, the drive elements of these devices from the pedal, the air distributor, the driving effectors mounted on the wheels, the adjusting levers, the expanding cams, and other similar elements. A failure of these elements is an emergency accident and requires not only a replacement or repair of this element, but also a change in its design or manufacturing technology if it is the reason for the failure.

The arrangement of the controls should ensure handling convenience and not create any danger for the driver and passengers in case of an emergency braking or accident.

The service brake system is required to act on all wheels and its action should be smooth. The distribution of this action should also be rational, i.e., in any wheeled vehicle at all admissible loads, the front wheels (the front group of wheels) should be blocked prior to the rear wheels (the rear group of wheels) when the ratio of the vehicle's deceleration to the gravitational acceleration (9.81 m/s^2) is within the limits of 0.15–0.30. This means that the advanced blocking of the front wheels is prescribed for HDWVs only on slippery roads.

Brake systems which have brake mechanisms mounted on the wheels should ensure a uniform distribution of the braking forces among the wheels of one axle, while the maximum deviation cannot exceed 15% of the largest braking force value. The wear of the brake mechanisms' friction surfaces should be compensated by manual or automatic regulation, so that the values of the gaps always remain within the preset limits.

The built-in sensor and signaling system for the technical condition of the brake control system should meet the requirements of OST 37.001.283, the pneumatic drive of the brake systems should meet GOST 4354, and the hydraulic drive should meet GOST 21181. The brake drive elements should also be reliably protected from external impacts and mechanical damage when the wheeled vehicle moves off-road.

Requirements for the exterior lighting system and signaling lights. In order to provide the driver with the required data about the visibility conditions of the road, it is necessary to equip the vehicle with four types of headlights. The installation of two to four high beam headlights and two low-beam headlights is mandatory; it is also recommended to install fog-lights and a high beam spotlight, especially in vehicles intended for long-distance hauls.

The minimum set of signaling lights mandatory for HDWVs includes direction indicators, a braking signal, dimensional marker lights, the license plate light, and the lights to mark a road train.

The arrangement, color, and visibility of external lights should meet the requirements of GOST 8769.

The international normative documents on external lights are Rule No. 48 of the UNECE; EU Directives 76/756, 80/233, 82/244, 83/276, 84/8, and 89/278; and the ISO-R-303. CIS countries use automobile lighting systems according to European safety standards; therefore, the requirements of state standards concerning the lighting equipment of vehicles practically coincide with the requirements of UNECE's Rule No. 48.

All vehicles should be equipped with European or American asymmetric systems of high and low beam headlights.

In recent years, automobile lighting systems have been utilizing lighting equipment with halogen lamps. The range of these headlights is much larger than common incandescent lamps and can reach up to 400 m. Headlights with halogen lamps are expedient for HDWVs used for long-distance hauls.

The light and color characteristics, norms, and test methods of signal lights are regulated by GOST 10984.

Requirements for the design's passive safety. Fulfilling the requirements related to a vehicle's cabin strength; the use of door locks, hinges, and controls; and using interior elements of a safe design are mandatory for improving the passive safety of freight vehicles of all load capacity classes.

When designing an injury-free cabin interior, the following should be envisaged:

- Placing control/measuring instruments, buttons, and handles in recesses of the instrument panel and placing window raiser handles and internal handles of the locking mechanisms inside recesses available in the internal door panels;
- Rounding the instrument panel's lower edge (the rounding radius should be at least 10 mm);
- Using a sun visor made of light-absorbing material and an injury-free, splinter-proof windscreen;
- Excluding dangerous irregularities or sharp edges in areas of a possible head blow;
- Lining interior cabin surfaces with soft panels, which quickly take the shape of the hitting body part, thus more uniformly distributing the shock load; and
- Equipping the steering column with an energy-absorbing device to protect the driver from injuries in case of a collision with other cars.

Statistics show that the main causes of severe traumas and lethal injuries to drivers and passengers of cargo vehicles at tipover and collisions are cabin deformation, colliding with elements of the interior equipment, and falling out of the cabin as a result of a spontaneous door opening.

The main document regulating the cabin strength requirements is UNECE Rule No. 29. It states that the cabin of a cargo vehicle with a full weight of more than 7 t should preserve the vital space after it has been affected by each of the following loads:

- A dynamic load (with an impact energy of 44.15 kJ) onto the front part of the cabin created by a freely suspended rectangular pendulum (2500 mm width, 800 mm height, 1500±250 kg weight) in the direction parallel to the vehicle's longitudinal vertical plane; the pendulum's center of gravity at the moment of impact should be in the vehicle's middle longitudinal cross-section, 150 mm above the center point of the human body's hip joint;
- A static load equal to half of the vehicle's full weight (but no more than 98 kN), applied uniformly to the cabin's roof;
- A static load equal to 1.962 kN per 1 t of cargo weight applied to the whole area of the cabin's back surface and created by a vertical solid slab moving in the direction parallel to the vehicle's longitudinal axis.

Russia operates under OST 37.001.221 which regulates the impact and strength properties of cargo vehicle cabins at tipover and frontal collisions; it also regulates the test methods aimed at defining these properties. It differs from UNECE's Rule No. 29 by the presence of vertical and lateral rigidity norms at impact loads and a more advanced method of estimating the cabin deformation. Sweden has the strictest requirements for the strength of truck cabins.

The following safety requirements are added for HDWVs intended for long hauls:

- A high stability against tipover when moving with a cargo in wind speeds of 20 m/s (with gusts of up to 25 m/s); in case of tipover, the liquid (fuel, oil, water, etc.) should not pour out of the tanks to avoid self-ignition; and

- The cabin's strength should be ensured against an overpressure in the air wave front of at least 0.03 MPa (0.3 kg/cm^2).

Requirements for ecological safety. In the recent years, all leading world companies, including the manufacturers of HDWVs and road trains, have been implementing programs aimed at improving the ecological parameters of their vehicles. This process was triggered by the UNECE Rules, which are continuously becoming stricter. Table 7.12 presents the international standards of the maximum permissible amounts of harmful substances emitted by the engine.

One of the major indicators describing the efficiency of a HDWV design is the level of external and internal noise. The international European norms on the allowable levels of a vehicle's external noise were first adopted in 1968 and then included into UNECE's Rule No. 6, and then into the national standards of the majority of European countries. In Russia, the restrictions on the admissible level of external and internal noise have been in effect since 1974. At present, the noise level is measured in decibels as per A scale (dBA) (GOST 17187, GOST 27435, and GOST 27436), which most fully corresponds to the human ear's hearing ability.

The real decrease in the noise level and its value measured in dBA are linked by a logarithmic, not arithmetic, dependence. For example, a 2 dB decrease in an 88 dB noise level according to the A scale corresponds to a 20% real decrease; a decrease of 6 dB equals a 50% real decrease.

To decrease the overall noise level of the vehicle by any value measured in dBA, it is necessary to reduce the level of each separate noise source by the same dBA value.

From 1974 to 1982, the EEC standard (EEC Directive 70/157) on the admissible external noise level was decreased from 91 to 86 dBA. In 1983, a still stricter norm (84 dBA) was adopted, which was introduced in 1987 for new models and since 1989 has been in effect for all existing vehicles.

Table 7.12 Limit values of harmful substances emitted by the engine, g/kW

Harmful substances	Rules of the UNECE		
	No. 49-02 A "*EURO*-I"	No. 49-02 E "*EURO*-II"	No. 49-02 "*EURO*-III"
	Introduction date		
	01.07.92	01.10.95	01.10.98
Carbon monoxide	4.5	4.0	Below 4.0
Hydrocarbons	1.1	1.1	Below 1.1
Nitrogen oxide	8.0	7.0	5.4
Solids (smoke)	0.36	0.15	Below 0.1

The overall noise level of modern HDWVs is mostly defined by the power unit's noise level, its supporting systems, and the transmission aggregates.

The steady decrease in the maximum allowable external noise level dictates the need to continuously improve the design of the power unit, its operation, and the gas exhaust and air intake systems, since they are the main sources of noise in HDWVs. A schematic example of the external noise field of the MAZ-543M chassis is presented in Fig. 7.6.

At a travel speed of 80 km/h or more, the noise level from the tires can exceed the noise level from other sources in the vehicle. The tire noise depends on the tread pattern, the degree of tire wear, the type of road surface, the travel speed, and the load on the tire. Reducing the tires' noise level without degrading their service life is one of the greatest challenges in conforming to the noise level regulations.

Within the last 20 years, the noise level in the cabin of HDWVs has considerably decreased (for example, from 77 to 73 dBA for road trains traveling at 80 km/h). The leading Western European companies-manufacturers of road trains have managed to reduce it to 65–68 dBA in their serially-made vehicles. This has been done by modifying the engine design, better cabin sealing, using linings with more efficient sound-absorbing materials, and filling the cavities with foaming polymers. An example of the admissible disturbances to oral communication (passengers to the driver) is presented in Fig. 7.7.

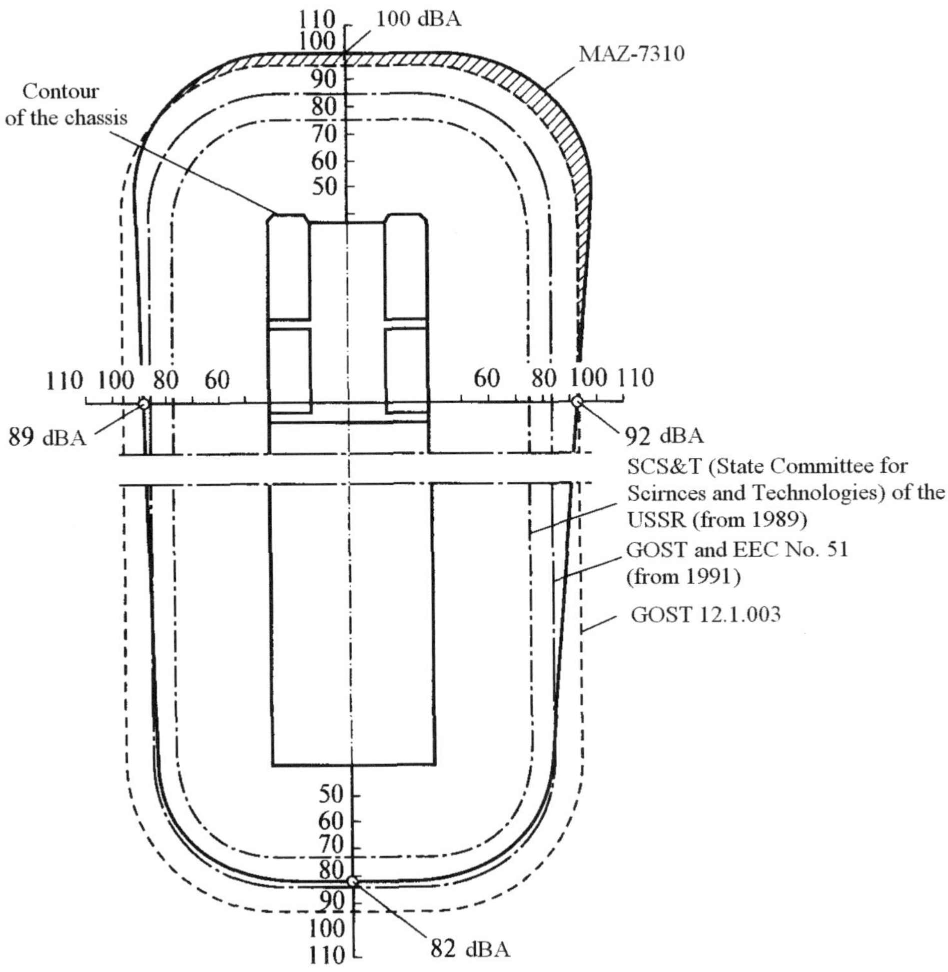

Fig. 7.6. External noise field of the MAZ-543M chassis

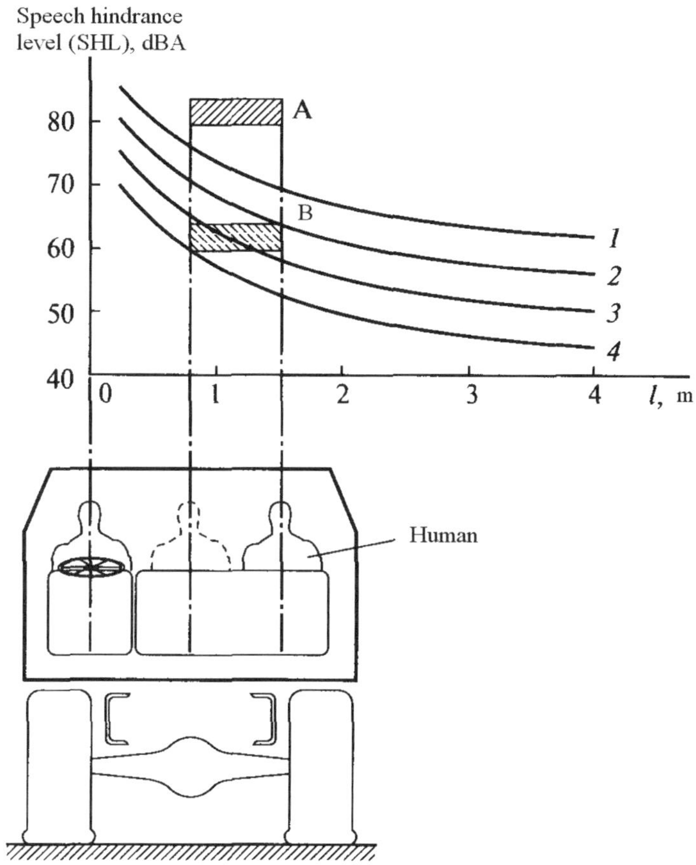

Fig. 7.7. Admissible levels of oral speech disturbances depending on the distance between the interlocutors at various voice strengths (passengers communicating with the driver): A—HDWVs; B—marine fleet norms for pilot houses; 1—shout; 2, 3, 4—loud, raised, and normal voice

The best way to decrease the power unit's noise level is to coat it with a soundproof encapsulation. However, this insulation hampers the heat removal from the engine; therefore, a corresponding increase in the cooling system's efficiency is needed. The use of electrochemical power units in vehicles is a promising option in reducing the noise level.

To decrease the noise level of the cooling system fans, the fan shaft's rotation frequency can be reduced, the respective impinging fan blade angle and fan jacket design selected, and its drive improved by using a hydraulic or pneumatic coupling which engages the fan when additional engine cooling is needed.

The noise level of the gas exhaust system can be reduced by installing mufflers with increased efficiency.

The noise level of the transmission depends on its type (mechanical, electric, or hydrostatic), the balancing degree of the cardan shafts, the wheels and tires, and the manufacturing class of various gearboxes.

Due to the high cost of the design-and-engineering projects aimed at solving ecological problems, the initial tryout and introduction of these designs are undertaken on the HDWVs which are most expensive and have fact payback (quarry dumpers and line road trains, operated in the two- and three-shift mode with an annual mileage of 100–120,000 km).

The main directions of improving the ecological properties of HDWVs are presented in Table 7.13.

7.6 Ergonomics

7.6.1 General provisions

The design features which ensure the ease and convenience of steering are characterized by their ergonomic parameters, which can be subdivided into four groups:

- The hygienic indicators, which reflects the degree to which the driver's workplace conforms to sanitary norms;
- The anthropological indicators, which describe the convenience of the driver's workplace;
- The physiological and psychological indicators, which define the degree to which the effects necessary to handle the controls and other impacts on the driver conform to the driver's psycho-physiological capabilities; and

Table 7.13 Basic trends in improving the ecological properties of HDWVs

Trend	Details
Decreasing fuel consumption	Nontraditional engines (including alternative fuels)
	Microprocessor-based control system of the power plant (power unit—transmission)
	Reducing the air resistance factor
	Independent automated engine heating systems (thermal conditioning systems)
	Optimizing the engine's thermal regulation parameters
	Optimizing the engine and transmission parameters
	Reducing losses in the engine and transmission units
Improving fuel combustion and better exhaust gas purification	Low-toxic operation with optimized (adaptive fuel equipment (r = 100 MPa) and recycled exhaust gas
	Neutralizing exhaust gas
	Particulate filter
	Pre-start engine warming (decreasing the emission of CO, CH, and solids at cool engine start)
	Intermediate air-to-air cooling
	Turbocompressor with regulated turbine geometry
	Parts with ceramic elements for thermal insulation of the combustion chamber and supercharging system
	Nontraditional engines (gas turbine, on alternative fuels)
Reducing noise and vibrations	Improving the silencing systems of the exhaust gas emission noise
	Encapsulating the chassis' engine and units
	Insulating the cabin against noise
	Increasing the gears' precision and cleanness
	Reducing air resistance
	Optimizing tread pattern
	Active suspension regulation

Table 7.13 Basic trends in improving the ecological properties of HDWVs	
Reducing the negative effects of wear-out products and POL	Asbestos-free friction pads
	Tubeless tires with improved tire life (1.5–2.0-fold decrease in ash dust emission)
	Anti-blocking and anti-skidding systems (reducing the wear of tires and brake laps)
	Brake-decelerator
	Pneumatic suspension (decreasing the road degradation intensity)
	"Eternal lubricants," excluding POL leakages
	Reducing tire wear due to an optimized steering system (StS) algorithm
Increasing environmental friendliness through technical improvements	Increasing the performance indicators (average travel speeds, load capacity, loading and unloading adaptability, etc.)
	Decreased frequency of maintenance
	Use of microprocessor diagnostics
Improving ecological aspects in HDWVs manufacturing	Improving manufacturing technologies
	Reducing unladen weight
	Using ecologically pure materials
Decreasing the negative effect on the support surface	Developing a complex methodology to evaluate HDWVs' impact on soil degradation and recovery.
	Creating automated systems for choosing and adjusting tire pressure depending on the vehicle's motion conditions
	Use of regulated and active wheel suspensions
	Optimizing the parameters of the wheeled mover

- A group of psychological indicators, which defines the design's suitability to use the driver's available skills and to form new ones with regards to perception and information processing.

The improvement of a vehicle's ergonomic indicators is paid great attention abroad, since it is connected to the driver's health, an increase in their labor productivity, and the vehicle's competitiveness. However, Russian-made HDWVs are notably inferior in these parameters to their foreign-made counterparts, and thus, have no sales markets abroad.

7.6.2 Hygienic indicators

One of such indicators is the internal noise level. Their requirements are considered above since this indicator also characterizes the HDWVs' ecological properties.

The vibration load level on the driver's workplace also refers to hygienic parameters. Vibrations excite the human nervous system, cause headaches, reduce visual acuity, raise fatigue, and deteriorate psychophysical and reflex activity.

In accordance with GOST 12.1.012, the amount of vibration affecting the person can be determined by one of following methods:

- A spectral analysis of the rated parameters, which are the average root-mean-square (RMS) values of the vibration velocity or vibration accelerations in the octave or one-third octave frequency bands, ranging from 0.7 to 90 Hz in the characteristic points of the vehicle, located in three mutually perpendicular planes;

- The integral vibration estimate by the rated parameters' frequency, which are the corrected vibration acceleration values; and
- The vibration doze, i.e., the maximum time a human is allowed in the workplace.

OST 37.001.291 establishes the technical standards of the maximum allowed vibrations on the truck drivers' seats and also in the characteristic points of the vehicle's sprung weight. It uses integrated vibration estimates, the corrected values of vertical and horizontal (longitudinal and lateral) vibration accelerations on the driver's seat, and the RMS values of vertical vibration accelerations with frequency ranges of 0.7–22.4 Hz in characteristic points of the vehicle's sprung weight. The driver's continuous working time is defined by the level of vibration load in his or her workplace.

The locations for measuring the vibration accelerations on the vehicle, the modes, and the test and measurement conditions are established by OST 37.001.275.

The international standard on traveling comfort (smoothness of travel) is ISO 2631-78 *Vibration transferred to the human body. Manual for assessing the influence on the human.*

The cabin's microclimate is defined by the combination of temperature, humidity, the speed of air blowing, and the air's chemical composition.

The cabin's air temperature is directly related to the external air temperature, the engine temperature, the cabin's heat insulation, and the heating and ventilation systems. The optimum temperature is 18–24°C. Above this range, the driver's attention weakens, his or her perception of the changing situation becomes poorer, and response time and fatigue increases. At 17°C and lower, the driver's body begins to cool. A temperature of 10°C is the minimum allowed. In low-temperature conditions, the performance efficiency of human muscles decreases and the constraint and discrepancy of human movements are observed. Excessively warm clothes hinder the driver's movements and warm footwear complicates foot-pedal control.

HDWV cabins which work in hot climates should be painted with light paint and equipped with air conditioners. Heat-absorbing glasses should also be used for glazing the cabin.

HDWV cabins intended for operation in low temperatures should have heat-insulated walls; double glazing with electric heating; and frost-proof sealing in piping, levers, pedals, and doors. Cabins should also be heated by a high-powered independent heater.

A human's thermoregulation depends on the air humidity and mobility. For the majority of people, a humidity of 30–70% is considered normal. A human perceives air flow at speeds above 0.25 m/s. Air flow speeds the cabin and passenger compartments should not exceed 1 m/s.

Proper temperature conditions in cabins are ensured by the heating, ventilation, and air-conditioning systems; their requirements are stipulated in Russia by OST 37.001.413.

The maximum allowable concentration of harmful emissions in the cabin is established by GOST 12.1.005, GOST B 21116, GOST 23185, and GOST B 23186.

The ergonomic indicators in cabins of foreign-made HDWVs, especially tow cars, are comparable to passenger cars. They are equipped with an independent automatic heater, an air conditioner, a refrigerator, a tachometer, a tachograph, a seat with electric heating and multiple adjustments, a heated rear-view mirror, and other accessories. The latest innovations include an air conditioning system with automatic control of the preset temperature and remote control of the right-hand mirror's position, and the right-hand window's opening. The cabins are quite spacious inside and allow for 95% of drivers to stand upright. The steering wheel has a large diameter (500–550 mm) and is a reminder to the driver that he or she is driving a heavy vehicle. Composite bituminous-and-felt flooring is used to improve the noise insulation. All cabins are dust-and-moisture-proof.

7.6.3 Anthropological indicators

One anthropological indicator is the driver's working posture. A physiologically correct working posture ensures the convenience of driving the vehicle, increases the accuracy and speed of the driver's locomotion actions, and allows the driver to work long continuous hours without significant fatigue, which is especially important for HDWV drivers. This is reached when the whole human body is in a position close to the condition of functional rest and his or her weight is uniformly distributed on the seating surface.

The workplace parameters in a cargo vehicle (the dimensions and adjustment of the seat and the arrangement of controls) are stipulated by OST 37.001.413. The characteristics of a working posture when driving are obtained by a special mannequin (dummy) (Fig. 7.8).

The upholstery of the driver's seat should be dense and rough, made of dense woolen and semi-woolen fabrics, leather, leather substitutes, and plastic materials. The material's air permeability should be at least 0.35 $cm^3/(cm^2 \times s)$. An example of a cabin layout is presented in Fig. 7.9.

7.6.4 Physiological indicators

These establish the efforts to be applied to the controls and are regulated by GOST 21398 (and were considered above).

7.6.5 Psychological indicators

The psychological indicators include the devices which ensure the internal visual presentation of information, including how the information is displayed, and the devices which improve visibility.

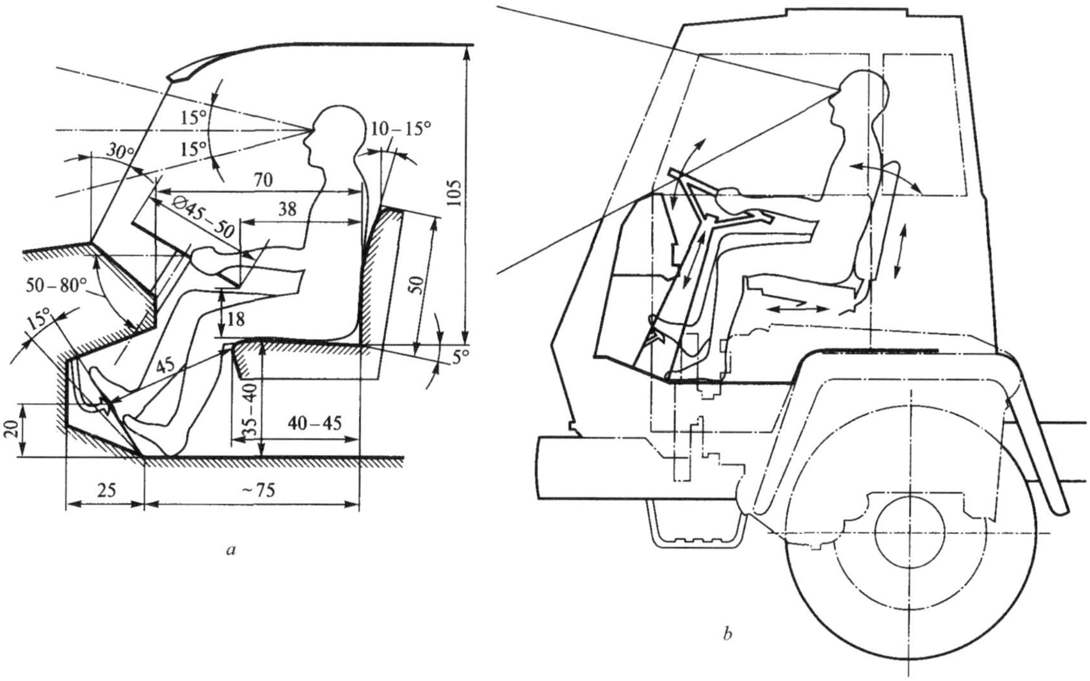

Fig. 7.8. Diagram of the driver's workplace: A—the dimensions; b—required displacements (adjustments) of the driver's seat and steering column

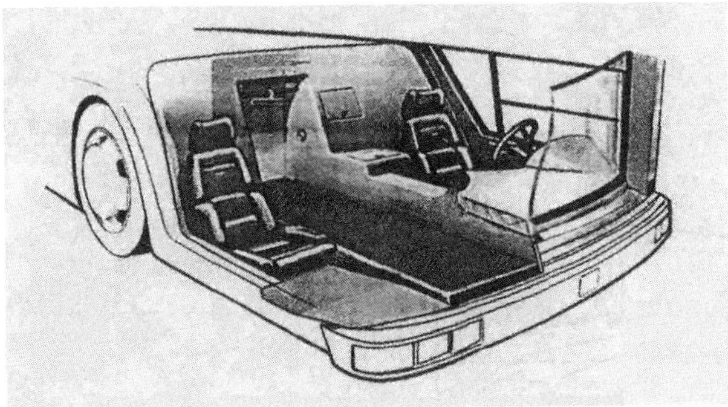

Fig. 7.9. Layout example of a HDWV cabin

In the process of driving a HDWV, the driver needs information about the status of technical systems, which should arrive in the coded form from the devices and indicators located on a special panel.

Since the design of a HDWV includes several unconventional systems, units, and assemblies, the flow of information addressed to the driver is much bigger. That is why the dashboard should have various indicators showing the condition of the equipment, the current active processes, the travel speed, etc. These devices should be designed according to the human ability to perceive information: they should ensure fast reading and correct (unequivocal) understanding by the driver of the dashboard's visual information. It should be remembered that about three quarters of all road accidents are caused by driver errors. When operating the equipment, the driver thus eliminates some indetermination in its functioning, i.e., he or she brings his or her information into the control system. The more sophisticated the task, the greater the indetermination that he or she has to eliminate (introduce more information into the system). To do so, the driver requires a sufficient amount of input data; however, the volume of this data should be as much as needed for decision-making in a specific situation.

OST 37.001.202 regulates the dashboard's placement in the vehicle cabin, the list of control and measuring instruments and signaling devices, as well as the principles of their arrangement on the dashboard.

The main requirement for the dashboard's configuration is to reduce the time needed by the driver to perceive the instruments' readings, which inform the driver about the status and provide the sufficient amount of data.

The visibility from the driver's seat defines his or her ability to obtain a complete perception and correct estimation of the road situation. The visibility parameters for cargo vehicles are regulated by OST 37.001.413.

To improve visibility, the driver's seat is brought closer to the front, the windshield area is increased, and a panoramic windshield is installed (see Fig. 7.8). However, because of the huge dimensions of HDWVs and their specific layouts together with the mounted equipment, these design measures are obviously not enough for estimating the road situation; therefore, it is necessary to equip HDWVs with television and radar systems that bring the information to the dashboard.

Irrespective of the meteorological conditions, the windshield wipers, the window spray, and the heating system are very important for ensuring proper visibility. The main requirement for the wipers is the ability to clear as much of the windshield as possible and ensure a high clearing quality at each travel of the wipers. The spray and heating system should prevent misting and frosting of all cleared windshield surfaces at outdoor temperatures reaching –45°C.

The driver often has to estimate the road situation behind the vehicle, especially when maneuvering at loading-unloading points or at construction sites. This is achieved by the rearview mirrors (television or radar systems), while the visibility depends on the shape of their reflective surfaces (convex or flat), the mirror's dimensions, and its location in relation to the driver's eyes. The visibility through the rearview mirrors should meet the requirements of OST 37.001.451 and UNECE's Rule No. 46.

The psychological indicators also include the vehicle's audio warning signals. At present, there are no international documents which standardize the audio warning devices and the values of their characteristics (acoustic intensity, intermittence, and tonal quality). Sound signals serve to provide the driver with elementary information as a warning. In particularly dangerous cases, an emergency light signal should be complemented with an intermittent audio signal (for example, the insufficient level of fluid in the brake system or air pressure in the tires, or a decrease in distance to obstacles below the critical value, etc.) [20].

The use of audio information allows for reducing the driver's visual load, improving the working conditions, and, increasing traffic safety. This is particularly important with the increase in devices displaying visual information and an increase in traffic density.

By summing up the ergonomic requirements, the following conclusion can be made. The improvement of ergonomic indicators in automobiles is a priority task abroad, in spite of the great resources needed to fulfill it. For this purpose, they permanently update or develop the respective normative documents and improve the legislative base. In Russia, this problem is traditionally almost ignored. Among the chief designers of HDWVs and scientists, only M.S. Vysotsky was the first to pay due attention to it [4].

Chapter 8
Main technical criteria and efficiency evaluation of HDWV solutions

8.1 Analysis of technical criteria of vehicle excellence

8.1.1 General provisions

The whole life cycle of a vehicle is accompanied by an evaluation of its quality and efficiency, that is, of its excellence. Efficiency (from the word "effect") is a generally adopted term in various areas and activities. However, each area or activity uses its own particular quantitative criteria. Currently, there are no generally adopted criteria to evaluate the efficiency of cargo vehicles; therefore, the works of many scientists are aimed at developing methods to evaluate the efficiency of an automobile in general and to make a comparative evaluation of a particular vehicle with various technical solutions used in designing its main units and systems.

One of the main stages in developing these methods is selecting and justifying the objective qualitative criteria, which is later used to build particular proposals for evaluation. In spite of the huge number of works, there is still no generally acknowledged methodology, built on the criterion (a complex of criteria), that is able to cover all automobile usages. That is why the authors of this book present below—prior to offering their own efficiency evaluation method for HDWV mobility— a brief analysis of the known problem-solving approaches from scientific publications.

In general, the aim of defining a system's efficiency is to optimize it for obtaining a new system, modernizing the existing one, or selecting a variant that is the closest to the optimal by a definite efficiency indicator.

A characteristic of complex systems is the necessity to evaluate them under several particular quality indicators: precision, reliability, cost, etc. According to the principle of uniqueness, the efficiency indicator of some system in general, as the optimality criterion, should be presented as one single indicator that includes all considered particular quality indicators. In general, the efficiency of the introduction or modernization of some system is evaluated by the following indicator:

$$E = \frac{B}{C},$$

where B is the effect, i.e., the value of the new or modernized system; and C is the expenses needed to design, introduce, and operate the new or modernized system.

Theoretically, the efficiency indicator of the new system E takes into account all the expenses and benefits from its commissioning: the labor input, satisfaction of the qualitatively new demands, quality improvement, etc. However, not all these indicators can be directly measured, that is why the idea of economic efficiency was offered for practical use [21].

If a complete effect from the introduction of a new system cannot be presented in monetary terms, the technical-economic efficiency is used, which takes into account—apart from the saving from the labor input—the measurable technical specifications of the new system:

$$E = F(y_1, y_2, \ldots, y_n) \quad (8.1)$$

where $y_1, y_2, \ldots, y_n$ are the measurable technical indicators, treated as particular quality indicators; and n is the number of accountable particular quality indicators, including the expenses $C = y_n$.

The particular quality indicators depend on the structural and design parameters, which can be measured in the process of designing and introducing the systems:

$$y = j_i(x_1, x_2, \ldots, x_m)$$

where $i = 1, \ldots, n$; $x_1, x_2, \ldots, x_m$ are the structural and designing parameters of the system and its elements.

If functions j_i and F are known, that is, expressed analytically, it is not difficult to define the efficiency indicator E, since the parameters $x_1, x_2, \ldots, x_m$ are known for every system variant. These parameters are the control inputs for optimizing the system in the design process, since designing and modernizing a system means selecting the structure and elements that ensure the maximum efficiency.

If the function F is unknown, the system's efficiency is evaluated by one—the most important—particular indicator (for example, by y_1), and restrictions are imposed on the other indicators to keep them within finite limits:

$$E = y_1, y_{li}, y_i, \ldots, y_{ui}, \quad i = 2, 3, \ldots, n$$

where y_{li} and y_{ui} are the lower and upper limits of the ith particular quality indicator. Depending on a particular quality indicator, one of the limits (upper or lower) can be left unlimited.

The evaluation of a system's efficiency by one particular indicator, with imposed limits on other particular indicators, has a drawback since the optimized problem solution or the choice of the practically optimal system variant to implement will be ambiguous (many-valued).

Several system variants with the same or approximately the same main particular quality indicator y_i, with essentially different other particular indicators which meet the restrictions, can be obtained. In this case, the system version closest to optimal cannot be confidently determined. Therefore, Chumakov and Serebryany [21] have offered a method for defining the efficiency and finding a practically optimal variant of a complex system, which assumes replacing the efficiency indicator function (8.1)—provided its analytical expression is unknown—with a linear function that includes all of the main particular quality indicators:

$$E = b_1y_1 + b_2y_2 + \ldots + b_ny_n$$

where $b_1, b_2, \ldots, b_n$ are the weight coefficients in the presence of restrictions y_{li}, y_i, and y_{ui}.

The linear form of a complex systems' efficiency indicator (criterion) is the simplest function, which takes into account all the main particular quality indicators. It is used when it is necessary to select—quickly and without special difficulties—a practically optimal variant out of all contending variants.

When using both the linear function (8.2) and any other function, the list of accounted (main) particular quality indicators y_i and their weight coefficients b_i must be defined.

8.1.2 Technical and economic criteria of a vehicle's efficiency

The usual practice is to start the justification of the criterion and particular parameters choice with a brief review of the existing formulas, which is used to define the respective values. First of all, which aims should be reached by means of this or that indicator need to be defined.

It is clear that for a designer, this criterion should be a tool for evaluating the excellence of the vehicle's design as a complex system (mainly, of the chassis, i.e., the power unit, transmission, and the mover). However, the customer, who may not understand the criterion at all, will choose a car based on his or her personal preferences. However, we should take as an axiom that the operation efficiency parameters are always derivatives of the vehicle excellence parameters, chosen and defined by the designer. Thus, the technical and economic analysis binds these two aspects together.

That is why M.I. Ipatov [22] has offered that—in order to reveal the most economically rational numerical values of design parameters—the technical and economic analysis methods should assume not only direct correlation between the vehicle's design and economic parameters, but also feedback, which makes it possible to optimize the created design. In this context, the following main principles of applied cybernetics plays a large role in the technical and economic analysis: studying the vehicle under design as a system, in which every element is a part of a greater ensemble; ensuring an optimal solution of multi-variant problems by choosing numerical characteristics of these elements under a certain criterion; and using specific cybernetic methods (simulation, feedback, etc.). The cybernetic approach allows for binding together four aspects of the analyzed system: technical, economic, production, and mathematical.

With the aim to conduct a systemic technical and economic analysis of automobiles, M.I. Ipatov [22] has offered a classification of parameters and indicators, which is notable for the following:

- All parameters and indicators are split into five groups: standardized, design, production, operational, and economic; when putting a "mixed" parameter or indicator (for example, the load capacity) into a certain group, the principle of priority is applied, that is, whether the parameters and indicators are primary or secondary in the design process;
- Within a classification group, parameters and indicators are differentiated by their nature (for example, design has the following subgroups: dimensional, weight, and power);
- By their usage character, all parameters and indicators are classified into groups preset by the higher organization or the customer; selected or obtained during the layout creation; or that are calculated;
- Based on the information available at the respective design stage, classification tables are compiled; moreover, a number of parameters and indicators, selected or calculated at one design stage, become the assigned or preset ones for the next design stage; and
- For the preliminary design stages (front-end engineering) and technical design, classification tables are compiled both for the automobile as a whole and for its main units and systems.

In his justification of this classification, M.I. Ipatov [22] gives special attention to ensuring— at early design stages—the sufficient calculation precision of the average travel speed of the designed and basic (compared as a prototype) automobiles or of the designed structural variants of automobiles when used in particular road conditions. These calculations are, in the opinion of the author, the starting and most important ones for defining the automobile's performance.

A.M. Ivanov [23] has introduced for the first time the notion of a vehicle's design (structural) efficiency, which is defined as the vehicle's ability, using a preset combination of technical parameters, to ensure the maximal haulage efficiency in particular operation conditions while observing the standard safety level.

This work also suggests using haulage profitability, P_{haul}, as the integral criterion of a vehicle's design efficiency. This technical and economic criterion allows for bringing all of the results and expenses into one single unit of measure under particular operating conditions.

The haulage profitability takes into account the maximum number of indicators and technical parameters associated with the automobile's design and reduces them to a single unit of measure. In particular, they include the following:

- m_{crg}: the mass of the cargo transported by the vehicle, in tons;
- v_{av}: the average speed en route, in km/h;
- Q_s: the average operational fuel consumption, in liters per 100 km;
- $T_{m\&r}$: the specific value of man-hours for technical maintenance and repairs, in man-hours per 1000 km;
- L_{mv}: the vehicle's service life (resource) until writing off or overhaul (major repairs), in km;
- n_{tire} and L_{tire}: the number of tires (in units) and their service life, in km;
- $t_{l\&u}$: the duration of loading and unloading in one travel, in hours;
- k_{rel}: the release coefficient;
- $k_{util.\,f}$: the travel utilization factor; and
- E_l: the technical and economic efficiency, corresponding to the lower level of particular quality indicators.

The dependence of the integral criterion of the design efficiency on technical indicators and parameters is as follows:

$$P_{haul} = \left(k_1 m_{crg} v_m K_{tuf} k_{rel}\right) / \left\{\left(k_2 k_{rel} K_{tuf} v_m k_3 k_{rel} K_{tuf} t_{np} v_m^2\right) \times \times \left(k_4 Q_s + k_5 T_{m\&r} + \frac{k_6 n_{tire}}{L_{tire}} + k_7\right)\left[\frac{k_8}{(1+E_1)^{L_{mv}/v_{av} k_9}} - 1\right] \times \times \left(t_{l\&u} v_{av} k_{exec} + k_{10}\right)\right\} - 1 \tag{8.3}$$

where k_1–k_{10} are the coefficients which take into account the transport and economic operating conditions.

The structural analysis of the expression (8.3) shows that the values v_{av} and Q_s are of general interest for both the designer and consumer. These vehicle parameters are the starting points for the consumer in choosing the automobile and for the designer in designing it. The other integral criterion components are those indicators and parameters that are preset by the Terms of Reference (TOR) or ensured by design methods and organizational measures.

The average travel speed, v_{av}, and the travel fuel consumption, Q_s, are, in their turn, derivatives of the energy excellence of the automobile as a whole and of its units and systems.

Selecting a criterion (or indicator) that allows for evaluating the energy excellence of a vehicle and its units and systems allows for being more reasonable in making optimal design decisions and promoting further improvement in the quality and efficiency of future automobiles. It is important that the evaluation criteria of the excellence of individual units and systems influence the design excellence criterion of the automobile as a whole (for example, v_{av} and Q_s).

8.1.3 Requirements for the technical criteria of vehicle efficiency

The automobile and tractor theories provide the criteria that allow for evaluating the design excellence of the vehicle as a whole and of its units and systems.

These criteria will be briefly considered from the viewpoint of the design excellence of the power unit-transmission-wheeled mover system and its influence on the vehicle's mobility evaluated.

The excellence of this system is understood as the degree to which the design ensures the distribution of torque (power) among the driving wheels according to the vehicle's layout, travel modes, and conditions.

These criteria should reflect the power change needed for the designed or evaluated automobile to move compared to some existing prototype or some other functionally identical (or ideal) automobile, which has the optimal transmission characteristics.

These criteria should be dimensionless, that is, independent of the load capacity, number of axles, type of tires, and other design characteristics.

Thus, the following requirements for such criteria can be formulated:

- They should allow for evaluating the transmission excellence from the viewpoint of ensuring the minimal power consumption necessary for the vehicle to move in different road conditions;
- The existing (evaluated) vehicle's transmission design excellence should be compared with the transmission of a functionally identical automobile, which ensures an optimal distribution of torque (moment) among the driving wheels, and with an automobile whose transmission does not meet the optimal power distribution requirement;
- The criteria should be quantitatively associated with the power (torque) change necessary for the vehicle to move; and
- The criteria should be dimensionless.

With the above considerations in mind, the technical criteria and indicators that are currently recommended in automobile and tractor theory will be briefly considered.

8.1.4 Evaluation of traction-and-speed properties of the vehicle

For the first time, the definition of a vehicle's traction properties was given in the works of Academician E.A. Chudakov, where these properties were understood as the vehicle's ability to move on different roads. It was assumed that traction properties could be fully evaluated by using the dynamic characteristic and the maximum and average travel speeds. Later, these criteria found their application and were further developed in several works [see References 22–51].

It was established that the traction-and-speed properties are essentially defining the utilization efficiency of automobiles. It follows from V.F. Platonov's [29] work that the efficiency should be understood as the ability of an automobile to perform haulage along the preset routes at the maximum possible speed. The speed properties, as indicated by V.F. Platonov [29], are usually evaluated through the average travel speed. Out of the many factors that influence the speed properties, the author highlights the specific power, stability and steering ability (controllability), braking properties, and smoothness of motion.

Out of these factors, Platonov [29] gives priority to specific power N_{spec}. It is proposed to evaluate the traction capacity by using power balance and dynamic factor equations.

To assess the vehicle's efficiency, Platonov offers two relative indicators of the efficiency and performance, which take into account the economic expenses (costs) or the vehicle's working parameters when performing a targeted job. The efficiency indicator is defined as the following ratio:

$$K_e = \frac{\sum \varphi_i \sum \Pi_{oi}}{\sum \Pi_i \sum \varphi_{oi}} \tag{8.4}$$

where $\Sigma\varphi_i$ is the level of complex functions, performed by the vehicle; $\Sigma\Pi_i$ is the reduced total costs; and $\Sigma\Pi_{oi}$ and $\Sigma\varphi_{oi}$ are characteristics of the basic model (prototype).

The performance (efficiency) indicator is equal to:

$$E = \frac{k_{crg}\, m_{crg}\, L\, k_{util.f}\, v_{av}\, TD\, \tau}{(L + k_{util.f}\, v_{av}\, t_{l\&u})}, \tag{8.5}$$

where m_{crg} and k_{crg} are the load capacity and its utilization factor; L is the vehicle mileage; $k_{util.f}$ is the utilization factor of the mileage; v_{av} is the average technical speed; T and D are the time of work per day and the number of working days in the period; τ is the vehicle's utilization factor during the considered period; and $t_{l\&u}$ is the loading and unloading time.

These indicators disclose the efficiency of using the vehicle en route; however, analysis of the parameters from equations (8.3)–(8.5) shows that these indicators are preferable for assessing the work efficiency of vehicles in transportation companies, not in the cases when HDWVs are used together with attached (installed) equipment.

A.A. Tokarev [16] studied the traction properties based on traction and power balances equations. Thus, for example, he offers the following power balance equation:

$$N_w = N_e\eta_N = (P_f \cos\alpha + G_{car} \sin\alpha + W_{air}v^2 + m_{car}\delta_M) v \cdot 10^{-3}, \quad (8.6)$$

where N_w is the power applied to the driving wheels; N_e is the efficient engine power; η_N is the power loss factor; P_f is the aggregate rolling resistance; G_{car} is the vehicle's weight; W_{air} is the air resistance factor; m_{car} is the vehicle's mass; α is the road ascending angle; v is the travel speed; and δ_M is the factor that takes the rotating masses into account.

In the above equation, the author has failed to take into account the power losses in the mover related to skidding. However, at an intensive acceleration from rest, circumferential forces are rather high on the driving wheels. They result in increased tangential tire deformations and, in certain cases, in the wheel sliding on the support surface, i.e., in increased power losses on mover skidding.

A.A. Tokarev believes that the main evaluation factors of traction properties are the maximum circumferential force, $P_{k\,max}$, on the driving wheels and the maximum dynamic factor, D_{max}; both are defined at a preset stable speed along a straight horizontal road having improved pavement. His indirect factors are the specific power, N_{spec}; the accelerating capacity ratio at the given travel speed, η; and the maximum slope conquered by the vehicle, α_{max}. The above author indicates the following as the main speed evaluation indicators: the maximum speed, V_{max}; the conventional maximum speed, $V_{c\,max}$, that the vehicle can gain on the limited travel section; the maximum (average) acceleration at speedup, j_{max}; the average travel speed on a typical route, V_{av}; and the average travel speed in the drive cycle, V_{avc}.

A.S. Litvinov and Ya.E. Farobin [31] defines traction-and-speed properties as a combination of properties that define the variety of travel speeds possible under the given engine parameters or given adhesion of the driving wheels with the road and the limit of vehicle speedup intensities at work in the traction mode under various road conditions. The traction mode is the mode when the power delivered from the engine to the driving wheels is sufficient to overcome the motion resistance.

These authors [31] have put forward the following list of the most usable and sufficient unit indicators to make a comparative evaluation of the traction-and-speed properties:

- The maximum speed;
- The rated maximum speed defined as the average speed at the final 400 m of a speedup on a 2000 m section with a full fuel supply and start of switching gears at the nominal crankshaft rotation frequency;
- The speedup time on the 400 and 1000 m sections;
- The acceleration time to the present speed;
- The speed characteristic of the speedup-runout;
- The speed characteristic of the speedup at the top gear;
- The speed characteristic on a road with a variable lateral profile;
- The minimum stable speed;
- The maximum overcome slope;
- The steady speed at long ascending grades;
- The acceleration at speedup;
- The traction force on the hook; and
- The length of dynamic climbing.

Some of these indicators have values prescribed by branch or state standards (known as OSTs and GOSTs), or their values can be defined experimentally or calculated by solving the force or power balance equations. For this purpose, for example, the power balance equation is written as follows:

$$m_a \delta_M \frac{\partial v}{\partial t} = P_w - P_d - P_{air},$$

where P_d is the road resistance force, which is equal to the sum of the climbing and rolling resistance forces; and P_{air} is the air resistance force. Then, the equation is transformed as follows:

$$m_a \delta_M \frac{\partial v}{\partial t} = a_i v^2 + b_i v + c_i,$$

where i is the gear number; a_i, b_i, c_i are the coefficients that change along with the transmission ratio and depend on the respective coefficients of the cubic trinomial (a, b, c), which approximates the dependence of the engine power N_e from the crankshaft rotation frequency n. The coefficients a, b, c are constant for the given engine.

The method offered by A.S. Litvinov and Ya.E. Farobin [31] allows for analyzing the traction-and-speed properties of the designed vehicle and comparing them with those of the prototype to study the efficiency of different design solutions aimed at improving the traction-and-speed properties.

However, the need to use novelties and innovations in vehicle designs (for example, a GTE, HST, ET, etc.) demands more efficient indicators for both evaluating the traction-and-speed properties of motor vehicles and defining the functional dependences of these properties from the design parameters.

In this context, many authors [see references 29, 32, 33, and others] use both the performance and the specific performance (referred to the average fuel consumption Q) as the main criterion of the vehicle's efficiency:

$$\Pi_{spp} = \frac{m_{crg}\, v}{Q}.$$

The specific performance (spp) parameter is very convenient, since it provides a complex evaluation not only of the vehicle's traction-and-speed properties and its fuel efficiency, but also the efficiency of the design as a whole. By its structure, this parameter is close to the efficient factor broadly discussed in technical literature.

The work of G. Wong [34] uses the specific transport efficiency, defined as follows:

$$\eta_{tr} = \frac{m_{crg} V}{N_{ex}},$$

where m_{crg} is the working load, V is the vehicle's speed, and N_{inp} is the input power.

The tractor theory [26] offers a range of indicators that allow for objectively evaluating the design excellence of a tractor and its individual units and systems.

8.1.5 Efficiency factor as a criterion of design excellence

One of the main indicators of a tractor's efficiency is the traction efficiency factor (η_T). It is measured (or calculated) for a steady motion along a horizontal surface without engaging the power take off (PTO) shaft according to the following formula:

$$\eta_T = \frac{N_{pn}}{N_e} \qquad (8.7)$$

where N_{pn} is the power necessary to overcome the traction resistance of the modular (unitized) equipment and N_e is the engine power.

A drawback of this indicator is the fact that when the vehicle moves without a trailer, the value of η_T will be close to zero.

Apart from coefficient η_T, the tractor theory uses the efficiency factors of the transmission (η_{te}), the internal losses in the mover (η_{ilw}), and the power losses on forming the rut ($\eta_{\text{col\&u}}$). The latter is defined by the formula:

$$\eta_{\text{col\&u}} = 1 - \frac{N_{\text{col\&u}}}{N_e}, \tag{8.8}$$

where $N_{\text{col\&u}}$ is the power needed to form the rut and N_e is the nominal engine power.

The drawback of this formula is that the efficiency factor varies depending on the tractor's loading mode. At increasing N_e (for example, when heavy equipment is attached) or at a slight increase of $N_{\text{col\&u}}$, the value of $\eta_{\text{col\&u}}$ will go down.

Of special interest are the transport and full efficiency factors offered by V.V. Vantsevich [25] and the coefficients, close by structure, offered by V.N. Shalyagin [27].

For defining the transport efficiency factor, V.V. Vantsevich and V.N. Shalyagain [25, 27] use the following formulas:

$$\eta_{te} = \frac{\sum_{i=1}^{m} N_{\varphi cari}}{\sum_{i=1}^{n} N_{wi}}, \tag{8.9}$$

where $\sum_{i=1}^{m} N_{\varphi\,car}$ is a part of the resistance power caused by the impact of the transported cargo; $\sum_{i=1}^{m} N_{wi}$ is the power delivered to the mover; m is the number of axles; and n is the number of driving axles.

If the adhesion conditions of the right and left wheels on each axle are different, the expression (8.8) will have the following form:

$$\eta_{te} = \sum_{i=1}^{m} N'_{\varphi cari} + \frac{\sum_{i=1}^{m} N''_{\varphi cari}}{\sum_{i=1}^{n} N'_{wi}} + \sum_{i=1}^{n} N''_{wi}, \tag{8.10}$$

where the indices (′) and (″) define the right and left wheels respectively.

If a vehicle is transporting not only the working load in the body, but also towing a trailer, the transport efficiency factor is expressed by the formula:

$$\eta_{te} = \sum_{i=1}^{m} N_{\varphi cari} + \frac{\sum_{i=1}^{c} \sum_{i=1}^{m_t} N_{\varphi \text{carin}}}{\sum_{i=1}^{n} N_{wi}}, \tag{8.11}$$

where $\sum_{i=1}^{m_{np}} N_{\varphi\text{carin}}$ is the additional resistance to rolling caused by the working load; m_t is the number of axles on one trailer; and c is the number of trailers.

If the vehicle is towing an empty trailer, it is proposed to use the traction efficiency factor from (8.7).

It follows from formulas (8.9)–(8.11) that the proposed indicators are the vehicle's real efficiency factors since by definition any efficiency factor is the useful-to-consumed power ratio. In general, its value is an indication of the design excellence. However, this factor is not directly correlated with the quantitative indicators of power and average travel speed variations, as well as with the fuel consumption. For example, with a 5% increase in the efficiency factor from (8.10), it is difficult to define the decrease in the power necessary for motion after design improvements are made or to what extent the given design improves v_{av} and decreases Q_s.

The efficiency factor value defined by formulas (8.10) and (8.11) also does not allow for evaluating the possibility and directions of further improvement to the transmission.

The use of the vehicle's reference (standard) efficiency factor as the main indicator of its design excellence was offered by M.I. Pogosbekov [28]:

$$h_{re} = \frac{f_{St}\, m_{car} \eta_T\, \eta_{Sl}}{f(G_{car} + G_{Ve}) + W_{air} v^2},$$

where f_{St} is the reference coefficient of rolling resistance; m_{car} is the mass of the cargo; η_T is the transmission's efficiency factor; η_{Sl} is the skidding efficiency factor; f is the actual coefficient of rolling resistance; $G_{car} = m_{carg}$ is the cargo weight; G_{Ve} is the vehicle's curb weight; and W_{air} is the air resistance force.

The formula offered by M.I. Pogosbekov [28] differs from the expressions (8.9)–(8.11) in that it, first, does not show the structure of the skidding efficiency factor; and second, it introduces the idea of the rolling resistance (f_{St}) reference coefficient and the actual rolling resistance (f) reference coefficient. As to the skidding efficiency factor, M.I. Pogosbekov believes that its value can be neglected when the vehicle moves along a hard road. However, it is well known that the application of a torque (or a longitudinal force) to the wheel always causes a change in the rolling radius, and, consequently of the skidding value. The latter is quite essential when driving on a dry road (especially when towing a trailer) and even more so if the road is wet or snowy (or if moving on the ground).

By introducing the f and f_{St} values, M.I. Pogosbekov asserts that f can be measured rather precisely by using strain-gauging. However, G.V. Zimelev called f a conditional value that cannot be directly measured. This viewpoint is reflected in GOST 17967. Only the rolling resistance coefficient in the driven mode can be defined with sufficient precision; its value can be found in all tables.

The exceptional role of the efficiency factor in studying the vehicle's properties is noted in the work by A.A. Tokarev [30].

According to this author, the vehicle's efficiency factor is as follows:

$$\eta_a = \frac{k_s\, m_{car} v^2}{\gamma Q_s H_u}, \tag{8.12}$$

where k_s is the proportion factor on the slowdown path; γ is the fuel density; Q_s is the fuel consumption; and H_u is the specific fuel thermal equivalent.

It is proposed to use equation (8.12) for analyzing not only functionally close vehicles, but also all vehicle types. Thus, the efficiency factor of gasoline non-four-and-four trucks, with a transported cargo mass of 0.8–6.0 tons, lies within 0.062–0.206. For full-drive trucks, the efficiency factor is within 0.040–0.117. The efficiency factor of passenger cars is mainly within 0.090–0.110.

Based on the conducted studies, A.A. Tokarev [30] makes the following conclusions: the efficiency factor is a complex indicator of the automobile's quality, making it possible to take into account all operational losses; and the efficiency factor is an important tool for choosing rational design parameters when designing and optimizing the vehicle.

When designing different units and systems and for a detailed study of the efficiency factor, the latter should be mathematically expressed as functionally dependent on the characteristics and parameters of these units and systems.

8.1.6 Q-factor (perfection factor) of the transmission

The maximum approximations of the above mentioned requirements include a transmission Q-factor (perfection factor) proposed by P.V. Aksyonov, B.N. Belousov, and A.F. Starikov [35].

Its structure is connected with the fact that in real conditions, each wheel is supplied with the power N_w, which differs from the optimal

$N_{w.op}$, providing minimal rolling resistance. The difference between them is shown as:

$$\Delta N = N_w - N_{w.op}.$$

Accounting for the fact that all wheels participate in the formation of the vehicle's performance properties, the transmission perfection factor can be presented in the following form:

$$K_q = 1 - \sqrt{\frac{\sum_{i=1}^{2n}\left(N_w - N_{w.op}\right)^2}{N_{0.op}}}, \quad (8.13)$$

where n is the number of axles; and $N_{0.op}$ is the optimal power, which should be applied by this transmission in given motion conditions.

Thus, an ideal transmission has $K_q = 1$. Real transmissions have a K_q that can change from zero in a fully differential transmission (friction in differentials is neglected) when one of the wheels loses contact with the road surface, to a negative value when the power circulation occurs in a fully blocked transmission.

In this transmission Q-factor (perfection factor) structure, it is difficult to take into account the effect of the engine's adaptability coefficient, to be more precise, to determine this effect in a quantitative form since it is possible only when evaluating the power balance of the engine-transmission-mover system. However, it does not exclude the possibility to use the idea embodied in the structure K_q when developing the coefficient that meets all of the requirements imposed on the criterion of system evaluations.

For this criterion-coefficient, which is conventionally called the optimal power distribution $K_{o.p.d}$ coefficient by Yu. Pirkovsky [54], it is necessary that its value quantitatively coincide with the quantitative change of the transmitted power of the considered transmission as compared to the transmission which ensures an optimal power distribution on the driving wheels. In the final form, according to Yu.V. Pirkovsky and S.B. Shukhman [52], the change in value of this coefficient should directly reflect the changes in average travel speed, v_{av}, and fuel consumption, Q_s.

When developing the analytical expression $K_{o.p.d}$, the separate (right and left) driving wheels of the axles are not considered, since in Yu.V. Pirkovsky and S.B. Shukhman's work [52] the assessment of the separate channel HST is important. If the separate wheels are considered, the analytical expression of the optimal power distribution coefficient will look as follows [54]:

$$K_{o.p.d} = \left[1 - \frac{N_{0\ w.op}}{N_{0\ w.actual}}\right] \cdot 100\%$$

$$= \left[1 - \frac{N_{0\ w.op}}{N'_{o.actual} + N_{en.actual}}\right] \cdot 100\%, \quad (8.14)$$

where: $N_{0.w.actul} = N'_{0w} + N_{en.actual}$ is the actual power delivered to the mover by the transmission (channel) whose perfection is being assessed; $N'_{oactual}$ is the power delivered to the mover, which is determined by the mover's power balance equation (the wheel) without taking into account the adaptability coefficient; and $N_{en.actual}$ is the power, which occurs as a result of the impact of the engine's adaptability coefficient.

The analysis of the expression (8.7) shows that the value $K_{o.p.d}$ is equal to the percentage power increase, which occurs as a result of improving the transmission scheme(or a single channel). According to the authors, this indicator can be used when comparing vehicles with equal weights, design parameters, and engines, but which have different transmission schemes, transfer functions (for example, inter-axle differential ratio), and a different layout. At the same time, when using $K_{o.p.d}$ to assess the mechanical transmission perfection, it is not necessary to take into account the distribution $N_{en.actual}$ on separate wheels or axles. If the transmission consists of different channels (e.g., ET or HST), then it is assumed that

$N_{en.actual}$ is distributed through the channels evenly (or in proportion to their load).

Based on the above materials, the main criteria for assessing the efficiency of adopted technical solutions on HDWVs can include the traction force, maximum and average travel speed, aggregated motion resistance coefficient, and fuel consumption. When solving specific issues (for example, evaluating transmission schemes), additional assessment criteria determined by the dependency equations (8.13) and (8.14) can be used. Namely these indicators determine the mobility of HDWVs as an efficiency criterion and they can create the basis for creating an assessment methodology of the SWC's mobility.

8.2 Methods for assessing vehicle designs

8.2.1 General provisions

Historically in Russia, an automobile was basically assessed in the design process from the viewpoint of the manufacturer, his interests, capabilities, and problems. This meant that the adopted assessment methods for the technical level of the specific property indicators in standard conditions failed to reflect the customer's interest and required him or her to conduct long and costly comparative tests of alternative vehicle samples in real operation conditions. Such an approach, when the sample is made virtually for "single" transportations is an unacceptable process for developing HDWVs. New economic conditions have also required creating a complex methodology for assessing the vehicle at all lifecycle stages in conditions close to the real operation.

Unfortunately, until now, the existing analysis methodology has failed to solve a number of problems: the analysis is conducted, as a rule, after the end of the design elaborations stage and, therefore, can only record the technical solution's design efficiency or inefficiency; analysis is not conducted at each stage and step of design; it applies very approximate and enlarged calculation methods; and, finally, there is no systemic approach in which the analysis would use the information on parameters and indicators corresponding to given design stages and which would select and study design parameters influencing the economic efficiency of the vehicles. In some cases, all of the above leads to significant errors in calculations and the necessity to re-develop or re-work the vehicle's design.

At the same time, all existing methodologies for evaluating criteria under the preset list were developed by separately considering the impact of the vehicle's sub-systems on its design efficiency by using numerous mathematical models that were virtually unconnected. Thus, a complex dynamic system like the driver-vehicle-road (Fig. 8.1) was assessed by a simple set of values of one or another design efficiency criterion, obtained as a result of multiple studies of unconnected mathematical models.

Based on the studies conducted by a group of scientists (A.F. Starikov, A.N. Parkhomenko, and V.G. Kornilov) under the leadership of B.N. Belousov, a complex mathematical model of the SWC's mobility was developed.

8.2.2 Overview of the complex mathematical model

In general, mobility (freedom of motion) (see Figs. 7.1 and 7.2) is a complex technical attribute of the equipment, including such important transportation consumer properties as the traction dynamics, passing ability, capability of stream crossing, reliability, adaptability to maintenance and repair, and others. This chapter uses a narrower definition of mobility based on a limited list of indicators and is understood as the ability of the SWC to ensure a quick and timely transportation of the installed equipment (cargo or load) along the preset (representative) route.

The comprehensive mathematical model presented in this book reflects the general framework of the existing common design solutions theory and represents the basis for the assessment technique

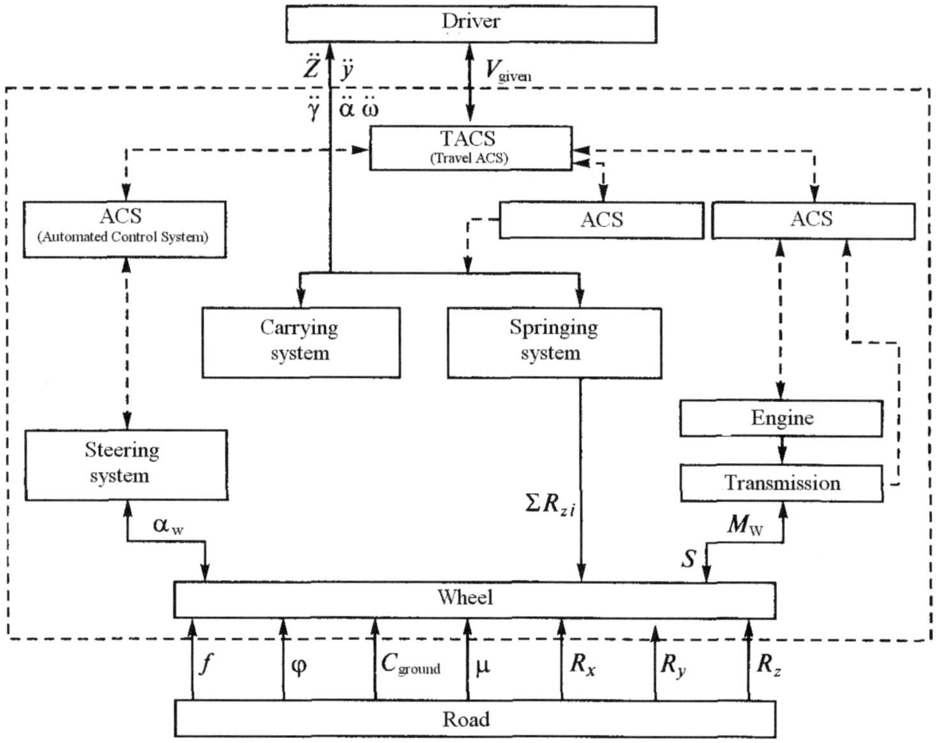

Fig. 8.1. Structural diagram of the dynamic systems: TACS—travel automated control system of the HDWV; ACS—Automated control system of a HDWV subsystem (component); $\ddot{Z}$, $\ddot{y}$, $\ddot{\gamma}$, $\ddot{\alpha}$, $\ddot{\omega}$ —The respective vertical, lateral, lateral- And longitudinal-angular accelerations of the SWC and its acceleration around the vertical axis Z; V_{given} — The SWC's preset travel speed; α_w—Turning angle of the controlled wheel; ΣR_{zi}—Sum of vertical reactions on the wheels; M_W—Torque on the wheels; S—Sliding of the wheel; f, φ, C_{ground}, μ, R_x, R_y, R_z—Parameters of external affections on the wheel (the values of the parameters indicated in the figure and their calculation process are given in the chapter)

of the SWC's efficiency of use (mobility) for special equipment at motion on the preset route.

This technique meets the following requirements:

- It allows for adequately, fully, and thoroughly describing the chassis' motion at each moment and at each point of the representative route;
- It is formed according to the adaptability principle;
- It takes into account the possibility of using it when conducting all forms of tests of the SWC's samples and also when justifying the choice of motion routes when organizing the real haulage; and
- In scientific studies, it is based on the initial data obtained in the course of tests and real operation of HDWVs.

To meet the above requirements, it is necessary to develop a route model that takes into account a variety of road conditions (see Chapter 6) occurring during operations and to develop a complex model of the chassis' motion, which would take into account basic properties of the SWC (traction dynamics, handling quality and stability, passing ability and soft driving, etc.).

In this technique, the maximum travel speed averaged for the representative route was selected as an assessment criterion of the SWC mobility.

The maximum average travel speed means the technically feasible (possible) average speed of covering the representative route by a SWC. At the same time, the influence of the driver's psychophysical features on the travel speed is not taken into account. Limits on the SWC's travel speed on the route corresponds to restrictions of the traction dynamics and passing ability, which ensure satisfactory indicators of handling quality, stability, and soft driving, regulated by normative and technical documents for transported equipment, chassis, and crew.

The route, which contains road conditions characteristic of the HDWV's real routes in the possible operation areas, is representative.

A route section is a piece of road where the values of longitudinal slope, road curvature in the plane view, and type of road surface are constant or vary by no more than a certain value.

The representative route includes a large number of sections in which the parameters correspond to real roads according to statistical data. The representative character is achieved by including a variety of road conditions, accounting for their specific content, and including a sufficient length. At the same time, the data presented in Section 6.3 are used for route modeling as well.

The methodology allows for varying the SWC's design parameters to select acceptable values. It is formed by modules based on the interconnectedness of particular mathematical models of the sub-systems and of the motion route. Particular mathematical models of the motion route also allows for varying the road surfaces' parameters. This is necessary for assessing the quality of existing and projected motion routes for correct selection and when developing proposals on additional engineering equipped for organizing the transportation of cargoes weighing 25 t or more.

Calculating the SWC's mobility is divided into two parts. The first part forms a representative route or presets some specific real route based on their measurements. The second part simulates the SWC's motion on the route (Fig. 8.2).

The representative route is characterized by the data from the road's longitudinal slope angle α_l, the road curvature K in plane, and the road surface type number $No._{t.s}$. Accordingly, with the help of initial data and mathematical models, probable implementations of the above parameters are developed. The implementations are then divided into separate sections, the parameters of which (α_l, K, $No._{t.s}$) are recorded in the representative route's data table.

Implementation of the road parameters means a sequence of parameter values (in particular, heights of the road's macro- and microprofile) for the points located in ascending order of the route's length coordinates with a fixed value of the step length.

Implementation of all road surface types is also calculated. Additional data for all surfaces use values of the rolling resistance coefficient f, the coefficient of tire adhesion with the road surface φ, the specific resistance of the ground to deformation C_{ground}, and the power indicator of the ground resistance μ. These additional implementations and data are used in the motion model according to the road surface's type number for the given section.

The structure of the representative route is shown in Fig. 8.3.

The SWC's motion is modeled with the help of mathematical models of traction dynamics, roadability, stability, passing ability, and soft driving. Analysis of the motion parameters is conducted in each subsequent section of the route. Following the results of the analysis, one or several of the above mathematical models is (are) included in the calculation.

The motion model also is comprised of a speed limitation module acting for the driver. In this book, this function means maintaining the maximum possible speed safe for the chassis

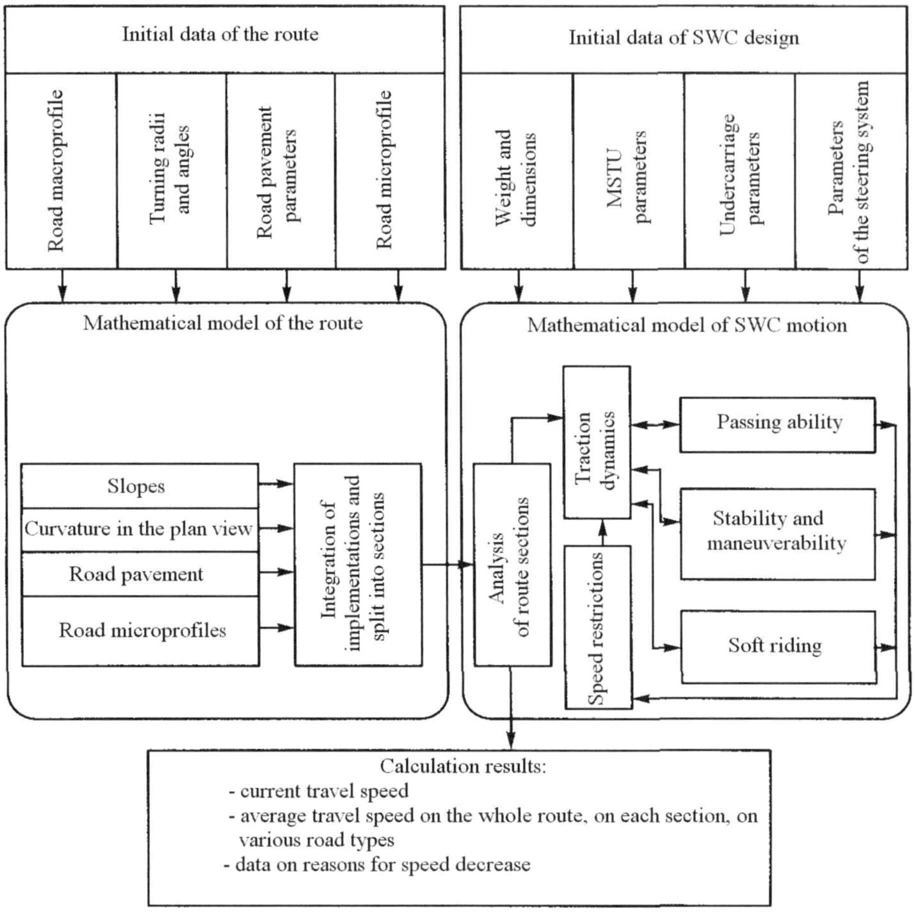

Fig. 8.2. Structural diagram of the sequence of calculating the SWC's mobility

and transported cargo in every moment of time. The module analyzed the SWC's achieved speed and decreased the speed when one of the criteria exceeded the permissible value.

The set of above technique components (see Fig. 8.2) allows for modeling the SWC's controlled motion on the representative route, with a maximal approximation of the processes to the real operation. This is the main difference between the proposed methodology for assessing the SWC's efficiency of use (mobility) and the earlier known mathematical models and techniques for assessing the vehicle dynamics.

8.3 Mathematical model of the SWC's representative route

8.3.1 Structure of the representative route

To simplify using the route data, their classification is developed in the mathematical model. In accordance with it, all route data are divided into four levels. The first level is comprised of the name of the geographical area in which the SWC's mobility assessment is conducted.

This work selected the western geographical area of Russia for illustration. Four typical terrain types are considered (see Fig. 8.3).

After geographical area, the second level indicates the specific contents of the road types

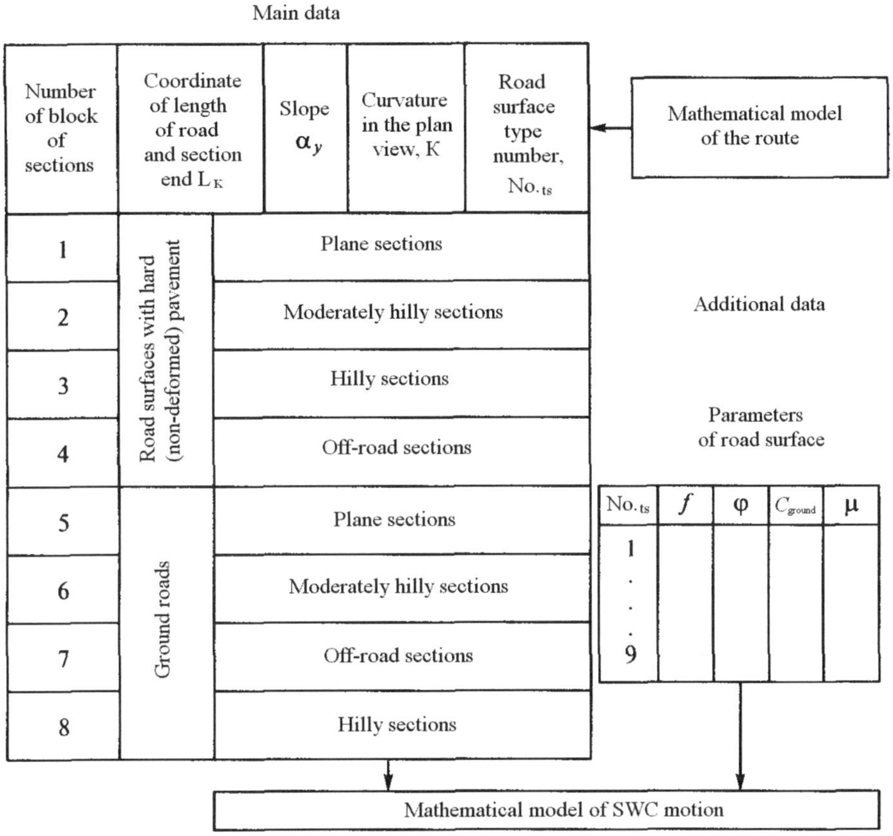

Fig. 8.3. Structure of the representative route

in the region. Likewise, in accordance with the data from studies of the existing roads, the specific contents of the road's hard and ground surfaces are calculated.

The third level includes statistical characteristics determining the occurrence probabilities of certain route parameters. According to the structure of the representative route, the following characteristics are determined as the statistical characteristics of the third level: spectral densities of the macroprofiles of roads in various terrains, the distribution of probable radii and angles of road turning, and the distribution of probable pavement types.

The fourth level is comprised of parameters measured directly on the route which affects the SWC's motion model and are considered below as basic and additional route parameters.

Data from the fourth level are calculated in the form of implementations with the help of statistical methods, that is, sets of parameter values for road points placed in the order of increasing route length coordinates with length step. They include implementations of the road's macroprofile heights (then converted into road slopes), road curvatures in the plane view, and types of road surfaces. The studies assume that these implementations are statistically independent of each other.

The route has a statistical (probable) nature, so it is necessary to meet the following requirements to ensure its representative character: the lengths of all slope implementations, curvatures in the plane view, and types of road surfaces should be large enough for statistical evaluation. This requirement determines the special transition order from one data level to another. Initially, the most laborious transition from the third to the fourth level is carried out according to the probabilities distribution, slope implementations, curvature in the plane view, and types of road surfaces. This allows for obtaining implementations of sufficient length by selecting the appropriate step on the basis of providing acceptable implementations precision. Thus, at the transition from the third to the fourth data level, the value necessary for statistical calculations is ensured for any route. All routes are formed from these implementations. After calculating the SWC's motion on the formed route and determining the average motion speeds, the parameters of the first two levels are taken into account: geographical area, specific contents of the road types, and specific contents of the road's hard and ground surfaces.

Road sections are the main route elements. The road parameters vary randomly from section to section according to the preset statistical dependences. To rationalize, these sections are united in eight blocks (see Fig. 8.3). The first four blocks describe road conditions on surfaces with hard pavements and the other four blocks describe road conditions on ground surfaces. Each block determines road sections for one relevant area out of four.

8.3.2 Longitudinal road gradients

The basic route parameters α_b, K, $No._{t.s}$ are determined as follows.

In accordance with the spectral densities of the macroprofiles heights [62] and with the help of the Fourier transformation for plain, moderately hilly (median terrain), hilly, and cross-country terrains, the appropriate implementations of the macroprofiles heights are determined, which are then calculated in angles of longitudinal road gradients.

The spectral densities of the road macroprofiles heights $K(\lambda)$, for the ith terrain, are defined by the following dependencies (Fig. 8.4):

$$\mathrm{K}(\lambda)_i = \frac{D_{1i}}{\lambda_i^2},\ \text{if}\ \lambda_i < \lambda_{\Gamma i}\ \text{(zone of medium irregularities);}$$

$$K(\lambda)_i = \frac{D_{2i}}{\lambda_i^4},\ \text{if}\ \lambda_i < \lambda_{\Gamma i}\ \text{(zone of small irregularities);}$$

$$D_{2i} = K(\lambda_{gi})\lambda_i^2,$$

where: λ_{gi} is the travel frequency in the boundaries of the medium and small irregularity zones; D_{1i}, D_{2i} are the spectral densities coefficients; and λ_i is the current travel frequency of the macroprofile heights.

The number of spectral density points:

$$\mathrm{K}_\lambda 2^{i_s+1},$$

where i_s is the spectrum window.

The minimum circular frequency of the macroprofile:

$$\lambda_{\min} = \frac{2\pi}{\ell_{\max}},$$

where $l_{\max}$ is the maximum length of the macroprofile wave.

Step of the macroprofile's circular frequency:

$$h_{\lambda i} = \frac{\lambda_{\max i} - \lambda_{\min}}{K_\lambda - 1},$$

where $\lambda_{\max i}$ is the maximum circular frequency of the macroprofile in the ith terrain.

The number of macroprofile height points on the implementation length:

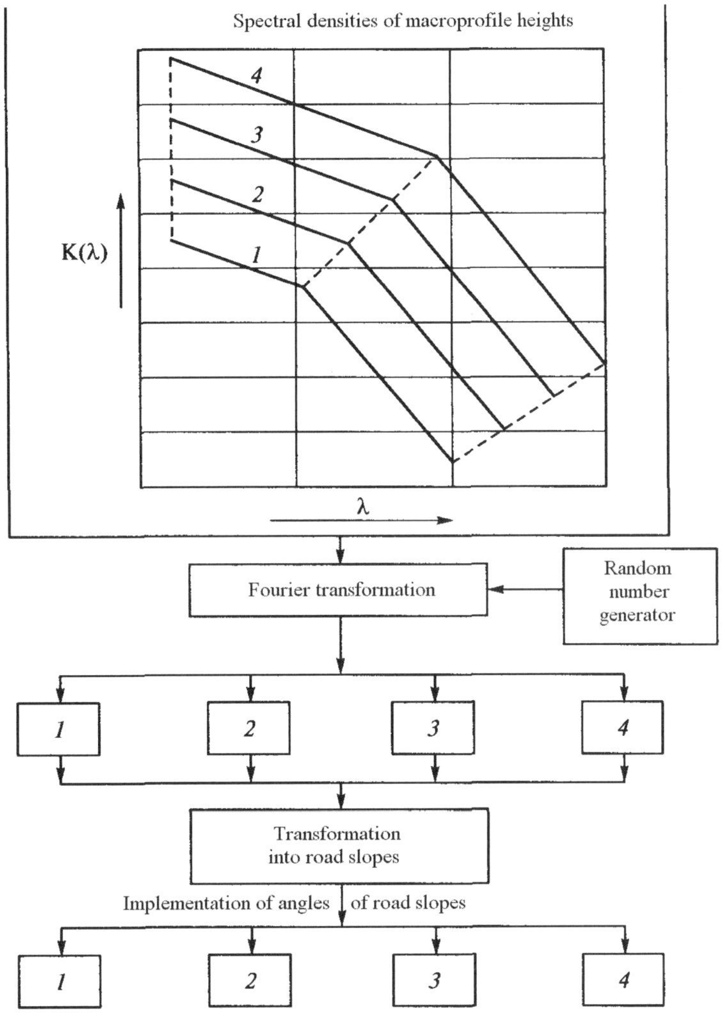

Fig. 8.4. Calculation diagram of longitudinal gradients of routes for the following localities: 1—Plain; 2—Moderately hilly; 3—Hilly; 4—Cross-country

$$K_\ell = \frac{1}{h_\ell},$$

where h_l is the step on implementation of the macroprofile,

$$h_\ell = \frac{1}{2^{i_s}(h_\ell / 2\pi)}.$$

Implementation of macroprofile heights in accordance with the Fourier transformation:

$$Z = \sum_j \sqrt{2K(\lambda_j)_i h_{\lambda i}} \cos\left(\left(\lambda_{\min} + (j-1)h_{\lambda i}\right) K\, h_\ell + \phi(j)\right),$$

where K is the number of microprofile height points at implementation for length; $\phi(j)$ is the phase randomly selected in the values range from $-\pi$ to π; j is the number of travel frequency points of spectral density; and i is the terrain number.

The angles of the longitudinal road gradients are calculated from the expression:

$$\alpha_{sr} = \frac{Z_{k+1} - Z_k}{h_\ell}.$$

The calculation diagram of the longitudinal gradients is presented in Fig. 8.4.

8.3.3 Road curvature in the plane view

The initial data for calculating the road curvature K in each terrain are the occurrence probability distributions of the radii, angles, and turning directions with specific values on the route. All values of these parameters are divided into intervals.

The number of radii K_{RL}, tuning angles $K_{\Theta L}$, and turning directions K_{NPL} (the values of which are within the Lth interval) are determined by the following dependences (curly brackets denotes the operation of finding the integral part):

$$K_{RL} = \{P_r(L)\mathrm{K}_{\text{turn}}\};$$
$$K_{\Theta \mathrm{L}} = \{P_\Theta(L)\mathrm{K}_{\text{turn}}\};$$
$$K_{NPL} = \{P_{NP}(L)\mathrm{K}_{\text{turn}}\},$$

where $P_R(L)$, $P_\Theta(L)$, $P_{NP}(L)$ are the occurrence probabilities of the radii, angles, and tuning directions within Lth interval of their values; Kturn is the number of turns on the route:

$$K_{\text{turn}} = K_n\, L_i,$$

where K_n is the number of turns per kilometer; and L_i is the route length in i^{th} terrain.

To form implementations of the radii, turning angles, and turning directions, the generator of random, evenly distributed numbers ψ within an interval from 0 to 1 and a digit filter for generating the distributions of the above parameters are used. In accordance with the current value ψ, the interval number n of the parameter values is defined.

When calculating the integral probability densities of the radii $F_R(n)$, turning angles $F_\theta(n)$, and turning direction $F_{NP}(n)$, the following dependencies are used:

$$F_R(n) = \sum_{n=1}^{n=\ell} P_R(n);$$

$$F_\theta(n) = \sum_{n=1}^{n=\ell} P_\theta(n);$$

$$F_{NP}(n) = \sum_{n=1}^{n=\ell} P_{NP}(n),$$

upon fulfillment of the condition $F_\Theta(n) > f$, $F_R(n) > f$, $F_{NP}(n) > f$, it is assumed that $l = n$.

The lengths of the straight sections (insertions) between turns are as follows:

$$L_{\text{insent}} = \frac{L - \sum_{j=1}^{K_{\text{turn}}} \frac{2\pi R_j}{360}\theta_j}{K_{\text{turn}}},$$

where R_j and θ_j is the radius and angle of the jth turning; L is the route length.

The calculation diagram of the route curvature in the plane view is presented in Fig. 8.5.

8.3.4 Types of pavements

The pavement types considered in the representative route, $\text{No.}_{\text{t.s}}$–No. 1–9 are as follows:

- 1 and 2 represent cement-concrete highway in satisfactory and worn-out condition;
- 3 and 4 represent cobblestone highway in satisfactory and worn-out condition;
- 5 and 6 represent crushed stone macadam in satisfactory and bumpy condition;
- 7 and 8 represent ground road in satisfactory and bumpy condition; and
- 9 represents off-road (terrain).

The distribution of pavement types on the route is equally probable and is defined with the help of expressions (8.12)–(8.14). Surfaces

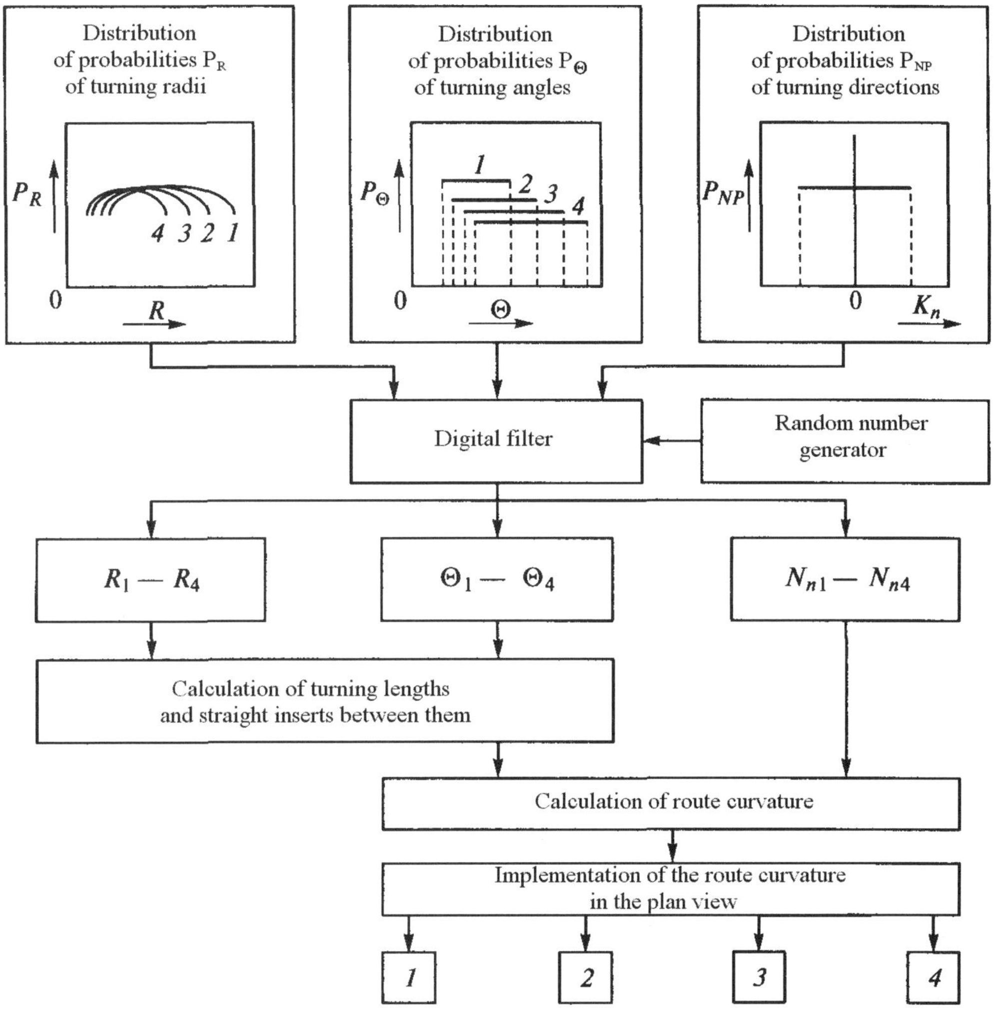

Fig. 8.5. Calculation diagram of the route curvature in the plane view for the localities: 1—Plain; 2—Moderately hilly; 3—Hilly; 4—Off-road

1–4 are assigned to hard road surfaces and 5–9 to ground road surfaces. Alternating road surface types is assumed as equally probable and formed with the help of a random numbers generator (Fig. 8.6). Road section lengths with various surfaces have an equal distribution of occurrence probabilities within the interval from 100 to 1000 meters.

Additional parameters of the route, except for mentioned coefficients f, ϕ, C_{ground}, μ, are implementations of the road's microprofile.

8.3.5 Road microprofiles

The road's microprofiles are calculated in the form of implementations for each assumed road surface type for the spectral densities of the microprofile's heights [62]. Depending on the travel frequency λ_i, the spectral densities $S(\lambda_i)$ are expressed by the following dependence

$$S(\lambda_i) = A\,\lambda_i^W ,$$

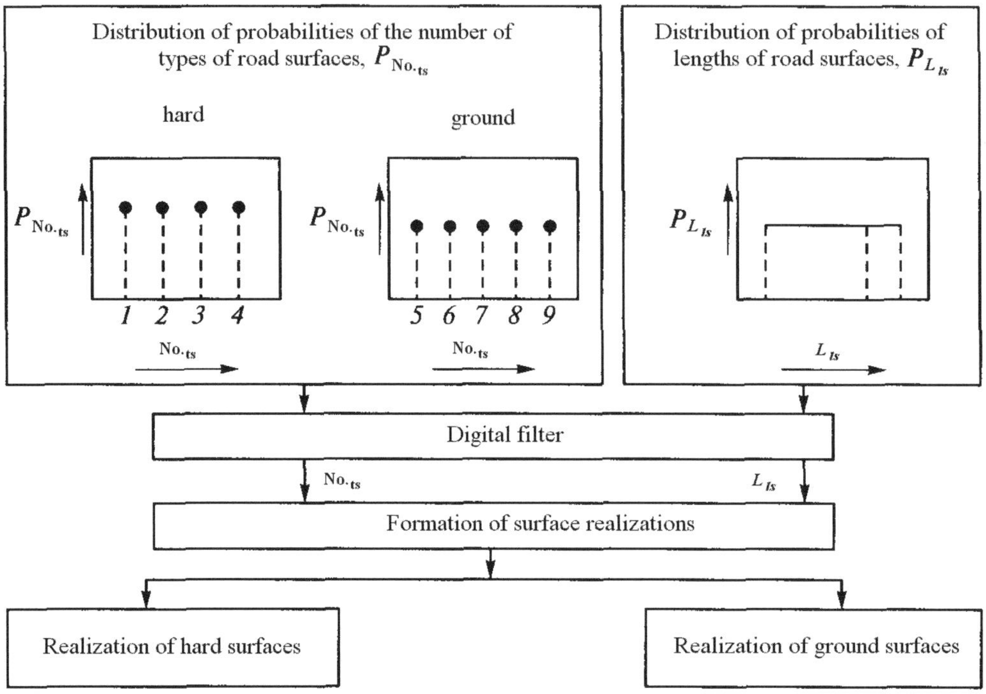

Fig. 8.6. Calculation diagram of the sequence of pavements on the route

where A and W are the constant parameters preset for each road surface type; λ are the numbers of the road surface types; and i is the section number.

The number of spectral density points is defined from the following expression:

$$K_\lambda = 2^{i_s - 1},$$

where i_s is the spectrum window.

The boundary values of the circular travel spectrum frequencies are

$$\lambda_{\min} = \sqrt{50A};$$

$$\lambda_{\max} = \lambda_{\min} + \Delta_\lambda,$$

where Δ_λ is the spectrum width preset in accordance with the recommendations [62].

The step of the circular travel frequency is defined from the expression:

$$h_\lambda = \frac{\lambda_{\max} - \lambda_{\min}}{K_\lambda - 1}.$$

The step of implementation on the road length is

$$h_\ell = \frac{1}{(h_\lambda / 2\pi)^{i_s}}.$$

The number of implementation points on the road length is:

$$K_\ell = \frac{1}{h_\ell}.$$

For calculating the implementations on the right and left road tracks, the correlation coefficient of parallel microprofile sections $\rho(\lambda_i)$ is used:

$$\rho(\lambda_i) = \frac{1}{1 + \left(\frac{B\lambda_i}{n_d}\right)^2},$$

where B is the SWC's track; and n_e is the road coefficient [62].

By means of the correlation coefficient, the spectral densities of the semi-sum $S_{zi}(\lambda_i)$ and semi-difference $S_{\psi i}(\lambda_i)$ of the tracks' microprofiles are calculated:

$$S_{zi}(\lambda_i) = S(\lambda_i)(1+\rho(\lambda_i));$$
$$S_{\psi i}(\lambda_i) = S(\lambda_i)(1-\rho(\lambda_i)).$$

From the spectral densities, implementations of the semi-sum Z_j and semi-difference ψ_j values of the tracks' microprofiles are calculated (with the help of the Fourier transformation):

$$Z_j = \sum \sqrt{2S_z(\lambda_i)h_\lambda} \cdot \cos\left((i\,h_\lambda)j\,h_\ell + \psi_1(i)\right);$$

$$\psi_j = \sum \sqrt{2S_\psi(\lambda_i)h_\lambda} \cdot \cos\left((i\,h_\lambda)j\,h_\ell + \psi_2(i)\right),$$

where i, j are the ordinal numbers of the travel frequency points and implementation points on the length; and $\psi_1(i)$, $\psi_2(i)$ are the harmonic component phases of implementations selected with the help of a random number generator in the interval from $-\pi$ to π.

Implementations of the microprofile heights on the right (Z_{rj}) and left (Z_{lj}) tracing ruts are:

$$Z_{rj} = (Z_j - M_z) - (\psi_j - M\psi);$$

$$Z_{lj} = (Z_j - M_z) - (\psi_j - M\psi),$$

where M_z and M_Ψ are the mathematical expectations of implementations of half-sum and half-difference in the microprofiles of the tracing ruts:

$$M_z = \sum_j^1 \frac{Z_j}{K_\ell};$$

$$M_\psi = \sum_j^1 \frac{\psi_j}{K_\ell}.$$

The calculation diagram of the route's microprofiles is presented in Fig. 8.7.

Thus, implementations united and divided into the sections of the route's main parameters—the longitudinal road gradient angles α_y, the route curvature K in the plane view, and road surface type numbers $No._{t.s}$—together with additional data—the roads' microprofile and coefficients f, φ, C_{ground}, μ—describe the road conditions and represent perturbations for the mathematical motion model.

8.3.6 Initial data of the representative route

Using the above approaches to form the SWC's motion conditions, a version of the representative route is further proposed.

Dividing the roads by types is an important issue when developing the route. In our opinion, the most reasonable division for multi-axle SWCs is the one in which category I-III roads with hard pavements makes up 15% of the road, ground roads of different conditions make up 65%, and off-road conditions make up 20%. Roads with hard pavement are divided into two types: asphalt-concrete highways and smoothly paved cobblestone roads in proportion of 2:3 according to *Dynamics of System "Road-Tire-Automobile-Driver"* [62]. Both types consist of satisfactory and worn-out sections. Ground roads are divided into two types: ground in satisfactory condition and worn-out ground in proportion of 3:1 in accordance with the above work [62]. Ground roads in satisfactory condition include crushed stone macadam roads with sections in satisfactory and worn-out conditions.

When modeling this representative route, it was taken into account that the highest impact on the SWC's average travel speed value along the entire route will be provided by the number and transition types from one homogeneous area (by pavement type) to another, and, therefore, the route was determined to make such transition occur by random distribution and, at the same time, have all probable transition options on the route.

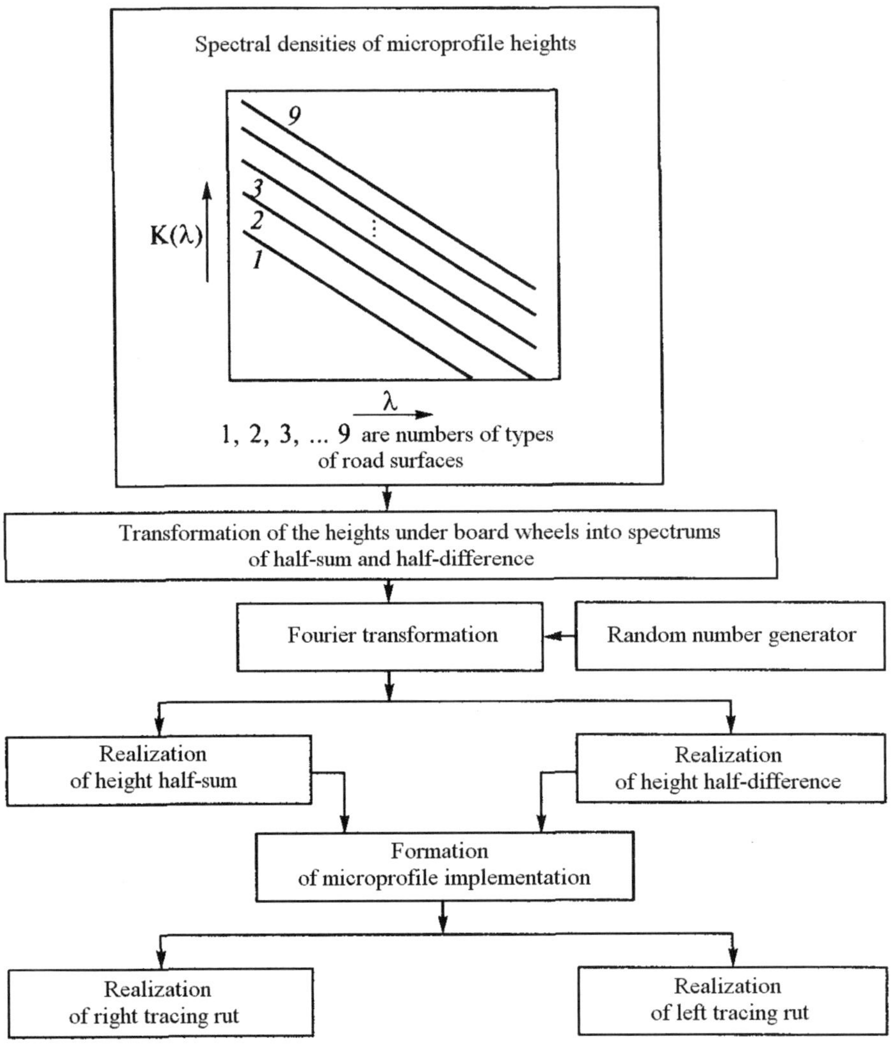

Fig. 8.7. Calculation diagram of the route's microprofiles

Some road condition parameters were not taken into account in the specific types of road pavements. For example, it was assumed that an asphalt-concrete highway in satisfactory condition was absolutely smooth (by microprofile) and non-deformable and that ground parameters affecting the passing ability of SWCs were preset only for motion on worn-out ground and terrain.

Dividing the route's longitudinal road gradients (macroprofile) was conducted as follows. It was assumed that the route is circular and laid in median terrain in conditions of western Russia, beginning and ending at the same level. In accordance with the statistical data, it was assumed that the maximum ascent (descent) in the route section depended on the pavement type and was 3° for asphalt, 5° for cobblestone roads, 7° for ground roads in different conditions, and 9° for terrain. At the same time, it was assumed that the maximum gradient angles occur in no more than 30% of the length of each pavement type, in other cases, the values of the angles were lower.

Dividing by the curve's radii value in the plane view and lengths of the turning sections was also conducted by statistical data for roads in western Russia. Each pavement had 90° turning with a 30 m radius for checking the chassis' maneuvering properties.

When introducing sections with low ground bearing capacities for checking the limits on passing ability, it was taken into account that the route would be laid on loose loam soil ($C_{ground} = 0.25$; $\mu = 0.5$), sand ($C_{ground} = 1.0$; $\mu = 0.25$), arable land ($C_{ground} = 0.5$; $\mu = 0.5$), and compacted grassed ground ($C_{ground} = 1.0$; $\mu = 1.0$).

In the microprofile description, the irregularity ordinate function depending on the distance traveled and defined through spectral densities for various pavement types was preset. At the same time, to obtain different ordinates under the wheels of the same axle, implementation of the microprofile was delivered under different sides of the chassis with random phase shift.

The results of preliminary studies revealed that assessment parameters depend on the lengths of implementations. At implementations of 20–25 km or more, the values of the assessment parameters stabilize. That is why the lengths of the representative macroprofile implementations and the route curvature in the plane view should be of at least 20 km.

On this basis, the total length of the route in the mathematical model was assumed to be 50 km.

The initial data for preparing the various options of the representative route are presented in Tables 8.1–8.6 [7, 62].

Table 8.1 Parameters of spectral densities of the macroprofile height

Parameter, m/rad	Locality			
	Plain	Moderately hilly	Hilly	Off-road
Dl	2.0·10-3	1.6·10-3	1.0·10-1	1.0
A	6.3·10-3	6.3·10-3	6.3·10-3	6.3·10-3
λΓi	6.3·10-3	4.0·10-2	2.5·10-2	1.6·10-2
λi	1.0·10-2	4.0·10-3	1.6·10-3	6.3·10-4

Table 8.2 Intervals of values of turning radii of the route in various localities

Locality	Intervals of values of turning radii, m								
	1	2	3	4	5	6	7	8	9
Plain	100–200 (150)	200–300 (250)	300–400 (350)	400–500 (450)	500–600 (550)	600–700 (650)	700–800 (750)	800–900 (850)	900–1000 (950)
Moderately hilly	48–96 (72)	96–144 (120)	144–192 (168)	192–240 (216)	240–288 (264)	288–345 (312)	345–384 (360)	384–432 (408)	432–480 (456)
Hilly	40–80 (60)	80–120 (100)	120–160 (140)	160–200 (180)	200–240 (220)	240–380 (260)	280–320 (300)	320–360 (340)	360–400 (380)
Off-road	24–48 (36)	48–72 (60)	72–96 (84)	96–120 (108)	120–144 (132)	144–168 (156)	168–192 (180)	192–216 (204)	216–240 (288)

Note. Average radii for each interval are presented in brackets.

Table 8.3 Probability distribution of turning radii

Probabilities of appearance of the turning radius within the intervals of Table 8.2								
1	2	3	4	5	6	7	8	9
0.041	0.095	0.094	0.106	0.174	0.155	0.090	1.106	0.140

Table 8.4 Turning angles of the route in various localities

Angle	Turning angles by types of locality, degrees:			
	Plain	Moderately hilly	Hilly	Off-road
Minimum	3	6	7.5	12.5
Maximum	18	36	45	75

Table 8.5 Parameters of spectral densities of microprofile height

Type of road pavement	Value	
	A. m/rad	*W*
Cement-concrete highway:		
in satisfactory condition	3.2·10-5	-2.0
worn-out	6.5·10-5	-2.0
Cobblestone highway:		
in satisfactory condition	3.5·10-4	-1.07
worn-out	2.2·10-3	-1.07
Crushed stone macadam:		
in satisfactory condition	2.5·10-4	-2.0
worn-out	5.0·10-4	-2.0
Ground road:		
in satisfactory condition	2.5·10-4	-2.0
worn-out	8.0·10-3	-2.12
Off-road (locality)	1.0·10-2	-2.0

Table 8.6 Parameters of road pavements

Type of road pavement	Value			
	f	φ	C_{ground}	μ
Cement-concrete highway:				
in satisfactory condition	0.020	0.75	–	–
worn-out	0.025	0.70	–	–
Cobblestone highway:				
in satisfactory condition	0.030	0.65	–	–
worn-out	0.050	0.65	–	–

Table 8.6 Parameters of road pavements

Crushed stone macadam:				
in satisfactory condition	0.025	0.60	–	–
worn-out	0.030	0.50	–	–
Ground road:				
in satisfactory condition	0.030	0.55	–	–
worn-out	0.100	0.45	1.0	1.0
Off-road (locality)	0.300	0.40	0.5	0.5

8.4 Mathematical model of the SWC's motion

The motion model takes into account the traction dynamics, stability and maneuverability, passing ability, and soft driving (travel smoothness). For the most complete description of a multi-axle chassis' motion, the current kinematical and dynamic chassis characteristics should be calculated with the help of numerical integration of differential equations, which describe the chassis' basic components, and their interaction with each other and the road surface.

Individual mathematical models of the multi-axle chassis' properties are involved either simultaneously or selectively in the calculation process depending on the road conditions: the presence of turns and the surface conditions. To do that, an analysis of the parameters of each subsequent route section is made. While on the route, the travel speed is kept at the maximum according to the SWC's technical capabilities and the traffic safety conditions.

In the mathematical model, control of the travel speed is carried out with the help of the speed limitation block. The data on the current values of the chassis' coordinates and their derivatives, and the power factors defined in the motion calculation, allow for determining the acceptable current travel speed for ensuring the safety of the chassis and transported cargo in every point of the route. In case of restrictions, the travel speed is reduced by forced braking with a fixed deceleration down to the speed at which there are no limitations.

The mathematical model has a block that controls the process of switching gears from the lowest to the highest and vice versa and a block that implements a simulation of blocking and unblocking the hydraulic torque converter.

The combination of above characteristics allows for simulating the controlled motion of the chassis on the representative route with the maximum similarity to the real operation process.

The following additional assumptions are made when developing the mathematical SWC motion model:

- The sprung weight of the chassis is a solid body with longitudinal and lateral planes of symmetry. The deformations of the sprung weight in torsion and bending are not taken into account;
- The springing system is symmetrical in relation to the longitudinal vertical plane. The characterization of the elastic suspension elements is piecewise-linear. The elastic characteristics of the tires are linear;
- The springing system "suspension-tire" is assumed to be independent torsion or candle type depending on the design sample under study;
- In the static position, the load is evenly distributed among the axles; and
- The imbalance and gyroscope torques of the rotating masses of the transmission and the engine are small.

The correctness of the above assumptions for SWCs has been confirmed by numerous experimental studies [1, 2].

The following differential equation of the longitudinal motion acceleration is the fundamental equation of dynamics that simulates the SWC's motion in the longitudinal direction:

$$\dot{V} = \frac{g\left(P_w - \sum P_{\text{srm}}\right)}{\delta_{rm}\, G}$$

where g is the gravity acceleration; P_w is the force on the wheels; ΣP_{srm} is the total force of the motion resistance; δ_{rm} is the coefficient accounting for the rotating masses; and G is the chassis' weight.

In general, the traction force on the wheels can be expressed as follows:

$$P_w = \frac{Mw}{r_w}, \tag{8.15}$$

where M is the engine's torque; w is the transmission's transfer function; and r_w is the radius of the driving wheel.

For the mechanical differential transmission, the expression (8.15) has the following form:

$$P_w = \frac{MU_g U_d U_b U_{wr} \eta_t}{r_w}, \tag{8.16}$$

where U_g, U_d, U_b, and U_{wr} are the transmission ratios of the gearbox, transfer (distribution) box, main gear, and wheeled reduction gearbox, respectively; and η_t is the transmission's efficiency factor.

The expression (8.16) is used given the assumption that it is true for all driving wheels. This precondition is reasonable for vehicles with driving wheels on two or three axles. However, for multi-axle vehicles, this expression calculates the chain of transmission parameters starting from the engine and moving to the most distant wheel or pair of wheels and the value P_w is valid only for them. All other driving wheels (for example, a 12×12 chassis has 10 such other wheels) receive the engine torque M, which differs from the one calculated in the expression (8.16). However, for comparative studies, the use of the expression (8.16) in the model is quite sufficient.

The maximum traction force on the wheels, limited by the adhesion of the driving wheels with the road, is as follows:

$$P_\phi = \phi\, G \cos \alpha_{sr}, \tag{8.17}$$

where: ϕ is the adhesion coefficient; and α_{sr} is the longitudinal road gradient angle.

The calculations use the smallest values defined from the expressions (8.16) and (8.17).

The total resistive force to the SWC motion is as follows:

$$\Sigma P_{\text{srm}} = P_\psi + P_w$$

where: $P_\psi = G(f \cos \alpha_{sr} + \sin \alpha_{sr})$ is the motion resistance force on the slope; f is the rolling resistance coefficient; $P_{\text{air}} = K_{\text{air}}\, F\, V_2$ is the air resistance force; F is the lateral cross-section area of the SWC; K_{air} is the air shape coefficient; and V is the travel speed.

The coefficient accounting for rotating masses is defined from the following formula:

$$\delta_{rm} = (1.03\ldots1.05) + (0.001\ldots0.002)\ U_t^2,$$

where U_t is the transmission's aggregate transmission ratio.

The external speed characteristic of the (diesel) engine is presented in the following form:

$$N_e = N_{\max}\left[a\frac{\omega_i}{\omega_N} + b\left(\frac{\omega_i}{\omega_N}\right)^2 + c\left(\frac{\omega_i}{\omega_N}\right)^3\right]$$

$$M_e = M_N \left[a + b\,\frac{\omega_i}{\omega_N} + c\left(\frac{\omega_i}{\omega_N}\right)^2 \right]$$

where N_e and M_e are the power and torque at the crankshaft's rotation rate ω_i; $N_{\max}$ and M_N are the maximum power and corresponding torque at the crankshaft's rotation rate ω_N; ω_i is the current rotation rate of the engine's crankshaft; ω_N is the crankshaft's rotation rate corresponding to $N_{\max}$; and *a*, *b*, and *c* are the external speed characteristics coefficients:

$$a = \frac{K_e^2 - K_a(2K_a - 1)}{(K_e - 1)^2}$$

$$b = \frac{2K_e^2 - K_a(K_a - 1)}{(K_e - 1)^2}$$

$$c = \frac{K_a - 1}{(K_e - 1)^2}$$

where: $K_e = \frac{\omega_{M\max}}{\omega_N}$ is the engine elasticity coefficient; $K_a = \frac{M_{\max}}{M_N}$ is the engine adaptability coefficient; and $\omega_{M\max}$ is the crankshaft's rotation rate corresponding to $M_{\max}$.

The free engine power, which determines the traction dynamic of SWCs, is defined from the following expression:

$$N_{fce} = N_e - N_{sc} - N_{ex} - N_{aps} - N_{æ}$$

where N_{sc}, N_{ex}, N_{aps}, and N_{ae} are the losses of in engine power in the cooling, exhaust, and air purification systems and to the auxiliary units' drive, respectively:

$$N_{sc} = (0.07...0.16)\,N_{\max}\left(\frac{\omega_i}{\omega_N}\right)^2;$$

$$N_{ex} = (0.03...0.04)\,N_e;$$

$$N_{aps} = (0.02...0.04)\,N_{\max}\left(\frac{\omega_i}{\omega_N}\right)^2;$$

$$N_{æ} = (0.03\ldots0.04)\,N_{max}.$$

For a mechanical transmission, the work of the clutch at starting the SWC from a stop is taken into account:

$$\dot{\omega}_1 = \frac{M_e(\omega_1) - M_a(t)}{J_1}, \text{ at } 0 \le \omega_2 \le \omega_1;$$

$$\dot{\omega}_2 = \frac{M_a(t)}{J_2}$$

$$M_a(t) = \begin{cases} \beta M_N \dfrac{t}{t_1} & \text{at } 0 \le t_1, \\ \beta M_N & \text{at } t \ge t_1; \end{cases}$$

where:

$$\dot{\omega}_1 = \dot{\omega}_2 = \frac{M_e(\omega_1)}{J_1 + J_2} \text{ at } \omega_1 = \omega_2,$$

where: ω_1, $\dot{\omega}_1$, ω_2, $\dot{\omega}_2$ are the angular velocities and accelerations of the engine's crankshaft and the clutch's driven shaft, respectively; J_1, J_2 are the equivalent inertia moments of the engine and transmission's rotating parts, respectively; $M_a(t)$ is the frictional torque of the clutch coupling; β is the margin coefficient of the clutch coupling; t_1 is the switching-off time of the clutch coupling; and *t* is current time.

When using a hydromechanical transmission, the joint work points of the engine and HTC are defined from the known dependences:

$$M_p = \alpha_p\,\omega_p^2; \qquad (8.18)$$

$$n_p = \frac{\omega_i}{U_{s.g}}; \quad (8.19)$$

$$M_p' = M_{fce}\, U_{s.g}\, \eta_{s.g}\,, \quad (8.20)$$

where M_p and M_p' are the torques on the HTC's pump wheel from the maximum and free engine torque, respectively; ω_p is the pump wheel's rotation rate; ω_i is the current rotation rate of the engine's crankshaft; M_{fce} is the engine's free torque; and $U_{s.g}$, $\eta_{s.g}$ are the ratios and efficiency factor of the matching gearbox (increasing gear).

From the condition of the joint work $M_p = M_p'$, the value ω_i is defined from formulas (8.18)–(8.20), and the output characteristic of the hydraulic torque converter M_T is defined:

$$\omega_{tw} = \omega_p U = \frac{\omega U}{U_{s.g}}$$

$$M_{tw} = M_w\, K_{tw} = M_{fce}(\omega_i)\, U_{s.g}\, \eta_{s.g}\, K_{tw}$$

where ω_{tw}, M_{tw} is the rotation rate and torque on the HTC's turbine wheel; U is the HTC's internal ratio; K_{tw} is the transformation coefficient; and M_w is the current value of the engine's torque.

The dependence $M_{tw}(\omega_{tw})$ represents the HTC's output characteristic and is used in formula (8.16) instead of the torque M for calculation of a P_w vehicle with an HTC.

Calculations for the indicated parameters of SWCs with nontraditional technical solutions, for example, with an electrical transmission and gas-turbine engine or other alternative power source, are presented in other chapters.

When moving on ground roads with low bearing capacities, the value of the rolling resistance coefficient is calculated from the known expression

$$f = \frac{\sum_{i=1}^{n} f_i}{n},$$

where $f_i = \dfrac{b_w\, C_{groundi}\, h_{gi}^{\mu+1}}{G_{wi}(\mu+1)}$ is the rolling resistance coefficient of the ith wheeled axle; b_w is the width of the wheel profile; $C_{groundi}$ is the specific ground resistance to deformation (soil settlement coefficient); μ is the degree indicator, which characterizes the change in ground resistance according to the deformation's depth; h_{gi} is the rut's depth after passage of the ith wheeled axle; G_{wi} is the normal load on the ith wheeled axle; and n is the number of axles.

The specific ground resistance to deformation goes up with every passage of the wheeled axle and is calculated as follows:

$$C_{groundi} = C_{g1} \prod_{k=1}^{i=1} (\Phi_k + 1)^{\mu}$$

where C_{g1}, $C_{groundi}$ are the specific deformation resistance under the first and i^{th} wheeled axles; and Φ_k is the ground compaction coefficient, which is numerically equal to the depth increment h_{gi} at passage of the ith axle.

The depth of the rut h_{gi} under the ith axle is calculated by using the successive approximations from the known dependence [52] as follows:

$$G_{wi} = C_{groundi}\, b_w \sqrt{2r_{ti}} \left(1 - \frac{\mu}{3}\right) h_{gi}^{\mu+5}$$

where: $r_{ti} = r_s\left(1 + \dfrac{h_{zi}}{h_{gi}}\right)$ is the radius of a rigid wheel, the curvature of which is equal to the deformable wheel; r_c is the statistical radius of the deformable wheel; $h_{zi} = \dfrac{G_{wi}}{C_{S.tir}}$ is the tire's radial deflection; and $C_{S'tir}$ is the tire's radial stiffness.

The depth of the track after passage of a multi-axle chassis is defined from the following expression:

$$h_{\text{pas}} = \sum_{i=1}^{n} h_{gi} \,.$$

At the mathematical description of the SWC's motion on deformable surfaces, a simple, common assessment model of the chassis' passing ability [52, 54], which is in good conformity with the experimental data, is used. Taking into account the representative nature of the route from the viewpoint of the distributions of the different support surfaces and the fact that the study of the chassis' passing ability is not the main calculation aim and is only used for comparative assessment with a better analogue, such approach can be considered as quite reasonable. At the same time, it should be noted that the modular approach of the considered technique, where each module describes a particular feature, allows for using any other, including specified (more detailed), mathematical assessment model of the passing ability, for example, when it is necessary to study the SWC's motion on any specific route characterized by a high specific content of deformable support surfaces.

For studies of turnability and motion stability, the chassis is represented as a single physical object in accordance with Fig. 8.8 and 8.9, moving in the longitudinal (coordinate X) and lateral (coordinate Y) directions with account of side rolls (coordinate γ) and rotating around the vertical axis Z. At the same time, the movable coordinate system is rigidly fixed to the vehicle's center of mass.

According to the calculation scheme (see Figs. 8.8 and 8.9), the SWC's motion along the curvilinear trajectory is described by the system of differential equations:

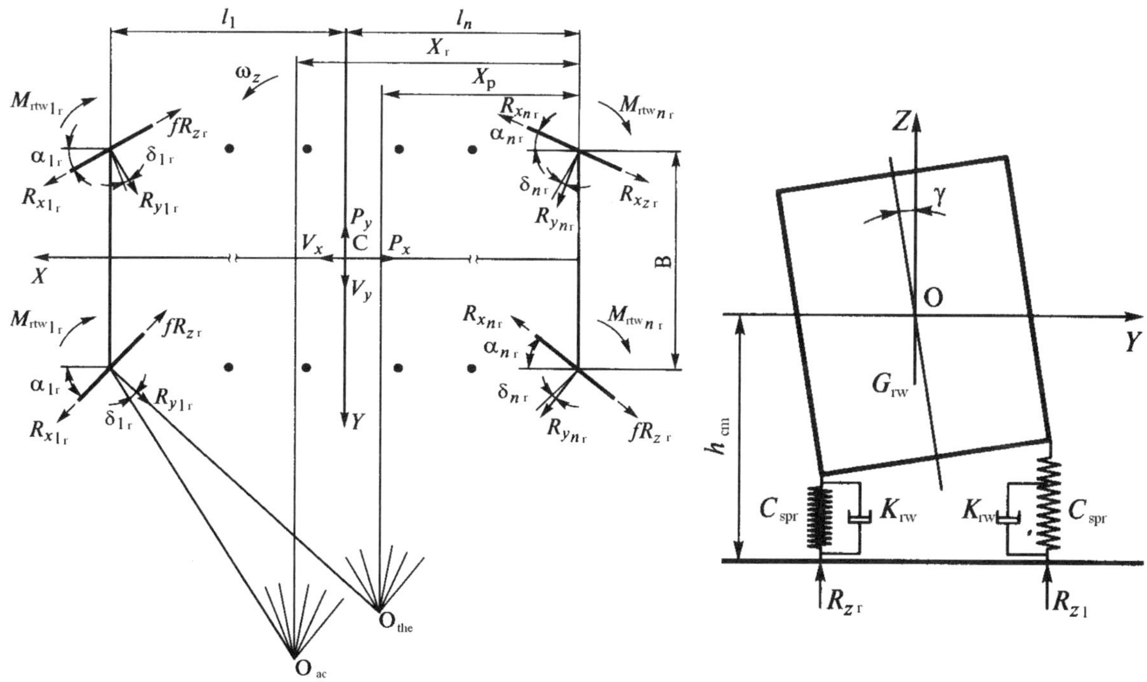

Fig. 8.8 Calculation diagram of the SWC's turnability: O_{ac} and O_{the}—Actual and theoretical turning poles; O—The chassis' center of mass

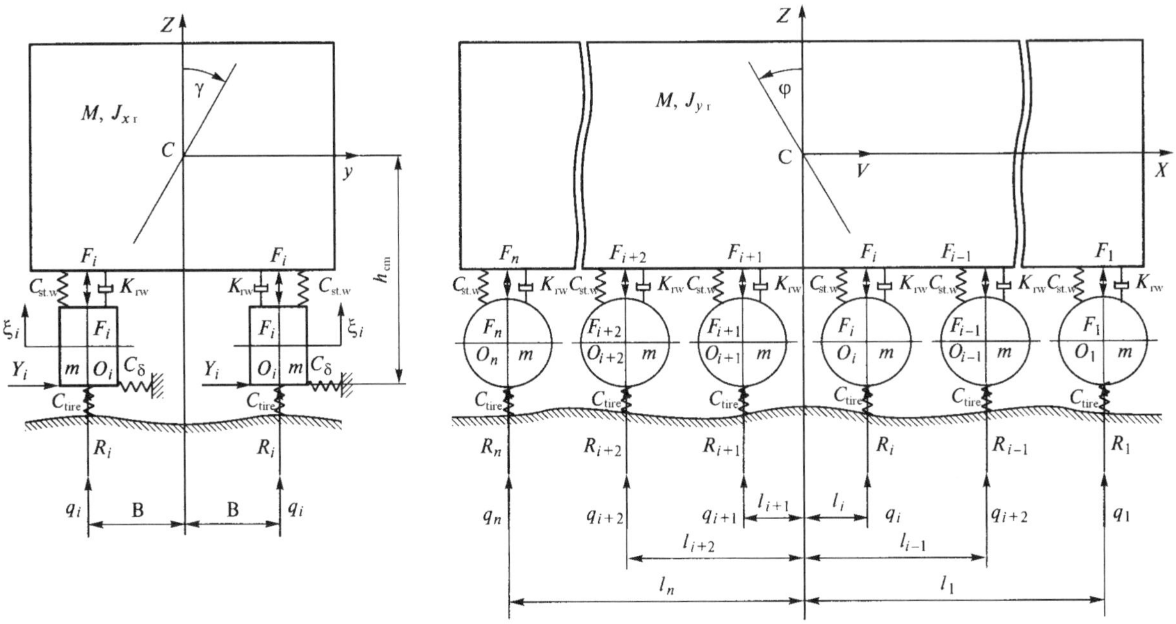

Fig. 8.9. Calculation diagram of the oscillations of an n-axle chassis: M and m—Sprung and non-sprung weights respectively; J_{xn} and J_{yn}—Inertia moments of the SWC's sprung weight; z, y, and x—Vertical, lateral, and longitudinal linear coordinates; γ, φ—Angular longitudinal and lateral displacement coordinates of the sprung weight; ξ_i—Vertical linear coordinate of the ith wheel; $C_{st.w}$—Stiffness of the wheel suspension; K_{rw}—Resistance coefficient of the wheel suspension's shock absorber; F_i— Suspension force; Y_i—Lateral force acting on the wheel; C_δ—Lateral tire stiffness; C_{tire}—Vertical tire stiffness; R_i—Vertical reaction on the wheel; q_i—Current coordinate of the microprofile under the ith wheel; l_i—Distance of the ith axle to the center of mass; and O_i—Center of mass of the ith non-sprung weight (of the wheel)

$$
\left.
\begin{aligned}
&\dot{V}_x = \frac{1}{M_g}\left[\sum_{i=1}^{n}\left(R_{xi\,r}\cos\alpha_{i\,r} + R_{xi\,\ell}\cos\alpha_{i\,\ell}\right) - \right.\\
&-\sum_{i=1}^{n}\left(R_{yi\,r}\sin\alpha_{i\,r} + R_{yi\,\ell}\sin\alpha_{i\,\ell}\right) - \\
&\left.-\sum_{i=1}^{n}\left(f\,R_{z\,r}\cos\alpha_{i\,r} + f\,R_{z\,\ell}\cos\alpha_{i\,\ell}\right) - P_{\text{more}}\right] + V_y\omega_z;\\
&\dot{V}_y = \frac{1}{M_g}\left[\sum_{i=1}^{n}\left(R_{yi\,r}\cos\alpha_{i\,r} + R_{yi\,\ell}\cos\alpha_{i\,\ell}\right) - \right.\\
&-\sum_{i=1}^{n}\left(R_{xi\,r}\sin\alpha_{i\,r} + R_{xi\,\ell}\sin\alpha_{i\,\ell}\right) - \\
&\left.-\sum_{i=1}^{n}\left(f\,R_{z\,r}\sin\alpha_{i\,r} + f\,R_{z\,\ell}\sin\alpha_{i\,\ell}\right)\right] - V_x\omega_z;\\
&\ddot{\gamma} = \frac{1}{J_x}\left\{h_{cm}\left[\sum_{i=1}^{n}\left(R_{xi\,r}\sin\alpha_{i\,r} + R_{xi\,\ell}\sin\alpha_{i\,\ell} + R_{yi\,r}\cos\alpha_{i\,r} + \right.\right.\right.\\
&\left.\left.\left.+R_{yi\,\ell}\cos\alpha_{i\,\ell} - f\,R_{z\,r}\sin\alpha_{i\,r} - f\,R_{z\,\ell}\sin\alpha_{i\,\ell}\right)\right] - \frac{B}{2}\sum_{i=1}^{n}\left(R_{z\,r} - R_{z\,\ell}\right)\right\};\\
&\dot{\omega}_z = \frac{1}{J_x}\left\{\frac{B}{2}\left[\sum_{i=1}^{n}\left(R_{xi\,r}\cos\alpha_{i\,r} - R_{xi\,\ell}\cos\alpha_{i\,\ell}\right) - \right.\right.\\
&\left.-\sum_{i=1}^{n}\left(R_{yi\,r}\sin\alpha_{i\,r} - R_{yi\,\ell}\sin\alpha_{i\,\ell}\right) - f\sum_{i=1}^{n}\left(R_{z\,r}\cos\alpha_{i\,rr} - R_{z\,\ell}\cos\alpha_{i\,\ell}\right)\right] + \\
&+\sum_{i=1}^{n}\ell_i\left(R_{xi\,r}\sin\alpha_{i\,r} + R_{xi\,\ell}\sin\alpha_{i\,\ell} - f\,R_{z\,r}\sin\alpha_{i\,r} - f\,R_{z\,\ell}\sin\alpha_{i\,\ell} + \right.\\
&\left.\left.+R_{y\,\ell}\cos\alpha_{i\,\ell} + R_{yi\,r}\cos\alpha_{i\,r}\right) - \sum_{i=1}^{n}\left(M_{rtwi\,r} + M_{rtwi\,\ell}\right)\right\},
\end{aligned}
\right\}
\tag{8.21}
$$

where $R_{xi\,(r,\,l)}$, $R_{yi\,(r,\,l)}$, $R_{z\,(r,\,l)}$ are the longitudinal, lateral, and vertical (normal) reactions in the contact pattern of the ith right (left) wheels with the support surface; M_g is the SWC's total weight; J_x, J_z are the inertia moments of the SWC relative to the longitudinal and vertical axles passing through the center of mass; V is the chassis' track, m; h_{cm} is the height of the center of mass, m; n is the number of axles; $\alpha_{i\,(r,\,l)}$ is the turning angle of the right (left) steered wheel; f is the motion resistance coefficient; $M_{rtwi\,(r,\,l)}$ is the moment of resistance to turning the wheels of the right and left sides; P_{more} is the additional external force caused by natural factors or their combination (air resistance, etc.) in the longitudinal direction at a curvilinear motion; V_x, V_y are longitudinal and lateral travel speeds of the SWC's center of mass; γ is the side roll angle; and ω_z is the SWC's angular travel speed along a curvilinear trajectory.

Vertical reactions, R_z, are obtained by solving the differential equations systems (8.26), presented below.

Lateral reactions, R_{yi}, are calculated with the help of the angular tire drift, δ_i, based on the theory of linear drift:

$$R_{yi\,(r,\,\ell)} = -K_u\,\delta_{i\,(r,\,\ell)}$$

where K_u is the resistance coefficient to lateral drift.

The lateral tire drift angles are defined as the angle between the wheel rotation plane and the wheel velocity vector from the following dependence:

$$\delta_i = \frac{V_e + \ell_i\,\omega_z + \dot{\gamma}\,h_{cm}}{V_x \pm 0.5B\,\omega_z} - \alpha_{wi}$$

where l_i is the distance from the HDWV's center of mass to the ith axle.

Longitudinal reactions on R_{xi} are defined at solution of equations of the traction dynamics of HDWV (8.16). Additional force of resistance to motion in longitudinal direction P_{more} at motion along a curvilinear trajectory is:

$$P_{more} = \sum_i^{2n} R_{yi}\,\delta_i\,.$$

The rotation resistance moment of the HDWV's right and left wheels when rolling along a curvilinear trajectory is determined from the following expression:

$$M_{rtwi\,(r,\,\ell)} = \frac{b_w}{3} R_{yi\,(r,\,\ell)} + \frac{2\phi\,R_{zi\,(r,\,\ell)}\,\ell_w}{2\pi\,(1+0.15\,R_{rti\,(r,\,\ell)} b_w)}$$

where b_K is the tire profile width; l_K is the tire print length; and $R_{rti\,(r,\,l)}$ is the rolling trajectory radius of the (right, left) i^{th} wheels.

The value R_{rti} is defined from the following geometric ratios:

$$R_{rti\,(r,\,\ell)} = \frac{R_{rtm} + 0.15B}{\cos(\alpha_{wi\,(r,\,\ell)} + b_i)}$$

where R_{trm} is the trajectory radius made by the turning pole and is calculated from the following formula:

$$R_{rtm} = \frac{L}{tg\left[\alpha_{cw1} + 0.5(\delta_{1\ell} + \delta_{1r})\right] - tg\left[0.5(\alpha_{r,\ell} + \alpha_{r,la}) + 0.5(\delta_{r,\ell} + \delta_{r,la})\right]}$$

where L is the HDWV's base; α_{cw1} is the turning angle of a conventional wheel; δ_{1l}, δ_{1r} are the drifting angles of the first axle's wheels; $\delta_{r,\,l}$, $\delta_{r,la}$, are the turning angles of the last axle's steered wheels; and $\alpha_{r,\,l}$, $\alpha_{r,la}$ are the drifting angles of the last axle's steered wheels.

The SWC's motion along a curvilinear trajectory with a given curvature is ensured by tracking the trajectory of a conventional wheel located in the middle of the ith wheeled axle. The turning angle of this wheel, α_{K1av}, is defined by solving the following differential equation:

$$\frac{d\alpha_{cw1}}{dt} = -K_b\left(\frac{dV_y}{dt} + V_x\omega_x + \ell_{cm}\frac{d\omega_z}{dt} + h_{cm}\frac{d^2\gamma}{dt^2} - V_x^2 K_o\right),$$

where K_b is the coefficient that takes into account the tracking accuracy of a given trajectory by the driver; l_{cm}, h_{cm} are the longitudinal and vertical location coordinates of the driver's working place; and K_o is the current curvature of the motion trajectory.

Depending on the value α_{cw1}, the turning angles of the right and left wheels of all wheeled axles are determined as follows:

$$\alpha_{wi\ (r,\ \ell)} = arctg\left[\frac{(\ell_i + |\ell_n| - X_p)\ tg\alpha_{cw1}}{\ell_i + |\ell_n| - X_p}\right],$$

where X_p is the distance from the last (rear) axle to the pole of the StS scheme in the longitudinal direction; and $|l_n|$ is the distance from the center of mass to the last axle.

The pole of the StS scheme is the projection of the turning pole on the vehicle's longitudinal axle when turning with the maximum possible angles of the steered wheels.

The calculation of all other parameters included in the system of differential equations (8.21) is thoroughly considered in *Scientific Approaches to Justification of Technical Appearance of Undercarriage of Future Multi-Axle Vehicles* [7]. It should be noted that among nontraditional technical solutions proposed for advanced SWCs, the all-wheel StS has been studied more completely and designs and operation algorithms of these StS systems were checked in the course of experimental tests on a full-scale, running model of the 79085 chassis (see Fig. 4.16.) with a 40 t load capacity [5, 7].

The turnability and stability mathematical model is used together with the main traction dynamics mathematical model and the undercarriage load and soft driving mathematical model in the presence of a turn in the current section.

For calculating the vertical loads in the undercarriage and soft driving, it is assumed that the sprung weight is a solid-state object, that the suspension characteristic is piecewise-linear, and that each non-sprung weight (wheels) has one vertical degree of freedom.

The system of differential equations composed according to the design scheme (see Fig. 8.9) has the following form:

$$\begin{cases} M\ddot{z} = \sum_{i=1}^{2n} F_i\ ; \\ M\ddot{Y} = \sum_{i=1}^{2n} y_i\ ; \\ J_{xn}\ddot{\gamma} = 0.5B\left(\sum_{i=1}^{n} F_{i\ell} - \sum_{i=1}^{2} F_{ir}\right) - h_{cm}\sum_{i=1}^{2n} y_i + M\ g\ h_c\ \gamma\ ; \\ J_{yn}\ddot{\phi} = \sum_{i=1}^{n} F_{ifa}\ell_i - \sum_{i=1}^{n} F_{ila}\ell_i\ ; \\ m\ddot{\xi} = R_{zi} - F_i\ , \end{cases} \tag{8.22}$$

where: M, M_g are the SWC's sprung and full weights; J_{xn}, J_{yn} are the inertia moments in the SWC's sprung part; z, y are the vertical and lateral linear coordinates; ϕ, γ are the angular longitudinal and lateral displacement coordinates of the SWC's sprung part; h_c is the height of the SWC's center of mass; ξ is the vertical linear displacements coordinate of the ith wheel; and y_i is the lateral force acting on the SWC. The forces acting in the suspension are defined by the following expression:

$$F_i = F_{pi} + F_{rwi} + F_{fri}, \tag{8.23}$$

where $F_{pi} = C_{pst.w}\,\Delta_i$ is the force determined by the characteristics of the suspension's elastic element (springs, torsion bar, etc.); $F_{rwi} = K_{rw}\dot{\Delta}_i$ is the force formed by the suspension's shock absorber; Δ_i, $\dot{\Delta}_i$ is the deflection and deflection speed in the suspension of the ith wheels; and $C_{st.w}$, K_{rw} are the spring stiffness and resistance coefficient of the shock absorber reduced to the wheel.

According to a well-known expression, the dry friction force is formed as follows:

$$F_{fri} = 0.11\left(\frac{M_g\ g}{2n}\right), \tag{8.24}$$

where n is the number of axles.

The vertical and lateral forces acting in the tires' contact point with the support surface is:

$$R_{zi} = R_{\text{tiri}} + R_{\text{tir.ri}}$$
$$Yi = C_\delta \, \Delta\delta i; \qquad (8.25)$$

where: $R_{\text{tiri}} = C_{\text{tir}} \Delta_{\text{tiri}}$ is the force formed by the tires' elastic properties; $R_{\text{tir.ri}} = K_{\text{r.tir}} \dot{\Delta}_{\text{tiri}}$ is the damping force in the tire; C_{tire}, C_δ are the radial and lateral tire stiffness; $K_{\text{.tire.ri}}$ is the damping coefficient in the tire of the ith wheel; and $\Delta\delta_i$ is the lateral tire deformation in the ith wheels.

In their turn, the force and displacement equations look as follows:

$$\begin{cases} F_i = C_{\text{st.w}} \, \Delta_{\text{st.wi}} + k_{r.w} \, \dot{\Delta}_{\text{rwi}} + 0.11 \dfrac{M_g \, g}{2\text{n}} \; ; \\ R_{zi} = C_{\text{tir}} \, \Delta_{\text{tiri}} + k_{\text{r.tir}} \, \dot{\Delta}_{\text{tiri}} \; ; \\ Y_i = C_\delta \, \Delta_{\delta i} \; ; \\ \Delta_{\text{i}} = \xi_i - z_i \pm 0.5B\,\gamma \pm \ell_i \, \varphi \; ; \\ \Delta_{\text{Hi}} = q_i - \xi_i \; ; \\ \Delta_\delta = h_{st}\gamma - y_i \; , \end{cases} \qquad (8.26)$$

where q_{i} is the current microprofile coordinate under the ith wheel.

It should be noted that the systems of equations (8.22) and (8.26) quite thoroughly and fully describe the undercarriage and soft driving loading concerning SWCs, which have a traditional springing "suspension-tire" system that is either independent torsion or candle type depending on the chassis' considered design sample. However, it is known that such a springing system does not fully correspond to the prospective requirements imposed on SWCs concerning satisfactory provisions of soft driving and vibration loads to the crew and transported equipment. That is why one of the nontraditional solutions for springing systems proposed for prospective SWCs is represented by, for example, a steered, long-stroke, C-shaped aircraft type suspension, the detailed description and mathematical modeling characteristics are considered in subsequent chapters.

The SWC's motion model includes speed limitations for accident prevention and safety of the transported cargo. At each calculation step, the following conditions are checked:

- The loss of passing ability because of the depth of the rut:

  $$h_{rd} < h_{\text{drd}}$$

  where h_{drd} is the allowable depth value of the rut;

- The loss of passing ability because of skidding:

  $$f \cos\alpha_y + \sin\alpha_y < \varphi$$

- The loss of passing ability because of insufficient engine power:

  $$f \cos\alpha_y + \sin\alpha_y < \frac{\sum_{1}^{2n} R_{xi\,(r,\,\ell)} 2n}{g}$$

- A failure to fit into the turn by the turning angle of the setting axle:

  $$\alpha_1 < \alpha_{1\text{rad}}$$

where $\alpha_{1\text{rad}}$ is the allowable turning angle value of the setting axle;

- A failure to fit into the turn by the speed of rotating the steering wheel:

  $$\dot{\alpha}_1 < \dot{\alpha}_{1\text{s}}$$

where $\dot{\alpha}_{1\text{s}}$ is the maximum allowable speed of the steering wheel turn;

- The chassis' lateral slipping at slipping of all axles:

  $$\sqrt{(2R_x)^2 + (R_{yi\,\ell} + R_{yi\,r})^2} < \frac{M_g \, g \, \phi \, \cos\alpha_y \cos\alpha_w}{n}$$

where α_w is the turning angle of the steered wheel;

- A dynamic tipover of the SWC at turning:

$$\frac{\gamma_{st}}{\gamma} < k_s$$

where γ_{st} is the lateral static stability angle; γ is the lateral roll angle in turning or at oscillations; and k_s is the marginal stability coefficient;

- A failure of the chassis to fit into the turn by the dimensional corridor:

$$H_c < H_{c.s}$$

- The lateral acceleration exceeds the allowable (threshold) value:

$$V_y < \dot{V}_{y\,s}$$

- A breakaway of the wheels from of the support surface:

$$\xi_i - q_i < \frac{M_g}{C_{\text{tir}}}$$

- An excess of the allowable lateral roll angle value :

$$\gamma < \gamma_{\text{more}}$$

- An excess of the allowable (threshold) vertical acceleration value:

$$\ddot{z} < \ddot{z}_{\text{more}}$$

- A failure of the chassis' wheel suspension:

$$\Delta_{\text{st.wi}} < \Delta_{\text{st.w.more}}$$

where $\Delta_{\text{st.w.more}}$ is the permissible bend of the suspension.

If at least one condition is not fulfilled, the chassis is forced to brake with deceleration $\dot{V}_x$ to the speed at which all the specified conditions are met.

Thus, the developed mathematical motion model of the chassis when used in conjunction with the route model (conditions) of the SWC's motion allows for determining and evaluating the parameters of a multi-axle chassis, its components, and different design solutions: with diesel engines, with mechanical and hydromechanical gear transmissions, with various StS combinations, and different springing system characteristics.

On the other hand, taking into account that a significant increase in the specifications level of advanced SWCs can only be achieved by using innovative technical solutions, for example, a GTE-generator power unit or another alternative energy source, electrical or hydrostatic transmission, carrying system of the all-wheel StS, and the long-stroke C-shaped suspension, the developed mathematical model can be regarded as a basic tool for studying the efficiency of using (mobility) the chassis.

Subsequent chapters present mathematical models that describe the functioning processes of the above mentioned nontraditional elements. These models, being relevant blocks (see Fig. 8.2), are part of the assessment technique for the efficiency of using SWCs and thus allows for analyzing the excellence of the adopted technical solutions.

8.5 Comparing SWC mobility on different routes

Current competition in the HDWV market increases requirements of performance properties and, in particular, readiness for the start of movement, passing ability, average travel speed on roads, and travel distance from the starting point.

Taking into account that the customer imposes stringent mobility requirements on SWCs, a comparative assessment of the traction-dynamic properties of existing chassis and chassis with advanced requirements was carried out for one version of continuous motion on the selected haulage route. The assessment used the methodology for assessing the SWC's efficiency (mobility) (hereinafter the Methodology) developed on the basis of material presented in previous sections of this chapter.

The assessment criteria include the average travel speed on all types of roads, the the vertical load dynamics coefficient in the undercarriage, and the distribution number of actuated limitations on the travel speed because of an imperfection in any system of the chassis.

The vertical load dynamics coefficient, K_d, characterizes a ratio, well-known among experts, between the real force acting on the undercarriage's motion process and the static vertical load on the wheel. K_d indirectly and quite precisely evaluates the undercarriage (axle formula, type of wheel suspension, wheel mover) and carrying system's perfection. This requires comparing the safety margin taken in calculations (generally, for SWCs, the coefficient is assumed as K_d = 2.0–3.0) with the real maximum value K_d determined on any chassis wheel in the course of its motion.

Analyzing the number of actuated limitations on the travel speed allows for revealing the systems and components that limit the maximum travel speed.

The characteristics of the road surfaces on the route and the list of existing restrictions are presented in Section 8.4. The results of the travel speed calculations and the K_d of the three SWC samples on the representative route are presented in Fig. 8.10. A comparison of the

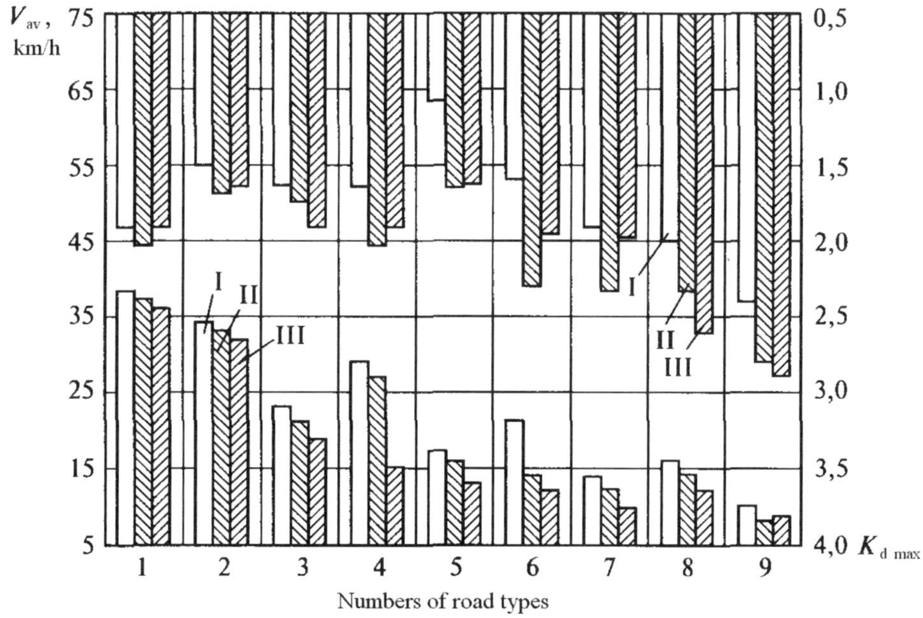

Fig. 8.10. Block diagram of the average travel speed and $K_{d\ max}$ on different roads for SWCs with a load capacity of 50–70 tons or more along a representative route: I—MAZ-547V; II—7917 chassis; III—7922 chassis road types: concrete highway (1—In satisfactory condition; 2—Worn-out); cobblestone highway (3—In satisfactory condition; 4—Worn-out); crushed stone macadam (5—In satisfactory condition; 6—Worn-out); ground road (7—In satisfactory condition; 8—Bumpy road); 9—Off-the-road

calculated data with experimental ones for one sample is presented in Table 8.7.

The main SWC systems that restrict the travel speed are as follows: the undercarriage, the StS, and the transmission.

The longitudinal slopes were assessed in motion by means of a gyroscope attached to the frame of the chassis. The longitudinal gradient of the frame during motion had two components: the gradient caused by oscillations of the chassis on its suspension and the one caused by the road gradient. The separation of the second component from the first was achieved by defining its mathematical expectation for each route section. Thus, we obtained the average values of the longitudinal gradient for each section.

Table 8.7 Average travel speed of SWCs along the representative route obtained by calculations and experimentally

Method of obtaining data	Average speed of SWCs on sections of typical route, km/h								
	1	2	3	4	5	6	7	8	9
Calculations	36.3	28.1	21.7	17.2	17.7	14.4	19.8	11.5	7.8
Experiment	28–34	17-25	19–23	16–20	-	15–19	21–25	11–16	-

The use of a HMT, a StS with controlled wheels of the front axle, a hydromechanical drive, and hydropneumatic suspensions allows for obtaining acceptable mobility and other specifications only in 8×8 SWCs weighing no more than 50 t. This is confirmed by numerous theoretical and experimental studies conducted with our participation and other scientists [5–7].

The studies have established that any further improvements in the traction-dynamic characteristics, stability, handling quality, and undercarriage properties of SWCs with a load capacity of 30–40 t and 70 t or more is only possible on the basis of non-traditional, completely new technical solutions of the HDWV, undercarriage, and StS.

Experimental studies to check the above Methodology were conducted using the prototype of a 4-axle BAZ-69502 model SWC with a gross weight of 30,780 kg on a section of the Moscow-Ryazan highway, on a ground road during autumn impassability, and using the SWC prototype as an equipment component in the Sverdlovsk Region.

The following basic parameters of the routes were defined experimentally: angles of the longitudinal slopes, the curvature of the turns in the horizontal plane, and the surface condition of the route sections (resistance to the chassis' motion).

The position of the chassis' frame on a horizontal surface was taken as a reference point in defining the longitudinal gradient. The lateral angle in this position was measured by an optical angle meter and taken into account in calculations of the route's longitudinal gradient.

The curvature of the turns in the plane view (the inverse value of the turning radius) was defined by measuring the angles of entering and quitting the turn and by the length of the chassis' motion trajectory on the turn. The angles in the plane were measured with the help of the course gyroscope fixed on the chassis' body. The trajectory length was defined by the readings of the calibration rpm sensor on the wheels. The sensor was calibrated on a reference 200-meter-long horizontal section of the ground route.

The chassis' motion resistance was assessed on three straight horizontal route sections by braking the chassis from the maximum possible travel speed under the given road conditions to a complete stop. The arising deceleration numerically corresponds to the motion resistance coefficient.

The deceleration was defined by the wheel rpm sensor at each current moment of the braking time.

The microprofile conditions of the road sections at the time of the experimental studies were assessed by measuring the maximum travels of the suspension with the help of displacement sensors. These travel values can be used later to monitor the road smoothness when comparing individual sections of the route with each other.

The suspension travels were also used to compare the microprofile defined during the experiment with the one calculated according to the Methodology. If the difference between the experimental and calculated suspension travels is essential (over 5–10%), then the amplitude value of the calculated spectral density in the computer-based microprofile should be corrected to eliminate the differences.

As a result of the above, the routes were divided into homogenous sections by the longitudinal gradient values, the curvature of the turns, and the surface condition. The average values of these parameters for each route section were defined.

It can be seen from the data in Table 8.7 that the experimental and theoretical results have a satisfactory match.

The Methodology results obtained by introducing it into the process of analyzing and examining the technical solutions of SWC designs offered by the industry and into the design process at the manufacturing enterprises have enabled us to build a complex evaluation algorithm for the influence of the general design solutions on the SWC's mobility within the complex "road-vehicle-driver" dynamic system.

The complex evaluation algorithm for the technical solutions is presented in Fig. 8.11. Its essence is as follows:

- A study of the external effects during the upcoming operation of SWCs and defining the basic performance properties, selecting a representative travel route, defining the main characteristics of the route, and correcting its mathematical description offered by the Methodology. Based on the defined basic performance properties of the SWC, the parameters of the acting restrictions are updated;
- Based on the design solutions proposed at the preliminary and technical project stages, the complex mathematical motion model is updated, a complex assessment of the influence of these solutions on the average travel speed on the representative route is conducted, and the design solutions that do not allow for achieving the chassis' preset main performance properties are defined;
- An extended and deep numerical study of each design solution that limits the performance properties of any individual chassis system is done by using a particular mathematical model; the rational parameters of the system that ensures achieving the basic performance properties are defined; and the technical capacity to realize the rational parameters of this chassis system is assessed based on the technical economic analysis of the level of the intended production facility; and
- A complex evaluation of the effects of the general design solutions with new parameters, obtained in the course of particular studies, on the SWC's average travel speed along the representative route is conducted and the final version of recommendations and proposals on the technical solutions offered in the Preliminary Design or in the Technical (Detailed) Design are elaborated.

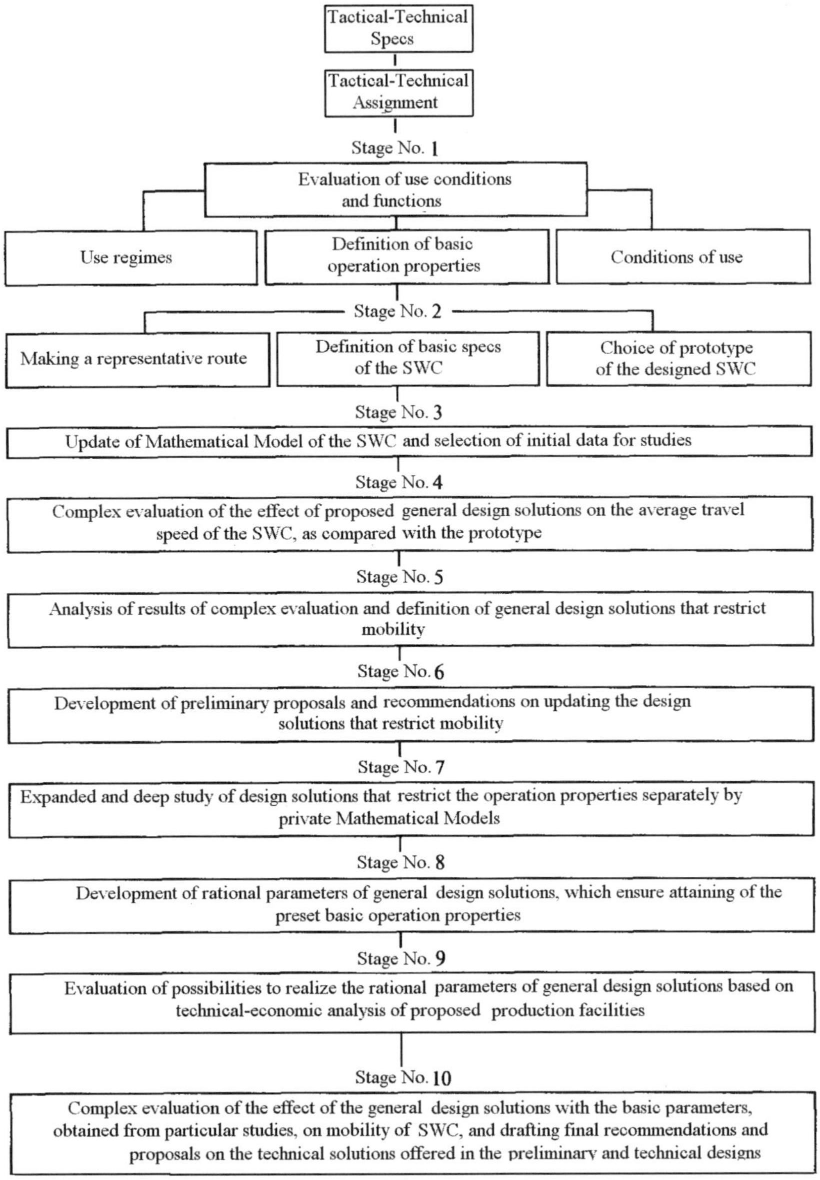

Fig. 8.11. Algorithm of using the assessment methodology of the efficiency (mobility) of the SWC

Part III

Fundamentals of HDWV theory and design

Chapter 9
Basic principles of HDWV general layout

9.1 Trends in overall layout

The general layout assumes defining the rational mutual positions, types, structures, and functions of the vehicle's main units, systems, cabin, and platform, which ensures the vehicle's functionality at the highest efficiency in the given operation conditions and the adaptability of the design to the manufacturing process.

The aim of the overall layout, as a design stage, is to define the main dimensions and the most beneficial mutual arrangement of the units and systems while keeping in mind the required performance properties, i.e., the correct choice of the layout parameters. In general, this is a multi-criteria problem.

Any vehicle design is based on the existing development trends of the vehicles in the given load capacity class. As a rule, the concept of a vehicle is implemented within the overall technical requirements and perspective ranges (series) of vehicle prototypes. This is especially characteristic for mass-produced vehicles.

According to V.V. Vantsevich, M.S.Vysotsky, and L.Kh. Gileles [15], the structure of the concept that reflects the basic principles of the systemic approach, is presented in Fig. 9.1.

For mass-produced vehicles, a huge work is underway to study the market structure and demand trends of vehicle consumers, the wanted range of modifications (Fig. 9.2) in basic models, and the short- and long-term development forecasts (of individual companies, the branch as a whole, etc.).

HDWVs are usually manufactured in small batches or even as individual items, according to particular one-time orders or for particular equipment or loads (cargos) to be transported. Thus, when designing the overall layout, it is important to account for particular (rather narrow) operational and usage conditions, as well as for past experience in designing similar vehicles.

It is well-known and generally adopted that HDWVs can be divided by their purpose into three main groups:

- Multi-axle special wheeled chassis (SWCs) intended for the installation and transportation of particular equipment (see Figs. 4.1 and 4.2);
- Quarry dumpers; and
- Multi-support vehicles (see Fig. 2.1).

Let us consider, as an example, the creation process of a prospective model series of quarry dumpers [11] as the most populous group of HDWVs.

At present, the designs of manufactured quarry dumpers meet, as a whole, the specific requirements of the mining industry and satisfy their numerous consumers as a technological means of transport. According to the overall layout, this group of HDWVs can be categorized into three types:

- Two-axle single-link dumpers intended for operation in open quarries (see Fig. I.1*c*) or in mines and tunnels;

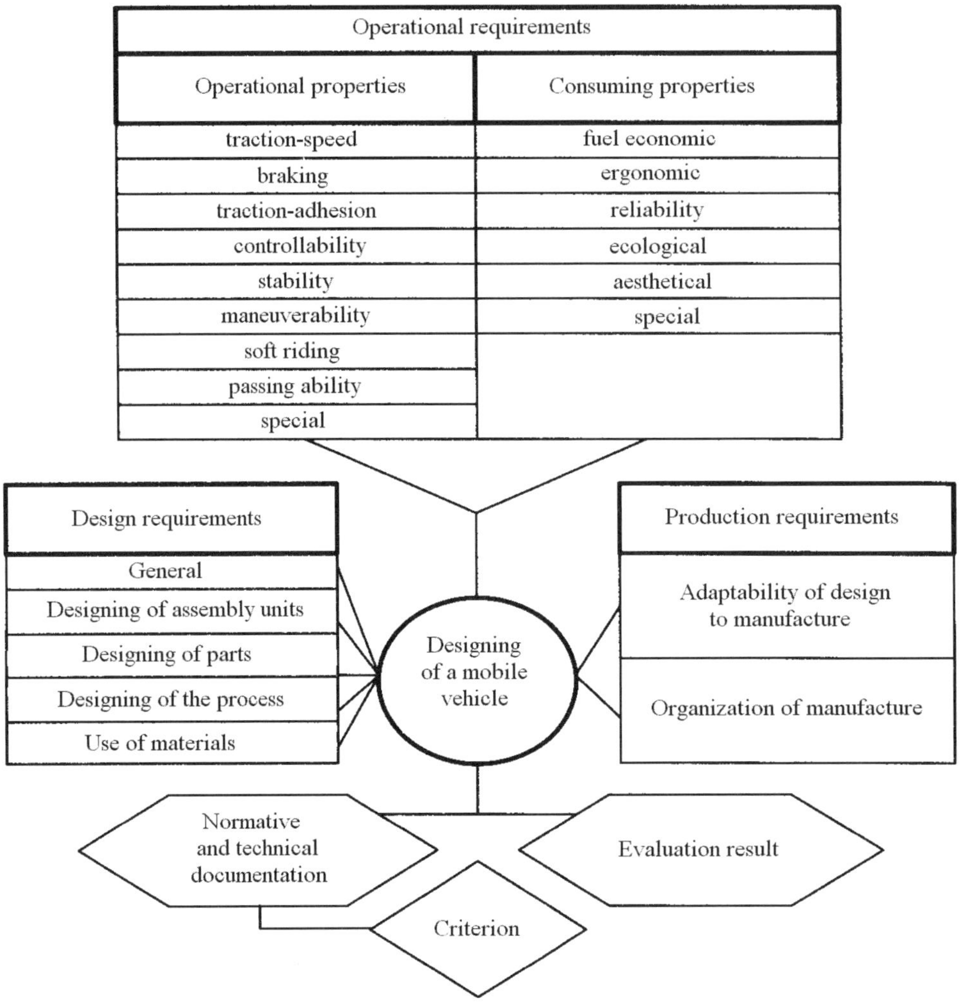

Fig. 9.1. Structural diagram of the process of creating a vehicle

- Three-axle single-link dumpers intended for operation on open sites (Fig. 9.3); and
- Three-axle articulated dumpers intended for broad use (Fig. 9.4).

Mining conditions are growing more and more complicated (quarries grow deeper, transportation loads and distances go up, etc.), while open developments move into areas of extremely cold or, on the contrary, extremely hot desert climates; as a rule, these areas are far away from industrially developed centers and have scant labor resources. This means that in the near future, more advanced extraction and transportation equipment and machinery capable not only of compensating for a certain decrease in performance due to the above problems, but also of raising its efficiency will be needed. In addition to better quarry dumpers that are efficient in new conditions and able to be used as trunk vehicles (for taking the rock from the excavator directly to the stockpile or mineral warehouse), vehicles fit for working under new cyclic-flow and other broadly implemented progressive extraction and transportation technologies are needed. In such cases, developers prefer using various vehicle combinations operated in the most rational way for each particular quarry zone. These systems

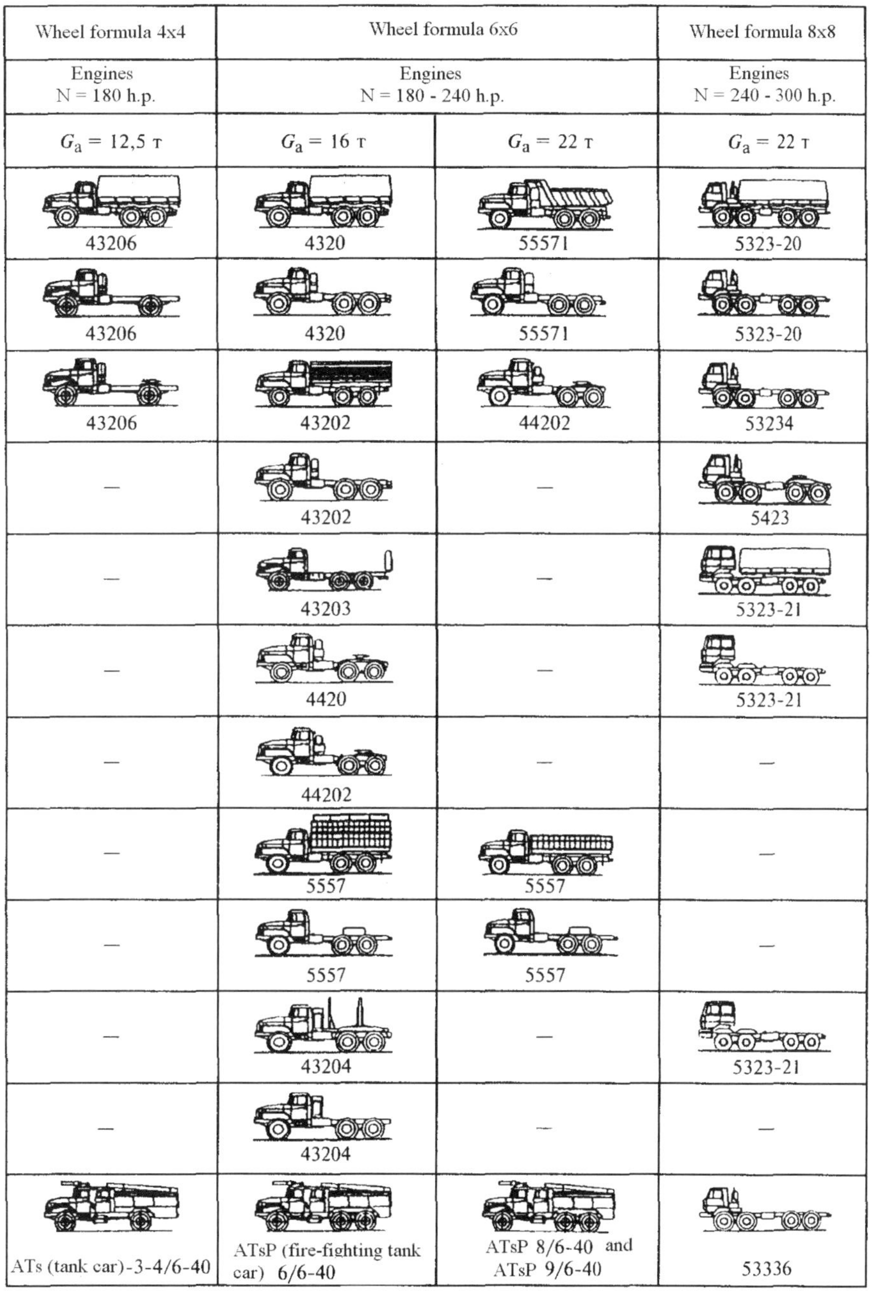

Fig. 9.2. Range (nomenclature) of modifications of vehicles made by the Open Joint-Stock Company (OJSC) "UralAZ"

use the motor transport as an assembly element (unit) in the transportation process, usually working on the bottom and lower levels (tiers) of a deep quarry and transporting the rock from the excavator to the reloading machines placed in the quarry. The route of these dumpers are not longer than 1.5 km, but transportation often goes along poor roads with irregularities, steep slopes,

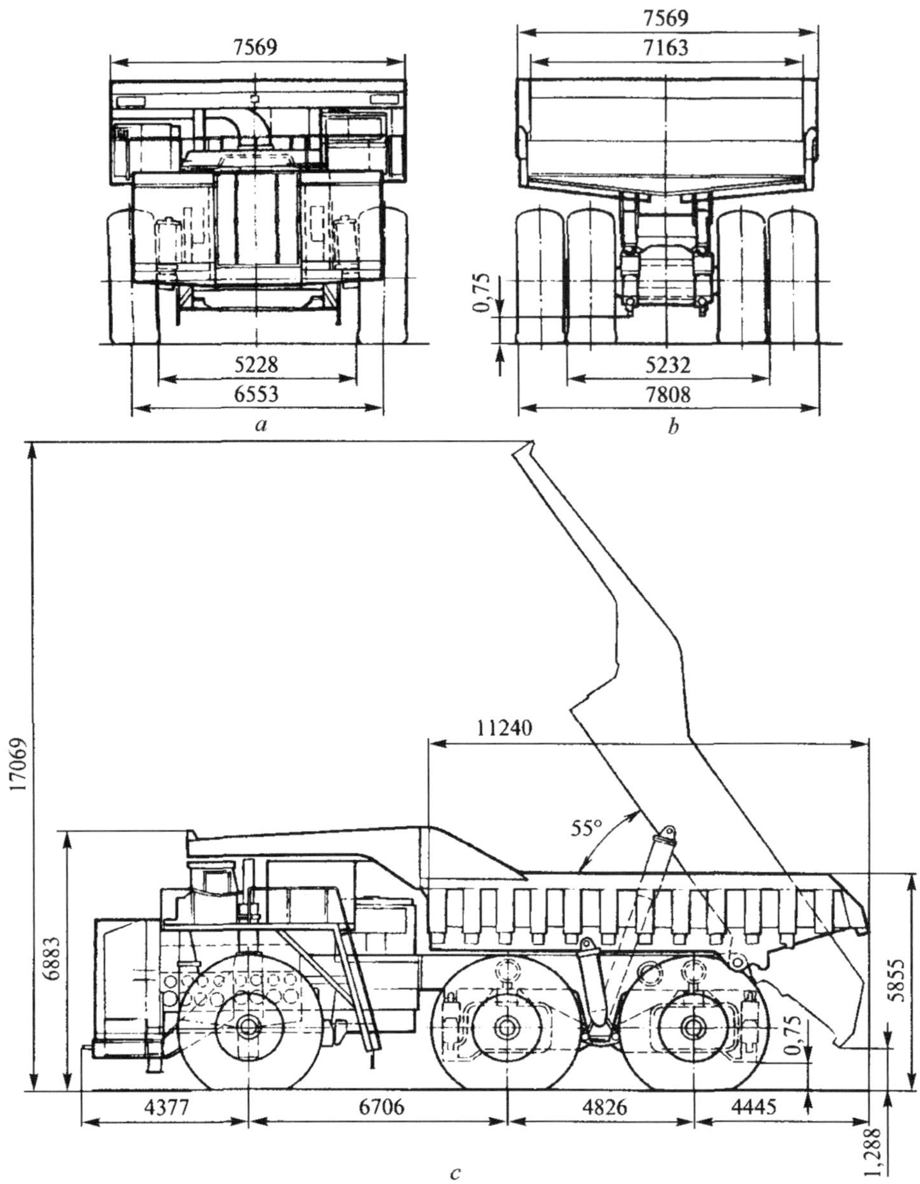

Fig. 9.3. Three-axle dumper: a—Front view; b—Back view; c—Side view

potholes, and are usually covered with dropped rock fragments.

The prototype range (series) (15–20 years ahead) designed with account of the above specificities (Table 9.1) includes two groups of quarry HDWVs: for trunk haulage and for the so-called "assembly" operations and delivery of the rock to the transshipment devices. The parametric series (with a coefficient of 1.5–1.7) is preserved here and covers six load capacity classes from 30 to 280 t (the dumper with a 250–280 t load capacity was excluded from Group 2 since the chances to use it in mines are limited).

The first four members of the parametric series are rational for this HDWV group and already have the traditional layout and wheel formula (4×2) for

Table 9.1 Prototypes of quarry dumpers

Parameter	Dumper load capacity class, t					
	30	50	80	120	180	250–280
Basic models						
Full weight, T	51	85	140	210	315	440–490
Wheel formula	4×2	4×2	4×2	4×2	4×4	4×4
Minimum standard mileage of main units till major repairs, thousand km	300	300	350	350	400	400
Dumpers for trunk quarry haulage (Group 1)						
Engine power, kW (hp)	346 (470)	562 (765)	948 (1290)	1264 (1720)	1900 (2580)	2940 (4000)
Specific power, kW/t	6.76	6.62	6.76	6.03	6.03	6.03–6.6
Maximum speed, km/h	50	50	50	50	45	40
Dumpers for assembly operations (Group 2)						
Engine power, kW (hp)	264 (360)	437 (595)	735 (1000)	958 (1300)	1470 (2000)	-
Specific power, kW/t	5.15	5.15	5.15	4.78	4.56	-
Maximum speed, km/h	40	40	40	40	40	-

Fig. 9.4. Articulated dumper

the Belarusian Automobile Works (BelAZ); they run on tubeless, radial, metal-cord tires sized 18.00–25, 24.00–35, 27.00–49, 33.00–52, and 40.00–57 accordingly (Table 9.2). The biggest dumper with a 250–280 t load capacity is an articulated 4×4 vehicle with twin wheels in the front and rear axles (tire 40.00–57). This layout (4×4) was used deliberately, since when using the 4×2 formula, we should have needed new and much bigger tires.

Accounting for the fact that the travel time of a loaded dumper uphill (ascend) in deep quarries is over 60% of the overall quarry cycle, the Group 1 series—used for trunk haulage (Table 9.3)—is based on engines with an increased power of 25–35% as compared to earlier dumpers. This allows for increasing their speed on long ascends at a slope of 8% up to 15.5–18.0 km/h and decreasing the time needed to get out of a 300 m deep quarry down to the time previously needed for a 200 m deep quarry.

It also seems very promising to use a part of the vehicle's power to overcome an increased—at least by 10%—limiting (maximum) gradient of quarry routes (which will result in a respective decrease in the route length), due to which the performance of the new dumpers at travel speeds of 13–15 km/h will improve by 23% as compared to today's BelAZ dumpers. In this case, the stripping volume can be essentially

Table 9.2 Specifications of tires

Parameter	Tire size (type)				
	18.00-25	24.00-35	27.00-49	33.00-51	40.00-57
Admissible load on tire, kg, at a maximum travel speed of 50 km/h:					
according to NTD	8650	14,800	24,600	32,500	50,500
according to Michelin	9250	18,500	27,250	38,750	60,000
Full weight of the vehicle, t, at equal load on tires:					
according to NTD	51.9	88.8	147.6	195.0	303
according to Michelin	55.5	111.0	163.5	232.5	360

Note: NTD – normative and technical documentation.

Table 9.3 Specifications of Group 1 dumpers

Parameter	Load capacity class, t					
	30	50	80	120	180	250-280
Weight, t:						
unladen vehicle	21	35	60	90	135	190-210
full weight	51	86	140	210	315	440-490
empty-weight-to-carrying-capacity ratio (area utilization)	0.7	0.7	0.75	0.75	0.75	0.76-0.75
Engine power:						
nominal, kW (hp)	346 (470)	563 (765)	949 (1290)	1266 (1720)	1899 (2580)	2944 (4000)
specific, kW/t (hp/t)	6.78 (9.21)	6.62 (9.0)	6.78 (9.22)	6.02 (8.18)	6.02 (8.19)	6.03-6.62
Transmission	HMT	HMT	ET	ET	ET	ET
Wheel formula	4×2	4×2	4×2	4×2	4×2	4×4
Tires	18.00-25	24.00-35	27.00-49	33.00-51	40.00-57	40.00-57

reduced (by 3.4–6.4 mln m^3), since under the increased slope in the quarry road, the pit itself can be made smaller.

The engines needed for Group 1 dumpers (Table 9.4) can be created by forcing (boosting) diesel engines: for 30 and 50 t load capacity classes, the 140×140 family of the Yaroslavl Engine Plant (YaMZ) can be used; for 80, 120, and 180 t, the ChN21/21 family of the Urals Turbine Engine Plant can be used; and for 250–280 t, the ChN 26/26 family of the Kolomna Locomotive Plant can be used. The specific power needed for these HDWVs—6.03–6.77 kW/t (8.2–9.2 hp/t)—can be ensured through the intermediate cooling of the supercharging air.

It is assumed that Group 2 dumpers—for assembly operations—will be equipped with smaller power units (a specific power of 4.78–5.52 kW/t, or 6.5–7.5 hp/t), i.e., with engines having a boosting level close to current serial

Table 9.4 Specifications of Group 1 engines

Parameter	Load capacity class of dumpers, t					
	30	50	80	120	180	250–280
Type*	8H	12H	6H	8H	12H	18H
Working volume, cm^3	1722	2564	4361	5815	8723	22080
Power, kW (hp):						
nominal	346 (470)	562 (765)	948 (1290)	1265 (1720)	1897 (2580)	2941 (4000)
in liters	20 (27.3)	21.9 (29.8)	21.9 (29.8)	21.8 (29.6)	21.8 (29.6)	17.8 (24.2)
cylinder	43 (58.5)	46.9 (63.7)	158 (215)	158 (215)	158 (215)	158 (215)
Cylinder dimension, mm	140×140	140×140	210×210	210×210	210×210	210×210
Average efficient pressure, MPa	1.148	1.148	1.74	1.74	1.74	1.6
Weight, kg	1300	1900	-	5800	8600	-
Specific weight, kg/kW (kg/hp)	3.76 (2.77)	3.37 (2.48)	-	4.57 (3.37)	4.52 (3.33)	-
Operating time till repairs, thousand motor-hours	10	10	16	16	16	24

*All engines have turbo-supercharge and intermediate cooling of the supercharging air.

BelAZ dumpers. To reduce the weight of these vehicles, it is expedient to use engines of the respective families from Group 1 with a lower cylinder count at the same liter capacity (18.4–22 kW/l, or 25–30 hp/l).

A new generation of traction electric equipment with AC synchronous generators and serially connected TEMs is used in dumpers of both groups. Both groups have models with engines working under the gas-diesel process.

Groups 1 and 2 differ in transmissions. The former assumes using along with the ET, a hydromechanical transmission with a broad gearbox range and automatic microprocessor control in the 80 and 120 t load capacity classes; while Group 2 uses transmissions that increase the traction moment on the driving wheels due to the decrease in the maximum travel speed down to 40 km/h.

All of the advanced models are traditionally equipped with fairly up-to-date aggregates for this type of HDWVs: high pressure pneumohydraulic suspensions, hydraulic dumping mechanism hoist, etc.; their bodies, frames, and other bearing elements are made of thermally strengthened sheet steel grades.

In order to achieve a rational utilization of the load capacity, it is proposed to create, for each class, three standard body sizes for transporting the overburden, the metallic and nonmetallic minerals, and coal.

New dumpers have progressive weight parameters. Thus, the empty-weight-to-carrying-capacity ratio (area utilization) of models with HMTs exceed 0.7, those with ETs exceed 0.75, and the 4×4 250–280 t models exceed 0.78–0.80.

The designs of diesel-generator aggregates of both series should have a united regulation of the engine and generator, microprocessor control systems and devices allowing for optimizing the work of other aggregates (brakes, StSs, and suspensions), built-in diagnostics systems, and modern information devices.

Special attention should be given to traffic safety; therefore, it is proposed to duplicate all the control system channels that affect safety and to equip all vehicles with devices aimed at improving visibility.

The solution of another major problem—reducing the negative impact on the environment—will be promoted by optimizing the diesel and transmission control, and, naturally, by equipping vehicles with neutralizers of exhaust gases, in particular, with three-component neutralizers with exhaust carbon filters.

The family of BelAZ dumpers assumes using the 6×4 wheel formula, though it was recognized as less efficient than the 4×4 formula. However, the 6×6 formula is globally used both in single-link and articulated models (Figs. 9.3 and 9.4).

An example of a single-link model is the Tereke three-axle quarry dumper made by General Motors (USA) and equipped with a 350 t load capacity diesel-electrical ETU (see Fig. 9.3). The dumper has 23,110 kg road weight, a 3,175,140 N maximum payload capacity, and can travel at a maximum speed of 45 km/h. When dumping, the maximum inclination of the platform is 55°, the lifting time is 28 s, and the lowering time is 30 s. The dumper is also equipped with a 16-cylinder V-shaped 16-645E4 diesel engine made by General Motors with turbo-supercharging and direct fuel injection. The engine cylinders' camber angle is 45°, the cylinder dimensions are 230×254 mm, and the working volume of a cylinder is 169.15 liters. The engine has a rated power of 2429 kW (3300 hp) at a rotation rate (frequency) of 1900 min^{-1}. The AR10 model AC generator with a built-in three-phase rectifier is driven by the engine. The dumper's driving wheels are rotated by D79CF model batch-manufactured electric motors built into each wheel. Both the traction motor and the cylinder are air-cooled.

The power aggregate, together with the cooling system and all auxiliary equipment, is mounted on a sub-frame and forms a power module. This module is located in an encapsulated compacted hood space where an overpressure of pre-filtered air is maintained which reduces the ambient noise level. Two six-section radiators, each cooled by a separate automatically started propeller fan, are used to cool the engine.

As a result of using the ET, the powertrain design is very simple. In order to improve maneuverability and reduce the turning radius, the rear axle wheels are steered. The wheel suspension is hydraulic and has a device to automatically level the rear axle load.

The independent front suspension, used as a guiding mechanism, has a turning swivel pin in combination with a shaking candle-type strut (post). A pneumatic rubber buffer (pillow) serves as an elastic element.

The two rear axles of the dumper are driven and the size 40.00–57 tires withstand a nominal load of 504 kN at a 4.9 kg/cm^2 air pressure. Each driving axle has two tires.

Since late 1970s, the articulated configuration has become popular in many countries, especially in dumpers (see Fig. 9.4). The load capacity of articulated dumpers is within 24–55 tons. As a rule, they have an articulated joint between the links with two degrees of freedom: in the horizontal (plan view) and vertical-lateral (displacement around the longitudinal axis of the vehicle) planes. The curvilinear motion is controlled based on the principle of forced folding of the links (jackknifing) in the plane view relative to each other by means of hydraulic cylinders.

Articulated vehicles are divided into two types based on their configurations—trailed and saddled. Trailed type vehicles (Fig. 9.5) can have 1-3 degrees of freedom in the link joint. In saddle type vehicles, the links are joined by a common

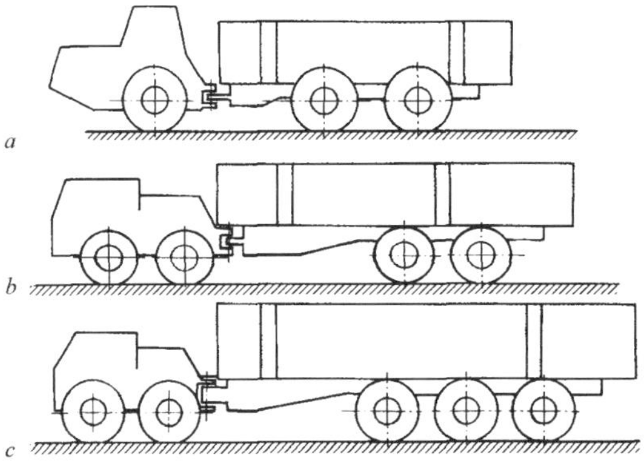

Fig. 9.5. Layouts of an articulated HDWV with a cargo link corresponding in dimensions and load capacity to the length and weight of the cargo: a—6×6 vehicle; b—8×8 vehicle; c—10×10 vehicle

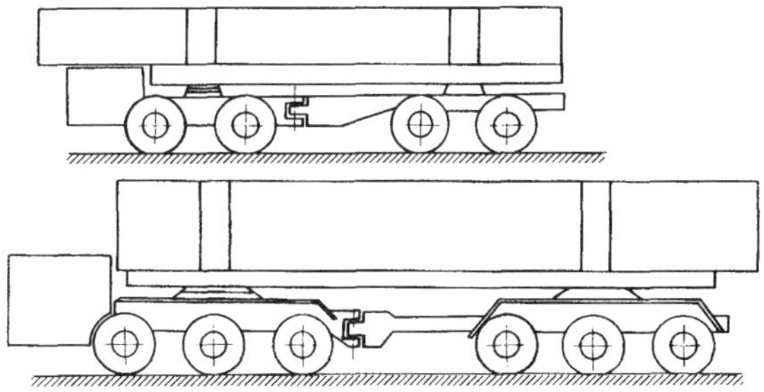

Fig. 9.6. Layouts of a saddle-type articulated vehicle with a common cargo platform for accommodating a monocargo: a—8×8 vehicle; b—12×12 vehicle

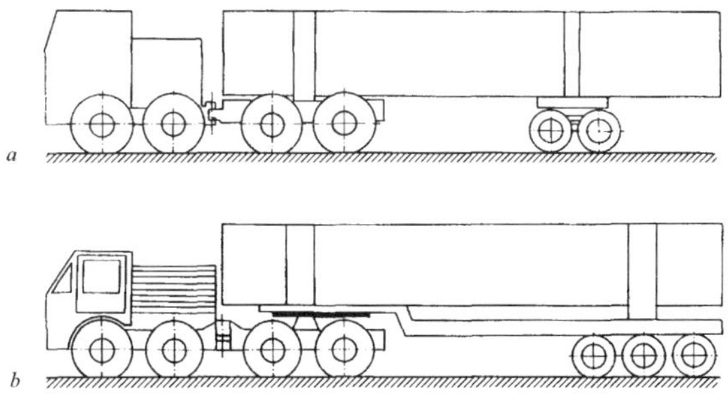

Fig. 9.7. Layouts of an articulated vehicle built as a road train: a—With a bolster-type (expandable) trailer; b—With a semi-trailer

cargo platform connected to the links by pivoting devices similar to truck bolsters used in road trains where tow cars are linked with semi-trailers (Figs. 9.6 and 9.7). The turning in both types is performed, as a rule, in a "forced" way by using a system of hydraulic cylinders, which turn the links in the horizontal plane (in plan view) up to 45° on each side. The pivoting joints can allow a relative displacement of links in the longitudinal- and lateral-vertical planes to an angle of up to 30°.

The layout configuration of the multi-axle single-link HDWV was developed in the 1960–70s and typical layouts are presented in Figs. 9.8, 9.9, 4.6, and 4.7. The overall layouts of heavy-duty vehicles with mounted equipment (e.g., crane trucks) are presented in Figs. 9.10–9.12.

Multi-support HDWVs manufactured by foreign companies can be divided by their purpose into five main subgroups:

- Those intended for use at metallurgy mills;
- Those intended for ship-building yards;
- Those intended for long-distant haulage;
- Those intended for transporting oversized metal structures and industrial installations; and
- Standard modular vehicles.

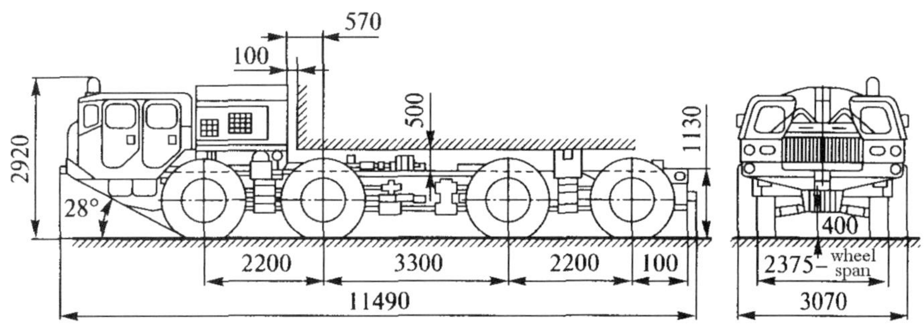

Fig. 9.8. Layout scheme of the MAZ-543 model chassis

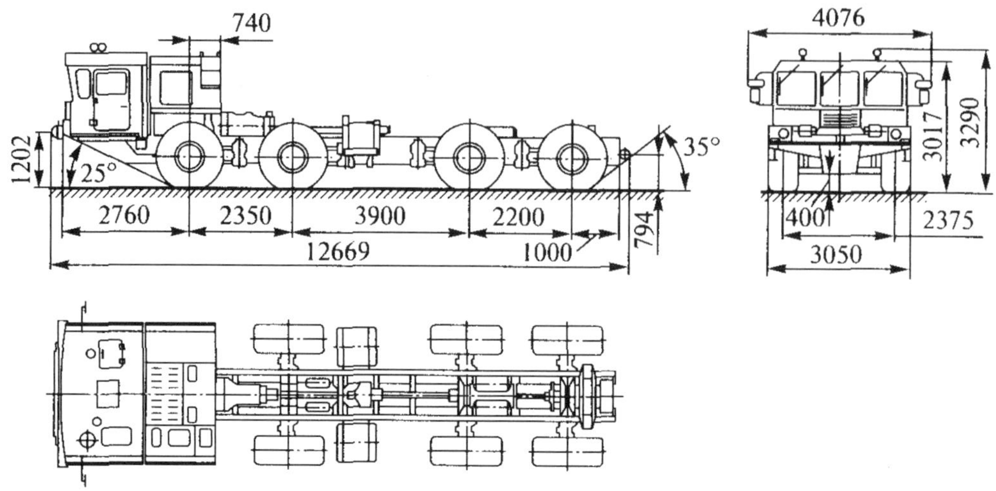

Fig. 9.9. Layout scheme of a new generation of 8×8 chassis

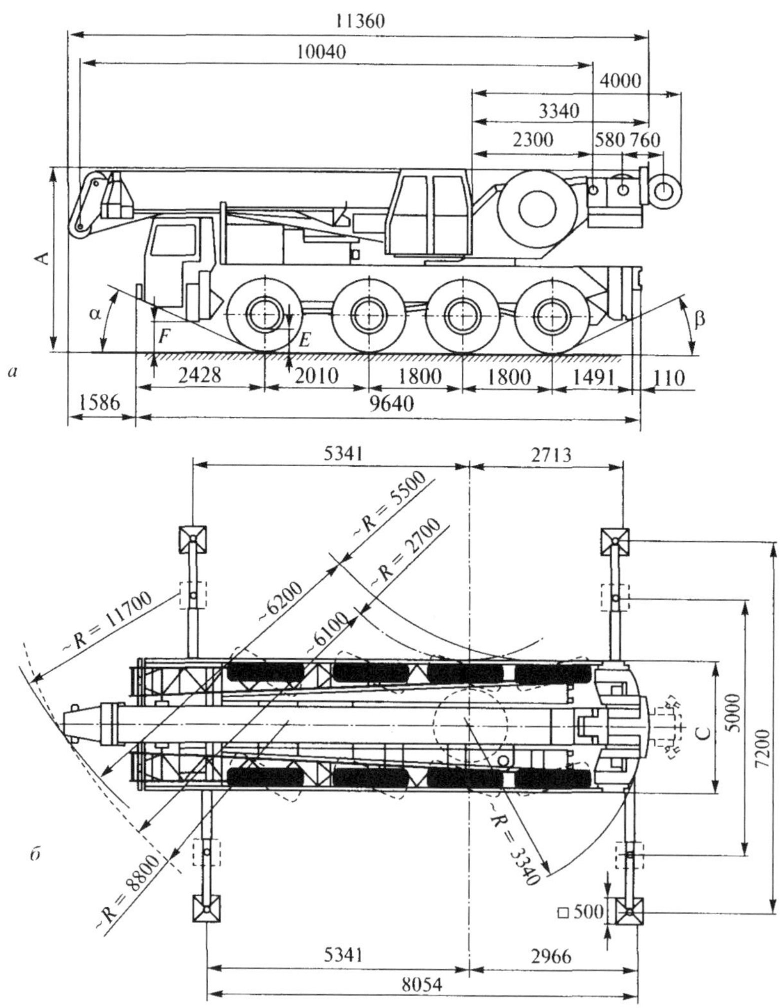

Fig. 9.10. Layout scheme of a crane truck on an 8×8 chassis with a 40 t gross (full) weight: a—Side view; b—Plan view. 16.00R25 tires; A = 3670 mm; C = 2700 mm; E = 330 mm; F = 460 mm; α = 25°; and β = 25°

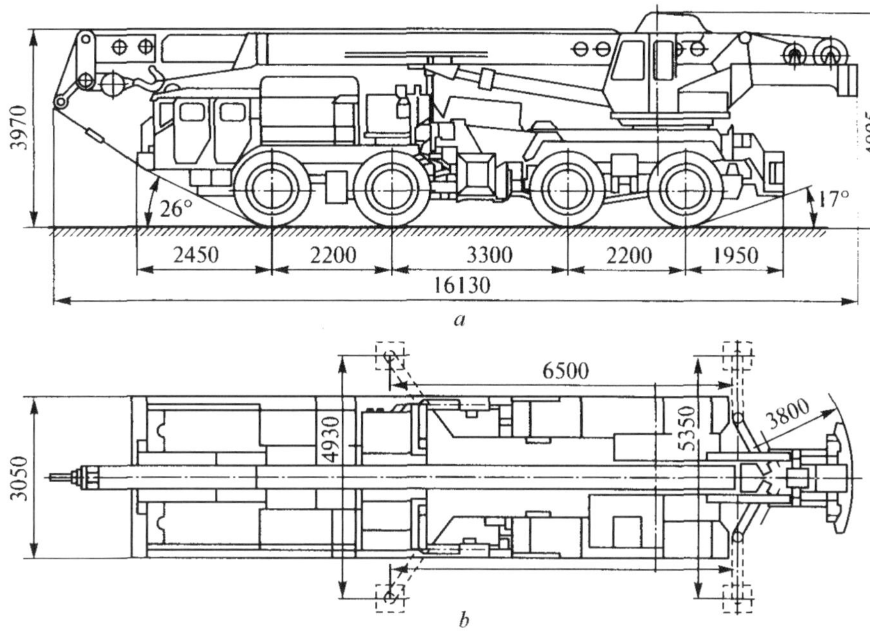

Fig. 9.11. Layout scheme of a crane truck based on the MAZ-73131 model SWC with a 43 t gross weight: a—Side view; b—Plan view

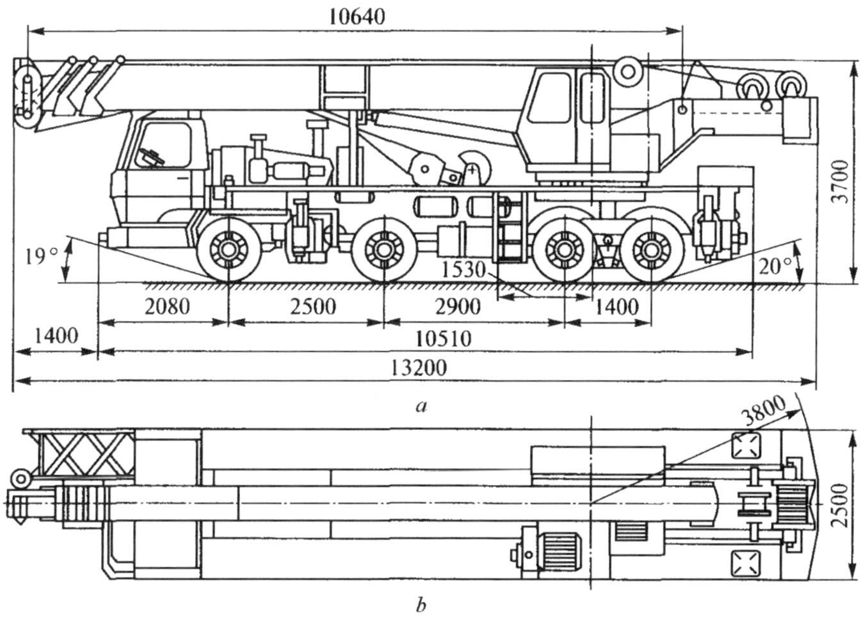

Fig. 9.12. Layout scheme of a crane truck based on the MAZ-6923 chassis with a 39 t gross weight: a—Side view; b—Plan view

All multi-support HDWVs are subdivided by their design solutions into semi-trailers working jointly with saddle-type vehicles, modular-type trailers working with ballast tow cars, and single-link multi-support transport vehicles.

By the load capacity of their individual modules, HDWVs form a unified series from 40 to 200 tons or more. They can also be also divided into two large groups. The first one includes HDWVs whose axial loads are within current legislative restrictions; therefore, they can be used on general-purpose roads (see Fig. 2.1). The second group includes HDWVs which exceed the restrictions and can only work off-road or on general-purpose roads with special permits.

HDWVs are manufactured mostly under individual orders at small factories by using flexible technologies. Their layouts are fairly diverse and all decisions on a particular vehicle are mainly defined by the nature of transportation and the characteristics of the cargo to be transported. The Cometto wheeled vehicle is an example (Fig. I.1*a*) and its overall layout is presented in Fig. 9.13.

Although companies abroad use different approaches to the design choice and layout options of particular units and aggregates, they all stick to a common approach to designing the vehicle as a whole, which is dictated by the transportation purpose. All multi-support vehicles are manufactured under the modular principle, which allows for configuring the vehicle for the required load capacity and overall dimensions.

Thus, the companies Nicolas, Cometto, and Kamag use the principle of a mono-block multi-support vehicle, whereas Scheuerle places traction hydraulic motors uniformly in each module of an articulated vehicle, while the control unit together with the operator's cabin is suspended at the front of the first module and the power unit is placed at the back of the last module. The support platform (see Fig. 2.2) is the base of the module, i.e., it is a lower level module. This achieves an absolute standardization in the series of manufactured vehicles.

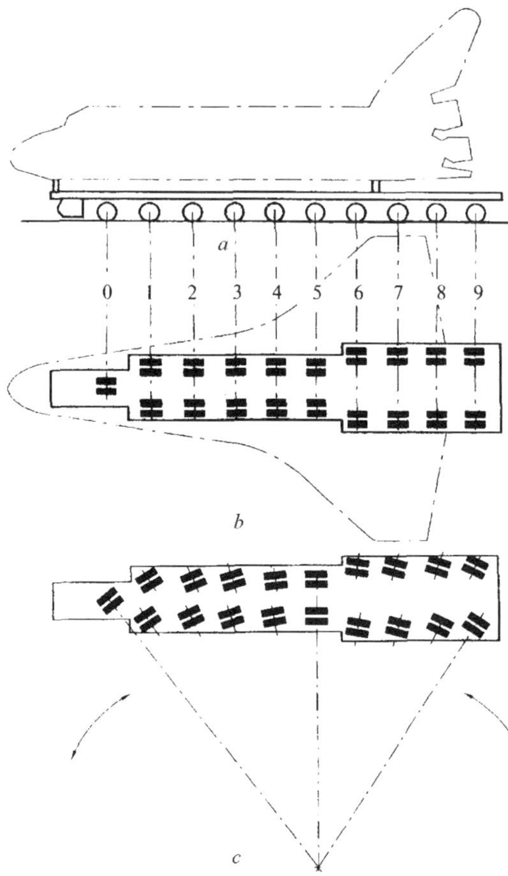

Fig. 9.13. Layout scheme of a multi-support vehicle for transporting an aircraft: a—Side view; b—Plan view; c—Turning diagram of steered wheels

When building heavy-duty transporters with large overall dimensions, the individual modules can be rigidly joined end to end to form a transporter with an integral bearing platform. It is also possible to place individual modules, or groups of modules with gaps between them, with the corresponding hydraulic loop connections of the suspensions and hydraulic systems of the power units. The gaps between modules or groups of modules can vary within a broad range with a definite assembly pitch in both the longitudinal and lateral directions. The value of this pitch depends on the software downloaded into the electronic system which controls the turning of the support bogies and mostly makes 0.1 m.

When assembling a multi-modular transporter, Cometto uses a definite number of active and passive (with no driving wheels) modules.

The active all-wheel-drive modules, each with 12 pneumatic-wheeled support bogies, are grouped into six axle rows and designated as A; the passive modules, also with 12 basic support bogies (as in Fig. 2.2), are designated as D.

The A group can be used either independently or in connection with the D modules. When assembling longitudinal lines of modules, each line should have at least one all-wheel-drive module. When maneuvering the HDWV, the turning (see Fig. 2.16) of each support bogie is controlled by electronic servo systems operated by computers mounted on each all-wheel-drive A module.

The multi-support A modules can form seven main configuration variants in which the turning of the wheels is controlled electronically (see Fig. 2.16). For this purpose, they use special control software stored in the computer memory for each layout version. Additional programs are available for layout schemes with both A and D modules, which are also stored in the computer memory of the A modules.

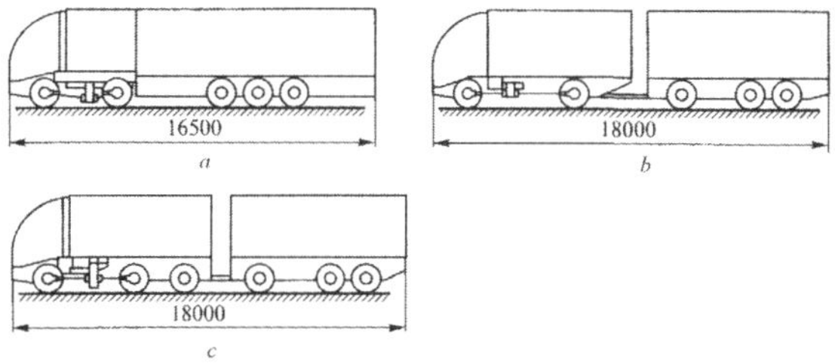

Fig. 9.14. Layouts of line-haul road trains with all-wheel-drive tow cars: a—Saddle type with a gross weight of 38–44 t; b and c—Trailer type with a gross weight of 38–44 and 50–52 t respectively

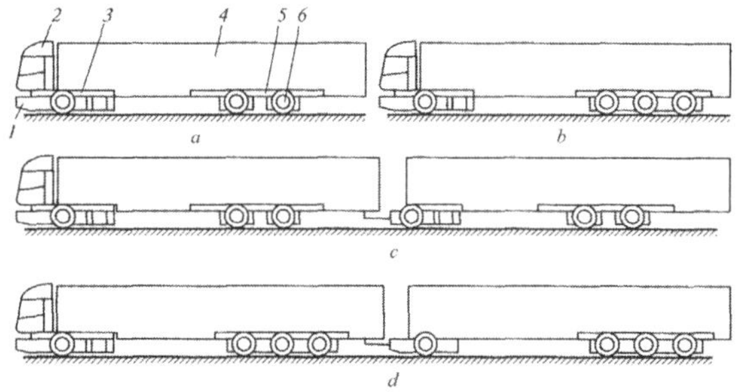

Fig. 9.15. Basic layouts of modular road trains: a and b—Composed of a single-axle tow car and a semi-trailer with a gross weight of 31 and 41 t respectively; c and d—A multi-link road train with a gross weight of 62 and 82 t respectively; typical modules: 1—Traction module; 2—Cabin; 3—Upper frame; 4—Cargo module; 5—Frame; 6—Pull-out bogie and driven axle

According to M.S. Vysotsky, L.Kh. Gileles, and S.G Khersonsky [4], essentially new layout schemes of line-haul road trains presented in Figs. 9.14–9.16 (see the general view in Fig. I.2*d*) are quite promising. Fig. 9.14 presents layout schemes of road trains with all-wheel-drive tow cars. At the same adhesive force ratio, compared to 6×4 tow cars, an all-wheel-drive 4×4 tow car has only four single-tire wheels or six twin-tire wheels instead of six and ten wheels, respectively, which reduces rolling resistance, fuel consumption, road weight, and operational expenses of the vehicle.

The "cabin on top of the engine" configuration complicates the possibility of designing 4×4 tow cars under the traditional layout because of the degraded lateral stability of the road train and a reduction in the useful volume of the cabin. That is why in 4×4 layout schemes (see Figs. I.2*d* and 9.14) the engine is placed horizontally between the axles. It also ensures the following:

- An approximately equal distribution of axial loads from both the road weight and the gross weight of the vehicle (with single-tire wheels);
- A decrease in the vertical coordinate of the center of mass because of a lower location of the engine;
- A simplified cabin design;

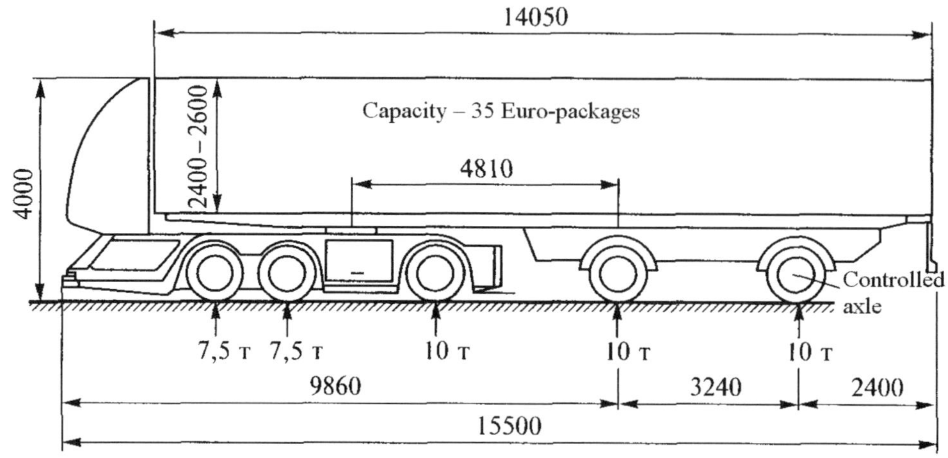

Fig. 9.16. Road train layout diagram

- A reduction in the internal noise level and an improved microclimate in the cabin; and
- Improved ergonomic indicators.

The almost equal distribution of axial loads also allows for using single wide-section tires with lower rolling resistance as compared to twin tires. The gross weight of the tow car can be increased by creating a three-axle vehicle by adding a third driving or driven axle (see Fig. 9.14*c*).

However, a disadvantage in placing the engine between axles is that it creates complicated access to it for maintenance.

An essentially new trend in line-haul road trains is based on using the modular principle in one of the MAZ models. The basic layout diagrams of modular road trains consisting of individual typified modules are shown in Fig. 9.15 and one implementation can also be found in Fig. I.2*d*.

Traction module *1* is comprised of an engine integrated with a gearbox, an articulated driving axle, and an independent pneumatic suspension of the driving wheels.

The upper frame is equipped with a coupling and fixing device intended for connecting a semi-trailer and forms the installation base for the cabin, fuel tank, power cylinders of the steering drive, and a swivel block.

The cargo module is a cargo platform with the bearing frame and is a replaceable body. A tank, a shipping container, etc., can also act as a cargo module.

Modular road trains have the following layout characteristics:

- The articulated modules can form a road train composed of a single-axle tow car (modules *1*, *2*, and *3*) with a traction module located under the cabin and a semi-trailer (modules *4*, *5*, and *6*);
- The articulated modules—the cabin, the upper frame, and the semi-trailer—form a united rigid system, which makes the road train behave as a front-driven vehicle;
- There is the possibility to distribute the traction modules along the whole length of the road train, including the trailing links; and
- There is the possibility to uniformly distribute the axial loads and use wide-section tires.

The main advantages of modular road trains as compared to traditional ones are as follows:

- It is possible to make a broad range of vehicles with a load capacity of 18–55 t for various needs on the basis of a small number of standardized modular links;

- It is possible to assemble—by selecting the corresponding number of modules (traction, cargo, and driving axles)—road trains highly adapted to particular haulage conditions with account of the required load capacity, cargo volume, and traction-and-speed properties;
- An increased load capacity of the road train as a result of placing the maximum permissible axial load onto the first steered axle;
- An increased cargo capacity because of the absence of a gap between the cabin and the front hull side of the platform; and
- A decrease in the fuel consumption as a result of reducing the aerodynamic resistance of the road train and replacing the twin tires with wide-section ones.

The drawbacks of modular road trains include the complexity of the control systems, the need to coordinate the work of several traction modules of a multi-link road train, and complicated access to the units, aggregates, and systems of the traction module.

9.2 Fundamentals of modular designing

The experience of the domestic (Soviet and Russian) and foreign motor industry has shown that, given the frequent turnover of manufactured articles dictated by the free market, standardization, unification, and aggregation (modularity principles or designing from building blocks) are the key conditions for reducing the lead time of new products and for applying the most efficient and economic manufacturing methods. As applied to transport vehicles, and to wheeled vehicles in particular, the modular design principle is an apt solution. This principle is key when creating families of dual-application vehicles.

A comparative analysis of the variants in different automobile technical systems and components when standardizing the modular automobile chassis is based—by analogy with radioelectronic equipment—on the general systems research theory. In general, such analysis is performed in three stages: substantial study, formalization, and quantification.

A substantial study assumes a general description of the vehicle and its needs; it also gives a qualitative description of the interrelation between the chassis modules.

Formalization assumes a schematic and mathematical description of a vehicle and the interrelation of its modules. The main formalization method is simulation, which can be illustrative, descriptive, quantitative, or qualitative.

Quantification means a quantitative expression of qualitative attributes. This is the most complex stage, consisting of establishing the requirements for the quality of modules, proceeding from the quality requirements for the HDWV (direct problem), or establishing the impact of the quality module indicators on the corresponding chassis specifications (inverse problem).

This book makes an attempt to consistently cover all three stages.

The term "unification" (from Latin *unio* – unity, *facere* – do) is interpreted in this chapter as the concept of limited variety. This is mostly justified in relation to HDWV groups such as SWCs and dumpers.

A correct description of the unification area is a necessary condition for the correct formulation of the problem of modular designing. If $n > 1$ ($n = 2, 3, \ldots$), the unification is an intersystem one; if $n = 1$, the unification is systemic, where n is the number of considered systems. Based on these assumptions, as they apply to SWCs and quarry dumpers and at the present economic development stage of the country, the unification, which defines the modular design, should be a systemic one. In other words, the intra-factory unification offers broad opportunities both for installing various equipment for SWCs and for the use of both HDWV groups to their purpose.

At present, three main trends in the world practice of modular design for different products can be distinguished:

- Using functional modules of relatively low complexity;
- Assembling prototypes from large, functionally complete units and aggregates; and
- Applying the modular principle at all levels of the structural hierarchy (modules of higher hierarchical levels are assembled from modules of lower levels).

As noted in Chapter 5, in our opinion, the hierarchical method of building systems is the most flexible and free from drawbacks. It assumes that the generic replacement elements are the modules of the first level, which, in their turn, are assembled to create larger modules of the second level, etc. Actually, the first level modules are the "building blocks" for creating modules of higher levels and of systems as a whole. All modules should have a uniform network of geometrical and connecting dimensions.

Under this design approach, when a product is a complex aggregate in which each function is performed by a separate module, the task of designing a product as a whole is subject to decomposition. This allows for paying more attention to the solution of each design subtask since the work is distributed among groups of designers and opens opportunities for modifying and modernizing the modular system at low expenditures of time and resources. A possible diagram of the hierarchical structure of modular designing is presented in Fig. 9.17 and its implementation in Figs. 9.18–9.20.

If a vehicle is modernized with the aim of improving its parameters and increasing its

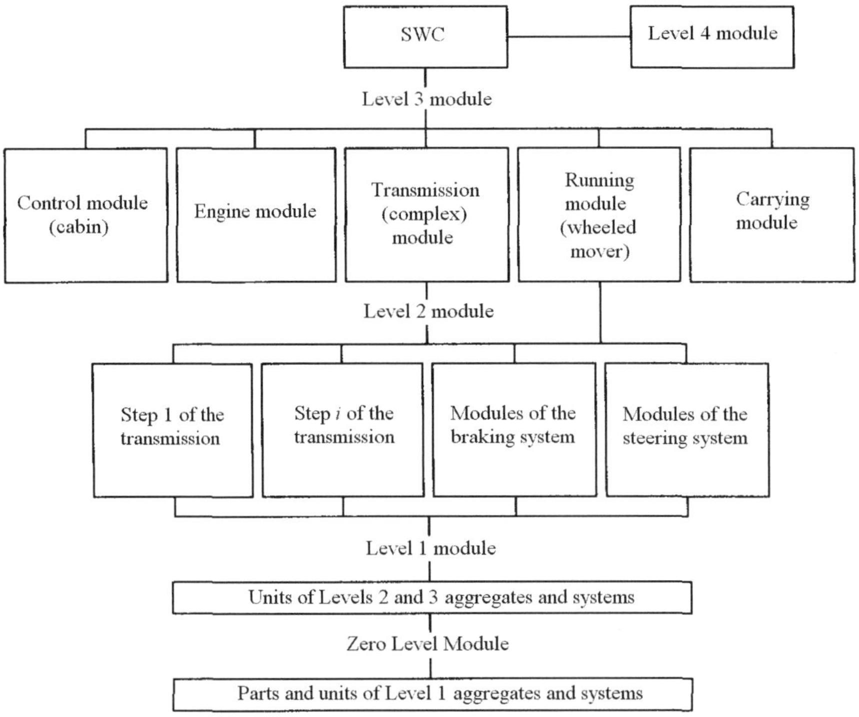

Fig. 9.17. Example of a hierarchical design structure of a mobile vehicle (like SWC)

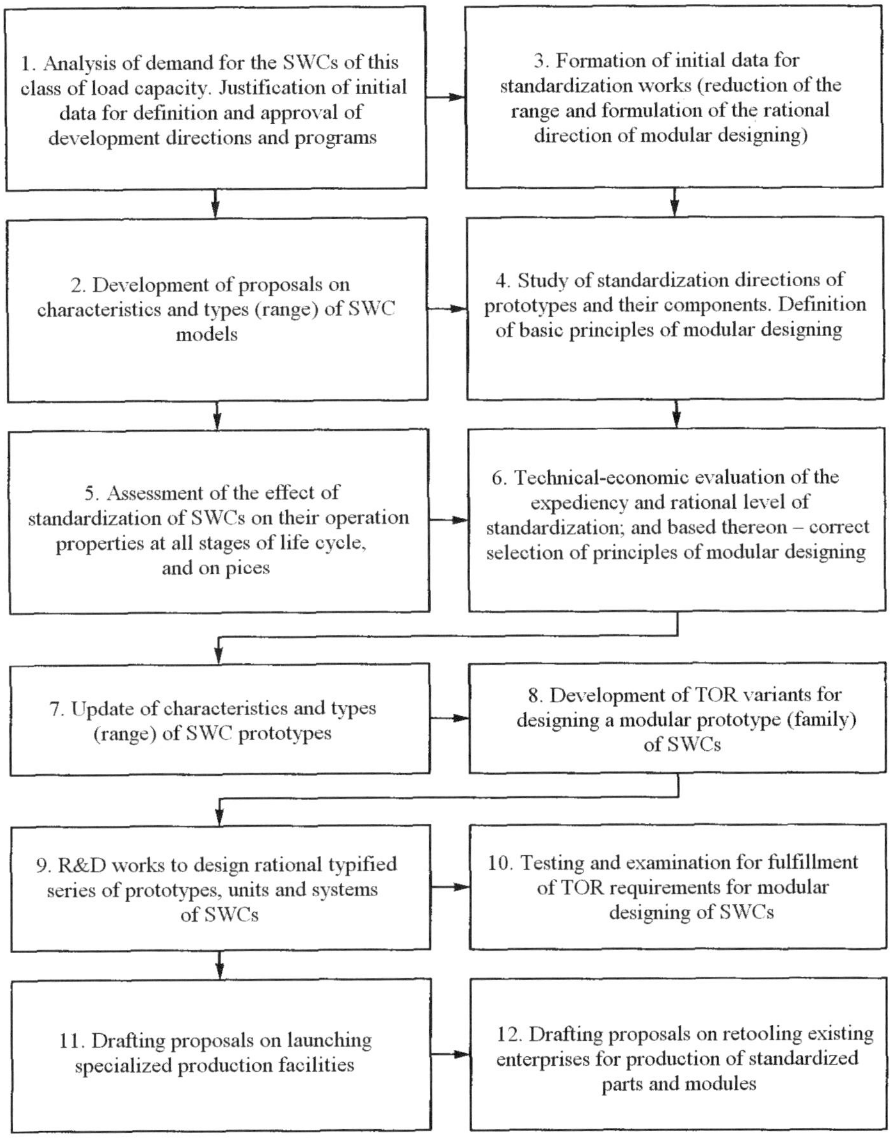

Fig. 9.18. Basic work stages of complex unification of SWCs

efficiency, its modification, on the other hand, is performed not so much to increase its efficiency, but to expand or change its capabilities by application. Product modifications can be recurrent when a primary modification is the basis for the secondary one, etc.

The modular principle allows for extending the maximum service life of products without their essential moral obsolescence. This is reached through a stage-by-stage modernization of the individual modules.

In general, all trends in modular designing are based on the general technical principle of aggregation, which assumes the following:

- Systems and others complex technical devices represent an aggregate consisting of several independent devices (modules);

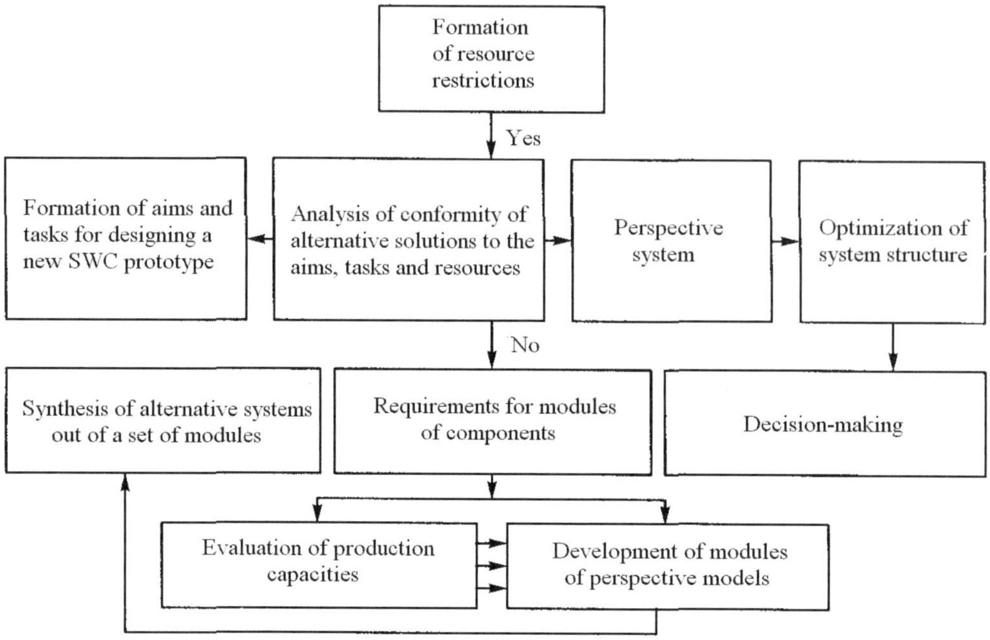

Fig. 9.19. Modular structure of SWCs

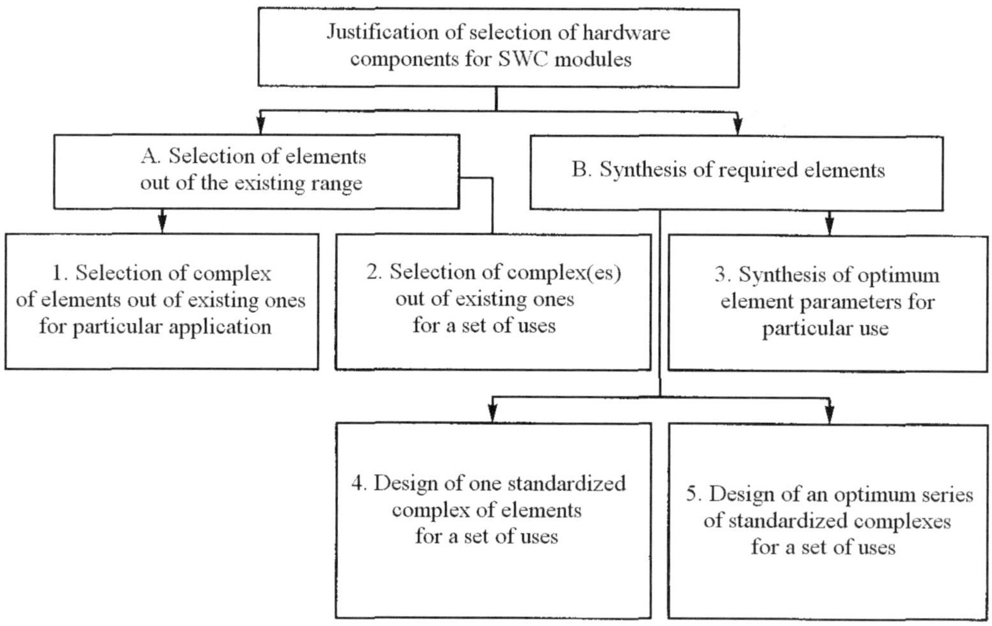

Fig. 9.20. Problem-solving structure for justifying the selection of the modules' element base

- The decomposition (split) into these devices (modules) is made in such a manner that each performs a definite function and is structurally and technically finalized;
- The types of devices are selected in such a way that it might be possible to assemble them into an aggregate with preset technical properties;
- The functional variety of the aggregates is reached by different combinations of devices (modules) and also by a possibility to build-up (scale-up) the aggregates structure in the process of using them; and
- The products of one and the same functional purpose must form a parametric series.

The basic precondition of aggregation (modularity) is module compatibility.
The concept of compatibility (energetic, informational, metrological, structural, and operational) is based on the consecutive unification and standardization of the modules' properties and parameters.

In the most general sense, the modular design principle assumes creating a wheeled vehicle as an aggregate of various independent and interconnected modules. These modules are the carrying structure, the cabin, the engine, separate aggregates of the transmission, and also the components of the springing, control, and braking systems, etc. (Figs. 9.21 and 9.22).

The advantage of the modular principle is the possibility to simultaneously work on many units (both designing and manufacturing), which reduces the design time of the entire vehicle. To reduce the amount of original documentation, the production lead time, and to facilitate vehicle maintenance, a decrease in the number of modules is needed. For this purpose, it is desirable to design aggregates and parts

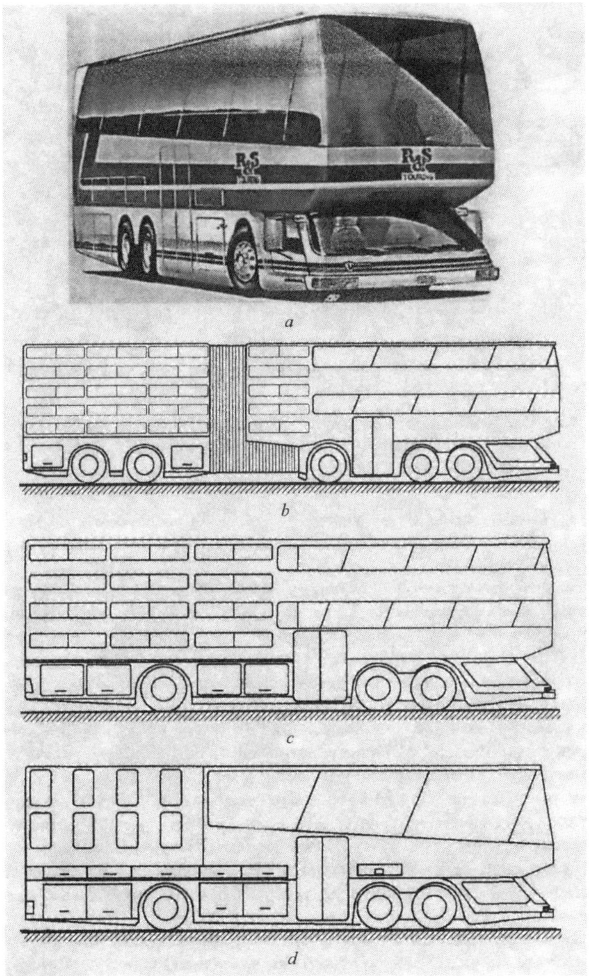

Fig. 9.21. Buses of extra large busloads: a—Outside appearance; b—Layout scheme of an articulated bus; c—Layout scheme with a 3-tier arrangement of passenger seats; d—Layout scheme with a 2-tier arrangement of passenger seats

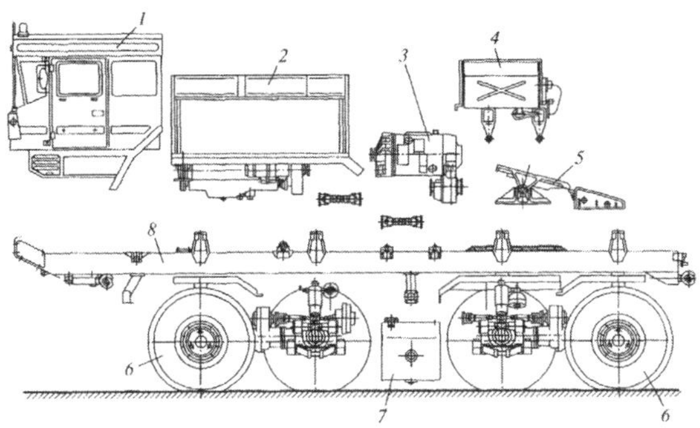

Fig. 9.22. Modular design of a tow car (hauler) proposed by KZKT. Modules: 1—Of control; 2—Engine; 3—Power; 4—Special equipment; 5—Coupling; 6—Traction; 7—Fuel; 8—Supporting frame

of the transmission, as well as of the springing and braking systems, identical for all wheels; however, this does not seem rational.

The modular principle allows for organizing the design and production processes in the simplest and cheapest way by using the same aggregates for the 6×6 and 8×8 vehicle families; however, it can decrease certain technical and tactical specifications. As an example, a hierarchical structure of a hypothetical SWC built from modules of different levels is presented in Fig. 9.17. In this case, the SWC itself forms—as a module of the fourth level—a part of the transport complex. A wheeled vehicle has several modules of the third level: a control module (a cabin), an engine module, a carrying module, a running module, and connecting control modules (for steering and braking), which are united into one assembly unit—a mobile SWC. It is then possible to add, for example, wheeled modules both with and without changing the dimensions of the carrying module, which can result in an increase in the length or width of the SWC, and, accordingly, the ability to transport cargoes of various sizes and weights. Any third level modules consist of the modules of lower levels. The lowest level is the level of constituent assembly units (components), which is accepted as the zero level and has no modular hierarchical structure.

Thus, all this confirms that an improvement in the design and production parameters is in a certain contradiction with the products' operational purpose.

The mainstream progress of the modular design principle as seen by foreign automobile companies is the possibility to make chassis from the minimal possible nomenclature of parts for further use, according to all the existing requirements in different load capacity classes. The result of the modular principle is a vehicle which corresponds to any customer's special requirements and thus guarantees its rational use, i.e., a profitable investment. For example, there is the possibility to equip a chassis with a gross weight of 36–40 t with engines of different power since the power requirement can vary depending on the use, e.g., for long-distance or intra-city hauls. However, some vehicle properties are degraded.

The basic work stages aimed at comprehensive modular designing are presented in Figs. 9.18–9.20; while examples of modular configurations of saddle and trailer road trains are presented in Fig. 9.15.

Since it is highly promising, the modular design principle is also used in creating passenger buses of large busloads (see Fig. 9.21) and in the production of parts irrespective of whether it is an engine, a gearbox, or a frame.

An example of a modular structure of a saddle wheeled tow car proposed by OJSC Rusich-KZKT is presented in Fig. 9.19.

The above ways of implementing the modular design principle are characteristic for traditional chassis designs, equipped, in particular, with mechanical and hydromechanical transmissions. If we identify the transmission as a module of the second level in a wheeled chassis module with an ET or an HST, then it can be assembled from the first level modules, which are identical independent loops (motors-in-wheel) of the wheeled module drive. In its turn, each loop of the wheel drive is assembled from the modules of the zero level, the assembly units (modules), which act as the basic design elements of an aggregate, system, or product as a whole.

It should be noted in conclusion that the modular design principle makes it possible (through replacing the individual modules) to adapt, at any moment, the vehicle to any modern market requirement and to the level of engineering development. This is an important step towards creating long-term vehicle models.

9.3 Theory for building HDWV families

In general, the task of optimizing (forming) the standardized series (families) of technical objects, including SWCs, adjoins establishing requirements for the main objects' parameter and defining the necessary quantity of their standardized series built according to this main parameter.

The standardized series (families) of mobile equipment are understood as sets (pluralities) of prototypes united by the general purpose and ranged by the value of the main parameter.

The research process assumes using a number of consistently interconnected initial concepts, which are specified by recommendations on the methods of building a design-unified series of machine-building products.

The interrelationship of the standardized and design-unified series is expressed through the fact that each product is included into the considered system of machines at the same time as a member of a standardized series and a basic model or modification of a design-unified series.

The designing of the interconnected standardized and design-unified series is performed on the basis of the objects (products) main parameter's united parametric series of values. The number of basic models of the products' standardized or design-unified series cannot exceed the number of members of the main parameter's parametric series of values.

In general, when building the parametric series of technical objects, we need to define the nomenclature of objects, the nomenclature (range) of their parameters, and the main parameter.

Defining the nomenclature of objects is performed with an account of the needs, technical capabilities, and technical and economic expediency of forming interconnected standardized and design-unified series. The nomenclature of objects includes all existing and (or) planned-for-production objects, the parameters of which essentially affect the technical and economic production efficiency and operation of the objects.

The choice of the nomenclature of the objects' parameters assumes identifying the parameters that define the most essential design, the technological and performance properties of the objects, and ensures a high level of standardization in their components. The list of objects' parameters and their components is composed, as a rule, with an account of the requirements grouped as follows: power parameters, weight parameters, kinematical parameters, dimensional parameters, performance parameters (parameters of profitability, reliability, etc.), and other parameters.

The main parameter is selected from a certain range of parameters. In doing so, the following requirements should be observed:

- The main parameter should be the same for the whole nomenclature of objects;
- The parametric series of its values is the basis for building all of the standardized series in the system of technical objects; and
- Each particular main parameter value remains constant for the objects included into a design-unified series.

The main parameter's parametric series of values is defined on the following basis:

- The range of values of the main parameter, i.e., defining the extreme members of the series, which satisfy the demands of the customer;
- The number of members of the parametric series is a rational quantity, which allows for matching together the demands of the consumer with the capabilities of the manufacturer under minimal production and operational expenses; and
- The frequency of the series, i.e., the intervals between the adjacent values of the main parameter in the parametric series.

Formulating the problem of optimizing the parametric series is as follows. One parameter is selected from all possible variants of the parametric series in which the values of the main parameter satisfy the consumer at the objects' maximum efficiency. An optimum parametric series of mechanical engineering items is understood as a series which meets the following requirements:

- The items with the values of the main parameter included in the series should be technically realizable and meet the demands of the consumers;
- The values of the main parameter creating a parametric series should be selected from the series of preferred numerals according to GOST 8032 (for any independent parameters); and
- The parametric series as a whole should meet the recommendations of the ISO, IEC, and other international organizations on standardization.

Various optimization criteria are used in the practice of forming parametric series. When forming the parametric series of mechanical engineering items, the cumulative costs of designing, manufacturing, and operating the objects are most often used as such criterion. In this case, the optimization criterion represents the dependence of the aggregate costs of designing, manufacturing, and operating versus the values of these parameters:

$$Q = Q_{dc} + Q_{mc} + Q_{co} = f(x_1, x_2, \dots, x_n),$$

where Q_{dc}, Q_{mc}, and Q_{co} are the cost of designing, the ex-factory price of the objects, and the operational expenses, respectively; $x_1, x_2, \dots, x_n$ are the values of the parameters of all members of the series.

The problem of optimizing a parametric series represents a problem of minimizing the aggregate expenses:

$$Q(x_1, x_2, \dots, x_n) \rightarrow \min.$$

The cost of designing a group of objects is defined by the number of developed standardized models and increases when the number of members in the parametric series grows. The cumulative design costs are defined by the expression:

$$Q_{\Sigma dc} = \sum_{i=1}^{n} q_{n_i},$$

where q_{n_i} is the cost of designing an item having the main parameter value equal to the value of the ith member of the series, $x = x_i$.

The design expenses (costs) include the cost of research and design (R&D) works, developing the design documentation, and manufacturing and testing the prototypes.

The aggregate (total) cost of the product is defined according to the formula:

$$Q_{mc} = K \sum_{i=1}^{n} q_{mc_i} p_i,$$

where K is the coefficient accounting for the reduction in costs with the increase in the scale of serial production. In accordance with GOST 20799:

$$K = 0.3 \frac{\Pi}{\Pi_o} < 1,$$

where Π_o is the initial number of the members in the series; q_{mc_i} is the production cost of one item; and p_i is the demand for the items.

Defining the price of the items takes into account the cost of the materials, energy, wages of the labor force, overhead costs, etc.

The operational expenses of products can increase with the reduction of the members in the series when items with a smaller parameter are replaced by items with a higher parameter through increasing the values of other indicators (weight, dimensions, energy intensity, etc.). At the same time, the nomenclature of the products which are in operation are reduced as a result of optimizing the parametric series. Under a preset demand for the products, if the series is reduced, the repeatability of the products is increased. In its turn, it will allow for decreasing the additional expenses due to reducing the amount of required spare parts, reducing the labor intensity of maintenance and repairs, increasing the reliability, and reducing expenses on training the service personnel.

The total operational expenses are defined according to the following formula:

$$Q_{co} = K_1 \sum_{i=1}^{n} q_{co_i} p_i \,,$$

where K_1 is the coefficient of the equivalent reduction in operational expenses accounting for the decrease in the nomenclature of products; and q_{co_i} means the operational expenses of the whole product with the parameter value $x = x_i$.

Operational expenses take into account the cost of spare parts, storage, repair and maintenance costs, and the salaries and training of the service personnel.

With an account of the total expenses at all stages of the product life cycle, the optimization criterion is defined by the following formula:

$$Q = \sum_{i=1}^{n} q_{n_i} + K_1 \sum_{i=1}^{n} q_{mc_i} p_i + K_1 \sum_{i=1}^{n} q_{co_i} p_i \,.$$

The cumulative (total) costs are calculated for the whole period of manufacturing and operating the product.

In certain cases not all costs are used as a criterion, but only the part that depends on changes to the optimization parameter. For example, at mass production, the design costs are sometimes neglected since they are small compared to the production operational expenses (the so-called Capex and Opex).

In other cases, in order to increase the scientific validity of the series, they add to the total expenses of the parametric series criterion; for example, the expenses from adapting the model based on discrepancies in the interests of consumers who seek to reduce the nomenclature because of their restricted material, labor, and financial resources.

Thus, for motor vehicles engaged in transportation, the total transportation costs can generally be roughly split into two components: the production and operational expenses (Capex and Opex).

Further, when defining the rational standardized series, only one group of HDWVs, namely, SWCs with a load capacity of 20 t or more, shall be considered as an example.

Forming highly unified families assumes establishing the requirements for the main parameter and defining the required number of standardized SWCs series, to meet, for example, the demand for the chassis intended for installing special equipment. Keep in mind that the number of standardized series should be minimal.

Let us consider the method of forming a standardized series of SWCs based on the implementation of design-and-layout schemes by means of a unified standardized series (USS) and aimed at ensuring the required mobility and load capacity.

The studies and accumulated design experience show that the basic models of SWCs can be created as a USS from uniform units and aggregates if they have the same axial load and differ only in the wheel formula. For practically all wheeled vehicles within the USS limits, the wheel formula is accepted as 6×6, 8×8, etc.; while the USS can include one to three standardized models. The gross weights of each USS standardized model are defined according to the following expression:

$$M_g = pn, \tag{9.1}$$

where M_g is the gross weight of the standardized model with the axle number n and p is the axial load in the given USS.

Defining the maximum axial load depends on the conditions of using the vehicle and the specific statistical load on the tire in its contact spot with the bearing (rolling) surface. Earlier, a maximum axial load of 10 t was set for SWCs, which was explained by the passing ability condition of military vehicles. However, in recent years, these vehicles are more often equipped with wide-section tires; as a result, the axial load went up to 15.0–16.0 t while preserving the acceptable specific pressure values in the contact spot. Thus, the maximum gross weight of an 8×8 wheeled vehicle can be 40.0–64.0 t.

The load capacity G_Γ of each model in the standardized series is defined by the weight utilization factor:

$$G_{Lc} = M_g K_{\text{u.m}} \tag{9.2}$$

Analysis of multiple study results shows that the alteration of $K_{\text{u.m}}$ can be described by a linear dependence in the function M_K of the type:

$$K_{\text{u.m}} = a + bM_g,$$

where a and b are the coefficients received after statistical processing of the information.

Further, the approach to selecting the standardized series is as follows. In the range of 10.0–16.0 t maximum axial load, 6×6 SWCs are taken as the basis for a standardized model. Their gradation is selected to meet the demand for load capacity in the economy, existing practice, etc. The most widespread load capacity classes in the world automotive experience are 15, 20, 25, 20–25, and 40 t.

If we take for a module a 4×4 vehicle with an axial load not exceeding 10.0 t, 6×6 and 8×8 models can be designed to increase the load capacity. There are examples of such vehicles in the domestic and foreign practice. In this case, the gross weight of each subsequent standardized model is found from the condition:

$$M_g = \frac{m_{g\,(4\times4)}}{2} n \tag{9.3}$$

After that, a possible load capacity G_Γ of the 6×6 and 8×8 vehicles with account of statistical data is defined as:

$$G_{Lc} = M_g K_{\text{u.m}} = M_g (a + b) \tag{9.4}$$

When building a USS system of the entire standardized series, the principle of load capacity equality in two base standardized models of adjacent USSs with different wheel formulas is used. The following condition is accepted in this approach:

$$G_n^{m+1} = G_{n+1}^{m} \tag{9.5}$$

where G_n^{m+1} is the load capacity of the base standardized model of (m + 1)-USS with the number of axles n; G_{n+1}^{m} is the load capacity of the base standardized model (m + 1)-USS with the number of axles n + 1; and m is the USS serial number.

Unified standardized series with a greater serial number is comprised of vehicles with a greater gross weight and axial load.

The use of this principle is explained by the fact that satisfaction of the consumer's needs (in this case in load capacity) can be provided by the chassis using various wheel formulas and, accordingly, various passing abilities and assembly properties accounting for the purpose of mounted equipment and conditions of its operational use. In the approach under consideration, the load capacity of a 8×8 chassis of the lower series is defined through the gross weight of the higher series model corresponding to it by load capacity:

$$M_{n+1}^{m} = M_{n}^{m+1} \frac{K_{u.m}}{K_{u.m}(n+1)} \quad (9.6)$$

or:

$$M_{8x8}^{m} = M_{8x8}^{m+1} \frac{K_{8.m\ 6x6}}{K_{8.m\ 8x8}} \qquad M_{6x6}^{m} = M_{4x4}^{m+1} \frac{K_{8.m\ 4x4}}{K_{8.m\ 6x6}} \quad (9.7)$$

Calculations according to formulas (9.1)–(9.7) allow for defining the gross weights and load capacities of all possible highly unified families (standardized models) of SWCs included in the USS system as automotive base chassis (ABC).

For SWCs, load capacity, maximum axial load, or maximum gross weight are taken as the main parameter.

The load capacity of models is defined by the customer's requirements. The model's maximum gross weight depends not only on the load capacity, but also on the road weight, which is defined by the vehicle's design (its materials consumption). In this context, it is expedient to take load capacity as the main parameter in SWCs standardized series.

Such standardized series may be corrected to meet specific customer's needs.

Table 9.5 and 9.6 present the calculation results on defining the possible standardized series of basic models of wheeled vehicles, their gross weights and load capacity based on military demand for SWCs, the conditions of their field use (for selecting the wheel formulas at identical or similar load capacities), and the expediency of applying various design-and-layout schemes with the use of unified units and aggregates.

In Table 9.5, calculations are made for SWCs equipped with 1500×600–635 VI-203 model tires providing an axial load of 11.5 t. In Table 9.6, the calculations are made using 1600×600–685 VI-178 AU model tires providing an axial load of 15.0 t.

The analysis of Table 9.5 from the point of allowable axial loads and gross weights of the chassis shows that in USS 1–3 it is expedient when forming the general SWC standardized series to limit the number of standardized models to three for 6×6 vehicles, two for 8×8 vehicles, and one for 12×12 vehicles. In its turn, it follows from Table 9.6 that within USS 1 it is sufficient to limit to one standardized model for 8×8, 12×12, and 16×16 vehicles. When accounting for equal load capacities in the 12×12 and 8×8 vehicles in Table 9.5 and 9.6 and considering real weight

Table 9.5 Main parameters of a standardized series of SWCs with an axial load of 11.5 t

USS #	6×6		8×8		12×12	
	M_g, t, less than 33.5	G_{Lc}, t, at least 14.0	M_g, t, less than 45.5	G_{Lc}, t, at least 22.0	M_g, t, less than 69.0	G_{Lc}, t, at least 35.0
1	29.0	14.0	43.0	22.0	69.0	35.0
2	32.0	17.0	45.0	24.0	-	-
3	33.0	18.0	-	-	-	-

Table 9.6 Main parameters of a standardized series of SWCs with an axial load of 15.0 t

USS #	6×6		8×8		12×12	
	M_g, t, less than 60.0	G_{Lc}, t, at least 35.0	M_g, t, less than 80.0	G_{Lc}, t, at least 55.0	M_g, t, less than 120.0	G_{Lc}, t, at least 85.0
1	60.0	35.0	80.0	55.0	120.0	80.0

and the dimensional parameters of mounted equipment, it is expedient to exclude 12×12 chassis with an axial load of 11.5 t from the USS.

Thus, the standardized series formed on the basis of using highly unified SWCs families can look like:

14.0–18.0, 22.0–24.0, 35.0, 55.0, and 80 t and over.

The series coefficient is equal to 1.57.

The existing standardized series of domestic HDWVs in the 22–24 and 80 t load capacity classes was formed as a result of applying the previous and existing series with account of the country's demand, the industry's capabilities, and corresponding to the theoretical series. Moreover, the START-2 (Strategic Arms Reduction Treaty) assumes reduction of the equipment, which uses the existing chassis from the above 35.0 and 55.0 t load capacity classes as the transport base.

It should be noted here that for maintaining the minimal operational expenses when installing and transporting mounted equipment (truck cranes, etc) in various road and climatic traffic conditions, it is necessary to have a sufficiently wide range of base item modifications for a corresponding standardized model, each of which would be adapted for the specific purpose.

9.4 Future transport-and-process wheeled vehicle

The consumer requirement of having the ability to drive at a high speed on all kinds of roads and terrains should serve as a starting point when developing possible designs of an advanced HDWV as a transport-and-technological vehicle. As shown by the experience of testing and operating multi-axle SWCs, the running gear concept of traditional HDWVs, which, being conservative today, has long satisfied the demand for consumer mobility properties and which at the initial creation stage of similar vehicles was quite reasonable, now has become the main limiting factor (see Chapter 4). Thus, the running gear of HDWVs in many respects defines the design of an advanced wheeled vehicle.

Accounting for the above main principles of modular design, one can say that the running gear of HDWVs can be considered as a support-running complex consisting of the following basic modules: the springing system control assembly of each wheel, the cross-axle connection systems, the elasto-damping system, the traction and braking module, the turning control system (steering module), the local power supply system, the local unified motion control system unit, etc.

The results of developing a new HDWV design on the basis of existing units and aggregates while providing for a certain unification with the existing vehicles are presented in Fig. 9.23 and 9.24.

A brief analysis of the data presented in Figs. 9.23 and 9.24 shows the following design drawbacks:

- The cantilever mounting of the wheels requires excessive reinforcement of their fixation elements to the lever system;

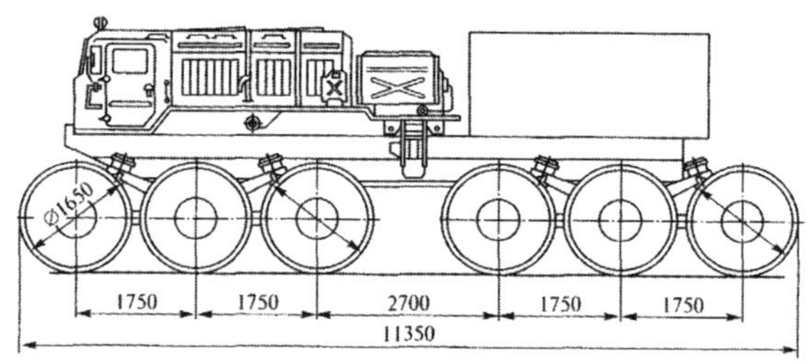

Fig. 9.23. Layout scheme of a HDWV with a three-support balance truck (version 1)

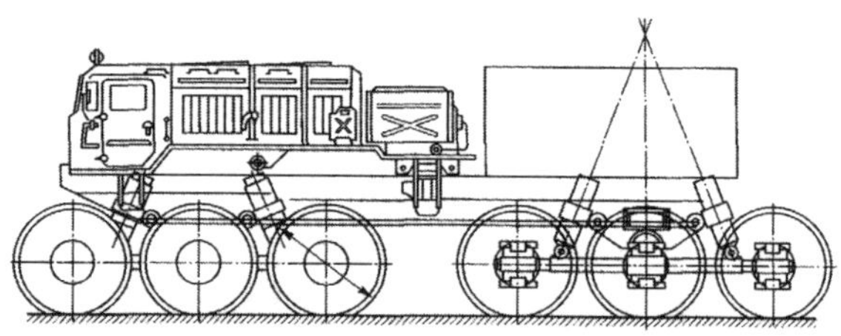

Fig. 9.24. Layout scheme of a HDWV with a three-support balance truck (version 2)

- The small recesses for the support-running modules do not allow for using spatial designs for the lever mechanisms and spaced-apart schemes for installing the bearings in link joints, which in combination with great static loads on the wheels and a high level of dynamic loads leads to oversized joint pins, as well as their bearings;
- The limited lateral overall dimension of the chassis does not allow the wheels to turn at greater angles, which drastically reduces the vehicle's maneuverability in a space-limited environment; and
- The lever elements occupy unreasonably large internal spaces of the vehicle, which complicates accommodating for the purpose designed equipment, displacing it high up in the vehicle.

All of these drawbacks are also distinctly revealed when attempting to compose possible designs of a vehicle with a three-link bogie of a similar type; even at limited vertical displacement of the wheels, the vehicle has an obviously irrational configuration because the entire empty space in the bottom of the body shell is occupied by the lever mechanisms of the support-running complex. The load ratio payback of a similar vehicle could hardly be at an acceptable level.

The results of this study (Fig. 9.25) allow for making a negative conclusion, not only in regard to this specific layout scheme, but also in relation to the attempts still made today to implement running gears by using lever-mechanical cross-axle systems of different complexity levels in such a complicated vehicle. It seems hardly possible to accommodate such systems in a real vehicle and to provide its necessary load ratio.

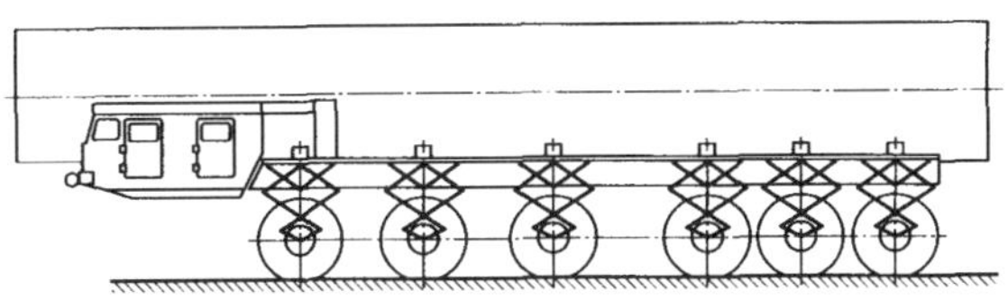

Fig. 9.25. Overall view of a chassis with a pantograph guidance mechanism of the long-stroke suspension

Designing support-running complexes on the basis of individual wheeled modules should be recognized as the most correct approach and the necessary characteristics of the complex will be reached by introducing controlled electrical and hydraulic cross-axle connections.

Extra long-stroke individual modules differ by the guidance mechanism diagram.

When developing a possible design of an advanced HDWV, the following versions have been considered:

- A damping strut with a pantograph guidance mechanism (see Fig. 9.25);
- A single-stage damping strut with a semi-levered type guidance mechanism and a two-wheeled one-axle bogie (Fig. 9.26);
- A single-stage telescopic damping strut with a two-wheeled one-axle bogie (Fig. 9.27);

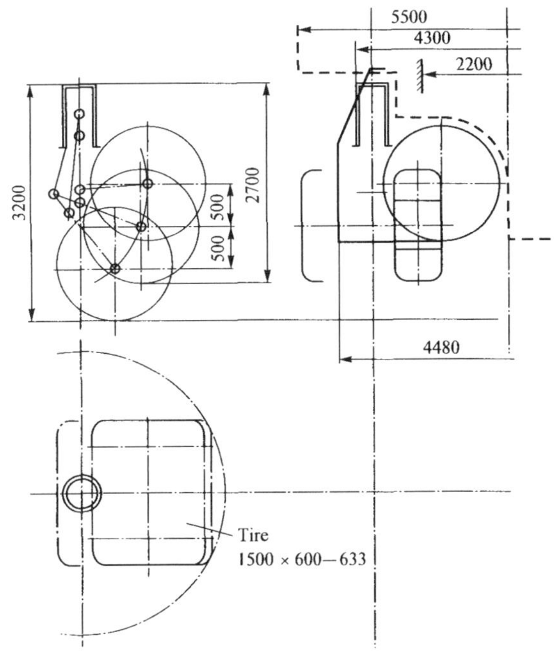

Fig. 9.26. Kinematics and layout restrictions of a single-stage, strut-mounted, extra long-stroke suspension of HDWVs

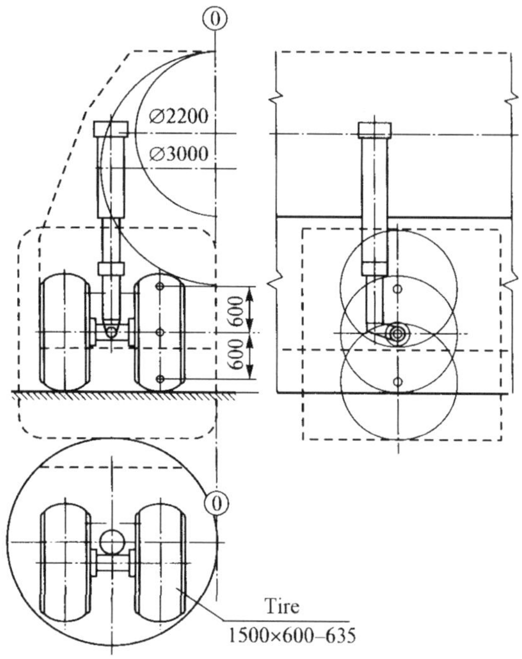

Fig. 9.27. Kinematics and layout restrictions of single-stage, semi-lever-strut-mounted, long-stroke suspension of HDWVs

- A two-step damping strut with a semi-levered type guidance mechanism and a two-wheeled one-axle bogie (Fig. 9.28);
- A two-step damping strut with a semi-levered type guidance mechanism with torque arms and a one-axle bogie; and
- A three-support center-point suspension with beamless connection of the elastic elements (Fig. 9.29).

The design development of an advanced HDWV, for the sake of an example, was carried out in relation to a pipe carrier intended for transporting various combinations of standard big-diameter pipes. A six-axle MAZ type chassis was taken as a prototype (Fig. 9.30). A comparative estimation of the profile passing ability and maneuverability parameters are presented in Figs. 9.31–9.34.

Assessment studies of the vehicle's longitudinal silhouette have shown that the fitting (negotiating) of an HDWV into the road profile can be ensured if the number of supports along the sideboard do not exceed six (a six-axle chassis). In this case, the rated load on the support-running module will be 9.5–10.0 kN. The acceptable fitting into the road profile can also only be provided by grouping the supports into conventional bogies shifted to the vehicle's ends. This coincides with the well-known requirement for the configuration of multi-axle chassis, which ensures suppression of the longitudinal-angular oscillations of the body; however, it creates certain problems in maintaining the due level of HDWV maneuverability.

The damping struts with a pantograph-type guidance mechanism can principally provide the required parameters of the wheel stroke and the use of the "wheel niche space" (see Fig. 9.25). At present, MSTU is engaged in deeper studies of such designs with the aim of defining their practical feasibility. That is why it is too early to make practical conclusions about the applicability of pantograph-type damping struts in the support-running complex of advanced HDWVs.

Single-stage damping struts with a lever- and semi-lever type guidance mechanism (see Fig. 9.26) fit into the acceptable lateral dimension of the vehicle at a compression stroke of no more than 600 mm. Here, preference should be given to the damping struts with reactive posts over traditional aviation semi-lever struts since the studies of the latter have revealed unacceptable overloads of the elastic element by reactive forces arising during the vehicle's braking and acceleration.

Complete satisfaction of the requirements can be ensured by two-step controlled damping (shock-absorbing) struts with a full wheel compression stroke of 1150–1200 mm, designed at MSTU (Figs. 9.35 and 9.36).

Structurally, these modules can be implemented in a one-wheel version and in a more traditional version with a one-axle two-wheel bogie (see Fig. 9.35*b*). Thus, we have synthesized the possible appearance of a gas pipe string carrier

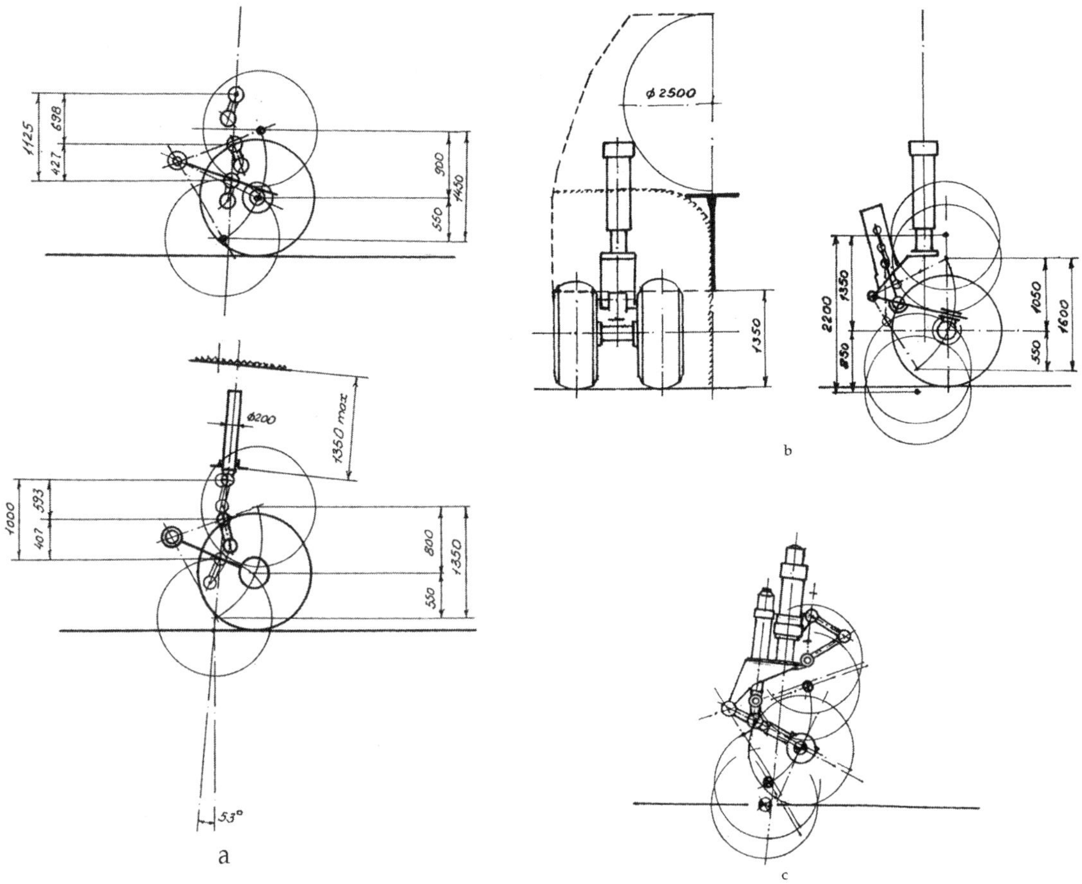

Fig. 9.28. Kinematics and layout restrictions of two-step damping strut with a semi-levered type guidance mechanism: a—Kinematics of the first step of the telescopic extra long-stroke suspension; b—Kinematics and layout capabilities of the suspension on the chassis; c—Suspension layout scheme

Fig. 9.29. Overall view of the chassis with a three-support, center-point suspension with beamless connection of the elastic elements

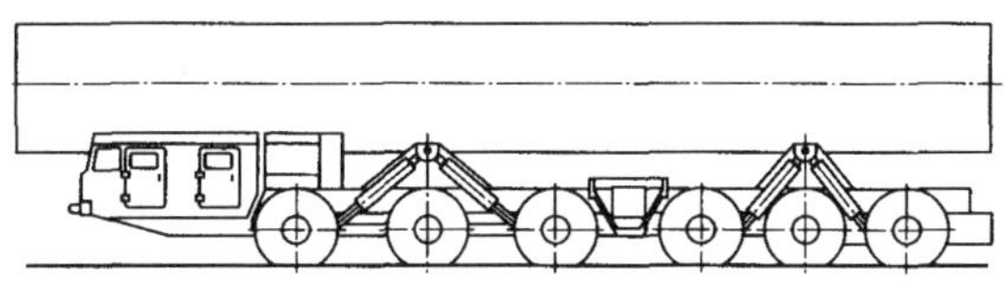

Fig. 9.30. Overall view of the chassis with the traditional suspension diagram with small wheel strokes

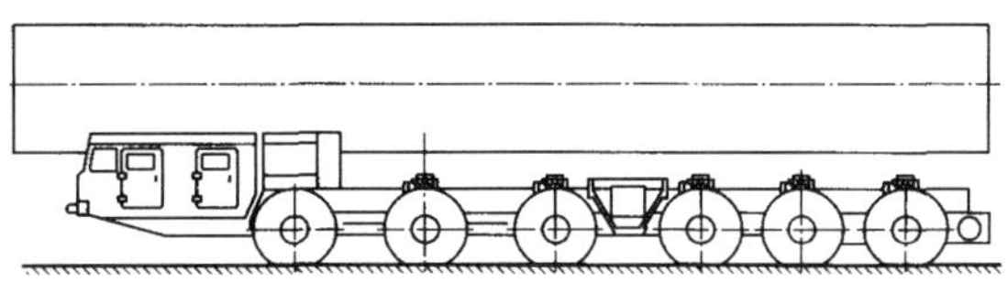

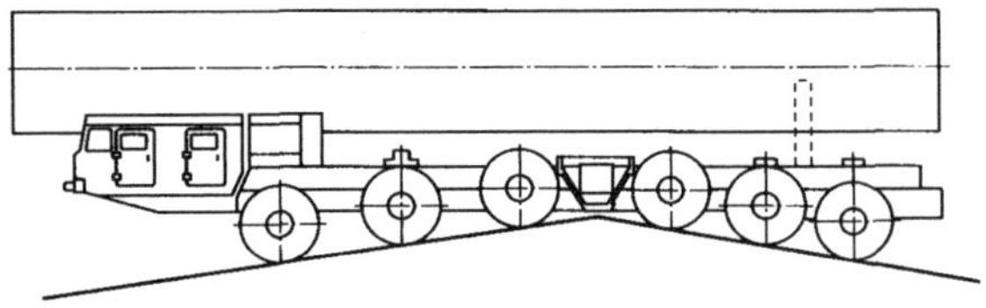

Fig. 9.31. Crossing a macro irregularity by a chassis with a suspension that ensures big vertical wheel strokes (travels)

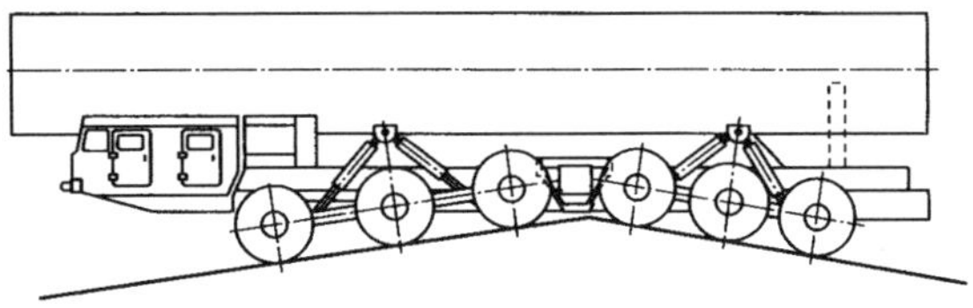

Fig. 9.32. Crossing a macro irregularity by a chassis with a three-support center-point suspension

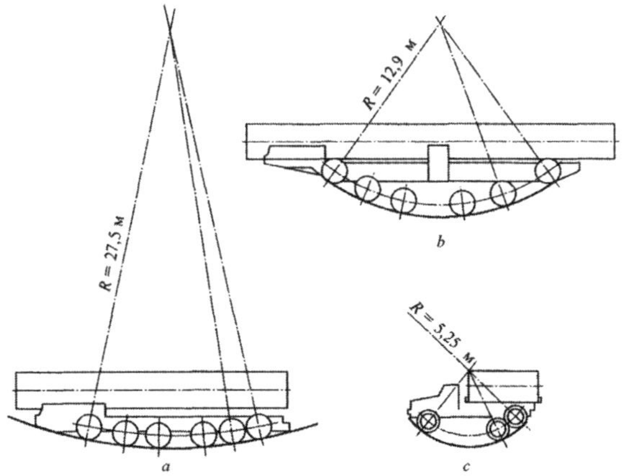

Fig. 9.33 Diagrams of wheeled chassis' fitting into the ground longitudinal profile folds: a—MAZ-type chassis prototype; b—Advanced HDWV with a long-stroke suspension; c—KrAZ-255B, a serial truck

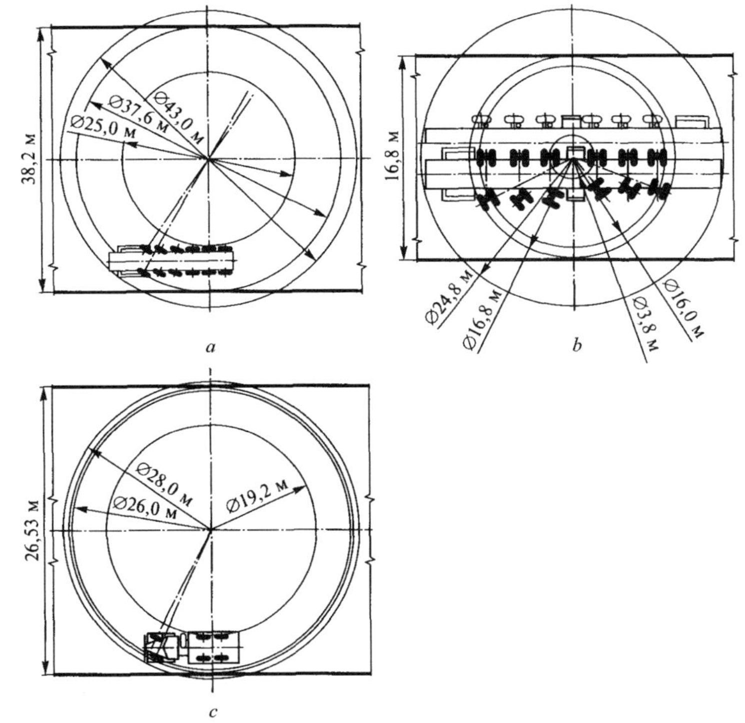

Fig. 9.34. Estimation of the wheeled chassis' maneuverability: a—MAZ-type chassis prototype; b—Advanced HDWV; c— KrAZ-255B, a serial truck

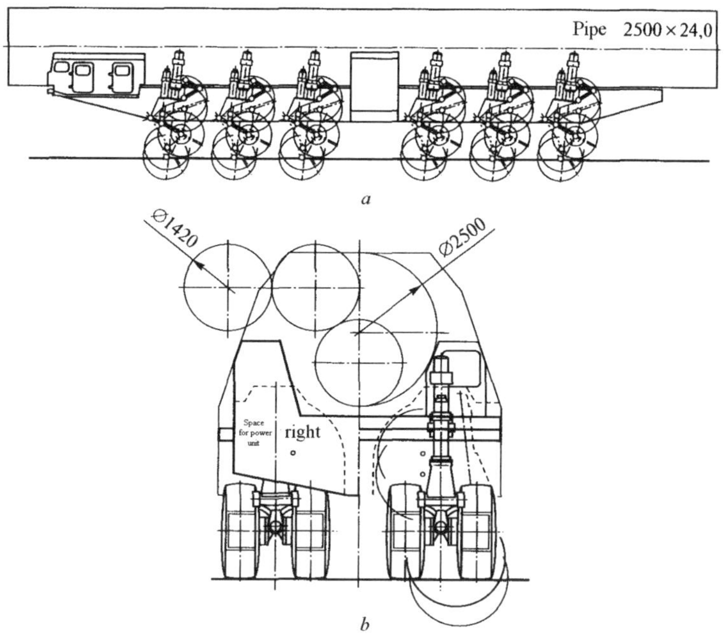

Fig. 9.35. High speed pipe carrier with a high passing ability and extra large wheel suspension strokes: a—Side view; b—Front view

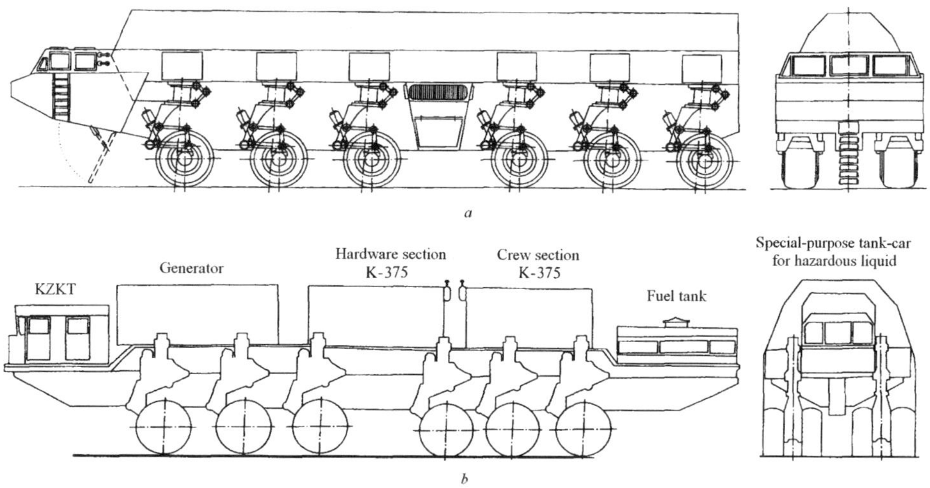

Fig. 9.36. Advanced gas pipe string carrier: a—Outside appearance; b—Layout scheme

with a six-axle chassis on the basis of a two-wheel, two-step damping strut-type support-running modules (see Figs. 9.35 and 9.36).

The analysis of this experimental design shows the following:

- The maximum allowed tire diameter for layout reasons is about 1600–1650 mm;
- The application of a one-axle two-wheel bogie requires a provision for the structural lateral skewing of the axle by at least 6° (12°

is better), which substantially complicates the chassis design and drastically increases the spaces squeezed by the module outside of the vehicle's body;

- The introduction of a bogie's axle skew requires a special device for damping the lateral axle oscillations and strong supports capable of sustaining the loads in case one of the bogie wheels hang;
- The design of the guidance device's lever mechanism in a two-wheel bogie is simpler, more compact, and readily producible, especially with the use of torque (reactive) arms;
- The second stage of the module can only be calculated to perceive the limit loads at suspension compression and its gas-hydraulic component can be made without splitters;
- The traction complex cannot be realized in principle on the basis of the mechanical transmission (a rational variant here is the electric motor-in-wheel drive design);
- A two-wheel bogie is convenient for designers of motors-in-wheel when organizing the assembly and maintenance procedures;
- The use of a two-wheel bogie eliminates the operational difficulties when replacing damaged tires since the wheel units have easy access from the outside and an autonomous system of bogie hanging;
- The traditional ladder-type spar (girder) framework is not suitable when designing a support-running complex based on the damping strut-type module;
- If necessary, an acceptable solution can be a truss or frame type spatial carrying system resting on a strong "keel girder," stringers for fixing the support-running modules, and load-bearing elements of the transverse framing; and
- From the viewpoint of the damping strut-type support-running complex design, the optimum solution is to use the carrying body (monocoque) of the respective shape.

When assessing the results of the above design of a damping strut with a two-wheel one-axle bogie, it should be recognized that the difficulties associated with a non-uniform load distribution among the bogie wheels during fast travel on the terrain and at crossing macro irregularities were not properly resolved. In this case, the HDWV version with a one-wheel two-step damping strut is most expedient.

The basic problem in designing such a damping strut is that at a tough restriction of the wheel diameter it is necessary to perceive the load at 9.5–10.0 kN and ensure an acceptable passing ability level of the vehicle at travel speeds of 70–80 km/h on a relatively hard road. At present, a solution of this task is theoretically possible only with the use of arch tires.

9.5 Weight equations of a multi-wheel vehicle and its elements

9.5.1 General equation of a transport vehicle weight in relative parameters

The gross (full) weight of any self-propelled vehicle or aggregate can be presented as the following equation:

$$M_{tw} = M_{fr} + M_{wu} + M_{g} + M_{st} + M_{\text{fuel}} + M_{sf} + M_{oe} + M_{p} + M_{\text{trans}} + M_{Lc} \tag{9.8}$$

where:

- M_g is the gross weight of the vehicle or aggregate with the load;
- M_{fr} is the weight of the framework (body) including the bearing elements (girders), lateral braces, bumpers, trailing gear, frame brackets, control cabins, and aerodynamic stabilizers;
- M_{wu} is the weight of the wheeled aggregates including the weight of the tires, tire rims,

wheel hubs with bearing units, a system of central tire pumping, and the air in the tires;

- M_{ws} is the weight of the springing system (shock absorbers) including the guiding and elastic elements and the system for regulating and stabilizing the body;
- M_{st} is the weight of the vehicle control elements including the steering system with the steering booster, the braking system with wheel brakes, air receivers, etc.;
- M_{fuel} is the weight of the fuel including the fuel tanks and fuel system equipment located outside the engine;
- M_{sf} is the weight of the system for fixing the vehicle including the jacks, support slabs, cross beams, and support mechanisms;
- M_{oe} is the weight of the auxiliary equipment including the electric equipment, hoists, and other unaccounted for weights;
- M_p is the weight of the power unit, its service systems, and cooling oil;
- M_{trans} is the weight of the transmission and its wheel control drives; and
- M_{Lc} is the payload weight including the body or platform and the technological equipment and technical systems for servicing the load.

For analysis convenience and for calculating the weight characteristics at the design stage, write equation (9.8) in relative parameters by dividing the left- and right-hand parts of the equation by the vehicle gross weight m_O. Then:

$$1 = \xi_{fr} + \xi_{wu} + \xi_{g} + \xi_{st} + \xi_{\text{fuel}} + \xi_{sf} + \xi_{oe} + \xi_{p} + \xi_{\text{tran==s}} + \xi_{Lc} \tag{9.9}$$

where $\xi_i = \dfrac{M_i}{M_{tw}}$ are the relative weights of the elements on the right-hand side of equation (9.8).

Based on the analysis of statistical data and experience in designing basic vehicles, all weight equation elements can be divided into three main groups.

The first group includes the elements whose weight does not depend or practically does not depend on the adopted power-to-weight ratio (power supply capacity) of the vehicle and is mainly defined by the level of passing ability, the allowable overloads, the materials used, and the design quality.

The second group includes the elements whose weights are essentially dependent on the vehicle's power-to-weight ratio.

The third group includes the elements which make the useful load of the vehicle, including the cargo (load) proper and the equipment for its mounting and servicing.

Let us now present the first group of elements as the following sum:

$$\sum_{1}^{n} \xi_{ki} = \xi_{fr} + \xi_{wu} + \xi_{g} + \xi_{st} + \xi_{fuel} + \xi_{sf} + \xi_{oe} \tag{9.10}$$

The second group of elements is expressed through specific parameters as follows:

$$\xi_p = m_p N_w \tag{9.11}$$

$$\xi_{\text{tras}} = m_{\text{trans}} N_w \tag{9.12}$$

where m_p and m_{trans} are the specific reduced weights of the power unit and transmission, respectively, and N_w is the specific efficient (implemented on the wheels) power-to-weight ratio of the vehicle.

By solving equation (9.9) relative to the payload weight ξ_{Lc} from equations (9.10), (9.11), and (9.12), we get:

$$\xi_{Lc} = 1 - \sum_{1}^{n} \xi_{ki} - (m_p + \mathrm{m}_{trans})\, N_w \tag{9.13}$$

When designing the basic vehicle for a transport or other mobile aggregate, the initial data should specify the full payload or its components and also the overall dimensions and positions of the centers of mass.

If the absolute weight of the payload is preset and there are dependencies for defining the components of equation (9.17) and the specific reduced weights in equations (9.11) and (9.12), then at the preset power-to-weight ratio N_w, the relative payload weight can be found, and, according to formula (9.13), the gross weight of the vehicle can be defined as follows:

$$m_{tw} = \frac{m_{Lc}}{\xi_{Lc}},$$

and then all of the design elements:

$$m_i = m_{tw}\xi_i \quad (9.14)$$

The accumulated statistical materials [3] on characteristics of individual design elements, the use of element models, and certain dependencies existing among the vehicle's parameters allow for deriving analytical expressions for calculating the relative weights of the vehicle's elements that are suitable for practical use at the design stage.

9.5.2 Definition of relative weights of aggregate elements

9.5.2.1 Definition of the relative weight of the frame

Let us use a beam resting on two supports and loaded with the equivalent uniformly distributed load of its own weight and the load of the elements located thereon (the sprung weights) as a model of a basic vehicle frame with any number of axles. Such a model is the simplest one, and, at the same time, is close enough to a real frame since the loading diagram of a real frame is always represented as a beam on supports (hanging up on two axles of a multi-axle vehicle). The traditional loading diagrams as a uniformly distributed load of the sprung weights are also close enough to the loading of a real frame.

The possible cases of loading the frame model (Fig. 9.37*g*) and the epures of the bending moments are shown in Fig. 9.37*a*, *b*. Case *a* corresponds to the limit loading, which is practically unachievable for a real frame because of the absence of overhangs (consoles). The maximum bending moment in the middle of the frame has the following value:

$$M_{\max} = \frac{q\,\ell_{fr}^2}{8},$$

where l_{fr} is the frame length and q is the load on the frame.

The same moment value, as a special case, will be observed in case the vehicle hangs on its middle axle or if the number of axles of the aggregate is odd and their spread is uniform.

Case *b* corresponds to the loading of the real frame. The bending moment on the supports or in the middle of the frame will be less than the maximum one, and, in general, depending on the ratio, l_k/l_{fr} it can be expressed by the following formula:

$$M = K_l q\, l_{fr}^2,$$

where K_l is the loading coefficient of the values of which, depending on the ratio, l_k/l_{fr} are presented in Fig. 9.37*e*.

The most typical cross-sections of frames are presented in Fig. 9.37*c* and *d*. The frame in Fig. 9.37*c* consists of two longitudinal girders (longerons) linked by cross-bars. The girders are made of standard channel beams, the h/b ratio of which varies within the range of 1.0–2.0. If at any h/b ratio we take the wall thickness δ and the length of the cross-section's center line as the same, which also means the same weight per unit of the girders' length, then the equatorial resisting moment W can be expressed by the following general formula:

$$W = \frac{\delta h}{3}(h+6b).$$

Then, the bending stress in the frame's weakest section will be:

$$\sigma = \frac{M}{W} = \frac{3K_l\, q\, \ell_{fr}^2}{\delta \mathrm{h}\,(\mathrm{h}+6\mathrm{b})}.$$

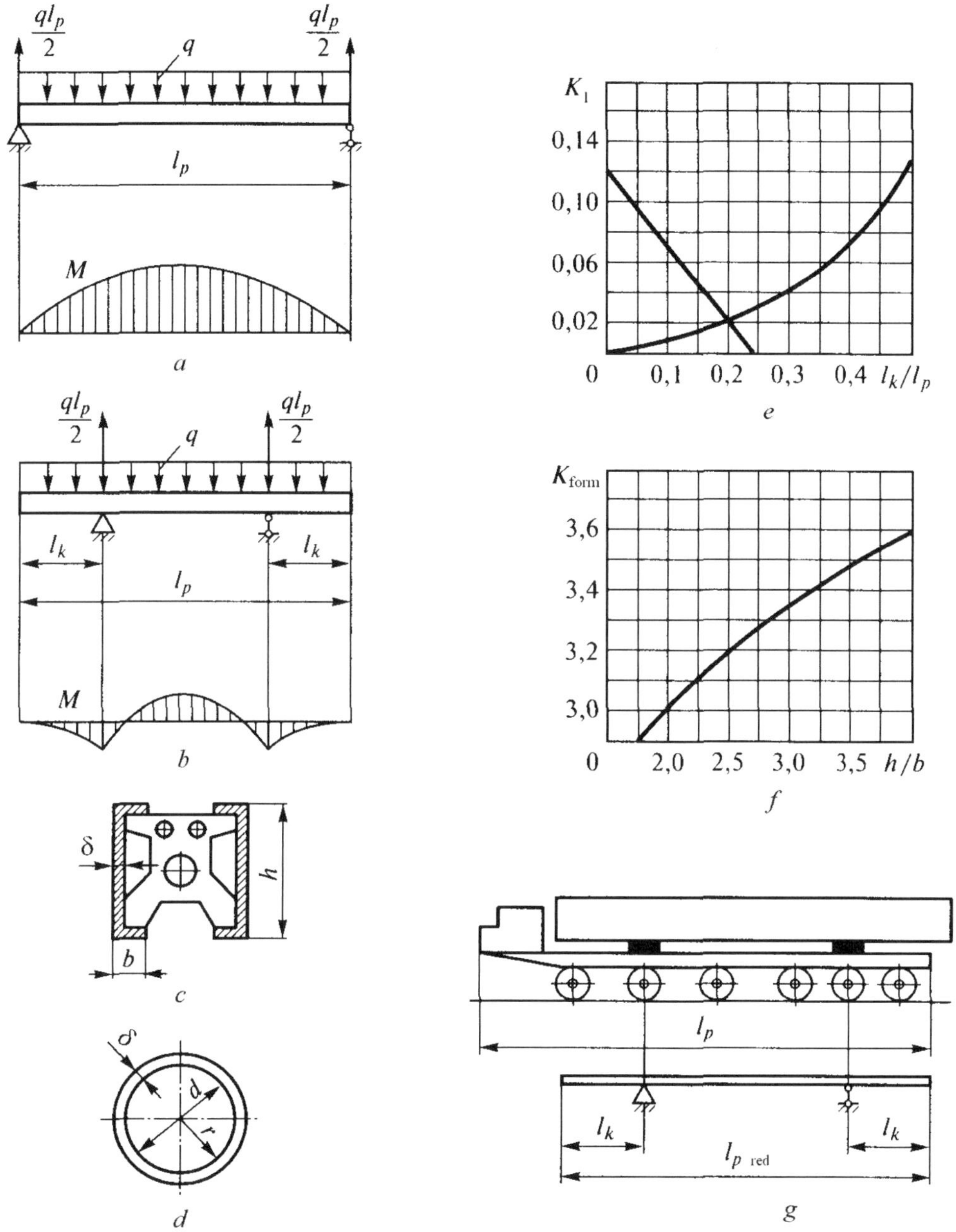

Fig. 9.37. Diagram for calculating the relative weight of the undercarriage (carrying system)

From which we get:

$$q = \frac{\sigma\delta h\,(h + 6b)}{3K_l\ \ell_{fr}^2}.$$

Thus, let us find an approximate full weight of the system:

$$m_{tr} = \frac{q\ell_{fr}}{g} = \frac{\sigma\delta h\,(h + 6b)}{3K_l\ g\ \ell_{fr}}$$

And frame weight:

$$m_{fr} = 2(h+2b)\,\delta\,\ell_{fr}\,\frac{\gamma}{g},$$

where h is the side rail height; b is the width of the side rail flange; γ is the specific weight of the frame material; and g is the gravity acceleration.

Then the relative frame model weight will be expressed by the formula:

$$\xi^*_{fr} = \frac{m_{fr}}{m_{tw}} = K_l\,\frac{6\,(h+2b)}{(h+6b)}\,\frac{\ell^2_{fr}\,\gamma}{\sigma\,h} = K_l K_{form}\,\frac{\ell^2_{fr}\,\gamma}{\sigma\,h}, \qquad (9.15)$$

where $K_{\text{form}} = \dfrac{6\,(h+2b)}{(\text{h}+6\text{b})}$ is the coefficient of the girders' cross-section shape depending on the h/b ratio defined by the diagram in Fig. 9.37*e*.

Considering that for the tubular frames (see Fig. 9.37*d*: the resisting moment)

$$W = \pi\delta r^2,$$

and:

$$m_{fr} = 2\pi\delta r^2\,\delta\ell_{fr}\,\frac{\gamma}{g},$$

we get in a similar way:

$$\xi^*_{fr} = \frac{m_{fr}}{m_{tw}} = \text{K}_l K_{form}\,\frac{\ell^2_{fr}\,\gamma}{\sigma\,d}, \qquad (9.16)$$

where d is the tubular frame diameter; r is the tubular frame radius; and K_{form} is the shape coefficient having a constant value for tubular frames of SWCs equal to 4.0.

If a tubular frame's diameter is taken as equal to the side rail height ($d = h$), then, according to formulas (9.15) and (9.16), the tubular frame weight will be 10–30% heavier than a straight (girder) frame in equal conditions. However, it should be noted that a girder frame requires cross-beams, the weight of which is comparable to the weight of longerons proper and is not considered by formula (9.15). Tubular frames do not require any cross-beams. We also note that open type longeron frames do not resist torsion as well as tubular ones. The polar moment of its resistance to torsion does not exceed 4% of the polar moment of resistance to torsion in a tubular frame. The torsion pressure in longeron frames can be twice as much or more than the bending pressure, thus requiring a greater cross-beam weight; therefore, with regard to reducing weight, tubular frames are more preferable than the open longeron frames.

Using statistical data [3], let us apply formulas (9.15) and (9.16) to real frames. Since loading in a real frame is only defined by the sprung weights, making:

$$m_{sm} = (0.78...0.9)\,m_{tw},$$

then formulas (9.15) and (9.16) should be complemented with ratios of the sprung and non-sprung weights:

$$K_m = \frac{m_{sm}}{m_{tw}} = 0.78\ldots0.9,$$

where large values relate to vehicles of rather low passing ability and mechanical drive of the wheels, and lesser values are for vehicles with high passing ability and motor-wheel drive (electrical or hydrostatic transmission).

The vehicle frame, according to formula (9.7), includes, in addition to the longerons, cross-beams, frame brackets, bumpers, and towing devices not considered in formulas (9.15) and (9.16), the weight of which can be 1.5–2 times greater than the weight of the longerons. That is why it is necessary to introduce the strength-weight ratio in these formulas:

$$K_k = \frac{m_{fr}}{m_{spar.f}} = 0.5\ldots2.0,$$

where lesser values refer to tubular frames and larger ones refer to open longeron frames.

The real frame is subject to greater dynamic loads, which need to be accounted for by the dynamic load factor:

$$K_d = 2.0\ldots2.5,$$

where lesser values are taken for speedy ($V \geq 60$ km/h) and highly passable vehicles.

Since the acting stresses must be less than allowable ones and the latter are defined by the strength standards as:

$$[\sigma] = \frac{\sigma_m}{n_m},$$

where $[\sigma]$ are allowable stresses; σ_m is the frame material yield point; and n_m is the strength margin taken within the limits of 1.4–2.0. Then, in formulas (9.15) and (9.16), instead of σ, the values of $\frac{\sigma_m}{n_m}$ should be placed.

The most common material for frame production is low-alloy steel grades 10ХСНД (σ_T = 400 MPa), 19ΦГС (σ_T = 430 MPa), 3OT (σ_T = 360 MPa), 14Г2С (σ_T = 390 MPa), 14Г2С (σ_T = 370 MPa), 25PC (σ_T = 280 MPa), etc.

To account for the nature of the concentrated forces application, the following coefficient is introduced in formulas (9.17):

$$K_{c.c} = 0.9\ldots1.05,$$

the lesser values of which refer to when the significant part of the load is concentrated above an axle (support) or above the group of axles (as, for example, in an attic), and greater ones to when the location of the concentrated forces is between the axles or at the frame ends.

In a real frame with identical loads on the wheel axles, the front and back overhangs, as a rule, are not equal in length, resulting from the non-uniform load distribution along the frame length. We shall consider this non-uniformity by using the reduced frame $l_{fr.p}$ for determining which front and back axle groups should be replaced with the equivalent front and back supports so that the reactions to them are identical. Then, in the layout drawing or loading diagram, the difference between the front and back overhangs should be subtracted from the length, l_{fr}, of a real frame. In other words, it is necessary to reduce the frame to the loading diagram, as shown in Fig. 9.37c. Now we finally have the formulas for calculating the relative weights of the longeron and tubular frames respectively:

$$\xi_{fr} = K_l K_{form} K_m K_k K_d K_{c.c}\, n_m \frac{\ell^2_{fr.p}\,\gamma}{\sigma_m\, h}; \qquad \xi_{fr} = K_l K_{form} K_m K_k K_d K_{c.c}\, n_m \frac{\ell^2_{fr.p}\,\gamma}{\sigma_m\, d}. \tag{9.17}$$

At preset K_e and n_m, the least frame weight may be obtained for a short-wheelbase vehicle with a tubular frame at l_k/l_{fr} = 0.21 and with use of light and strong alloys. Thus, for example, if duralumin (σ_T = 200 MPa) is used instead of grade 10ХСНД steel, then, with all other conditions being equal, the weight of such frame will be less by about 25%; if titanium alloys are used then the weight will be reduced by 40%. However, the decision to use these materials should only be made on the basis of a thorough technical and economic analysis.

In the majority of implemented multi-axle base heavy duty vehicle designs, the relative frame weight without control cabins and facing elements makes in average:

$$\xi_{fr} = 0.07\ldots0.104.$$

The frame height h is taken, as a rule, within the limits

$$h = (0.06\ldots0.07)l_{fr}.$$

The relative frame weight is calculated after developing the vehicle's design-layout scheme. The weight of the cabin and facing elements should be set in absolute values for greater calculation accuracy and included in the payload weight in calculations of the elements' relative weights within the limits

$$m_{in} = 900\ldots1200 \text{ kg}.$$

9.5.2.2 Definition of relative weights of wheeled units (wheeled running gear)

The wheeled running gear includes the wheel hubs with bearing units and elements of the central pumping system, tire rims, and tires. In-built wheel-hub drives relate to the transmission and the wheel brakes and brake drums relate to the brakes of the vehicle's control system; as a result they are not accounted for in the wheeled running gear weight.

Processing large statistical data has shown that the relative wheel hub weights may be set within the limits:

$$\xi_{wh} = 0.02\ldots0.03,$$

where greater values should be taken for high-speed vehicles and lesser values for slow-acting ones ($V_{max} \leq 50$).

The relative weights of the tire rims are also quite stable:

$$\xi_{wr} = 0.012\ldots0.02\,.$$

The relative tire weights to a great extent depend on the desired (designed) level of the passing ability defined by the minimum specific ground pressure, q_{min}. This dependence is sufficiently approximated by the empirical formula:

$$\xi_{tir} = \frac{0.004}{q_{min}}\sqrt[3]{\frac{q_{min}}{0.05}}\,.$$

The relative weight of the wheeled running gear is defined as

$$\xi_{wu} = \xi_{tir} + \xi_{wr} + \xi_{wh}\,.$$

Fig. 9.38 presents the dependence of the wheeled running gear's relative weight on the specific ground pressure constructed according to the above formulas and the statistical data of implemented designs. It can be seen from the diagram that reducing the specific ground pressure, for example, from q_{min} = 0.02 MPa to

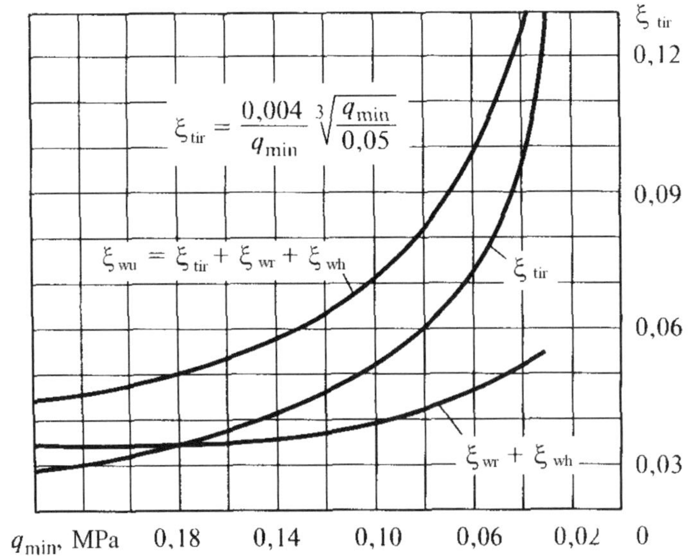

Fig. 9.38. Dependence of the wheeled running's gear relative weight on the specific ground pressure

q_{min} = 0.05 MPa, results in doubling the wheeled running gear's relative weight, which, at the vehicle's fixed weight, will lead to a considerable reduction in its load capacity. Further, at wheel loads of $G_{lw} = 5{\cdot}10^4$ H, its diameter increases from D_{tir} = 0.95 m at q_{min} = 0.02 MPa to D_{tir} = 2.1 m at q_{min} = 0.05 MPa, which will also reduce the vehicle's layout capabilities. Therefore, improving the passing ability of wheeled vehicles to account for the reduction in the specific ground pressure represents an intractable problem.

For vehicles with a high passing ability ($q_{min} \approx$ 0.2–0.1 MPa), the wheeled running gear's relative weight may be

$$\xi_{wu} = 0.070\ldots0.10.$$

9.5.2.3 Definition of the springing system relative weights

The springing system includes elastic, guiding, and damping elements of the wheel suspension, as well as the body regulation and stabilization system. The main influences on the springing system's relative frequency according to the results of research are the exertion of the full rated wheel stroke, suspension type, and the availability of the body regulation and stabilization system. The springing system's relative weight is defined as:

$$\xi_s = K_{rw} h_w + \xi_{m.r.s} + \xi_{s.h} \quad (9.18)$$

where K_{rw} is the coefficient depending on the elastic element type (for leaf springs K_{rw} = 0.07–0.08, for torsion springs K_{rw} = 0.05–0.06, for hydropneumatic springs K_{rw} = 0.04–0.05); h_w is the full wheel stroke; $\xi_{m.r.s}$ is the relative weight of the shock absorbers' guiding elements (for leaf spring suspensions $\xi_{m.r.s}$ = 014–0.02, for torsion spring suspensions $\xi_{m.r.s}$ = 0.012–0.016, for hydropneumatic spring suspensions $\xi_{m.r.s}$ = 0.01–0.015); and $\xi_{s.h}$ is the relative weight of the body regulation and stabilization systems if they are provided for by the design (for suspensions with metal elastic elements, $\xi_{s.h}$ = 0.01; for suspensions with hydropneumatic springs, $\xi_{s.h}$ = 0.005).

For vehicles with a high passing ability ($h_w \geq 0.3$ m), the relative weight of the springing system without regulation and stabilization systems may be

$$\xi_s \approx 0.025\ldots0.05\,.$$

9.5.2.4 Definition of the relative weights of the vehicle's control elements

Control elements include the steering system with a steering booster and elements of the braking system with wheel brakes. The main influences on the control elements' weight are exerted by the number of controlled wheels, and, more exactly, by the vehicle's weight falling on the steered wheels, and by the transmission type. The relative weight of the control elements is defined as:

$$\xi_{st} = 0.026\frac{m_{st}}{m_{tw}} + \xi_b \quad (9.19)$$

where m_{st} is the vehicle's weight falling on the steered wheels; m_{tw} is the vehicle's (aggregate's) gross weight; and ξ_b is the relative weight of the elements of the braking system with wheel brakes, according to statistical data ξ_b = 0.015–0.023.

In a vehicle equipped with a hydrostatic transmission with two-gear wheel-hub drives, the braking function can be fulfilled by the brakes of the wheel-hub drive's switching gears, then the brakes' weight is accounted for by the specific weight of the wheel-hub drive and in formula (9.19) the value ξ_b = 0.0.

When the load is equal on all axles, the steering system's relative weight is defined through the ratio of the number of steered wheels n_{st} to the total number of the vehicle's wheels n_Σ:

$$\xi_{st} = 0.026\frac{n_{st}}{n_\Sigma}\,.$$

9.5.2.5 Definition of the fuel relative weight

The fuel relative weight includes the weight of the fuel, fuel tanks, and fuel equipment located outside the engine. The fuel relative weight is defined according to the formula:

$$\xi_{fuel} = K_{\sigma} \frac{g_e \, D \, S_m}{360 \, \eta_p \eta_{st} \eta_{trans}} \tag{9.20}$$

where K_{σ} is the coefficient accounting for the weight of the tanks, K_{σ} = 1.1–1.2; D is the ability rating of the vehicle for which the cruising fuel range is defined, D = 0.35–0.045; S_m is the cruising fuel range, $S_m \geq 500$ km; and η_p, η_{st}, and η_{trans} are the coefficients accounting for the power take-off of the engine auxiliaries, control systems, and transmission efficiency factor, respectively.

As can be seen from formula (9.20), the relative weight of the fuel does not depend on the specific power-to-weight ratio of the vehicle and is only defined by the engine type, the motion resistance, the preset mileage of the vehicle without refueling, and the general losses in the power unit and transmission.

The specific fuel consumption for diesel engines is g_E = 0.224–0.244 kg+t / (kW+h) and for carburetor engines, g_E = 0.30–0.32 kg+t / (kW+h).

9.5.2.6 Relative weight of the supports

If the vehicle is equipped with suspension support, the weight of the supports depends only on the gross weight of the vehicle and, according to statistical data [3], with account of the cross-beams, leveling jacks, and drives can be

$$\xi_{sf} = 0.020\ldots0.032,$$

where greater values refer to outriggers.

9.5.2.7 Relative weight of the auxiliary equipment

The auxiliary equipment includes the electrical equipment of the vehicle's thermostatic system with battery pack; individual sets of spare parts, tools, and accessories; self-recovery winches; and other non-specified equipment. According to statistical data [3] for existing designs of multi-axle vehicles, the relative weight of such equipment can be

$$\xi_{oc} = 0.010\ldots0.015.$$

9.5.2.8 Relative weight of the power unit

The power unit includes the traction motor or motors (if more than one), engine mounting fixtures on a frame, and the systems serving the engine, including the motor oil and cooling liquid. The relative weight of the power unit depends on the vehicle's specific power-to-weight ratio and the specific reduced weight of the power unit and, therefore, may be defined according to the formula:

$$\xi_p = m_p N_w \,.$$

The specific reduced weight of the power unit is defined as:

$$m_p = K_p \frac{m^1{}_p + m^2{}_p}{\eta_p \, \eta_{st} \, \eta_{trans}} \tag{9.21}$$

where K_p is the coefficient accounting for the weight of the fixtures used to attach the engine to the vehicle's frame (K_p = 1.05–1.1); $m^1{}_p$ is the engine's specific weight by the nameplate data as delivered by the manufacturer (for carburetor engines, $m^1{}_p$ =1.1–1.8 kg/kW; for turbocharged diesel engines, $m^1{}_p$ = 2.3–3.0 kg/kW; for long-running diesel engines, $m^1{}_p$ = 3.75–4.6 kg/kW; and for automobile gas-turbine engines (AGTEs) with heat exchanger, $m^1{}_p$ =0.8–1.2 kg/kW with account of its servicing systems); $m^2{}_p$ is the specific weight of the engine systems for heavy multi-axle vehicles, according to statistical data, $m^2{}_p$ = 2.1–2.4 kg/kW; η_e, η_y are the coefficients accounting for the power take-off for the respective needs of the engine and control system ($\eta_p \eta_{st}$ = 0.80–0.85); and η_{trans} is the transmission's efficiency factor defined by the transmission type and number of driving axles.

For heavy multi-axle vehicles with hydromechanical transmissions and boosted diesel engines, the power unit's specific reduced weight can amount to m_p = 9–12 kg/kW; for the same vehicles with AGTEs, m_p = 3–4 kg/kW, i.e., three times less (better). That is why the use of AGTEs in heavy-duty vehicles is a promising development trend in basic vehicles.

9.5.2.9 *Relative weight of the transmission*

The transmission includes all of the transmitting (delivering) and transforming (converting) elements, the distribution and wheel reduction gearboxes, and also the transmission control, oil, and lubrication systems which service its elements. The relative weight of the transmission, like the relative weight of the power unit, depends on the vehicle's specific efficient power-to-weight ratio and the transmission's specific reduced weight, $m_{y.Tp}$, and is defined according to the following formula:

$$\xi_{trans} = m_{trans} N_w .$$

The specific reduced weight of the transmission depends on the transmission type and the number of driving axles. The average statistical data for the existing vehicles give the following values of the specific reduced weight of the transmission: the mechanical transmission of a vehicle with one driving axle, m_{trans} = 4–6 kg/kW; the mechanical transmission of the three-axle all-wheel-drive ZIL, Ural, KrAZ model vehicles, m_{trans} = 8–10 kg/kW; the hydromechanical transmission of the multi-axle MAZ-537 and MAZ-543 model vehicles, m_{trans} = 18–20 kg/kW; the hydrostatic transmission, irrespective of the number of driving axles, with a two-gear reduction gearbox for vehicles with a high passing ability, m_{trans} = 10–12 kg/kW; the DC electric transmission, irrespective of the number of driving axles, m_{trans} = 20–30 kg/kW; and the AC electric transmission, irrespective of the number of driving axles, m_{trans} = 18–25 kg/kW.

As can be seen from the above average statistical data, it is expedient to use hydrostatic transmissions (HSTs) in multi-axle vehicles with high passing abilities. They are especially beneficial in multi-drive vehicles and aggregates, which have, apart from the wheel drives, many special drives, for example, the load lifting drives, suspension supports, winches, etc., and also for engineering vehicles, where all of these mechanisms can be driven by the pumps of the transmission.

For more accurate calculations of the specific reduced weight of the hydrostatic transmission, measured in kg/kW, the following formula can be used:

$$m_{trans} = \frac{m_{w.g}}{\eta_{w.g}} + \frac{m_{a.g}}{\eta_{a.g}\eta_{h.p}\eta_{h.m}\eta_{w.g}} + \frac{m_{h.p}d_{h.p}K_N}{\eta_{h.p}\eta_{h.m}\eta_{w.g}} + \frac{m_{h.m}d_{h.p}d_{h.m}K_N}{\eta_{h.p}\eta_{w.g}} \cdot$$
$$+ m_{fl} + m_{a.m} + m_{p.l} \qquad (9.22)$$

where

- $M_{w.g}$ is the specific weight of the wheel gearboxes ($m_{w.g}$ = 2.8–3.1 kg/kW);
- $M_{a.g}$ is the specific weight of the distribution reduction gears ($m_{a.g}$ = 0.4–0.6 kg/kW);
- $M_{h.p}$ is the specific weight of the pumps ($m_{h.p}$ = 0.38–0.42 kg/kW);
- $M_{h.m}$ is the specific weight of the motors ($m_{h.m}$ = 0.65–0.75 kg/kW);
- $m_{f.l}$ is the specific weight of the working fluid ($m_{f.l}$ = 1.0–1.6 kg/kW);
- $m_{a.m}$ is the specific weight of the control equipment ($m_{a.m}$ = 0.4–0.6 kg/kW);
- $m_{p.l}$ is the specific weight of the power pipelines (for the steel grade 1X18H9T (σ_{cp} = 550 MPa) at a fluid flow rate in the pipes V = 6–8 m/s and an average distance from the power unit to the wheels of l = 5 m, $m_{p.l}$ = 1.0–1.5 kg/kW);
- $d_{h.p}$ and $d_{h.m}$ are regulation ranges of the pressure and specific volume, respectively;
- K_N is the coefficient of the hydraulic machines' power reserve, which depends on the required transmission service life and the pressure regulation range; it is defined from the diagram in Fig. 9.39 as the ratio of the maximum pressure in the hydraulic

machines, Δp_{max}, to the calculated pressure, $\Delta p_{max\,p}$, for the preset service life T and the pressure regulation range $d_{h.p}$; and

- $\eta_{a.g}$, $\eta_{h.p}$, $\eta_{h.m}$, and $\eta_{w.g}$ are the efficiency factors of the distribution reduction in the gears, pumps, motors, and wheel gearboxes, respectively.

Fig. 9.39 presents the dependence of the hydrostatic transmission's specific reduced weight on the regulation range and the maximum rated pressure in the system for the hydromechanical transmissions, built according to formula (9.22). The initial data are put on the field of the diagram. At a maximum pressure in the system of Dp_{max} = 40 MPa, the transmission ensures the service life $T \approx$ 300 h and at Dp_{max} = 30 MPa, $T \approx$ 800 h.

At the required regulation range of the gear, $d_{h.p} - d_{h.m}$ = 4–5, which in cases of two-gear wheel gearboxes in motors-in-wheel ensures the general regulation range, d_O = 16–20, the transmission's specific reduced weight will not exceed m_{trans} = 10.5–13.0 kg/kW, which is 1.5–1.7 times less than for hydromechanical transmissions and twice as small for electrical transmissions.

The relative weights of the power unit and the transmission depend on the vehicle's preset specific efficient power-to-weight ratio, N_w, which, according to the horsepower formula, defines the maximum travel speed, V_{max}, at the preset ability rating (dynamic factor) D_{min}:

$$N_w = \frac{g\,D_{min} V_{max}}{3600} \qquad (9.23)$$

where g is the gravity acceleration.

In order to ensure a maximum speed of V_{max} = 90 km/h at D_{min} = 0.03, the required power-to-weight ratio, according to formula (9.23), is N_w = 0.00735 kW/kg = 7.35 kW/t. The relative weight of the power unit and the transmission for this case can make:

- For vehicles with hydromechanical transmissions:

$$\xi_{trans} + \xi_p = (m_p + m_{trans})N_w$$
$$= (12 + 20) \cdot 0.00735 = 0.235 \text{ ; and}$$

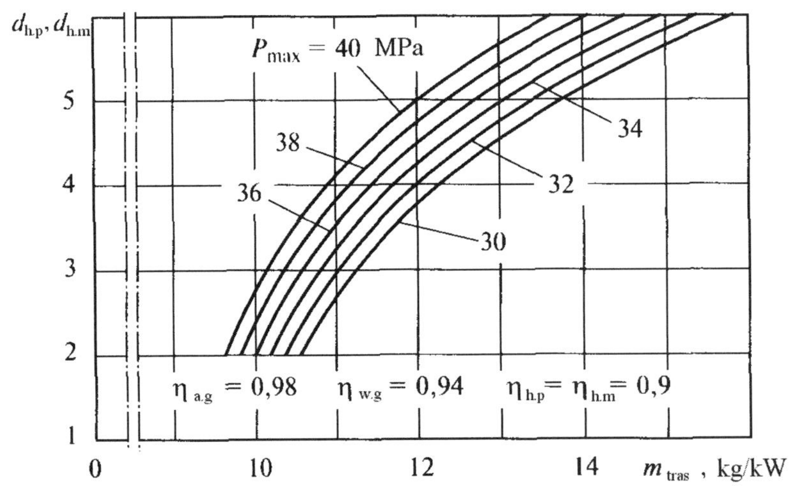

Fig. 9.39. The specific reduced weight of the hydrostatic transmission versus the regulation range and the maximum pressure: $d_{h.p}$ = 2.0; $m_{w.g}$ = 3.0 kg/kW; $m_{a.g}$ = 0.5 kg/kW; $m_{f.l}$ = 1.3 kg/kW; $m_{a.m}$ = 0.5 kg/kW; $m_{p.l}$ = 1.2 kg/kW; $m_{h.p}$ = 0.7 kg/kW; $m_{h.m}$ = 0.4 kg/kW

- For vehicles with hydrostatic transmissions:

$$\xi_{\text{trans}} + \xi_p = (m_p + m_{\text{trans}})N_w$$
$$= (12 + 12) \cdot 0.0735 = 0.176$$

If we take into account that in the existing vehicles, $\sum_1^n \xi_{ki} \approx 0.3$ (see formula (9.20)), then the relative payload weight will be within the limits of $\xi_{Lc} = 0.465\text{–}0.524$ and the vehicle's capacity utilization factor will be close to one, i.e., the vehicle will have its payload roughly equal to the weight of the vehicle itself.

9.6 Calculating the weight of wheeled vehicles

9.6.1 Designing and analyzing the design-layout schemes of basic vehicles for transport aggregates

HDWVs intended for mounting special equipment designed for various purposes (hereafter referred to as transport aggregates), can be designed and built as self-propelled vehicles or road trains consisting of a tow car (hauler) with a semi-trailer, a trailer, or several trailers. The layout selection of the transport aggregate and its basic vehicle is defined by the purpose of the aggregate, the full weight and overall dimensions of the transported cargo, the requirements for the traction-dynamic properties, and the level of its passing ability. However, it is not always clear which aggregate layout will turn out to be the most efficient or expedient. That is why sketches of several layouts are examined at the design stage and only after making a proper analysis can the choice of the optimal solution be made and its design started.

When starting the design of a wheeled vehicle intended to carry special equipment, the designer is guided by the technical requirements developed by the customer's engineers, which reflect the technical and techno-economic requirements of the transport aggregate's main purpose. Based on these requirements, the designer makes preliminary calculations and starts the design process.

The design process is a creative process, based, apart from the designer's technical skills and knowledge, on his or her intuition, i.e., his or her ability to choose the best solution out of the variety of possible design alternatives. One should not think, however, that the designer is provided with a complete freedom of intuition. It is restricted and limited, not only by the requirements of the customer's TOR (Terms of Reference), but also by the technical requirements of rational designing, which will finally define the degree of perfection (excellence) or the drawbacks of the finished product.

The rational requirements can be general, particular (specific), or special. The general requirements assume that the design should be simple, reliable, survivable, compact, small in dimensions and weight, and have a simplicity in service, repair, unification, and efficiency. Special requirements reflect peculiarities of the main technical applicability of the vehicle. The technical assignment for designing a basic vehicle for special aggregates with high passing ability and load capacity should specify its purpose, peculiarities, and operation conditions, as well as the general characteristics of the aggregate or a set of aggregates and initial data.

The following initial data is necessary for the design:

- The weight of the basic payload m_{Lc};
- The weight of the special equipment m_{ae};
- The overall dimensions, volume, and centers of mass of the cargo and special equipment (outline drawing);
- The minimal specific ground pressure q;
- The cruising fuel range S_m;
- The total warranty mileage for the whole service life S and the distribution of road

conditions as a percentage of the total mileage;

- The dimensional or road legislative restrictions; and
- The traction-dynamic qualities and other specific parameters.

The design-layout schemes of a wheeled vehicle or an aggregate are developed after the approximate calculation of its gross weight on the basis of the prototype is known. For example, it is recognized that heavy multi-axle vehicles with a high passing ability have a load capacity factor roughly equal to one.

The load capacity (useful load) factor is usually the ratio of the useful load weight to the basic vehicle weight:

$$K_g = \frac{M_{Lc}}{M_g} \tag{9.24}$$

where M_{Lc} is the payload weight including the actually transported cargo, special equipment, and container (body, cargo platform, hoppers, etc.); and M_g is the road weight of a wheeled vehicle completely equipped and tanked.

To determine the gross payload weight, statistical data on implemented designs are used. Thus, for example:

- The weight of sided platforms is 15–20% of the cargo weight transported on it;
- The weight of an air-tight closed refrigerator body or bodies for special purposes to accommodate equipment and people is 20–25% of the cargo weight transported in it;
- The weight of the tanks for fuel or other liquid components can be 25–30% of the transported component's weight;
- The weight of the hoppers for equipment located along the vehicle's boards is 10–15% of the weight of the equipment placed in them;
- The weight of independent electric power sources (diesel-electric sets) not including the fuel reserve is an of average 40–60 kg/kW;
- The weight of refrigerating units varies within the limits of 60–100 kg/kW of the refrigerating capacity; and
- The weight of a cargo boom with a hydraulic lift cylinder is at least 4–5% of the hoisted cargo's weight.

Thus, the payload weight is defined by summing up the weight of the cargo, special equipment, electric power sources, and fuel reserve for the aggregate's preset independent operation. If the weight is placed in a body, the body's weight is added and the aggregate's nominal gross payload M_{Lc} is received. Then, given the vehicle's load capacity factor K_g from formula (9.24), the basic vehicle weight is defined and the aggregate's gross weight $M_{tw} = M_g + M_{Lc}$ in a first approximation is defined. In presetting the load capacity factor, the passing ability level and the vehicle's preset traction-dynamic properties as compared to the prototype are accounted for. The higher the passing ability level and the dynamic properties preset by the technical assignment, the lower the preset load capacity factor should be.

The following stage is to determine the number of axles in view of the allowable axial load so as not to exceed the permissible load (for example, 10 t per axle) under the road legislation.

A 15 t or more axle load is sometimes provided for extra heavy aggregates of unit production. By dividing the aggregate's gross weight M_{tw} by the axial load M_{axis}, the minimum axle number is found:

$$n = \frac{M_{tw}}{M_{axis}}.$$

Further, the load per one wheel is defined as:

$$G_w = \frac{M_{axis}}{2} q,$$

and the tire type is selected accounting for the preset passing ability and the distribution of road conditions according to the assignment, the loading capacity diagram of the preset tires is constructed, and the wheels' diameter D_{tir} is defined at the preset specific ground pressure q. Then, the basic vehicle design-layout schemes are developed accounting for the rational designing requirements, the unification of wheeled vehicles, the service convenience of the mechanisms and systems, the compactability, etc. Here it is better to bear in mind that the overall dimensions of traction engines (diesels) of transport aggregates irrespective of their typical power is $N_{E\,min}$ = 150 kW; $N_{E\,max}$ = 600 kW traction engines are approximately 1 m wide, 1.0–1.1 m high, and differ only in length: 1.1–1.2 m for smaller power and 1.4–1.6 m for greater power diesels.

Systems serving the engine require space along the vehicle's boards and in its free places approximately equal to the engine volume. Fuel reserves for 500 km are approximately 2–3% of the aggregate's gross weight and about 2 m^3 of fuel tank volume is required for one ton of fuel. The vehicle's road clearance at a static load should be at least 0.5 m. The front and rear angles of the vehicle's overhang should be at least 25–30° according to the design drawing; and the vehicle's frame height should be at least 7–8% of its actual length. Control cabins should be at least: 1.6–1.8 m in length; 1.5–1.7 m in height; and 0.7–0.8 m in width (for a single-seat cabin). The cabin doors should be at least 60–70 cm wide and at least 1.2–1.3 m high. The inside height of the cabin body should be at least 1.8–1.9 m. The non-steered wheels (no more than three axles) should be mounted together having a maximum gap between the wheels of 5–10 cm. The steered wheels should have a gap of at least 30–40 cm. Zero bulkiness may also be applied to heavy wheeled vehicles. For a more efficient use of the lateral dimensions of railway load platforms, it is expedient to use a wheel suspension with a controlled variable clearance. In this case, the vehicle's board height can be greater by the value of the clearance, because the suspension is adjustable.

Let us consider the preliminary calculations procedure and defining the aggregate and its basic vehicle's general design-layout scheme with reference to a specific example [3]. The development of the design-layout scheme of a mobile automotive shop (MAS) according to the technical assignment and initial data is required.

The MAS is intended for repairs and maintenance of automobiles and other wheeled machinery in field conditions during the operation. The MAS crew consists of five persons. The stand-alone selector with regard to fuel reserves, water, and rations should be at least 10 days. The electric power consumption of machine tools, mechanisms, and equipment is estimated as 20 kW. The back-up electric power supply is doubled. All workshop equipment should be accommodated in an air-tight, heat-insulated body with a total space of 100 cu m. The weight of machine tools and other technological equipment with spare parts and materials is estimated as M_{Lc} = 20,000 kg.

The aggregate should have high passing ability and traction-dynamic properties expressed as follows:

- The maximum travel speed, $V_{max} \geq 60$ km/h;
- The minimum ability rating (dynamic factor), $D_{min} \geq 0.03$;
- The maximum ability rating, $D_{max} \geq 0.5$;
- The minimum specific ground pressure, $q_{min} \leq 0.1$ MPa;
- The cruising fuel range, $S_m \geq 500$ km; and
- The total mileage for the whole service life, $S \geq 20{,}000$ km.

The aggregate is supposed to operate in mixed road conditions (50% of mileage on hard roads and 50% on field roads). It should have zero bulkiness without dismantling and have supports for jacking and leveling on site.

It is allowed to vary the body's dimensions depending on the adopted aggregate's design diagram, and to modify its layout and the center of mass position, provided the total preset volume of $V_{K.y.3} = 100$ cu m is preserved.

First, let us estimate the aggregate's total payload weight according to these data. The diesel-electric's set specific weight is taken as $m_p = 60$ kg/kW. Then, at power value $N_{e.u} = 20$ kW and double back-up, we get the weight of the independent electric power sources:

$$m_{e.u} = 2N_{e.u}m_p = 2 \cdot 20 \cdot 60 = 2400 \text{ kg}.$$

The specific fuel consumption of a diesel-electric set can be taken as $g_e = 0.224$ kg/(kW·h). Then, accounting for a general efficiency factor $\eta_\Sigma = 0.75$, the following amount of fuel will be consumed in 1 hour:

$$m_{fuel} = \frac{N_{e.u}g_e}{\eta_\Sigma} = \frac{20 \cdot 0.224}{0.75} = 6 \text{ kg}.$$

Accounting for the stand-alone operation period of 10 days (T = 240 h) and the weight of fuel tanks, the total fuel system weight is defined as:

$$m_{\Sigma T} = 1.2Tm_{fuel} = 1.2 \cdot 240 \cdot 6 \approx 1700 \text{ kg}.$$

It is assumed that the body is equipped with a cooler-heater with a cooling efficiency of 4 kW. Thus, the weight of the air-conditioning system will be:

$$m_{air\ con} \approx 400 \text{ kg}.$$

The crew personnel weight with full equipage is:

$$m_{crew} \approx 400 \text{ kg}.$$

The weight of water and rations for the crew members for a stand-alone operation period of 10 days and daily per capita rate of 20 kg will be:

$$m_{food} \approx 1000 \text{ kg}.$$

The weight of the cabin's domestic equipment (tables, bunk beds, chairs, wash sink, stove, etc.) is:

$$m_{kitchen} \approx 400 \text{ kg}.$$

The weight of the aggregate field camouflage set is:

$$m_{camouflage\ cover} \approx 300 \text{ kg}.$$

Since the weight of the control cabins does not practically depend on the weight and overall dimensions of the aggregate, let us preset their weight complete with cabin equipment as:

$$m_{cabin} \approx 1200 \text{ kg}.$$

Then, the full (rated) weight of the transported cargo will be:

$$\begin{aligned} m^1_{Lc} &= m_{Lc} + m_{e.u} + m_{\Sigma T} + m_{air.con} + m_{food} + m_{food} + \\ &\quad m_{crew} + m_{kitchen} + m_{camouflage.cover} + m_{cabin} \\ &\approx 20{,}000 + 2400 + 1700 + 400 + 500 + 1000 \\ &\quad + 400 + 300 + 1200 \\ &\approx 28{,}200 \text{ kg}. \end{aligned}$$

Let us take the body weight to be equal to 23% of the transported cargo, which is:

$$m_{body} = 0.23\, m_{tw} \approx 0.23 \cdot 0.23 \cdot 28{,}200 \approx 6{,}500 \text{ kg},$$

And define in a first approximation the full (rated) weight of the basic load:

$$M_{tw} = m_{body} + m^1_{Lc} = 28{,}200 + 6500 = 34{,}700 \text{ kg}.$$

Given a load capacity factor of $K_g = 1.3$, the basic vehicle weight is defined (see formula (9.24)) as:

$$M_{base\ car} = \frac{M_{tw}}{K_g} = \frac{34{,}700}{1.3} \approx 26{,}700 \text{ kg}.$$

The gross weight of the aggregate in this case is:

$$M_O = M_{base\ car} + M_{tw} = 34{,}700 + 26{,}700 = 61{,}400 \text{ kg}.$$

Having preset the vehicle's weight on one axle to 10 t, we get a six-axle vehicle.

A wide-section tire with controlled air pressure may be proposed for the aggregate designed, and for load $G_K = 50000$ N ($m_{axle\ load}/2 \approx 5000$ kg) at the preset q_{min} = 0.1 MPa, the tire diameter is calculated as D_{tir} = 1.5 m.

Now we can start to develop and define the possible design-layout schemes of the basic vehicle and the aggregate as a whole. In doing so, we shall be geared to use diesel engines, hydrostatic transmissions and hydropneumatic wheel suspensions.

The creation of the aggregate is possible according to the three following basic diagrams [3]:

- A self-propelled aggregate on a unified six-axle chassis;
- A road train consisting of a four-axle tow car and a two-axle semi-trailer; and
- A four-axle hull-type vehicle with a two-axle power trailer that is not detachable from the basic vehicle.

Aggregates developed according to these three design-layout schemes and accounting for the above recommendations are shown in Fig. 4.40.

Fig. 9.40*a* presents the diagram of the aggregate on a self-propelled six-axle chassis with three steered front axles and three non-steered rear axles. The aggregate's power unit consists of two diesel engines located in bins at the frame boards between the rear and front wheel units. The aggregate's control cabin and the front suspension support are mounted in a front part of a longeron frame and the cross-beam with two back supports is mounted in the rear part of a frame. The road clearance is taken to be equal to 0.5 m, the height of the frame longerons to be 1.0 m, and the frame length to be 17 m. The wheels' working stroke is taken as 0.3 m. The wheel suspension is variable by means of hydropneumatic springs and lateral guiding elements which provide lowering by 0.3 m in rail

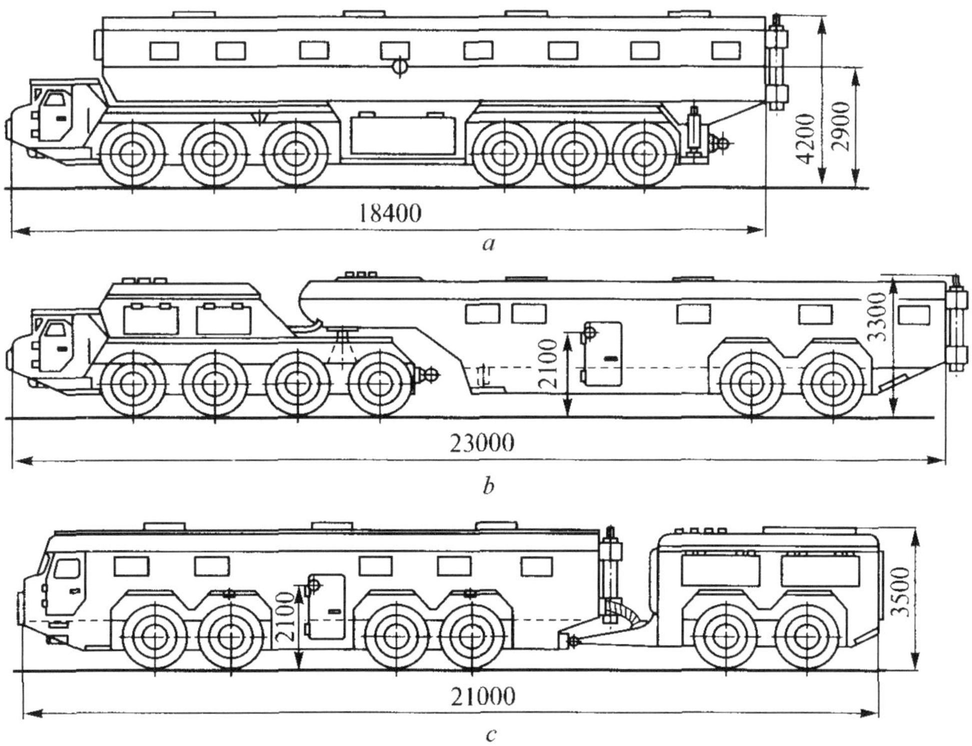

Fig. 9.40. Design-layout schemes of a transport aggregate on the basis of: a—A six-axle chassis; b—A saddle road train; c—A trailing road-train

transportation, therefore the aggregate's board height to the bevel edge is 2.9 m. The body's loading height is equal to 2 m.

The body's lateral section area, with its 2.2 m height and 3.4 m width accounting for the zero bulkiness, outlines to 6.4 sq.m. To get the body's preset volume (V_{body} = 100 cu m), a length of nearly 16 m is required and, if the body is accommodated on the chassis directly behind the control cabin, the aggregate's overall length will be 18.4 m. At that length, the body's center of mass with equipment will be nearly 2.9 m high and the aggregate's traveling position height will be 4.2 m, which, accounting for the ability to lower the suspension by 0.3 m, allows for providing transportation for the aggregate by rail without dismantling the body.

The main drawbacks of the aggregate on a self-propelled chassis are the high position of the body's mass center with equipment, which can result in an insufficient lateral stability when the aggregate moves; an outline of the body's cross-section inconvenient for layout (the height of the body's board up to the bevel edge is 0.9 m); and inconvenient to service and dismantle the power unit and its systems located under the body's bottom between the wheels in bins and engines accessed through hatch covers on one side.

Fig. 9.40*b* presents a diagram of the road train type aggregate consisting of a four-axle tow car and a two-axle semi-trailer with a bearing body (the semi-trailer has no frame). The overall length of the semi-trailer accounting for the projection of a saddle-type coupling unit is 16 m with the length of the main body part being 12 m. The total body space is about 110 cubic meters with an overall traveling height of 3.3 m. Accounting for the layout of the wheel suspensions, the body's floor is 0.7 m above its bottom so that the body's internal height is 2.0 m with almost no bevel edges of the boards. The body's cross-section outlines are more adapted to accommodate the internal equipment and service personnel. The body's loading height is 1.2 m, which is much more convenient than the aggregate in Fig. 9.40*a*.

The four-axle tow car has a uniform arrangement of the wheels along its length, two front axles are steered, the frame is longeron type, and the control cabin is located in its front part. The power unit and systems servicing it are placed on a frame of the tow car behind the control cabin in a special bin providing access to the engine and its systems from practically all sides. The semi-trailer and the tow car's centers of mass are approximately 2 m above the road surface, which provides for high stability for the moving aggregate. Using the bearing body layout allows for reducing the aggregate's gross weight as compared to the design in Fig. 9.40*a*.

The basic drawbacks of the diagram is in the greater overall length (23 m), which is 4.6 m longer than the self-propelled aggregate in Fig. 9.40*a*, a wider turning corridor and greater turning radius, and complicated reverse maneuvering.

Fig. 9.40*c* presents the diagram of the same aggregate in the form of a hull-type, four-axle chassis with two steered front axles and a two-axle power trailer with engines of the power unit and systems serving it, independent electric power sources, and a cooler-heater unit. The space in the four-axle chassis is approximately 100 cu m and the control cabin is combined with the hull of the main body (bus-like layout). This allows for considerably reducing the gross weight of the aggregate. The aggregate's high maneuverability, low position of the mass centers, and the convenience of servicing the power unit and its systems are preserved. The overall length of the road train is 21.0 m, which is only 2.6 m more than in Fig. 9.40*a*.

The difficulty in reverse maneuvering should be referred to the drawbacks of the design in Fig. 9.40*b*.

Thus, the developed versions of the aggregate's diagrams and their brief qualitative analysis provide the basis for selecting calculation model in Fig. 9.40*c*. The preference may be given to any diagram presented in Fig. 9.40 for other conditions and other vehicle applications.

9.6.1 Calculation of the aggregate's weight at the design stage

After defining the aggregate's overall design-layout solution resulting from the preliminary (approximate) calculations and making the outline drawing, we can start a precise calculation of the aggregate's weight characteristics at the preset traction-dynamic properties and the passing ability level on soft soils (grounds) using the formulas and recommendations of point 9.4. Let us consider the process of these calculations for the three aggregate diagrams (see Fig. 9.40) according to the above initial data.

9.6.1.1 Self-propelled six-axle chassis (see Fig. 9.40a)

According to the recommendations for formula (9.17) and by using the outline drawing, the reduced frame length l_{fr} = 15.3 m and the ratio l_k/l_{fr} = 1.6/15.3 = 0.105 (at hanging on the extreme axles), which in Fig. 9.37*e* gives K_l = 0.07. If we take the ratio for the frame h/b = 4.0 in accordance with Fig. 9.37*f*, then we get K_{form} = 3.6. Let us then set the values for the following parameters as: K_m = 0.9; K_k = 2.0; K_d = 2.2; $K_{\text{C.C}}$ = 1.05; and n_T = 2.0. Then, for 10ХСНД steel grade, we set: σ_T = 400 MPa; γ = 78,000 N/m^3; and h = 1.0 m.

Thus, let us define the relative weight of the frame according to formula (9.17) as follows:

$$\xi_{fr} = K_l K_{\text{form}} K_m K_d K_{\text{C.C}}\, n_m \frac{\ell_{fr}^2 \gamma}{\sigma_m} =$$

$$0.07 \cdot 3.6 \cdot 0.9 \cdot 2 \cdot 2.2 \cdot 1.05 \cdot 2.0 \frac{15.3^2 \cdot 78{,}000}{400 \cdot 10^6 \cdot 1.0} = 0.0715.$$

The relative weight of the wheeled aggregates for $q_{\min}$ = 0.1 MPa is defined according to the diagram shown in Fig. 9.38 as follows:

$$\xi_{wu} = 0.0885.$$

Let us adopt these values for the following parameters: K_e = 0.04, h_K = 0.5 m, $\xi_{\text{Н.э}}$ = 0.010, $\xi_{\text{рЕГ}}$ = 0.005, and, according to formula (9.18) we get the relative weight of the springing system as:

$$\xi_s = K_d h_k + \xi_{\text{m.r.s}} + \xi_{s.h.} = 0.04 \cdot 0.5 + 0.01 + 0.005 = 0.035$$

Let us set $\xi_{\text{ТОрМ}}$ = 0.0 (accounted for in the two-gear wheel gearboxes). According to the diagram in Fig. 9.40*a*, having defined the ratio $\frac{m_{st}}{m_g}$ according to formula (9.19), let us find the relative weight of the control system:

$$\xi_{st} = 0.026 \frac{m_{st}}{m_g} + \xi_b = 0.026 \cdot 0.5 + 0.0 = 0.013.$$

Let us adopt these values for the following parameters: K_σ = 1.2; g_1 = 0.23; D = 0.04; S_m = 500 km; $\eta_d \eta_{st}$ = 0.85; and η_{trans} = 0.7 and find the relative weight of the fuel and tanks according to formula (9.20):

$$\xi_{fuel} = K_\sigma \frac{g_1\, D\, S_m}{360\, \eta_d \eta_{st} \eta_{trans}} = 1.2 \frac{0.23 \cdot 0.04 \cdot 500}{360 \cdot 0.85 \cdot 0.7} \approx 0.027.$$

Let us also set the relative weight of the rigidly fixed supports as ξ_{sf} = 0.02 and the relative weight of the auxiliary equipment as ξ_{oe} = 0.01. Then, the relative weight of the designed assembly according to formula (9.18) will be:

$$\sum_1^n \xi_{ki} = \xi_{fr} + \xi_{wu} + \xi_g + \xi_{st} + \xi_{fuel} + \xi_{sf} + \xi_{oe} + \xi_p =$$

$$= 1.0715 + 0.0885 + 0.035 + 0.013 + 0.027 + 0.02 + 0.01$$

$$= 0.265.$$

According to formula (9.49), we get:

$$N_w = \frac{g\, D_{\min} V_{\max}}{3600} = \frac{9.8 \cdot 0.03 \cdot 360}{3600} = 0.0049\text{, kW/kg.}$$

At a general regulation range of

$d = \frac{D_{\max}}{D_{\min}} = \frac{0,5}{0,03} = 16,7$ and a two-gear reduction box, $d_u = \sqrt{d} = 4.1$ let us find the transmission's specific reduced weight according to the diagram in Fig. 9.39:

$$m_{\text{trans}} = 12 \text{ kg/kW}.$$

By setting $K_{c.c} = 1.1$; $m^1_p = 4.6$ kg/kW; and $m^2_p = 2.1$ kg/kW, we get the power unit's specific reduced weight according to formula (9.21):

$$m_p = K_d \frac{m^1_p + m^2_p}{\eta_d \, \eta_{st} \, \eta_{trans}} = 1.1 \frac{4.6 + 2.1}{0.85 \cdot 0.7} = 12.3 \text{ kg/kW}.$$

The relative payload weight is defined according to formula (9.12) as follows:

$$\xi_{tw} = 1 - \sum_1^n \xi_{ki} - (m_p + m_{trans}) N_w$$
$$= 1 - 0.625 - (12.35 + 12) \cdot 0.0049 = 0.616$$

Thus, the aggregate gross (full) weight is:

$$M_o = \frac{m_{tw}}{\xi_{tw}} = \frac{34,700}{0.616} = 56,500 \text{ kg},$$

of which the body's weight with equipment according to the initial data and calculations is $m_p = 33{,}500$ kg and the basic vehicle weight is $m_{\text{b.M}} = 23{,}000$ kg. The components of the basic vehicle will have the following values according to formula (9.14):

$$m_{\text{cabin}} = 1200 \text{ kg};$$
$$m_{fr} = 4050 \text{ kg};$$
$$m_{wu} = 5000 \text{ kg};$$
$$m_g = 2050 \text{ kg};$$
$$m_{st} = 740 \text{ kg};$$
$$m_{\text{fuel}} = 1520 \text{ kg};$$
$$m_{sf} = 1130 \text{ kg};$$
$$m_{oe} = 570 \text{ kg};$$
$$m_p = 3400 \text{ kg};$$
$$m_{\text{trans}} = 3340 \text{ kg}.$$

The vehicle's load capacity factor will be:

$$K_g = \frac{M_{Lc}}{M_g} = \frac{33,500}{23,000} = 1.46 \, .$$

The load per one wheel is $G_w = 46{,}000$ N, which at $D_{\text{tir}} = 1.5$ corresponds to $q_{\min} = 0.09$ MPa; thus, the vehicle's passing ability will be somewhat higher than the preset level.

After selecting the engines and transmission elements, the weight of the power unit and transmission can be specified and corrected, which will not essentially change the aggregate's gross weight in case of a corrected definition of the specific weights of its components.

9.6.1.2 Road train with a two-axle semi-trailer (see Fig. 9.40b)

The calculation of the weight characteristics of this construction is made for the semi-trailer first, and then for the tow car. Since the body of the semi-trailer has the same volume as the body of the design in Fig. 9.40*a*, we can take the same full weight for it as:

$$M_{Lc} = 33{,}500 \text{ kg}.$$

To define the gross weight of the semi-trailer, we need to take into account the weight of the wheel units, wheel suspensions, and supports. If we assume that the body's load is distributed equally among the rear wheel unit and the saddle coupling, then the weight equation in relative values will be expressed by the following formula:

$$\xi_{Lc} = 1 - 0.5(\xi_{wu} + \xi_g) - \xi_{sf}.$$

Since the passing ability characteristics and the types and strokes of the suspensions remain the same as for the design in Fig. 9.40*a*, we can assume the following:

$$\xi_{wu} = 0.0885; \xi_g = 0.035; \xi_{sf} = 0.02.$$

The payload relative weight will be defined as:

$$\xi_{Lc} = 1 - 0.5(\xi_{wu} + \xi_g)$$
$$= 1 - 0.5(0.0885 + 0.035) - 0.02$$
$$= 0.8565,$$

And the full semi-trailer weight will be:

$$m_s = \frac{M_{Lc}}{\xi_{Lc}} = \frac{33,500}{0.8565} = 39,100 \text{ kg}.$$

According to the adopted distribution, the semi-trailer's load on the tow car's saddle coupling unit will be half of the weight of the semi-trailer and will be the tow car's payload needed for calculating its weight:

$$m_{Lc.t} = \frac{m_s}{2} = 19,550 \text{ kg}.$$

The tow car's payload should also include the weight of the control cabin (m_{cabin} = 1200 kg), as well as the weights of the power unit, the transmission pumps, and the saddle coupling unit ($m_{\text{t.p}}$ = 1500 kg). Finally, the tow car's payload should be complemented with the weight of the power unit and transmission needed to convey the weight falling on the rear wheel unit of the semi-trailer with the same dynamic properties as for the gross weight of the tow car. According to calculations (see Fig. 9.40*a*), $\xi_p + \xi_{\text{tras}} = 0.119$. Then the weight of the power unit and transmission for the drive of the semi-trailer's rear wheel unit will be:

$$m_{p.s} + m_{\text{act}.s} = \frac{m_{nn}}{2}(\xi_p + \xi_{trans})$$
$$= 19,550 \cdot 0.119 = 2340 \text{ kg}.$$

The full conventional (rated) payload of the tow car is:

$$m_{\Sigma\text{Lc.t}} = m_{\text{Lc.t}} + m_{\text{cabin}} + m_{\text{t.p}} + (m_{\text{p.s}} + m_{\text{act.s}})$$
$$= 19,500 + 1200 + 1500 + 2340 = 24,590 \text{ kg}.$$

As applied to the tow car design presented in Fig. 9.40*b*, let us define the relative weight of the frame as:

$$\xi_{fr} = K_l K_{form} K_m K_k K_d K_{c.c} n_m \frac{\ell^2_{fr.s}\gamma}{\sigma_m h} =$$

$$0.07 \cdot 3.6 \cdot 0.9 \cdot 2 \cdot 2.2 \cdot 1.0 \cdot 2.0 \cdot \frac{7,6^2 \cdot 78\ 000}{40 \cdot 10^6 \cdot 0,8} = 0.035.$$

Let us adopt the following values of the parameters:

$$\xi_{wu} = 0.0885;$$
$$\xi_g = 0.035;$$
$$\xi_{st} = 0.013;$$
$$\xi_{\text{fuel}} = 0.04;$$
$$\xi_{oe} = 0.01;$$
$$m_p = 12,3 \text{ kg/kW};$$
$$m_{\text{trans}} = 12,0 \text{ kg/kW};$$
$$N_w = 0.0049 \text{ kW/kg}.$$

Then, the relative weight of the whole structure will be:

$$\sum_1^n \xi_{ki} = \xi_{fr} + \xi_{wu} + \xi_g + \xi_{st} + \xi_{fuel} + \xi_{oe}$$
$$= 0.035 + 0.0885 + 0.035 + 0.013$$
$$+ 0.040 + 0.01 = 0.2215.$$

The relative payload weight is:

$$\xi_{Lc} = 1 - \sum_1^n \xi_{ki} - (m_p + m_{trans})N_w$$
$$= 1 - 0.2215 - (12.3 + 12) \cdot 0.0049$$
$$= 0.6595$$

Now, the tow car's full weight is defined as:

$$M_{o\Sigma} = M_{o.s}\frac{m_{\Sigma Lc.t}}{2} = 37{,}200 + 19{,}550 = 56{,}750 \text{ kg}$$

And the gross weight of the whole aggregate (road train) as:

$$M_{o.t} = \frac{m_{\Sigma Lc.t}}{\xi_{Lc}} = \frac{24,590}{0.6595} = 37,200 \text{ kg},$$

which is 250 kg more than for the design in Fig. 9.40*a*.

The full weight of the tow car without any load on the saddle coupling unit is as follows:

$$M_t = M_{o.t} - M_{Lc.t} = 37{,}200 - 19{,}550 = 17{,}650 \text{ kg}.$$

In this case, the components of the tow car will have the following weights:

$$m_{cabin} = 1200 \text{ kg};$$
$$m_{fr} = 1300 \text{ kg};$$
$$m_{wu} = 3280 \text{ kg};$$
$$m_g = 1300 \text{ kg};$$
$$m_{st} = 480 \text{ kg};$$
$$m_{fuel} = 1490 \text{ kg};$$
$$m_{t.p} = 1500 \text{ kg};$$
$$m_{oe} = 370 \text{ kg};$$
$$m_p + m_{Ttrans} = 6730 \text{ kg}.$$

We see from the calculations that the weight parameters of the road train are not inferior to the design of the self-propelled chassis, in spite of the presence of the additional weight from the power unit bin and the saddle coupling unit (not accounting for the semi-trailer's weight).

9.6.1.3 Road train with a two-axle trailer (see Fig. 9.40c)

The calculation of the weight characteristics for this layout is similar to that of Fig. 9.40*b*, with the only difference being that we initially define the weight of the basic vehicle and then the weight of the powered trailer. When calculating the layout, it is expedient to place the electric power independent sources ($m_{e.u}$), the cooler-heater ($m_{air.con}$), and the fuel reserve for the independent sources (m_{fuel}) on the trailer and not in the body of the basic vehicle. Then, without the above weights and accounting for the control cabin, the weight of the tow car's body will be:

$$M_b = 30{,}200 \text{ kg}.$$

The relative payload weight will be:

$$\begin{aligned}\xi_{Lc} &= 1 - \xi_{wu} - \xi_g - \xi_{st} - \xi_{sf} - \xi_{oe} \\ &= 1 - 0.0885 - 0.035 - 0.013 - 0.02 - 0.01 \\ &= 0.8335,\end{aligned}$$

and the full weight of the tow car will be:

$$\xi_{Lc} = 1 - \xi_{wu} - \xi_g - \xi_{st} - \xi_{sf} - \xi_{oe} = \text{kg}.$$

Its constituent weights will have the following values:

$$M_{wu} = 3250 \text{ kg};$$
$$m_g = 1280 \text{ kg};$$
$$m_{st} = 480 \text{ kg};$$
$$m_{osf} = 730 \text{ kg};$$
$$m_{oe} = 360 \text{ kg}.$$

In order to achieve the preset traction-dynamic properties of the tow car, the required weights of the power unit, transmission, and fuel will be as follows:

$$m_p + m_{trans} = M_{o.t}(\xi_p + \xi_{trans}) = 36{,}300 \cdot 0.119 = 4320 \text{ kg};$$
$$m_{fuel} = M_{o.t}\xi_{fuel} = 36{,}300 \cdot 0.027 = 980 \text{ kg}.$$

Then, the rated payload of the trailer will be:

$$\begin{aligned}m_{Lc.s} &= (m_{p.s} + m_{trans.s}) + m_{fuel} + m_{eu} + m_{\Sigma fuel} + m_{air.con} \\ &= 4320 + 980 + 2400 + 1700 + 400 = 9800 \text{ kg}.\end{aligned}$$

Thus, we have the following for the trailer:

$$\sum_{1}^{n} \xi_{ki} = \xi_p + \xi_{wu} + \xi_g + \xi_{oe}$$
$$= 0.035 + 0.0885 + 0.035 + 0.027 + 0.02$$
$$= 0.2055.$$
$$\xi_{Lc} = 1 - \sum_{1}^{n} \xi_{ki} - (m_p + m_{trans})N_w$$
$$= 1 - 0.2055 - 0.119 = 0.6755$$

The full weight of the trailer is:

$$M_{o.s} = \frac{m_{Lc.s}}{\xi_{Lc}} = \frac{9800}{0.6755} = 14,500 \text{ kg},$$

and the gross weight of the aggregate is:

$$M_{o\Sigma} = M_{o.t} - M_{o.s} = 36{,}300 - 14{,}500 = 50{,}800 \text{ kg},$$

which is 5700 kg less than for the design in Fig. 9.40*a*, which is caused by the presence of the carrying body in the design and by smaller frame lengths.

Thus, the calculation of the weight characteristics shows a considerable advantage to the aggregate in Fig. 9.40*c* as compared to the designs in Fig. 9.40*a* and *b*.

9.6.1.4 Definition of the aggregate's general center of mass and the center of the sprung weights

After defining the gross weight of the aggregate and its components, it is now necessary to find the aggregate's center of mass position as a whole and the center of its sprung weights (masses). We need the position of the general center of mass for defining the loads on the wheels and calculating the motion stability and soft driving (traveling comfort) parameters.

To calculate the center of mass position, we need to have the aggregate's general design-layout scheme made on a scale. This scheme is used to place—after placing (for) the aggregate's component weights—the center of mass positions of all its components as presented in Fig. 9.41.

The following legend is used in Fig. 9.41: m_{oe} is the weight of the auxiliary equipment, including the spare parts, tools, and accessories (SPTA); m_{cabin} is the weight of the cabin; $m_{p.}$ is the weight of the power unit; m_{st} is the weight of the vehicle's control elements; m_g is the weight of the springing system (shock absorbers); $m_{wu.}$ is the weight of the wheel units including the weight of the tires; m_{trasp} is the weight of the transmission; m_{fr} is the weight of the body frame; m_o is the vehicle's gross weight; m_{fuel} is the weight of the fuel with the fuel tanks; and m_{Lc} is the payload weight, including the body or the platform.

Then, we set the coordinate system *XY*. We can choose the coordinate axes and the origin at random; however, it is better to align the *X* axis with the surface of the wheels and the *Y* axis with the front plane of the aggregate as shown in Fig. 9.41. In this case, the aggregate's general center of mass is found

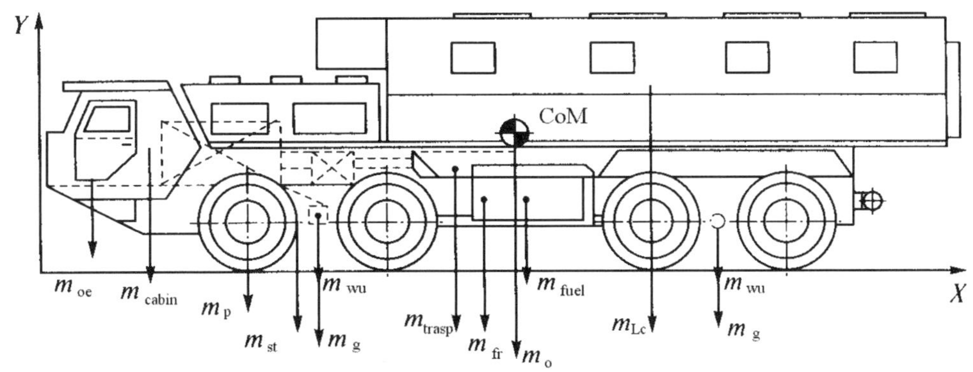

Fig. 9.41. Diagram of a wheeled vehicle for calculating the center of mass

as the application center of the resulting force from all forces of the elements' weights. Two equations of the moments in relation to each coordinate axis are compiled for this purpose:

$$X_{c.m} = \frac{\sum_{1}^{n} m_i \, x_i}{m_{>}};$$

$$Y_{c.m} = \frac{\sum_{1}^{n} m_i \, y_i}{m_{>}},$$

where x_i and y_i are the coordinates for the aggregate's elementary weights according to the design drawing and m_i are the elementary weights of the aggregate's units.

In order to improve the vehicle's passing ability and stability, we need to position its center of mass as low as possible and at a position along the aggregate's length where the load is the same on all wheels. An overloading of the rear wheels relative to the front ones should be no more than 5–10%.

In order to calculate the lateral stability and soft driving, we need to know the center of the aggregate's sprung weights. Therefore, to finalize the layout and ensure a uniform loading, we need to calculate the sprung weights' center by using the same formulas. To do so, the torques m_{ixi} and m_{iyi} of the non-sprung weights are excluded from them, i.e., the weights of the wheel units, half of the weight of the springing system, and part of the elements of the transmission and brakes mounted in the wheels; while the value of the aggregate's sprung weight m_{sp} is put instead of the weight m_o.

After defining the centers of the sprung weights, we define the inertia moments of the sprung weight relative to the longitudinal and lateral axes passing through the center of mass. These inertia moments are necessary to calculate the parameters of soft driving.

Chapter 10
Power units of wheeled vehicles

10.1 General characterization of power units

Choosing the power units (PUs) in the process of designing HDWVs is an important task, which in general requires considering the main initial data on the existent and perspective energy generation systems. Such findings are also necessary as the starting data for computer-aided designing (CAD); that is why we have summed up the main data on PUs borrowed from different literature sources.

The term power unit is understood as a thermal engine (or another source of energy) and its servicing systems are: the fuel and air supply, the cooling and lubrication, the startup, and the heating and ignition systems.

The basic part of any PU is the engine where the energy released at the combustion of fuel turns into mechanical energy on the exit (output) shaft of the engine.

The engine and its external systems is one of the main units of the HDWV's motor-transmission installation (system), which forms—by its units, systems, and communications—an essential part of the vehicle's road weight and affects its traction-dynamic, economic, and other characteristics.

That is why the choice of the engine type and its characteristics is a very important stage in the process of designing a vehicle.

Diesel engines (a class in the scope of internal combustion engines (ICEs)) and gas-turbine engines (GTEs) are used as part of PUs in HDWVs. Carburetor (petrol) ICEs cannot be used in HDWVs because of their limited power output; thus, they are excluded from review in this book.

When choosing an engine for a HDWV, its external characteristics, overall dimensions, weight, and reliability should be taken into account.

The external characteristics of diesel and gas-turbine engines are presented in Figs. 10.1 and 10.2. The curves represent the dependencies of the main engine parameters (power N_e, turning moment (torque) M_{et}, and the hourly Q_{hr} and specific g_e fuel consumption) on the engine's crankshaft rpm ω at full fuel supply.

The following rotation modes of the engine's crankshaft are distinguished in the engine characteristics:

- $\omega_{\min}$ – the minimum stable rpm of the engine;
- ω_M – the rpm corresponding to the maximum torque;
- ω_{ge} – the rpm corresponding to the minimum specific fuel consumption;
- ω_N – the rpm corresponding to the maximum engine power; and
- $\omega_{\max}$ – the maximum engine rpm.

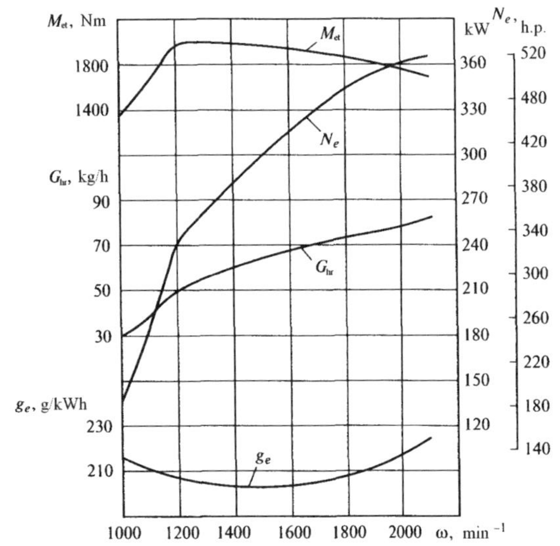

Fig. 10.1. External characteristic of the YaMZ-846.10 engine model

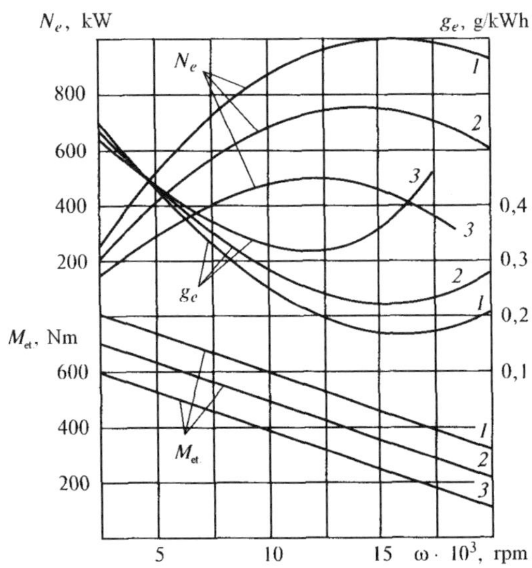

Fig. 10.2. External characteristic of a two-shaft GTE: 1—At the rated mode of compressor operation; 2 And 3—At the non-rated modes

The change character of the torque curve is expressed through the adaptability coefficient K_M (torque coefficient), which is understood as the ratio of the maximum engine torque to the torque at the maximum power.

Despite the continuous development and upgrading of piston ICEs and their broad use in transport means, they are still far from ideal thermal machines and have a number of inherent defects, the most essential of which are as follows [55]:

- Low efficiency factor—in diesel engines it does not exceed 0.4–0.41;
- The cycling (intermittent) operation mode;
- The presence of a complex crank-and-rod mechanism;
- High requirements for the fuel properties and quality;
- Insufficient adaptability to external loads;
- Contamination of the environment with exhaust gases; and
- Problems with starting the engine in cold weather.

For these reasons, the piston (reciprocating) engines do not yet fully meet the requirements for PUs in HDWVs. That is why the engineering society is constantly working toward designing new engines which are free, to a certain extent, from the drawbacks of piston ICEs; the automobile gas-turbine engines (GTEs) are the most promising option.

Automobile GTEs (hereafter, AGTEs), as the most promising engine for HDWV PUs have a number of essential advantages over ICEs, namely:

- A direct and continuous operation cycle, which ensures a simple design and drastically reduces the overall dimensions and weight;
- The absence of the cooling system and its units and parts;
- High starting qualities at any ambient air temperature;
- High adaptability to external loads and a linear descending dependence of the torque versus the rpm;
- Low requirements for fuel quality; and
- Small overall dimensions and weight.

However, they also have a number of disadvantages, including:

- Low overall efficiency;
- Poor operation at transition operation modes;
- High cost of used materials; and
- High weight and overall dimensions of heat exchangers and air cleaners.

Another more distant perspective in developing PUs is designing non-mechanical converters of the fuel's chemical energy into electricity, the so-called fuel elements (combustion cells), which are today at the stage of theoretical studies and experimental verification.

The comparative characteristics of different engine types are presented in Table 10.1.

base (dimensions, number and location of cylinders, etc.). It allows for designing highly efficient diesel engine modifications, which, while differing in the number of cylinders, body part configurations, and in the absence or availability of forced boosting units and boosting air coolers, can be used as part of power units for military vehicles, automobile and marine transport means, etc., while maintaining a high level of standardization (unification) (up to 80–85%).

The following can be highlighted among the reasons that contributed to the development of multi-purpose transport diesel engines: on the one hand, the presence of the accumulated scientific background, which in combination with the high level of engine-building technologies allowed for solving several functional tasks on the basis of one universal engine configuration;

Table 10.1 Comparative characteristic of engines

Engine type	Specific weight, kg/kW	Specific fuel consumption, kg/kW-h	Effective efficiency factor	Adaptability coefficient	Assessment of starting at low temperatures
Carburetor	1.1–2.0	0.3–0.36	0.3–0.35	1.25–1.35	Satisfactory
Diesel	2.3–5.0	0.22–0.25	0.35–0.41	1.05–1.115	Bad
AGTE	1.0–1.4	0.35–0.45	0.25–0.30	2.0–2.5	Good

10.2 Engine development trends

In recent years, the world engine industry has exhibited a trend toward designing multifunctional engines and an intention to reduce the number of specialized designs [55].

The idea of multifunctional engines is based on the current scientific-technical background in engine-building, newly developed high-strength and fire-resistant materials, and advanced technologies, which resulted in manufacturing the main engine elements with high strength and thermal margins without any significant increase in their weight-dimensional indicators and preserving the permanent general configuration

and, on the other hand, a reduction in the manufacturing of individual (customized) engines, which was becoming unprofitable. It is clear that reserves of industrial capacities are indirectly created, capable of reorienting the production of the required number of engines for specific technical purposes within the shortest lead time (1.5–2.0 months).

In order to have the required energy property values, it is not necessary to design and build any special high-energy power units for HDWVs, it is enough to upgrade (modernize) the existing models of basic diesel engines. At the same time, the main trend is to use turbo charging (boosting), which, apart from the power increase, allows for meeting the environmental requirements, improving the utilization of exhaust gases, and increasing the average effective pressure of the engine operation to the value, which guarantees the required level of reliability and endurance in severe operation conditions [18, 54].

The best foreign-made automobile diesel engines have a liter capacity of about 23–50 kW/l and a minimum specific fuel consumption of 192–202 g/kW×h. These indicators are reached by increasing the efficiency factor of the operation cycle through increasing the maximum pressure of the fuel combustion; diesel engines with a power of 200–450 kW have an average effective pressure of up to 1.6 MPa [53, 55].

The most advanced Russian-made engines have a specific fuel consumption of 203–217 g/kW×h, which is 5–12% worse than in foreign-made analogs.

The inferior fuel efficiency of Russian-made diesel engines, as compared to foreign ones, is caused by the lower intra-cylinder pressure of the working medium in the course of doing the effective work, and, accordingly, by the lower values of the average effective pressure. Thus, the average effective pressure in engines without supercharging is 0.65–0.69 MPa, while diesel engines with gas-turbine supercharging have the values of 0.85–0.95 MPa. Only advanced diesel engines of the YaMZ-840 series have an average effective pressure comparable to foreign analogs. For example, in the YaMZ-846 it is 1.22 MPa. In this context, the liter capacities of Russia-made (hereinafter referred to as "domestic") engines are within the limits of 13–23 kW/l, which is somewhat lower than the liter capacities of the best foreign analogs.

The increase in fuel efficiency for modern engines is significantly contributed to by the introduction of new turbo superchargers (TSCs) [56] with an efficiency factor of 50–60%, a raise in future TSCs of up to 70%, and the increase in the supercharging level of up to 3–3.5.

The use of TSCs with a high efficiency factor and a flat characterisric curve allow for raising the torque backup to 20–23%, the adaptability coefficient up to two, and overcoming the smoke emissions at low rpms of the crankshaft, an essential drawback of supercharged diesel engines.

In order to improve fuel efficiency and reduce the amount of toxic gas emissions, several companies are trending toward implementing electronics in the control systems of diesel engines (Ford, Caterpillar, Cummins, Toyota, Ricardo, etc.) [56].

Using electronics to control the fuel supply according to the load, rpm, temperature, and air weight at the input decreases the specific fuel consumption by 15%.

At present, only a complex system is able to solve this task, therefore, foreign companies pay great attention to selecting the system functions that yield the most significant effect resulting from the compromise between the fuel efficiency and the vehicle's dynamic properties, reliability, and system costs.

Thus, Cummins designed the XAP-1000 power unit with the XAV-28 diesel engine for a fourth generation advanced tank. The XAV-28 diesel engine is considered a basic one for the development of 4-, 6-, 8-, and 12-cylinder serially produced engines with a power range of 150–1490 kW. A turbo-supercharger with elements of variable flow geometry is applied in the engine and there is also possibility to disconnect one of its cylinder blocks. It is fully controlled by an onboard computer.

Nowadays, the XAV-28 diesel engine is the only purely military diesel engine equipped with such a high level of computer automation, which testifies to its qualitative advantage over other analogs.

Currently, domestic engine-builders are mainly oriented toward the use of mechanical engine control, although works on implementing electronic systems are under way.

An analysis of the conditions and developments of diesel engine-building shows that traditional hydromechanical systems of the fuel supply fail to provide hazardous waste reduction at the level of Euro II requirements. Radical improvement in engine characteristics up to the Euro III level is only possible by using electronic control systems of the fuel supply with increased injection in every cycle of every cylinder. Despite a 25–30% rise in the cost of the vehicle's power unit due to using electronic control systems, engine-making companies find these costs inevitable since constantly toughening control of toxic emissions remains a main factor.

The use of electronic systems for domestic, serially produced engines is only achieved through a complex adaptation to current engine configurations and fuel equipment. That is why for diesel engines, for example, YaMZ, the only possible option to reduce toxic emissions is to use high pressure fuel equipment (HPFE), which is developed on vehicle diesel engines of foreign production and facilitates the implementation of the electronic systems [56].

This HPFE (of the "column" type) represents individual high pressure pumps for every injector ("column") with a common drive and regulation. This fuel equipment configuration, at the expense of reducing the length of the fuel high pressure pipeline, allows for providing an essential decrease in the pressure line volume and, subsequently, an increase in the fuel supply accuracy, stability of its characteristics, and controllability. The increase in the maximum and average fuel injection pressures also provides for quality mixture formation and improving the diesel engine's fuel efficiency. In addition, the designed HPFE configuration promotes an easier adaptation of fuel supply electronic control systems in the future.

Thus, the development of engine and fuel equipment electronic control systems has become the main development concept for leading engine-producing companies of the world and is supported by their various governments. They are in demand for engines of military vehicles due to the flexibility of electronic control systems and the ability to adapt to operation conditions and locate and make up for engine defects.

Reducing the weight-dimension indicators of diesel engines is performed by using new materials having a low specific weight and high physical and mechanical properties (high-duty cast iron and steel, aluminum alloys, plastics, composite materials, and ceramics) And also due to design solutions and the development of production technologies.

The specific weight of the best foreign engine models produced by leading companies such as Caterpillar, Cummins, and Comatsu is 1.2–4.3 kg/kW.

At present, domestic engines have a specific weight of 3.2–5.2 kg/kW, which is 1.2–2.6 times greater than the indicators of foreign analogs.

Along with reducing toxic emissions and fuel consumption, special attention is given to improving the reliability of diesel engines. The service life till capital repairs of foreign-made automobile diesel engines has reached 600–800 thousand km. They are also now guaranteeing a service life of up to 1 million km for advanced automobile engines. The reliability analysis of domestic and foreign analogs shows a significant lag in domestic automobile diesel engines by 1.6–3 times on the average. The comparative service life data of domestic and foreign diesel engines with a power of over 298 kW (400 hp) is presented in Table 10.2.

Thus, for the main characteristics of domestic engines to comply with the level of the leading world companies, the following is necessary:

- Introducing supercharging with boost air cooling;
- Developing and launching TSCs with a high efficiency factor;
- Using fuel equipment with a high fuel injection energy and developing its control systems;

Table 10.2 Comparison of domestic and foreign diesel engines with a power of over 298 kW

Characteristics of engines	OM501LA (Mercedes-Benz, Germany)	OM501LA (Mercedes-Benz, Germany)	D12C (Volvo, Sweden)	DSC14-09 (Scania, Sweden)	BF8M-1015C (Doyle, Germany)	N5000E (Cummins, USA)	Ser. 55 (DDA, USA)
Cylinder arrangement and number	6V	8V	6P	8V	6V	6P	6P
Cylinder capability, l	12.0	16.0	10.0	14.2	11.2	14.0	12.0
Power, kW (hp)	Up to 313 (426)	Up to 423 (571)	340 (460)	368 (500)	400 (544)	375 (507)	272 (369)
Service life before capital repairs, thousand km (machine hours)	1000.0	1000.0	1000.0	800.0	800.0	over 1000.0	1200.0

Characteristics of engines	YaMZ-8401	YaMZ-8424	YaMZ-846	YaMZ-847	YaMZ-7511	B-46-5MC	B-85	D12-525
Cylinder arrangement and number	12V	8V	8V	12V	8V	12V	12V	12V
Cylinder capability, l	25.9	17.2	17.2	25.9	14.9	38.9	38.9	38.2
Power, kW (hp)	478 (650)	308 (420)	368 (500)	625 (850)	295 (400)	522 (710)	618 (840)	386 (525)
Service life before capital repairs, thousand km (machine hours)	1500.0 (10000)	600.0 (-)	- (7500)	- (5000)	800.0 (-)	40.0 (-)	- (700)	40.0 (1000.0)

- Reducing weight and dimension indicators by applying advanced technologies and using new materials and design solutions;
- Developing and implementing electronic control systems; and
- Developing engines that comply with ecological standards.

10.3 Fundamentals of HDWV engine designs

10.3.1 State of development of domestic diesel engines

The traction-dynamic properties of HDWVs are defined by the value of the maximum turning torque of PUs and the progressive characteristic of its course, estimated by the adaptability coefficient. At the same time, PUs must have low weight and dimension indicators, which ensure fitting into the specified volume of all systems, and reasonable performance indicators. At present, a power unit fully complying with all requirements is not yet designed. That is why selecting the PU type is usually carried out according to its capability to meet the technical requirements imposed on the vehicle.

Developing SWCs towards load capacity improvement and the increase in the number of axles have required a high aggregate capacity of PUs and have toughened, in terms of configuration, the requirements for their weight and dimensions.

The problem of designing PUs to meet the modern requirements imposed on HDWVs is considered in the following aspects:

- The development and design of diesel engines with a high aggregate capacity;
- The application of multi-engine units to use with serially produced diesel engines; and
- The development and design of automobile gas-turbine engines.

The solution to this problem is based on improving the reciprocating engines of different types with a gradual transition to the highly accelerated, turbo-piston diesel engines that have recently emerged.

Over the last 10–15 years, there has been a significant increase in the technical level of domestic diesel engines used as power units of automotive transport vehicles (ATVs), including SWCs. The dynamics of the change in main diesel engine indicators is presented in Table 10.3.

In order to meet the advanced requirements of HDWVs, including SWCs, there is a demand for diesel engines (with an operating volume range of 6–27 l) with a liter capacity of up to 30 kW/l (41 hp/l)–35 kW/l (47.6 hp/l).

Diesel engines of advanced vehicles must have the energy data presented in Table 10.4.

Tank type diesel engines were used as PUs of earlier designed SWCs having a load capacity of 12, 20, and 50 t. At the time of design, they satisfied the chassis designers with their power level (from 220.6–522 kW), weight indicators (a specific weight from 2.49–3.89 kg/kW), and fuel consumption (a specific fuel consumption from 217.6–239.4 g/kW×h. The use of the mentioned tank type diesel engine series was explained by the absence, at that time, of automobile diesel engines of corresponding unit capacities.

Basic indicators of domestic diesel tank engines are presented in Table 10.5.

Table 10.3 The dynamics of the change in main engine indicators within the 1980–1993 period

Diesel	Power, kW	Specific fuel consumption, g/kW-h	Liter capacity, kW/l	Specific weight, kg/kW
Natural aspiration:				
1980	154–386	223–226	9.9–14.2	3.9–7.5
1993	125–213	218–224	14.3–14.9	4.6–5.4
Forced aspiration:				
1980	221–478	224–227	12.3–14.9	3.2–4.9
1993	85–625	196–215	17.6–24.2	3.1–5.3

Note. The analysis used the characteristics of the following engines:
In 1980, diesel engines: YaMZ-238D, YaMZ-238M, YaMZ-238N, KamAZ-74, D12A-525, D12AH-650, and V-58-7;
In 1993, diesel engines: GAZ-544.10, GAZ-5424, ZIL-645, Ural-744, Ural-745, KamAZ-7403, KamAZ-7406, KamAZ-7482, YaMZ-8424, YaMZ-846, YaMZ-847, and YaMZ-8401.

Table 10.4 Rational power property parameters of advanced diesel engines

Parameter	Multi-purpose vehicles with a load capacity of		Multi-axle wheeled chassis and haulers
	up to 1.5 t	over 1.5 t	
Power range, kW	70–150	100–400	350–800
Operation type	Diesel (gasoline injection)	Diesel	Diesel

Table 10.4 Rational power property parameters of advanced diesel engines			
Configuration	4P, 6P	6P, 8V	8V, 12V
Cooling system type	Liquid Air	Liquid Air	Liquid
Average effective pressure, MPa	1.2–1.25	1.38–1.45	1.35–1.55
Liter power, kW/l	25–32	25–29	25–28

Table 10.5 Basic indicators of diesel tank engines used in SWCs

Engine model	Power, kW	Specific weight, kg/kW	Liter power, kW/l	Specific fuel consumption, g/kW-h
5D20K-300	220.6	3.02	13.87	224.4
UTD-25	294.0	2.72	13.87	224.4
D12A-525A	386.0	3.89	9.95	239.4
D12AN-650	477.9	3.24	12.32	217.6
V-58-7	522.0	2.49	13.45	217.6

One should note that these diesel tank engines are upgraded diesel engines designed as early as the 1930s, and their characteristics do not meet the requirements of modern engine-building.

A high forcing level of the diesel tank engine is used to achieve the required power level, which leads to a rapid depreciation of the cylinder piston group resulting in a high operational consumption of motor oil and low reliability.

All of the above has resulted in the current situation, where designers of SWCs are refusing to use diesel tank engines as PUs and tend to use automobile diesel engines.

Modern engine-building adopts the following principles of automobile diesel engine design:

- Designing diesel engines in the form of a unified series, including models with a diverse number of cylinders, which ensures the interchangeability of common rapidly wearing parts;
- Using turbo supercharging and intermediate cooling of boost air, which allows for considerably expanding the power range in every series and a resulting decrease in the number of original engines in the approved engine range for SWCs;
- A wide unification of engines, which are the basis for designing diesel engine modifications for SWCs; and
- Implementing new configuration solutions accepted in modern engine-building, which ensures a high competitiveness compared to the best serially produced foreign analogs (individual heads of cylinders, valve gears with four valves, cylinder heads with spiral inlet, regulated cooling system fans using hydraulic clutches, high pressure fuel pumps with high energy fuel injection, etc.).

At present, the Joint-Stock Company (JSC) Autodiesel (known as YaMZ) is completing the design of a new advanced series of diesel engines with turbo charging and boost air cooling with a power ranging from 325 to 650 kW (from 450 to 850 hp); their basic characteristics are presented in Table 10.6.

Basic improvements to PUs of perspective SWCs should be aimed at using automobile diesel

Table 10.6 Diesel engine series of the Yaroslavl Engine Plant

Parameter	Diesel model					
	846	8461	8462	849	847	8471
Consumer plant	MZKT	BAZ	BAZ	KrAZ	M(K)ZKT	IZTM
Cylinder arrangement and number	V8 90	V8 90	V8 90	V8 90	V12 90	V12 90
Cylinder capability, l	17.24	17.24	17.24	17.24	25.86	25.86
Rated power, kW (hp)	373 (500)	373 (500)	335 (450)	335 (450)	596 (800)	596 (800)
Nominal rpm of the crankshaft, minutes^{-1}	2100	2100	2100	2100	2100	2100
Maximum torque M_e, N×m	2000	2000	1800	1800	3150	3150
Rpm of the crankshaft at M_e, minutes^{-1}	1200–1400	1200–1400	1200–1400	1200–1400	1300–1500	1300–1500
Weight, kg	1440	1460	1460	1430	2000	2000
Specific weight, kg/kW (kg/hp)	2.15 (2.88)	2.18 (2.92)	2.41 (3.24)	2.37 (3.18)	1.86 (2.50)	1.86 (2.50)
Specific effective fuel consumption, g/kW×h (g/(h.p.×h)):						
minimal	112 (150)	112 (150)	112 (150)	112 (150)	112 (150)	112 (150)
nominal	125 (168)	125 (168)	125 (168)	125 (168)	125 (168)	125 (168)
Oil consumption for burn-out loss, %	0.4	0.4	0.4	0.4	0.4	0.4
Service life, hours	7500	7500	7500	7500	5000	5000
Overall dimensions, mm:						
length	1231	1341	1341	1341	1594	1594
width	1178	1166	1166	1166	1070	1070
height	1220	1126	1126	1126	1220	1220
Fan drive clutch (frictional)	No	Yes	Yes	Yes	Yes	Yes
Pre-heating system (pre-start hot oil pumping)	Yes	Yes	Yes	Yes	Yes	Yes
Back-up starting system (pneumatic starter)	Yes	Yes	Yes	Yes	Yes	Yes

engines with turbo charging and boost air cooling; improving the energetic properties of diesel engines by increasing the forcing level of the operation course and the energy conversion efficiency in all performance mode scopes; and, due to the increase in resistance of the power unit parts to thermal and mechanical loads, ensuring high reliability and long service life of the engines.

Together with the increase in the engines' technical characteristics, it is necessary to improve their traction characteristics and ensure the reliability of the main parameters at motion in highlands, hot-desert areas, and the Far North. The mass and dimension parameters of domestic diesel engines must also be reduced since they do not meet modern requirements as compared to foreign analogs. The service life and reliability of domestic diesel engines are inferior to the best foreign samples. The tasks related to ensuring the multi-fuel properties, the air-tightness of diesel engines at stream crossings, etc. are not yet solved.

Despite the number of positive properties, domestic diesel automobile engines have some drawbacks. Diesel automobile engines are characterized by their insufficient forcing level on the maximum turning torque and low adaptability coefficient values equal to 1.15–1.2. That is why when improving the traction-dynamic characteristics of SWCs, designers have to use hydrodynamic and ranged gears in combination with multi-stage mechanical gearboxes, which complicates transmission, decreases the chassis' reliability, and increases service and repair complexity.

Another significant drawback is that domestic diesel automobile engines are formed with priority given to the highest fuel efficiency. Thus, the value of the minimum specific fuel consumption without calculating the nature of the fuel efficiency change in all performance scopes of the PU's operation modes is still used for its estimation. The minimum specific fuel consumption is positioned in the mode closest to the maximum torque at its lowest forcing on the absolute value. This engine mode is not widely used in SWC operation and is applied only shortly for passing high road irregularities. That is why improving the fuel efficiency in that mode alone cannot positively affect the chassis' operational fuel consumption.

Thus, the engine's operation range at the minimum specific fuel consumption (dashed zone in Fig. 10.3) is just 25% of the PU's overall operation scope in a vehicle.

At the same time, the diesel engine's low forcing level on the turning torque affects its traction characteristics and limits the SWC's mobility (see diagram in Fig. 10.3).

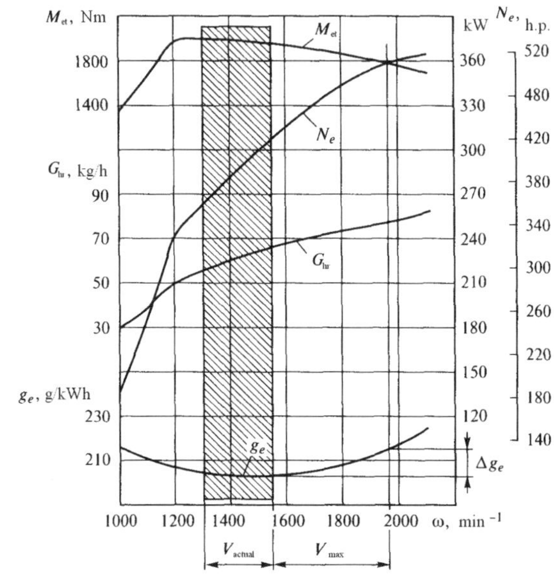

Fig. 10.3. Example of irrational use of the YaMZ-846.10 engine characteristics in the 7930 chassis

10.3.2 Power-to-weight ratio of SWCs

The defining parameter for evaluating a vehicle's power-to-weight ratio is the chassis' specific power. To achieve the high traction-dynamic indicators of HDWVs having a gross weight of 40–80 t, the specific power must be equal to 8.1–8.8 kW/t (11–12 hp/t). It may be performed, for example, on an 80 t load capacity chassis using 970–1060 kW (1390–1440 hp) engines.

At the same time, with an increase in the gross weight of modern SWCs there is a tendency to decrease their specific power. Table 10.7 displays the relative value of this indicator for some samples of serial and perspective SWCs. Indicators of an 8×8 chassis are taken as 100%.

Table 10.7 Ratio of the gross weight and the specific power of SWC			
Type of chassis, wheel formula	Gross weight of the SWCs, %	PU Power, %	Specific power of the SWC, %
MAZ-7911, 8×8	100	100	100
MAZ-7929, 10×8	150	90	59
MAZ-547V, 12×12	210	124	59
MAZ-7917, 14×12	260	135	52
MZKT-79221, 16×16	300	152	52

The analysis of data in Table 10.7 shows that with an increase in the gross weight of SWCs by 1.5–3.0 times, the power of the power unit increases by 1.24–1.52 times and the specific power decreases by 1.6–2.0 times.

The specific power indicators reflect a qualitative demonstration of the decrease in the power-to-weight ratio in a chassis with an increase in the load capacity and number of axles. By using absolute values, the insufficiency in automobile engine power can be better traced. It is assumed that on a 20 t load capacity chassis, instead of real diesel engines, 368 kW automobile diesel engines of the YaMZ-840–YaMZ-846 series are installed; on an 80 t load capacity chassis, the 625 kW YaMZ-847 is used. The power-to-weight ratio is considered adequate when the SWC's specific power is equal to a minimum of 8.1 kW/t. Data on the power-to-weight ratio level of chassis with YaMZ-840 series automobile diesel engines are presented in Table 10.8.

Data in Table 10.8 confirms the previous conclusion that using automobile diesel engines ensures a sufficient power-to-weight ratio level and, as result, high traction-dynamic properties for the 20 t load capacity 8×8 chassis. However, the specific power of SWCs with a gross weight of 80 t or more and a 12×12 or more wheel formula is considerably lower than the required power-to-weight ratio level.

It should be noted here that in domestic engine-building with the existing level of technology, quality of materials, and motor oils, the maximum liter capacity level is 25 kW/l. The working volume (liter capacity) of the YaMZ-847 engine is 25.85 l. Simple calculations show that with these mass and dimension parameters, the reserves of further power increases in the 840 series of automobile diesel engines are practically exhausted.

This can be explained by the fact that it is necessary to increase the cylinder diameter and piston stroke for an increase in the working volume. There is a known limit on the average piston speed (not more than 10–13 m/s), which restricts the operability of piston rings and cylinder sleeves. An excess in the average piston

Table 10.8 Power-to-weight ratio level of the chassis with YaMZ-840 series diesel engine				
Type of chassis, wheel formula	Gross weight, t	Diesel power, kW		Specific power, kW/t
		YaMZ-846	YaMZ-847	
MAZ-7911, 8×8	41.4	368	-	8.89
MAZ-7930, 8×8	45.0	368	-	8.18
MAZ-7929, 10×8	60.0	368	-	6.13
MAZ-547V, 12×12	82.5	-	625	7.58
MAZ-7917, 14×12	105	-	625	5.95
MZKT-79221, 16×16	120	-	625	5.20

speed over specified values will inevitably lead to rapid wear of the sleeve assembly, high motor oil consumption, and, as a result, a reduction in the engine service life to diesel tank engine levels.

That is why increasing the piston stroke at a specified average speed reduces rpm and a growth in the cylinder diameter (piston weight) increases the dynamic loads that require reinforcing engine parts. Thus, further power increases cause a drastic raise in the volume and weight of diesel engine PUs.

Due to the above reasons, it is not deemed possible to design an automobile diesel engine of the required power for HDWVs at domestic engine-making plants now or in the near future.

Moreover, high powered diesel engines (1100 kW) used on 7904 and 7906 chassis are not designed for wheeled vehicles; they refer to industrial (ship) ones and cannot be applied in modern SWCs due to unacceptable mass and dimension parameter values (for example, the M351 diesel engine weighs 7220 kg, has a volume of 5.83 m^3, and a specific weight of 8.4 kg/kW).

10.3.3 New power units

Along with the studies on diesel engine vehicles taking into account their development tendencies considered in item 10.2 of this chapter, the search for other solutions enabling designers to create highly efficient PUs for transport and combat vehicles still continues abroad.

Foreign experts believe that an adiabatic engine has potential advantages due to its ability to adapt to the operational conditions of the transportation vehicle.

The adiabatic engine work process, as is known, is understood as the process with a minimal heat exchange with the environment owing to the thermal insulation of the combustion chamber and output line. In general, the work process of an adiabatic engine is characterized by the formula:

$$PV^k = \text{const},$$

where P and V are the pressure and working gas volume; and k is the adiabatic indicator.

The full operation cycle of a adiabatic engine is calculated according to the following dependence:

$$L = \frac{R}{k-1}(T_3 - T_4 - T_1 + T_2),$$

where R is the absolute gas constant; k is the adiabatic index (for air, $k = 1.4$); T_1 is the ambient temperature; $T_2 = \frac{P_2 V_2}{R}$ is the temperature at the end of compression; P_2 is the gas pressure after the adiabatic compression with parameters P_1 and V_1; P_1 is the gas pressure before compression; V_1 is the gas pressure before compression; V_2 is the gas pressure at the end of compression; T_3 is the gas temperature after the heat supply; and T_4 is the gas temperature after the adiabatic expansion.

While working toward improving the adiabatic engine, American experts came to the conclusion that the ongoing design studies cannot exhaust all of its progress opportunities. They believe that the creation of so-called diesel engines with minimal losses is one of the promising strategies. It is planned to obtain an essential reduction of losses due to a set of design measures—the use of gas bearings, pistons without rings, dry ceramic bearings with a low friction coefficient, and solid lubricants. It is supposed that this engine can adopt most of the parts and units mastered earlier in a traditional adiabatic engine and achieve—by decreasing the friction—a specific fuel consumption around 154 g/kW×h (the effective efficiency factor = 54%), while at the same time abandoning the engine oiling system.

It is known that current engine designers are actively working toward improving the characteristics of high-powered automobile diesel engines, including designing turbo-compound engines. In particular, an improvement in the power, fuel efficiency, and ecological parameters can be achieved by solutions such as:

- Turbo charging (boosting) and intermediate cooling of the boost air;
- Turbo compounding (using a power turbine connected via a gearbox with the engine crankshaft);
- Combined supercharging (a turbocharger with an inertial charging of the intake pipelines);
- Regulated supercharging (removing a part of the air after the compressor and a part of the gas before the turbine);
- A two-stage boosting (to increase the engine's throttle response and provide an additional reduction in the effluent gas toxicity due to shortening the transient processes);
- Using fuel equipment with a high injection energy and variable supply start depending on the crankshaft rpm with adaptive microprocessor control, etc.

Diagrams of some of the above-mentioned design solutions are shown in Fig. 10.4.

Studies show that at the current stage of engine-building the only alternative engine type capable of meeting the requirements imposed on PUs of a vehicle with a gross weight of 80 t or more related to the power, weight, and overall dimensions is the automobile gas-turbine engine (AGTE).

The investment scales allocated in most developed industrial countries on designing a AGTE, creating and using the ceramic materials in them, and the results achieved so far allow for expecting a technological breakthrough in the world engine-building.

Most of the AGTEs currently designed are equipped with a free power turbine and a regulated nozzle unit (RNU). They are available with various types of heat exchanger or without it. The engines are designed in either two- or three-shaft versions (Fig. 10.5).

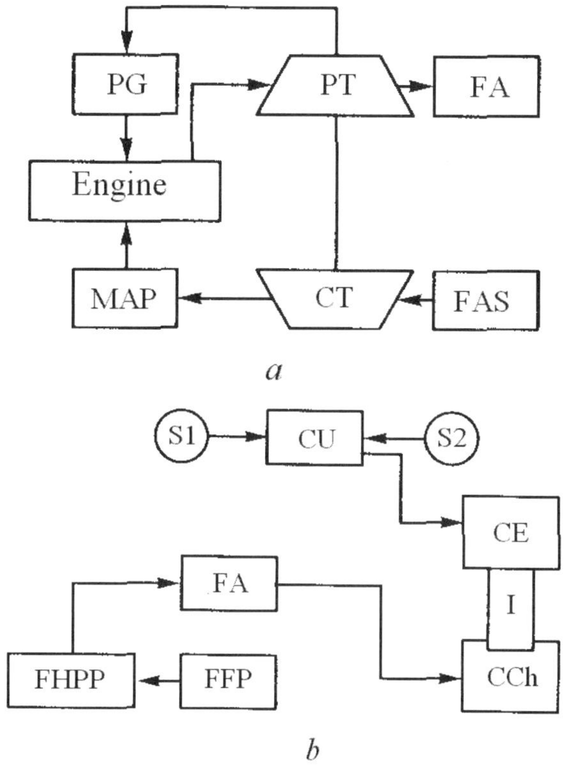

Fig. 10.4. Development diagrams of diesel engines: A—basic diagram of a turbo-compound diesel engine (YaMZ); PT—Power turbine; CT—Compressor turbine; FA—Removal of free air; FAS—Supply of free air to the compressor turbine; MAP—Main air pump; PG—Power gearbox; b—Basic diagram of accumulator fuel feeding systems (Common Rail type); FHPP—Fuel high pressure pump; FFP—Fuel feed-up pump; FA—Fuel accumulator; I—Injector; CE—Controlled electromagnet; CU—Control unit; S1—Rotation speed sensor of the engine's crankshaft; S2—Diesel engine load cell; CCh—Combustion chamber

The main problem faced by AGTE designers is in improving the engine's efficiency. In order to achieve it, they regulate the passage cross sections in the engine's air-gas channel by turning the compressor's directing blades and the power turbine's nozzle blades (the RNU) in the partial load and idle running modes. Using the RNU not only allows for improving

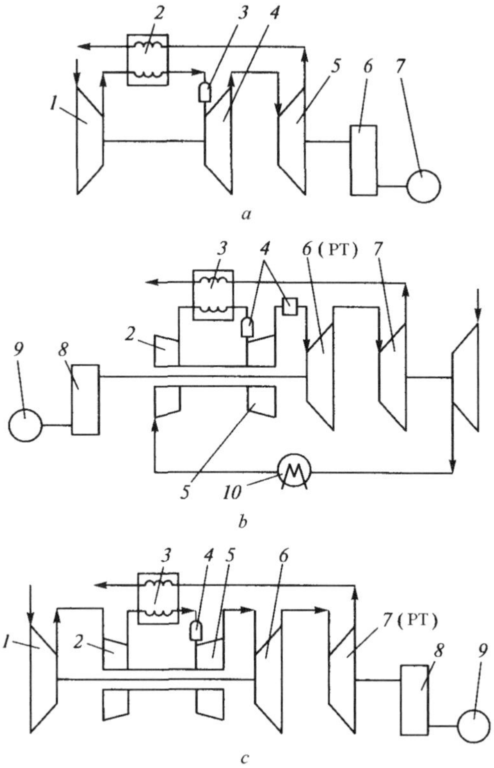

Fig. 10.5. Diagrams of AGTEs: a—Two-shaft diagram: 1—Compressor; 2—Heat exchanger; 3—Combustion chamber; 4—Compressor turbine; 5—Power turbine; 6—Gearbox; 7—Load; b and c—Three-shaft diagram: 1—Low pressure compressor; 2—High pressure compressor; 3—Heat exchanger; 4—Combustion chamber; 5—High pressure turbine; 6—Middle pressure turbine; 7—Low pressure turbine; 8—Gearbox; 9—Load; 10—Air cooler; PT—Power turbine

the efficiency in partial load modes, but it also improves the engine's adaptability factor and makes it possible to brake by using the engine. However, the introduction of the RNU results in a significant increase in the cost of the AGTE, complicates its design and operation, and can reduce the efficiency factor of the power turbine in the rated mode because of leaks through the gaps in the RNU and as a result of changing the approach angle when turning the blades.

The development of the AGTE in Russia can be illustrated by the following examples. The Yaroslavl Engine Works (YaMZ) have created an AGTE with a power of 735–1100 kW. The engine was built according to a two-shaft layout; its specific fuel consumption in trial conditions was 381–408 g/kW×h.

Another domestic engine-making company has offered a multi-fuel AGTE for ground vehicles with high passing ability. It is designed according to a three-shaft layout without a heat exchanger. The 919 kW GTD-1250A engine model in trial conditions had a specific fuel consumption of 306 g/kW×h, a weight of 800 kg, and overall dimensions of 1929×1210×932 mm.

These GTEs were tested on some SWC models. These SWC models were also used to test military tank GTEs—735 kW GTD-1000A models, 882 kW GTD-1000TFM models, and 919 kW GTD-1250A models. All engines weighed between 800–960 kg (1300–1500 kg with their auxiliary systems) and the specific fuel consumption in the trial conditions was 306 g/kW×h.

The main parameters of some domestic AGTEs are presented in Table 10.9.

Analysis of the data in Tables 10.3, 10.6, and 10.9 shows that only the perspective models of AGTEs that are at the technical design stage (GAZ-9031 and GTD-KOMB) have a fuel efficiency close to the corresponding parameters of tank and automobile diesel engines. Thus, the main obstacles to a broad use of AGTEs are their low fuel efficiency as compared with diesel engines, the high cost of prototypes, and a need of essential investments for retooling the production facilities to manufacture them. This issue was considered in more detail in point 10.2.

Studies have confirmed the fundamental prospects of using a high powered gas-turbine electric power plant with an electric transmission as the power unit of a SWC. While studying the possibilities and prospects of using GTEs as SWC power units, MSTU has developed (up to the level of technical proposals) a highly efficient, open cycle, 1600 kW, transport gas-turbine unit (GTU).

Table 10.9 Basic indicators of domestic AGTE

Engine model	Power, kW	Specific weight, kg/kW	Overall power, kW/m³	Specific fuel consumption, g/kW×h
Prototype GTD-NAMI	258	2.90	235	258
Prototype GAZ-902.10	310	2.74	230	289
Project GTD-KOMB	400	-	154	219
Project GAZ-9031	480	1.98	390	230
Prototype YaMZ-E901	735	2.04	285	381
Prototype GTD-1000TFM	882	1.47	530	306
Project YaMZ-E904.10	1100	1.00	423	277

The diagram of this GTU is presented in Fig. 10.6.

The gas-turbine unit consists of a high-pressure turbo compressor (mounted on the same shaft as the collet electric generator), a low-pressure turbo compressor, a gas recuperator, an intermediate cooler, an intermediary heater, and a forcing (boosting) system.

The unit operates in the open cycle mode and its working medium is air.

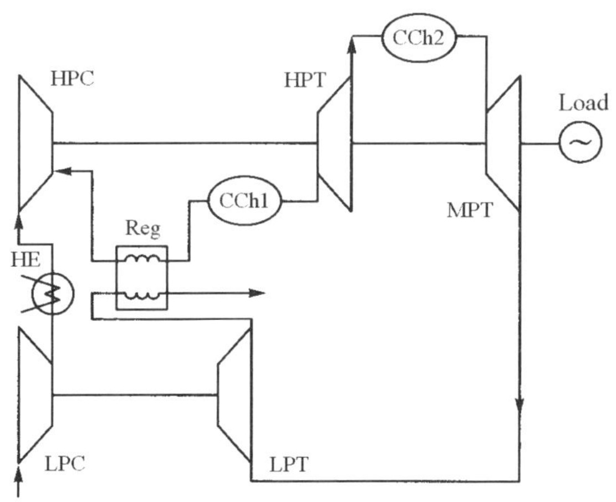

Fig. 10.6. Diagram of a 1600 kW gas-turbine PU for perspective HDWVs: HPC—High pressure compressor; HPT—High pressure turbine; CCh1 and CCh2—Combustion chambers; LPC—Low pressure compressor; LPT—Low pressure turbine; PTMPT—Medium pressure turbine, H_l—Load (generator); Reg—Regulator; HE—Heat exchange

The high-pressure turbo compressor consists of a centrifugal compressor (designated as HPC in the diagram) and two axial turbines of high and medium pressure. In front of the high-pressure turbine (HPT), there is the main combustion chamber (CCh1) operating under the principle of micro-torch combustion. The nozzle unit of the HPT is also the device for the fuel distribution of the intermediate heating chamber (CCh2) working under the same micro-torch principle.

The rotation rate of the gas-turbine unit is defined by the electric generator characteristic and is 24,000 rpm.

The shaft rotates on two supporting and one thrust gas bearings. Due to a relatively low gas temperature, T_Γ = 1173 K, no cooling of the working and nozzle blades is provided.

The configuration design of the main GTU components is presented in Figs. 10.7 and 10.8.

Thus, by their technical and economic parameters (traction properties, specific weight and power-to-size ratio, endurance and reliability due to the absence of the cooling system) AGTEs have some advantages over diesel engines. Comparability of specific fuel consumption is possible at high power levels; therefore GTE use is justified specifically on HDWVs.

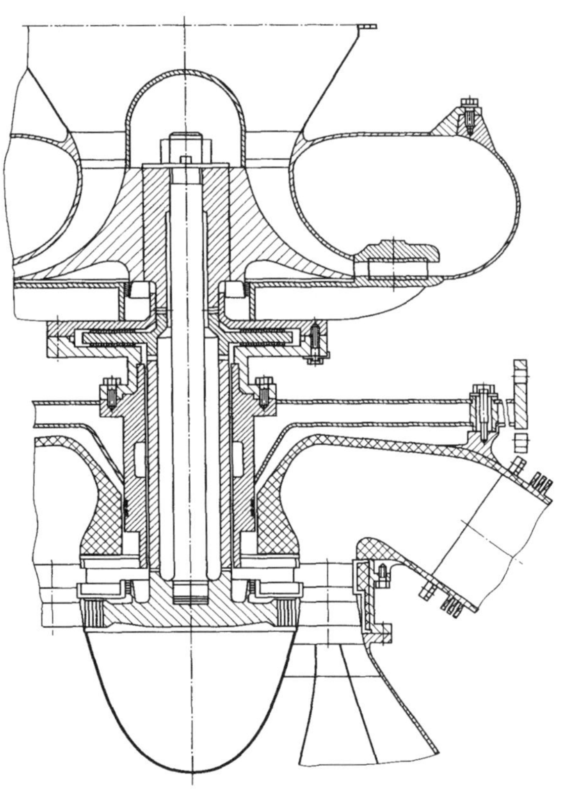

Fig. 10.7. Low pressure block of the GTU

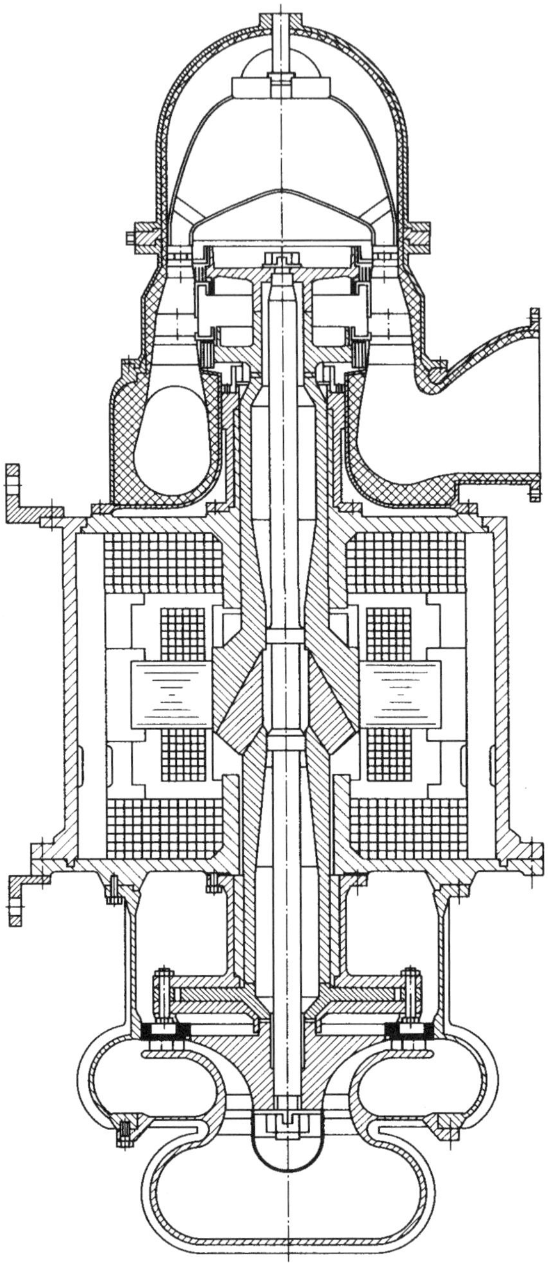

Fig. 10.8. High pressure block of the GTU

10.4 Features of mathematical modeling a diesel engine and gas-turbine power plants for HDWVs

For use as a part of the general mathematical model (MM) (see Chapter 8, Figs. 8.1 and 8.2) of a SWC, which allows for studying its mobility parameters by PC simulation, a model of the diesel engine is needed.[1] In the selected approach, the engine is represented as a separate element (block) of the general system (a SWC) and is connected to it by a set of variables (Fig. 10.9). The task of a diesel engine's MM may be formulated as a determination of the functional relations (transfer functions) connecting incoming and outgoing data. Since the model must be universal, i.e., allow for designing basic diesel engine models

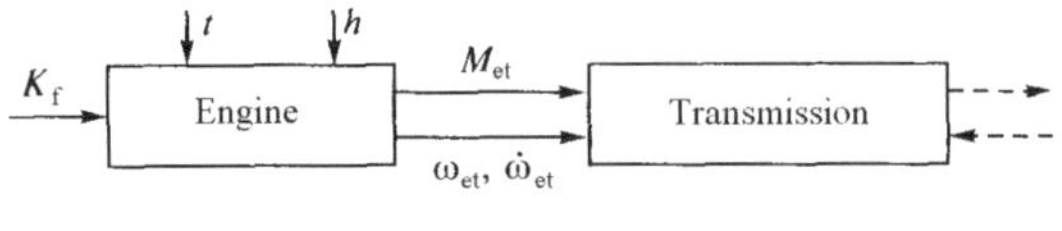

Fig. 10.9. General structural layout of the variable interrelations in a diesel engine–transmission system

1 The approaches mentioned in work of A.V. Denisov [57] were used in the development of the MM of gas turbine engines.

installed on SWCs, the MM is complemented with an additional set of equations for the preliminary calculation of the transfer function coefficients on specified basic indicators of technical characteristics of a particular designed engine. The indicators for diesel engines are as follows:

- N_r is the rated (maximum) power;
- M_{max} is the maximum torque;

The rpm of the crankshaft:

- ω_N – at rated power;
- ω_M – at maximum rotation torque;
- ω_{idle} – at maximum idle rpm;
- ω_{min} – at minimum steady rpm; and
- ω_e – at the current rpm.

The main input variable determining the engine's operation model is the position of the fuel supply's control body—the fuel-feed pedal or high-pressure fuel pump (HPFP) control rack.

In Fig. 10.9, this variable (the fuel supply coefficient) is indicated as K_f. The range of its alteration in relative units is from 0 (pedal eased) to 1 (pedal fully pressed). In the design, this signal can be formed as a separate block ("driver," "speed set unit," etc.) under any specified rule, including K_f = const.

The power flow of the diesel engine is characterized by two interdependent factors—power and speed. The power factor—the engine's turning torque (M_{et})—is taken as the output variable, which is the result of calculating the diesel engine's MM on every stage of simulation. The speed factor—the diesel engine's crankshaft rpm ω_e—is taken as the input variable. That is why the second input variable is the crankshaft's current rpm ω_{et}. To estimate the diesel engine's dynamic properties that determine its inertia, one more value is needed—the derivative of the crankshaft's rpm $\dot{\omega}_{et}$.

It is known that a diesel engine's operation indicators to a great extent depend not only on its operation mode (K_f, ω_{et}, M_{et}), but also on its operating conditions. These are the ambient air parameters (pressure, density, temperature, humidity, chemical composition, dust content, etc.) and the properties of the fuel and oil used. The engine's technical condition, its spatial position in the vehicle, and some other factors may also have an essential effect. In this model, only the influence of the ambient air temperature (t) and the altitude above sea level (h) are considered, on which pressure (P) and air density (ρ_o) are defined.

The torque taken from the diesel engine's flywheel and transferred into the "transmission" block is defined according to the formula:

$$M_{et} = (M^1{}_{et} - M_j)\, k^1{}_d \tag{10.1}$$

where $M^1{}_{et}$ is the torque developed by the diesel engine at steady run of the specified crankshaft's rpm in normal conditions; $M_j = \frac{\dot{\omega}_{et}}{j_{et}}$ is the diesel engine's inertia torque; jet is the inertia moment of the diesel engine's mobile parts; and $k^1{}_d$ is the coefficient of the engine indicators reduction depending on the ambient conditions:

$$k^1{}_d = \left(\frac{100}{P}\right)^{0.65} + \left(\frac{t+273}{298}\right)^{0.5} \tag{10.2}$$

The density and ambient air pressure is adjusted depending on the temperature and the altitude above sea level:

$$\begin{aligned} \rho_o &= 1{,}2928 - 0{,}0047536\, t + 0{,}000015738\, t^2\ ; \\ P &= e^{4{,}618333} - \frac{\rho_o \cdot 9{,}81\, h}{101325}\ . \end{aligned} \tag{10.3}$$

The engine torque at steady run is taken as one of the values:

$$\begin{aligned} {}_{et.ex} &= k_{01} + k_{02}\,\omega_{et} + k_{03}\,\omega_{et}^2\ ; \\ {}_{et.r} &= \left(\omega_{min} + (\omega_{idle} - \omega_{min})\ {}_f - \omega_{et}\right) k_{04}\ ; \\ {}_{et.b} &= k_{05}\,\omega_{et}\ , \end{aligned} \tag{10.4}$$

where $M_{et.ex}$, $M_{et.r}$, and $M_{et.b}$ is the engine torque according to the external speed, regulatory, and braking characteristics, respectively (Fig. 10.9); and k_{01}, …, k_{05} are the coefficients of the approximation characteristics:

$$\begin{aligned} k_{01} &= \frac{(9570N_r - M_{\max}\,\omega_N)\,\omega_M^2}{\omega_N(\omega_N-\omega_M)^2} + M_{\max}\;; \\ k_{02} &= \frac{2n_M(M_{\max}\,\omega_N - 9570N_r)}{\omega_N(\omega_N-\omega_M)}\;; \\ k_{03} &= \frac{9570N_r - M_{\max}\,\omega_N}{\omega_N(\omega_N-\omega_M)^2}\;; \\ k_{04} &= \frac{9570N_r}{\omega_N(\omega_{idle}-\omega_N)}\;; \\ k_{05} &= \frac{-2871N_r}{\omega_N^2}\;, \end{aligned} \qquad (10.5)$$

where n_M is the engine crankshaft's rpm corresponding to the maximum torque.

It was assumed when setting the diesel engine's braking characteristic (coefficient k_{05}) that it is linear and that for A forced running (motoring), i.e., at the engine's operation in the brake mode at the rated crankshaft rpm, the power spent is approximately equal to 28–35% of N_r. These assumptions are based on the available experimental data. [57].

The criterion for choosing M_{et} from formula (10.4) is the torque mark on the regulatory characteristic of $M_{et.r}$, which definitively indicates the diesel engine's operation mode: the brake mode ($M_{et.r} < 0$) or the motoring mode ($M_{et.r} \geq 0$).

Thus, using formulas (10.1)–(10.5), the diesel engine's torque at the input to the transmission is defined according to the following algorithm:

$$\text{at } M_{reg} \geq 0 \;\; M^1_{et} = \min\{M_{et.ex}, M_{et.r}\},$$

$$\text{at } M_{reg} < 0 \;\; M^1_{et} = \max\{M_{et/r}, M_{et.b}\}.$$

It is known that in the presence of a load on the GTE shaft, the differential equation of torques is presented in the following form:

$$T\frac{d\omega}{dt} = \bar{M}_{GTE} - \bar{M}_c \qquad (10.6)$$

where T is the mechanical time constant of the engine, reduced to its output shaft; $\bar{M}_{GTE} = \dfrac{M_{GTE}}{M_{N_{GTE}}}$; and $\bar{M}_c = \dfrac{M_c}{M_{N_{GTE}}}$; where M_c is the relative value of the corresponding engine torque and the resistive torque equal to the ratio of the current value to the nominal engine torque.

Formula (10.6) displays the necessary description of the GTE torque characteristics and the general resistive torque for transient modes.

The theory of vane-type devices states that the turning torque course of a GTE in operational modes is fairly accurately defined by the relation:

$$M_{GTE} = M_0\left[K_\mu - \frac{(K_\mu - 1)\omega}{\omega_0}\right] \qquad (10.7)$$

where K_μ is a constant coefficient, equal to the ratio of the maximum turning torque to the torque value corresponding to the maximum efficiency factor of the free turbine; ω is the current value of the free turbine rpm; and M_0 and ω_0 are the torque and rpm of the free turbine at its maximum efficiency factor value (along the line of optimal fuel efficiency), respectively.

In view of the fact that reducing the free turbine rpm at constant pressure and stable initial parameters causes the gas consumption through the stage to increase due to the reduced reactivity level, the coefficient K_μ subsequently increases (proportionally to the increase in gas consumption) as well.

Considering this, formula (10.7) may be presented in the following form:

$$M_{GTE} = M_0\left[k^0\, K_\mu - \frac{(k\, K_\mu - 1)\omega}{\omega_0}\right] \qquad (10.8)$$

where k^0 is the coefficient accounting for K_μ's growth when reducing the free turbine rpm.

Subsequently, the values of M_0 and ω_0 are interrelated at a certain assumption of the following relation:

$$M_0 = a\,\omega_0^2 \tag{10.9}$$

where $a = \dfrac{M_N}{\omega_{0N}}$ is a constant coefficient equal to the ratio of the turning torque to the free turbine's rpm in the normal operation mode.

However, it is difficult to determine M_0 and ω_0 by the regular methods, not only in transition, but also in the engine's steady operation modes. That is why when designing this mathematical model, the results of experimental studies were used [57] where values of M_0 and ω_0 were defined in operational conditions with the braked free turbine.

In the braked free turbine, the torque depends on the turbo compressor's operational speed mode and on the position of the free engine's controlled nozzle unit. Then the M_0 torque is presented as follows:

$$M_0 = \frac{M^1{}_o}{k^0\ K_\mu} \tag{10.10}$$

where $M^1{}_0$ is the torque at the braked free turbine and a certain position of PCA.

In joint solution of equations (10.8)–(10.10), the GTE torque is presented in the following form:

$$M_{GTE} = M^1{}_o - \frac{(k^0\ K_\mu - 1)\sqrt{k^0\ K_\mu M_N}}{k^0\ K_\mu} - \sqrt{M^1{}_o \frac{\omega}{\omega_{0N}}}\ .$$

After replacing the absolute values of $M^1{}_0$, ω, and M_{GTE} by their relative values in accordance with the formulas:

$$\bar{M}_o^1 = \frac{M^1{}_o}{M_N};$$

$$\bar{M}_{GTE} = \frac{M_{GTE}}{M_N};$$

$$\varpi = \frac{\omega}{\omega_{0N}};$$

the formula (10.13) is presented in the transformed form:

$$\bar{M}_{GTE} = \bar{M}_o^1 - \frac{k^0\ K_\mu - 1}{\sqrt{k^0\ K_\mu}} - \sqrt{\bar{M}_o^1\ \varpi}\ .$$

In the engine's variable operation modes, $\bar{M}_o^1$ is defined by the relation:

$$\bar{M}_o^1 = \bar{M}_o^2\, \bar{f}(\alpha_o)\,,$$

where $\bar{M}_o^2$ is the relative value of the dynamic rotation torque at the braked free turbine and nominal angular position of PCA; and $\bar{f}(\alpha_o)$ is the relative alteration of the turning torque depending on the angular position of PCA.

The values of $\bar{M}_o^2$ and $\bar{f}(\alpha_o)$ are defined on the basis of experimental relations obtained directly in operational conditions. For convenience, the characteristics of the turning torque at power surge and dumping are presented in differential form and the $\bar{f}(\alpha_o)$ function is approximated by the third grade polynomial and defined in relation to the turbo compressor rpm.

At a power surge, the GTE's dynamic torque at the braked free turbine is described by the exponential relation of the kind:

$$\bar{f}(\alpha_o) \tag{10.11}$$

where $\bar{M}_{ct}$ is the GTE torque at the power surge and the braked free turbine in the end of transition course; $\bar{M}_{ct0}$ is the GTE torque at the braked free turbine in the beginning of the transition course; and z is a constant coefficient, varied depending on the turbo compressor's response time.

The value of $\bar{M}_{ct}$ depends on the turbo compressor's rpm, the nominal angular position of PCA, and is defined by the formula:

$$\bar{M}_{ct} = f(\omega_{\text{turboch}}) \tag{10.12}$$

Graphically, the dependence (10.12) defined in the course of experimental studies in operational conditions is presented in Fig. 10.10.

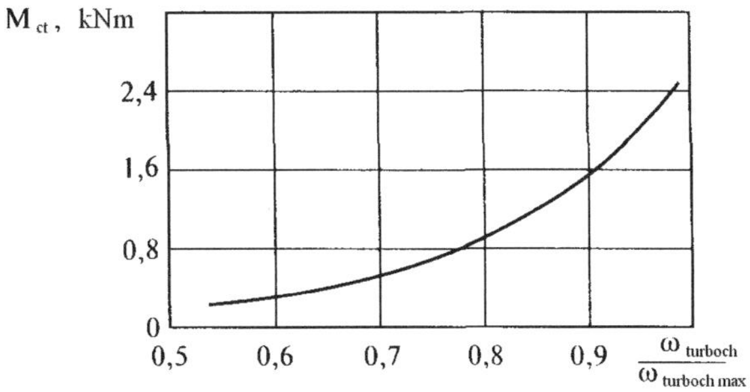

Fig. 10.10. The turning torque dependence of a GTE at the braked free turbine on the turbo compressor's rpm at the nominal angular position of PCA

The differential equation (10.11) will take the form:

$$\frac{d\bar{M}^{H}_{ctd}}{dt} = z\,\bar{M}_{ct} - z\,\bar{M}_{ctd}\,.$$

Accounting for the operational course of the GTE's torque in the transition mode, the value of the constant coefficient z is defined by the dependence:

$$z = \frac{3}{t_1},$$

where t_1 is the GTE turbo compressor's response period.

The GTE torque at the braked free turbine in the transition process, when the load is dropped, is defined by the following dependence:

$$\bar{M}^{c}_{ctd} = \bar{M}_{ct0} + (\bar{M}_{ct} - \bar{M}_{ct0})e^{-z_1 t},$$

where $\bar{M}_{ct0}$ is the GTE torque at the power dump and the braked free turbine at the end of the transition course; $\bar{M}_{ct0}$ is the GTE torque at the braked free turbine at the beginning of the transition course; and z_1 is a constant coefficient.

The differential equation (10.20) takes the following form:

$$\frac{d\bar{M}^{c}_{ctd}}{dt} = z_1\,\bar{M}_{ct0} - z_1\,\bar{M}^{c}_{ctd}\,.$$

Thus, the MM obtained supplements the SWC's motion and maneuvering models in regards to the mathematical modeling the diesel engine and gas-turbine PU. The diesel engine's turning torques and the GTE calculated by the above formulas are output characteristics of various PU types and are taken into account as the input turning torque in transparent (mechanical or hydromechanical) and non-transparent (electrical) transmissions of SWCs in the efficiency assessment of the technical solutions proposed by the industry at the SWC design stage, which was described in Chapter 8.

10.5 State of things with electrochemical engines

As was mentioned above, the capabilities of the domestic industry to design a high capacity (over 625 kW) automobile diesel engine are practically exhausted, while the widespread introduction of automobile gas-turbine engines is inhibited by the low fuel efficiency compared to diesel engines, high costs of prototypes, and the demand for essential investments for switching over to AGTE production.

However, the most preferred transmission type for a multi-axle special wheel chassis with a gross weight of 80 t or more is the electric transmission (ET). Low efficiency coefficient values of the power transmission unit consisting primarily of a thermal engine (diesel engine or GTE) and an electric transmission have been indicated as the main disadvantage of implemented designs of a SWC with an ET.

This may be explained by the necessity for double energy conversion—chemical fuel energy into mechanical energy of the primary engine's output shaft and then mechanical energy into electrical through an electric generator; therefore, we shall study different alternative sources of electrical energy.

Since the mid-1970s, there has been an active search worldwide for alternative air-independent energy units (EUs) with electrochemical generators (ECGs). In Russia, such works started in the early 1970s.

The performance of EUs with ECGs is based on the principle of directly converting the chemical energy produced in the reaction of the fuel and oxidant into electrical energy without an intermediate conversion of the chemical energy into thermal energy and the corresponding losses inherent in all thermal engines and defined by the Carnot principle. The electrical energy conversion reaction occurs in a fuel element.

Power units with low temperature fuel elements comprise of a cryogenic tank, a protection block and control system, an electrochemical generator, a power supply system switchgear, pumps of the thermostatic system, a heat exchanger, a reaction products collecting system tank, intermetallic tanks, and fire and explosion safety equipment systems. A basic diagram of an EU with ECGs is presented in Fig. 10.11.

The advantages of EUs with ECGs are as follows:

- A high efficiency factor (70–75% as compared to 20–30% for traditional power units);
- Absolute ecological cleanliness;
- Low noise;
- Low temperature level and a considerably lower quantity of subtracted heat as compared to traditional PUs; and
- A high energy input.

At present, within 2–4 years, it will be possible to develop, produce, and supply (depending on the

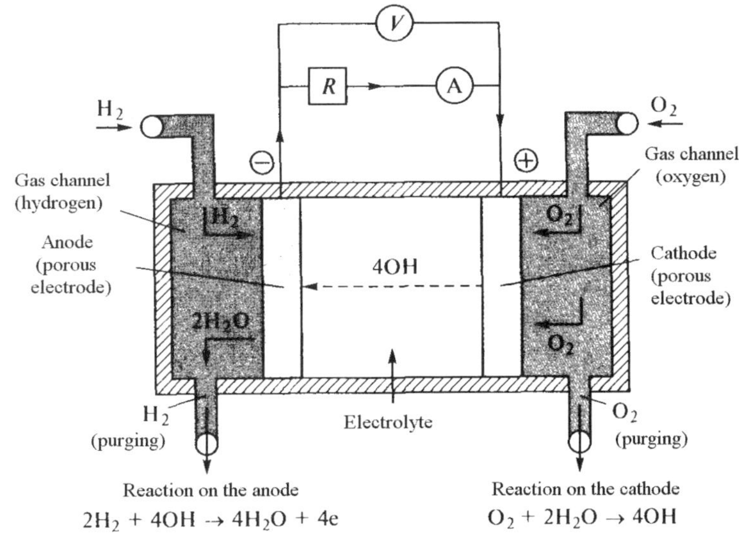

Fig. 10.11. Basic diagram of an EU with ECGs

power level and energy input) EUs with ECGs with the following indicators:

- A power of 10–600 kW (peak power of up to 4000 kW);
- An energy input of 100–100,000 kW/h; and
- A specific energy input of 150–200 W×h/kg (200–250 W×h/hp).

Power units with ECGs can be used for both marine and terrain facilities.

In addition to non-traditional electric energy sources with low temperature fuel elements, there are sources related to a category of energy systems: metal–air (iron–air, zinc–air, and aluminum–air).

Work on aluminum–air systems were launched abroad around the early 1980s. These systems relate to semi-fuel elements (SFE) since they are designed for mechanical recharging (installing aluminum plate sets). At the same time, they do not need to be charged from an external electrical network compared to traditional accumulator batteries (ACBs).

Let us briefly consider the operation principle of EUs in the form of aluminum–air SFEs.

Anodic plates from low aluminum alloys and porous air cathodes (from carbon and Teflon with an accelerator) are installed into the reactor (cell unit). During the operation, a mixture of cleaned air and water vapor passes through the air electrode and an alkaline electrolyte is continuously pumped through the gaps between electrodes. Aluminum, water, and aerial oxygen are consumed in the operation.

The resultant is a residue of aluminum hydrate, which is washed out of the cell by the electrolyte flow; after the wash and forthcoming drying in the EU sections, the residue is fed into the collecting tank in the form of a fine powder.

Aluminum hydrate is a ready semi-finished product for the industrial production of aluminum in the regular production cycle.

That is why aluminum electrodes serve as a renewable stage between the industrial source of electric energy and the electromobile.

Although the operational principle of EUs on aluminum–air SFEs is experimentally proven, there are still a number of engineering problems to be solved:

- There is a side (non-current generating) reaction of a simple aluminum dilution in the alkaline electrolyte; therefore, it is necessary to alloy plates for forming a special oxide film that inhibits this reaction (for example, 0.04% gallium, 0.4% iron, and 0.04–0.8% magnesium). This technological operation is considered expensive and complex and is currently not developed;
- The efficiency and reliability of the air electrode's performance requires further improvement since it mainly limits the system's power;
- The cell configuration development is required for convenient and fast replacement of the aluminum anodes And also for the removal of aluminum hydrate;
- The system's operational characteristics in winter conditions are not defined; and
- It is required to design complex experimental EUs, test them, and further optimize.

Except for the above-mentioned problems, there is one more essential drawback in such systems: the direct use of the regenerative braking energy for the electromobile in this system is excluded.

The main developer of aluminum–air type EUs is the Lawrence Livermore National Laboratory (LLNL; USA). A number of specialized companies in the U.S. are also involved in the work for solving particular matters.

After studying models and full-sized cells of aluminum–air type EUs and after further calculations and design, it was defined that the system is designed for the following operational modes:

- The aluminum plates' initial reserve is 56 kg, i.e., with an efficiency of 36.0 t×km/kg of aluminum, the fuel distance of the electromobile before anodes replacement is approximately 1600 km;
- Every 400 km, a short break is needed to refuel the 23 liters of water (tap water is suitable) and remove the 41 kg of aluminum hydrate (approximately 10 liters by volume); and
- Replacing the anode plates set at a service station after 1600 km (regular stations' network is used) is performed within 20–30 minutes.

After the first two years of development, the efficiency of the air electrode's power is defined. It is 54 t×km/kg of aluminum, which is an achievable indicator in the future. At the same time, the system's efficiency factor including the current costs of electric energy for aluminum melting is about 40%.

Based on the above-mentioned data, one can conclude that using EUs with ECGs and metal–air energy systems as power units in HDWVs designed for mounting traditional types of equipment is not likely to be rational up to 2015 because the improvement perspectives of specified power diesel engines are far from being exhausted.

Chapter 11
Fundamental theory and calculation of power transmission from engine to wheeled mover

11.1 General overview, classification, and requirements for transmissions

A deep study of any system is based on the analysis of the subject from the outside and inside: from the outside it is a study of the interaction between the system and the environment; from the inside it is a study of the interactions between and within the system's structural components.

A systemic approach considers three main transmission structures as a system, namely: the output of the system and its input, the output of the preceding subsystem and the input into the following one, and the output of the preceding system and its input into several subsequent ones. There is no doubt that HDWV transmissions belong to the third system type. The place of the transmission in the "driver-vehicle-road" system is presented in Fig. 8.1.

The character of the internal interaction between the structural components of different transmission types was presented earlier in Figs. 4.6, 4.12, 4.13, and 4.14*a*; that of the hydrostatic transmission (HST) is presented in Fig. 11.1.

The theory and calculation of HDWV transmissions is based on the main provisions of the knowledge areas that support the design of the transmission's components; for example, for the mechanical transmission it is the area of classic mechanics; for the electric transmission, it is the area of electric engineering; for HSTs, it is the area of hydrodynamics, etc. Thus, the main task of the vehicle theory in this regard is to devise the laws, regularities, and algorithms from which the transmission should function, while ensuring the utmost utilization of the power unit's energy and the maximum achievement of the HDWV's consumer properties. The major problem is in ensuring the rational distribution of power (torque) from the power unit (PU) among the vehicle's wheels. The consideration of this problem, i.e., the interaction regularities of the system's main structural components is the topic of this chapter.

The transmissions used in HDWVs are quite different and can be classified in many ways: by the principle of operation or regulation, or by the type of transformer used, and so on.

By the operation principle, transmissions are divided into mechanical, hydraulic, electric, and combined transmissions; by the regulation principle, they are divided into regulated and automated (automatic) transmissions; and by the type of the transformer used, they are divided into friction-gear fixed-ratio (stepwise) transmissions, friction continuous (with variators (stepless speed regulators)), pulse-type, hydrodynamic (hydromechanical), hydrostatic, electric (with motors-in-wheel), electromechanical, and electrohydraulic transmissions.

When designing a wheeled vehicle, and, in particular, its transmission, the parameters

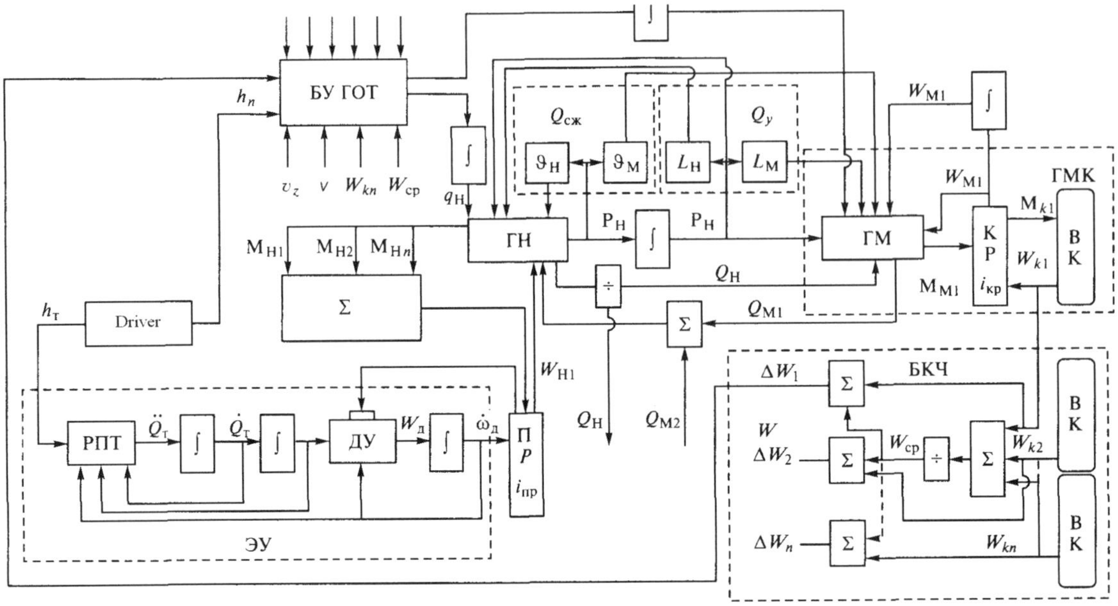

Fig. 11.1. Flow-chart of the main interconnections existing in the HST: PST—Fuel supply regulator; BU—HST control unit; D—Diesel power plant; HM—Hydraulic wheel motor; IG—Intermediate gearbox; FCU—Frequency control unit; HP—Hydraulic pump; DW—Driving wheel; WG—Wheel gearbox; HM—Hydraulic motor; PP—Power unit

which satisfy the technical and consumer requirements are initially defined; then, the design is optimized under the main requirement. Once the requirements for load capacity, traction-dynamic properties, passing ability, soft driving, etc., are met, the main requirement can be the weight of the vehicle, which depends on the transmission type. For example, Tables 4.3 and 4.4 have presented the weight of the different transmission types used in HDWVs. The tables reveal the obvious advantages of the hydrostatic transmission (HST).

In order to advance the HDWV motion theory in some new direction, we need to establish the interrelations of the vehicle parameters and systems with its performance properties and to justify the evaluation indicators under which we plan to operate the systems and the vehicle as a whole. For example, a review of the numerous studies performed in the area of wheeled vehicle motion theory allows us to identify a group of performance properties greatly affected by the systems used to bring the power to the driving wheels. For HDWVs, these are the traction-dynamic properties, maneuverability, and motion stability (road-holding).

The traction-dynamic properties create the defining vehicle efficiency factors. These properties are especially relevant for HDWVs. Under any evaluation criteria of the transportation work performed by the vehicle (in tons, t/km, etc.), experts need to use its main traction-dynamic properties (the maximum and average travel speed en route, specific traction force, and the acceleration characteristic) in order to make a comparison,.

That is why in modern HDWVs any increase in their load capacity as a way to maximize their efficiency is accompanied by a mandatory ensuring of the required traction-dynamic properties.

The generally adopted evaluation criteria that define the parameters of a vehicle's traction-dynamic properties are the engine's specific power and the average travel speed in various road conditions. Table 10.7 has presented information on the dynamic change in the specific power in various SWCs, which have practically the same traction-dynamic property indicators. For special-purpose multi-axle HDWVs, with an increase in the vehicle's full weight, there is a trend toward decreasing the specific power. The data of Table 10.7 indicates that it is practically impossible to further improve the traction-dynamic properties of HDWVs only by increasing the engine power.

The specific power is a conventional indicator. It completely disregards the nature of the external and specific characteristics of the engine and transmission, which are essentially affecting the engine power's utilization efficiency.

The character of the engine's external and specific characteristics affects engine adaptability. In different countries, engine adaptability is assessed by different parameters:

- By the torque (turning moment) adaptability coefficient (or adjustability factor):

$$K_M = \frac{M_{et\max}}{M_{Ne}};$$

- By the rpm adaptability coefficient:

$$K_\omega = \frac{\omega_{Ne}}{\omega_{M_{et\max}}};$$

and

- By the integral adaptability coefficient:

$$K_o = K_M K_\omega,$$

where: M_{Ne} is the engine torque corresponding to the maximum engine power; ω_{Ne} is the crankshaft's nominal rotation rate corresponding to the maximum engine power; $M_{\text{et max}}$ is the maximum engine torque; and $\omega_{M_{et\max}}$ is the crankshaft's rotation rate corresponding to the maximum engine torque.

The parameters of the vehicle's average speed and fuel consumption, which have the same specific power, notably depend on the engine adaptability and transmission characteristics. That is why in the opinion of British, Italian, and Danish experts, the main precondition for improving a vehicle's traction-dynamic properties, especially in HDWVs, is to increase the traction force of the driving wheels, which depends on the engine torque and transmission characteristics.

British specialists believe [34] that it is more reasonable to equip heavy-duty vehicles and road trains with relatively slow and highly economical engines, which ensure a specific power of at least 4.41 kW/t (6.0 hp/t) (power was measured according to the British standard) and have a high adaptability coefficient.

Specialists from Mack Trucks (USA) believe that the most suitable parameter for defining the performance of a vehicle is not the maximum, but the average power generated by the engine within its rpm working range. This parameter takes into account both the engine adaptability and the number of steps in the transmission.

While advancing this idea still further, French companies engaged in manufacturing cargo vehicles have suggested evaluating the engine power with the help of the so-called "equivalent power."

The equivalent power defines the engine's utilized power, measured at the average rpm n of the crankshaft caused by the presence of a particular transmission.

The expression for defining the ω_m is as follows [34]:

$$\omega_m = 0.5\,\omega_{Ne}\left(1 + \sqrt[n-2]{\frac{u_i}{u_2}}\right),$$

where: n is the number of vehicle gearbox stages; and u_2 and u_i are the transmission ratios of the second and the highest steps of the gearbox respectively.

Specialists at the Yaroslavl Engine Factory have conducted experimental studies on the operation modes of the YaMZ-236 engine model in a MAZ-500 truck's motion on roads with asphalt-concrete pavement outside a suburban zone (Operation Category 1), in urban conditions (Category 2), and the operation modes of the YaMZ-238 and YaMZ-240N engine models installed on the KrAZ-256B and BelAZ-548A dumpers, respectively, during their work in quarries (Category 3).

The following average travel speeds were obtained: 30–32 km/h during the tests of the MAZ-500 truck in urban conditions; 45–50 km/h during the tests of the MAZ-500 in conditions of intercity haulage; and 21–23 km/h during the work of the loaded BelAZ-548A dumper in a quarry.

Analysis of the data shows that the dominant working regimes of the YaMZ-236 engine in a MAZ-500 in urban conditions fall within a narrow rpm range (970–1170 rpm) and a broad loading range—from the full fuel supply to the full braking mode by means of the engine.

When a single MAZ-500 truck moves on the highway, the YaMZ-236 engine is mainly working within a narrow loading and a broad range of speed regimes.

The working regimes of the YaMZ-238 engine when the KrAZ-256B dumper works for a full cycle in the quarry are the same as when the MAZ-500 moves on a highway. The YaMZ-238 engine works mainly within the external speed characteristic zone, i.e., within a narrow loading and a broad speed range. However, since the longitudinal slopes of the roads in the quarry and the relative loading of the KrAZ-256B dumper are higher than those of the MAZ-500, the operation regimes of the former within the external speed characteristic zone at moving with a load against a slope become dominant.

The dominant working modes of the YaMZ-240N engine installed on the BelAZ-548A dumper in the whole operation cycle of the quarry, and, especially when moving fully loaded uphill, are the modes within the external speed characteristic, i.e., the YaMZ-240N engine works within a narrow loading range. Thus, 58.1% of the time it works under a torque range of 1560–1800 Nm (the torque at its maximum power is 1665 Nm) when the BelAZ-548A dumper performs its full cycle and 81.84% of the time it works within the same torque range when loaded and moving uphill.

However, if the dominant operating modes of the YaMZ-236 engine when the MAZ-500 moves on a highway and the dominant operating modes of the YaMZ-238 engine when the KrAZ-256B dumper works in the quarry are spread within a rather broad speed range, the dominant operating modes of the YaMZ-240N engine are within a narrow loading range and a narrow speed range.

Thus, when the BelAZ-548A dumper works for a full cycle, 49.8% of the time the YaMZ-240N engine works within a rotation rate range of 1770–1970 rpm and 8.3% of the time within a rotation rate range of 1970–2170 rpm; when moving fully loaded uphill within the same rotation rate ranges, it works for 62.5% and 19.34% of the time respectively.

Thus, the engine-transmission units (ETUs) of the KrAZ-256B and BelAZ-548A dumpers ensure during their work in a quarry that the engine operations within the rational (from the viewpoint of the engine manufacturer) loading and speed ranges; however, they fail to ensure the main thing (from the viewpoint of the consumer): a high average travel speed (35–40 km/h for KrAZ and 20–25 km/h for BelAZ), although their specific power is at the level of European standards (at least 6.0 hp/t).

HDWV design experience and the calculations made show that in order to achieve acceptable traction-dynamic properties and passing ability, the power plant should be able to yield at least 800 kW and have the external speed characteristic of ensuring highly efficient operation within a narrow speed and loading range. At present, Russia has no diesel automobiles of that power. The high powered diesel engines used today in quarry dumpers and SWC prototypes made by BelAZ and MZKT are industrial engines—they are designed for operation in stationary modes and cannot be used in future HDWV models because of their inadequate dimensions and weight.

As noted earlier, the only alternative engine currently capable of meeting the requirements for power, weight, and dimensions for HDWV power units is the automobile gas-turbine engine [57].

Thus, it is practically impossible to ensure the required traction-dynamic properties of HDWVs by only increasing the power of the power unit. It is necessary to optimize the power distribution and transmission type for every HDWV group depending on their function, conditions of use, or special application.

An operational characteristic of the working conditions for transmissions used in special-purpose multi-axle and multi-support HDWVs (with 4–12 axles) is the fact that the number of possible combinations for the operation modes of the driving wheels at every specific moment is essentially higher than in HDWVs with 2–3 axles. This is especially evident during curvilinear motion, motion on deformable grounds, and motion on bumpy bearing surfaces when some wheels work in the driving mode and others in the driven, free, or neutral mode. It is even possible that some wheels may switchover into the braking mode. Moreover, the probability of power and speed asymmetry increases with the increase in the number of axles (wheels) and overall dimensions.

A.A. Shimkov [18] has conventionally classified HDWVs into four groups: light vehicles with a full weight of up to 20 t, average vehicles with a full weight of 20–40 t, heavy vehicles with a full weight of 40–100 t, and super-heavy (oversized) vehicles with a full weight of over 100 t. Based on the results of studies and tests and an analysis of the technical solutions of ETUs in wheeled vehicles weighing 20–100 t, Shimkov [18] has developed a methodology for designing and selecting ETUs based on the hydromechanical transmission (HMT). As it was already mentioned, the use of these transmissions in HDWVs manufactured by MAZ and BelAZ has taken these vehicles to a new qualitative technical level; however, it has not solved the problem of individual power delivery to the wheel depending on the particular rolling conditions. Additionally, an increase in the total weight of HDWVs up to 100 t or more with six or more axles has made it necessary to use two parallel power flows with a separate HMT in each axles. This creates a cumbersome mechanical section of the transmission, and, consequently, huge problems with the vehicles' layout (for example, in chassis models 7904 and 7906).

The increase in the HDWV's load capacity and axle number leads to an increase in the variety of possible mechanical transmission designs, which is associated with the presence or absence of the differentials in the points where the power flow is split.

The need to deliver a huge power (over 800 kW) and the absence of a proper HMT capable of providing it resulted in using, for example, in the 7904 and 7906 chassis models, two batch-produced HMTs intended for a power of up to 600 kW. This, in turn, required installing an additional summing up (totaling) gearbox unit. However, the parallel installation of the two HMTs and the combination of their power flows in one gearbox practically excludes any blocking of the hydraulic transformers. The impossibility to block these transformers results in an increased fuel consumption and decreased traction-dynamic properties when moving on highways; it also moves the special requirements for the cooling system, forcing the design of—in very dense layout conditions—complicated systems aimed at removing heat and diffusion. The composition of the transmissions in the 7904, 7906, and 7907 chassis models are presented in Table 11.1.

Analysis of the data presented in Table 11.1 shows that the transmission in the six-axle 7904 chassis is comprised of 24 main units and the eight-axle 7906 chassis is comprised of 32 main units with 45 cardan shafts used to drive them. Nine units in the 7904 model and 13 units in the 7906 model are equipped with differential mechanisms (inter-bogie and inter-axle differentials and freewheel clutches), which has notably complicated the design of the units and transmission of the entire chassis.

The 800 kW power transfer through the transmission causes the necessity to install many heavy units, which results in a great increase in the transmission's overall weight. Thus, for example, the weight of the matching gearbox in the 7906 chassis is 665 kg, the weight of the HMT is 1100 kg, the weight of the summing up gearbox is 710 kg, and the weight of the cardan shafts is 3150 kg. Thus, the total weight of the transmission units in the 7906 chassis is 17.6 t and the specific weight of the transmission is 39.4 kg/kW.

The use of two HMTs and large number of mechanical gearboxes in 12×12 and 16×16 HDWVs decreases the transmission's efficiency factor. It is theoretically defined that the total efficiency factor of the transmission in the nominal operation mode is 0.54 on the 7904 chassis and 0.52 on the 7906 chassis, which indicates a great decrease in the transmission units' power.

The structural complexity of the above mechanical (hydromechanical) transmissions for HDWVs has resulted in the situation where flexible transmission types, non-traditional for the motor industry, are designed and manufactured both in Russia and abroad.

The electric transmission (ET) is of the greatest interest in this context. The main advantages of the ETs are their remoteness and flexibility. Their flexibility makes it easier to solve the problem of transferring power from the driving engine to the driving wheels of HDWVs and to the trailed links. This is especially important for HDWVs with significant lengths (28–33 m) and a large number of driving wheels (12 and more).The ET can also ensure engine operation in the quasi-stationary mode.

Designing ETs of wheeled vehicles is carried out in three main directions: direct current (DC), alternating current (AC), and AC-DC transmissions.

Table 11.1 Composition of the transmissions used in the 7904, 7906, and 7907 chassis

Unit name	Quantity of the transmission units in the chassis model		
	7904 (12×12)	7906 (16×16)	7907 (24×24)
Matching gearbox	1	1	1
Hydromechanical transmission	2	2	-
Summing up (totaling) gearbox	1	1	-
Distributing gearbox	2	4	-
Axle gearbox	6	8	-
Cardan shaft	32	45	24
Wheel transmission	12	16	24
Reducing gearbox of the electric motor	-	-	24
Traction electric motor	-	-	24
Power generator	-	-	1

The DC electric transmission is used in BelAZ-549 quarry dumpers with a load capacity of 75 t and the AC/DC in the BelAZ-7519 with a load capacity of 180 t. MZKT has tested prototypes of chassis with an AC ET. The main parameters of the design-layout solutions for HDWVs with an electric transmission are presented in Table 4.4.

The important design-layout indicators of ETUs are their weight and dimensions. The total weight, the load-carrying capacity, the axle load, the fuel efficiency, and the dynamic qualities of the chassis and other features depend on the weight of the ETU. The dimensions influence the configuration and size of the HDWV and the ease of placing the transported cargoes and equipment. This increases the importance of evaluating the size and weight parameters when choosing the appropriate ETU for multi-axle wheeled vehicles and, primarily, for HDWVs.

A comparative evaluation based on the methodology presented by A.A. Shimkov [18] of a diesel engine and HMT (prototypes 1 and 2) ETU and a GTE with an ET (prototypes 3 and 4) ETU is given in Table 11.2. The main data are the results of comparative tests on prototypes of the 547V and 547E chassis (prototypes 2 and 3, respectively) and prototypes of the 7906 and 7907 chassis (prototypes 1 and 4, respectively). The ETU parameters in prototypes 1 and 2 were taken as 100% and the relative percentage of the ETU parameters was defined (parameter of prototype 3 relative to prototype 1, parameter of prototype 4 relative to prototype 2).

The data presented in Table 11.2 shows that the ETUs of prototypes 1 and 3 have approximately the same dimensions, while their weights differ by almost 10%. However, the more than 70% difference in the power on the driving wheels (because of the higher powered engine and the higher efficiency factors of the transmission) results in the fact that specific indicators of ETUs consisting of a GTE and an ET (prototype 3) are significantly higher than those of vehicles equipped with a diesel engine and an HMT.

Table 11.2 Absolute and specific indicators of the ETUs in multi-axle HDWVs

Parameter,%	Prototypes with ETU comprising:			
	Diesel engines and HMT		GTE and ET	
	No. 1	No. 2	No. 3	No. 4
	Wheel formula			
	12×12	16×16	12×12	24×24
Full weight	100.0	100.0	101.2	100.0
Engine power	100.0	100.0	153.8	80.0
Total power on the wheels	100.0	100.0	172.1	76.0
Weight of the ETU	100.0	100.0	90.6	87.0
Dimensions of the ETU	100.0	100.0	99.4	103.0
Weight of the transmission	100.0	100.0	91.9	113.9
Dimensions of the transmission	100.0	100.0	91.3	113.3
Specific weight of the ETU	100.0	100.0	53.0	114.5
Volume power of the ETU	100.0	100.0	172.0	75.8
Specific weight of the transmission	100.0	100.0	53.7	151.3

A comparison of the ETU parameters in prototypes 2 and 4 shows that the weight of the ETU in the 7906 chassis (prototype 2) is 13% higher than that of the 7907 chassis (prototype 4), while their dimensions are approximately equal. Under the same load capacity and full weight of the prototypes, the 7907 chassis has four (eight) additional axles (wheels). This situation is mainly due to the low specific size and weight parameters of the diesel power unit with objective systems of the 7906 chassis, which are 4.1 times higher than the corresponding weight parameters of the GTE and twice as high as its size parameters. In general, the ETU weight accounting for 1 kW of the wheeled power in the 7907 chassis is 8.5% lower, while the dimensions are 8.3% higher than those in the 7906 chassis.

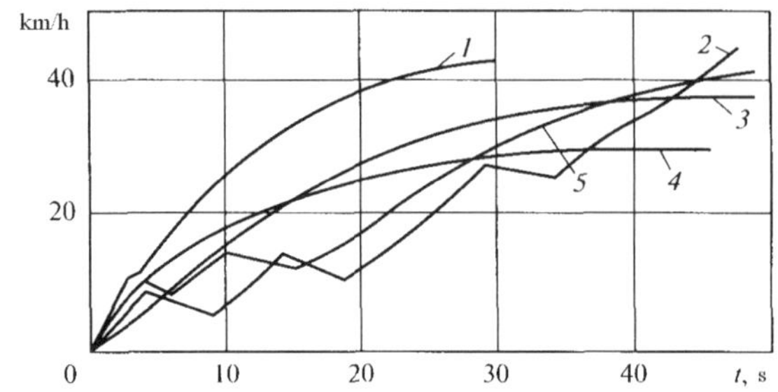

Fig. 11.2. Acceleration speed characteristic versus time in HDWVs with different ETU types: 1—547E chassis (GTE + ET); 2—547D chassis (E + mechanical gearbox); 3—7906 chassis (E + HMT); 4—7907 chassis (GTE + EMT); 5—7917 chassis (E + HMT); Δ*Pgen*—Power losses in the generator

It should be noted here that reserves for improving the weight and size parameters and increasing the efficiency factor of the 7907 chassis (prototype 4) look highly promising. Thus, in particular, installing traction electric motors in the wheel hubs (using motors-in-wheel, see Fig. 4.14), not inside the chassis frame, will allow for giving up 24 angular gearboxes and 24 cardan shafts, whose full weight and dimensions are 5.5 t and 4.3 m^3, respectively. At the same time, it will allow for increasing the ET's efficiency factor up to 0.58–0.60.

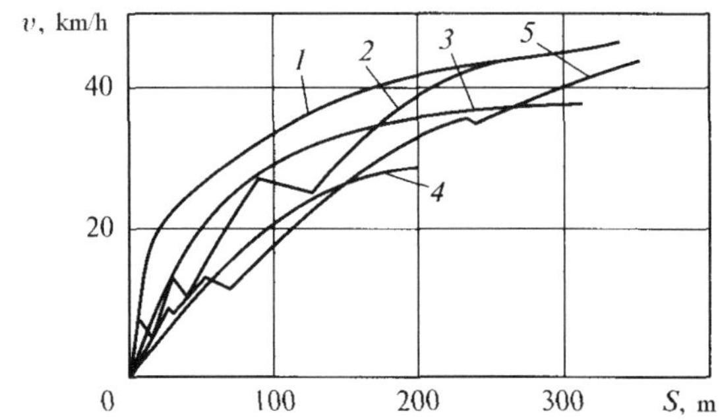

Fig. 11.3. Acceleration speed characteristic versus travel of HDWVs with different ETU types: 1—547E chassis (GTE + ET); 2—547 chassis (E + mechanical gearbox); 3—7906 chassis (E + HMT); 4—7907 chassis (GTE + EMT); 5—7917 chassis (E + HMT); Δ*PE*—Power losses in the electric motor

The standard acceleration characteristic of HDWVs with different ETUs is presented in Figs. 11.2 and 11.3.

Analysis of these figures shows that the traction-speed properties of the prototypes with the ET are at the same level as the indicators of the prototypes with diesel engine and HMT ETUs with the same load capacity and axle number.

The basic obstacle impeding a broader use of ETs in wheeled vehicles is the absence of the duly justified and mastered technical requirements for this transmission type. In addition, the designing and producing of ETs for wheeled vehicles are covered by different branches of the science and engineering industries (the motor industry and the electric engineering industry). As a rule, specialists from these spheres are not competent enough and skilled in special issues of the other sphere to address the tasks with due qualification. The issues of the design and calculation methods for the main parameters of traction electric motors, generators, and their regulation systems as applied to ETs of wheeled vehicles are insufficiently developed and poorly covered by technical literature.

Additionally, a number of the essential drawbacks were revealed while designing and manufacturing electric transmissions for SWCs:

- The works to create ETs were conducted mainly toward adapting the existing models of the electric units, which had been designed without due account of the chassis' character and operating modes; and
- The technical requirements for chassis with ETs were formulated according to the principle of achieving "at least the same" indicators as the chassis with same load capacity equipped with an HMT.

These drawbacks are best illustrated in the 7907 chassis (see the ETU diagram in Fig. 4.13). Figs. 11.4–11.15 present the dependencies of the HDWV's performance properties with the ETU diagrams presented in Fig. 4.13. When designing HDWVs with ETs, the data presented in Figs. 11.4–11.15 are basic. If the designed performance properties are lower than the basic ones, the use of the ET is inefficient.

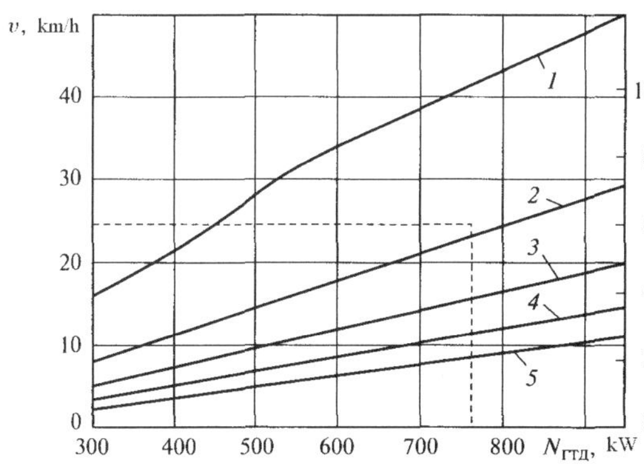

Fig. 11.4. Travel speed of a 24×24 chassis versus the GTE power and motion resistance ψ: 1—ψ = 0.02; 2—ψ = 0.04; 3—ψ = 0.06; 4—ψ = 0.08; 5—ψ = 0.10. For the curves in the transmission ratio of the wheel gearbox, *Uw.g.* = 67.815; - - - - - the maximum GTE power in the object

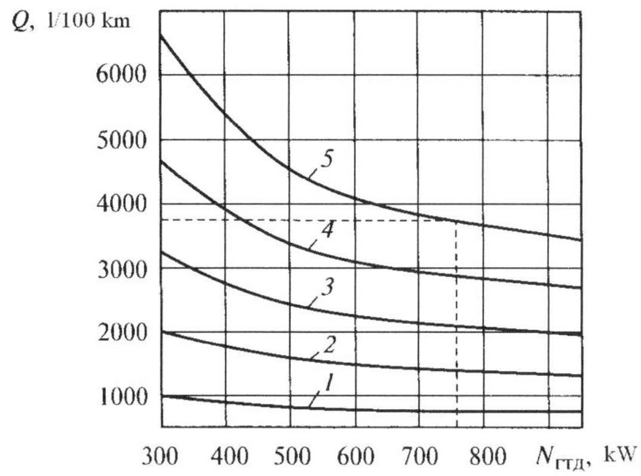

Fig. 11.5. Fuel consumption in a 24×24 chassis versus the GTE power and motion resistance ψ: 1—ψ = 0.02; 2—ψ = 0.04; 3—ψ = 0.06; 4—ψ = 0.08; 5—ψ = 0.10. For the curves the transmission ratio of the wheel gearbox, *Uw.g.* = 67.815; - - - - - the maximum GTE power in the object

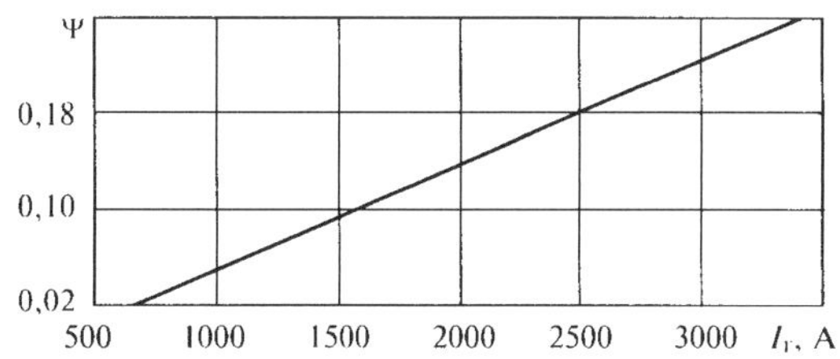

Fig. 11.6. Motion resistance ψ conquered by a 24×24 chassis versus the phase current *Ig* of the VSG-625 generator: ----—phase current limitation in the VSG-625 generator

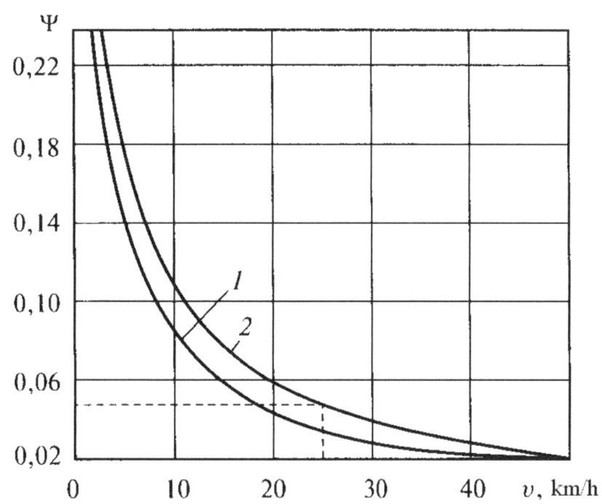

Fig. 11.7. Traction-speed characteristic of a 24×24 chassis: 1— GTE power in the 759 kW object; 2—GTE power in the 950 kW object; - - - - limitation by the generator's phase current to 2500 A

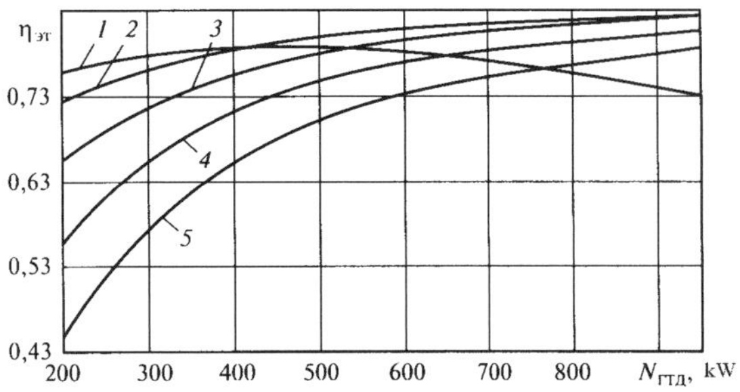

Fig. 11.8. Efficiency factor η_{ET} of the 24×24 chassis' ET on the GTE power in the object: 1—ψ = 0.02; 2—ψ = 0.04; 3—ψ = 0.06; 4—ψ = 0.08; 5—ψ = 0.10

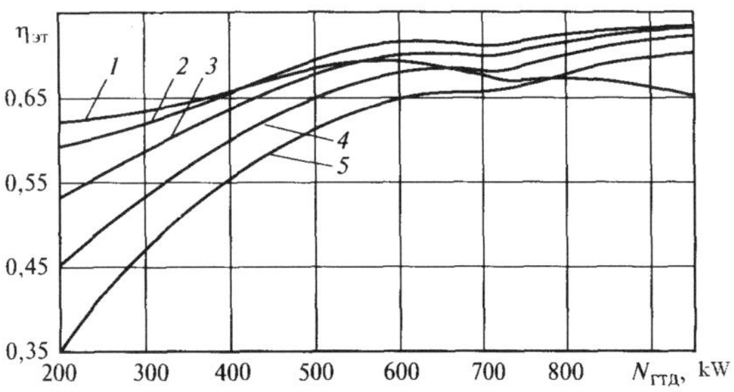

Fig. 11.9. Efficiency factor η_{EMT} of the electromechanical transmission in the 24×24 chassis versus the GTE power in the object: 1—ψ = 0.02; 2—ψ = 0.04; 3—ψ = 0.06; 4—ψ = 0.08; 5—ψ = 0.10

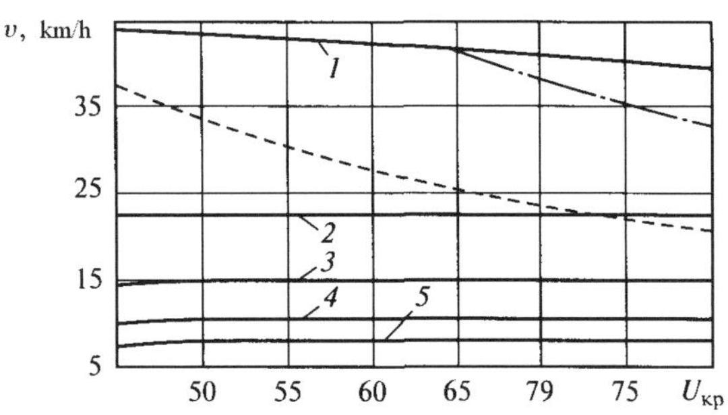

Fig. 11.10. Maximum speed v of the 24×24 chassis versus the transmission ratio of the wheel gearboxes $U_{K.P}$ at the full load of the 759 kW GTE in the object: 1—ψ = 0.02; 2—ψ = 0.04; 3—ψ = 0.06; 4—ψ = 0.08; 5—ψ = 0.10; - - - - - limitation of the electric motors' rotation rate to 6000 rpm; -- -- ---—limitation of the electric motors' rotation rate to 9600 rpm

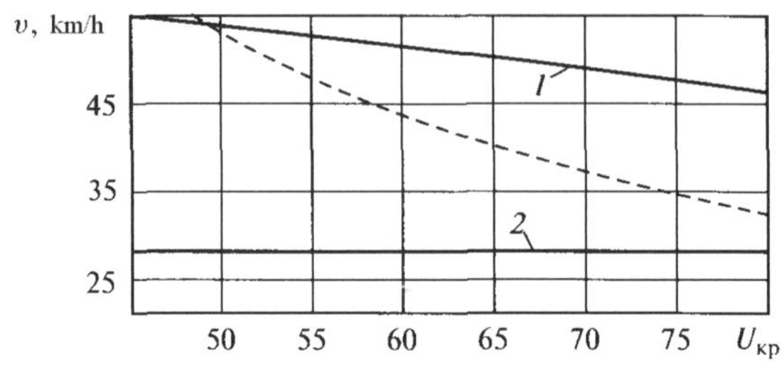

Fig. 11.11. Maximum speed v of the 24×24 chassis versus the transmission ratio of the wheel gearbox $U_{K.P}$ at the full load of the 9509 kW GTE in the object: 1—ψ = 0.02; 2—ψ = 0.04; - - - - - limitation of the electric motors' rotation to 9600 rpm

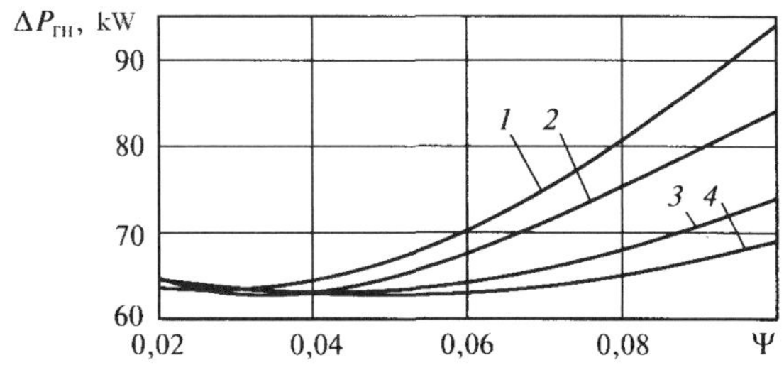

Fig. 11.12. Power losses in the generator ΔP_{ge} versus the motion resistance Ψ and transmission ratio U_{wg} 1—U_{wg} = 45.21; 2—U_{wg} = 52.8; 3—U_{wg} = 67.815; 4—U_{wg} = 80.0; N_{GTE}—759 kW

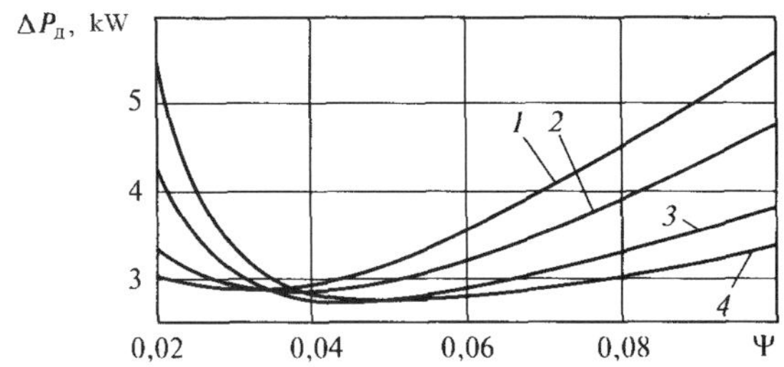

Fig. 11.13. Power losses in the generator ΔP_{ge} versus the motion resistance ψ and the transmission ratio of the wheel gearbox U_{wg}: 1—U_{wg} = 45.21; 2—U_{wg} = 52.8; 3—U_{wg} = 67.815; 4—U_{wg} = 80.0; N_{GTE}—759 kW

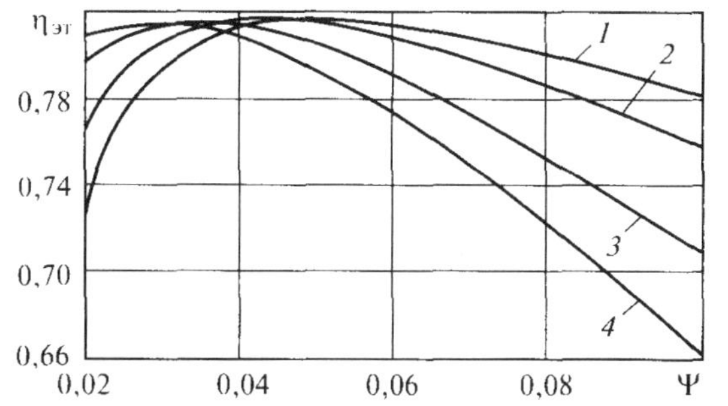

Fig. 11.14. Efficiency factor η_{ET} of the electric transmission of the 24×24 chassis versus the motion resistance ψ and the transmission ratio of the wheel gearbox U_{wg}: 1—U_{wg} = 80.0; 2—U_{wg} = 67.815; 3—U_{wg} = 52.8; 4—U_{wg} = 45.21; N_{GTE}—759 kW

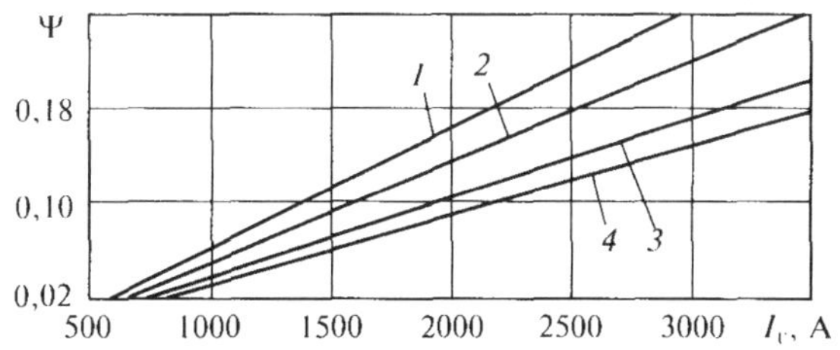

Fig. 11.15. Motion resistance ψ conquered by the 24×24 chassis versus the generator's phase current *Ig* and the transmission ratio of the wheel gearbox U_{wg}: 1—U_{wg} = 80.0; 2—U_{wg} = 67.815; 3—U_{wg} = 52.8; 4—U_{wg} = 45.21

Thus, the common deficiencies in the electric transmissions used in the above prototypes are as follows:

- Poor designs of the generator and traction electric motors, which resulted in restricting the maximum speed and free drawbar capacity;
- Insufficient reliability of the ET elements because of their structural imperfections and the absence of requirements for them; and
- Poor automatic control systems which prevent using the main advantages of the ET, including individual control of the wheel drives depending on road conditions and motion modes.

Analysis of these and other drawbacks indicates the insufficient depth of the theoretical studies for designing automobile ETs for SWCs while accounting for their construction and conditions of use. The traditional approach to designing transmissions for chassis of this class used in the 7904 and 7906 chassis demonstrates once again that it should be rejected.

An interesting new direction in the designs of traction electric drives for wheeled vehicles was the R&D work launched in 1990 at Kamsk Automobile Works (known as KamAZ) aimed at creating an advanced, multi-function, ecologically friendly vehicle by the year 2000. The designers used an ETU comprising of an automobile gas-turbine engine and an AC electric transmission with asynchronous traction electric motors. Within the framework of this R&D work, specialists from MSTU and the "VNIITransmash" (Scientific and Research Institute for Transport Machine-Building), with the authors' direct participation, created a prototype of the "Tabun" vehicle on the basis of a 4×2 KamAZ vehicle. It was used to run tests for developing the technical requirements for the regulation and control system of the advanced traction electric drive relating to the range of vehicles in the "Tabun" project. The main specifications of the running prototype are presented in Table 11.3, while its layout is presented in Fig. 11.16. The data in Table 11.3 indicates that the running "Tabun" prototype was developed to make the switch from the

Fig. 11.16. "Tabun" project running prototype

Table 11.3 Specifications of the running "Tabun" prototype

Parameter	Value
Wheel formula	4×2
Maximum speed (limited by the gearboxes used), km/h	55
GTE:	
nominal power, kW (hp)	154 (210)
rotation rate of the output shaft, rpm	8000
Generator:	
type	synchronous
nominal power, kW (hp)	130 (177)
rotation rate, rpm	8000
Traction electric motor:	
nominal power, kW (hp)	44 (60)
maximum rotation rate, rpm	5500
Overall dimensions, mm:	
length	6000
width	3330
height	2960
Wheelbase, mm	3600
Wheel span, mm	2980

Note. The regulation and control system of the traction electric drive: rpm governor of the traction electric motors—transistor-based by the generator's excitation current; the motion control system—stepless from the drive pedal detectors.

customary approach (used by MZKT and NIIKE (The Scientific and Research Institute of the Complete Electric Drive)) to the design of ETs for wheeled vehicles, especially of the automated control systems. The rotation rate (rpm) of the traction electric motors was varied by a transistor regulator on the basis of the pulse-width modulation principle instead of the generally adopted thyristor-based frequency converters.

Designers of automobile transmissions continue to pay great attention to the application of the hydrostatic drive (HSD). Here, it is important to identify the following three main directions in the research: first, the use of adjustable hydraulic machines in the main or parallel power flow for the stepless change of the total transmission ratio; second, ensuring the drive of some axles, engaged only in special driving conditions; and, third, creating multi-channel drive systems with individual power supply to all driving wheels (axles)—hydrostatic transmissions (HSTs).

Transmissions created within the first and second directions are combined and represent a complex of interconnected hydraulic machines and mechanical units. Using HSTs only for the stepless range variation in the general transmission ratio (first direction) or to drive some of the driving axles (second direction) prevent any essential increase in the wheeled vehicle indicators since the main obstacle is the huge number of mechanical assemblies and units such as onboard, central, and wheel gearboxes; distribution boxes; and cardan transmissions. The power losses and irrational power distribution in this mechanical part of the multi-drive transmission is one of the main factors that restrains the qualitative increase in the most important properties of HDWVs: the traction-dynamic properties, the passing ability, and others.

That is why the third direction, using hydrostatic drives (HSDs) in transmissions of multi-axle vehicles, is currently the direction of highest interest. Accounting for the design features, specific use conditions, and vehicle requirements, this direction is realized in several multi-axle HDWVs: loaders, construction-engineering machines, and other types of vehicles.

The application analysis of a HSD in HDWV transmissions shows that the most reasonable application is in vehicles where criteria such as the ease of control and driving, the possibility of individual stepless power regulation driven to the wheels, automation, and remote control dominate over the price and design simplicity.

The main obstacles to the widespread application of HSTs in vehicle designs are the same as noted during the analysis of the ET. By their functional capabilities, both HSTs and ETs can realize—rather flexibly and precisely—the same algorithms of the power transfer from the power-generating unit to the vehicle's wheeled mover. If we imagine a transmission as a "black box," then the task of the vehicle theory is to search, develop, and justify the rational transformation regularities and the most effective individual ways to deliver power from the power-generating unit to the wheeled movers depending on the operation conditions of each wheel of the multi-axle vehicle. The task of designing a specific flexible transmission, operating under a preset algorithm, will be imposed on specialists of the corresponding fields of knowledge.

Thus, further progress in HDWVs has to do with designing flexible ETUs based on using new methods of power transfer and distribution. The power plant (unit) of the ETU should be adapted as much as possible—by all of its external speed characteristics—to work within narrow speed and load ranges, while the transmission should ensure a rational, individual power transfer to each wheel depending on its travel conditions for the maximum utilization of the torque. The main task of the vehicle theory in this process is to create the laws (operation algorithms) of the power transfer from the engine to the wheeled mover and develop the general technical requirements (GTRs) for flexible ETUs on the basis of those laws. The design embodiment of these GTRs is impossible without integrating the corresponding scientific and technical fields.

A vehicle designer should address the issues of selecting the correct transmission type, which realizes the preset regularities (ET, HST, etc.), including the economic, mass-dimensional, or other criteria of a particular vehicle based on its intended use and definite operation conditions.

Each HDWV group and vehicle type has particularities defined by their intended use and operational conditions. That is why one cannot formulate requirements for transmissions that would fit all vehicle types and not be in contradiction with each other. We can only set several general requirements, which refer to essentially two features of the vehicle: its ability to carry cargoes at the maximum speed and develop the maximum traction force at the minimum speed.

HDWV transmissions should meet the following general requirements:

1. Have a nontransparent characteristic or, in other words, the load variations on the wheels should not influence the engine's operation mode, i.e., the engine should operate at a continuous rpm at constant power;
2. Continuously vary the power flow factors in accordance with changes in the motion conditions and conditions of the wheel-support surface interaction;
3. Ensure an automated (automatic) and, if needed, forced regulation;

4. Have a sufficiently high efficiency factor in all operation modes. If that is impossible for all modes, then, the highest values of the efficiency factor should cover the most frequently used operating modes;
5. Have the least possible overall dimensions and weight;
6. Be reliable and have a long service life; and
7. Ensure an easy branching of the power flow among the driving wheels and auxiliary drives.

The mathematical model (MM) of the hydrostatic transmission describes the dynamic interacting processes of the elements and subsystems, which are represented in the HST functional diagram (see Fig. 11.1).

The HST MM generically defines the loading (the forces and moments affecting the input shaft of the hydraulic pump and hydraulic motor) and balance equations of the flow rates defining the volume of the working fluid passing through the cross-sections of the hydraulic machines.

The HST input characteristic is the moment on the shaft of the ith hydraulic pump M_{Hi}, which can be calculated according to the load equation:

$$M_{pui} = P_{pu}\frac{q_{pu}}{2\pi} - U_{gear}\omega_{pu}b_{pu} - U_{dear}\dot{\omega}_{pu}J_{pu}\ ,$$

where: P_{pu} is the pumping mains pressure; q_{pu} is the pump's specific volume; U_{gear} is the transmission ratio of the intermediate gearbox; ω_{pu} is the rpm of the hydraulic pump's output shaft; and b_{pu} is the coefficient accounting for the mechanical energy losses in the pump and gearbox:

$$b_{pu} = 0.06 \cdot 10^5\ q_{pu}/2\pi;$$

J_{pu} is the aggregate inertia moment of the pump's rotating elements:

$$J_{pu} = 0.03(q_{pu})^{4/3} \cdot 10^5.$$

The fluid pressure differential in the pressure mains P_{pu} is defined by transforming the differential equation of the flow rate balances, which takes into account the leakages and compression losses:

$$\dot{P}_{pu} = \left(U_{gear}\omega_{pu}\frac{q_{pu}}{2\pi} - \sum_{i=1}^{2n} Q_{hm_i} - P_{pu}L_{pu}\right)\frac{E_g}{V_{pu}}\ ,$$

where: Q_{hm_i} is the supply of the axle's i^{th} hydraulic motor; L_{pu} is the leakage coefficient related to the pressure mains and the pump unit:

$$L_{pu} = 0.105 \cdot 10^6 q_{pu}/2\pi;$$

E_g is the specific volume modulus of the volumetric elasticity; and V_{pu} is the maximum volume of the pressure mains, including the pump unit chambers.

The moment on the shaft of the ith hydraulic motor M_{Mi}, as an output characteristic of the hydraulic drive, is calculated from the loading equation, which takes into account the mechanical losses and the inertia component of the rotating elements in the corresponding machines:

$$M_{hmi} = P_{pu}\frac{q_{hm}}{2\pi} - \omega_{hm}U_{wg}b_{hm} - \dot{\omega}_{hm}J_{hm}U_{wg}\ ,$$

where: q_{hm} is the specific volume of the hydraulic motor; ω_{hm} is the rpm of the hydraulic motor shaft; U_{wg} is the transmission ratio of the wheel gearbox; b_{hm} is the coefficient of the hydraulic motor's mechanical losses:

$$b_{hm} = 0.09 \cdot 10^5 q_{hm}/2\pi;$$

and J_{hm} is the aggregate inertia moment of the rotating elements in the hydraulic motor and the wheel gearbox:

$$J_{hm} = 0.06(q_{hm})^{4/3} \cdot 10^5.$$

The supply of one of the hydraulic motor axles Q_{hmi} is defined by the consumption balance equation including the leakage and compression components:

$$Q_{hmi} = \omega_{hm} U_{wg} q_{hm} + \dot{P}_{pu} \frac{V_{hm}}{E_g} + P_{pu} L_{hm} ,$$

where: V_{hm} is the maximum volume of the motor units' chambers; and L_{hm} is the leakage coefficient related to the hydraulic motors:

$$L_{hm} = 0.05 \cdot 10^6 q_{hm} / 2\pi .$$

The main variables in the HST MM are the specific volumes of the hydraulic machines ($q_{pu(hm)}$). Variations in these values allows for organizing the pressure distribution in the mains, which causes rearrangement of the power constituent in the HST-moments. At the same time, variation in the specific volumes influencing the fluid consumption results in a change in the power flows' speed component in the transmission.

Thus, the presented MM fully describes the physical processes taking place in the hydraulic drive when it is operating as a transmission. Variation in the definite MM parameters allows for organizing an efficient power distribution in the transmission.

The preliminary control equations include the following:

1. The operation algorithms of the pumping unit responsible for accelerating the wheeled vehicle and for tracing the changes in the linear speed irrespective of the individual wheels' rolling conditions:

$$\dot{q}_{pui} = \left(\left(q_{pu\max} - (v_z - v_{ki})K_r\right) - K_c \dot{q}_{pui}\right) / T_c ,$$

   where v_z and v_{ki} are the vehicle's linear speed and the preset level of its value; K_r is the control coefficient; K_c is the conformity coefficient; and T_c is the channel response time. K_r, K_c, and T_{BC} are defined by the hydraulic drive characteristics under development and its control system properties.

2. The hydraulic motor unit's operation algorithms, which are defined by the motion modes and the operational conditions of the each wheel of the vehicle:

   1) $q_{hmi} = q_{hm2} = q_{\text{hmad}}$;

   2) $q_{hmi} = \left(q_{hm\max} - (q_{hm.ad} - K_c)\right)\sqrt{\left|M_{w.ave} / M_{wi}\right|}$

   3) $q_{hmi} = \left(q_{hm\max} - (q_{hm.ad} - K_c)\right)\dfrac{\omega_{w.ave}}{\omega_{wi}}$;

   4) $q_{hmi} = \left(q_{hm.\max} - (q_{hm.ad} - K_c)\right)\dfrac{N_{w.ave}}{N_{wi}}$;

   5) $\ddot{q}_{hmi} = \left((\dot{\omega}_{wi} K_{c1} - \dot{v}_{wi} K_{c2}) q_{m\max} - \dot{q}_{mi}\right) / T_c$,

   where: $q_{\text{hm.ad.}}$ is the hydraulic motor setting volume; $q_{\text{hm max}}$ is the hydraulic motor's maximum volume; M_{wi} is the torque on the ith wheel; $M_{\text{w.ave.}}$ is the average torque on the wheels; N_{wi} is power on the ith wheel; $N_{\text{w.ave.}}$ is the average power on the wheels; ω_{wi} is the rpm of the ith wheel; $\omega_{\text{w.ave}}$ is the average rotation rate of the wheels; and K_{c1} and K_{c2} are the coefficients of the concordance to the motion modes.

A calculation experiment was conducted to estimate the preliminary algorithms. A 6×6 wheeled vehicle with an HST of the presented diagram performed speeding-up, turning, and coming out of the turn on asphalt pavement. The following parameters, which were also used in the full-scale experiment, were chosen as the initial data for the HST:

- P_{sp} is the charge pressure and is equal to 2,500,000 N/m (25 bars = 2.5 MPa);
- $P_{\max}$ is the maximum pressure in the hydraulic drive and is equal to 4,500,000 N/m (450 bars = 45 MPa);

- q_{0hm} is the initial volume of the hydraulic motor;
- $q_{hm.ad}$ is the setting volume of the hydraulic motor;
- $q_{0hm} = q_{hm.ad} = 110 \times 10^{-6}$ m^3/rev.;
- $q_{ym\,min}$ is the minimum volume of the hydraulic motor and is equal to 33 10^{-6} m^3/rev.;
- $q_{ym\,max}$ is the maximum volume of the hydraulic motor and is equal to 160 10^{-6} m^3/rev.;
- q_{0H} is the pump initial volume and is equal to 0;
- q_{0hm} is the maximum volume of the pump and is equal to 125 10^{-6} m^3/rev.;
- K_r is the proportionality factor of the pump and is equal to 0.897 m^3/s;
- V_{pu} is the maximum volume of the pressure mains, including the pump unit chambers, and is equal to 7.0 10^{-4} m^3;
- V_{hm} is the maximum volume of the motor unit and is equal to 3.2 10^{-4} m^3;
- J_{pu} is the aggregate inertia moment of the pump's rotational elements and is equal to 0.0193 N m s^2;
- J_{hm} is the aggregate inertia moment of the hydraulic motor's rotational elements and the wheel gearbox and is equal to 0.18H N m s^2;
- v_z is the preset (required) linear speed of the vehicle;
- v_{pu} is the HST pump part elasticity criterion; and
- v_{hm} is the HST motor part elasticity criterion and

$$v_{pu(hm)} = V_{pu(hm)}/E_g,\ \text{m}^5/\text{N}.$$

The obtained mathematical model adds to the vehicle's efficiency (moveability) evaluation method as per the HST description and significantly expands the application range.

Further work in this direction should define the method of selecting rational algorithms under certain criteria and formulating proposals on the effective use of controlled HSTs in HDWVs.

11.2 Trends in designs of mechanical transmissions

The use of mechanical transmissions in HDWVs is still in demand by the industry, so the authors are including a brief analysis of their state-of-the-art design based on the materials of famous authors.

11.2.1 Gearbox units

To provide minimal performance costs during transport of cargoes in diverse motion conditions, it's necessary to have an extended range of wheeled vehicles, which can be accommodated for specific aims. This requires producing engines of different powers and the corresponding transmission units.

To avoid developing new transmission units for every new or updated engine, different combinations of mechanical transmission units are used: auxiliary transmissions, distribution gearboxes, driving axles, and multiple-speed gearboxes.

In the currently produced vehicles of Kamsk (KamAZ-5320, -53212, -5410, -54112, -55102, etc.), Kremenchug (KrAZ-260, -260B, -255B1, etc.) and Minsk Automobile Works (MAZ-6422, -5432, -5335, -53352, etc.), the so-called point-to-point gearboxes (GBs) are used, consisting of a gearbox and a divider. They provide two spans of the speed ratios with maximum kinematical ranges of 9.65 (KamAZ) and 10.89 (KrAZ and MAZ). However, it is obvious that these kinematical ranges are insufficient since for heavy-duty vehicles there is a global tendency to use gearboxes with an even bigger range.

For example, they are planning to use a 12-range gearbox on promising ZIL vehicles and on KamAZ, MAZ, and KrAZ vehicles with a 12–15 range. On the English 3-11.4×2 117L25 cargo vehicles with a load capacity of 11.6 t, a 12.5-range GB and a divider with the ratios of 4.1 and 1.0 are installed. Gearboxes with a range of ≈ 12.5 and a divider with two ratio variants are installed on the following vehicles: the American "Autocar-64F" with a load capacity of 18 t (the divider ratios are 3.38:1.0 and 1.0:0.87); the German-Italian "190.42 Turbostar" and "190.42 RT Turbostar" with a load capacity of 11.5 t (ratios of 3.8:1.0 and 1.0 and 1.16:1.0); Japanese CXZ180T and CXZ18Y with a load capacity of 17.6 and 18 t (ratios of 3.38:1.0 and 1.16:1.0).

Abroad, the direction related to using automatic gearboxes is under development. Alongside shaft (pinion) gearboxes, planetary ones are being used extensively [58].

The American company Allison, specialized in producing hydromechanical gears for cargo vehicles and buses, as well as for HDWVs, has started to produce a new generation of the planetary gearboxes. Its specialists are making attempts at developing a standard series, which differ by the number of planetary gears and control elements.

The theoretical foundations of mechanical transmission development and the diagram synthesis of the drive to the driving axles and wheels are covered extensively in several studies [14, 25, 54, 58, 59, etc.]. These studies contain a defined technical overview and the basic requirements for the parameters of this type of transmission, which easily fit the in the adopted diagram (see Fig. 8.1). However, they don't cover the opportunities for developing planetary gearboxes. That is why we shall address this issue more carefully.

An even bigger problem is represented by the development of standard planetary gearboxes under kinematical range and generating in them the required transmission ratios for vehicles of all types and classes. But this problem is also solvable, which is proved in the considered study by Yu.A. Stepanov [63], with a two-level planetary gearbox with three degrees of freedom.

The kinematical and power flow diagrams of this gearbox are represented in Fig. 11.17*a*

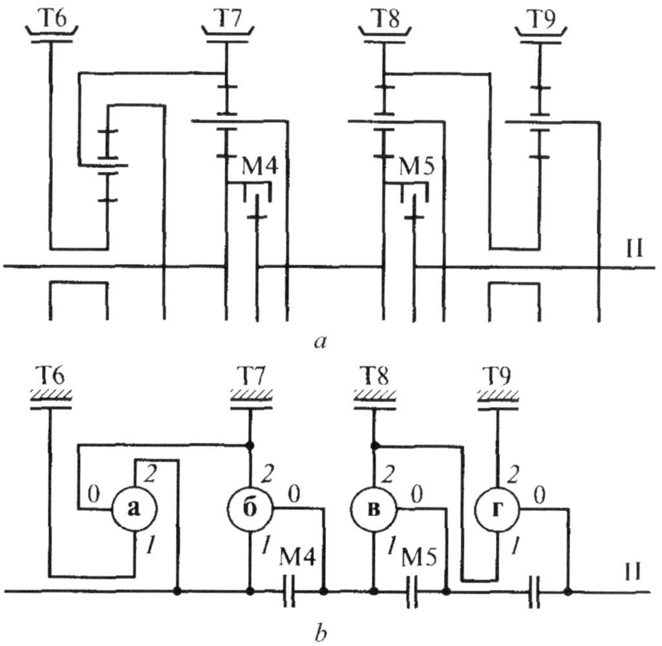

Fig. 11.17. Planetary gearbox: a—Kinematical diagram; b—Power flow diagram

(see Reference [2]). In Fig. 11.17*b*, straight-line planetary gears **a–d** are marked with circles, the upper vertical lines *2* are marked flows of the epicycle gear wheels, the lower lines *1* are noted flows of the sun pinions, and the horizontal lines *0* as a carrier are noted. Control elements (joints M4 and M5 and brakes T6, T7, T8, T9) are shown with two short parallel lines. Thus, the gearbox consists of the body, input and output shafts **I** and **II**, four straight-line planetary gears, four brakes, and two joints.

At engagement of clutches and brakes according to the control elements corresponding to the respective gear, the planetary gears are connected in a certain kinematical power flow diagram; as a result, the rotation from input shaft **I** to output shaft **II** is transferred with the transmission ratio corresponding to the diagram.

The gears operate in each specific case as follows:

At the first gear engagement (Fig. 11.18), brakes T7 and T8 are applied; as a result, the epicycle pinions 2 of the second **b** and the third **c** planetary rows are restrained. These rows operate under load in first gear and the first **a** and fourth **d** rows are idle. As shown in the power flow diagram, the transmission ratio of the gearbox in this case is equal to the product of the transmission ratios in the **b** and **c** planetary rows. However, under the theory of A.S. Antonov [2], in each planetary row at the restrained epicycle pinion, the transmission ratios of 2.44–5.88 are realized. Thus, with this gearbox, one can obtain transmission ratios of 5.95–34.6 in first gear.

At the second gear engagement (Fig. 11.19), brakes T6 and T8 are applied and sun pinion *1* of the first row **a** and epicycle pinion 2 of the third planetary row **c** are restrained. In this gear, three planetary rows operate under load: **a**, **b**, and **c.** Row **d** is idle.

At the third gear engagement (Fig. 11.20) with brake T7 and joint M8, epicycle pinion 2 of the second planetary row **b** is restrained and row **c** is blocked with the joint M5. In this gear, only planetary row **b** operates under load; rows **a**, **b**, and **c** are idle.

At the fourth gear engagement (Fig. 11.21), the joint M4 and brake T8 are on, planetary row **b** is blocked, and epicycle pinion *2* of planetary row **c** is restrained. In this gear, only planetary row c operates; rows **a**, **b**, and **d** are idle.

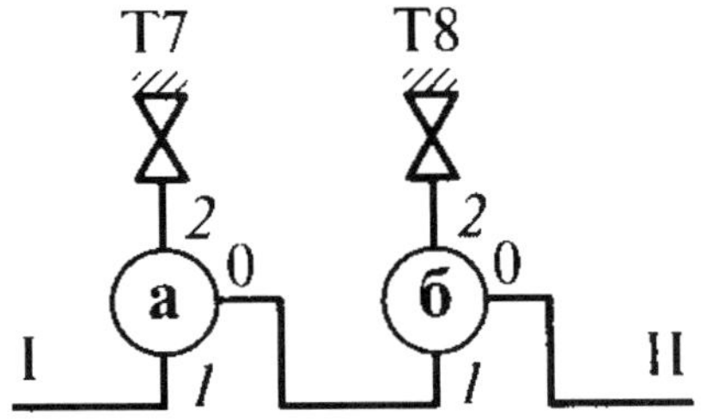

Fig. 11.18. Diagram of the power flow startup in a planetary gearbox at the first gear engagement

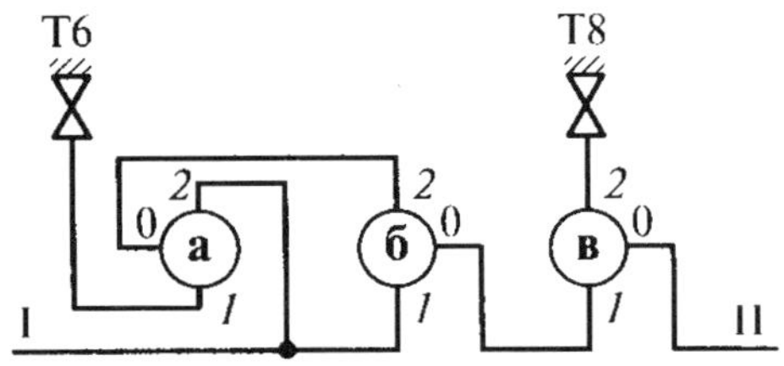

Fig. 11.19. Diagram of the power flow startup in a planetary gearbox at the second gear engagement

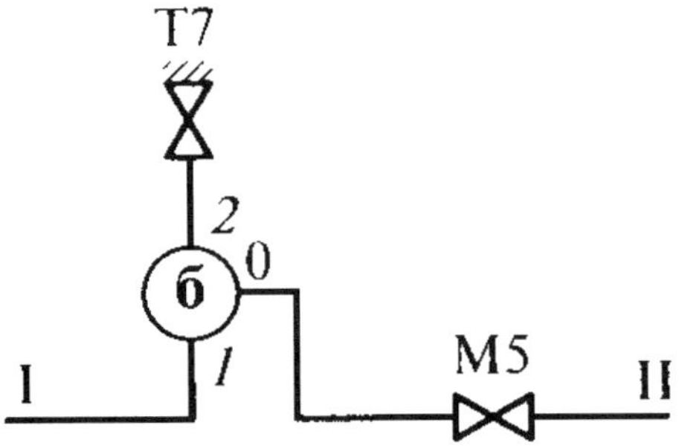

Fig. 11.20. Diagram of the power flow startup in a planetary gearbox at the third gear engagement

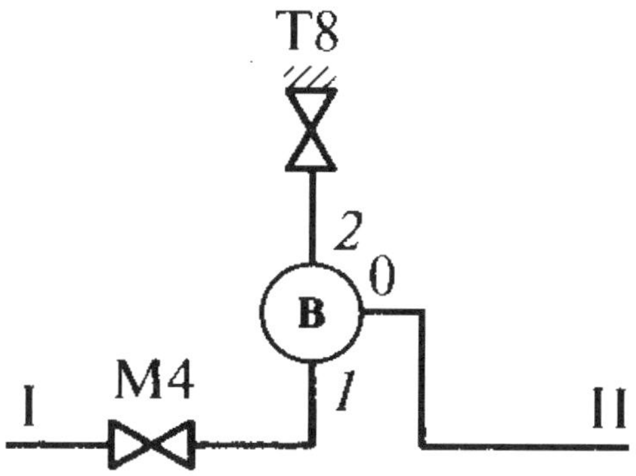

Fig. 11.21. Diagram of the power flow startup in a planetary gearbox at the fourth gear engagement

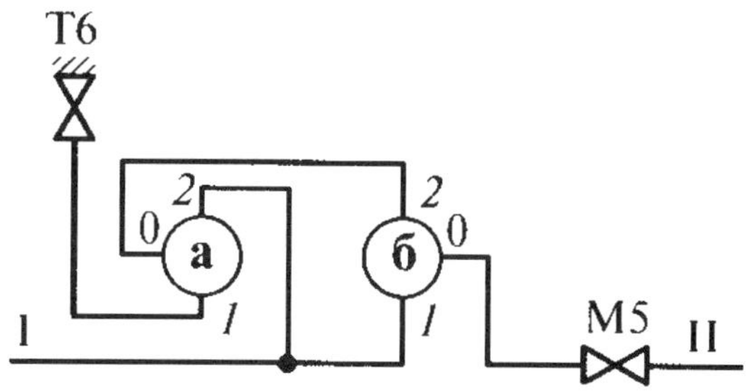

Fig. 11.22. Diagram of the power flow startup in a planetary gearbox at the fifth gear engagement

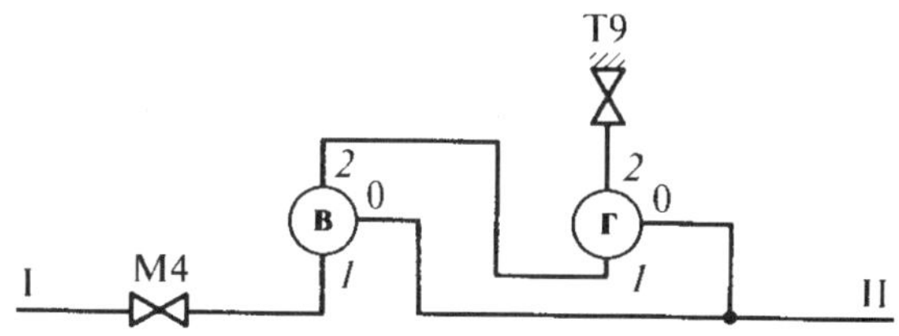

Fig. 11.23. Diagram of the power flow startup in a planetary gearbox at reverse gear engagement

In fifth gear (Fig. 11.22), the brake T6 and joint M5 are on, sun pinion *1* of planetary row **a** is restrained, and planetary row **b** is blocked. In this gear, planetary rows **a** and **b** operate; rows **c** and **d** are idle.

If joints M4 and M5 are powered on in the gearbox power flow diagram (Fig. 11.19), planetary rows **b** and **c** are blocked and the gearbox gives a direct gear.

In reverse gear (Fig. 11.23), joint M4 and brake T9 are on, planetary row **b** is blocked, and the epicycle pinion of planetary row **d** is restrained. Planetary rows **c** and **d** operate under load; rows **a** and **b** are idle.

Unfortunately, the transmission ratios produced in the gearbox in gears **I** and **II** are interconnected and one should not randomly select a ratio from within this range. That is why in the studies, the gearbox was selected and common ratios between gears were recommended. In particular, it was specified that for a range up to and equaling 15–16 (which meets the requirements for vehicles and tracked vehicles of all types and classes), the ratio between first and second gears should not exceed 2.5, i.e., it is quite acceptable for modern vehicles. In the process of a further kinematical range increase, the first transmission gives off to a second one. Therefore, the first transmission separates out into a retarding and the second will be first, etc.

However, this is not a drawback of the gearbox, but one of its advantages, as this significantly expands the usage areas of the gearbox.

Since connections between planetary rows are GB constituents and the set-out of the control elements were optimized during the studies, Yu.A. Stepanov [63] was able to not only make a compact and producible gearbox design, but to precondition it with a large (34.6) kinematical range alongside the transmission ratios family as well.

Hydromechanical automatic and semi-automatic gearboxes are designed based on the modular concept.

The examined transmission diagrams belong to the third type of basic complex network structures. Variants of this diagram can also be used as a module in principally new transmission diagrams.

It should be noted that planetary gearboxes are very often used as part of a torque converter (TC).

11.2.2 Hydraulic torque converter

The theory of calculating hydraulic torque converters (HTC) (transformers) is well developed; however, to receive a complete picture of the transmission development concept, the most interesting, in our opinion, common regularities of this theory and examples of its realization are cited in this work.

At HTC calculation, the consumption determination for each transmission ratio value $i = \omega_{tw}/\omega_{pu}$ has spread, which leads to minimal mistakes since the pressure balances experimental coefficients (losses for friction λ_{fric}, for impact φ, for flow deflection) and geometrical parameters of the working wheels [59, 60] are considered completely. A.N. Narbut's outlet characteristics optimization [60] is based on the suggestions of Yu.N. Laptev and A.S. Gordeev and have the following features:

1. The efficiency factor values at the predicted (non-impact) condition are defined from pressure balances when the transmission ratios' extreme HTC values disagree ($i^* \neq i\eta \neq iq$), i.e., disagreement of the non-impact inlet conditions in all working wheels (i^*), the maximum efficiency factor ($i\eta$), and the deflection extremum iq. That case is general and is comprised of the particular case of the transmission ratios extremums agreement; moreover, as the studies have shown, the last one is not acceptable for an automobile HTC with a centripetal turbine.

   The agreed notations match the following HTC parameters:

   - i is the HTC transmission ratio;
   - i^* is the HTC transmission ratio at disagreement of the non-impact inlet conditions in all working wheels;
   - $i\eta$ is the HTC transmission ratio at the maximum efficiency factor; and
   - iq is the HTC transmission ratio at maximum working fluid consumption.

2. Introducing the condition solution according to paragraph one ensures obtaining for the given i^*, consumption coefficient q^*, and pressure coefficient h_{pu^*}, the highest of all probable efficiency factors and blade angles matching these parameters, i.e. the optimal combination of all geometrical parameters (blade angles) at the same time. Thus, for an m-wheeled HTC, optimization is held at the same time at 2m geometrical parameters.

   The assignment of the values i^*, q^*, and h_{pu^*} clearly corresponds to the assignment of the outlet characteristics parameters and the energy-output ratio $\lambda_{pu^*} \cdot 10^4 = Aq_* h_{pu^*}$, the transparency $\Pi_* = \lambda_{pu.0}/\lambda_{pu^*}$, and the transformation coefficient Ki = K0 (0: condition $i = 0$, m: condition with $Ki = 1$).

   In the text of paragraph two, the following agreed notations are used:

   - q^* is the working fluid's consumption coefficient at i^* in the HTC;
   - h_{pu^*} is the working fluid's pressure coefficient in the HTC at i^*;
   - Π_* is the HTC's transparency at i^*;
   - Π is the general HTC transparency;
   - K is the HTC's transformation coefficient;
   - λ_{pu0} and λ_{pu1} are the HTC's transparency coefficients at K0 and $K = 1$ respectively; and
   - A is the constant coefficient.

3. Friction losses are defined according to A.N. Narbut's work [60]; moreover, for the common three-wheeled HTC, loss coefficients for friction are accepted as $\xi_{pu} = \xi_{tw} = \xi_{\text{reac}} = \xi$ (pu: pump, tw: turbine, reac: reactor) and $\xi = \lambda_{\text{fric}} L/8R_{\text{gave}} \approx 0.16$ (at friction coefficient $\lambda_{\text{fric}} = 0.08$). Validity of this consumption is confirmed by numerous experiments [60].

4. For determining the highest efficiency factor at predicted non-impact conditions, the equation of the pressure balances, written for

non-impact conditions, transforms into the following quadratic equation:

$$A_1\lambda_*^2 + B_1\lambda_* + C = 0,$$

where λ^* is the energy-output ratio; and $A1$, $B1$, and C are constants of the equation.

5. Transforming the pressure balance equation is done in the following way. For an m-wheeled HTC, each angle is figured out by one of the chosen $2m - (m + 2) = m - 2$ angles since the number of angles ($2m$) exceeds the number of available ($m + 2$) angle relation equations (m equations of the losses for impact and the equations of the pressure at the pump and turbine). When inserting the obtained formulas in the pressure balance equation at the non-impact condition ($m - 2$), the formula found is $ctg\beta_n = f(i_*, q_*, h_{pu*}, \lambda_*)$. By substituting the ctg, β_n are eliminated from the pressure balance equation and it is worked out relative to the efficiency factor (η^*).
6. Calculation is held in three steps: a) assuming the values i, i^* (all linear dimensions are known, i.e., relative radius of the inlet and outlet of the blades of the turbine and pump wheels and channel width coefficient for each variant) λ^* is defined; b) all characteristics of the HTC [60] are calculated in functional relation from I; c) blades angles are defined, moreover blades angles and characteristics may be calculated independently each from another.

According to A.N. Narbut's work [60] for a three-wheeled HTC, calculation of the HTC and blade angle characteristics is held in an electronic digital computer.

Examples of the graphical interpretation of Narbut's calculations [60] are represented in Figs. 11.24–11.27. The analysis of the represented data allows for making the following conclusions.

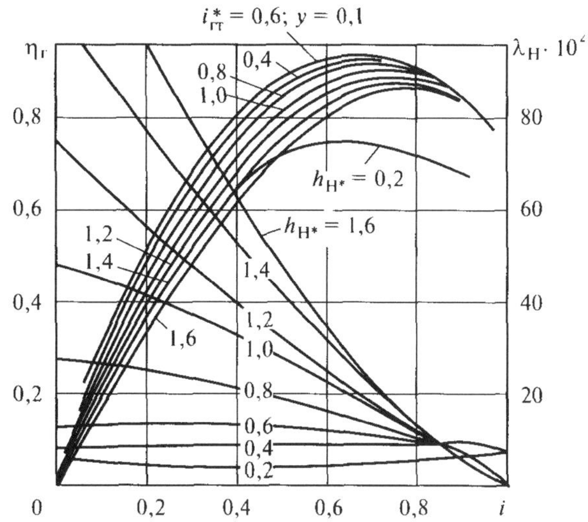

Fig. 11.24. Dimensionless characteristics of the HTC: $r2T$ = 0.56; i^* = 0.6; Y = 0.1

The consumption coefficient q* essentially influences the energy output ratio since λH^* is proportional to q^*. At that value of q^*, it doesn't influence the transparency, if the condition dislodgement iN when $K = 1$ is not taken into consideration because of the efficiency factor reduction at increasing q^*. The higher q^*, the lower the efficiency factor and K0 (Figs. 11.26 and 11.27), but i and h_{pu}^* also influence the maximum efficiency factor values (Fig. 11.26). Thus, the maximum allowable q^* value is defined by the value of the allowable decrease in the maximum efficiency factor and may be accepted as equal to approximately 0.4 at $0.4 \le i^* \le 0.8$ and $\xi = 0.16$.

The pressure coefficient h_{pu}^* is distinct from the consumption coefficient's influences on not only the energy-output ratio but also on the transparency (Fig. 11.27): with h_{pu}^* increasing, Π and the high efficiency factor zone (Fig. 11.27) are also increasing; with h_{pu}^* increasing, the active diameter values D0.75 are increasing, which hardly influences the maximum efficiency factor value. However, at low h_{pu}^* values (usually at $h_{pu}^* = 0.4$–0.6 if the turbine is centripetal and $h_{pu}^* = 0.6$–0.8 if the turbine is axial) an intense decrease in the efficiency factor is observed. Taking into consideration that fact and also the

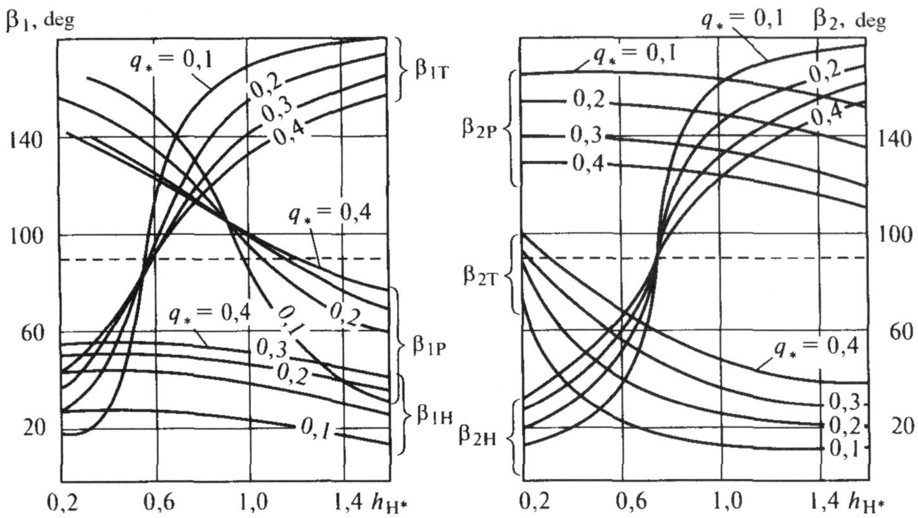

Fig. 11.25. Value relation of the blade angles β1 and β2 from *h**, *r2T*, and *i**

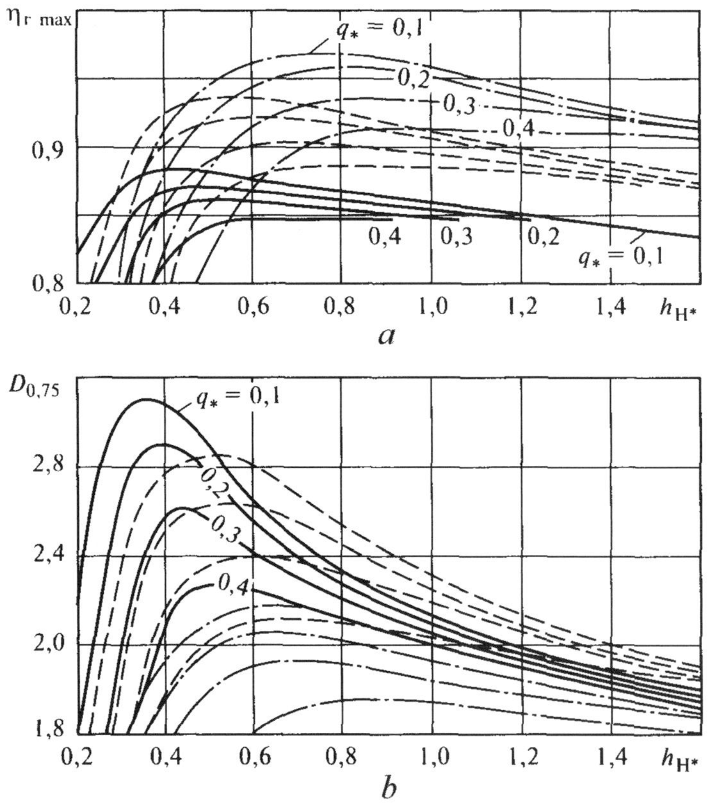

Fig. 11.26. Relation of the efficiency factor *h*г max (a) and D0.75 (б) from the pressure coefficient h_{pu}*: ----r2T = 0.56; i* = 0.4; – – – ––r2T = 0.6; *i** = 0.8; -·-·-—*r2T* = 0.8; *i** = 0.8

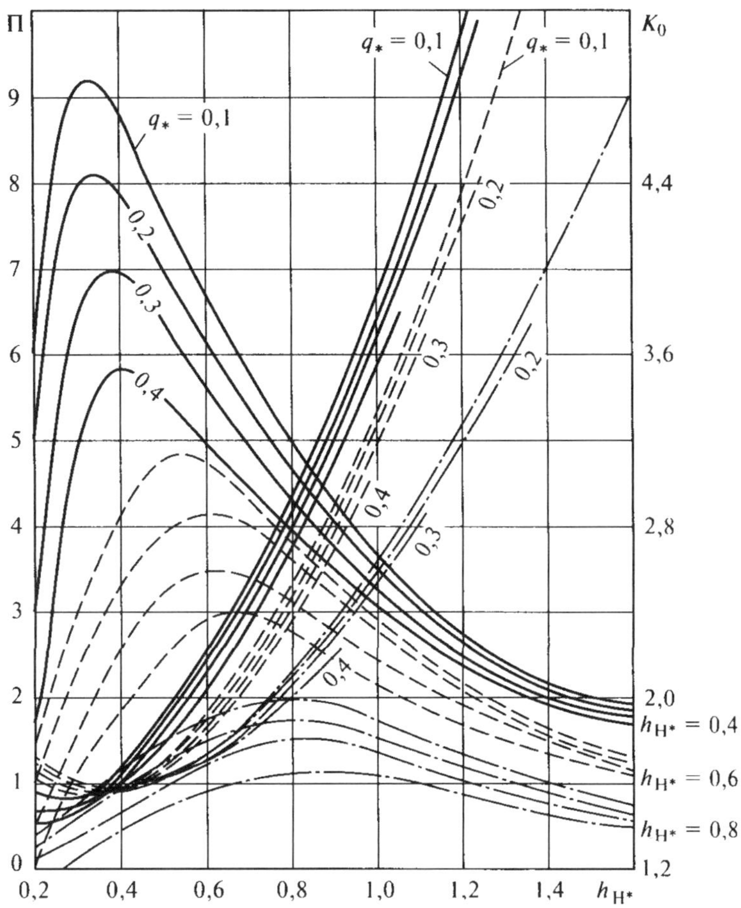

Fig. 11.27. Relation of the transparency Π and transformation coefficient K0 from the pressure coefficient h_{pu}*: ---—*r2T* = 0.56; *i** = 0.4; – – – – —*r2T* = 0.6; *i** = 0.8; -.-.-—*r2T* = 0.8; *i** = 0.8

necessity to limit the transparency with assumed values, the allowable h_{pu}* limits may be set. For example, for the HTC of road-building vehicles Π < 1,6 and for an automobile HTC, high Π values are allowable and sometimes even necessary.

The calculating condition *i** at some decrease in the maximum efficiency factor ensures a widening of the high efficiency factor zone and K0 if the turbine is centripetal and the dislodgement of this zone at K0 in the direction of the lower transmission ratios if the turbine is axial. Other characteristic parameters depend on *i** in essentially lower degrees.

Blade angles (Fig. 11.25) are approximate to 90° when *q** increases;, moreover, at all h_{pu}* angles (β1H and β2H < 90°and β2P > 90°) and graphs of the blade angles (β1P, β1T, and β2H) are disposed on both sides from 90° and have a maximum density value of nearly 90°. Outlet angles and the β1H angle change slightly when the calculating condition *i** changes.

Studies have shown that present-day HTCs have characteristic parameters close to extreme and at the achieved resistance coefficient ξ, obvious increases in the maximum efficiency factor and diapason of the high efficiency factor are not expected. However, there are some unimproved opportunities, for example, simplifying the rotary-vain systems and on the energy-output ratio's changing diapason for newly developed parametric rows.

An example that confirms that conclusion may be represented by studies held by NAMI and BelAZ on choosing the parameters of the normalized HTC for the BelAZ series of vehicles [59].

For BelAZ vehicles with a load capacity of 27–65 t and 265–382 kW (360–520 hp) YaMZ engines, the complex, single-staged, four-wheeled HTC BelAZ-540 with an active diameter of 466 mm (Da = 466 mm) (Table 11.4 and Fig. 11.28) was chosen. To use that HTC with the YaMZ engines, a conforming mechanical transmission that increases the hydraulic transmission's weight, complicates its construction, and decreases the efficiency factor by 3–4%, was installed in front of the engine.

In this context, for future hydromechanical transmissions of the BelAZ series, it was suggested to develop a range of normalized HTCs, which have the same circulation circle and basic parts (nave, free-wheel clutch, etc.) but different blade inlet angles of the pump wheel and second reactor wheel in order to achieve the appropriate load properties [59]. In this case, conforming transmission may be excluded. It is also reasonable to use HTCs with a wider higher efficiency factor zone and those with a higher maximum efficiency factor value.

From the HTC series developed by NAMI, four-wheeled, single-staged, LG-340-35 HTCs mostly satisfied the requirements and working conditions in BelAZ vehicles (Table 11.4 and Fig. 11.29).

Foreign HTCs used in heavy vehicles commonly have an efficiency factor of 0.87–0.90 and a transformation coefficient K = 1.8–2.0.

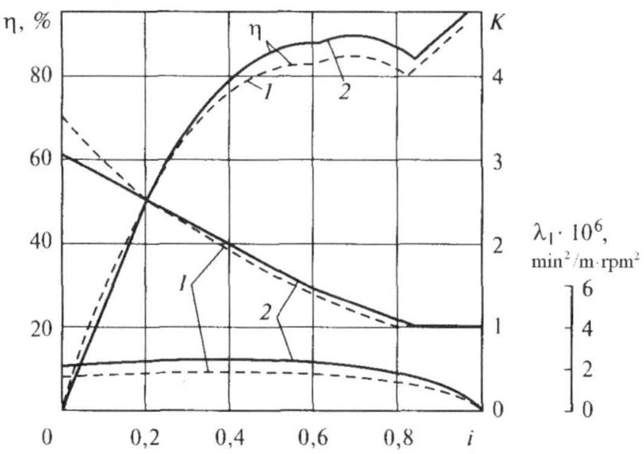

Fig. 11.28. External characteristic of the HTC: 1—BelAZ-540; 2—LG-470-23

Table 11.4

Hydraulic torque converter	Transformation coefficient at i = 0, KO	Transformation coefficient at efficiency factor = 80%, KO	Maximum efficiency factor	Transmission ratio at maximum efficiency factor, i	Transmission ratio hydraulic clutch, i_{hm}	Maximum torque coefficient	Transparency coefficient	Reverse transparency coefficient
BelAZ-540	3.5	1.8	0.85	0.7	0.83	1.9	1.45	1.12
LG-340-35	3.1	2.0	0.90	0.7	0.85	2.2	1.37	1.11
LG -340-36	2.9	1.9	0.89	0.7	0.85	3.2	1.63	1.03
LG -349-36B	2.7	1.9	0.90	0.7	0.86	3.2	1.63	1.00
LG -340-38	2.9	1.9	0.89	0.7	0.85	2.0	1.40	1.05

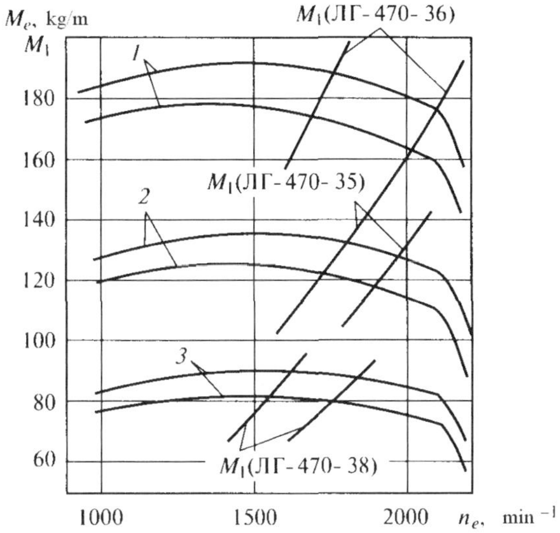

Fig. 11.29. Combined work characteristic of the LG-470 HTC with different engines: 1—YaMZ-240N; 2—YaMZ-240; 3—YaMZ-238

For cargo transport (CT) developed by NAMI, the notation system is as follows: after the letters LG (cast CT), the CT's active diameter appears and then the model's serial number.

The comparison of the CT characteristics given in Fig. 11.29 allows for concluding that the efficiency factor of the LG-340-35 hydraulic torque converter is 4–6% higher in the main working area With an increase in the given CT's dimensions from D_a = 340 mm to D_a = 440–470 mm, there can be an expected increase in the efficiency factor only by 1–15% and K_0 by 5–8%.

Studies have shown that for the YaMZ-240 engine, it is better to use the CT LG-470-35 (Fig. 11.29).

Engine characteristics in Fig. 10.1 are shown by two curves. The top curve of the torque corresponds to development testing and the bottom curve to the real operating conditions, including the power takeoff for the drive of the engine boosters and hydraulic mechanical gearbox. According to BelAZ, this loss of power is 10–12%.

The dependency $M_1 = f(\Pi_e)$ characterizing the joint operation of the engine with the CT is presented in the form of a band limited by two curves, the left of which corresponds to $i = 0$ and the right to $i = i_{hm}$, i.e., the transfer relationship under which transition to the hydraulic clutch mode takes place.

When selecting the operating modes of the engine and CT, the following circumstances were taken into consideration.

As is known, transforming characteristics of the low-transparent, complex, hydraulic torque converters Π_1 = 1.3–1.5 possess a higher completeness and have a more expanded high efficiency factor zone as compared with transparent CT Π_1 = 1.7–2.0. In this context, applying a CT with a low-transparent characteristic contributes to improving the vehicle's traction-dynamic qualities and also allows for reducing the range of the mechanical reduction gearbox installed behind the hydraulic torque converter. The latter is very important for heavy vehicles since in this case it makes the operating conditions of the frictions perceptibly easier when shifting the gear and also reduces the relative speeds in the rotating friction elements.

The joint operation mode of the engine and CT is selected on the basis of compromise. This is connected with the fact that the conditions appropriate for the maximum use of the engine power are not appropriate for the conditions under which good fuel efficiency is gained. Thus, for example, for a low-transparent CT with the aim of maximizing the engine power use at the CT mode, its active diameter should be selected in such a way that $n_{10} \approx n_{pu}$ (n_{10}: engine rotations under full fuel feed at the mode $i = 0$, n_{pu}: rotations corresponding to the maximum engine power).

The analysis of the developed designs allows for making the conclusion that usually for heavy-duty vehicles with diesel engines, n_{10} = (0.75–0.85) n_{pu}.

For the YaMZ-240 engine it was assumed reasonable to accept $n_{10} = 0.8\ n_{pu}$ and in accordance with this condition, CT LG-470-35 (see Fig. 11.29) was designed; for the YaMZ-240N engine, CT LG-470-36 differs from LG-470-35 only by the blade extension angles of the pump wheels. The experimental characteristic of the prototype (LG-340-38) is presented in Fig. 11.30 and typical parameters in Table 11.4.

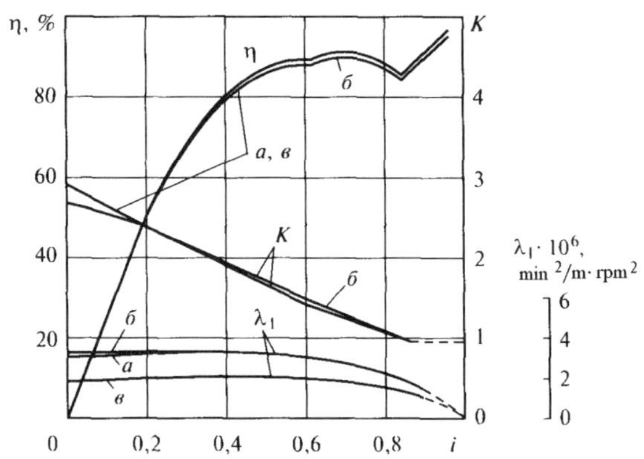

Fig. 11.30. Comparison of the CT's external characteristics: a—LG-470; b—LG-470-36B; in—LG-340-38; i—CT reduction ratio

The above materials show that due to the relevant selection of the geometrical parameters of the operating wheels it is possible to change the CT's load properties.

Fig. 11.31 presents the design characteristic of the BelAZ-540 vehicle with a three-speed gearbox. This figure shows the change in the CT's efficiency factor in transmissions depending on speed.

As is clear from Fig. 11.31, the LG-470-35 has a 4–6% higher efficiency factor in the main operating range at travel speeds from 4 to 50 km/h of the hydraulic torque converter. Correspondingly, while applying the given CT, the fuel consumption reduces, which is illustrated by the curves characterizing the given fuel consumption that is defined according to the formula:

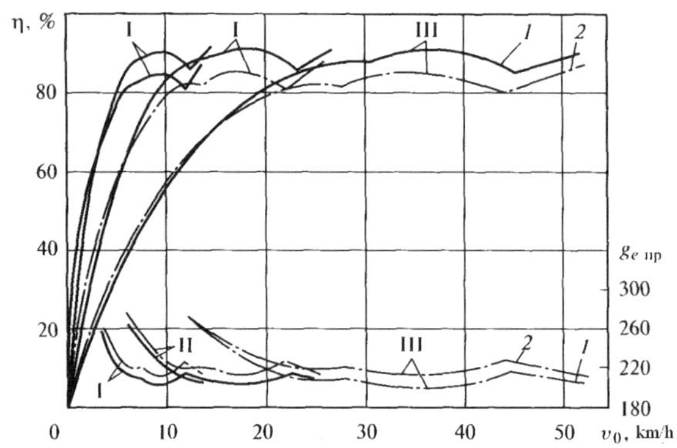

Fig. 11.31. Change in the CT's efficiency factor and specific fuel consumption q_{enp}: 1— G-430-35; 2— BelAZ-540 (transmissions I–III)

$$g_{en} = \frac{g_e}{\mathrm{COP}}.$$

Under a reduced fuel consumption rate it is meant fuel consumption of the system "engine – CT" reduced to output shaft of the CT.

If you consider that the transmission efficiency factor will additionally increase by 3–4% due to eliminating the mechanical matching of the transmission, it is possible to expect that, as a whole in the main operating range, the efficiency factor of the hydromechanical gearbox with a CT LG-470-35 will be 6–10% higher.

Fig. 3.2 shows the design of a unitized CT developed jointly by NAMI and BelAZ. The presented CT allows for implementing the recommended characteristics.

Thus, due to the appropriate selection of the pump wheel blade shapes, the reactor with the same circulating circle, and preservation of the main parts, the unitized CT LG-470 can be used for operation with the 265–382 kW (from 360–520 hp) YaMZ engines. In this the module design principle can be seen but at the same time, due to the increase in the CT's efficiency factor and elimination of the mechanical matching transmission, the HMT efficiency factor in the main operating zone can increase by more than 6–10%. Meanwhile, a 3–4% decrease in mechanical reduction gearbox can be seen.

11.3 Brakes-retarders of transmission type

Specific operating conditions of HDWVs working in hard road conditions, in mountain country, at construction sites, or in the metal mining industry cause a considerable load on their brake system. Movement of the vehicles with cargo on long and steep descents leads to overheating and wear of the frictional brake mechanisms. That is why in accordance with Regulations No. 13 of the UNECE, secondary brake systems or retarders are mounted on these vehicles. Their main function is to maintain a safe travel speed while descending and decelerating before a stop. Secondary brake systems should provide descending HDWVs a speed of 30 ± 2 km/h along a 7%-slope with a length of 6 km. Deceleration j_M provided by this system should be more than 0.6–2.0 m/s^2.

Retarders have the ability to relieve service brakes, increase their durability, and raise vehicle productivity by increasing the average travel speed. According to the American firm "Case," the annual savings from using power-intensive retarders on log trucks and vehicles operating in the metal mining industry can be 540 dollars per one vehicle due only to reducing the wear of the service brakes.

As a secondary brake system, an engine and electrodynamic and hydrodynamic retarders are used. In the retarder mode, special cooled friction multiple-disk brakes running on oil are also used. Such devices are mounted, for example, on the 769 and 773 "Caterpillar" vehicle-dumper models with a load capacity of 35 and 50 t, respectively.

An engine retarder ensures an artificial increasing of the engine shaft's rotation resistance moment in the forced idling mode, as a rule, by overlapping the engine's exhaust pipe and switching off the fuel supply. Such retarders are mainly used in vehicle-dumpers made in Europe, for example, in the "Aveling Bartford" (England) with load capacities from 10 to 35 t.

However, the efficiency of such retarders is rather low since it is limited by the pressure value in the pipe, which is defined by the spring rate of the exhaust valves. In addition, some portion of the compressed air energy in the engine cylinders is lost in the expansion stroke when the intake and exhaust valves are closed.

The widespread use of these retarders is prevented by the necessity to create a special design for each engine model and in some cases to make design changes to the engine. Also, high noisiness is noted in engines with a retarder.

Hydrodynamic and electrodynamic retarders, however, have become widespread.

Hydrodynamic retarders are installed, as a rule, in the transmission. The American company Allison specializes in manufacturing hydraulic transmissions for vehicles with 200 to 1000 hp engines for different applications and produces blade hydraulic retarders (Fig. 11.32) that can be installed between the hydraulic torque converter

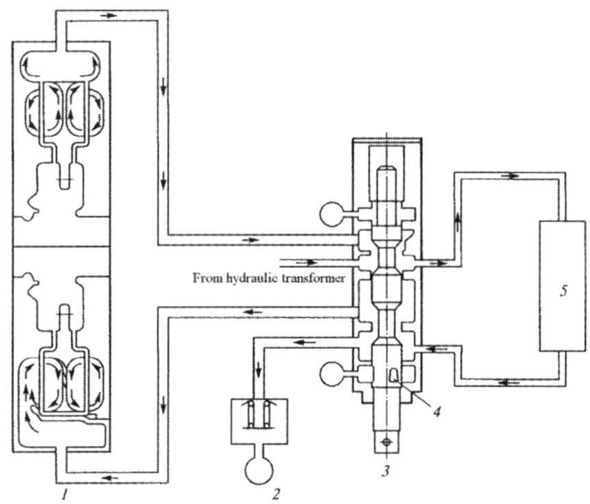

Fig. 11.32. Control circuit of the Allison hydrodynamic retarder: 1—Hydraulic retarder; 2— Discharge valve from the hydraulic torque converter; 3—Retarder control valve; 4—Slots from the three sides; 5—Heat exchanger

and planetary gearbox. The retarder rotor is attached on the hydraulic torque converter's turbine shaft. The working cavity is formed by the duplex hydraulic clutch and, when the retarder is switched on, it is filled with oil going through the control valves from the hydraulic mainline of the hydraulic transmission. The retarder's switching time is 2.4 s and has a working cavity volume of 3.8 l.

The pressure created in the peripheral part of the working cavity due to the action of the centrifugal force when the rotor is revolving ensures the fluid circulates through the heat exchanger attached to the engine cooling system and through the retarder along the closed loop. The maximum pressure in the retarder's working cavity of 3.8 kgf/cm^2 when the control valve is completely switched on is set with the help of the discharge valve from the hydraulic torque converter. The braking torque is adjusted by changing the amount of liquid filling the working cavity by moving the spool valve, through the slots of which part of the fluid is dumped from the retarder's closed loop into the pan.

The maximum size of the hydraulic retarder (by diameter) when installing it in the CLBT-750 is 431 mm and its weight is 90 kg, which is 20% of the hydraulic transmission's dry weight. The hydraulic retarder ensures maintaining a speed of 25 km/h when an equipped vehicle with a gross weight of 59 t moves on a 5% slope with the switched direct gear in the transmission.

At present, Allison is installing hydraulic retarders in many hydraulic transmission models used mainly for dumpers, building, and road machines.

A characteristic of Caterpillar's (USA) "Brake Saver" hydraulic retarder (Fig. 11.33) is that it is placed in the engine's flywheel housing (4) and its blade rotor is attached directly on the engine crankshaft between the back base bearing and the flywheel. Oil from the engine's lubricating system is used as the power fluid in the retarder. The oil from the pump (3) goes to the restrictor

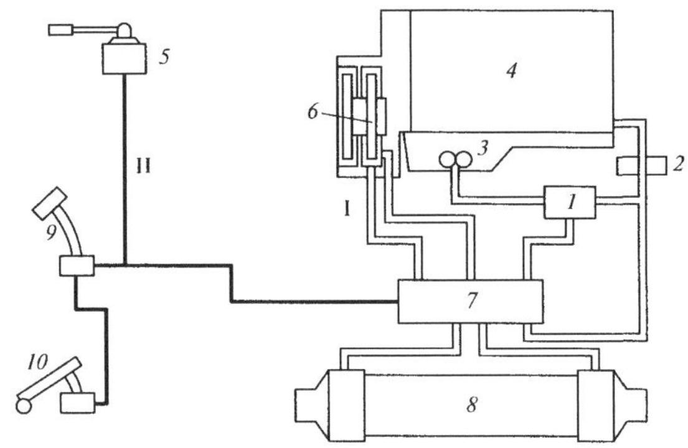

Fig. 11.33. Diagram of the "Brake Saver" hydraulic retarder

valve (*1*), which, in case of a pressure drop in the engine's lubricating system, overlaps the oil's access to the valve (*7*) of the hydraulic retarder control (*6*) and guides it through the filter (2) into the engine. When the retarder is switched on, the pumping action of its blade rotor circulates the oil along the closed loop between the retarder's working cavity and the heat exchanger (*8*).

Pneumatic retarder control is carried out with the help of the manual control crane (*5*) and the electropneumatic valve switching off the retarder automatically by pressing the clutch pedal (*9*) or the acceleration pedal (*10*).

Under the pressure from the air supplied through hole *1* (Fig. 11.34), the spool control valve is shifted to the right and the oil from the pump goes through holes *2* and *3* to the retarder's working cavity. From the incoming side of the retarder, oil is supplied to the heat exchanger through channels *4* and *5*, from which a part of it again returns to the retarder's suction cavity through channels *6* and *3*, while the other part is supplied to the oil filter and then into the engine's lubricating system through channel *7*.

The retarder's braking efficiency is controlled automatically by changing the amount of oil filling the working cavity under the changing air pressure supplied to the left butt end of the spool. For this, the incoming side of the retarder communicates with the internal cavity of the

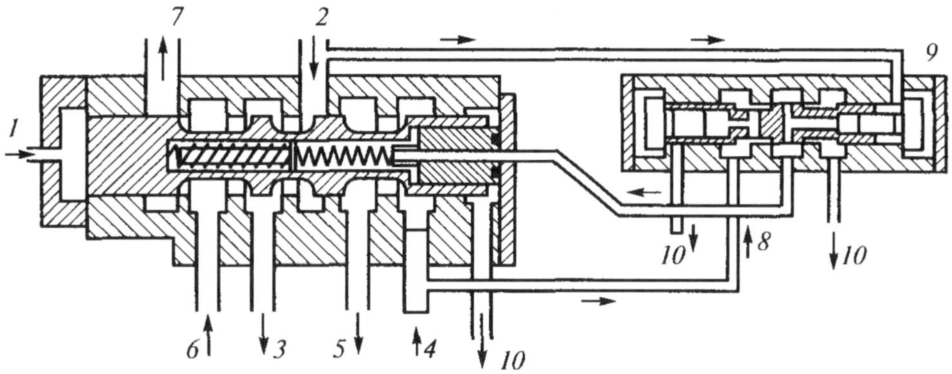

Fig. 11.34. Section drawing of the "Brake Saver" hydraulic retarder

spool through channels *4* and *8*, differential valve *9*, and channel *10* and the piston butt end of valve *9* is under pressure from the oil supplied to the retarder. When the number of rotor revolutions increases and the oil pressure in channel *4* is raised, the piston of valve *9* is shifted to the right and the oil pressure forces the spool control valve to the left blocking hole *2*. Thus, the degree of the retarder's filling goes down and the braking torque is restricted. The oil is discharged into the engine pan through channel *10*.

Combining the retarder with the engine can be used in vehicles equipped with transmissions of any type. With this version it is also possible to achieve the minimum lengths of the hydraulic mainlines in the control and cooling systems. By customer request, the "Brake Saver" retarder can be equipped on 1693T and 1693TA Caterpillar engines with a power of 325 and 375 hp, respectively, which can be mounted on cargo vehicles, log trucks, and vehicle-dumpers.

The high energy content of modern hydraulic retarders allows them to be controllable. Companies producing hydraulic retarders fix, as a rule, their maximum power so it does not exceed the maximum power of the engine to avoid overheating the service fluid and water in the engine.

Analysis of patent materials from Caterpillar, for example, over the past 10 years, shows that the company develops hydraulic retarders even for cargo vehicles with a diesel-electric power unit, where rheostat braking of the electric motors is possible. If in such transmissions there is an effective hydraulic retarder linked with the engine shaft, it is possible to use the effect of the braking by the engine to switch off the fuel supply. In this case, there is no need for the engine to drive the boosters when braking and fuel consumption goes down.

Electrodynamic retarders are included in the regeneration system of electric transmission energy and discussed in detail in special literature, for example, on electric railway locomotives.

11.4 Main principles of power distribution in transmissions of modern and future HDWVs

11.4.1 General provisions

Until recently, designers of drives for vehicle movers have widely used traditional methods of designing, testing, and studying mechanical and hydromechanical transmissions, which are rather simple and efficient for vehicles with a small number of driving axles.

In the simplest case of distributing power among two wheeled movers, four groups of design and construction solutions are applied. The drive of two-axle movers can be independent, blocked, differential, and with the use of different types of clutches.

With the increase in the number of driving wheels, the number of transmission layout solutions goes up in the geometric progression.

That is why scientific and technical theoretical justification of the general principles used for distributing power among the driving wheels while applying the most advanced systems for driving the mover is so important and urgent.

The problem of advancing the power transmission theory to the wheeled mover, as a constituent part of the vehicle theory, has always been in the spotlight of experts. In the process of improving vehicle designs, it has become a consistent necessity for a deep scientific consideration of theoretical aspects of building and using the mechanical, hydro- and electromechanical, hydrostatic, electric, and combined transmissions. Here, one of the most important issues has always been the efficient distribution of the power generated by the PU among the individual wheels.

The numerous studies performed in this area have established, in particular, that in different motion conditions and modes (solid or deformable support surface; acceleration, steady motion, or braking; travel along a trajectory with a small or large curvature; the presence or absence of a trailing load; etc.), the ratios of the torques delivered to the wheels of different boards (sides) and axles should be different. It follows from here that the best performance parameters of the vehicle—the traction-dynamic, fuel-economic, passing ability, stability, etc.—in constantly varying travel conditions can be achieved only under the appropriate continuous regulation of the power delivered to the wheeled mover depending on its rolling conditions.

Possibilities to improve the main performance indicators of SWCs through improving the regulation laws of their transmissions have been mentioned by authors of many theoretical works and confirmed in a number of experimental studies and tests of chassis prototypes with electric transmissions (ETs).

Along with that, analysis of domestic and foreign vehicles with traction electric drives show that to date there has been no uniform approach to selecting criteria of assessing the transmission diagrams and laws of controlling the power transfer to the driving wheels. Work on ETs and HSTs, as noted above, is carried out mainly in the direction of adapting the existing structures of electric vehicles and their control systems, which are designed without considering the nature and operating modes of SWCs. That is why the major advantage of these transmissions—the ability to individually distribute the power to the wheeled movers depending on road conditions and motion modes—is almost never used.

Thus, the problem of creating and using advanced transmissions in SWCs assumes in its first approximation the answers to the following questions:

- What are the general principles of the power distribution for wheeled movers to ensure the specified performance and consumer properties of the chassis?
- What kind of interpolation formula reflects the law of efficient (optimal) distribution of the power through wheeled movers of SWCs?
- What criterion should be used when assessing transmission diagrams?
- What scientific and methodological basis of the already known ones should be selected when developing the theoretical foundations of the quantitative and qualitative assessment methods of the transmission diagrams?

For specifying the main problems considered in this chapter, a number of common assumptions and restrictions are taken.

First, it is assumed that basic design and constructional solutions, which affect selecting the number of axles, placement of the axles on the basis and selection of the StS, and significantly influencing the transmission's operation, are considered given and are not variable factors.

Thus, for four- and eight-axle SWCs the following parameters of the basic design and constructional solution characteristics are assumed:

Wheel formula	8×8	16×16
Axle formula	2-2	2-2-2-2
Formula of the handling quality	12-00	123-00-678
Formula of the drive of the axles	1234	12345678

Second, studies are carried out mainly in relation to non-transparent (ET, HST) transmissions. For transparent transmissions, theoretical foundations of the assessment methods are discussed in general terms. The operation mode and transmission characteristic parameters are considered given and are appropriate for advanced SWCs.

Third, the processes occurring in the transmission at the first stage of the studies are not considered (ET and HST—a "black box" with input and output). Parameters of the tire characteristics are constant and fit advanced SWCs tires.

Fourth, patterns of the quantitative and qualitative change in the performance properties in these studies are established only in relation to the chassis' traction-dynamic properties. At the same time, the following principles of the power distribution to the wheels are considered:

- Equality of torques in drives of all wheels—fully differential transmission;
- Equality in rotation frequencies of all driving wheels—fully blocked transmission;
- Equality of power—products of the power (points) and speed (rotational rate) factors; and
- Equality of absolute slides in the contact spot of the wheeled mover with the support surface.

With account of the made assumptions and restrictions, the particular tasks of this chapter are as follows:

- To formulate and justify the basic power distribution principles among the driving wheels of a SWC;
- To derive the complex assessment criterion of the transmission diagrams; and
- To justify and derive the theoretical foundations of the basic methods of analyzing the power distribution in the transmission on the basis of the power flow theory.

11.4.2 Initial provisions for formulating basic principles of power distribution among wheeled movers

The vehicle motion theory and practice showed that with sufficient engine power high performance properties are ensured provided there is an adequate reserve of adhesion in the wheels with the support surface on which the chassis moves. Mathematically, this can be written down with the known dependency of the following kind:

$$G_g\varphi > G_{car}(f \pm i) \pm m_{car} j \frac{\delta - 1}{1-\sigma} + P_w + P_{tr} = \sum_{i=1}^{2n} P_{wi},$$

where G_g and G_{car} are the adhesion and gross weight of the vehicle respectively; $f \pm i = \psi$ is the total coefficient of the vehicle's motion resistance including the longitudinal slope of the road; m_{car} is the weight of the vehicle; j is the faster acceleration, deceleration; P_w, P_{tr} and $\sum_{i=1}^{2n} P_{wi}$ are, respectively, the air resistance forces, hook resistance, and full traction force value on the wheels for the engine; n is the number of the axles; δ is the rotational inertia coefficient; and σ is the sliding coefficient. Mathematically, in basic terms, the rational law of the power distribution for wheeled movers can be presented in the following expression:

$$N_{wi} = N_{fwi} + k\frac{N_{srm}}{2n_k} \leq N_{\text{slip max}} \qquad (11.1)$$

where N_{wi} is the instantaneous power supplied to the ith wheel; N_{fwi} is the instantaneous power of the rolling resistance of the ith wheel; N_{srm} is the instantaneous power of the external resistances to vehicle motion; n_k is the number of driving axles (for SWCs $n_k = n$); and k is the correction factor taking into account the interaction conditions of the wheeled mover with the support surface.

The expression (11.1) is taken as a first prime proposition when formulating basic principles of the power distribution for wheels and the law of the rational power distribution.

As a second prime proposition, the known pattern in the power change of the rolling resistance of the wheel depending on its rolling mode is taken.

It is theoretically and experimentally proven that there is the least rolling resistance in a wheel that is not loaded with longitudinal and lateral forces and moves only under the influence of the torque equal to the rolling resistance moment. This mode used to be called the free rolling mode [52].

The third prime proposition is the known general pattern of the change in the traction force at the wheel depending on sliding. For many types of ground and support surfaces, the maximum traction force corresponds approximately to 30–40% of the wheel slide coefficient. At higher values of the sliding coefficient, traction forces begin to fall especially sharply—for example, on ice and wet roads.

The fourth prime proposition is an assertion that, when moving, each wheel of a multi-axle SWC is in complex interaction with the support surface. Moreover, all wheels, as a rule, have different conditions and outcomes of this interaction, especially when there is a sharp change in the conditions when the wheeled mover is on deformable and sealing grounds along one track. When wheels are passing on such grounds, their mechanical properties change and that is why the parameters of the following wheels' interaction with the ground are different. Relying on all four prime propositions, it is possible to formulate in general terms the basic principles and the law of rational power distribution for wheeled movers, which the selected transmission diagrams should be appropriate for.

The transmission layout and its functioning can be considered rational under the following conditions:

- First, at any time, under any conditions and motion modes, the power supply to each wheeled mover must be equal to the power of the rolling resistance, i.e., transmission diagrams should provide rolling of the wheels in the free mode;
- Second, the power required to overcome external motion resistances (acceleration, hook traction, grade climbing, air resistance, resistance at curvilinear motion, etc.), should be evenly distributed by transmission diagrams for all wheeled movers with account of the correction factor for the interaction conditions of the wheels with the support surface; and
- Third, the total power supplied to the wheeled mover should not exceed the maximum power corresponding to a specific value of the wheel sliding both in the traction and braking motion modes.

As a result of the studies, it is suggested to consider that $k \cong 1$ at rectilinear motion of the chassis on solid roads and on incompressible grounds when the interaction conditions of the wheels are approximately equal.

The main theme of further discussions in this chapter is to disclose constituents of the expression (11.1) applied to the transmission diagrams under study and developing the general method of its analysis.

11.5 Optimization of the power transfer to the wheeled mover

11.5.1 Main method of analysis

Among actually made designs of modern SWC transmissions, there has not been a single transmission created that would provide the rational law of power distribution to wheeled movers according to the expression (11.1).

In this context, there is a need to define the degree of appropriateness of one or another transmission diagram to the rational law. To solve this problem, there is a necessity for a basic methodology for analyzing the transmission's operation in the current multi-axle chassis.

In n-axle chassis there can be $2n$ points, which are supplied with power to drive the wheeled movers. Consequently, a transmission of a multi-axle chassis is a complex system transmitting and converting different kinds of energy. At the same time, SWC transmissions can be viewed as systems of the power servo drives with multi-channel actuating elements.

Using the known vehicle theory research techniques in this case is rather difficult. New techniques and a new theoretical basis are urgent. In this work, the power flow theory developed for studies and calculations of track-type power trains and wheeled vehicles by Professor A.S. Antonov [2] are taken as a theoretical basis. The propositions of A.S. Antonov's power flow theory were developed most fully in the works of Yu.A. Stepanov and P.V. Aksyonov and are used by the authors.

The methodology of the power flow theory is based on the transition from real complex physical systems to a design scheme, which makes it possible to reveal internal links and patterns that govern the functioning of the single elements in the system. The advantage of the power flow theory is that its methods and principles allow us to consider the operation of any transmission and its individual elements irrespective of the type of mechanisms, their design, and forms of the converted and transmitted energy.

As a consequence of using the power flow theory, it is possible to assume that the patterns obtained in this chapter can be applied to all types of multi-axle chassis transmissions.

According to the power flow theory, all varieties of the converting devices involved in the transmission fall into three types of nodal points (NP): branching, kinetic, and integrated. In a branching NP, transformation of only the power factors in the power flow takes place. In kinetic NP, the speed factors are transformed. An integrated NP is comprised of kinetic and branching NP, which is why simultaneous transformation of the power and speed factors occurs in it. These types of NP are interlocked with one another by directed energy fluxes (linkage fluxes). In addition to linkage fluxes, the transmitting system also has accumulating and scattered fluxes.

Any transmission is presented by a set of NP, linkage (directed), accumulating, and scattered fluxes. Geometrical representation of such a set gives a power flow (PF) circuit, which is the starting point and basis of the power transmission studies.

In the power flow diagram of a multi-axle chassis transmission, the main component is the wheel represented in the form of the integrated NP. The diagram of the forces affecting the wheel during its rolling and its integrated NP are presented in Fig. 11.35.

The nodal point of the wheel has the following flows:

- The rotary flow at the speed of the ω_w and power M_{et} ($M_{E.mw}$) factors;
- The rotary reactive flow with power factor $M_R = M_f + M_t$;

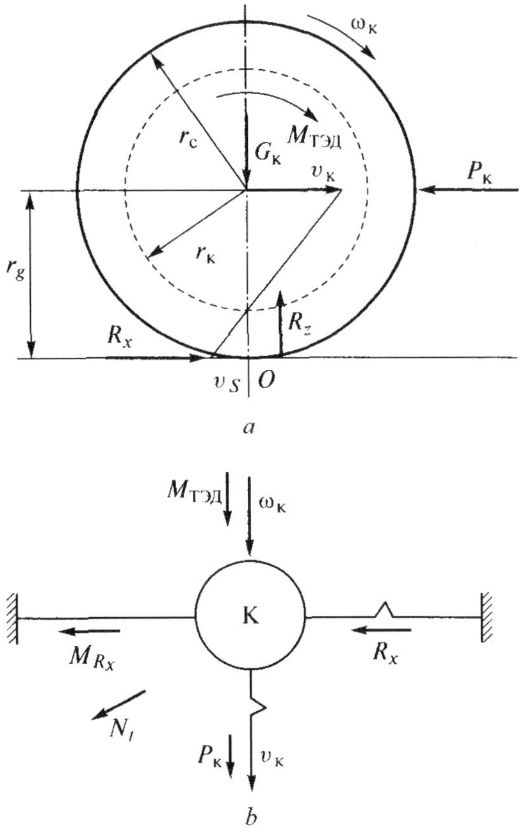

Fig. 11.35. Calculation diagram of the wheel motion in plane of its rotation: a—Diagram of the forces applied to the elastic wheel; b—Integrated common point of the elastic wheel

- The progressive active flow of the wheel axle at the speed of the v_w and power P_w factors;
- The progressive reactive flow of the wheel axle at the speed of the factor $R_x = P_t$; and
- The dissipative (scattered) power flow N_t.

By using the flow direction in the NP, it is easy to determine the wheel's operating mode. Thus, the braking wheel has a flow direction opposite to the ones shown in Fig. 11.35 (except for N_t). The driven wheel has one active progressive flow from the frame to the wheel and also reactive and scattered fluxes. A similar approach is used for representing the wheel in free and neutral rolling modes.

When the wheel is rolling in the general case and is additionally affected by lateral force and yaw moments, the wheel's integrated NP turns into a *vector* nodal point. The PF represented in Fig. 11.35 turns into *vector fluxes*, which mathematically are described by vector values. It is important to note that the number of PF and geometrical images are not changed.

The analysis of all possible power flow circuits in existing multi-axle chassis transmissions allowed for revealing common properties of these circuits. Thus, it was found that, accounting for the occurrence of relative power flows through the ground and support system, power flows of multi-axle chassis are closed loops. Such closed loops are the loops between the wheels of one axle (left and right) and between the axles. The number of single closed loops is equal to $(2n - 1)$. The proof this theorem follows is from the diagrams in Fig. 11.36, which shows the closed loops of the PF in transmissions of three-axle (*b*) and *n*-axle (*c*) all-wheel drive vehicles.

For SWCs with a transparent transmission circuit (mechanical, hydromechanical) see Fig. 11.36*b*.

1, 2, 3, …, *n* closed loops are formed by the left and right-hand wheels interlocked between one another through the axle drive gear and semi-axles. The relative power flow through the ground is indicated by a dotted line. These loops are called cross-axle; in each vehicle their number is equal to the number of axles.

The fourth loop is formed by the front axle wheels, cardan shafts, transfer box mechanisms, and wheels of the second axle. The fifth loop is similarly formed by wheels of the second and the third axles. The fourth and fifth loops are called inter-axle loops. In any vehicle, the number of inter-axle loops is one less than the number of axles, i.e., $n - 1$.

Thus, the total number of the loops is $n + n - 1 = 2n - 1$. For a three-axle vehicle, the number of loops is five. In side transmission diagrams, the *n*-axle vehicle also has $2n - 1$ loops, $2(n - 1)$ of

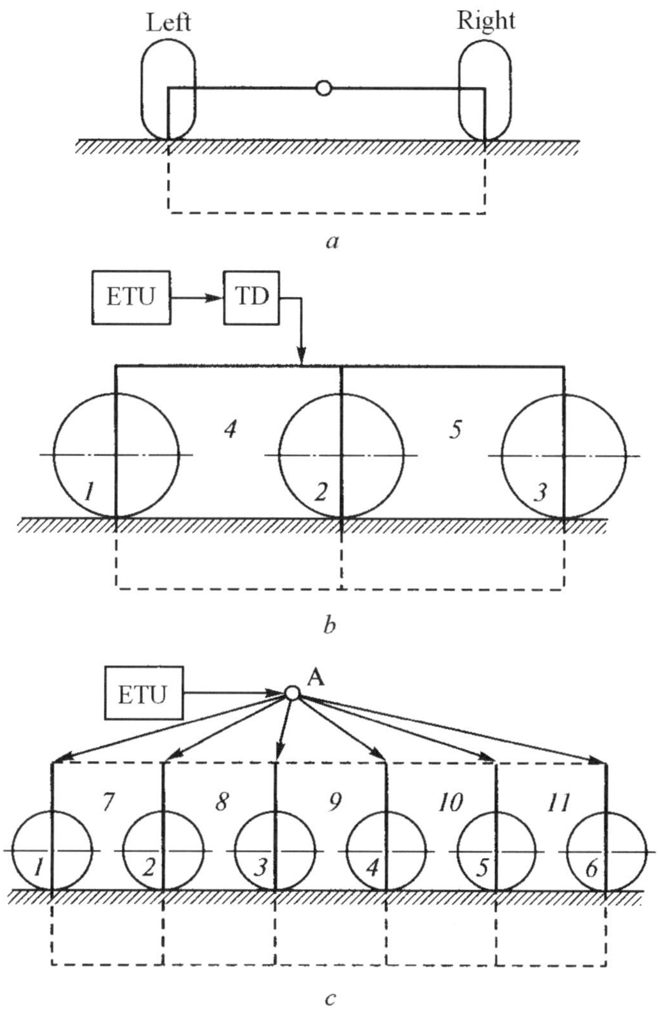

Fig. 11.36. Diagram of the closed loops in a transmission of an n-axle vehicle: a—Closed cross-axle loop of the lateral-vertical plane of the vehicle; b and c—Closed cross-axle loops of the three-axle and n-axle vehicle, respectively, in the longitudinal-vertical plane

which are along the sides of the vehicle and one is inter-side, i.e., $2(n - 1) + 1 = 2n - 1$.

For SWCs with non-transparent transmission diagrams (electric with motors-in-wheel and hydrostatic, see Fig. 11.36*c*), such loops form relative power flows connecting the wheels along the axles and sides through the support surface, suspension, and frame. The priniciples of the power distribution for wheels in such a loop are discussed in this chapter.

In accordance with the above, the theorem important for calculations is formed: the transmission of any multi-axle vehicle can be decomposed into single interconnected loops, the number of which is one less than the doubled number of the vehicle's axles.

The theorem allows for the total calculation of any transmission to be reduced to a successive calculation of the power distribution by elements of a single loop, similar to the calculation according to the method of the final elements. In this case, the final elements are closed loops of the transmission.

The common property of all closed loops under consideration is the presence of a large number of elastic ties in them. The chief of them are elastic ties to the deformable tires with the ground and the elastic ties to the mechanical elements of the loop (shafts, pinions, clutches, etc.).

A detailed kinematical and power analysis of the various types of closed loops (blocked, differential, with clutches of various design performance) is carried out in the works of P.V. Aksyonov and Yu.V. Pirkovskiy.

Thus, it is possible to formulate the following sequence for solving the problem of power distribution among the wheels:

1. Collect and specify the initial data for calculation concerning the parameters of the chassis characteristics, the transmission design solutions, the external conditions, and the motion modes. Reduce all initial data into a table.
2. Construct the power flow circuit of the transmission and decompose it into designed closed loops.
3. Determine the total power of the chassis' external and internal motion resistances according to known dependencies in the vehicle theory. Calculate the total power distribution for the designed closed loops.
4. Calculate the power distribution for the wheels of the each closed loop according to the methods of this chapter.
5. Compare the results obtained with the rational law of power distribution. Assess the conformity of the entire transmission to the rational law (11.1).

11.5.2 Criterion for complex assessment of transmission diagrams

The idea of working out a new criterion for assessing the transmission diagrams is based on the rational law hypothesis for the traction power distribution to the wheeled movers, the implementation of which ensures high performance properties in the chassis. In accordance with this hypothesis, it is assumed that in any infinitesimal time interval there is a rational value of the power $N_{0i\,\text{rational}}$ supplied to the ith wheel. It allows the wheeled mover to perform work under minimal loads and rolling power losses and also gives the wheel sufficient traction force with the support surface allowing it to create traction, stability, and handling motion qualities. At the same time, it is assumed that all wheeled movers operating totally in the rational power mode ensures high performance properties in the given conditions and motion modes.

It is obvious that the real instantaneous power N_{0ip} supplied to the wheeled mover at a given moment and in the same conditions will differ by some value of the rational power. That is why it is possible to put down as follows:

$$\Delta N = N_{0i\,\text{rational}} - N_{0ip} \quad (11.2)$$

It would make sense to assume that the deviation of real power from the rational one is a basis for assessing the transmission diagrams.

It is easy to prove that in a variety of operating conditions and at various transmission diagrams the difference (11.2) can be either positive or negative. That is why the standard deviation should be taken as a general criterion. Taking into consideration the fact that when forming the chassis' performance properties under discussion all of its wheeled movers and transmission elements take part, a general criterion can be presented as follows:

$$K_Q = 1 - \frac{\sqrt{\sum_{i=1}^{2n} (N_{0i\,\text{rational}} - N_{0ip})^2}}{\text{N}_{0a}} \quad (11.3)$$

where K_Q is the general complex assessment criterion of the transmission diagrams called the layout merit (perfection) coefficient; and

N_{0a} is the total traction power that should be supplied from the engine to the transmission and distributed to the wheeled movers in set conditions and motion modes.

The analysis of the expression (11.3) allows for determining the limits and conditions of the change K_Q.

Thus, at equal actual and rational power, $K_Q = 1$. This case is appropriate for an ideal transmission, which rationally distributes power to the wheels. Consequently, the highest indicator of the transmission diagrams' merit is 1, or $K_Q = 100\%$. With respect to 100%, all other transmission diagrams are assessed.

If the transmission fails to supply traction power to the wheeled movers, then $N_{0ip} = 0$ and $K_Q \cong 0$, i.e. the transmission has a zero merit coefficient for the calculation conditions. Such a case is possible, for example, in a fully differential transmission when the wheels of one axle or a single wheel breaks away from the support surface and traction power is virtually not supplied to the wheeled movers and the chassis loses mobility.

In fully blocked power distribution layouts when moving along hard surface roads, K_Q can be negative because of the power circulation in the transmission.

Thus, the analysis of the assessment criterion for the transmission diagrams shows the possibility of its use in carrying out calculations.

11.5.3 Mathematical model of the rational power supplied to the wheeled mover of an all-wheel chassis

A separate consideration of the input power's impact on traction-dynamic properties, fuel efficiency, handling quality, motion stability, and reliability allows for making some generalizations and conclusions.

First, it was found that an excess of input power, as well as its lack, has a negative impact on the indicators of the each vehicle property.

Second, each property has its own strictly defined quantity of rational power supplied to the wheeled mover, which ensures the best indicator of this property. Thus, for example, for ensuring a high passing ability by reducing the depth of the rut made by the mover, the power supplied to the wheel should be equal to the rolling resistance power of the wheels in the free mode. If there is a necessity to provide the maximum traction force, the input power should be maximal but not higher than the adhesion power of the wheels with the support surface.

Third, the rational power value depends on the capabilities of the wheeled mover to use the input power. For example, when the wheeled mover breaks away from the support surface, the rational power should be zero.

On the basis of these three generalizations, it is possible to create a mathematical expression for defining the rational power that should be supplied to the wheeled mover at any time and in set external motion conditions as follows:

$$N_{0i\varphi} \geq N_{0i\ \text{rational}} = \left(N^{c}_{fKi} + \frac{N'_{\text{car}}}{n} k_i \right) \leq N_{\text{tr max}},$$

or in the expanded form for the rigid support surface:

$$R_{zi}\ \varphi\ V_{\text{car}} \geq N_{0i.\text{rational}} = \left(G_w\ f^{c}_{wi}\ V_{\text{car}} + \frac{N'_{\text{car}}}{n} k_i \right) \leq N_{\text{tr max}} \tag{11.4}$$

where $N_{0i\varphi}$ is the adhesion power of the wheeled mover with the support surface; $N_{0i\ \text{rational}}$ is the rational power, which should be supplied to the chassis' i^{th} wheeled mover; N^{c}_{fwi} is the wheeled mover's free rolling mode power; and N'_{car} is the traction power necessary for overcoming the chassis' external motion resistance forces (air resistances, climbing, acceleration, curvilinear motion):

$$N'_{car} = N_{et.w} - \sum_{i=1}^{n} N^{c}_{fwi} \approx N_{et.w} - N_{srm.car};$$

where n is the number of chassis wheels; and k_i is the correction factor taking into account the capability of the wheeled mover to use power in set conditions and depending on the wheeled mover's adhesion properties:

$$k_i = \frac{\varphi_{wi}\, R_{zi}}{\varphi_{\max} \dfrac{G_{car}}{n}};$$

where: $N_{\mathrm{et.w}}$ is the power unit's traction power in the wheeled movers; φ_{wi} is the instantaneous adhesion factor of the ith wheeled mover with the support surface in the set motion conditions; $\varphi_{\max}$ is the maximum adhesion factor of the tires, which was obtained on a dry asphalt-concrete road at full skidding; f^{c}_{wi} is the rolling resistance coefficient of the wheeled mover in the free mode; V_{car} is the linear speed of the wheeled mover; $N_{\mathrm{tr\,max}}$ is the power corresponding to the maximum wheel sliding; R_{zi} is the vertical reaction at the ith wheel; G_K is the vertical load on the wheel in statics; and G_{car} is the weight of the vehicle.

From the expression (11.4) it follows that the rational power is equal to the free mode power of the wheeled mover's rolling and part of the chassis' total traction power going to this wheeled mover. Meanwhile, it is necessary to observe two restrictions: non-excess in the adhesion power of the wheeled mover with the support surface and the sliding power (skidding) of the wheeled mover in set motion conditions.

When moving on deformable sealing grounds, the expression of the free mode power of the wheeled mover's rolling is as follows [52]:

$$N^{c}_{fwi} = \frac{(2G^2_{wi})^{\frac{\mu+1}{2\mu+1}} \left(1 - \frac{2}{3}\mu + \frac{1}{6}\mu^2\right) \omega_{Ki}\, r^{c}_{Ki}}{(r^{c}_{wi})^{\frac{\mu+1}{2\mu+1}} (C_{groundi} B)^{\frac{1}{2\mu+1}} \left(2 + \frac{13}{15}\mu + \frac{1}{5}\mu^2\right)^{\frac{2\mu+2}{2\mu+1}}},$$

where $C_{\mathrm{ground}i}$ and μ are the ground characteristic parameters; ω_{wi} is the angular velocity of the ith wheel; r^{c}_{wi} is the rolling radius of the wheeled mover in the free mode; *In* is the width of the wheeled mover; and G_{wi} is the vertical load on the ith wheel in motion.

The calculations should take into account changes in the ground characteristic parameters during the second pass of the wheeled mover along the same track. During the second pass, rolling resistance and gain of the rut depth goes down as compared to their values during the pass of the first wheel. In this study, it is assumed that μ does not change during the second pass and that $C_{\Gamma i}$, changes according to the following dependency:

$$C_{groud.i} = C_{groud.i-1}\, h^{\mu}_{i-1},$$

where $C_{\mathrm{ground}.i\text{-}1}$ is the parameter value of the rut ground characteristic during the previous pass of the wheeled mover; i is the number of passes; and $h_{i\text{-}1}$ is the numerical value of the rut depth during the previous pass.

The numerical value of the rut depth during the ith pass of the wheeled mover is defined according to the well-known dependency:

$$h_i = \left[\frac{2G_{wi}}{C_{groud.i}\, B \sqrt{2r^{c}_{wi}} \left(2 - \frac{13}{15}\mu + \frac{1}{5}\mu^2\right)}\right]^{\frac{1}{\mu+0.5}}.$$

It can be assumed that by applying the rational law of supplying power to all wheeled movers of the SWC, high indicators of the chassis' most important performance properties listed above will be obtained. At the same time, the resulting dependency of the determining rational power supplied to the wheel can be taken as a basis when developing an algorithm for the functioning of the transmission's automatic control system and also when creating automatic power distribution systems for wheeled movers in any multi-axle SWC, especially with non-transparent transmissions.

11.5.4 Mathematical models for calculating the real power supplied to the wheeled mover in the non-transparent transmission loop

When finding the merit coefficient of the transmission diagrams according to the suggested dependence, it is necessary to know the real power supplied to the wheel under set conditions of the given layout being assessed.

For transparent (mechanical, hydromechanical) transmissions, there are several methods to determine this power. But for the development of multi-axle SWCs, it is more important to work out theoretical foundations of the power distribution in non-transparent (electric or hydrostatic) transmissions, than the mathematical model of calculating the real power supplied to the wheel in the transparent transmission loop, which in this case is not considered.

In the study of a specific chassis' transparent transmission, the mathematical model described in Chapter 8 can be an example.

As the main tool to ground the general power distribution principles for driving wheels of SWCs with non-transparent transmissions, the efficiency (mobility) assessment method of the usage area was applied that was presented in Chapter 8. Mathematical models for transparent ETUs were created; in accordance with the accepted methodology for making it possible to consider ET with individual power distribution to the wheels, improvement of the equation:

$$\eta_{tr} = \frac{m_{Lc} V}{N_{\text{input}}} \tag{11.5}$$

was fulfilled and the systems of equations (8.7)–(8.9).

At the same time, mathematical models that describe the processes occurring in the main constituent ET parts (generator, traction electric motors, frequency converters, etc.) with varying degrees of detail have been worked out by experts in the field of electric engineering, which increases the reliability of the results obtained.

For the present study, a mathematical model is used, which, with the help of vehicle theory methods with satisfactory accuracy, describes the dynamic properties of SWCs with ETs and individual power distribution and takes into account the relations of the main chassis parameters. Meanwhile, in accordance with the made assumption, electric transmissions are regarded as a "black box" and characteristics of transmitting and transforming energy processes that occur in electric machines when functioning are not taken into account.

The basis of the improvement is the mathematical description of the multi-axle SWC (see Fig. 8.9), which allows us to investigate the influence of the above-suggested general principles (algorithms) for power distribution in wheeled movers on the chassis' traction-dynamic properties in various motion conditions.

The motion equation of any SWC, including those having electric transmissions with individual power distribution to the wheeled movers, in the general form is represented as a slightly transformed dependency (11.5):

$$\frac{dV}{dt} = \frac{R_x - F_f - P_N - P_{air}}{\delta\, M_{car}},$$

where R_x is the total longitudinal reaction in the wheel's contact spot with the support surface; F_f is the rolling resistance force of the chassis; P_N is the weight component of the chassis when ascending (descending); P_{air} is the air resistance force; δ is the rotational inertia coefficient; and M_{car} is the gross weight of the SWC.

The calculation of the total traction force P_w for all wheels as defined by the total longitudinal reaction R_x for ETs differ significantly from the P_w calculation for SWCs with transparent transmissions and is explained when considering motors-in-wheel.

The chassis' rolling resistance force, weight component when climbing (descending), and the air resistance are calculated by the known vehicle theory methods and are distributed to the wheels in accordance with the above-stated analysis method of the transmission diagrams.

In mathematical modeling of the block "Engine" (see Fig. 8.1), the following considerations were taken into account. One of the positive properties of an ET is its non-transparency when any change in the load on the wheels does not affect the operating mode of the power unit. That is why the primary engine, as a rule, operates in a rather narrow load and speed range and the ACS of the electric transmission ensures a constant-power mode in compliance with the conditions:

$$M_{et}\omega^{v} = const \tag{11.6}$$

where M_{et} is the torque of the primary engine; ω is the rotation rate of the engine's output shaft; and ν is the rigidity coefficient.

Diesel and automobile gas-turbine engines with a high modular power, which notably differ in torque character and power characteristics, can be used as the primary engines in SWCs with ETs. In this case, there is no need to take into account the internal processes of the energy conversion in the various types of power units. The issues of a joint operation of the AGTE and ET are discussed in detail in Chapter 10. Additionally, the task of this study is the comparative assessment and justification of the general power distribution principles for the driving wheels on the basis of determining the maximum possible parameters, especially the traction-dynamic characteristic of SWCs in various road conditions, the achievement of which is only possible at full use of the PUs power. That is why the condition (11.6) can be represented as:-

$$M_{et}\omega^{v} = N_{et}\eta_{tras} \tag{11.7}$$

where N_{et} is the maximum PU power in the object conditions, including the losses; and η_{tras} is the transmission's efficiency factor (in this study it is equal to 0.7).

As an assumption, it is suggested that at mathematical modeling of the chassis' motion in case any restrictions may occur (method condition in Chapter 8, for example, restriction for lateral or vertical acceleration), the transmission input power is not supplied (the driver removes the fuel supply) until the restrictions are removed and then gradually increased to the maximum.

The most important element of the electric transmission is the electric motor-in-wheel (EMW) which, in turn, combines the electric motor and wheel with the elastic tire connected through a reduction gearbox. For obtaining a mathematical model of the motor-in-wheel block (see Fig. 8.1), the plane-parallel wheel motion in its plane rotation is considered. Calculation schemes for this case are presented in Fig. 11.35, where the legend is as follows: $M_{E.mw}$ is the torque supplied to the wheel from the block ETU in accordance with the taken power distribution principles; G_w is the vertical load on the wheel; r_{free} is the free radius of the wheel; P_w is the traction force; ω_w and v_w are the angular and linear velocities of the wheel, respectively; v_s is the linear skidding speed of the wheels on the support surface; point *0* is the wheel's instantaneous rotation center; and R_x and R_z are the longitudinal and vertical (normal) reaction in the wheel's contact spot with the support surface.

Based on the calculation scheme (see Fig. 11.35) and the approach taken earlier (omitting the preliminary computation), the mathematical model of the wheel in its final form is presented in the following system of equations:

$$\begin{cases} \dfrac{d\omega_{wi}}{dt} = \dfrac{1}{J_w}(M_{E.mwi} - M_{fi} - M_{Ji}) \ ; \\ \dfrac{dr_{di}}{dt} = \dfrac{1}{T_{tir}}(r_{free} - C_{s.tir}R_{zi} - \lambda R_{xi} - r_{xi}) \ ; \\ M_{fi} = r_{wi}(R_{zi}f - R_{xi}\delta_i) \ ; \\ M_{Ji} = J_w \dfrac{d\omega_{wi}}{dt} \ ; \\ \mathrm{r}_{st} = \mathrm{r}_{free} - R_{zi}/C_{s.tir} \ ; \\ r_{wi} = \mathrm{r}_{st} - \lambda R_{xi} \ ; \\ R_{xi} = (M_{E.mwi} - M_{Ji})/r_{wi} \ , \end{cases}$$

where $C_{s.tir}$ and λ are the radial and tangential stiffness of the tires, respectively; J_w is the inertia moment of the wheel; M_{Ji} is the inertia moment depending on the changes in the angular velocity of the wheel; r_{di} is the dynamic radius of the wheel; r_{wi} is the rolling radius of the wheel; T_{tir} is the time constant of the tires; M_f is the rolling resistance moment; r_{st} is the static radius of the wheel; f is the rolling resistance coefficient; and δ_i is the sliding of the ith wheel.

In the system of equations (11.8), the value of the wheel's rolling radius is applied without taking into account skidding. Along with that, it is known that between the longitudinal reaction R_{xi} and skidding S_i there exists a nonlinear dependence, whose form is presented in Fig. 11.37. From the data in Fig. 11.37, it follows that the longitudinal reaction R_{xi}, and, accordingly, the wheel's traction force P_{wi}, reach the maximum value at a certain S_{limit}, the value of which depends on many factors and, as is commonly believed, is within the limits 0.15–0.4.

In the present model for determining the skidding value when it does not reach the limiting value, the dependency between the longitudinal reaction R_{xi} and skidding S_i (Fig. 11.37) is approximated by the following expression:

$$S_i = \left(\frac{R_{xi}}{R_{zi}\phi_i} \right)^k (\varphi_i^{1/3} - \varphi_i) \qquad (11.8)$$

where φ_i is the wheel's adhesion factor with the support surface; and k is the power indicator.

The value S_{limit} is also calculated depending on the adhesion factor with the support surface according to the empirical dependence

$$S_{limit} = \varphi_i^{1/3} - \varphi_i .$$

With account of the skidding, the wheel's actual angular velocity ω_w and the increment $d\omega_w$ for calculating the inertia moment are defined by the following dependencies:

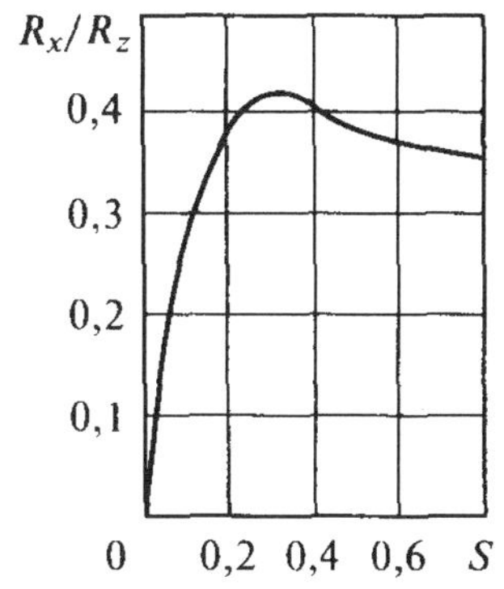

Fig. 11.37. Specific traction force R_x/R_z versus sliding S

$$\omega_{wi} = v_{wi}/[r_{wi}(1 - S_i)];$$
$$d\omega_{wi} = (\omega_{wi} - \omega'_{wi})/dt \qquad (11.9)$$

where ω_{wi} is the wheel's angular velocity at the previous calculation stage when implementing the model on a PC (personal computer).

The wheel's rectilinear speed at curvilinear motion is calculated from the velocity vector diagram by the following expression:

$$v_{wi} = v\left(R \pm \frac{B}{2} \right) \Big/ (\cos \alpha_i R) \qquad (11.10)$$

where v is the linear speed of the chassis; R is the turning radius of the chassis (the distance from the turning pole to the StS pole); B is the wheelspan of the chassis; and α_i is the turning angle of the ith wheel of the jth axle.

The wheel's rectilinear speed v_{wi} at rectilinear motion is equal to the chassis' linear speed.

The rate of the longitudinal reaction cannot be exceed, except for defined restrictions for skidding and the maximum adhesion force, which is why the obligatory fulfillment of the condition:

$$R_{xi} \le R_{zi}\varphi_i, \qquad (11.11)$$

is introduced when solving the system of equations (11.8).

If the sliding S_i defined from the expression (11.8) exceeds the value S_{limit} (drop-down curve branch in Fig. 11.37), then correction of the longitudinal reaction value is performed according to the formula:

$$R_{xi} = R_{zi}\,\phi\left[\frac{1-(S_i - S_{limit})}{1-S_{limit}}\right].$$

Also, when calculating S_i, it is necessary to fulfill the condition

$$S_i \leq 1.$$

When a chassis moves over deformable ground (soil), the coefficients f (rolling resistance) and φ (the wheel's adhesion with the support surface) vary, as well as the parameters C_{ground} and μ of the ground characteristic, as it has already been adopted above, depending on the parameter n (the number of wheel passes over the same track). The behavior of coefficients f_i and φ_i are presented in Fig. 11.38. That is why, unlike the methodology presented in Chapter 8, the coefficients f_i and φ_i in the given model can be different under each wheel that additionally increases the kinematical and power mismatch in the closed-loop circuits (in drives) of individual chassis wheels.

The change in the rolling resistance coefficient (see Fig. 11.38) depending on the number of passes is calculated by the following formula:

$$f_i = b_w C_{groud} R_{xi} = R_{zi}\,\varphi\left[\frac{1-(S_i - S_{limit})}{1-S_{limit}}\right]_i h_i^{\mu+1} \Big/ G_{wi}(\mu+1)\;,$$

where b_w is the sectional wheel width; $C_{groundi}$ is the ground resistivity to the indentation (ground settlement coefficient); G_{wi} is the normal load on the ith wheel; and h_i is the track depth of the ith wheel.

The ground resistivity to indentation goes up with each pass of the wheel axle and is calculated by the following formula:

$$C_{3i} = C_{31}\prod_{K=1}^{i=1}(\Phi_K + 1)^{\mu}\,60^{\circ}\,,$$

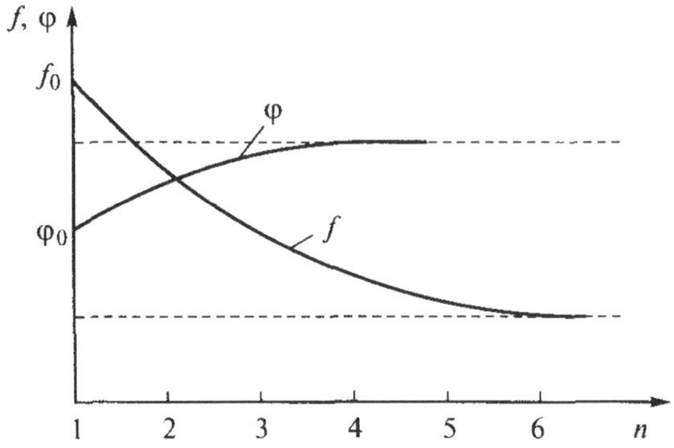

Fig. 11.38. The behavior of adhesion *j* and rolling resistance *f* coefficients depending on the number of passes of the wheeled mover *n* along one track of the deformable support surface

where $C_{ground1}$ and $C_{groundi}$ are the ground resistivities to indentation on the first and i^{th} wheel axles; and Φ_K is the consolidation coefficient, numerically equal to the incrementing track depth h_i while the ith axle passes.

There are no analytical formulas for calculating the φ_i coefficient depending on the number of passes, but the results of numerous experimental studies that the authors were involved in show that φ_i goes up by an average of 15–20% while the wheels of axles 1–4 pass gradually along the deformable ground and remains practically permanent. That is why the diagram of the changing adhesion coefficient φ_i (see Fig. 11.38) in the current model is approximated by the third degree polynomial.

Thus, for solving the system of equations (11.8), all variables are specified, except $M_{E.mw}$. When calculating $M_{E.mw}$, the following shall be taken into account.

The electric transmission in these studies is considered as some idealized power distribution system, which is lacking influence on the output data of its components, for example, generator, TEM, ACS, etc. A conditional description of the transmission in this model lies in the assumption that the specified variant of the power distribution to the wheels is realized at any time irrespective of the simulated chassis conditions and motion modes. That is why for an ETU block (see Fig. 8.1) performing ET functions with individual power distribution to the wheels, the output parameter is the torque $M_{\text{E.mw}i}$ in the separate wheel drives, composed in accordance with the chosen algorithm.

For the specification of the moments $M_{\text{E.mw}i}$ (power factors) characterizing the wheeled mover operating, the account of the linear v_{wi} and angular ω_{wi} velocity of the ith wheel's motion (velocity factors) shall be required, as well as the skidding S_i in the wheel's contact area with the support surface.

Taking into account formulas (11.7)–(11.10), the formation of the torques according to the chosen principals (algorithms) of power distribution to the wheels is performed as follows.

For the principal of the moment equality in a fully differential transmission, $M_{\text{E.mw}i}$ is calculated with the help of the following system of formulas:

$$\begin{cases} \omega_{w.ave} = \dfrac{\sum_{i=1}^{2n} \omega_{wi}}{2n}; \\ M_{o.tras} = \dfrac{N_{et}\eta_{tras}}{\omega_{w.ave}}; \\ {}_{\text{E.mwi}} = \dfrac{M_{o.tras}}{2n}, \end{cases} \tag{11.12}$$

where $\omega_{\text{w.ave}}$ is the average angular velocity of the chassis' wheels; and $M_{\text{o.tras}}$ is the general moment at the transmission's input taken from the power unit.

For the equality principle of the power, $M_{\text{E.mw}i}$ is calculated using the following system of formulas:

$$\begin{cases} N_{wi} = \dfrac{N_{et}\eta_{tras}}{2n}; \\ M_{E.mwi} = \dfrac{N_{wi}}{\omega_{wi}}, \end{cases} \tag{11.13}$$

where N_w is the power applied to the ith wheel.

For the equality principle of the rotation frequency, $M_{\text{E.mw}i}$ is calculated using the following system of formulas:

$$\begin{cases} \omega_{w.ave} = \dfrac{\sum_{i=1}^{2n} \omega_{wi}}{2n}; \\ M_{\omega i} = \dfrac{\omega_{w.ave} - \omega_{wi}}{\omega_{wi}} M'_{E.mwi}; \\ M_{o.tras} = \dfrac{N_{et}\eta_{tras}}{\omega_{w.ave}}; \\ {}_{\text{E.mwi}} = \dfrac{M_{o.tras}}{2n} + M_{wi}, \end{cases} \tag{11.14}$$

where $M_{\omega i}$ is the torque increment, which shall be applied to the ith wheel for its angular velocity to be equal to $\omega_{\text{w.ave}}$; and $M'_{E.mwi}$ is the current (momentary) torque value applied to the ith wheel.

For the equality principle of the slides, $M_{\text{E.mw}i}$ is calculated using the following system of formulas:

$$\begin{cases} S_{ave} = \dfrac{\sum_{i=1}^{2n} S_i}{2n}; \\ M_{\text{S}i} = (S_{ave} - S_i)M'_{E.mwi}/S_i \quad \text{при} \;\; S_{ave} < S_{\text{limit}}; \\ M_{\text{S}i} = (S_{\text{limit}} - S_i)M'_{E.mwi}/S_i \quad \text{при} \;\; S_{ave} > S_{\text{limit}}; \\ \omega_{w.ave} = \dfrac{\sum_{i=1}^{2n} \omega_{wi}}{2n}; \\ M_{o.tras} = \dfrac{N_{et}\eta_{tras}}{\omega_{w.ave}}; \\ M_{\text{E.mwi}} = \dfrac{M_{o.tras}}{2n} + M_{Si}, \end{cases} \tag{11.15}$$

where M_{Si} is the torque increment, which shall be applied to the ith wheel for its skidding to be equal to S_{ave} or S_{limit}.

A detailed description of the system equation derivation for (11.12)–(11.15) is not given, since it is justified in the previous material of this chapter.

Thus, the mathematical model of the electric (hydrostatic) power transmission system for HDWVs is obtained with different principles (algorithms) of the power distribution to the wheels of SWCs, which, as part of the complex motion MM (see Chapter 8), allows for tackling the assigned tasks.

11.5.5 Analysis of non-transparent transmissions with different principles of power distribution among the wheels of SWCs

HDWVs with ETs using different principles of the power distribution to the wheels; 8×8 and 16×16 wheel formulas; and mass, dimension, and geometrical parameters which correspond to actual SWCs have been selected as objects for theoretical studies. The justification for this approach is confirmed by technical project data from the 16×16 MZKT-7923 chassis equipped with ETUs composed of GTE and electric transmissions. The mass and dimension parameters of the MZKT-7923 chassis are similar to the corresponding figures, which were selected as study objects for a diesel engine and HMT chassis.

At simulation modeling of the motion conditions of a chassis equipped with a ET with different principles of power distribution among the wheels (simulating the block "road," see Fig. 11.1) and for creating the maximum kinematical mismatch with the aim to reveal the advantages and disadvantages of different power distribution principles, the following was realized out of the representative route offered in Chapters 6 and 8:

- A straight Category 1 road section having a length sufficient for reaching the maximum travel speed used for accelerating the chassis;
- Rectilinear road sections with a length of up to 3000.0 m with different types of support surfaces;
- Rectilinear road sections with a length of up to 1000.0 m with different support surfaces and an ascent angle from 2° to the maximum possible under the terms of adhesion and engine power;
- Curvilinear sections with a length of up to 400.0 m with different support surfaces and a turning radius from 100 m down to the minimum possible for the given chassis; and
- Road sections with different support surface properties along the boards of the chassis.

The type of support surface in all areas is defined by the values of the coefficients of the rolling resistance and tire traction with the road surface. Bumpy ground roads and locality are additionally defined by the ground resistivity to indentation value C_{ground} and the degree of the ground resistivity value μ.

With the aim of defining the limiting average travel speed, the chassis' motion is performed with an ET on the representative route with a length of 10 km (see Chapter 8), including all the above support surface areas. Meanwhile, different principles of power distribution to the wheeled movers are alternatively studied.

After passing the representative route, the PC screen displays as a result histogram of the limiting average travel speeds and mean square values of the transmission's Q-factor coefficient on various types of support surfaces and also the chassis' common limiting average speed for the whole route. The road surface parameters are equally defined for each area of the route with different conditions under the SWC's boards within the changing f, φ, C_{ground}, and μ coefficients which are experimentally defined for the given road type and listed in various literature [4, 6, 12, and 29]. Various types of support surface characteristics, offered in Table 11.5, are taken from the studies on SWCs-7930, -79221, and the KamAZ vehicle from *All-Drive Motor Vehicles* [29].

At mathematical modeling of the motion conditions of a chassis equipped a ET with different distribution principles of the power to the wheeled movers, it should be taken into account that in the absence of any differences

Table 11.5 Various types of support surfaces used at simulation of representative route

Type of the support surface	Board characteristics					
	f		φ		C_{ground}	μ
	right	left	right	left		
Cement-concrete highway:						
satisfactory condition (type 1)	0.02	0.03	0.75	0.70	–	–
worn-out (type 2)	0.25	0.03	0.65	0.60	–	–
Cobblestone highway:						
satisfactory condition (type 3)	0.03	0.04	0.65	0.60	–	–
worn-out (type 4)	0.04	0.05	0.60	0.55	–	–
Crushed stone, macadam:						
satisfactory condition (type 5)	0.03	0.035	0.60	0.50	–	–
worn-out (type 6)	0.035	0.45	0.50	0.40	–	–
Ground road:						
satisfactory condition (type 7)	0.03	0.035	0.45	0.40	–	–
worn-out (type 8)	0.035	0.045	0.40	0.35	1.0	1.0
Off-road (locality):						
(type 9)	0.15	0.20	0.40	0.30	0.5	0.5

in the motion conditions and operation of the driving wheels, the distribution principles of the power to the wheels justified above will work identically on any assessment figures, since in this case there is no kinematical and, as a consequence, power disagreement in loading the separate wheel drives.

That is why a MM performance check is performed with motion conditions modeling the typical exploitation of SWCs. The chassis' curvilinear motion on a turn with a bending radius in the plane view of 55, 45, 35, 25, and 15 m is considered. Deformable ground with different carrying capacities and adhesion properties alongside the chassis is taken as the support surface. The support surface properties are defined by the following coefficients: $\varphi_r = 0.4$; $\varphi_l = 0.3$; $f_r = 0.15$; $f_l = 0.2$; $C_{ground.r} = 1$; $C_{ground.l} = 0.5$; $\mu_r = 1$; and $\mu_l = 0.5$. Indices "r" and "l" represent the right and left board of the chassis, respectively.

To define the maximum possible travel speed and the maximum ascent angle in areas with different types of support surfaces, motion of the chassis is performed, whereby the final speed values obtained on each section is initial for the next area with more severe motion conditions (passing to a greater lead angle value or a smaller turning radius).

The preliminary analysis of the MM results showed that the maximum possible travel speed and the power expenses in the percentage to slips equality algorithm are similar for all of the chassis selected as study objects. That is why in further material of the chapter, the motion modeling results of the 16×16 79221 chassis, which is the most representative, is used.

The exception belongs to the studies on defining the maximum ascent angle, the data on which are presented for the 8×8 7930 chassis. This is because the ascent angle conquered by the 79221chassis has a value which is not enough for any essential redistribution of the vertical reactions due to its insufficient engine power. This results in practically similar values of the ascent angles for all distribution principles of the power among the wheeled movers of the SWC.

Thus, for theoretical studies:

- We have chosen the KamAZ-6350, MZKT-7930, and MZKT-79221 chassis models whose basic specifications corresponds to real 8×8 and 16×16 chassis models;
- We have formed the chassis' motion conditions (sections of the support surface with different characteristics specified by f, φ, C_T, and μ of the turning radius and ascent angle); and
- The methodology was designed, which established the succession of assessments on the influence of different principles of power distribution among the wheeled movers on the SWC's traction-dynamic properties.

The results of the calculation diagram for the transmission Q-factor of the KamAZ-6350 vehicle are presented in Table 11.6.

The analysis of the data in Table 11.6 shows that the transmission layout of the KamAZ-6350 (8×8) vehicle has unsatisfactory indicators of the Q-factor at linear speeds without load and at curvilinear speeds with a minimum radius with and without load. Low K_Q figures point at an exceptional inequality in the power distribution to the wheels and at turning when there is a circulating (parasitic) power between the vehicle's axles in the front truck.

The main reason for the low indicators in the transmission merit diagram for the KamAZ-6350 vehicle is the accepted constructive solutions: the foremost blocked connection between the vehicle's axles.

Serious degradation of the K_Q value in the transmission layout is most noticeable when the vehicle is without a load. An unfavorable distribution of the vehicle's full weight along axles in tension should be considered the main reason for this. At the accepted layout, the admitted vehicle weight without load falls on the front axles and the power distribution to the axles remains permanent, which results in a great departure of the real input power from the rationally required one. To decrease the

Table 11.6 Transmission Q-factor K_Q (8×8) of the KamAZ-6350 vehicle

No. n/n	Motion mode	K_Q value of the design			
		blocked		differential	
		with load	unloaded	with load	unloaded
1	Rectilinear motion (heavy duty), v = 60 km/h	0.882	0.541	0.997	0.914
2	Rectilinear motion uphill, i = 0.05, v = 20 km/h	0.974	0.913	0.995	0.896
3	Rectilinear acceleration at i = 0.05, g to v = 20 km/h	0.891	0.639	0.996	0.967
4	Curvilinear motion on asphalt road with R = 14.5 m, v = 10 km/h	0.209	−0.879	0.996	0.869
5	Curvilinear motion on dry sand with R = 14.5 m, v = 5 km/h	0.772	0.775	0.902	0.792
6	Curvilinear motion on loam with R = 14.5 m, v = 5 km/h	0.844	0.745	0.837	0.766
7	Curvilinear motion along the tillage with R = 14.5 m, v = 5 km/h	0.854	0.742	0.872	0.795
8	Curvilinear motion on wet ground road with R = 14.5 m, v = 5 km/h	0.839	0.741	0.884	0.834

effect of this mismatch, the following design modifications are offered:

- An improved vehicle layout;
- Using an inter-bogie differential with a variable internal transmission ratio—one for the motion with the load and the other for the unloaded vehicle's motion; and
- Eliminating the blocked link of the axles in the bogies.

The calculation data from the transmission Q-factor diagram of the MZKT-7930 chassis with different distribution principles of the power to the wheeled movers are presented in Table 11.7.

The analysis of the data in Table 11.7 shows the advantages of the transmission diagram when using the slide equality and torque equality principles over the other analyzed principles. Thereby both of the mentioned power distribution principles provide good transmission adaptability (adaptation) to changing environment conditions and motion modes. Also, in contrast to the data on the KamAZ-6350 vehicle with a mechanical transmission, the 7930 (8×8) chassis for all diagram variants has a smoother changing (absence of jumps) Q-factor both on hard roads and in a field. This is explained by the effect of the input power adjustability on the wheel depending on the influence of the support surface. The qualitative character of forming the transmission K_Q diagram with different power distribution principles for the 79221(16×16) chassis is similar to the 7930 chassis.

The comparison of the data from Tables 11.6 and 11.7 suggests that the regulations of the coefficient parameters in the transmission diagram which merit formation while moving along roads with a hard surface differ from the same ones while moving on deformable grounds. The interconnection between K_Q, the motion mode, and the support surface characteristic parameters are worthy of individual profound studies.

Table 11.8 lists the results from calculating the limit average travel speed for the 79221 chassis model with different variants of ET control on a standard route formed by conditions approaching the real conditions of using a HDWV of the given load capacity class.

As the analysis in Table 11.8 shows, the limiting average travel speed of the slide equality principle is 3–10% higher than the toque equality and power equality algorithms depending on the type of support surface and is 11.5–13% higher than rpms equality algorithm.

Insignificant differences in the limiting parameters of the average travel speed for various power distribution principles must be explained

Table 11.7 Q-factor K_Q of the 7930 (8×8) chassis

Rectilinear motion at maximum possible speed with restrictions on sections	Q-factor for equality of			
	slides	torques	rpms	powers
1	0.980	0.961	0.856	0.957
2	0.975	0.957	0.848	0.953
3	0.979	0.959	0.842	0.958
4	0.974	0.955	0.840	0.952
5	0.980	0.960	0.850	0.954
6	0.978	0.956	0.843	0.952
7	0.977	0.956	0.843	0.952
8	0.900	0.869	0.792	0.858

Table 11.8 Limiting average travel speed of the 79221 (16×16) chassis along the standard route

Power distribution principles	Average travel speed of, km/h, on types areas									
	1	2	3	4	5	6	7	8	9	Whole route
Equality of slides	54.5	39.1	27.5	27.2	23.0	22.0	18.4	12.5	8.3	16.2
Equality of moments	52.9	37.9	26.4	25.9	21.9	20.8	17.6	11.5	7.6	15.4
Equality of rpms	47.8	34.9	24.7	24.4	20.8	20.0	16.3	11.1	7.4	14.5
Equality of powers	52.3	37.2	26.2	25.7	21.7	20.6	14.4	11.4	7.5	15.2

by the fact that parameters of the representative route (curvature radius in the plane view of at least 30 m, the maximum ascent (descent) angles are no more than 9°, etc.) in motion conditions do not provide significant asymmetric loading on the separate wheel drives.

With the aim to reveal the utmost efficiency assessment when comparing various algorithms of the power distribution among the wheeled movers, Figs. 11.39–11.42 present calculation results for tougher road conditions than the representative route. The chassis moved along a curvilinear trajectory in the plan view with a curvature radius from 30 to 17 m (the minimum turning radius for the 79221 chassis model) with the following support surface characteristic parameters: $f_r = 0.025$; $\varphi_r = 0.65$; $f_l = 0.2$; $\varphi_l = 0.3$; $C_{\text{ground}\,l} = 0.5$; and $\mu_l = 0.5$.

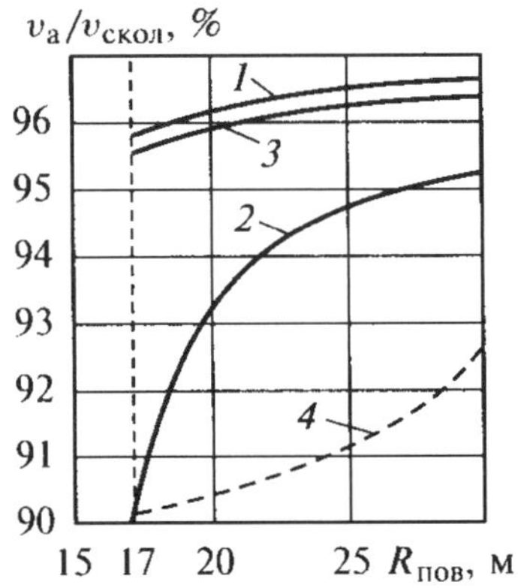

Fig. 11.39. Relative maximum speed possible for the curvilinear motion of the 79221 model at $N_{\text{o.tras}}$ = 588 kW (800 hp): 1—Equality of torques; 2— Quality of rpms; 3—Equality of powers; 4—Chassis with HMT

In Fig. 11.39, the chassis' relative maximum possible travel speed with the equality of slides algorithm was taken as 100%. In this case, the chassis' speed reduction with the equality of torques and the equality of powers algorithms is 9.0–11.7%; for the chassis with the equality of rpms algorithm, the speed reduction is 13.7–18.3%. Thus, with a growth in the curvature, the advantages of the chassis with the equality of slides algorithm by the maximum possible travel speed criterion becomes stronger.

The power input dependency on motion at the preset speed has a similar effect. In Fig. 11.40, the power input on the chassis' motion with the equality of slides algorithm is taken as 100%. The power input in Fig. 11.40 is expressed through the ratio of the power spent on the curvilinear motion (N_{cm}) to the power required for the rectilinear motion of the same chassis. The power input increase with the growth of the curvature for the chassis with the equality of torques and equality of powers algorithms is 8.5–11.5% and for the chassis with the equality of rpms algorithm it is 8.5–14.5%.

The reasons for the motion power input increase can be explained by considering the criterion characterizing the power input on sliding. Fig. 11.41 presents the dependencies of the power

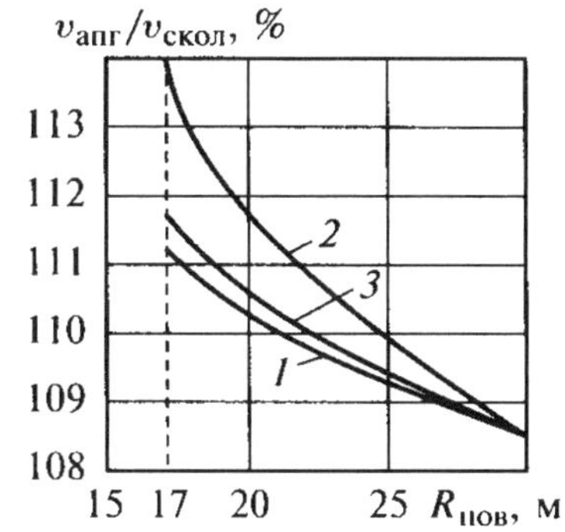

Fig. 11.40. Relative power input on the curvilinear motion of the 79221 model at a speed of the v = 2 m/s: 1—Equality of torques; 2—Equality of rpms; 3—Equality of powers

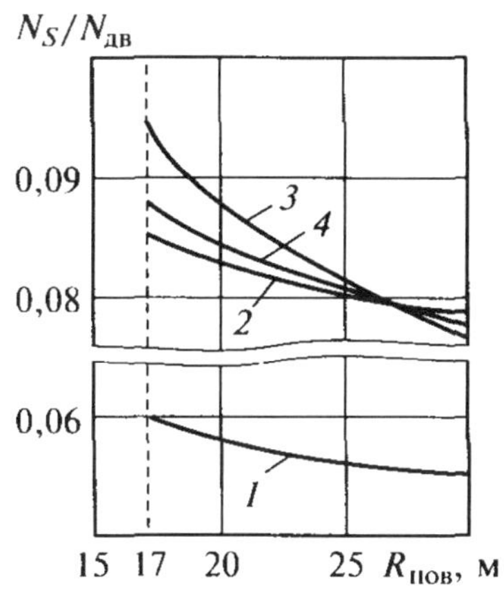

Fig. 11.41. Relative power input on sliding during curvilinear motion of the 79221 model at a speed of the v = 2 m/s: 1—Equality of slides; 2—Equality of torques; 3—Equality of rpms; 4—Equality of powers

input on sliding with a decrease in the chassis' turning radius. For the chassis with the equality of slides principle, the power input value on sliding is 5.6–6.0% of the total power from the transmission input and practically does not change when the turning radius decreases. The chassis with other power input algorithms have greater absolute values on sliding (for example, the chassis with the equality of rpms algorithm has a value of 8.0–9.5%); and, with an increase in the curvature, they intensely increase at the expense of the kinematical mismatch.

The combined analysis of the data in Figs. 11.39–11.41 allows for making an important conclusion: that the main advantage of the chassis with the equality of slides principle lies in the realization of the higher longitudinal reaction R_x in the wheel-contact surface-contact area (in all motion conditions and modes), which also means a higher pulling power P_w at a lower aggregate motion power input.

Fig. 11.42 shows the changes in the maximum conquerable ascent angle α for the 7930 chassis versus the adhesion factor φ on a rectilinear section of the deformable ground surface with the following support surface characteristic parameters: f = 0.15; C_{ground} = 0.5; and μ = 0.5 (wet loam). Pursuant to Fig. 11.42 and under these conditions, the drawbacks typical for the chassis with the equality of torques principle are most clearly apparent.

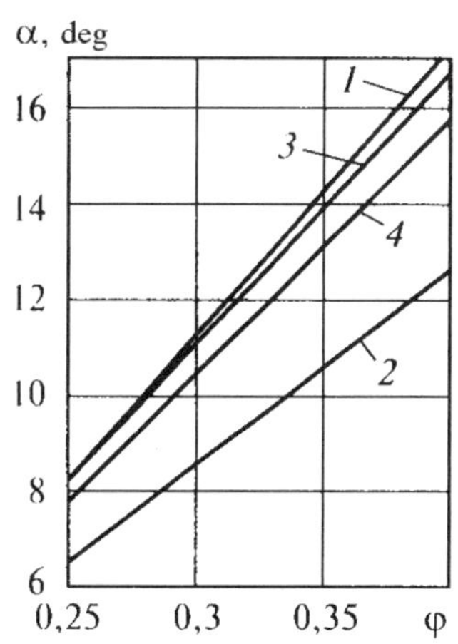

Fig. 11.42. Maximum ascent angle a conquered by the 7930 chassis on a deformable ground surface: 1— Quality of slides; 2—Equality of torques; 3—Equality of rpms; 4—Equality of powers

An increase in the ascent angle leads to a decrease in the vertical response of the first axle and violation of the condition (11.11). In this case, the sliding exceeds the value of the S_{limit}, while the differential links of the wheels and axles finally results in the loss of passing ability by adhesion. That is why a chassis with the equality of torques principle is inferior by the maximum ascent angle value to the chassis with all other power distribution variants, for example, to those with the equality of slides principle by 28.5–33.6%.

Thus, the developed mathematical model of the controlled motion of a HDWV with an ET has allowed for making a theoretical evaluation of the efficiency of the various power distribution principles. The most preferable of the above principles is the equality of slides. At the same time, it should be noted that this principle will ensure similar benefits in HDWVs with hydrostatic transmissions (HSTs).

With an increase in the asymmetry of the individual wheel drives, which is typical for actual SWC operating conditions, the ET with the above power distribution principles will experience a stronger manifestation of the drawbacks typical for traditional transmissions. First, the reduction in adhesive properties of the support surface degrades the chassis' mobility with the differential drive down to its complete loss if there is an engine of sufficient power. Second, the kinematical asymmetry at the curvilinear motion leads to a substantial increase in the chassis' power input with a locked drive.

The driving wheels' equality of slides principle excludes any loss of the HDWV's mobility at significant skidding (even when one or more wheels hang) and also prevents any circulation of the power in the transmission at the chassis' curvilinear motion. Finally, it leads to an increase in the travel speed by 9.0–18.3%, the maximum possible in various road conditions, and of the limit average travel speed on the representative route by 5.3–11.6%. At the same time, the power input on sliding goes down 43–75% and the ascent angle goes up 2–33.6% as compared to other power distribution principles.

Thus, without taking into account the technical difficulties in implementing the ET design, the considered power distribution principles can be classified by the degree of their impact on the SCW's traction-dynamic properties in the following order:

- The equality of wheel slides;
- The equality of torques on the wheels;
- The equality of powers in the wheel drives; and
- The equality of wheel rotation frequencies (rpms).

11.6 Promising electric transmissions

11.6.1 Theoretical backgrounds for ET designs

As noted above, an important advantage of the ET is the possibility to individually control the operation modes of each wheeled mover in the SWC in accordance with their motion conditions and support surface state. It is known that such a control allows for improving the traction-dynamic properties, the passing ability, and the mobility of the chassis as a whole as a transport base (platform) for special units of the mounted equipment.

The problem of organizing this type of control is in the development of the appropriate functioning principles of the ET and the formation on this basis of the requirements for the ETU components and for the structure and principles of the traction electric drive as a whole. This is also true for the hydrostatic transmission.

Baseline data for the development of ETs or HSTs should rely on science-based principles and algorithms for the change in the power and kinematical parameters in ETU sections under various HDWV motion modes and conditions.

This problem can and should be solved by experts in the area of vehicle theory with the help of simulating the ETU operation. At the same time, for the successful use of the simulation results to develop requirements for ETs and their creation, the simulation should be based on theoretical foundations and analogues (substitution diagrams) understandable by experts of the related fields of knowledge.

This section has made an attempt of building—based on the approaches outlined in *Development Trends in the Theory and Practice of Delivering Power to Movers of Multi-Axle Full-Drive Vehicles* [6]—a model of a chassis with an ET in which the description of the real processes occurring in the electromechanical devices was replaced (wherever possible) by the description of similar mechanical devices. The main condition for substitution was to select the analogs whose principles are close to electromechanical devices.

This allows, first, for getting away from the consideration of specific and highly complex processes and interaction patterns of various electric values (current, voltage, active and reactive resistances, etc.), the study of which is within the competence of specialists in the areas of electromechanics and automatics. Second, it allows for simplifying the mathematical model and focusing on the development of the ET's operation principles from the vehicle theory standpoint.

When developing the calculation diagram (substitution diagram) and mathematical model at this stage of the study, the following assumptions were made:

- The chassis moves only in the forward direction on a flat surface;
- The uphill motion is simulated by the climbing resistance force as a component of the chassis' weight with account of the load redistribution among the axles;
- The change in the wheel rolling conditions is simulated by the rolling resistance and adhesion factor coefficients; and
- The ET control system responds only to motion resistance and adhesive properties of the support surface.

Accounting for the above and the assumptions made, the substitution (equivalent) diagram of the ETU with an ET has the form as proposed by V. Kornilov, Candidate of the Technical Sciences, and is shown in Fig. 11.43.

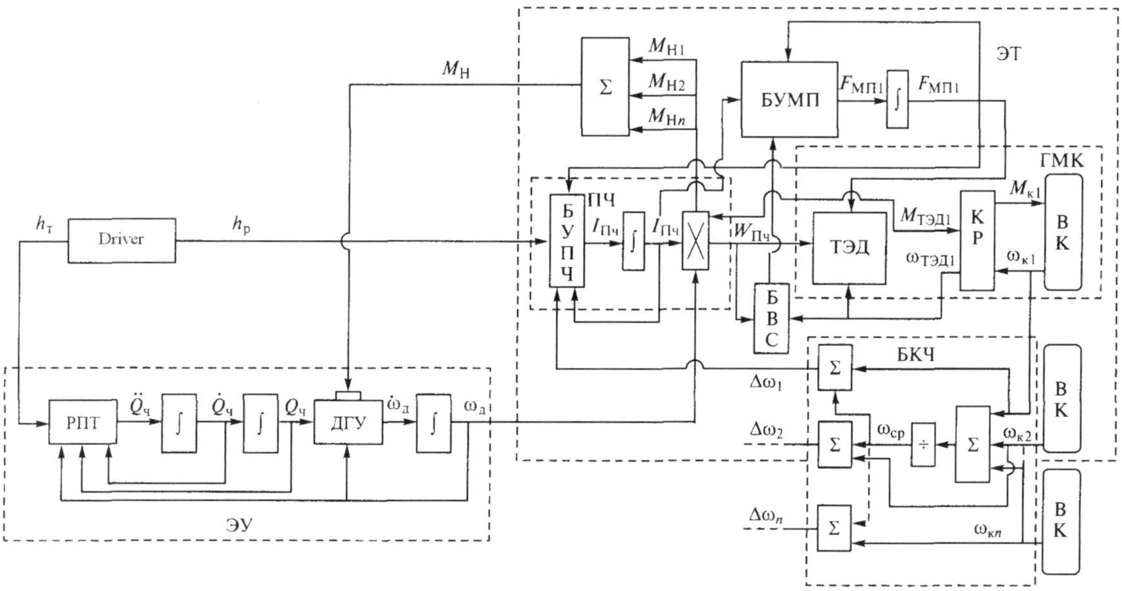

Fig. 11.43. Calculation diagram electric transmission: FSR—Fuel supply regulator; DGU—Diesel-generator unit; FCU—Frequency converter unit; ET—Electric transmission; FCU— Frequency control unit; MFCU— Magnetic flow control unit; SCU—Slide calculation unit; TEM—Traction electric motor; WG—Wheel gearbox; DW—Driving wheel

In general, the diagram includes two basic parts: the power unit (PU) and the electric transmission (ET).

The PU input parameter is the displacement of the fuel supply setting control (h_f) and the output parameter is the rotation rate (rpm) of the PU output shaft (ω_{em}). The load torque (M_l), which depends on the ET's operating mode, is used as feedback.

The internal structure of the PU is presented by the diesel-generator unit (DGU) and the fuel supply regulator (FSR).

The input signal for the FSR is the displacement of the fuel supply control (h_f) and the output signal is the hourly fuel flow rate (Q_h). The rotation rate of the DGU's output shaft (ω_{em}), the hourly fuel flow rate (Q_h), the speed of its variation ($\dot{Q}_h$), and the regulator decay characteristic (ξ_p) are used as feedbacks.

For the DGU, the input parameter is the fuel consumption (Q_h) and the output parameter is the output shaft's angular rotation rate (ω_{em}). The load torque (M_l) from the ET, which represents the sum of the traction electric motors' (TEM) torque, is used as feedback.

Structurally, the ET consists of the electric motor-in-wheel (EMW), the frequency converter unit (CU), the magnetic flow control unit (MFCU), the frequency control unit (FCU), and the mechanical sliding calculation unit in the TEM (MSC). The main objects of the control in the ET are the traction electric motors (TEM).

The output parameter of the TEM is the electromechanical torque ($M_{\text{E.mw}}$), which is supplied to the driving wheels (DW) through the transmission ratio of the wheel reduction gearbox (WG). The feedback is performed through the angular speed of the BK (ω_w) with account of the WG transmission ratio.

The frequency converter unit (FCU), which has at its input the displacement of the frequency control (h_c), the rotation rate (rpm) of the DGU's output shaft (ω_{em}), the magnetic flow ($F_{\text{m.f}}$), and the error signal ($\Delta\omega_w$) from the frequency control unit (FCU), generates an output signal that determines the magnitude of the magnetic field W_{FCU} of the TEM stator.

The relationship between h_f and h_p as the function $h_f = f(h_w)$ or $h_w = f(h_f)$ is quite complex and is conditioned by a program joint regulation and control. This work does not consider this relationship.

The input parameter for the magnetic flow control unit (MFCU) is the transmission ratio (FCU) (I_{FCU}) and the sliding error signal in the TEM (ΔS_{TEM}) is calculated by the TEM sliding calculation unit (SCU) relative to the critical sliding of the traction electric motor ($S_{\text{TEM W.g}}$).

At the output, the MFCU generates a signal defining the magnetic flow volume of the TEM, which is simultaneously used for correcting the transmission ratio (FCU).

In accordance with the substitution diagram, the work of the DGU can be described by the first order differential equation as follows:

$$\dot{\omega}_{dgu} = \frac{Q_h - K_{qt} - M_l K_a - \omega_{dgu}}{T_{dgu}}, \qquad (11.16)$$

where ω_{dgu} is the angular rotation rate of the DGU output shaft; Q_h is the hourly fuel flow rate; M_l is the load torque at the DGU shaft; T_{dgu} is the DGU time constant; K_{qt} is the proportionality factor between the fuel supply and DGU shaft's rotation rate; and K_a is the proportionality factor between the load torque and DGU shaft's rotation rate.

The load torque M_l is defined as the sum of the traction electric motor ($M_{\text{E.mw}}$) torques with account of the transmission ratio U_{FCU} of the frequency converter unit:

$$M_l = \sum_{i=1}^{2n} (\mathrm{M}_{E.mwi} U_{\text{FCU}\,i}) \,.$$

In accordance with the substitution diagram, the fuel supply to the DGU (Q_h) can be defined by solving the second order differential equation, which has the following form:

$$\ddot{Q}_h = \frac{h_f K_1 - \omega_{dgu} K_2 - 2\xi_h T_h \dot{Q}_h - Q_h}{T_h^2} \quad (11.17)$$

where h_f is the parameter defining the engine fuel pump's rack position; ξ_h is the decay factor of the fuel supply regulator; K_1 is the proportionality factor between the control unit position and the fuel supply; K_2 is the proportionality factor between the DGU shaft's rotation rate and the fuel supply; and T_h is the time constant of the fuel supply regulator.

The proportionality factors of the various values in (11.16) and (11.17) are defined form the following formulas.

The coefficient K_1 can be defined from the following correlation:

$$K_1 = (Qh_{\max} - Q_{h\,\min})/(h_{f\max} - h_{f\min}),$$

and the coefficient K_2 from the following dependency:

$$K_2 = (\omega_{\text{dgu max}} - \omega_{\text{dgu min}})/(Q_{h\max} - Q_{h\min}).$$

The coefficient K_a depends on the correlation between the maximum rotation rate of the DGU shaft $\omega_{\text{dgu max}}$ and the maximum load torque $M_{l\max}$, as follows:

$$K_a = \omega_{\text{dgu max}}/M_{1\max},$$

and the coefficient K_{qt} is the inverse value of the coefficient K_1:

$$K_{qt} = 1/K_1.$$

Thus, the equations (11.16) and (11.17) describe the work of the PU depending on the driver's impact on the fuel supply regulator (h_h), load torque M_l of the DGU shaft, and the system's internal operation parameters.

As noted above, the main object of ET control is the TEM. For an ET on AC, it is recommended to use asynchronous traction motors (AM).

It is obvious that in order to obtain the required law of change in the AM output characteristic (electromechanical torque), it is necessary to change the state of the stator's magnetic field and the volume of the magnetic flow. In this case, for steady work of the AM, the mechanical sliding in the TEM ($S_{\text{E.mw}}$) should not exceed its critical value ($S_{\text{E.mw. W.g K.P}}$) except for when the AM starts. In this case, the mechanical sliding in the TEM (S_{TEM}) is the ratio of the instantaneous rotor rotation rate, defined by electric quantities and supplied to the TEM and the load on the wheel defined by the conditions of its interaction with the support surface, to the calculated value of the instantaneous rotor rotation rate, defined only by the electric quantities supplied to the TEM.

The analysis of the AM operation principles, the nature of the changes in its output characteristic, and the methods of its control allows for presenting an asynchronous engine in the form of a unit comprising of internal and external plugs of the cylinder form. The annular gap between those plugs is filled in with a medium that has a moment of angular resistance. Then the torque from the external (driving) plug to the internal one will be transmitted through the elastic medium regardless of its nature.

Drawing an analogy, we can assume that the external plug rotation represents the stator's magnetic field and the internal plug rotation is similar to the AM rotor rotation. In this case, the elastic properties of the medium are simulated by the volume of the magnetic flow. In the absence of a load applied to the internal plug, the maximum possible value of its rotation rate will be defined by the magnetic flow properties (mechanical sliding is equal to 0); in the presence of a load, the maximum value of its rotation rate is defined by the load value and the magnetic flow properties. Then the operation of the TEM power control system drive can be arranged in such a manner that by regulating the electric parameter values applied to the TEM, it will provide equal mechanical sliding of the TEM to all wheels of the vehicle.

This substitution allows for describing the work of the AM in a generalized form, without affecting the processes and patterns that lie in the area of electrotechnics.

Based on the above and in accordance with the substitution diagram (Fig. 11.43), the AM's output shaft torque, including the damping properties of the elastic medium (in this case, magnetic flow), can be written in the following form:

$$M_{TEM} = F_{M.f}(W_{\mathrm{FCU}} - \omega_w U_{w.g}) - (\dot{W}_{\mathrm{FCU}} - \dot{\omega}_w U_{w.g})\mathrm{K}_g,$$

where $F_{\mathrm{M.f}}$ is the magnetic flow; and W_{FCU} is the rotation rate of the stator's magnetic flow defined by the frequency converter unit, that is:

$$W_{\mathrm{FCU}} = U_{\mathrm{FCU}}\omega_{\mathrm{dgu}},$$

where ω_w is the wheeled mover's rotation rate; $U_{\mathrm{w.g}}$ is the transmission ratio of the wheel reduction gearbox; and K_g is the decay factor of the magnetic flow.

Knowing the value of the electric motor's electromagnetic torque, we can define the driving wheel's angular speed with account of the transmission ratio of the wheel reduction gearbox:

$$\dot{\omega}_w = \frac{M_{TEM}U_{w.g} - R_x r_w}{J_w},$$

where R_x is the longitudinal reaction in the wheel's support surface contact area; r_w is the radius of the wheel's rolling when loaded; and J_w is the wheel's inertia moment with account of the rotating parts of the electric motor and wheel reduction gearbox.

The longitudinal reaction can be defined from the following formula:

$$R_x = \frac{\omega_w r_w - v}{\lambda_{tir}} - (\dot{\omega}_w r_w - \dot{v})\mathrm{K}_{tir},$$

where v is the chassis' longitudinal travel speed; λ_{tir} is the tangential tire elasticity; and K_{tir} is the tire damping coefficient.

With account of the assumptions made, the rectilinear motion equation can be written in the following form:

$$\dot{v} = \frac{1}{m_{car}}\left(\sum_{i=1}^{2n} R_{xi} - \sum_{i=1}^{2n} R_{zi}f_i - F_\alpha - F_{air}\right),$$

where R_{xi} is the longitudinal reaction of the ith wheeled mover; R_{zi} is the nominal normal reaction in the contact area pattern with the support surface of the ith wheeled mover; f_i is the rolling resistance coefficient of the ith wheeled mover; F_α is the resistance to ascend force; m_{car} is the chassis' weight; and F_{air} is the air resistance force.

The vertical reactions R_{zi} of the air resistance force and the resistance to ascend force are defined by known vehicle theory methods.

The chassis' travel speed control in conditions of the PU's constant rotation numbers (non-transparent transmission) is performed by regulating the frequency converter's transmission ratio by exposing the appropriate control unit in a manner similar to the driver shifting gears in the gearbox. The considered transmission lets us do this steplessly, that is, the driver sets the control unit position based on the desired travel speed. In accordance with that, the transmission ratio's control circuit fulfills a specified transmission ratio to yield for a specified motion mode.

The operation of such an ET circuit should meet the following requirements.

First, it is necessary to consider the possible difference in the driving wheels' angular velocities, the appearance of which may be caused by skidding in the individual wheels, the motion in a circular path or support surface relief.

Second, it is necessary to coordinate its work with the magnetic flow control circuit for the same purposes and for retaining the mechanical sliding value in the TEM in the subcritical zone.

The ET magnetic flow control circuit should ensure a smooth decrease in this value depending on the transmission ratio with account of the electric sliding in the TEM.

Following the above considerations, the joint work of the ET circuits connected with control of the frequency converter and magnetic flow transmission ratios can be presented in the form of the first order differential equation systems as follows:

$$\dot{I}_{FCU} = \frac{h_{FCU} K_{FCU} - F_{M.f} K_{M.f} - \Delta\omega_w K_\omega - I_{FCU}}{T_{FCU}} \quad (11.18)$$

$$\dot{F}_{M.f} = \frac{I_{FCU} K_F - (S_{TEM\,w.g} - S)\, K_S - F_{M.f}}{T_{M.f}} \quad (11.19)$$

where h_{FCU} is the frequency converter's control unit position; K_{FCU} is the proportionality factor between the frequency converter's control unit position and its transmission ratio; $K_{M.f}$ is the proportionality factor between the magnetic flow's volume and the transmission ratio; $\Delta\omega_w$ is the difference between the wheel's rotation rate and the average rotation rate of the wheeled movers; K_ω is the proportionality factor between $\Delta\omega_w$ and the transmission ratio; K_f is the proportionality factor between the frequency converter's transmission ratio and the magnetic flow's volume; K_S is the correction coefficient of the magnetic flow's volume on mechanical sliding in the TEM; T_{FCU} and $T_{M.f}$ are the time constants of the frequency converter and magnetic flow's control circuits; and I_{FCU} is the frequency converter's current.

The proportionality coefficients in (11.18) and (11.19) can be defined from the variable values associated with them, by successive approximation methods, or by prescribing the boundary conditions of the described system's operation in equilibrium.

It is clear from the presented equations of the ET's functioning that the driver has the leading role in controlling the chassis' travel speed and that other functions associated with adjusting the transmission ratio and electric quantities are carried out automatically. Moreover, a significant place in this process is given to the value and nature of the mechanical sliding changes in the TEM. This parameter is an indicator of the working conditions in the TEM, which almost instantly respond to the wheeled mover's working mode, and thus, is indirectly connected with its sliding in the contact area pattern with the support surface. This is very important since it was justified earlier that for an ET, the most appropriate regulation of the power distribution to the wheeled movers is the one based on the equality of slides condition in the contact area pattern with the support surface. However, at present, to measure this parameter with the required accuracy is difficult, and, at the same time, the mechanical sliding in the TEM, by contrast, is easily defined. That is why the mechanical sliding in the TEM in conjunction with other parameters can be used for developing algorithms of the ET work as a whole.

On the basis of the developed mathematical model, the software for studying the processes occurring in the ET in various chassis conditions and motion modes was developed.

Examples of such studies are shown in Figs. 11.44–11.46.

Fig. 11.44 shows the variation of the studied parameters in the process of accelerating the chassis from rest to a specified speed. This process is characterized by a significant resistance in the initial moment and a gradual decrease in the motion resistance due to the forces of inertia.

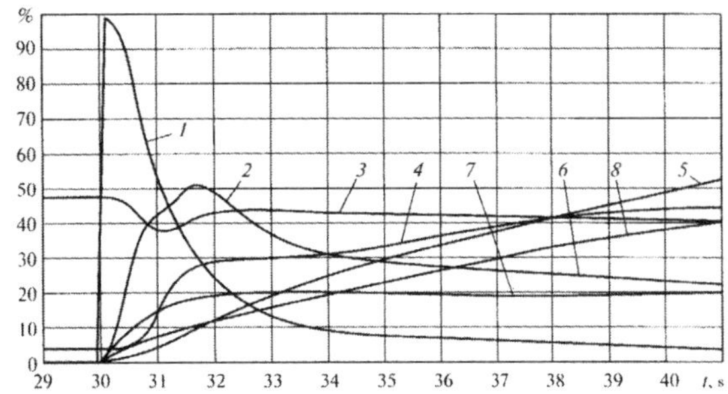

Fig. 11.44. Nature of the change in the TEM parameters at acceleration of a chassis with an ET: 1—Mechanical sliding in the TEM; 2— TEM torque; 3—Rotation rate of the DGU output shaft; 4—Fuel consumption of the primary engine; 5—The travel speed of the chassis; 6—Transmission ratio of the frequency converter; 7—Magnetic flow in the TEM; 8—Rotation rate of the driving wheel

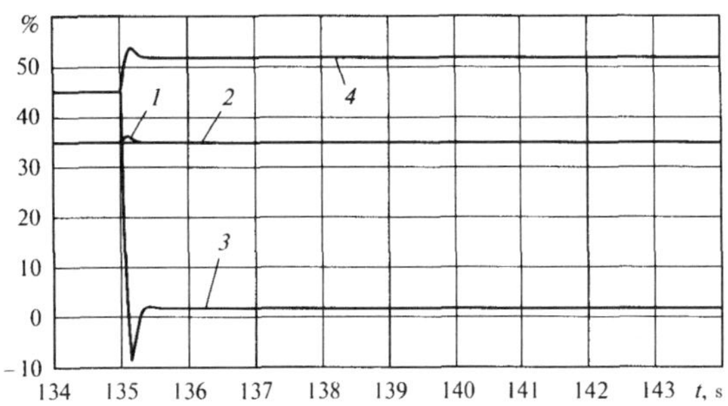

Fig. 11.45. Reaction of the ET to a step change in the adhesion coefficients under the first right-hand wheel (with φ = 0.7 to φ = 0.05) at the chassis' ascending motion: 1—Rotation rate of the first right-hand wheel; 2—Rotation rate of the first left-hand wheel; 3—Torque of the right-hand TEM; 4—Torque of the left-hand TEM

It is clear from the data shown in Fig. 11.44 that at the resting chassis' starting torque, the mechanical sliding in the TEM has its minimum value ($S_{TEM} \approx 0$) (curve *1*); however, rapid growth in the magnetic flow (curve *7*) and transmission ratio of the frequency converter (curve *6*) leads to a change in the mechanical sliding to its limits, allowing the TEM to work in the stable zone ($S_{TEM} < 0.12$).

The torque of the TEM (curve *2*) due to rapid growth of the magnetic flow and the

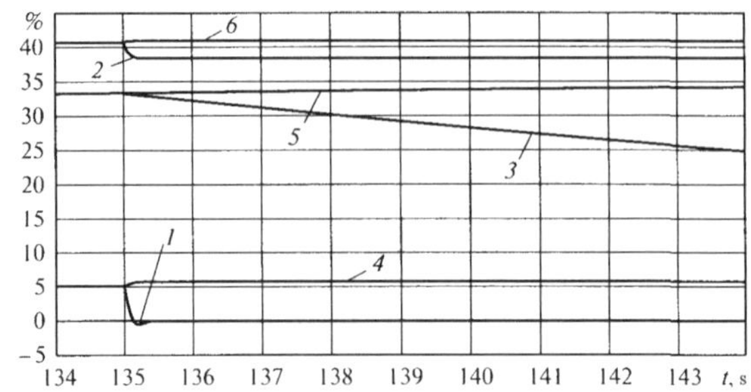

Fig. 11.46. Reaction of the ET to a step change in the adhesion coefficients under the first right-hand wheel (with φ = 0.7 to φ = 0.05) at the chassis' ascending motion: 1—Mechanical sliding in the first right-hand TEM; 2—Frequency converter's transmission ratio in the first right-hand TEM; 3—Magnetic flow on the first right-hand TEM; 4—Mechanical sliding in the first left-hand TEM; 5—Frequency converter's transmission ratio in the first left-hand TEM; 6—Magnetic flow on the first left-hand TEM

transmission ratio increases sharply in the early acceleration and gradually decreases with decreasing of the force of the inertia and increase of the travel speed (curve *5*) to the specified value.

The rotation rate of the DGU output shaft (curve *3*) has a distinctive "drawdown" in the beginning of the acceleration process because of the heavy loads; however, an automatic increase in the fuel supply (curve *4*) ensures the output of the DGU to the specified mode.

Fig. 11.45 presents the ET's reaction to a step change in the adhesion factor under the first right-hand wheeled mover from the maximum value $\varphi = 0.7$ to $\varphi = 0.05$ (simulation of the sliding surface, for example, ice) in the chassis' ascending motion. The presented data shows that the torque in the right-hand TEM drastically and significantly decreases (curve *3*). Simultaneously, the torque is redistributed to other wheeled movers to maintain the specified travel speed.

Fig. 11.46 (curve *4*) shows that the torque on the left front wheeled mover increases by some value. The torques on the other wheeled movers increased by the same value.

For traditional transmission types(for example, a mechanical transmission with symmetric differentials) in such conditions we should expect a "spin-up" of the wheeled mover, which occurs on support surfaces with low adhesive properties and a decrease in the torque on the rest of the wheeled movers; however, it does not happen in ETs. The angular velocity of the right and left-hand wheeled movers (curves 1 and 2) remain the same.

The described reaction of the ET on the changed conditions was caused by the joint work of the transmission ratio and magnetic flows' control circuits. Fig. 11.46 shows that the sliding of the right-hand TEM, whose wheel is on the sliding surface, drastically changed (curve *1*). This caused a decrease in the transmission ratio (curve *2*) and magnetic flow (curve *3*) and finally led to a significant reduction in the torque supplied to the right-hand wheel, which then operated in unfavorable conditions. At the same time, a slight increase in the sliding on the left-hand TEM (curve *4*) caused a corresponding increase of its magnetic flow (curve *6*) and transmission ratio (curve *5*).

The software developed on the basis of the mathematical model, which adequately reflects the qualitative aspect of the processes occurring in the real chassis with an ET in various motion conditions; the results of the studies conducted with its use; and the results obtained in the course of developing the methods for the complex evaluation of the transmission schemes' perfection, served as the basis for developing the science-based requirements for the design, principles, and algorithms of the electric transmission's functions.

11.6.2 Design of prospective automobile traction electric motors

On the basis of this and many other studies in the area of electrotechnics, the Scientific and Production Center of Special Machine-Building of the N.E. Bauman Moscow State Technical University (MSTU) together with the Joint-Stock Company (JSC) "Aviaelektromash" developed, manufactured, and tested in site conditions the experimental model of the 12×12 547V chassis and an automobile traction electric motor whose design took into account achievements from previous studies on the issue.

The result of the collaboration was the AD60 asynchronous electric motor and the RD60 step-down reduction gearbox with a 60 kW nominal power to the asynchronous electric motor's output shaft. The design process managed to combine them structurally into a single package. Consequently, from a constructive point of view, the asynchronous electric motor and reduction gearbox can be considered as a single traction electric mechanism for motors-in-wheel.

In addition, when assembling the electric mechanism, its fitness condition in the wheel reduction gearbox of the 547V chassis demanded installing standard automobile components (axle swivel and brake caliper). Therefore, in its final form, the traction electric mechanism of the motors-in-wheel represented a unit, including the traction electric mechanism and automobile components. The layout of the traction electric mechanism without a brake caliper is presented in Fig. 11.47.

According to the results of the preliminary theoretical studies, the traction electric motor type was clearly defined as a three-phase asynchronous electric motor with a wide frequency range of the three-phase supply voltage (from 40 to 1200 Hz); therefore, selecting the asynchronous electric motor's performance version was defined by the conditions of its use by electromechanical characteristics, control system, cooling method, and the installation specifics at the site.

The AD60 design was based on preset external and internal diameters, capability, torque, maximum starting moment, and maximum efficiency factor. Additional parameters were a

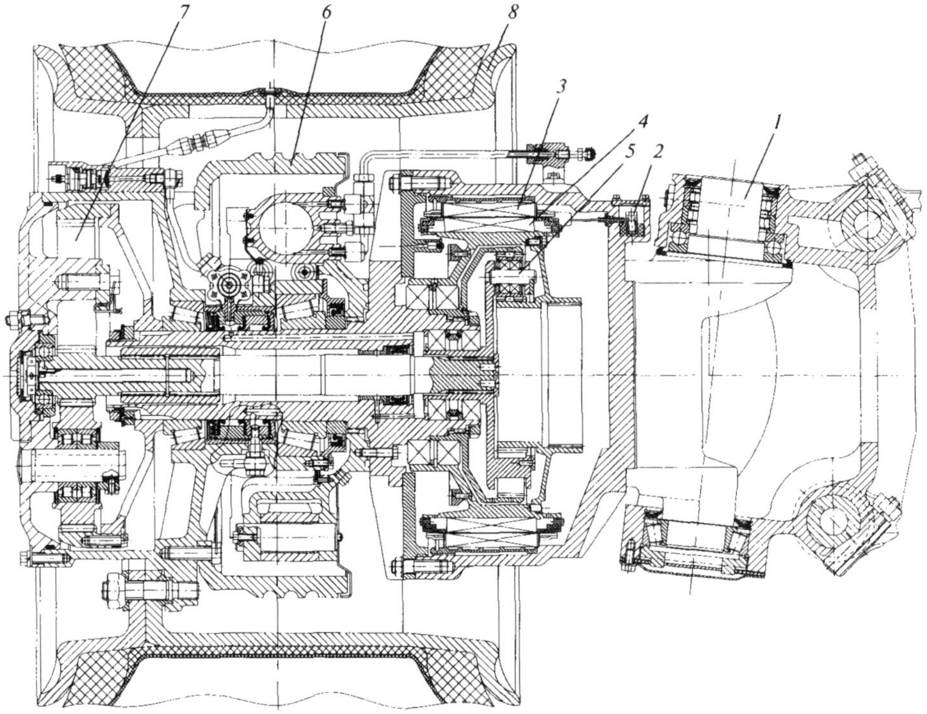

Fig. 11.47. Assembled layout of the traction electric mechanism with a wheel reduction gearbox and axle swivel: 1—Axle swivel; 2—Joint box; 3—Stator; 4—Rotor; 5—Planetary gearbox; 6—Brake drum; 7—Wheel gearbox; 8— Wheel tread

range of the possible supply voltage, the power unit's (diesel-generator) operating frequency, and the performance of the cooling system. The power characteristics of the thermal engine and electric power source were not preset, so it was assumed that the electric power source has infinite power.

Following the results of the conducted calculations on the parameters, an asynchronous electric motor with 15 pole pairs was developed; at feeding with a 40 Hz three-phase AC voltage, it allowed for obtaining a synchronous speed of the magnetic field rotation of 167 rpm, and, at a 1200 Hz three-phase AC voltage, a synchronous speed of the magnetic field rotation of 5000 rpm.

The calculated parameters of the AD60 are shown in the classical substitution diagram presented in Fig. 11.48. The values of the active and inductive resistances are given in absolute

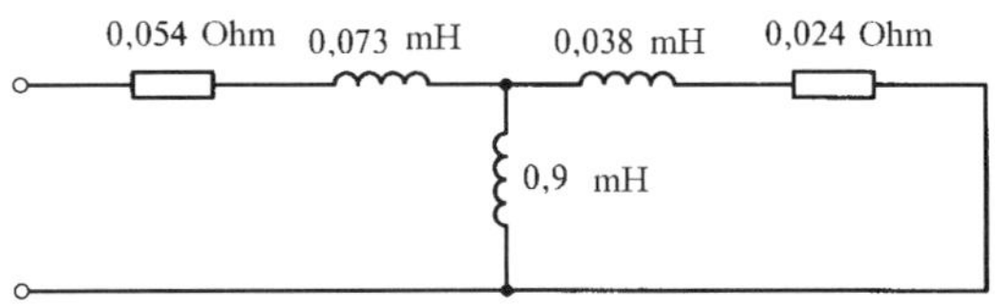

Fig. 11.48. Classic substitution diagram of the traction electric motor

units. Structurally, the AD60 has a conventional asynchronous design of the electric motor.

The design principles for high-performance engines allow for asserting that as the weight reduction and efficiency increase, the engine's active part should have a doughnut form: the stator and rotor are placed on the engine's circumference, while the central part not involved in forming the electromagnetic torque bears no active materials.

The external diameter of the stator core is 420 mm. Grooves, in which copper conductors of the stator with rectangular cross sections are stacked, are semi-closed and have a 3.85×8.8 mm rectangular shape. Fig. 11.49 shows the stator magnetic core with the winding compacted in the electric mechanism case.

Fig. 11.49. A stator magnetic core with the winding compacted in the electric mechanism case

The AD60 rotor is designed in accordance with the squirrel cage circuit and made of rectangular copper rods for improving the starting characteristics. The dimensions of the groove in the rotor core is 2.65×5.45 mm.

The external diameter of the AD60 rotor is equal to 383 mm. It relies on a fixed support with the help of bearings located at one side of the rotor. Since the rotor has the doughnut form, the central part inside it is occupied by a reduction gearbox. The rotor has one support, which is located on one side; therefore, to exclude its beats, two bearings are mounted on the support next to each other. The working gap between the rotor and stator is 0.5 mm.

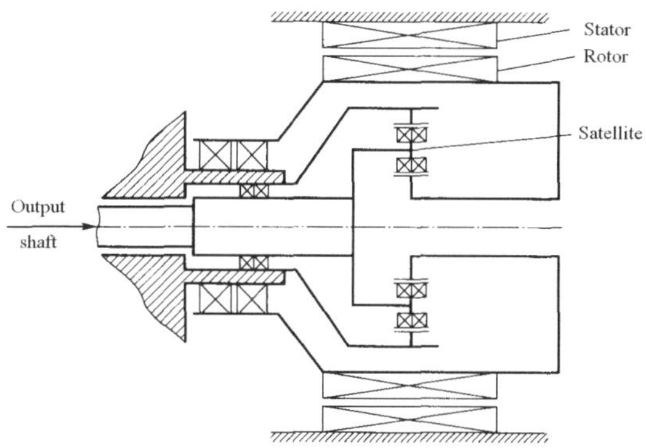

Fig. 11.50. The kinematical diagram of the "AD60 rotor-RD60 gearbox-output shaft"

The kinematical "AD60 rotor- RD60 gearbox-output shaft" layout is presented in Fig. 11.50. The RD60 reduction gearbox is planetary type with eight satellites. The gearbox's gear ratio is 2.628.

The AD60 rotor and RD60 reduction gearbox are cooled by IPM-10 grade oil. The oil's flow rate through the electric motor is defined by its heat generation and is about 2 l/min per 1 kW of the losses. The oil pressure at the inlet to the electric motor should be at least 3–4 atm.

The cooling oil enters the electric motor's body through a sealing union, passes through the spiral channels above the stator magnetic core pack, and then is supplied—through the metering orifices—into the cavity of the rotor hub. Under the influence of centrifugal forces, the IMP-10 oil washes the hub under the rotor pack and the toothed wheels of the reduction gearbox and is then drained by the rotor faces to the frontal parts of the stator winding. From the drain cavity, the oil is pumped out and through the filter into the heat exchanger from which it is fed back into the electric motor by the forcing pump. The pressure in the electric motor cavity does not exceed—due to a discharge pump with excessive capacity—the ambient pressure, and, consequently, no enhanced air tightness is required.

To control the oil temperature, the electric mechanism is equipped with a temperature detector. If the oil temperature exceeds 150°C, the detector sends a signal to the external circuit instructing it to turn off the power supply.

The rotation rate of the AD60 rotor is measured by the DSSP-5 rpm sensor. The sensor responds to changes in the magnetic field created by the sensor itself. The magnetic field changes due to the influence of a special toothed wheel's rotation connected to the AD60 rotor. The alternating "tooth-dent" changes the magnetic flow in the sensor and induces the EMS (electromotive force) in its output winding with a frequency proportional to the AD60 rotor's rpm.

Structurally, the motor-wheel consists of a wheel reduction gearbox from a 547 chassis and the TEM described above. The wheel reduction gearbox has a modified driving shaft and a hub as compared to the standard 547V reduction gearbox. As for the rest, the design of the reduction gearboxes is the same.

The control assembly process of the motors-in-wheel is presented in Fig. 11.51.

Fig. 11.51. Assembly process of the motor-in-wheel at the site

Chapter 12
Oscillations in a multi-axle wheeled chassis with resilient tires

12.1 Characteristics of vertical- and longitudinal-angular oscillations

Oscillations that arise in HDWVs, which have huge inertial masses and rather high travel speeds, are a defining factor to be accounted for in calculating and designing the undercarriages and secondary springing. That is why the regularities of the oscillatory processes in this group of machines require a separate detailed consideration.

Different variants of simplified mathematic models of a multi-axle wheeled vehicle's spatial movements, adapted to address various general and specific problems, were considered earlier in Chapter 8.

When considering the oscillatory movements of a multi-axle chassis' body, the processes under study are usually simplified by splitting them into weakly interconnected phenomena.

Traditionally, an automobile is designed to endure its stable longitudinal oscillations and to minimize the longitudinal and especially lateral linear oscillations. By using a number of special measures, it is possible to suppress the chassis' spontaneous turns in the horizontal plane (the so-called *"wandering"* along the course).

If these design measures are successful, the sprung (cushioned) body of the chassis will make—in its motion—only the linear vertical and lateral, longitudinal-angular, and lateral-angular oscillations. It is believed that the sprung body of a multi-axle chassis can be considered as a dynamic system whose motion is described by the following generalized coordinates (Figs. 12.1 and 12.2):

- The vertical displacements of the vehicle's body on the resilient-damping suspension elements (generalized coordinate z);
- The longitudinal-angular displacements or the longitudinal-angular oscillations, i.e., the turns of the body relative to the center of mass or its lateral axis y (generalized coordinate θ);
- The lateral-angular displacements or lateral-angular oscillations, i.e., the turns of the body relative to the longitudinal axis x (generalized coordinate γ).

In the general mathematical model of the body's motion, this schematization means using—in the implicit form— the assumption that the body's angular displacements are insignificant, which allows us to linearize the system of its general kinematical equations.

Indeed, let us suppose that:

$$\begin{cases} \varphi = \Phi + \Delta\varphi \\ \eta = E + \Delta\eta \\ \gamma = \Gamma + \Delta\gamma \end{cases} \tag{12.1}$$

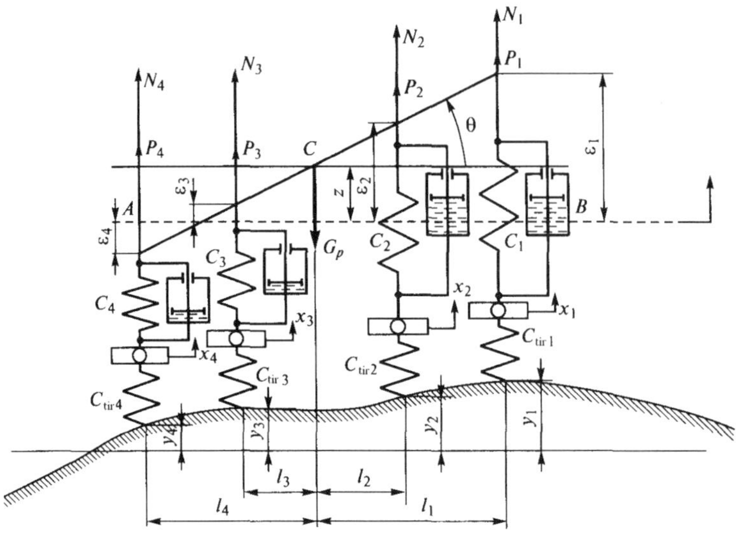

Fig. 12.1. Springing diagram of the multi-axle chassis' body (longitudinal plane)

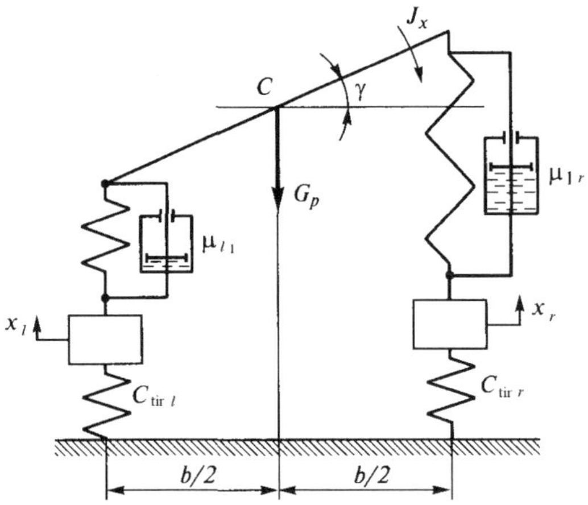

Fig. 12.2. Springing diagram of the multi-axle chassis' body (lateral plane)

where $(\varphi, \eta, \gamma)^T$ are the Krylov angles describing the rotational motions of the SWC body relative to its center of mass and that:

$$\begin{cases} \Delta\varphi = 0[\varepsilon] \\ \Delta\eta = 0[\varepsilon] \\ \Delta\gamma = 0[\varepsilon] \end{cases} \tag{12.2}$$

where $0[\varepsilon]$ is the value of the first order of smallness.

Also, let us introduce the notion of the "evolutional rotational motion of the body" at the angular rate of $(P, Q, R)^T$, $\Omega_b = (P, Q, R)^T$, set by the projections onto the axes of the bound root-mean-square (RMS) coordinate system.

In this case, the system of kinematical equations describing the motion of the bound RMS coordinate system relative to the normal earth-based coordinate system (CK0—zero coordinate system) can be written down in the following form:

$$\frac{d}{dt}\begin{pmatrix}\Delta\varphi\\ \Delta\eta\\ \Delta\gamma\end{pmatrix}=\begin{pmatrix}0 & \dfrac{\sin(\Gamma)}{\cos(E)} & \dfrac{\cos(\Gamma)}{\cos(E)}\\ 0 & \cos(\Gamma) & -\sin(\Gamma)\\ 1 & \tan(\Gamma)\sin(E) & \tan(\Gamma)\cos(E)\end{pmatrix}\begin{pmatrix}\Delta p\\ \Delta q\\ \Delta r\end{pmatrix}+$$

$$\left[+\begin{pmatrix}0 & \dfrac{\sin(\Gamma)\sin(E)}{cos^2(E)} & -\dfrac{\cos(\Gamma)\sin(\Gamma)}{cos^2(\Gamma)}\\ 0 & 0 & 0\\ 1 & \tan(\Gamma)\cos(\Gamma) & -\tan(\Gamma)\sin(\Gamma)\end{pmatrix}\begin{pmatrix}P\\ Q\\ R\end{pmatrix}\Delta\eta\right]+0\left[\varepsilon^2\right] \tag{12.3}$$

If we assume, as usual, that the oscillation parameters are first order values of smallness and that the evolutional motion is independent from the oscillations with an accuracy of up to the second order values of smallness, then we obtain from the above formula:

$$\frac{d}{dt}\begin{pmatrix}\Phi\\ E\\ \Gamma\end{pmatrix}=\begin{pmatrix}0 & \dfrac{\sin(\Gamma)}{\cos(E)} & \dfrac{\cos(\Gamma)}{\cos(E)}\\ 0 & \cos(\Gamma) & -\sin\Gamma\\ 1 & \tan(\Gamma)\sin(E) & \tan(\Gamma)\cos(E)\end{pmatrix}\begin{pmatrix}P\\ Q\\ R\end{pmatrix} \tag{12.4}$$

and:

$$\frac{d}{dt}\begin{pmatrix}\Delta\varphi\\ \Delta\eta\\ \Delta\gamma\end{pmatrix}=$$

$$\begin{pmatrix}0 & \dfrac{\sin(\Gamma)}{\cos(E)} & \dfrac{\cos(\Gamma)}{\cos(E)}\\ 0 & \cos(\Gamma) & -\sin\Gamma\\ 1 & \tan(\Gamma)\sin(E) & \tan(\Gamma)\cos(E)\end{pmatrix}\begin{pmatrix}\Delta p\\ \Delta q\\ \Delta r\end{pmatrix}+$$

$$+\left[\begin{pmatrix}0 & \dfrac{\cos(\Gamma)}{\cos(E)} & \dfrac{\cos(\Gamma)}{\cos(E)}\\ 0 & -\sin(\Gamma) & -\cos(\Gamma)\\ 1 & \dfrac{\sin(E)}{cos^2(\Gamma)} & \dfrac{\cos(E)}{cos^2(\Gamma)}\end{pmatrix}\begin{pmatrix}P\\ Q\\ R\end{pmatrix}\Delta\gamma+\right.$$

$$\left.+\begin{pmatrix}0 & \dfrac{\sin(\Gamma)\sin(E)}{cos^2(E)} & -\dfrac{\cos(\Gamma)\sin(E)}{cos^2(E)}\\ 0 & 0 & 0\\ 1 & \tan(\Gamma)\cos(E) & -\tan(\Gamma)\sin(E)\end{pmatrix}\begin{pmatrix}P\\ Q\\ R\end{pmatrix}\Delta\eta+0\left[\varepsilon^2\right].\right] \tag{12.5}$$

The formula (12.4) presents the kinematical equations for the evolutionary motion of the chassis' body. On the other hand, the formula (12.5) shows that in general, even linearization of the kinematical equations gives no grounds for a doubtless use of the hypothesis about the independence of the longitudinal-angular and lateral-angular oscillations from each other, as it is adopted in the vehicle theory and offered for use in most guidelines on assessing the vehicle's soft riding and handling quality.

If we exclude from consideration the oscillations at the chassis' curvilinear motion ($R = 0$), the case of the chassis' motion over longitudinal irregularities comparable with its wheelbase or "macro irregularities" ($Q = 0$), and the cases of lateral tipover (tripping) ($P = 0$), then:

$$\frac{d}{dt}\begin{pmatrix}\Delta\varphi\\ \Delta\eta\\ \Delta\gamma\end{pmatrix}=\begin{pmatrix}0 & \dfrac{\sin(\Gamma)}{\cos(E)} & \dfrac{\cos(\Gamma)}{\cos(E)}\\ 0 & \cos(\Gamma) & -\sin\Gamma\\ 1 & \tan(\Gamma)\sin(E) & \tan(\Gamma)\cos(E)\end{pmatrix}\begin{pmatrix}\Delta p\\ \Delta q\\ \Delta r\end{pmatrix}+0\left[\varepsilon^2\right]. \tag{12.6}$$

Finally, if we consider only the chassis' plane motion, which has no static roll and pitch, then:

$$\frac{d}{dt}\begin{pmatrix}\Delta\varphi \\ \Delta\eta \\ \Delta\gamma\end{pmatrix} = \begin{pmatrix}0 & 0 & 1 \\ 0 & 1 & 0 \\ 1 & 0 & 0\end{pmatrix}\begin{pmatrix}\Delta p \\ \Delta q \\ \Delta r\end{pmatrix} + 0\left[\varepsilon^2\right]. \quad (12.7)$$

This means that at fulfillment of the above restrictions, the angular rate projections of the body's perturbed motion onto the axes of the bound coordinate system are equal to the time derivatives of the perturbed part of the respective turning angles of the coordinate axes, which, in their turn, can be considered independent from each other. At these conditions, the calculation diagrams presented in Figs. 12.1 and 12.2 can be used to describe the oscillatory motion of the vehicle's body, which is usually done.

To define the position in the space of non-sprung masses reduced to the multi-support vehicle's wheels, we need to introduce 2*n* of generalized coordinates (*n* is the number of wheels of one board with an in-line linear support-running complex). Then, the multi-axle chassis is simulated as a dynamic system with $2n + 3$ degrees of freedom.

A number of different particular cases may exist, allowing for simplifying the model further.

Thus, for example, at a completely symmetrical system, the vertical, longitudinal-angular, and lateral-angular oscillations will not depend on each other, while the system of motion equations can be split into three simpler systems.

If a sprung system is symmetrical only relative to one main axis of inertia (most often to the vehicle's longitudinal axis), then the oscillations in the longitudinal plane will not depend on the oscillations in the lateral plane. This means that the chassis can be simulated separately in the both the longitudinal (dynamic system with three degrees of freedom) and lateral (dynamic system with two degrees of freedom) planes.

When simulating the motion of ordinary vehicles, the oscillatory system simplification is often brought practically to the limit.

Let us assume that ρ is the inertia radius of the vehicle in the longitudinal plane, l_1 is the distance from the center of mass to the first axle, and l_2 is the distance from the center of mass to the second axle. It is easy to show that at fulfillment of the condition:

$$\rho^2 = l_1 l_2. \quad (12.8)$$

The oscillations of the front and rear sections of the vehicle in the longitudinal plane can be considered independent and simulated separately by means of two so called "two-mass systems," the analysis of which is the foundation for the traditional springing theory of the vehicle. This assumption is as a whole acceptable for traditional vehicles, since most passenger cars and two- and three-axle cargo vehicles with a balancing lever bogie are characterized by the following values of the sprung masses distribution coefficient:

$$\varepsilon_y = \frac{\rho^2}{l_1 l_2} = 0.8\ldots1.2 \quad (12.9)$$

Multi-axle wheeled vehicle are essentially different from traditional ones in their geometry, weight report elements, and carrying system design. As a result, the traditional two-mass model proves to be absolutely unacceptable to evaluate the soft driving of such vehicles.

The reason is in the following circumstances.

First, multi-axle wheeled vehicles (up to the 10×10 wheel formula) are prone to intensive longitudinal-angular oscillations (*"galloping"*). The driver of such a vehicle and some sensitive equipment elements are located, as a rule, on the frame overhang at the vehicle's extremities; therefore, they are subjected to additional accelerations arising from the body's longitudinal swinging.

Second, the unsprung masses of the wheels are usually about 30% of the vehicle's full weight (see Chapter 9). If the vehicle has two axles, the weight of each axle's units will be about

15% of its full weight. This means that any axle oscillation should essentially represent the character of oscillations of the vehicle as a whole, especially if the automobile has "continuous" (unsplit) driving axles, which is typical for multi-purpose vehicles of increased passing ability.

As a rule, multi-axle vehicles have independent wheel suspensions. In a four-axle vehicle, the unsprung weight of one wheeled unit will be 3.8–4.0% of the vehicle's full weight. Naturally, the influence of their oscillations on those of the whole vehicle will be much smaller. The relative unsprung weight of the support-running module of a six-axle wheeled chassis will be still smaller, around 2.5–3.0%. Thus, in many cases we can disregard the influence of the oscillations of unsprung masses (of the wheels) on the oscillations of the vehicle's sprung mass (of the body).

Third, the dimensions of SWCs are usually much bigger than the dimensions of the obstacles they have to overcome. On the other hand, multi-axle chassis are usually equipped with large-diameter tires (1600–3000 mm). The contact spot dimensions and the profile height of these tires are much bigger than the height of typical road irregularities.

Finally, special chassis, as a rule, move in groups on permanent or specially prepared routes. In these cases, they form a quite definite and non-occasional character in the road's longitudinal profile. This means that oscillations in such vehicles should be considered under the influence of other affections than those experienced by the wheels of ordinary vehicles.

By means of the calculation diagrams presented in Figs. 12.1 and 12.2, we can build differential equations of the body's rotational motion (twist) with a symmetrical inertia tensor at the location of its main central axes with the axes of the bound RMS coordinate system. In traditional vehicles this condition is usually observed, although not completely; however, in HDWVs it is not always observed. Thus, for example, studies of heavy-duty dumpers have shown that this condition is quite essentially broken.

In designing ordinary vehicles, great attention is given to vertical oscillations of characteristic points and to longitudinal-angular oscillations. The lateral-angular oscillations, in the opinion of many experts, are not as important since they are not especially dangerous for an ordinary vehicle. However, multi-axle wheeled chassis are notable for a rather high center of mass and an almost always limited width. In this context, even minor roll angles may become quite dangerous from the lateral stability viewpoint.

On the other hand, a long vehicle length decreases the probability for a wheel of one of the boards to get on an irregularity that can cause a total roll of the body close to its value of the static tipover threshold. Such road conditions, as a rule, are easily predictable. In addition, a long chassis frame has a higher twisting compliance (up to 7–8° for the MAZ-547V), which affects the SWC's lateral-angular oscillations. This characteristic of long-base multi-axle chassis is arousing interest to study the lateral-angular oscillations arising when the vehicle passes through asymmetrical short and high irregularities. These oscillations have been scantily studied so far.

Finally, SWCs are often equipped with a long cabin, which occupies the first row of the front overhang's second and third tiers. Such chassis have long and narrow frames with low twisting stiffness. Due to this fact, the accelerations perceived in the driver's workplace are defined not only by the parameters of the wheels' springing system, but also by the cabin's lateral-angular oscillations suspended to the end of a long compliant frame. This characteristic of multi-axle chassis is outside the calculation schemes presented in Figs. 12.1 and 12.2.

Thus, a two-mass model is inapplicable even for rough estimates of a multi-axle wheeled chassis' soft driving (except for the very initial and approximate calculations). This model is well-known and there is no reason to describe it.

A simple model (see Figs. 12.1 and 12.2), in spite of its limitations and drawbacks, is broadly used at the initial design stages of wheeled chassis. Consequently, it is reasonable to consider in more detail its basic variant, different simplifications, and particular cases.

At the first step of building a model, the characteristics of all elements are usually taken as linear. The motion of the components of this system is considered within the normal earth CK0 coordinate system.

In order to stick to the generally adopted notations used in the most popular literature, we shall use in this chapter the legend and coordinate system presented in Figs. 12.1 and 12.2 (although sometimes it is not quite convenient).

The displacement of the chassis' center of mass (coordinate z) is measured against the position of its static balance. The displacement of the wheels' center of mass (coordinate x) is also measured against the position of its static equilibrium. It is assumed that the center of the wheels coincides with the center of mass. The protrusions and depressions of the road are set by the coordinate y, which is measured relative to the line that represents the mathematical expectation of the road's microprofile. Let us note here that the road's microprofile is practically always viewed as a random function with a constant mathematical expectation (unfortunately, this condition is usually not directly specified).

The motion equations of the system presented in Fig. 12.1 can be obtained differently. Let us take the simplest variant—the Koenig theorem about the motion of a system's center of mass, the d'Alembert principle, and the equations of a solid body's rotational motion. The Lagrange theorem is also often used for this purpose; however, in this case it looks like an unnecessary complication.

12.1.1 Oscillations of an asymmetric vehicle

Let us consider a multi-support vehicle which has a non-symmetrical suspension with a shock absorber under the following assumptions:

- All the resilient and damping (cushioning) elements have a linear characteristic, in other words, the rigidities of all resilient elements and the resistance coefficients of all damping elements, reduced to the wheels' centers, are constant (but not equal);
- The resilient characteristics of the tires are linear, i.e., their rigidities are constant (but also not equal);
- The internal friction losses in the tires are so small that they can be neglected;
- All of the energy losses in the suspension can be accounted for by selecting a respective resistance coefficient of the damping element (of the shock absorber);
- All of the shock absorbers are double-acting with equal resistance coefficients both at the direct and reverse travel of the wheels;
- The wheels leave the ground during the oscillatory processes;
- The resistance coefficient is constant;
- The suspension is symmetrical relative to the vehicle's longitudinal axis (the axes of the wheels of the left-hand board coincide with the axes of the respective wheels of the right-hand board);
- The characteristics of the tires and suspension elements of the right-hand and left-hand boards are the same; and
- The road profile under the right-hand and left-hand wheels is the same.

We omit the intermediary results, which refer to theoretical mechanics, and present below the system of differential equations describing the motion of a multi-axle vehicle in the longitudinal plane:

$$\begin{cases} M_{sp}\ddot{z} + 2\dot{z}\sum_{i=1}^{n}\mu_i + 2z\sum_{i=1}^{n}C_{st.wi} + 2\dot{\theta}\sum_{i=1}^{n}\mu_i\ell_i + 2\theta\sum_{i=1}^{n}C_{st.wi}\ell_i = \\ = 2\sum_{i=1}^{n}(\mu_i x_i + C_{st.wi}x_i)\ ; \\ \mathrm{I_y}\ddot{\theta} + 2\dot{\theta}\sum_{i=1}^{n}\mu_i\ell_i^2 + 2\theta\sum_{i=1}^{n}C_{st.wi}\ell_i^2 + 2\dot{z}\sum_{i=1}^{n}\mu_i\ell_i + 2z\sum_{i=1}^{n}C_{st.wi}\ell_i = \\ = 2\sum_{i=1}^{n}(\mu_i\dot{x}_i + C_{st.wi}x_i)\ell_i\ ; \\ \sum_{i=1}^{n}\left[m_i\ddot{x}_i + \mu_{tiri}\dot{x}_i + (C_i + C_{tiri})x_i\right] - \dot{z}\sum_{i=1}^{n}\mu_i - z\sum_{i=1}^{n}C_{st.wi} - \dot{\theta}\sum_{i=1}^{n}\mu_i\ell_i - \\ - -\theta\sum_{i=1}^{n}C_{st.wi}\ell_i = \sum_{i=1}^{n}(C_{tiri}y_i)\ , \end{cases} \tag{12.10}$$

where: M_{sp} is the body's weight (the weight of the chassis' so-called "sprung parts"); μ_i is the resistance coefficient of the shock absorbers in the suspension of the ith axle; $C_{st.wi}$ is the stiffness of the resilient element in the suspension of the ith axle; I_y is the body's inertia moment relative to the lateral axis passing through its center of mass; m_i is the weight of the wheels on the ith axle; C_{tiri} is the tire stiffness of the wheels on the ith axle; μ_{tiri} is the tire damping coefficient of the wheels on the ith axle; z_i is the road profile's vertical coordinate in the contact point of the wheels on the ith axle with the road; θ is the body's damping angle (pitch angle); z is the vertical displacements of the body's center of mass; and x_i is the vertical displacements of the wheels' center on the ith axle.

This system of equations shows that an excitation in the multi-axle chassis' vertical oscillations is surely accompanied by an appearance of its longitudinal-angular oscillations and oscillations of the wheels.

Two approaches exist to analyze oscillations of multi-axle wheeled vehicles.

The first (earliest) one is based on applying operational calculus (usually in the variant of a one-sided Laplase transformation) that allows passing over in the equations in (12.10) to images and later uses the frequency methods for analyzing linear dynamic systems. This approach has become traditional in the vehicle motion theory.

The second approach is developing the appearance context and broad spread object simulation systems, which allow executing—rather easily and quickly—imitational simulation of dynamic systems (an example of such an environment is the package SIMULINK, expanding the popular environment MATLAB). In recent times, the second approach is more and more popular.

Nevertheless, the traditional direction in the studies of multi-axle chassis oscillations has not lost its value since it allows for rather quickly obtaining the needed assessments of the created SWC's efficiency and ensuring good visual expression, reliability, and a high level of generalized results. This direction has also accumulated huge calculation and experimental material, which allows for conducting—quickly and reliably—an analysis of the proposed technical solutions and selected vehicle parameters, which is, as shown in Chapter 5, a major advantage, which is especially clear at the early design stages when a great number of reviewed variants demand a quick and fundamental decision by neglecting insignificant particulars.

At implementing the first, traditional, approach in equations (12.10), we pass over to images by Laplace (the parameter of a one-sided Laplace transformation is usually designated by the letter *s*). For reducing the writings, let us introduce the following legend:

$$a_{11} = \sum\mu_i;\ a_{10} = \sum C_i;\ a_{21} = \sum\mu_i l_i;\ a_{20} = \sum C_i l_i;$$

$$b_{11} = \sum\mu_i l_i; b_{10} = \sum C_i l_i; b_{21} = \sum\mu_i l_i^2; b_{20} = \sum C_i l_i^2; \tag{12.11}$$

$$a_{21} = b_{11}, a_{20} = b_{10}$$

Then, the system of equations (12.10) is brought to the following form:

$$\begin{cases}(M_{sp}\,s^2+2a_{11}s+2a_{10})\tilde{z}+2(b_{11}s+b_{10})\theta=2\sum_{i=1}^{n}(\mu_i s+C_{st.wi})\tilde{x}_i\ ;\\(I_y s^2+b_{21}s+b_{20})\tilde{\theta}+2(a_{21}s+a_{20})\tilde{z}=2\sum_{i=1}^{n}(\mu_i s+C_{st.wi})\ell_i\tilde{x}_i\ ;\\\sum_{i=1}^{n}(m_i s^2+\mu_i s+C_{npi})\tilde{x}_i-(a_{11}s+a_{10})\tilde{z}-(b_{11}s+b_{10})\tilde{\theta}=\sum_{i=1}^{n}(C_{tiri}\tilde{y}_i)\ .\end{cases}\qquad(12.12)$$

In the system (12.12), a wavy line over the variables defines their images according to Laplace.

Now, it is necessary to set some rules for the signs of the parameters included into the formulas (12.12). These rules are also traditional, although not quite correct. Thus, those distances from the center of mass to the axles of the wheels, which at a positive angle θ value displace the system element from the stable equilibrium position in the positive direction are treated as positive (which is defined by the orientation of the earth CK0 coordinate system's axes). Thus, in the wheels whose axles are located in front of the vehicle's inertia center, the *li* values are positive, while in the wheels whose axles are behind the inertia center have negative *li* values.

If a sprung system is symmetrical relative to the longitudinal and lateral axes, then:

$$b_{11}=\sum\mu_i l_i=\mu\sum l_i=0\ ;$$

$$b_{10}=\sum_{i=1}^{n}C_{st.wi}\ell_i=C_{st.w}\sum_{i=1}^{n}\ell_i=0$$

$$a_{21}=\sum\mu_i l_i=\mu\sum l_i=0;$$

$$a_{20}=\sum_{i=1}^{n}C_{st.wi}\ell_i=C_{st.w}\sum_{i=1}^{n}\ell_i=0\ ;$$

$$\sum_{i=1}^{n}(\mu_i\ell_i)=\sum_{i=1}^{n}(C_{st.wi}\ell_i)=0\ .\qquad(12.13)$$

In this case, the equations in (12.12) will look as follows:

$$\begin{cases}(M_{sp}\,s^2+2a_{11}s+2a_{10})\tilde{z}=2\sum_{i=1}^{n}(\mu_i s+C_{st.wi})\tilde{x}_i\ ;\\(I_y s^2+b_{21}s+b_{20})\tilde{\theta}2\sum_{i=1}^{n}(\mu_i s+C_{st.wi})\ell_i\tilde{x}_i\ ;\\\sum_{i=1}^{n}(m_i s^2+\mu_i s+C_{pi})\tilde{x}_i-(a_{11}s+a_{10})\tilde{z}=\sum_{i=1}^{n}(C_{tiri}\tilde{y}_i)\ .\end{cases}\qquad(12.14)$$

where: C_{pi} is the reduced stiffness in the springing system of the wheels on the ith axle.

Let us note here that in a symmetrical multi-axle vehicle with elastic tires, the connection of the vertical and longitudinal-angular oscillations is preserved, although it is essentially weaker and is executed through oscillations of non-sprung masses.

Following the well-known provisions of the springing system theory, let us introduce the following legend:

$$d_{1z}(s)=M_{sp}s^2+2a_{11}s+2a_{10};$$
$$d_{1\theta}(s)=(b_{11}s+b_{10});$$
$$d_{2z}(s)=(a_{21}s+a_{20});$$
$$d_{2\theta}(s)=I_y s^2+2b_{21}s+2b_{20};$$
$$d_{3z}(s)=a_{11}s+a_{10}.\qquad(12.15)$$

In a non-symmetrical system, $d_{1\theta}(s)=d_{2z}(s)$.

Let us connect the images of displacements $x_i(t)$ with the help of the "retardation" effect in time by introducing the following parameter:

$$\tau_i=\frac{l_1-l_i}{V},$$

where v is the chassis' travel speed.

Let us apply the same method to the image of the microprofile coordinates under each of the wheels $y_i(t)$. Then, following the well-known provisions, let us introduce the following legend:

$$K_1(s) = 2\sum_{i=1}^{n}(\mu_i s + C_{\text{st.wi}})\exp(-\tau_i s)\ ;$$
$$K_2(s) = 2\sum_{i=1}^{n}(\mu_i s + C_{\text{st.wi}})\ell_i \exp(-\tau_i s)\ ;$$
$$K_{tir}(s) = \sum_{i=1}^{n}(C_{\text{tiri}})\exp(-\tau_i s)\ ;$$
$$K_n(s) = \sum_{i=1}^{n}(m_i s^2 + \mu_i s + C_{pi})\exp(-\tau_i s)\ . \quad (12.16)$$

With account of (12.15) and (12.16), the equations in (12.12) will look as follows:

$$\begin{cases} d_{1z}(s)\tilde{z} + 2d_{10}(s)\tilde{\theta} = K_1(s)\tilde{x}\ ; \\ d_{2\theta}(s)\tilde{\theta} + 2d_{2z}(s)\tilde{z} = K_2(s)\tilde{x}\ ; \\ K_n(s)\tilde{x} - d_{3z}(s)\tilde{z} - d_{1\theta}(s)\tilde{\theta} = K_{tir}(s)\tilde{y}\ . \end{cases} \quad (12.17)$$

By dividing the left- and right-hand parts of each of the system equations in (12.17) by the image of the microprofile coordinates under the front wheel of the chassis, $y(s) = y_1(s)$, we obtain the following:

$$\begin{cases} d_{1z}(s)W_z(s) + 2d_{1\theta}(s)W_\theta(s) - K_1(s)W_x(s) = 0\ ; \\ 2d_{2z}(s)W_z(s) + d_{2\theta}(s)W_\theta(s) - K_2(s)W_x(s) = 0\ ; \\ d_{3z}(s)W_z(s) + d_{1\theta}(s)W_\theta(s) - K_n(s)W_x(s) = -K_{tir}(s)\ , \end{cases} \quad (12.18)$$

where, in accordance with the work of A.A. Silaev [64]:

$W_z(s) = z(s)/y(s)$ is the transfer function from the road to the vertical displacement of the chassis' center of mass; $W_\theta(s) = \theta(s)/y(s)$ is the transfer function from the road to the longitudinal-angular displacement of the chassis' body ("pitch"); and $W_x(s) = x(s)/y(s)$ is the transfer function from the road to the linear vertical displacement of the center of the first front wheel.

The expressions (12.18) can be considered as a system of linear equations relative to the transfer functions. Its determinant is defined by the formula (with account of the fact that $d_{1\theta}(s) = d_{2z}(s)$:

$$\Delta_w = \begin{vmatrix} d_{1z} & 2d_{1\Theta}(s) & -K_1(s) \\ 2d_{2z} & d_{2\Theta}(s) & -K_2(s) \\ d_{3z} & d_{1\Theta}(s) & -K_n(s) \end{vmatrix} =$$

$$= -K_1(s)\left[2d_{2z}^2(s) - d_{3z}(s)d_{2\Theta}(s)\right] + K_2(s)d_{2z}$$

$$(s)d_{2\Theta}(s)\ -\ 2d_{3z}(s) - K_n(s)\left[d_{1z}(s)d_{2\Theta}(s) - 4d_{2z}^2(s)\right]. \quad (12.19)$$

Respectively, the transfer functions are defined by the following expressions:

$$W_z(s) = \frac{1}{\Delta_W}\begin{vmatrix} 0 & 2d_{1\theta}(s) & -K_1(s) \\ 0 & d_{2\theta}(s) & -K_2(s) \\ -K_{tir}(s) & d_{1\theta}(s) & -K_n(s) \end{vmatrix} = \frac{K_{tir}(s)[2d_{2z}(s)\mathrm{K}_2(s) - d_{2\theta}(s)\mathrm{K}_1(s)]}{\Delta_W} \quad (12.20)$$

$$W_\theta(s) = \frac{1}{\Delta_W}\begin{vmatrix} d_{1z}(s) & 0 & -K_1(s) \\ 2d_{2z}(s) & 0 & -K_2(s) \\ d_{3z}(s) & -K_{tir}(s) & -K_n(s) \end{vmatrix} = \frac{K_{tir}(s)[2d_{2z}(s)\mathrm{K}_1(s) - d_{1z}(s)\mathrm{K}_2(s)]}{\Delta_W} \quad (12.21)$$

$$W_x(s) = \frac{1}{\Delta_W}\begin{vmatrix} d_{1z}(s) & 2d_{1\theta}(s) & 0 \\ 2d_{2z}(s) & d_{2\theta}(s) & 0 \\ d_{3z}(s) & d_{1\theta}(s) & -K_{tir}(s) \end{vmatrix} = \frac{K_{tir}(s)[4d_{2z}^2(s)\mathrm{K}_1(s) - d_{1z}(s)d_{2\theta}(s)]}{\Delta_W} \quad (12.22)$$

By setting now in formulas (12.20)–(12.22) $s = j\omega$, we obtain the system's amplitude-phase-frequency characteristics (APFC) (harmonic locus), where j is an imaginary unit.

To define the linear accelerations, which appear in the body's center of mass, the linear accelerations of the wheels, and the angular accelerations of the body, we can use the transfer functions and their respective APFCs obtained by multiplying formulas (12.20)–(12.22) by s^2.

Apart from the vertical (transfer) displacements, each component of the vehicle's sprung system executes, together with the center of mass, a relative motion at longitudinal-angular and lateral-angular oscillations. This phenomenon should be taken into account when assessing accelerations in the points that do not coincide with the vehicle's center of mass.

12.1.2 Oscillations of a symmetrical vehicle

A completely symmetrical sprung system has $d_{1\theta}(s) = d_{2z}(s) = 0$.

Then, from (12.19) we have:

$$\Delta_W = K_1(s)[d_{3z}(s)d_{2\theta}(s)] - K_n(s)[d_{1z}(s)d_{2\theta}(s)] \quad (12.23)$$

It follows from formulas (12.21) and (12.22) that:

$$W_z(s) = \frac{K_{tir}(s)\, K_1(s)}{K_n(s)d_{1z}(s) - K_1(s)d_{3z}(s)} \quad (12.24)$$

$$W_\theta(s) = \frac{d_{1z}(s)}{d_{2\theta}(s)}\ \frac{K_{tir}(s)\, K_2(s)}{K_n(s)d_{1z}(s) - K_1(s)d_{3z}(s)} \quad (12.25)$$

Consequently, the APFCs of the vertical and longitudinal-angular oscillations of a symmetrical vehicle are:

$$W_z(j\omega) = \frac{K_{tir}(j\omega)\, K_1(j\omega)}{K_n(j\omega)d_{1z}(j\omega) - K_1(j\omega)d_{3z}(j\omega)} \quad (12.26)$$

$$W_\theta(j\omega) = \frac{d_{1z}(j\omega)}{d_{2\theta}(j\omega)}\ \frac{K_{tir}(j\omega)\, K_2(j\omega)}{K_n(j\omega)d_{1z}(j\omega) - K_1(j\omega)d_{3z}(j\omega)} \quad (12.27)$$

The parameters of formulas (12.26) and (12.27) are defined with the help of the following expressions:

$$d_{1z}(j\omega) = (2a_{10} - M_{sp}\omega^2) + 2a_{11}\omega j;$$

$$d_{2\theta}(j\omega) = (2b_{20} - I_y\omega^2) + 2b_{21}\omega j;$$

$$d_{3z}(j\omega) = a_{10} + a_{11}\omega j.$$

In this case, by virtue of the system's complete symmetrical nature and the complete identical nature of all the wheels and suspensions:

$$a_{11} = \sum \mu_i = n\mu;$$

$$a_{10} = \sum C_i = nC;$$

$$b_{21} = \sum \mu_i l_i^2 = \mu \sum l_i^2;$$

$$b_{20} = \sum_{i=1}^{n} C_{st.wi}\ell_i^2 = C_{st.w}\sum_{i=1}^{n} \ell_i^2;$$

$$K_1(j\omega) = 2\sum_{i=1}^{n}(j\mu_i\omega + C_{st.wi})\exp(-j\tau_i\omega) =$$

$$= 2(j\mu\omega + C_{st.wi})\sum_{i=1}^{n}\exp(-j\tau_i\omega);$$

$$K_2(j\omega) = 2\sum_{i=1}^{n}(j\mu_i\omega + C_{st.wi})\ell_i\exp(-j\tau_i\omega) =$$

$$= 2(j\mu\omega + C_{st.wi})\sum_{i=1}^{n}\ell_i\exp(-j\tau_i\omega);$$

$$K_{tir}(j\omega) = \sum_{i=1}^{n}(C_{tiri})\exp(-j\tau_i\omega) = C_{tir}\sum_{i=1}^{n}\exp(-j\tau_i\omega);$$

$$K_n(j\omega) = \sum_{i=1}^{n}(-m_i\omega^2 + j\mu_i\omega + C_{pi})\exp(-j\tau_i\omega) =$$

$$= [(C_p - m\omega^2) + j(\mu\omega)]\sum_{i=1}^{n}\exp(-j\tau_i\omega),$$

where τ_1 is the time of the action delay.

Thus:

$$d_{1z}(j\omega) = (2nc - M_{sp}\omega^2) + (2n\mu\omega)j;$$

$$d_{2\Theta}(j\omega) = \left(2c\sum l_i^2 - I_y\omega^2\right) + \left(2\mu\omega\sum l_i^2\right)j;$$

$$d_{3z}(j\omega) = n[c + (\mu\omega)j].$$

With account of the above transformations, formulas (12.26) and (12.27) can be presented as follows:

$$W_z(j\omega)=\frac{2C_{tir}(j\mu\omega+c)\left[\sum_{i=1}^{n}\exp(-j\tau_i\omega)\right]}{[(c-m\omega^2)+j(\mu\omega)]\,d_{1z}(j\omega)-2(j\mu\omega+c)\,d_{3z}(j\omega)} \quad (12.28)$$

$$W_\theta(j\omega)=\frac{d_{1z}(j\omega)}{d_{2\theta}(j\omega)}\,\frac{2C_{tir}(j\mu\omega+c)\left[\sum_{i=1}^{n}\ell_i\exp(-j\tau_i\omega)\right]}{[(c-m\omega^2)+j(\mu\omega)]\,d_{1z}(j\omega)-2(j\mu\omega+c)\,d_{3z}(j\omega)} \quad (12.29)$$

It is seen from expressions (12.28) and (12.29) that the influence of the axle multiplicity factor on the APFCs is accounted for by the two following complex coefficients:

$$\overline{C_z}(j\omega)=\sum\exp(-j\tau_i\omega);\overline{C_\Theta}(j\omega)=\Sigma l_i\exp(-j\tau_i\omega).$$

Occasionally, these coefficients are treated as the accounting coefficients of the non-simultaneous affection character on the wheels at the vertical and longitudinal-angular oscillations of the chassis' body, or, more briefly, the coefficients of the non-simultaneous character on the vertical and longitudinal-angular oscillations. They take into account the obvious fact that the same single road irregularity is affecting the wheels successively as they reach this irregularity, i.e., not at the same time.

It is more convenient to present the non-simultaneous character coefficients in the following indicative form:

$$\overline{C_z}(j\omega)=C_z(\omega)\exp[-j\Phi_z\omega],$$

$$C_z(\omega)=\sqrt{\left[\sum\cos(\tau_i\omega)\right]^2+\left[\sum\sin(\tau_i\omega)\right]^2}, \quad (12.30)$$

$$\Phi_z(\omega)=arctg\left(\sum\sin(\tau_i\omega)\Big/\sum\cos(\tau_i\omega)\right);$$

$$\overline{C_\Theta}(j\omega)=C_\Theta(\omega)exp\left[-j\Phi_\Theta(\omega)\right],$$

$$C_\Theta(\omega)=\sqrt{\left[\sum l_i\cos(\tau_i\omega)\right]^2+\left[\sum l_i\sin(\tau_i\omega)\right]^2},$$

$$\Phi_\theta(\omega)=arctg\left(\sum_{i=1}^{n}\ell_i\sin(\tau_i\omega)\Big/\sum_{i=1}^{n}\ell_i\cos(\tau_i\omega)\right). \quad (12.31)$$

In calculations they usually use the amplitude-frequency characteristics, which are modules of the complex functions (12.28) and (12.29) instead of the complex functions themselves. This means that the modules of the complex non-simultaneous character coefficients (12.30) and (12.31) are of special interest.

At $\omega = 0$, the values of the non-simultaneous character coefficients for the vertical oscillations are different from the unit, which is not convenient when making comparative evaluations. The formula (12.31) should be "normalized" by dividing it by the coefficient value in point $\omega = 0$

$$C_{Hz}(\omega)=\frac{\sqrt{\left[\sum_{i=1}^{n}\cos(\tau_i\omega)\right]^2+\left[\sum_{i=1}^{n}\sin(\tau_i\omega)\right]^2}}{n} \quad (12.32)$$

It is obvious that the value of the second coefficient in point $\omega = 0$ for a completely symmetrical system is zero. In order to bring this coefficient to a dimensionless form, it is necessary to refer its value to the value of some characteristic dimension of the vehicle, for example, the wheelbase L of the wheeled chassis:

$$C_{H\theta}(\omega)=\frac{\sqrt{\left[\sum_{i=1}^{n}\ell_i\cos(\tau_i\omega)\right]^2+\left[\sum_{i=1}^{n}\ell_i\sin(\tau_i\omega)\right]^2}}{L} \quad (12.33)$$

The dependence of the normalized non-simultaneous affection character coefficients from the frequency for various chassis travel speeds (10, 20, 30, and 40 km/h) is presented in Figs. 12.3 and 12.4. We have used the layout of a four-axle SWC as an example, which is close by the dimensions and axle layout to the 7908 chassis, but which was deliberately brought to the symmetrical variant (l_1 = 4.2 m; l_2 = 1.4 m; l_1 = –1.4 m; l_4 = –4.2 m; and L = 8.4 m).

With an increase in the travel speed, the time τ_i of the affection delay goes down, which is why the growth rate of the $\tau_i\,\omega$ in formulas (12.32) and (12.33) decreases and all of the dependences somewhat stretch to the right.

At $\omega = 0$, the normalized non-simultaneous affection character coefficient for the vertical displacements of a completely symmetrical system is equal to one (unit) and does not depend on the chassis' travel speed. The normalized non-simultaneous affection character coefficient for the longitudinal-angular oscillations of a completely symmetrical system at $\omega = 0$ also does not depend on the travel speed and is equal to zero. Its maximum values also do not depend on the travel speed, although they are reached at different frequencies (the higher the travel speed, the higher the frequency corresponding to the extreme value).

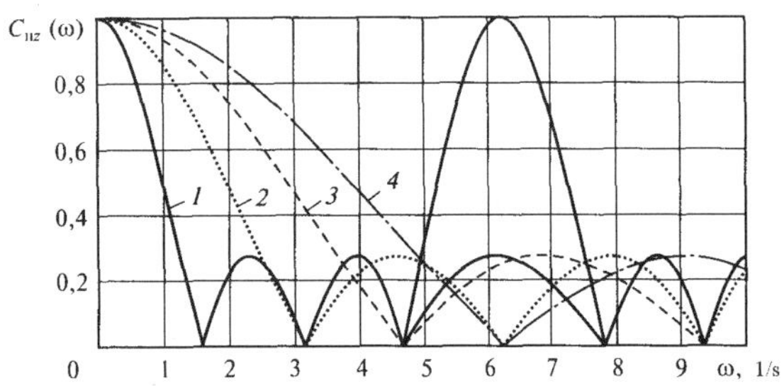

Fig. 12.3. Normalized non-simultaneous affection character coefficient for the vertical oscillations versus the affection frequency and versus the travel speed of a completely symmetrical MAZ-7908 chassis at a travel speed of: 1—10 km/h; 2—20 km/h; 3—30 km/h; 4—40 km/h

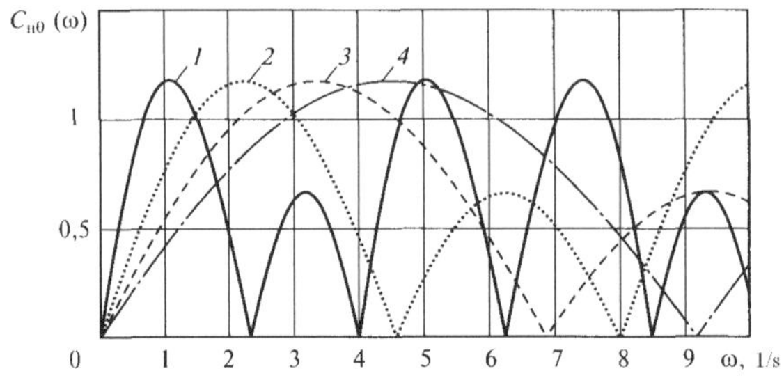

Fig. 12.4. Normalized non-simultaneous affection character coefficient for the longitudinal-angular oscillations versus the affection frequency and versus the travel speed of a completely symmetrical MAZ-7908 chassis at a travel speed of: 1—10 km/h; 2—20 km/h; 3—30 km/h; 4—40 km/h

12.1.3 Non-symmetrical springing system

In most cases, multi-axle wheeled chassis have non-symmetrical sprung systems. Most often the suspension and tire characteristics are the same for all wheels. In this case, formulas from (12.11) look as follows:

$$a_{11} = n\mu;$$

$$a_{10} = nC_i;$$

$$a_{21} = \mu \sum_i \ell_i\,;$$

$$a_{20} = C_i \sum_i \ell_i\,;$$

$$b_{11} = \mu \sum_i \ell_i\,;$$

$$b_{10} = C_i \sum_i \ell_i\,;$$

$$b_{21} = \mu \sum_i \ell_i^2\,;$$

$$b_{20} = C_i \sum_i \ell_i^2\,;$$

$$a_{21} = b_{11};$$

$$a_{20} = b_{10} \qquad (12.34)$$

If we introduce the following notation: $d_0(s) = C_i + \mu s$, $v = m/M_{sp}$ and define the body's inertia radius as $R_{uy}^2 = I_y/M_{sp}$, then the transfer functions acquire the following form:

$$\Delta_W = -\left\{M_{sp}^3[vR_{uy}^2]s^6 + M_{sp}^2\left[2v(\sum_i \ell_i^2)+(1+2nv)R_{uy}^2\right]d_0 s^4 + \right.$$

$$+2M_{sp}\left[(1+2nv)(\sum_i \ell_i^2)-2v(\sum_i \ell_i)^2\right]d_0 s^2\Bigg\}\sum_{i=1}^{n}\exp(-\tau_i s)+$$

$$+\left\{2\left[M_{sp}(\sum_i \ell_i)\right]d_0^2 s^2\right\}\sum_{i=1}^{n}\ell_i\exp(-\tau_i s)\,;(12.35)$$

$$\Delta_W W_z(s) = 2d_0 C_H\left\{2d_0(\sum_i \ell_i)\sum_{i=1}^{n}\ell_i\exp(-\tau_i s)-\right.$$

$$\left.-\left(M_{sp}R_{uy}^2 s^2 + 2d_0\sum_i \ell_i^2\right)\sum_{i=1}^{n}\exp(-\tau_i s)\right\}\sum_{i=1}^{n}\exp(-\tau_i s)\,; \quad (12.36)$$

$$\Delta_W W_\theta(s) = 2d_0 C_H\left\{2d_0(\sum_i \ell_i)\sum_{i=1}^{n}\exp(-\tau_i s)-\right.$$

$$\left.-\left(M_{sp}s^2 + 2nd_0\right)\sum_{i=1}^{n}\ell_i\exp(-\tau_i s)\right\}\sum_{i=1}^{n}\exp(-\tau_i s)\,; \quad (12.37)$$

$$\Delta_W W_x(s) = -C_{tir}\left\{M_{sp}^2 R_{uy}^2 s^4 + 2M_{sp}\left[\sum_i \ell_i^2 + 2nR_{uy}^2\right]d_0 s^2 + \right.$$

$$\left. + 4d_0^2\left[n\sum_i \ell_i^2 - (\sum_i \ell_i)^2\right]\right\}\sum_{i=1}^{n}\exp(-\tau_i s) \quad (12.38)$$

Thus, the denominator of the transfer functions and, respectively, the formulas for defining the APFC of a non-symmetrical sprung system is comprised of summands with two types of sums, which define the non-simultaneous affection character on the chassis' wheels. This means that applying the non-simultaneous affection character coefficients for the vertical and longitudinal-angular oscillations for systems of this type is, strictly speaking, unjustified.

12.1.4 Symmetrical analog of a non-symmetrical springing system

If we mirror the front part of a support-running complex relative to the vertical axis passing through the chassis' center of mass, we get a "symmetrical analog" of a non-symmetrical sprung system. Let us define ℓ_i^* as the wheel center coordinates of the analog's layout and $\Delta\ell_i = \ell_i - \ell_i^*$, then:

$$\sum_{i=1}^{n}\ell_i = \sum_{i=1}^{n}(\ell_i^* + \Delta\ell_i) = \sum_{i=n/2}^{n}\Delta\ell_i = \varepsilon \quad (12.39)$$

and formula (12.35) can be transformed as:

$$\Delta_W/M_{sp} = -\left\{M_{sp}^2[vR_{uy}^2]s^6 + M_\kappa\left[2v(\sum_i \ell_i^2)+(1+2nv)R_{uy}^2\right]d_0 s^4 + \right.$$

$$\left.+2(1+2nv)(\sum_i \ell_i^2)d_0^2 s^2\right\}\sum_{i=1}^{n}\exp(-\tau_i s)+$$

$$+2d_0^2 s^2\left\{\sum_{i=1}^{n}\ell_i\exp(-\tau_i s) - 2v\varepsilon\sum_{i=1}^{n}\ell_i\exp(-\tau_i s)\right\}\varepsilon\,. \quad (12.40)$$

For most SWCs, the value of ε is small, for example, for the four-axle MAZ-7908 chassis $|\varepsilon/L| = 0.051$, and for a six-axle MAZ-7916 chassis $|\varepsilon/L| = 0.151$. This means that in the first approximation, with the help of non-simultaneous affection character coefficients, we can evaluate the oscillation characteristics of a multi-axle wheeled chassis with a non-symmetrical springing system.

An idea about the change in the dependence of the non-simultaneous character coefficients on the affection frequency, when the design of the sprung system deviates from the symmetrical one, can be obtained from Figs. 12.5 and 12.6, where we see the curves of these functions for a real non-symmetrical wheeled MAZ-7908 chassis and for its symmetrical analog (the travel speed is 20 km/h).

With an increase in the frequency, the value of the normalized non-simultaneous affection

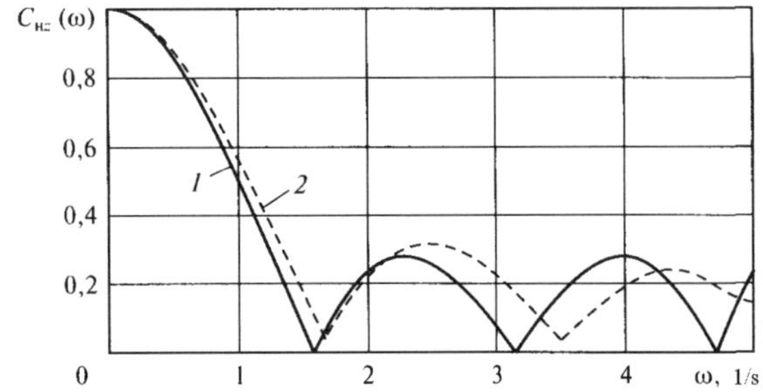

Fig. 12.5. Normalized non-simultaneous affection character coefficient for vertical oscillations versus the affection frequency for a real non-symmetrical MAZ-7908 chassis and its symmetrical analog at a travel speed of 20 km/h: 1—Symmetrical chassis; 2—Non-symmetrical chassis

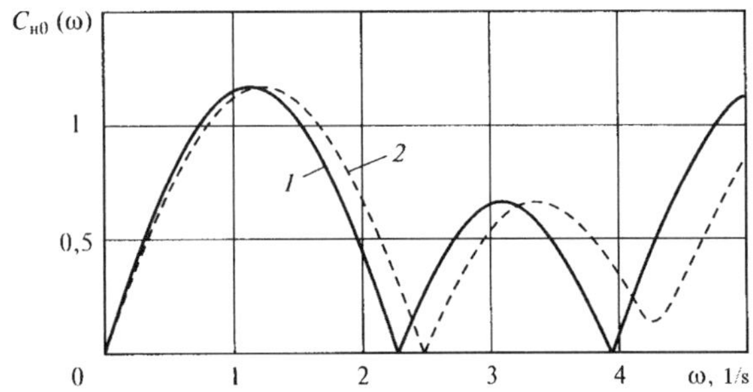

Fig. 12.6. Normalized non-simultaneous affection character coefficient for longitudinal-angular oscillations versus the affection frequency for the real non-symmetrical MAZ-7908 chassis and its symmetrical analog at a travel speed of 20 km/h: 1—Symmetrical chassis; 2—Non-symmetrical chassis

character coefficient for vertical oscillations decreases. This means that when a chassis moves over short irregularities, the effect of the multi-axle factor for vertical oscillations of the center of mass is expressed stronger than at its motion over long irregularities.

At high motion speeds, the affection delay is insignificant and the normalized non-simultaneous coefficient for vertical oscillations decreases slower with the increase in the affection frequency, i.e., at an increase in the chassis' travel speed, the amplitude of the vertical oscillations in its center of mass increases and decreases when the speed is reduced. This regularity is explained by the fact that at an increase in the travel speed, the affection delay on the following wheels decreases and the affection on all wheels becomes more simultaneous. It is natural that the displacement of the center of mass at a simultaneous displacement of the wheels in one direction will be much higher than at displacement of only one group of wheels when all the remaining wheels do not change their positions.

The amplitude of the longitudinal-angular oscillations decrease with the increase in the travel speed, since at better simultaneous affection characters on the wheels, the angular displacement of the body goes down.

Figs. 12.3–12.6 show that the frequency characteristics of multi-axle wheeled chassis are essentially different from those of ordinary vehicles. It looks like the non-simultaneous affection character coefficients somehow modulate these functions, making them pulsing ones; while the inter-resonance zone, nicely expressed in the frequency characteristics of double-mass sprung systems, practically disappears here.

In general, the formulas describing the frequency characteristics of multi-axle sprung systems are easily obtained from formulas (12.15) and (12.19–12.22) by substituting $j\omega$ instead of argument s. Formulas (12.28) and (12.29) define the frequency characteristics of a completely symmetrical springing system.

Of most interest is the intermediate case: a non-symmetrical springing system with practically similar wheels and resilient-damping elements in the suspensions of each wheel.

In the traditional approach, we need to transform the formulas defining the APFCs of a multi-axle chassis' oscillations into a clearly algebraic form. To do so, we isolate the real and imaginary parts in the numerator and denominator and then exclude the complex value in the denominator.

Then, we convert the obtained expression in such way as to identify in the numerator and denominator the characteristic semi- and full nominal structures.

These actions as applied to the formulas describing the transfer functions of oscillations in a multi-axle wheeled chassis are extremely cumbersome, while the results are bulky and unfit for elementary analysis. It is preferable to make a direct calculation with subsequent evaluation of the respective functions.

A detailed analysis of the oscillation characteristics in a multi-axle wheeled chassis with various springing system designs is presented in the next chapter. Here, we consider only the basic feature of frequency oscillation characteristics in SWCs, which should give a general idea about their distinctions from the respective characteristics of ordinary vehicles.

As an example, we have taken a typical four-axle wheeled MAZ-7310 chassis and its conventional symmetrical MAZ-7310C analog. The parameters used in the calculations are presented in Table 12.1.

The APFC character of vertical and longitudinal-angular oscillations is illustrated by the MAZ-7310C symmetrical chassis-analog, which moves at a speed of 20 km/h (ω = 0–10 1/s) (Figs. 12.7 and 12.8). For a real non-symmetrical chassis, these curves are so intricate that it is not convenient to use them in practice; therefore, we do not consider them here.

In this respect, the amplitude-frequency characteristics (AFCs) are more indicative and are more often used at practical evaluations and calculations. Their distinctions from AFCs of ordinary vehicles and the influence of multiple

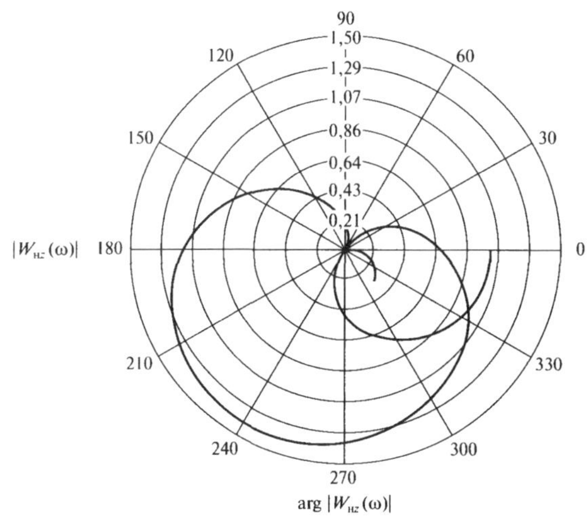

Fig. 12.7. Normalized amplitude-phase-frequency characteristic (APFC) of vertical oscillations in a symmetrical MAZ-7310C four-axle chassis-analog

Table 12.1 Parameters of the MAZ-7310 chassis and its symmetrical MAZ-7310C analog

Parameter	Symbol	MAZ-7310	MAZ-7310C
Full weight, kg	M_{sp}	3.85×10⁴	3.85×10⁴
Inertia radius, m	R_{uy}	3.624	3.624
Relative un-sprung mass	N	0.145	0.145
Coordinates of wheel centers, m	l_1	4.15	4.15
	l_2	1.95	1.95
	l_3	-1.35	-1.95
	l_4	-3.55	-4.15
Rigidity of resilient element of suspension, N/m	C	2.77×10^5	2.77×10^5
Rigidity of tires, N/m	C_{tir}	5.00×10^5	5.00×10^5
Resistance coefficient of damping element, N×s/m	μ	1.00×10^4	1.00×10^4
Number of axles	n	4	4

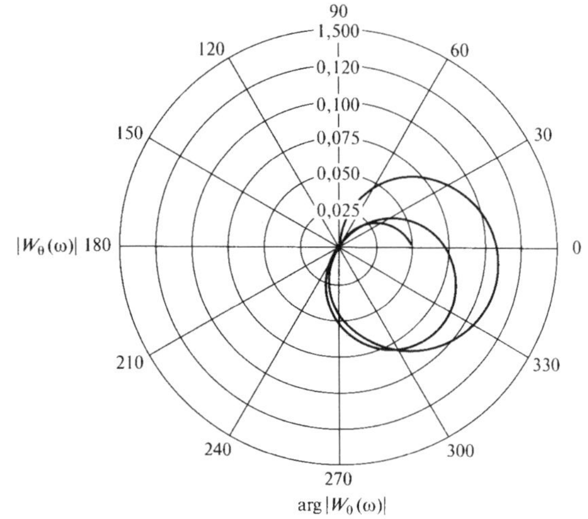

Fig. 12.8. Amplitude-phase-frequency characteristic (APFC) of longitudinal-angular oscillations in a symmetrical MAZ-7310C four-axle chassis-analog

axles on chassis oscillations are nicely seen at comparing their vertical and longitudinal-angular oscillations with the wheel oscillations of a real MAZ-7310 chassis (Figs. 12.12–12.14) and its symmetrical MAZ-7310C analog (Figs. 12.9–12.11).

Figs. 12.9 and 12.10 show a characteristic pulsation in the curves caused by the non-simultaneous road affection on different wheels of the chassis. The very shape of the curves is totally different from AFCs of usual vehicles, which is especially vivid in AFCs of the body's longitudinal-angular oscillations. The intricate shape of the curves also has to do with the fact

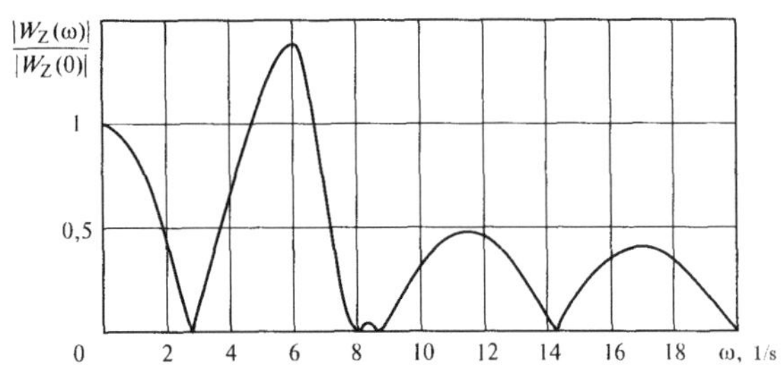

Fig. 12.9. Normalized amplitude-frequency characteristic of vertical oscillations in a symmetrical MAZ-7310C chassis-analog (at a travel speed of 20 km/h)

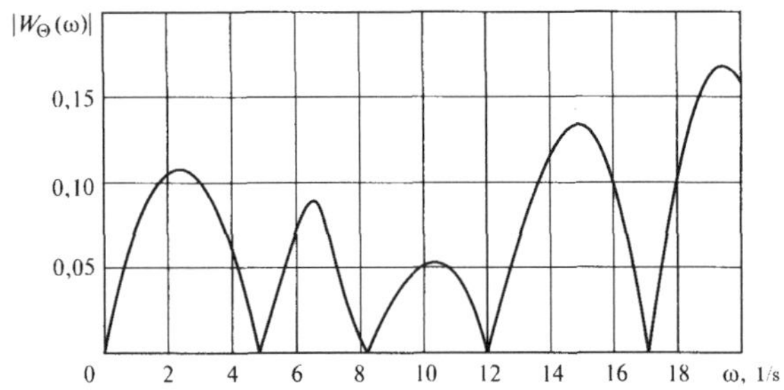

Fig. 12.10. Amplitude-frequency characteristic of longitudinal-angular oscillations in a symmetrical MAZ-7310C chassis-analog (at a travel speed of 20 km/h)

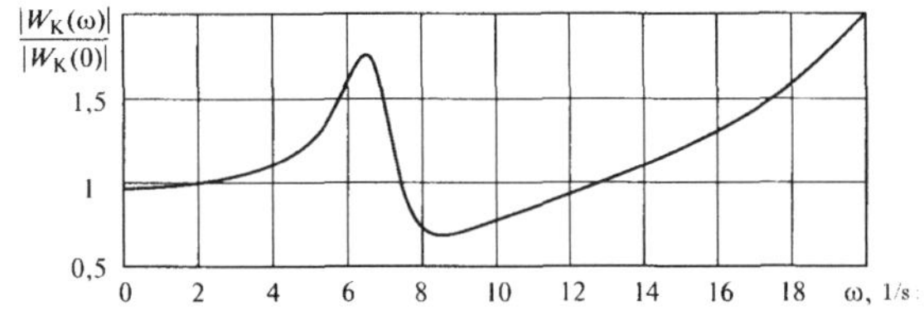

Fig. 12.11. Normalized amplitude-frequency characteristic of vertical wheel oscillations in a symmetrical MAZ-7310C chassis-analog (at a travel speed of 20 km/h)

that apart from the non-simultaneous road affection character, an interrelation of the body's vertical and longitudinal-angular oscillations is present here. At the same time, the AFC of the wheels' oscillations is shaped quite traditionally, since in symmetrical sprung systems, the body's oscillations have practically no influence on oscillations of the chassis' wheels.

A design deviation in the sprung system from symmetry radically complicates the oscillatory processes (Figs. 12.12–12.14). First of all, it is manifested through an essential strengthening in the interrelation between the oscillations of the body and the wheels, as well as of the vertical and longitudinal-angular oscillations. A practical disappearance of the so-called "inter-resonance zones" is noted, which served as the basis for many methods of automated control over parameters of soft driving. This is an indication of the fact that the traditional methods aimed at improving soft driving, designed for ordinary

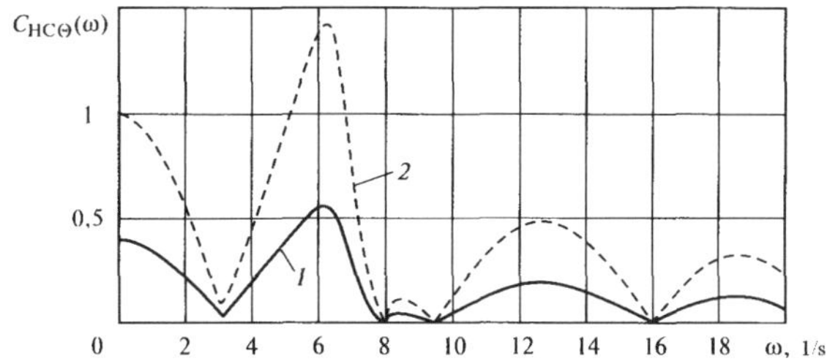

Fig. 12.12. Normalized and non-normalized amplitude-frequency characteristic of vertical oscillations in a MAZ-7310C chassis (at a travel speed of 20 km/h): 1—Symmetrical chassis; 2—Non-symmetrical chassis

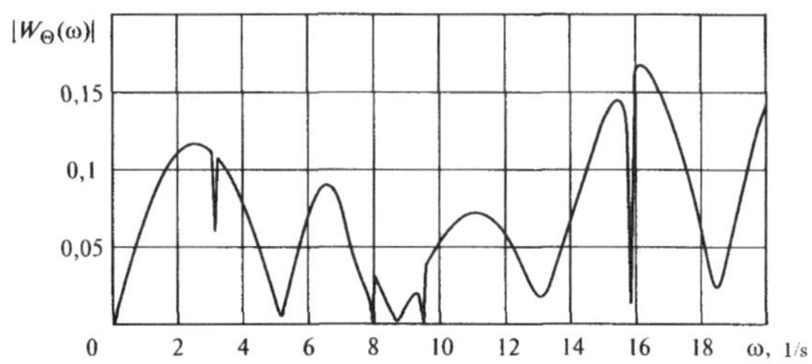

Fig. 12.13. Amplitude-frequency characteristic of longitudinal-angular oscillations in the MAZ-7310 chassis (at a travel speed of 20 km/h)

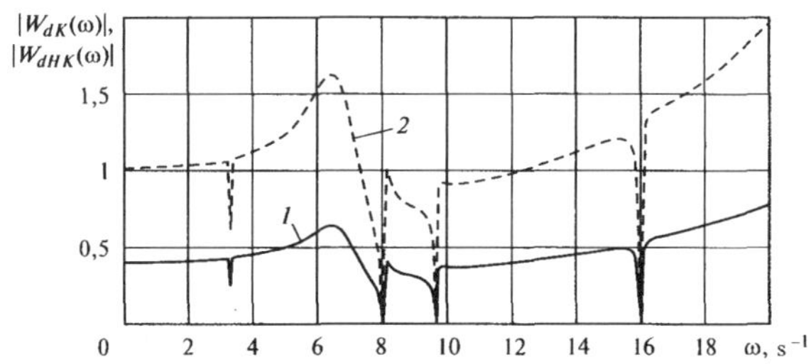

Fig. 12.14. Normalized and non-normalized amplitude-frequency characteristic of vertical wheel oscillations in the MAZ-7310 chassis (at a travel speed of 20 km/h): 1—½$W_{dK}(\omega)$½; 2—½$W_{dHK}(\omega)$½

vehicles, can be inefficient for multi-axle wheeled chassis (the next chapter will deal with these issues in more detail).

12.2 Characteristics of lateral-angular oscillations

In the first approximation we can believe that lateral-angular oscillations of a wheeled vehicle with resilient tires are observed in those cases when the road microprofile under the right- and left-hand wheels is different. In reality, a more complex body's motion is possible, but at the initial design phases we can use a simplified model, which considers the longitudinal-angular and lateral-angular oscillations as independent from each other. This means that it is possible to use a flat (in the lateral plane) calculation diagram and the respective motion equations.

The motion equations, like in the case of longitudinal-angular oscillations, can be obtained by any method; however, the easiest is to build them by using the d'Alembert principle:

$$\begin{cases} \sum_{i=1}^{n} [m_i \ddot{x}_{ir} + \mu_i \dot{x}_{ir} + (C_{st.w} + C_{tir}) x_{ir}] - \sum_{i=1}^{n} [\mu_i \dot{z} + C_{st.wi} z] - \\ \quad - \sum_{i=1}^{n} [\mu_i \dot{\theta} + C_{st.wi} \theta] \ell_i = \sum_{i=1}^{n} C_{tir_i} y_{ir} \; ; \\ \sum_{i=1}^{n} [m_i \ddot{x}_{il} + \mu_i \dot{x}_{il} + (C_{st.w} + C_{tir}) x_{il}] - \sum_{i=1}^{n} [\mu_i \dot{z} + C_{st.wi} z] - \\ \quad - \sum_{i=1}^{n} [\mu_i \dot{\theta} + C_{st.wi} \theta] \ell_i = \sum_{i=1}^{n} C_{tir_i} y_{il} \; ; \\ I_x \left[\ddot{\psi} + 0.5B^2 \sum_{i=1}^{n} (\mu_i \dot{\psi} + C_{st.wi} \psi) \right] = 0.5B^2 \sum_{i=1}^{n} [\mu_i (\dot{x}_{il} - \dot{x}_{ir}) + \\ \quad + C_{st.wi} (x_{il} - x_{ir})] \; , \end{cases} \qquad (12.41)$$

where the indices "*l*" and "*r*" indicate the left- and right-hand boards respectively.

In Laplace images we have:

$$\begin{cases} \sum_{i=1}^{n}[m_i s^2+\mu_i s+(C_{st.w}+C_{tir})]\tilde{x}_{ir}-\tilde{z}\sum_{i=1}^{n}[\mu_i s+C_{st.wi}]- \\ \quad -\tilde{\theta}\sum_{i=1}^{n}[\mu_i s+C_{st.wi}]\ell_i=\sum_{i=1}^{n}C_{tir_i}y_{ir}\ ; \\ \sum_{i=1}^{n}[m_i s^2+\mu_i s+(C_p+C_{tir})]\tilde{x}_{il}-\tilde{z}\sum_{i=1}^{n}[\mu_i s+C_i]- \\ \quad -\tilde{\theta}\sum_{i=1}^{n}[\mu_i s+C_i]\ell_i=\sum_{i=1}^{n}C_{tir_i}y_{il}\ ; \\ \mathrm{I}_x\tilde{\psi}\left[s^2+0.5B^2\sum_{i=1}^{n}(\mu_i s+C)\right]=0.5B^2\sum_{i=1}^{n}(\mu_i s+C_i)(\tilde{x}_{il}-\tilde{\bar{x}}_{in})\ . \end{cases} \quad (12.42)$$

By extracting the second equation from the first one in (12.42), we have:

$$\begin{cases} \sum_{i=1}^{n}[m_i s^2+\mu_i s+(C_{st.w}+C_{tir})](\tilde{x}_{ir}-\tilde{x}_{il})=\sum_{i=1}^{n}C_{tir_i}(\tilde{y}_{ir}-\tilde{y}_{il})\ ; \\ \mathrm{I}_x\tilde{\psi}\left[s^2+0.5B^2\sum_{i=1}^{n}(\mu_i s+C)\right]=0.5B^2\sum_{i=1}^{n}(\mu_i s+C_{st.wi})(\tilde{x}_{il}-\tilde{\bar{x}}_{ir})\ . \end{cases} \quad (12.43)$$

With account of the "delay" effect:

$$\begin{cases} (\tilde{x}_r-\tilde{x}_l)\sum_{i=1}^{n}[m_i s^2+\mu_i s+(C_{st.w}+C_{tir})]\exp(-\tau_i s)= \\ \quad =(\tilde{y}_r-\tilde{y}_l)\sum_{i=1}^{n}C_{tir_i}\exp(-\tau_i s)\ ; \\ \mathrm{I}_x\tilde{\psi}\left[s^2+0.5B^2\sum_{i=1}^{n}(\mu_i s+C_{st.w})\right]= \\ \quad =0.5B^2(\tilde{x}_r-\tilde{x}_l)\sum_{i=1}^{n}(\mu_i s+C_{st.wi})\exp(-\tau_i s) \end{cases} \quad (12.44)$$

Thus, we get:

$$\begin{cases} (\tilde{x}_r-\tilde{x}_l)\mathrm{K}_{st.w}(s)=(\tilde{y}_r-\tilde{y}_l)\mathrm{K}_{tir}(s)\ ; \\ \mathrm{I}_x\tilde{\psi}\left[s^2+0,5B^2\sum_{i=1}^{n}(\mu_i s+C)\right]=0,5B^2(\tilde{x}_n-\tilde{x}_l)\mathrm{K}_1(s)\ . \end{cases} \quad (12.45)$$

If we consider the height difference in the irregularities between the right- and left-hand chassis boards as an input affection (Δy) and exclude from (12.42) the difference in the displacements of the wheels on the right- and left-hand boards, we can build the following function:

$$\mathrm{W}_{\psi}(s)=\frac{\tilde{\psi}}{\Delta\tilde{y}}=\frac{K_{tir}(s)\,\mathrm{K}_1(s)\,\mathrm{B}^2}{I_x K_n(s)\left[2s^2+B^2\sum_{i=1}^{n}(\mu_i s+C)\right]} \quad (12.46)$$

If we use the motion of the inertia radius in the lateral plane, then expression (12.46) will take the form:

$$\mathrm{W}_{\psi}(s)=\frac{K_{tir}(s)\,\mathrm{K}_1(s)}{M_x K_n(s)\left[2s^2+B^2\sum_{i=1}^{n}(\mu_i s+C_{st.wi})\right]}\ \frac{B^2}{R_{ux}^2} \quad (12.47)$$

Let us treat this function as the transfer function describing the lateral-angular body oscillations in a multi-axle wheeled chassis.

A sprung system having the same tires and resilient-damping suspension elements will be described by a simpler expression for the transfer function:

$$\mathrm{W}_{\psi}(s)=\frac{K_{tir}(s)\,\mathrm{K}_1(s)}{M_x K_n(s)\,[2s^2+nB^2(\mu s+C)]}\ \frac{B^2}{R_{ux}^2} \quad (12.48)$$

The formula structure in (12.48) shows that all of the basic conclusions obtained by analyzing the body's longitudinal-angular oscillations as a whole are applicable to its lateral-angular oscillations in a multi-axle wheeled chassis (there are essential distinctions from the body's motion character in an ordinary vehicle, in pulsations, etc.). For illustration, Figs. 12.15 and 12.16 show

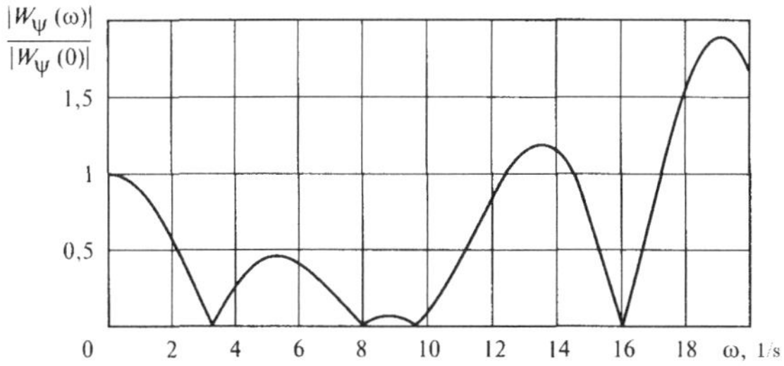

Fig. 12.15. Normalized AFC of lateral-angular oscillations in a MAZ-7310 four-axle wheeled chassis

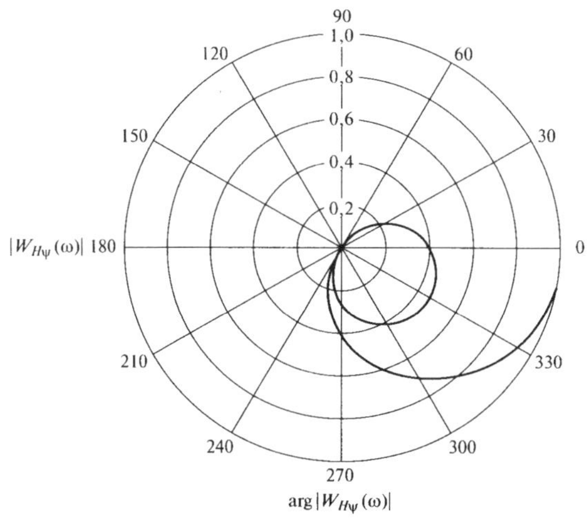

Fig. 12.16. Normalized amplitude-phase-frequency characteristic (APFC) of lateral-angular oscillations in a MAZ-7310 four-axle wheeled chassis

the AFCs of lateral-angular oscillations in a four-axle wheeled MAZ-7310 chassis.

When describing and studying the vertical and angular oscillations of a typical multi-axle wheeled chassis, it is reasonable to set the input affection by using the "averaged" road profile, which essentially simplifies the calculations. At the same time, there are quite a few multi-support wheeled vehicles in which the axles of the right-hand wheels do not coincide with the left-hand ones. In such cases the description of the support surface's microprofile becomes essentially more complicated.

12.3 Frequency characteristics of oscillation velocities

When designing motor vehicles, one should know the speed of the oscillation processes arising in the chassis' motion since it affects the crew's fatigue (along with the amplitude of oscillations); in addition, the speed of the oscillation processes is used in calculations of the suspension's damping elements (in particular, it is necessary to know the most probable speed range of the resilient-damping element's rod under the preset motion conditions).

Let us define the transfer function of the speed of the sprung system's oscillations as the ratio of the Laplace transformations' speed to the similar Laplace transformations in relation to the affection delivered to the vehicle by the road. Then, to obtain it, it is enough to multiply the expression for the transfer function of the sprung system's oscillations by the parameter of the Laplace transformation:

$$W_{\dot z}(s) = s\frac{K_{tir}(s)\,[2d_{2z}(s)\,K_2(s) - d_{2\theta}(s)\,K_1(s)]}{\Delta_W}$$

$$W_{\dot\theta}(s) = s\frac{K_{tir}(s)\,[2d_{2z}(s)\,K_1(s) - d_{1z}(s)\,K_2(s)]}{\Delta_W}$$

$$W_{\dot x}(s) = s\frac{K_{tir}(s)\,[4d_{2z}^2(s) - d_{1z}(s)\,d_{2\theta}(s)]}{\Delta_W}$$

$$W_{\dot\psi}(s) = s\frac{K_{tir}(s)\,K_1(s)}{M_x K_n(s)\,[2s^2 + nB^2(\mu s) + c]}\,\frac{B^2}{R_{ux}^2} \tag{12.49}$$

We do not present further transformations here since they are of routine character, but allow for obtaining a set of APFCs and AFCs for the speeds of the vertical, longitudinal-angular, and lateral-angular oscillation processes of a special wheeled chassis. Figs. 12.17–12.20 present such characteristics for the MAZ-7310 chassis.

12.4 Frequency characteristics of oscillation accelerations

When designing a motor vehicle it is important to know the acceleration of the oscillation processes that take place at the chassis' motion, especially the acceleration of vertical oscillations in the body's center of mass, since this parameter is most frequently regulated by various normative documents. In addition, accelerations in various points of the chassis' body allow for obtaining an idea about the loads acting on the SWC elements installed in these places; these loads were used at strength calculations.

Let us define the transfer function of accelerations in the sprung system's oscillations as a ratio of a Laplace-transformed acceleration to the Laplace-transformed road affection on the vehicle. Then,

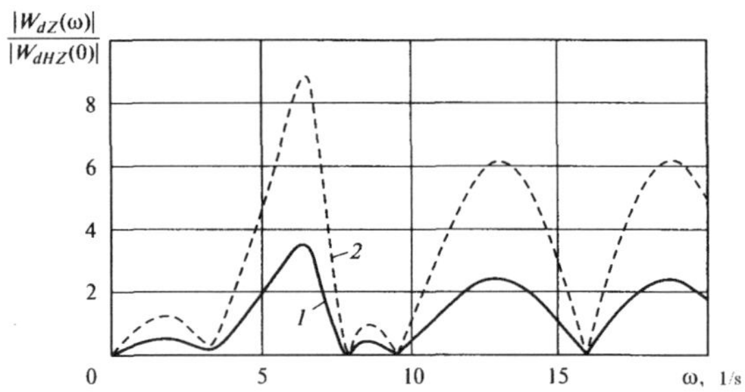

Fig. 12.17. Amplitude-frequency characteristic of the speed of the center of mass's vertical oscillations in the MAZ-7310 chassis (at a speed of 20 km/h): 1—$/W_{ddZ}(\omega)/$; 2—$/W_{ddHZ}(\omega)/$

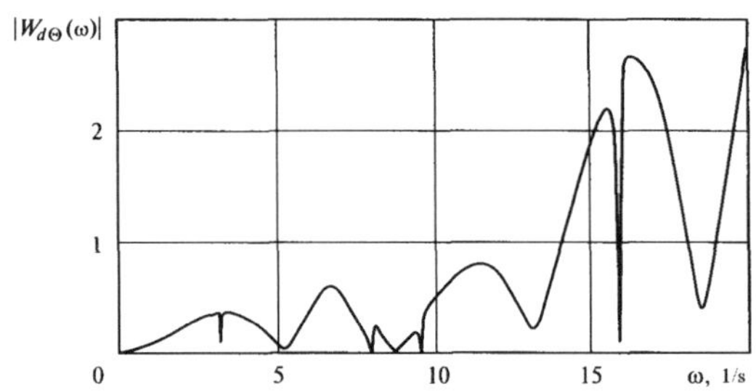

Fig. 12.18. Amplitude-frequency characteristic of the angular rate of the center of mass's longitudinal-angular oscillations in the MAZ-7310 chassis (at a speed of 20 km/h)

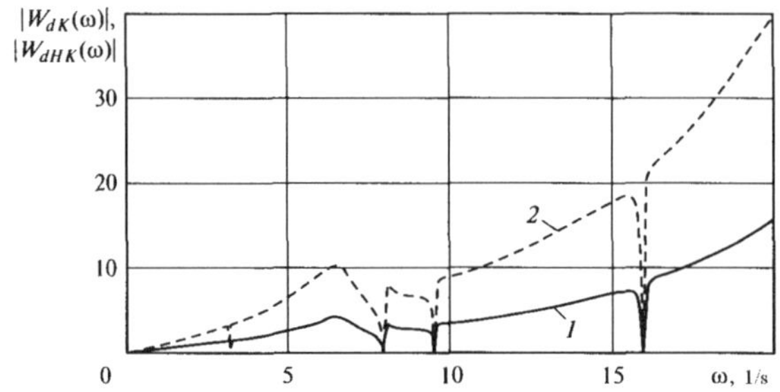

Fig. 12.19. Amplitude-frequency characteristic of the speed of the non-sprung masses' vertical oscillations in the MAZ-7310 chassis (at a speed of 20 km/h): 1—$/W_{d K}(\omega)/$; 2–$/W_{dHK}(\omega)/$

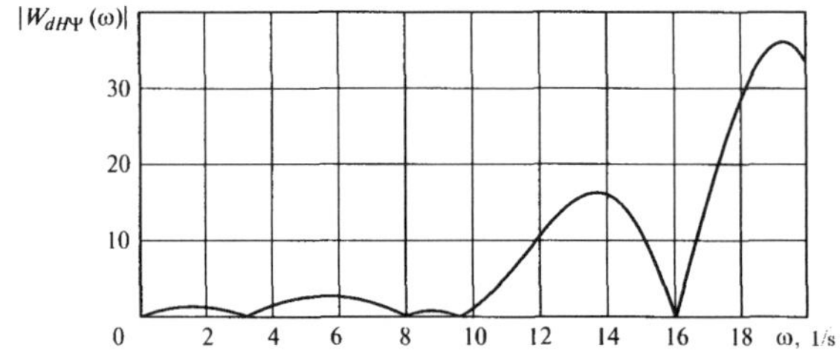

Fig. 12.20. Amplitude-frequency characteristic of the angular rate of the center of mass's lateral-angular oscillations in the MAZ-7310 chassis (at a speed of 20 km/h)

in order to obtain it, it is enough to multiply the expression for the transfer function of the sprung system's oscillations by the squared parameter of the Laplace transformation:

$$W_{\ddot{z}}(s)= s^2 \frac{K_{tir}(s)\,[2d_{2z}(s)\,K_2(s)- d_{2\theta}(s)\,K_1(s)]}{\Delta_W};$$

$$W_{\ddot{\theta}}(s)= s^2 \frac{K_{tir}(s)\,[2d_{2z}(s)\,K_1(s)- d_{1z}(s)\,K_2(s)]}{\Delta_W};$$

$$W_{\ddot{x}}(s)= s^2 \frac{K_{tir}(s)\,[4d_{2z}^2(s)- d_{1z}(s)\,d_{2\theta}(s)]}{\Delta_W};$$

$$W_{\ddot{\psi}}(s) = s^2 \frac{K_{tir}(s)\,K_1(s)}{M_x K_n(s)\,[2s^2+nB^2(\mu s)+c]}\,\frac{B^2}{R_{ux}^2} \tag{12.50}$$

Examples of AFCs for the MAZ-7310 chassis are presented in Figs. 12.21–12.24.

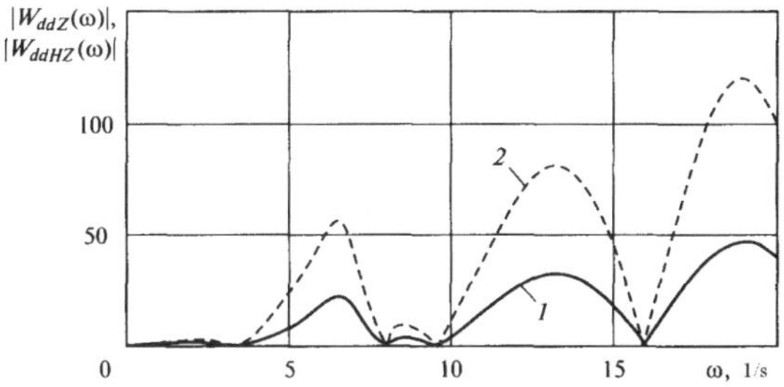

Fig. 12.21. Amplitude-frequency characteristic of accelerations of the center of mass's vertical oscillations in the MAZ-7310 chassis (at a speed of 20 km/h): 1—/$W_{ddZ}(\omega)$/; 2—/$W_{ddHZ}(\omega)$/

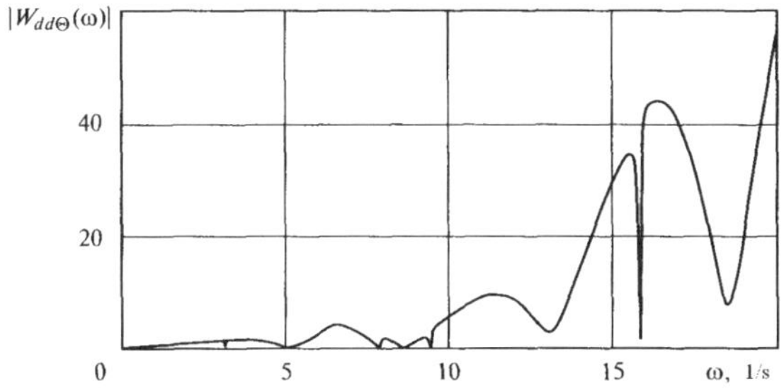

Fig. 12.22. Amplitude-frequency characteristic of the angular acceleration of the center of mass's longitudinal-angular oscillations in the MAZ-7310 chassis (at a speed of 20 km/h)

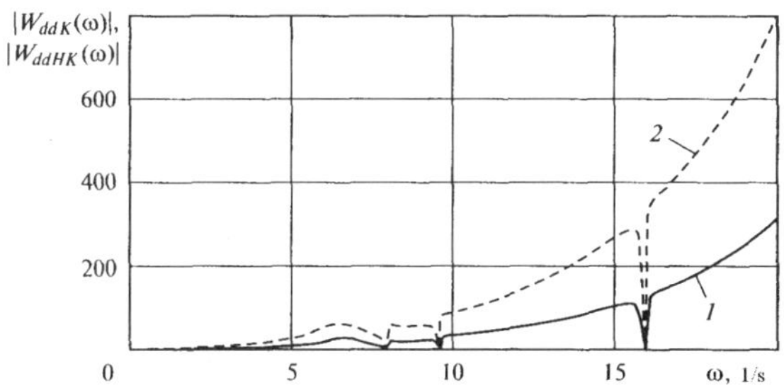

Fig. 12.23. Amplitude-frequency characteristic of accelerations of the non-sprung masses' vertical oscillations in the MAZ-7310 chassis (at a speed of 20 km/h): 1—/$W_{ddK}(\omega)$/; 2—/$W_{ddHK}(\omega)$/

For building a frequency characteristic of accelerations of a random point in the body, preset in the bound RMS system by coordinates (x_{a3}, y_{a3}, 0), the following formula is used:

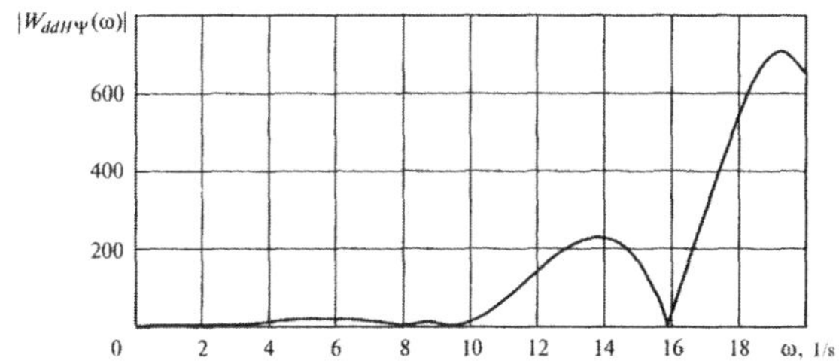

Fig. 12.24. Amplitude-frequency characteristic of angular acceleration of the center of mass's lateral-angular oscillations in the MAZ-7310 chassis (at a speed of 20 km/h):

$$W_{\ddot{z}A}(j\omega) = W_{\ddot{z}}(j\omega) + W_{\ddot{\theta}}(j\omega)\, x_{A3} + W_{\ddot{\psi}}(j\omega)\, y_{A3} \tag{12.51}$$

The data in Fig. 12.25 shows that accelerations at the driver's workplace in a typical multi-axle wheeled chassis have a very complex character for the following reasons:

- The mutual influence of vertical, longitudinal-angular, and lateral-angular oscillations of the carrying system;
- The remoteness of the driver's workplace from the chassis' center of mass;
- An essential width of the chassis and specificities of transported cargoes, which results, in combination, in an essential

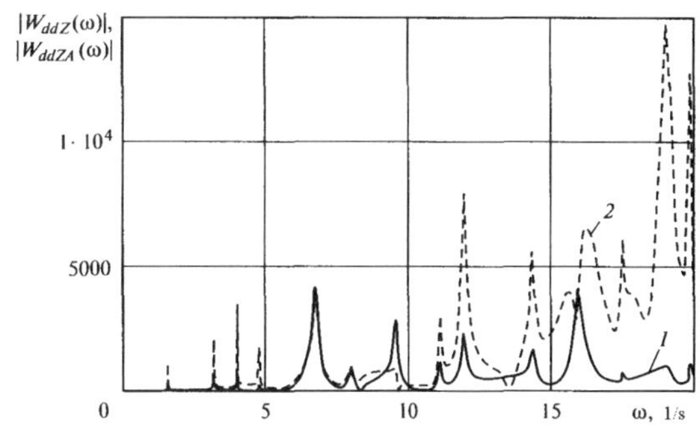

Fig. 12.25. Amplitude-frequency characteristic of accelerations of the vertical oscillations at the driver's workplace in the MAZ-7310 chassis (at a speed of 20 km/h): 1—/$W_{ddK}(\omega)$/; 2—/$W_{ddHK}(\omega)$/

displacement of the driver to the vehicle's outer board; and

- The general influence of multiple axles.

In addition, in a number of cases when assessing the ergonomic properties of the driver and crew's workplaces in a multi-axle wheeled chassis, one has to take into account additional displacements in the location area of workplaces, which have to do with the compliance of the vehicle's carrying system. This additionally complicates the problem of an adequate assessment of the accelerations present at the driver and crew's workplaces, as well as of the accelerations in the locations of the most critical components of the transported equipment or cargo.

12.5 Spectral analysis of oscillations

In the considered models of oscillations in a multi-axle wheeled chassis which moves on a bumpy road, the suspension system is treated as a linear open system of automatic regulation, which is under affection of a stationary random perturbation.

It is known that the energy output signal's amplitude spectrum in such a system can be obtained by multiplying the input affection's energy spectrum by the squared function, setting the respective amplitude-frequency characteristic.

In the considered problem, the input affection is the support surface's microprofile (in a particular case, of the road pavement). The general descriptions and characteristics of typical road pavements can be found in Chapters 6 and 8. For an elementary analysis and preliminary evaluation of a SWC's oscillations, we can use the recommendations from the work of A.A. Silaev [64], which offers the following approximation of the microprofile's spectral density, applicable for most road and ground conditions:

$$\Phi(\omega) = \frac{2\alpha v[(\alpha^2+\beta^2)v^2+\omega^2]}{\omega^4+2(\alpha^2-\beta^2)v^2\omega^2+(\alpha^2+\beta^2)^2v^4} \quad (12.52)$$

The parameters α and β are defined for a travel speed of 1.0 m/s. Usually, these parameters are within the following limits: α = 0.014–0.111 1/m and β = 0.025–0.140 1/m (the unit of measure of products αv and βv is 1/s when the speed is measured in m/s). If the travel speed is set in km/h, as is usually done when describing transport vehicles, then the numerical values and units of measure of these parameters should be corrected accordingly.

When we selected the illustrative materials for this section, we assumed that the chassis moves on ground roads characterized by the following spectral density parameters: α = 0.017 h/(s×km) and β = 0.024 h/(s×km). The curves of the microprofile's spectral densities corresponding to these parameters and travel speeds of 10, 20, 30, and 40 km/h, are presented in Fig. 12.26.

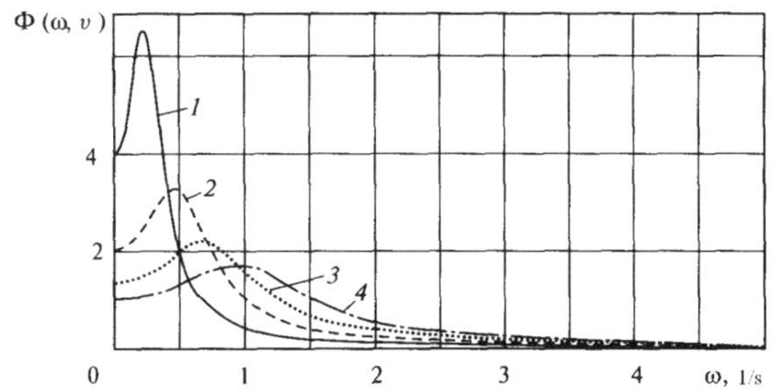

Fig. 12.26. Curves of microprofile densities corresponding to parameters α = 0.017 h/(s·km) and β = 0.024 h/(s·km) at a travel speed of: 1—10 km/h; 2—20 km/h; 3—30 km/h; 4—40 km/h

The spectral density of amplitudes and accelerations of longitudinal-angular oscillations of the multi-axle wheeled chassis' carrying system is defined from the following formulas:

$$S_\theta(\omega) = |W_\theta(j\omega)|^2\,\Phi(\omega) \quad (12.53)$$

$$S_{\ddot{\theta}}(\omega) = \left|W_{\ddot{\theta}}(j\omega)\right|^2\Phi(\omega) = \omega^4\left|W_\theta(j\omega)\right|^2\Phi(\omega) \quad (12.54)$$

An idea about the character of dependences can be obtained if we consider the motion of a four-axle wheeled MAZ-7310 chassis model on a bumpy ground road in speed intervals from 10 to 40 km/h.

We prepare the initial data for making a spectral analysis of the center of mass's vertical oscillations in a multi-axle wheeled chassis in a similar way.

The spectral density of amplitudes and accelerations of the carrying system's vertical oscillations in a multi-axle wheeled chassis is defined from the following formulas:

$$S_Z(\omega) = |W_Z(j\omega)|^2 \Phi(\omega) \quad (12.55)$$

$$S_{\ddot{z}}(\omega) = |W_{\ddot{z}}(j\omega)|^2 \Phi(\omega) = \omega^4 |W_z(j\omega)|^2 \Phi(\omega) \quad (12.56)$$

The character of dependences is also illustrated with the example of the four-axle wheeled MAZ-7310 chassis' motion along a bumpy ground road within a speed range of 10–40 km/h. It allows for comparing the processes of the carrying system's vertical and longitudinal-angular oscillations of the same chassis in the same motion conditions (Figs. 12.27–12.29).

For making a spectral analysis of the carrying system's lateral-angular oscillations we also need—apart from the earlier information—the

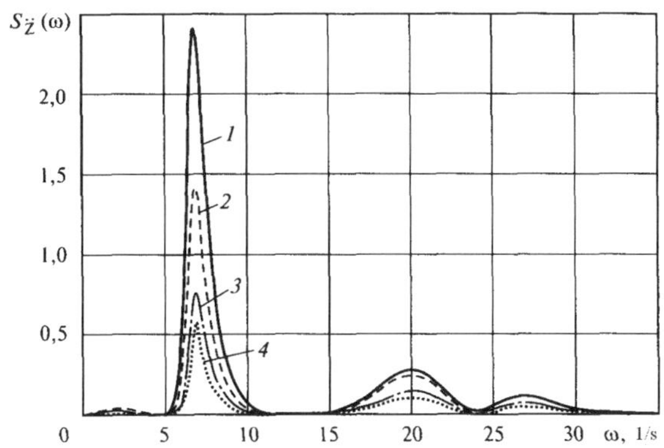

Fig. 12.27. Spectral density of accelerations of the carrying system's vertical oscillations in the MAZ-7310 multi-axle wheeled chassis at a travel speed of: 1—10 km/h; 2—20 km/h; 3—30 km/h; 4—40 km/h

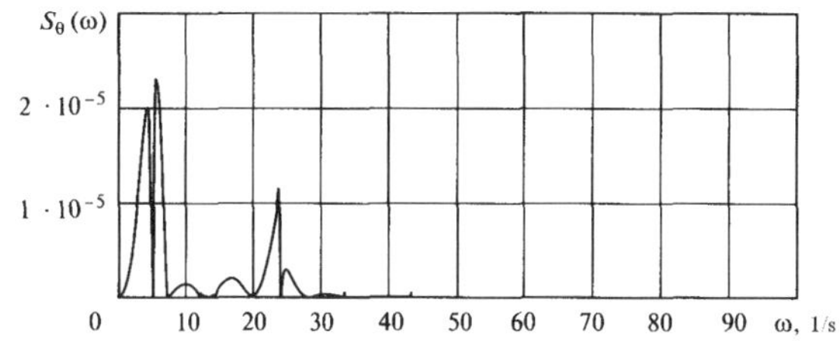

Fig. 12.28. Spectral density of amplitudes of the carrying system's longitudinal-angular oscillations in the MAZ-7310 multi-axle wheeled chassis

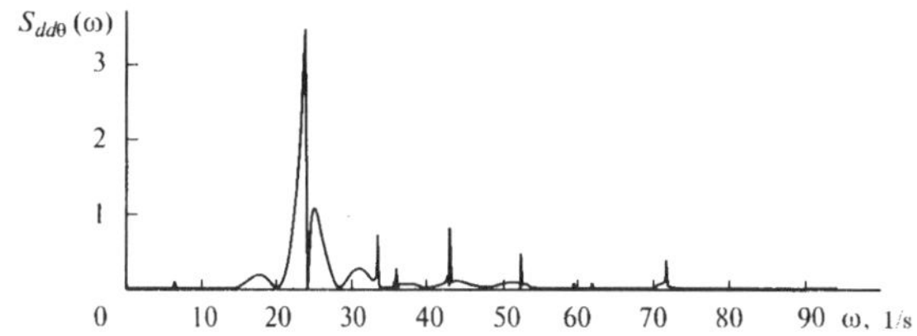

Fig. 12.29. Spectral density of accelerations of the carrying system's longitudinal-angular oscillations in the MAZ-7310 multi-axle wheeled chassis

data about the spectral density of the input affection, which gives rise to the chassis' oscillations along the roll angle.

Microprofile studies of the roads intended for HDWV traffic are very rarely done (except for the longitudinal microprofile of ground roads). The characteristics of the lateral-angular oscillations of multi-axle wheeled chassis in the lateral plane are described in *Spectral Theory of Vehicle Springing* [64]. In particular, it is shown that at initial design stages, we can use the spectral density of the input affection with a single structure; it is enough to take $\beta = 0.112$ 1/s (0.031 h/(s×km)). In other words, the lateral microprofile of ground roads usually has rather small and short irregularities.

After defining the structure and numerical values of the input affection parameters, the calculation of the carrying system's lateral-angular oscillations is rather simple.

The spectral density of amplitudes and accelerations of the carrying system's lateral-angular oscillations is defined from the following formulas:

$$S_\psi(\omega) = |W_\psi(j\omega)|^2 \Phi_\psi(\omega). \qquad (12.57)$$

$$S_{\ddot\psi}(\omega) = \left|W_{\ddot\psi}(j\omega)\right|^2 \Phi_\psi(\omega) = \omega^4 \left|W_\psi(j\omega)\right|^2 \Phi_\psi(\omega) \qquad (12.58)$$

For the most widespread road types, we can take $\alpha = 0.15$ 1/s (0.042 h/(s×km)).

Examples of dependences are presented in Figs. 12.30–12.31.

It was not the aim of this chapter to make a comprehensive analysis of the oscillation characteristics in multi-axle wheeled chassis, their evaluation, and ways to improve specifications. These issues are considered in detail in the section dealing with calculating and designing springing systems and their elements. For the purpose of a primary analysis of design solutions and evaluating the difference in oscillations of multi-axle wheeled chassis from oscillations of ordinary vehicles, it is enough to demonstrate that the above results allow for solving the following problems:

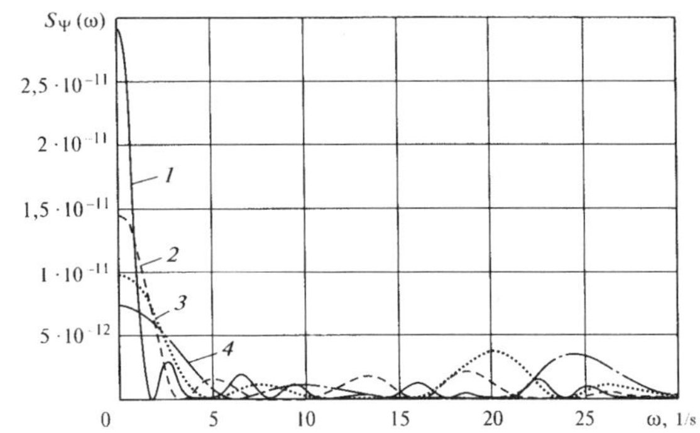

Fig. 12.30. Spectral density of amplitudes of the carrying system's lateral-angular oscillations in the MAZ-7310 multi-axle wheeled chassis at a travel speed of: 1—10 km/h; 2—20 km/h; 3—30 km/h; 4—40 km/h

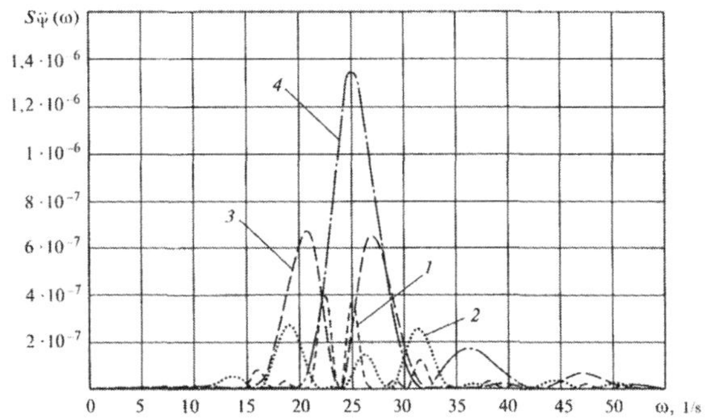

Fig. 12.31. Spectral density of accelerations of the carrying system's lateral-angular oscillations in the MAZ-7310 multi-axle wheeled chassis at a travel speed of: 1—10 km/h; 2—20 km/h; 3—30 km/h; 4—40 km/h

A. Evaluation of the amplitude and acceleration dispersions (RMS deviations)of the carrying system's vertical, longitudinal-angular, and lateral-angular oscillations, as well as the displacements and accelerations at the driver's workplace, at the crew seats, and in the equipment component's attachment points.

For example, the amplitude dispersion of the carrying system's longitudinal-angular oscillations looks as follows:

$$D[\theta(t)] = \frac{1}{\pi}\int_0^\infty S_\theta(\omega)\,d\omega \qquad (12.59)$$

In a similar way, we can define the amplitude dispersions for other types of oscillations.

B. Evaluation of the probability of oscillations with amplitudes or accelerations exceeding the preset limits. This evaluation is made when designing the multi-axle wheeled chassis intended for transporting certain types of equipment sensitive to dynamic loads. If we assume that the distribution of road irregularities is normally centered, then the distribution of amplitudes of the carrying system's oscillations will also normally be centered. This allows for building a probability distribution function of the amplitudes of oscillations and accelerations in different points of the carrying system and evaluating the probability of surpassing certain definite (preset) thresholds by these values.

Motion characteristics of multi-axle wheeled chassis are most vividly manifested during the analysis of vertical accelerations in different points of the carrying system. The same methodology allows for making a full analysis of the carrying system's oscillations.

The dispersion of vertical accelerations in the center of mass of the MAZ-7310 chassis, which is moving at a speed of 30 km/h along different ground roads, has the following values defined under formula (12.59):

$$D_{\ddot{z}} = (1.682 \quad 1.181 \quad 0.404 \quad 0.643)^{\mathrm{T}} \ \mathrm{m^2/s^4}.$$

Respectively, the RMS deviation of vertical accelerations is:

$$\sigma_{\ddot{z}} = 1.297 \quad 1.087 \quad 0.636 \quad 0.802)^{\mathrm{T}} \ \mathrm{m/s^2}$$

or RMS deviation in shares of g:

$$\sigma_{g\ddot{z}} = (0.132 \quad 0{,}111 \quad 0.065 \quad 0{,}082)^{\mathrm{T}}$$

The distribution of vertical accelerations in the MAZ-7310 chassis' center of mass is normal, in accordance with the adopted assumptions (Fig. 12.32), and is defined according to the formula:

$$f(\ddot{z}_g) = \frac{1}{\sigma_{g\ddot{z}}\sqrt{2\pi}} \exp\left(\frac{-\ddot{z}_g^2}{2\sigma_{g\ddot{z}}^2}\right) \qquad (12.60)$$

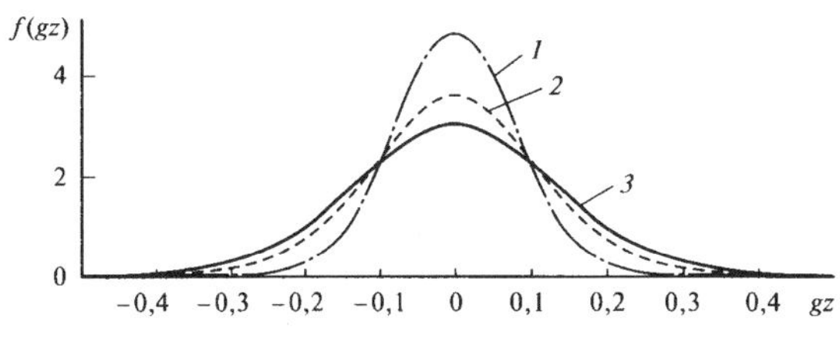

Fig. 12.32. Distribution of vertical accelerations in the MAZ-7310 chassis' center of mass moving along various ground roads at a speed of 30 km/h: 1, 2, 3—Ground roads of various quality

The terms of reference (ToR), or the technical assignment for designing a chassis, most often prescribe a certain level of maximum admitted accelerations in certain characteristic points of the body. For example, for one type of special equipment it is extremely undesirable if the linear displacement has an acceleration of over 0.7 g. We can define the ratio between the maximum permissible and RMS acceleration in a certain point at motion in the preset mode as:

$$K_n = \frac{\ddot{z}g \ \max}{\sigma_{g\ddot{z}}} \qquad (12.61)$$

In accordance with the Gaussian inequality, the appearance probability of accelerations exceeding the preset maximum level is defined as:

$$P = \frac{\exp\left(\frac{-K_n}{2}\right)}{K_n\sqrt{2\pi}} \qquad (12.62)$$

In the considered example:

$$K_n = (5.294 \quad 6.318 \quad 10.798 \quad 8.567)^{\mathrm{T}};$$
$$P = (5.34 \cdot 10^{-3} \quad 2.68 \cdot 10^{-3} \quad 1.67 \cdot 10^{-4} \quad 6.43 \cdot 10^{-4})^{\mathrm{T}}.$$

Thus, the appearance probability of high accelerations in the MAZ-7310 chassis' center of mass that moves slowly on ordinary ground roads is rather small.

If we now calculate the appearance probability of such vertical accelerations on the longitudinal axis of the vehicle in its nose (at the level of the driver's workplace), it will be much higher:

$$P = (0.446 \quad 0.442 \quad 0.209 \quad 0.291)^{\mathrm{T}}.$$

Thus, multi-axle wheeled chassis are characterized by essential crew and equipment overloads located on the nose and back overhangs of the vehicle's frame (because of the carrying console's large length). High accelerations in these zones can appear when the vehicle moves on rather good roads at moderate speed or when accelerations in the center of mass are at an acceptable level. This is caused by a high appearance probability of resonance high-frequency oscillations of non-sprung masses (wheels) and the carrying frame console.

With an increase in the multi-axle wheeled chassis' travel speed, RMS amplitude values of longitudinal-angular oscillations usually go up at the beginning, reach their maximum within a rather narrow speed range, and then continuously go down. Despite a large difference in the dispersion of irregularities on various roads, the RMS amplitude values only differ a little (when dispersions differ by 2.5 times, the RMS amplitude values of longitudinal-angular oscillations differ by only 20%).

Earlier, we gave an example of defining the displacements and accelerations of a point located on the chassis' longitudinal axis and not coinciding with its center of mass. It is important to have a model that allow us to assess the objects' oscillatory motion parameters located in the random point of the carrying system (usually, at initial design stages, they confine themselves to the points located in the plane (*x*3, *y*3) of the bound RMS coordinate system). If we set the position of point A by coordinates in the RMS system (*x*A, *y*A), it is obvious that the vertical displacements of this point are defined as:

$$Z_A(t) = z(t) + \theta(t) \cdot x_A + \psi(t) \cdot y_A \quad (12.63)$$

It follows from (12.63) that:

$$\tilde{z}_A(j\omega) = \tilde{z}(j\omega) + \tilde{\theta}(j\omega) \cdot x_A + \tilde{\psi}(j\omega) \cdot y_A \quad (12.64)$$

In assessment calculations we can use a single input affection to form both the lateral-angular and longitudinal-angular oscillations. If this simplification is valid, then:

$$S_{zA}(\omega) = |W_z(j\omega) + W_\theta(j\omega) \cdot x_A + W_\psi(j\omega) \cdot y_A|^2 \, \Phi(\omega) \quad (12.65)$$

and:

$$S_{\ddot{z}A}(\omega) = \omega^4 \left| W_z(j\omega) + W_\theta(j\omega) \cdot x_A + W_\psi(j\omega) \cdot y_A \right|^2 \Phi(\omega) \quad (12.66)$$

Then, we apply the whole described spectral analysis process of the object's oscillatory motion fixed in point *A* of the chassis' carrying system.

12.6 Spatial models of oscillations

In a number of cases, applying simplified models is insufficient for an adequate description of the oscillations in multi-axle wheeled chassis since such models disregard the resilient properties of the vehicle's carrying system. At the same time, experimental data show that the frame's elasticity (resilience or compliance) can essentially affect the evaluation results on soft driving even of ordinary long-wheel-base vehicles, which have a frame-based design; and, first of all, it refers to the compliance or yield of the carrying system to twisting. Thus, for example, Fig. 12.33 [65] shows RMS values of the

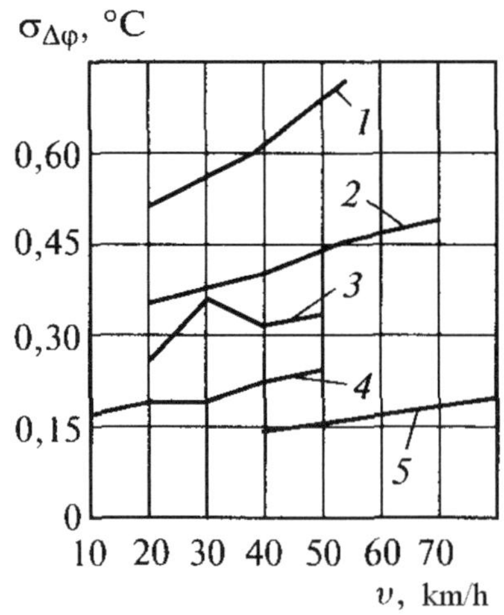

Fig. 12.33. RMS deviations of the frame twisting angle versus the travel speed for the ZIL-130 vehicle with a load of 4.0 t at movement on different roads: 1—Ground road; 2—Cobble road in satisfactory condition; 3—Worn-out cement concrete pavement; 4—Smooth cement concrete pavement; 5—Smooth asphalt concrete pavement

frame's twisting angle versus the travel speed for a ZIL-130 short-base vehicle. It is seen that even at a small frame length, the twisting angles can reach one degree. According to Fig. 12.33, we can calculate that the frame of the ZIL-130 vehicle is twisted—with a probability of 10^{-3}—by the following angles:

- At motion on ground roads by 5.5°;
- At motion on cobble roads by 5.5°; and
- At motion on roads with cement concrete pavement by 0.835°.

It is explained by the fact that the carrying system's resistance to twisting is often comparable with the suspension's lateral angular rigidity. The frames of typical multi-axle wheeled chassis are long at the limited lateral cross-section dimensions. Also, the cabins for the driver and crew are mounted on the frame's long front overhangs. As a result, the angular displacements in the cabins' attachment points are essentially dependent on the frame's angular twisting (oscillations), especially at motion on bumpy ground roads with large irregularities, i.e., in the typical operation conditions of multi-axle wheeled chassis.

The carrying system's twisting stiffness (rigidity) also reflects the soft driving assessment and, especially, the evaluation of the vibration load on the crew, although this influence is essentially weaker (Fig. 12.34).

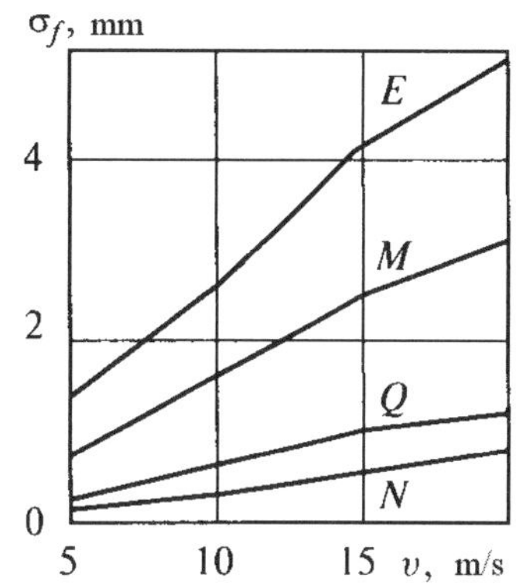

Fig. 12.34. RMS deviations of beam bending in different points on the ZIL-130's frame versus the travel speed (motion with a load of 4.0 t on a cobble road in satisfactory condition): E—Front bumper; N—Base middle; M—Rear overhang; Q—Point at the distance of 0.62 m to the left of the rear suspension's axle

In most studies it is considered admissible to use—for assessing soft driving—the chassis model made up of concentrated masses linked by resilient and shock-absorbing elements because soft driving is assessed within the band of low frequencies: 0.7–22.5 Hz (a discrete model). This allows for using—at the initial design stages—the spectral analysis method for a multi-axle chassis' oscillations.

When building a discrete model, which accounts for the frame's compliance, the above-described system of assumptions (see point 12.1) is usually used. The fastening of the main units to the frame is considered rigid (Fig. 12.35). At the initial design stages of a multi-axle chassis, the simplest rod-based model of the frame is the most expedient one [65]. When defining the parameters of the simulating rod, one should keep in mind that under the available experimental data the carrying system's resistance to twisting in a cargo vehicle is 30–50% higher than the stiffness of a separate frame.

The chassis' carrying system is simulated by a rod resilient to twisting and bending in the vertical plane with a variable unit mass and stiffness. The rod's bending in the horizontal plane is disregarded. The distribution of masses and rigidities along the rod's length should correspond to the distribution of these parameters in the chassis' simulated carrying system.

At the rod's bending, we neglect the turning inertia of its cross sections and shear deformations, while at twisting we believe that the rod's lateral cross-sections remain plane and not deformed. It is natural that the rod's resilient oscillations are considered small, while the dependences between deformations and power factors are linear. Then, the differential equations of the rod's oscillations are as follows:

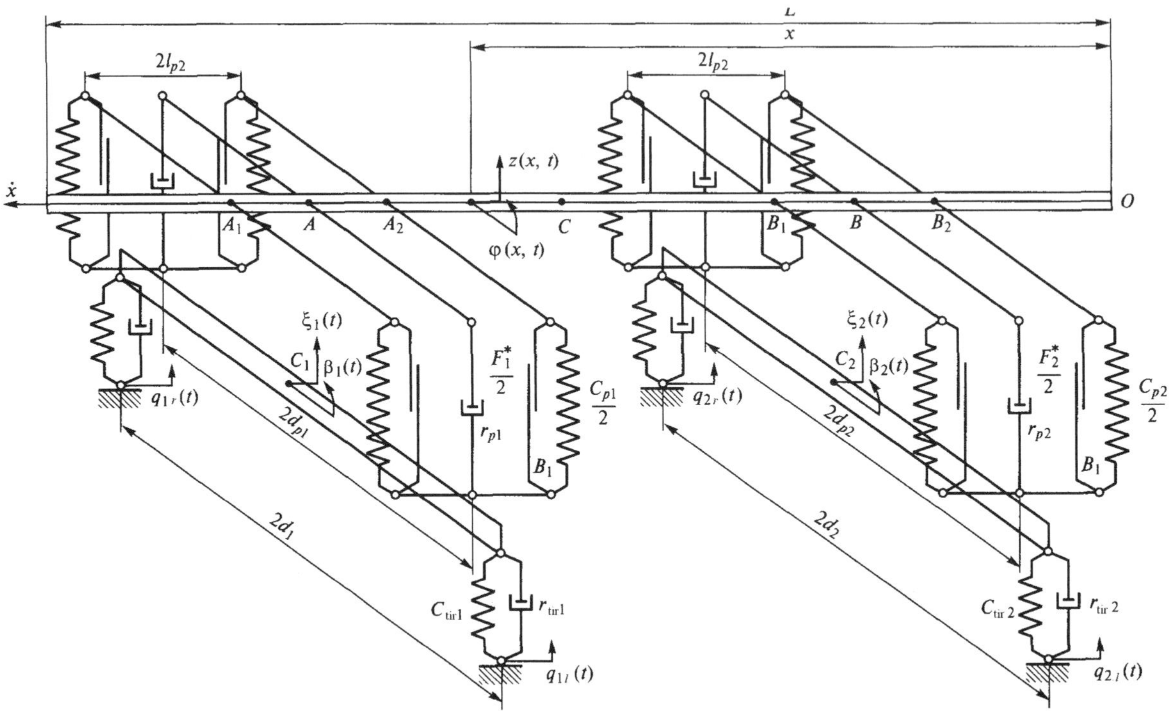

Fig. 12.35. Calculation diagram of a multi-axle chassis' oscillations with the carrying system resilient to bending and twisting and a bridge-type layout of the support-running complex

$$\begin{cases} C_\varphi \dfrac{\partial \varphi(x,t)}{\partial x} = M_{sp}(x,t)\ ; \\ EI_z(x,t)\dfrac{\partial^2 z(x,t)}{\partial x^2} = M_u(x,t)\ , \end{cases} \tag{12.67}$$

where: $C_\varphi = C_\varphi(x)$ is the rod's resistance to twisting; $EI_z(x)$ is the rod's resistance to bending; $M_k(x,t)$ is the torque in cross-section with the current coordinate x; $M_u(x,t)$ is the bending moment in cross-section with the current coordinate x; $z(x,t)$ are the displacements of the rod's rigidity axis caused by its bending; and $\varphi(x,t)$ is the turning angle of the rod's respective cross section.

At this design stage, it is recommended to account for the internal friction forces in the carrying system with the help of the Voigt hypothesis (the hypothesis of viscous friction). It is considered that the additional strengths arising in the rod at deformations from the forces of non-resilient resistance are proportional to the speed of deformations, while the distribution of friction in the rod corresponds to the distribution of its rigidities:

$$\begin{cases} C_\varphi \dfrac{\partial \varphi(x,t)}{\partial x} + k_\varphi \dfrac{\partial}{\partial t}\left(C_\varphi \dfrac{\partial \varphi(x,t)}{\partial x}\right) = M_{sp}(x,t)\ ; \\ EI_z(x,t)\dfrac{\partial^2 z(x,t)}{\partial x^2} + k_z \dfrac{\partial}{\partial t}\left(EI_z(x,t)\dfrac{\partial^2 z(x,t)}{\partial x^2}\right) = M_u(x,t)\ , \end{cases} \tag{12.68}$$

where k_φ and k_z are the coefficients characterizing the viscous friction at twisting and bending of the rod.

The legend for Fig. 12.35 is as follows:

- x is the current coordinate, which is setting the rod's cross-section position and simulating the carrying system (the start of counting coincides with the frame's rear perpendicular);

- $z(x,t)$ are the vertical displacements of the rod's rigidity axis points;
- $\varphi(x,t)$ are the turn angles of the rod's cross-sections relative to the longitudinal axis;
- $\xi_i(t)$ is the vertical displacement of the axle's center of mass ($i = 1$ corresponds to the front axle, $i = n$ corresponds to the rear axle);
- $\beta_i(t)$ is the angle of the axle's lateral turn;
- C is the carrying system's center of mass;
- C_i is the axle's center of mass;
- A is the point of the rod corresponding to the axis of the front axle;
- A_1 is the point of the rod corresponding to the nose bracket's attachment point to the spring of the ith axle;
- A_2 is the point of the rod corresponding to the back bracket's attachment place to the spring of the ith axle;
- $2l_{pi}$ is the distance between the axles of the spring brackets of the ith axle;
- L is the length of the carrying system;
- $2d_i$ is the wheelspan of the ith axle;
- $2d_{pi}$ is the distance between the springs of the ith axle;
- $C_{st.wi}$ is the stiffness of the spring;
- r_{pi} is the shock absorber's reduced resistance coefficient;
- F_i^* is the dry friction force in the suspension of the ith axle;
- C_{tiri} is the radial stiffness of the tires on the ith axle;
- R_{pi} is the viscous friction coefficient of the tires on the ith axle;
- $q_{il}(t)$ are the ordinates of the road microprofile under the left-hand wheel of the ith axle; and
- $q_{ir}(t)$ are the ordinates of the road microprofile under the right-hand wheel of the ith axle.

The viscous friction hypothesis fails to take into account the independent character of the absorption coefficient from the frequency of oscillations. Different ways to improve the models are known: the Bokk correction ([65, 66]), the Sorokin hypothesis ([65, 67]), etc. At earlier design stages, the decisive aim is to reduce the calculation volume under the acceptable precision of results. The practice has shown that based on the viscous friction hypothesis we can obtain rather accurate results.

When compiling the differential equations of the rod's oscillations, which simulate the carrying system, the generalized coordinates are measured from the centers of the corresponding masses under non-deformed resilient elements of the suspension and non-deformed carrying system.

The methodology of building models is described in detail in *Calculation of Operational Motion Parameters of [an] Automobile and Train* [65].

Since the bending moment in the rod's cross-sections, the shearing force, and the per-unit load are interlinked by the following relations:

$$\begin{cases} \dfrac{\partial M_u(x,t)}{\partial x} = Q(x,t)\ ; \\ \dfrac{\partial Q(x,t)}{\partial x} = q(x,t)\ , \end{cases} \tag{12.69}$$

then:

$$\begin{cases} \dfrac{\partial^2}{\partial x^2}\left(c_\phi \dfrac{d\phi(x,t)}{\partial x}\right) + k_\phi \dfrac{\partial^2}{\partial t\,\partial x}\left(c_\phi \dfrac{d\phi(x,t)}{\partial x}\right) = -m_k(x,t)\ ; \\ \dfrac{\partial^2}{\partial x^2}\left(EI_z(x,t)\dfrac{d^2 z(x,t)}{\partial x^2}\right) k_z + \dfrac{\partial^3}{\partial t\,\partial x^2}\left(EI_z(x,t)\dfrac{d^2 z(x,t)}{\partial x^2}\right) = q(x,t)\ , \end{cases} \tag{12.70}$$

where $m_k(x,t)$ is the per-unit moment acting on the rod.

At oscillations, the rod is loaded by the inertia forces and concentrated forces in the attachment places of the suspension's resilient and damping elements:

$$\begin{cases} q(x,t) = -m_0(x)\dfrac{\partial^2 z}{\partial t^2} + \sum\limits_{q=1}^{m} F_q \delta(x - x_q) - m_0(x) g \ ; \\ m_k(x,t) = -I_\phi(x)\dfrac{\partial^2 \phi}{\partial t^2} + \sum\limits_{q=1}^{m} F_q d_q \delta(x - x_q) \ , \end{cases} \tag{12.71}$$

where: $m_0(x)$ is the rod's unit mass; F_q is the concentrated force applied in the rod's cross-section with the coordinate x_q at the distance d_q from the rod's axis; and $I_\varphi(x)$ is the rod's unit inertia moment.

The rod's unit mass and unit inertia moment are calculated with account of the concentrated masses of the aggregates fixed in the points with coordinates x_j:

$$\begin{cases} m_0(x) = m_0'(x) + \sum\limits_{j=1}^{k} m_j \delta(x - x_j) \ ; \\ I_\phi(x) = I_\phi'(x) + \sum\limits_{j=1}^{k} I_j \delta(x - x_j) \ . \end{cases} \tag{12.72}$$

Formulas (12.69) and (12.70) allow for compiling the differential equations of the resilient rod's oscillations simulating the chassis' carrying system:

$$\begin{cases} \dfrac{\partial^2}{\partial x^2}\left(C_\phi \dfrac{d\phi(x,t)}{\partial x} \right) + k_\phi \dfrac{\partial^2}{\partial t \, \partial x}\left(C_\phi \dfrac{d\phi(x,t)}{\partial x} \right) = \\ \qquad = I_\phi(x)\dfrac{\partial^2 \phi}{\partial t^2} - \sum\limits_{q=1}^{m} F_q d_q \delta(x - x_q) \ ; \\ \dfrac{\partial^2}{\partial x^2}\left(EI_z(x,t)\dfrac{d^2 z(x,t)}{\partial x^2} \right) k_z + \dfrac{\partial^3}{\partial t \, \partial x^2}\left(EI_z(x,t)\dfrac{d^2 z(x,t)}{\partial x^2} \right) = \\ \qquad = -m_0(x)\dfrac{\partial^2 z}{\partial t^2} + \sum\limits_{q=1}^{m} F_q \delta(x - x_q) - m_0(x) g \ , \end{cases} \tag{12.73}$$

In accordance with Fig. 12.35, it is assumed that external forces act only in the cross-sections corresponding to the location of the axles and the points where they are fixed; therefore:

$$\begin{cases} \sum\limits_{q=1}^{m} F_q \delta(x - x_q) = \sum\limits_{i=1}^{n} (F_{li} + F_{ri}) \, \delta(x - x_i) \ ; \\ \sum\limits_{q=1}^{m} F_q d_q \delta(x - x_q) = \sum\limits_{i=1}^{n} (F_{li} - F_{ri}) \, d_i \delta(x - x_i) \ , \end{cases} \tag{12.74}$$

By using formulas (12.73) and (12.74), it is easy to obtain the differential equations of the rod's oscillations simulating a multi-axle wheeled chassis' carrying system:

$$\begin{cases} I_\phi(x)\dfrac{\partial^2 \varphi}{\partial t^2} - \dfrac{\partial^2}{\partial x^2}\left(C_\phi \dfrac{d\varphi(x,t)}{\partial x} \right) - k_\phi \dfrac{\partial^2}{\partial t \, \partial x}\left(C_\phi \dfrac{d\varphi(x,t)}{\partial x} \right) = \\ \qquad = \sum\limits_{i=1}^{n} (F_{li} - F_{ri}) \, d_i \delta(x - x_i) \ ; \\ \dfrac{\partial^2}{\partial x^2}\left(EI_z(x,t)\dfrac{d^2 z(x,t)}{\partial x^2} \right) k_z + \dfrac{\partial^3}{\partial t \, \partial x^2}\left(EI_z(x,t)\dfrac{d^2 z(x,t)}{\partial x^2} \right) + \\ \qquad + m_0(x)\dfrac{\partial^2 z(x,t)}{\partial t^2} = \sum\limits_{i=1}^{n} (F_{li} + F_{ri}) \, \delta(x - x_i) - m_0(x) g \ , \end{cases} \tag{12.75}$$

The surges (jumps) in the shearing forces and torques in the suspension's attachment points are accounted for by the structure of equations in (12.75). It is also necessary to take into account the conditions at the ends of the rod (bending and twisting moments and the shearing forces on the free ends of the rod should be absent—equal to zero):

$$\begin{cases} EI_z \left.\dfrac{\partial^2 z}{\partial x^2}\right|_{x=0,\ x=L} = 0 \ ; \\ \left.\dfrac{\partial}{\partial x}\left[EI_z \dfrac{\partial^2 z}{\partial x^2} \right]\right|_{x=0,\ x=L} = 0 \ ; \\ C_\phi \left.\dfrac{\partial \phi}{\partial x}\right|_{x=0,\ x=L} = 0 \ . \end{cases} \tag{12.76}$$

In order to preserve the general nature of the models describing with different approximation degrees the oscillations of a wheeled chassis, *Calculation of Operational Motion Parameters of [an] Automobile and Train* [65] suggests to consider only low-frequency oscillations and replace the differential equations in private derivatives by a system of ordinary differential equations by using the Bubnov-Galerkin method [65, 68].

The solutions of the equations in (12.75) are sought for in the following form:

$$\begin{cases} z(x,t) = \sum_{j=-1}^{N} f_j(x)\,\theta_j(t)\ ; \\ \phi(x,t) = \sum_{k=0}^{N_1} \psi_k(x)\,\phi_k(t)\ . \end{cases} \tag{12.77}$$

The forms of the bending and twisting oscillations of the rod with free ends, which has the same distribution of masses and rigidities as the rod simulating the carrying system, are used as coordinate functions. This means that the coordinate functions are solutions of the following boundary problems:

$$\begin{cases} \frac{\partial^2}{\partial x^2}\left[EI_z \frac{\partial^2 f(x)}{\partial x^2}\right] - m_0(x)\,p^2 f(x)\ ; \\ EI_z \left.\frac{\partial^2 f(x)}{\partial x^2}\right|_{x=0,\ x=L} = 0\ ; \\ \left.\frac{\partial}{\partial x}\left[EI_z \frac{\partial^2 f(x)}{\partial x^2}\right]\right|_{x=0,\ x=L} = 0\ . \end{cases} \tag{12.78}$$

$$\begin{cases} \frac{\partial}{\partial x}\left[C_\phi \frac{\partial \psi(x)}{\partial x}\right] + I_\phi(x)\,p^2 \psi(x)\ ; \\ C_\phi \left.\frac{\partial \psi(x)}{\partial x}\right|_{x=0,\ x=L} = 0\ , \end{cases} \tag{12.79}$$

where p is the frequency of the free rod's oscillations.

The solutions of the boundary problems in (12.78) and (12.79) are satisfied by orthogonality conditions [65, 68]:

$$\begin{cases} \int_0^L m_0(x)\,f_i(x)\,f_k(x)\,dx = 0\ , \quad j \neq k\ ; \\ \int_0^L I_\phi(x)\,\psi_i(x)\,\psi_k(x)\,dx = 0\ , \quad j \neq k \end{cases} \tag{12.80}$$

or:

$$\begin{cases} \int_0^L EI_z(x) \frac{d^2 f_i(x)}{dx^2} \frac{d^2 f_k(x)}{dx^2}\,dx = 0\ , \quad j \neq k\ ; \\ \int_0^L C_\phi(x) \frac{d\psi_j(x)}{dx} \frac{d\psi_k(x)}{dx}\,dx = 0\ , \quad j \neq k\ . \end{cases} \tag{12.81}$$

It is natural that the solutions of the boundary problems in (12.78) and (12.79) are the functions describing the free rod's oscillations as a solid body (at $p = 0$, it is checked by the direct substitution):

$$\begin{cases} f_{-1}(x) = 1\ ; \\ f_0(x) = x - x_C\ ; \\ \psi_0(x) = 1\ , \end{cases} \tag{12.82}$$

where x_C is the coordinate of the rod's center of mass of the rod.

Then, following *Calculation of Operational Motion Parameters of [an] Automobile and Train* [65], the following transformations are made:

- The expressions in (12.77), which define the adopted solution structures, are substituted into the initial system of differential equations in (12.75);
- The right- and left-hand parts of the first equation of the obtained system are multiplied by $f_j(x)$; and
- Integration is made within the limit from 0 to L:

$$\int_0^L f_j(x)\frac{\partial^2}{\partial x^2}\left(EI_z(x,t)\frac{\partial^2\left[\sum_{k=-1}^{N} f_k(x)\,\theta_k(t)\right]}{\partial x^2}\right) dx +$$

$$+\int_0^L k_z\, f_j(x)\frac{\partial^3}{\partial t\,\partial x^2}\left(EI_z(x,t)\frac{\partial^2\left[\sum_{k=-1}^{N} f_k(x)\,\theta_k(t)\right]}{\partial x^2}\right)+$$

$$+\int_0^L m_0\, f_j(x)\frac{\partial^2\left[\sum_{k=-1}^{N} f_k(x)\,\theta_k(t)\right]}{\partial t^2} =$$

$$= \int_0^L f_j(x)\sum_{i=1}^{n}(F_{li}+F_{ri})\,\delta(x-x_i) - \int_0^L m_0(x)\,f_j(x)\,g\ , \qquad (12.83)$$

where $j = -1.0\ldots N$.

It is clear that by changing the integration and summation order, formula (12.83) can be presented as follows:

$$\sum_{k=-1}^{N}\left\{\theta_k(t)\int_0^L f_j(x)\frac{\partial^2}{\partial x^2}\left(EI_z(x,t)\frac{\partial^2 f_k(x)}{\partial x^2}\right)dx + \right.$$
$$+k_z\dot{\theta}_k(t)\int_0^L f_j(x)\frac{\partial^2}{\partial x^2}\left(EI_z(x,t)\frac{\partial^2 f_k(x)}{\partial x^2}\right)dx +$$
$$+\ddot{\theta}_k(t)\int_0^L m_0(x)\,f_j(x)\,f_k(x)\,dx =$$
$$\left. = \sum_{i=1}^{n}(F_{li}+F_{ri})\int_0^L f_j(x)\,\delta(x-x_i)\,dx - g\int_0^L m_0(x)\,f_j(x)\,dx\right\}. \qquad (12.84)$$

The integrals, which form part of equation (12.84), can be transformed as follows.

The first integral is integrated twice in parts and presented as:

$$\int_0^L f_j(x)\frac{\partial^2}{\partial x^2}\left(EI_z(x,t)\frac{\partial^2 f_k(x)}{\partial x^2}\right)dx = \left[f_j(x)\frac{\partial^2}{\partial x^2}\left(EI_z(x,t)\frac{\partial^2 f_k(x)}{\partial x^2}\right)\right]_0^L -$$

$$-\int_0^L\left[\frac{\partial}{\partial x}\left(EI_z(x,t)\frac{\partial^2 f_k(x)}{\partial x^2}\right)\right]\cdot\frac{\partial f_i(x)}{\partial x}\,dx = \left[f_j(x)\frac{\partial^2}{\partial x^2}\left(EI_z(x,t)\frac{\partial^2 f_k(x)}{\partial x^2}\right)\right]_0^L -$$

$$-\left[\frac{\partial}{\partial x}\left(EI_z(x,t)\frac{\partial^2 f_k(x)}{\partial x^2}\right)\right]_0^L + \int_0^L EI_z(x,t)\frac{\partial^2 f_k(x)}{\partial x^2}\,\frac{\partial f_i(x)}{\partial x^2}\,dx\ . \qquad (12.85)$$

Then, we take the boundary conditions into account (12.76):

$$\int_0^L f_j(x)\frac{\partial^2}{\partial x^2}\left(EI_z(x,t)\frac{\partial^2 f_k(x)}{\partial x^2}\right)dx = \int_0^L EI_z(x,t)\frac{\partial^2 f_k(x)}{\partial x^2}\,\frac{\partial^2 f_i(x)}{\partial x^2}\,dx \qquad (12.86)$$

And we account for the orthogonality conditions and introduce the abridged symbol for the first integral:

$$\begin{cases} C_{jk} = \int_0^L EI_z(x,t)\dfrac{\partial^2 f_k(x)}{\partial x^2}\,\dfrac{\partial^2 f_j(x)}{\partial x^2}\,dx = 0\ , \quad \mathrm{j}\neq\mathrm{k}\ ; \\ C_{jj} = \int_0^L EI_z(x,t)\left[\dfrac{\partial^2 f_i(x)}{\partial x^2}\right]^2 dx\ . \end{cases} \qquad (12.87)$$

It is obvious that the transformation results of the second integral differ from those of the first one only by the multiplier k_z. By introducing the notation $b_{jj} = k_z C_{jj}$, we pass over to the transformations of the third integral.

The condition of orthogonality (12.80) can be directly applied to it; therefore, by introducing the symbol a_{jk}, we immediately obtain:

$$\begin{cases} a_{jk} = \int_0^L m_0(x)\,f_j(x)\,f_k(x)\,dx = 0\ , \quad j\neq k\ ; \\ a_{jj} = \int_0^L m_0(x)\,f_j(x)^2\,dx\ . \end{cases} \qquad (12.88)$$

The fourth integral is transformed using the definition and properties of the δ-function:

$$\int_0^L f_j(x)\,\delta(x-x_i)\,dx = f_i(x_i)\,. \qquad (12.89)$$

Finally, the last integral is designated as G_{jk} and, with the help of the orthogonality condition (12.80), we have:

$$G_{jk} = g\int_0^L m_0(x)\,f_j(x)dx = \begin{cases} 0,\ j\,, & k \neq -1\,; \\ M_n g,\ j\,, & k = -1\,. \end{cases} \qquad (12.90)$$

As a result of the transformations, the system of equations in (12.84) takes the following form:

$$\sum_{k=-1}^{N}\left[\,C_{jj}\theta_k(t)+ b_{jj}\dot\theta_k(t)+ a_{jj}\ddot\theta_k(t)\right]= \sum_{i=1}^{n}(F_{li}+ F_{ri})\,f_i(x_i)- G_{jj} \qquad (12.91)$$

Finally, we have:

$$C_{jj}\theta_j(t)+ b_{jj}\dot\theta_j(t)+ a_{jj}\ddot\theta_j(t)= \sum_{i=1}^{n}(F_{li}+ F_{ri})\,f_i(x_i)- G_{jj} \qquad (12.92)$$

where $j = -1, 0, 1, \ldots, N$.

Let us now make the same transformations with the first equation of the system in (12.75).

The result of substituting the function in (12.77), multiplication by $\psi_j(x)$ and integration looks as follows:

$$\int_0^L \psi_j(x)\,I_\phi(x)\frac{\partial^2\left[\sum_{k=0}^{N_1}\psi_k(x)\,\phi_k(t)\right]}{\partial t^2}$$

$$dx - \int_0^L \psi_j(x)\frac{\partial^2}{\partial x^2}\;\frac{\partial\left[\sum_{k=0}^{N_1}\psi_k(x)\,\phi_k(t)\right]}{\partial x}\;dx -$$

$$-\int_0^L k_\phi\psi_j(x)\frac{\partial^2}{\partial t\,\partial x}\left(C_\phi\frac{\partial\left[\sum_{k=0}^{N_1}\psi_k(x)\,\phi_k(t)\right]}{\partial x}\right)dx =$$

$$= \int_0^L \psi_j(x)\sum_{i=1}^{n}(F_{li}- F_{ri})\,d_i\delta(x- x_i)\,dx\,. \qquad (12.93)$$

By changing the summation and integration order, we obtain the following:

$$\sum_{k=0}^{N_1}\left\{\ddot\phi_k(t)\int_0^L I_\phi(x)\psi_j(x)\psi_k(x)dx - \phi_k(t)\int_0^L \psi_j(x)\frac{\partial^2}{\partial x^2}\left(C_\phi\frac{\partial\psi_k(x)}{\partial x}\right)dx -\right.$$

$$\left.-k_\phi\dot\phi_k(t)\int_0^L \psi_j(x)\frac{\partial}{\partial x}\left(C_\phi\frac{\partial\psi_k(x)}{\partial x}\right)dx\right\} =$$

$$= \sum_{i=1}^{n}(F_{li}- F_{ri})\int_0^L \psi_j(x)\,d_i\,\delta(x- x_i)\,dx\,. \qquad (12.94)$$

The transformation result of the first integral of the formula in (12.94) by the orthogonality condition looks as follows:

$$\begin{cases} a'_{jk} = \int_0^L I_\phi(x)\,\psi_j(x)\,\psi_k(x)\,dx = 0\,, & j \neq k\,; \\ a'_{kk} = \int_0^L I_\phi(x)\,\psi_k(x)^2\,dx = 0\,, & j = k\,. \end{cases} \qquad (12.95)$$

As a result of transforming the second integral of the formula in (12.94), we have:

$$\int_0^L \frac{\partial^2}{\partial x^2}\left(C_\phi\frac{\partial\psi_k(x)}{\partial x}\right)\psi_j(x)\,dx = \frac{\partial}{\partial x}\left(C_\phi\frac{\partial\psi_k(x)}{\partial x}\right)\psi_j(x)\Bigg|_0^L -$$

$$-\int_0^L \frac{\partial}{\partial x}\left(C_\phi\frac{\partial\psi_k(x)}{\partial x}\right)\frac{\partial\psi_j(x)}{\partial x}dx = \frac{\partial}{\partial x}\left(C_\phi\frac{\partial\psi_k(x)}{\partial x}\right)\psi_j(x)\Bigg|_0^L -$$

$$-\left(C_\phi\frac{\partial\psi_k(x)}{\partial x}\right)\frac{\partial\psi_j(x)}{\partial x}\Bigg|_0^L + \int_0^L\left(C_\phi\frac{\partial\psi_k(x)}{\partial x}\right)\frac{\partial\psi_j(x)}{\partial x}dx \qquad (12.96)$$

With account of the boundary conditions and the use of the orthogonality condition, formula (12.96) takes the following form:

$$\begin{cases} C'_{jk} = \int_0^L \psi_j(x) \frac{\partial^2}{\partial x^2}\left(C_\varphi \frac{\partial \psi_k(x)}{\partial x}\right) dx = \\ \quad = \int_0^L \frac{\partial\, \psi_j(x)}{\partial x}\left(C_\varphi \frac{\partial \psi_k(x)}{\partial x}\right) dx = 0\ , \quad j \neq k; \\ C'_{kk} = \int_0^L C_\varphi \left(\frac{\partial \psi_k(x)}{\partial x}\right)^2 dx = 0, \quad j = k \end{cases} \tag{12.97}$$

The transformation result of the free member is:

$$\sum_{i=1}^{n} (F_{li} - F_{ri}) \int_0^L \psi_k(x)\, d_i\, \delta(x - x_i)\, dx = \sum_{i=1}^{n} (F_{li} - F_{ri})\, d_i\, \psi_k(x_i)\, . \tag{12.98}$$

By transforming the first equation of the system in (12.93), we obtain:

$$a'_{kk}\varphi_k(t) - C'_{kk}\varphi_k(t) - b'_{kk}\varphi_k(t) = \sum_{i=1}^{n} (F_{li} - F_{ri}) d_i \psi_k(x_i) \tag{12.99}$$

where $k = 0, 1, \ldots, N_1$, $b_{kk} = k_\varphi C_{kk}$.

For $j = -1$, with account of the fact that in accordance with (12.82) $f_{-1} = 1$ and by introducing $\theta_{-1} = z_0$, we get the following out of (12.92):

$$\begin{cases} M_n \ddot{z}_0(t) = \sum_{i=1}^{n} (F_{li} + F_{ri}) - M_n g\ ; \\ \mathrm{M_n} = \int_0^L m_0(x) dx\ . \end{cases} \tag{12.100}$$

For $j = 0$, with account of the fact that in accordance with (12.82) $f_0(x) = (x - x_c)$ and by introducing $\theta_0 = \alpha$, we have the following from (12.92):

$$\begin{cases} I_{yy}\ddot{\alpha}(t) = \sum_{i=1}^{n} (F_{li} + F_{ri})(x_i - x_C)\ ; \\ \mathrm{I_{yy}} = \int_0^L (x_i - x_C)^2 m_0(x) dx\ . \end{cases} \tag{12.101}$$

And for $k = 0$, with account of the fact that $\psi_0(x) = 1$, we have:

$$\begin{cases} I_{xx}\varphi_k(t) = \sum_{i=1}^{n} (F_{li} + F_{ri})\, d_i\ ; \\ \mathrm{I_{xx}} = \int_0^L I_\varphi(x)\, dx\ . \end{cases} \tag{12.102}$$

For the values of j and k different from one, we obtain a system of ordinary differential equations:

$$\begin{cases} a_{jj}\ddot{\theta}_j(t) + b_{jj}\dot{\theta}_j(t) + c_{jj}\theta_j(t) = \\ \quad = \sum_{i=1}^{n} (F_{li} + F_{ri})\, f_i(x_i) - G_{jj}\ , \quad j = 1, \ldots, N; \\ a'_{kk}\, \ddot{\varphi}_k\, \varphi_k(t) + c'_{kk}\, \dot{\varphi}_k(t) - b'_{kk}\varphi_k(t) = \\ \quad = \sum_{i=1}^{n} (F_{li} - F_{ri})\, d_i \psi_k(x_i)\ , \quad k = 1, \ldots, N_1. \end{cases} \tag{12.103}$$

Equations (12.100–12.103) describe oscillations of a multi-axle wheeled chassis' carrying system as discrete systems. The first group of system equations (12.103) describes the carrying system's bending oscillations, while the second group describes its twisting oscillations.

In the process of deriving these equations, we did not use any assumptions about the design and make of the springing system, while the axle diagram of the support-running complex was used only as an example. This means that the obtained model is applicable for studying oscillations of multi-axle wheeled chassis with any support-running complex.

12.7 Spatial model of oscillations in a multi-axle saddle road train

Chapters 1–4 showed that one of the rather broadly popular models of heavy-duty multi-axle vehicles is a multi-link road train and most often the two-link saddle layout is used. This type of transporter is not described by the earlier considered models; however, when designing a specialized calculation sequence, one can use as a whole the earlier adopted system of assumptions and restrictions (Fig. 12.36).

In particular, it is admitted that the road train is symmetrical relative to its longitudinal plane and moves rectilinearly at a constant speed; we can neglect the influence of tangential reactions on the links' vertical oscillations. Oscillations of the road train's links are considered small and the influence of the lateral reactions on the links' vertical oscillations is disregarded.

The carrying systems of the links are simulated—same as in point 12.6—by the rods, which have variable rigidities, unit inertia moments and masses, and workings for the twisting and bending in the vertical plane. The bending in the horizontal plane and the longitudinal compressions of the rods is not taken into account.

As earlier, we assume that the concentrated masses are rigidly connected to the frames of the road train's links.

When simulating the bend of the rods, we disregard the inertia of the cross-sections' turn and the shear deformation; and we also assume that the rod's cross-section points, which are on the axis of its rigidity, move vertically inline. When simulating the twisting of the rods, we disregard the phenomenon of the cross-sections' warping (twist).

The friction in the carrying system is accounted for under the Voigt hypothesis (viscous friction hypothesis).

It is considered that the drawbar has its resilience in the vertical and longitudinal directions and also at relative lateral bends of the links' carrying systems (it is assumed, however, that under

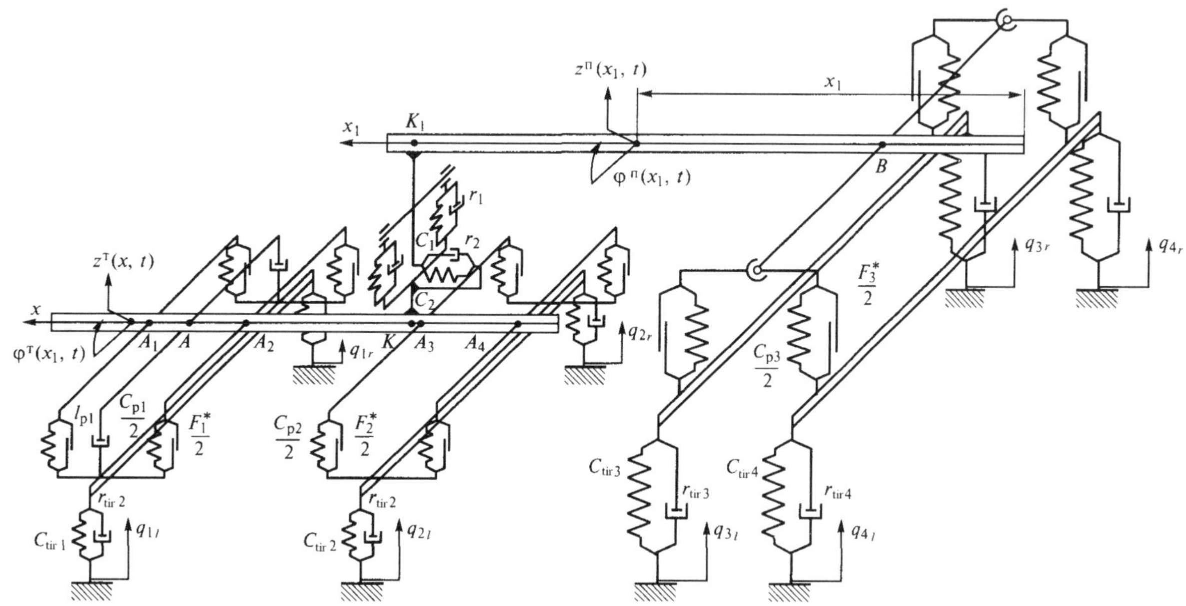

Fig. 12.36. Spatial model of oscillations in a saddle road train

relative turns of the links in the longitudinal plane, no resilient forces appear, but that viscous or dry friction is taking place).

Finally, all of the generalized coordinates are measured from the position of the system's static equilibrium.

Fig. 12.36 uses the following legend for a saddle drawbar: C_1 is the vertical stiffness; r_1 is the viscous friction coefficient (in the vertical direction); C_2 is the longitudinal stiffness; r_2 is the viscous friction coefficient (in the longitudinal direction); and the other elements are the same as in Fig. 12.35.

Given that all of the above conditions are met, the differential equation of the rod's bending oscillations simulating the tow car's carrying system has the following form (see also the formulas in (12.75)):

$$\frac{\partial^2}{\partial x^2}\left(EI_{zm}(x,t)\frac{\partial^2 z_m(x,t)}{\partial x^2}\right)+ k_{zm}\frac{\partial^2}{\partial t\,\partial x^2}\left(EI_{zm}(x,t)\frac{\partial^2 z_m(x,t)}{\partial x^2}\right)+$$

$$+m_m(x)\frac{\partial^2 z_m(x,t)}{\partial t^2}= \sum_{i=1}^{n}(F_{li}+F_{ri})\,\delta(x-x_i)-F_{Kx}hK\frac{d}{dx}\delta(x-x_i)-$$

$$-F_{Kx}\delta(x-x_K)+M_{sp}\frac{d}{dx}\delta(x-x_K) \qquad (12.104)$$

The equation of the rod's twisting oscillations simulating the tow car's carrying system is as follows:

$$I_{\phi m}(x)\frac{\partial^2\phi_m(x,t)}{\partial t^2}-\frac{\partial^2}{\partial x^2}\left(C_{\phi m}\frac{\partial\phi_m(x,t)}{\partial x}\right)-$$

$$-k_{\phi m}\frac{\partial^2}{\partial t\,\partial x^2}\left(C_{\phi m}\frac{\partial\phi_m(x,t)}{\partial x}\right)=$$

$$=\sum_{i=1}^{n}(F_{li}-F_{ri})\,d_i\,\delta(x-x_i)-M^{*}_{sp}\,\delta(x-x_K)+m_n g\,h(x)\,\varphi_m(x,t) \qquad (12.105)$$

Formulas (12.104) and (12.105) use the following new elements (in these formulas, the rod simulates the tow car's carrying system):

- $EI_{zm}(x)$ is the rod's bending stiffness;
- $C\varphi(x)$ is the rod's twisting stiffness;
- $z_m(x, t)$ is the vertical displacement of the rod's rigidity axis;
- $\varphi_m(x, t)$ is the lateral inclination angle of the rod's cross-section;
- k_{zm} is the coefficient characterizing the internal friction in the rod (at its bending);
- $k_{\varphi m}$ is the coefficient characterizing the internal friction in the rod (at its twisting);
- $m_m(x)$ is the rod's unit mass (including the concentrated masses fixed on the tow car's frame);
- $I_{\varphi m}(x)$ is the rod's unit inertia moment (including the inertia moments of the aggregates fixed to the tow car's frame);
- F_{Kx} is the longitudinal component of the reaction in the saddle-drawbar;
- F_{Kz} is the vertical component of the reaction in the saddle-drawbar;
- M^{*}_{sp} is the total moment arising in the saddle-drawbar at the relative lateral turn of the carrying systems of the tow car and semi-trailer;
- h_K is the distance (vertically) from the saddle-drawbar's axis to the rod's rigidity axis; and
- $h(x)$ is the distance (vertically) from the rod's rigidity axis to the gravity centers of its cross-sections.

The differential equations of the bending and twisting oscillations of the rod simulating the carrying system of a semi-trailer can be written as follows:

$$\frac{\partial^2}{\partial x_1^2}\left(EI_{zn}(x_1,t)\frac{\partial^2 z_n(x_1,t)}{\partial x_1^2}\right)+$$

$$+k_{zn}\frac{\partial^3}{\partial t\,\partial x_1^2}\left(EI_{zn}(x_1,t)\frac{\partial^2 z_n(x_1,t)}{\partial x_1^2}\right)+$$

$$+m_n(x_1)\frac{\partial^2 z_n(x_1,t)}{\partial t^2}=\sum_{i=n+1}^{m}(F_{li}+F_{ri})\,\delta(x_1-x_{1i})+F_{Kx}h_{K1}$$

$$- F_{Kz}\ \delta(x_1 - x_{K1}) - M_{sp}\frac{d}{dx_1}\delta(x_1 - x_{1K1}) \qquad (12.106)$$

$$I_{\phi n}(x_1)\frac{\partial^2\phi_n(x_1,t)}{\partial t^2} - \frac{\partial^2}{\partial x_1^2}\left(C_{\phi n}\frac{\partial\phi_n(x_1,t)}{\partial x_1}\right) -$$

$$- k_{\phi n}\frac{\partial^2}{\partial t\ \partial x_1^2}\left(C_{\phi n}\frac{\partial\phi_n(x_1,t)}{\partial x}\right) =$$

$$= \sum_{i=n}^{m}(F_{li} - F_{ri})\ d_i\ \delta(x_1 - x_{1i}) + M_{sp}^{*}\ \delta(x_1 - x_{1K1}) +$$

$$+ m_n g\ h_1(x_1)\ \varphi_m(x_1,t) \qquad (12.107)$$

Formulas (12.106) and (12.107) use the following additional designations:

- the index "n" marks the parameters characterizing the semi-trailer's carrying system;
- x_1 is the longitudinal coordinate describing the position of the elements in the semi-trailer's carrying system;
- h_{K1} is the distance from the saddle-drawbar's axis to the rigidity axis of the rod simulating the semi-trailer's carrying system; and
- $h_1(x_1)$ is the distance (vertically) between the cross-sections' center of gravity in the rod simulating the semi-trailer's carrying system and its rigidity axis.

The forces acting in the saddle-drawbar are defined by the following relations:

$$M_{sp} = r_k\left(\left.\frac{\partial^2 z_m}{\partial t\ \partial x}\right|_{x=x_K} - \left.\frac{\partial^2 z_n}{\partial t\ \partial x_1}\right|_{x_1=x_{1K1}}\right) \qquad (12.108)$$

$$M_{sp}^{*} = 2r_1\ d_C^2\left(\left.\frac{\partial\varphi_m}{\partial t}\right|_{x=x_K} - \left.\frac{\partial\varphi_n}{\partial t}\right|_{x_1=x_{1K1}}\right) +$$

$$+ 2c_1\ d_C^2\left[\phi_m(x_K) - \phi_n(x_{1_{K1}})\right] \qquad (12.109)$$

$$F_{Kz} = 2c_1[z_m(x_K,t) - z_n(x_{1K1},t)] +$$

$$+ 2r_1[\dot{z}_m(x_K,t) - \dot{z}_n(x_{1K1},t)] \qquad (12.110)$$

$$F_{Kx} = C_2\left[\left(x_0 - h_K\left.\frac{\partial z_m}{\partial x}\right|_{x=x_K}\right) - \left(x_{01} - h_{K1}\left.\frac{\partial z_n}{\partial x_1}\right|_{x_1=x_{1K1}}\right)\right] +$$

$$+ r_2\left[\left(\dot{x}_0 - h_K\left.\frac{\partial^2 z_m}{\partial t\ \partial x}\right|_{x=x_K}\right) - \left(\dot{x}_{01} - h_{K1}\left.\frac{\partial^2 z_n}{\partial t\ \partial x_1}\right|_{x_1=x_{1K1}}\right)\right]$$

$$(12.111)$$

Formulas (12.108)–(12.111) use the following legend:

- r_K is the viscous friction coefficient;
- C_1 is the stiffness defining the vertical elastic forces acting in saddle-drawbar;
- r_1 is the viscous friction coefficient defining the vertical friction forces acting in saddle-drawbar;
- $2d_C$ is the distance between the saddle-drawbar's supports;
- C_2 is the stiffness defining the longitudinal elastic forces acting in the saddle-drawbar;
- r_2 is the viscous friction coefficient defining the longitudinal friction forces acting in the saddle-drawbar;
- x_0 is the longitudinal displacement of the tow car's center of mass; and
- x_{01} is the longitudinal displacement of the semi-trailer's center of mass.

In order to use the Bubnov-Galerkin method, the displacements of the cross-sections of the rods simulating their carrying systems are presented as follows:

$$z_m(x,t) \approx \sum_{j=-1}^{n_1}\theta_{mj}(t)f_{mj}(x) =$$

$$= z_{m0} + (x - x_C)\alpha_m + \sum_{j=1}^{n_1}\theta_{mj}(t)f_{mj}(x) \qquad (12.112)$$

$$\phi_m(x,t) \approx \sum_{k=0}^{n_2} \Phi_{mk}(x)\phi_{mk}(t) = \phi_{m0} + \sum_{k=1}^{n_2} \Phi_{mk}(x)\phi_{mk}(t) \tag{12.113}$$

$$z_n(x_1,t) \approx \sum_{j=-1}^{n_3} \theta_{nj}(t) f_{nj}(x_1) = z_{n01} + (x_1 - x_{1C1})\alpha_n + \sum_{j=1}^{n_3} \theta_{nj}(t) f_{nj}(x_1) \tag{12.114}$$

$$\phi_n(x_1,t) \approx \sum_{k=0}^{n_4} \Phi_{nk}(x_1)\phi_{nk}(t) = \phi_{n0} + \sum_{k=1}^{n_4} \Phi_{nk}(x_1)\phi_{nk}(t) \tag{12.115}$$

Formulas (12.112)–(12.115) use the following additional symbols:

- x_C is the center of mass coordinate for the rod simulating the tow car's carrying system;
- x_{C1} is the center of mass coordinate for the rod simulating the semi-trailer's carrying system;
- Φ_{mk} is the function describing the distribution of the relative turn angles of the tow car's frame along its length; and
- Φ_{nk} is the function describing the distribution of the cross-section's relative turn angles of the semi-trailer's frame along its length.

Same as before for the coordinate functions $f_j(x)$ and $\Phi_j(x)$ ($j = -1, 0, \ldots; k = 0, 1, \ldots$), we take the forms of the bending and twisting oscillations of the free rods simulating the carrying systems of the tow car and semi-trailer, which are defined through solving the following boundary problems:

$$\begin{cases} [EI_{zm}(x)f_m''(x)]'' - p^2 m_m f_m(x) = 0\ ; \\ EI_{zm}(x)f_m''(x)\big|_{x=0,\ x=L} = 0\ ; \\ [EI_{zm}(x)f_m''(x)]'\big|_{x=0,\ x=L} = 0\ ; \\ [EI_{zn}(x_1)f_n''(x_1)]'' - p^2 m_n f_n(x_1) = 0\ ; \\ EI_{zn}(x_1)f_n''(x_1)\big|_{x=0,\ x=L_1} = 0\ ; \\ [EI_{zn}(x_1)f_n''(x_1)]'\big|_{x=0,\ x=L_1} = 0\ ; \\ [C_{\phi m}(x)\Phi_m'(x)] + p^2 I_{\phi m}(x)\Phi_m(x) = 0\ ; \\ C_{\phi m}(x)\Phi_m'(x)\big|_{x=0,\ x=L} = 0\ ; \\ [C_{\phi n}(x_1)\Phi_n'(x_1)] + p^2 I_{\phi n}(x_1)\Phi_n(x_1) = 0\ ; \\ C_{\phi n}(x)\Phi_n'(x)\big|_{x=0,\ x=L_1} = 0\ . \end{cases} \tag{12.116}$$

The coordinate functions should satisfy the conditions of orthogonality at $j \neq k$:

$$\begin{cases} [EI_{zm}(x)f_m''(x)]'' - p^2 m_m f_m(x) = 0\ ; \\ EI_{zm}(x)f_m''(x)\big|_{x=0,\ x=L} = 0\ ; \\ [EI_{zm}(x)f_m''(x)]'\big|_{x=0,\ x=L} = 0\ ; \\ [EI_{zn}(x_1)f_n''(x_1)]'' - p^2 m_n f_n(x_1) = 0\ ; \\ EI_{zn}(x_1)f_n''(x_1)\big|_{x=0,\ x=L_1} = 0\ ; \\ [EI_{zn}(x_1)f_n''(x_1)]'\big|_{x=0,\ x=L_1} = 0\ ; \\ [C_{\phi m}(x)\Phi_m'(x)] + p^2 I_{\phi m}(x)\Phi_m(x) = 0\ ; \\ C_{\phi m}(x)\Phi_m'(x)\big|_{x=0,\ x=L} = 0\ ; \\ [C_{\phi n}(x_1)\Phi_n'(x_1)] + p^2 I_{\phi n}(x_1)\Phi_n(x_1) = 0\ ; \\ C_{\phi n}(x)\Phi_n'(x)\big|_{x=0,\ x=L_1} = 0\ . \end{cases} \tag{12.117}$$

$$\begin{cases} \int_0^L EI_{zm}(x)\dfrac{d^2 f_{jm}(x)}{dx^2}\ \dfrac{d^2 f_{km}(x)}{dx^2}dx = 0\ ; \\ \int_0^L C_{\phi m}(x)\dfrac{d\Phi_{jm}(x)}{dx}\ \dfrac{d\Phi_{km}(x)}{dx}dx = 0\ ; \\ \int_0^{L_1} EI_{zn}(x_1)\dfrac{d^2 f_{jn}(x_1)}{dx_1^2}\ \dfrac{d^2 f_{kn}(x_1)}{dx_1^2}dx_1 = 0\ ; \\ \int_0^{L_1} C_{\phi n}(x_1)\dfrac{d\Phi_{jn}(x_1)}{dx_1}\ \dfrac{d\Phi_{kn}(x_1)}{dx_1}dx_1 = 0\ . \end{cases} \tag{12.118}$$

While acting under the earlier described sequence, we can obtain a system of ordinary differential equations describing the oscillations of the tow car and semi-trailer's sprung masses with account of their carrying systems' resilience. These equations are complemented naturally by the equations of the oscillations in the tow car and semi-trailer's non-sprung masses; the methodology for deriving them is not different from the universally known one, while the form and structure are defined by particular design versions of the tow car and semi-trailer's support-running complexes.

The system of differential equations describing the oscillations in the tow car and semi-trailer's sprung masses with account of the carrying systems' resilience is as follows:

$$\begin{cases} \mathrm{M_m}\ddot{z}_{0m}(t)=\sum\limits_{i=n}^{m}(F_{li}-F_{ri})-F_{Kz}\ ; \\ M_m=\int\limits_0^L m_m(x)dx\ . \end{cases} \qquad (12.119)$$

$$\begin{cases} \mathrm{I_{yym}}\ddot{\alpha}_m(t)=\sum\limits_{i=1}^{n}(F_{li}-F_{ri})(x_i-x_C)-F_{Kz}(x_K-x_C)+F_{Kx}\mathrm{h}\ -M_{sp}\ ; \\ \mathrm{I_{yym}}=\int\limits_0^L(x_i-x_C)^2 m_m(x)dx\ . \end{cases} \qquad (12.120)$$

The system of equations in (12.116)–(12.120) is complemented by the equations of longitudinal oscillations in the tow car and semi-trailer; when deriving the latter, it is reasonable to assume that because the tangential reactions in the center of mass are so small, the whole system moves practically uniformly. This allows for considering only one generalized coordinate $x=(x_{()}-x_{(\Pi)})$ instead of two.

In this case the differential equation of the tow car and semi-trailer's longitudinal oscillations is as follows:

$$\frac{\partial^2}{\partial t^2}\overline{x}-F_{Kx}\left[\frac{1}{M_m+M_1+M_2}+\frac{1}{M_n+M_3+M_4}\right] \qquad (12.121)$$

where: M_m+M_1+M_2 is the full mass of the tow car and M_n+M_3+M_4 is the full mass of the semi-trailer.

The longitudinal accelerations of the tow car and semi-trailer's center of mass are defined from the following formulas:

$$\begin{cases} \dfrac{\partial^2}{\partial t^2}x_0=\dfrac{\partial^2}{\partial t^2}\overline{x}\left[\dfrac{M_m+\sum\limits_{i=1}^{n}M_i}{M_0}\right] \\ \dfrac{\partial^2}{\partial t^2}x_{01}=\dfrac{\partial^2}{\partial t^2}\overline{x}\left[\dfrac{M_n+\sum\limits_{i=n+1}^{m}M_i}{M_0}\right] \end{cases} \qquad (12.122)$$

where $M_0=M_n+\sum\limits_{i=1}^{m}M_i$ is the full mass of the road train.

Chapter 13
Fundamental theory and design of springing systems in HDWV wheels

13.1 Function, classification, and requirements for wheel suspensions

The basis of any vehicle's springing system (SpS) is a suspension. Springing systems are traditionally divided into primary and secondary suspensions. The springing systems allow for implementing the recommendations and meeting the requirements for the vehicle's oscillations and the vibration loads on the crew, which were considered in Chapter 12.

A suspension is a combination of devices and parts which connect the wheels with the frame or the vehicle's body (primary springing), or with the frame and the seats of the driver and passengers and with the frame and the body of the mounted equipment (secondary springing).

Springing systems are designed because of the need to ensure an acceptable level of soft (comfortable) driving, vibration loads on the crew, and continuous contact of the wheels with the road.

The suspension should ensure the required soft driving, which is understood as the vehicle's ability to move at the prescribed speed, guarantee the good condition and preservation of cargoes and the normal operation of mechanisms, and keep the vibration loads on the crew within the limit which does not cause any harmful effect on their physiological condition.

The suspension components are divided into three groups by their function: a resilient element, a guiding device, and a damping device (a shock absorber); in active and controlled suspensions the fourth element is added—an automated control system.

Suspensions are subdivided by the link's character (for example, of the wheels with the frame) into dependent, balancing (group-type), independent, and support-types.

The traditional designs of springing systems may include, as a resilient element, a leaf spring, a torsion bar, or a coil spring, and, as the damping element, a hydraulic shock absorber. Basically, they allow for satisfying the soft driving requirements for the existing models of multi-target vehicles and SWCs with load capacities of 14–18 t and 22–24 t. Along with that, the longstanding operation experience of vehicles with traditional springing systems in real conditions shows that such suspensions are essentially decreasing the vehicle's performance properties when traveling on Categories IV and V roads and in off-road conditions. Thus, according to certain data, the average travel speed decreases 2–3 times when moving on ground roads, the fuel consumption per unit of travel goes up 1.5–2 times, and the prime cost of transportation goes up 2–2.5 times as compared to the motion on roads covered with some modern pavement.

Springing systems of existing HDWVs are designed as independent with hydropneumatic elements, which are connected, as a rule, along the board by hydraulic balancing links. Such a design ensures satisfactory soft driving for SWCs of this class (the levels of maximum accelerations of transported cargoes does not exceed 1 g) and the load on the undercarriage. This is explained by the tough requirements for the motion routes and low speeds of SWC travel (no more than 30–40 km/h).

The less stringent requirements for the routes and higher travel speeds of current HDWVs dictate the need to radically change the wheel suspension's design in the near future, since the existing SpSs fail to resolve the following technical contradiction. On one hand, the high loading of the carrying system (frame) and undercarriage (axial static loads are over 150 kN) and the restricted ability to overcome unit obstacles (macro irregularities) cause a need to use "softer" long-stroke suspensions than the existing ones. On the other hand, the requirement to ensure a high lateral stability does not allow for using such "softer" suspensions. This technical contradiction can be resolved by absolutely new SpS designs: active or passive regulated (controlled) springing systems.

The theoretical aspects of active vibration-protected systems for a general case of springing a solid body and for a number of particular cases, for example, springing the suspension of the driver's seat, were studied by many scientists: V.Ya. Balagula, M.M. Gaitsgori, M.D. Genkin, V.V. Yablonskiy, S.A. Eliseev, S.V. Inosov, A.V. Sinyov, F.A. Furman, and others. The active springing of wheels as it applied to vehicles was considered in works by V.I. Koltsov, R.I. Furunzhiev, A.N. Ostanin, V.D. Sharapov and a number of authors abroad: H.B. Satton, A.G. Thompson, and others. All of the researchers have noted that active SpSs are most efficient from the viewpoint of soft driving.

Suspensions are treated as active when they execute a complex of technical measures for a partial or full compensation of deviations from the preset values of the current vertical surface forces, acting through the suspension elements on the body of a vehicle that moves on an irregular road.

Active suspensions can be of two essentially different types. Suspensions of the first type have no passive resilient-damping elements. The vehicle's body is linked to the wheels by non-resilient links, which are able to change the suspension stroke in accordance with a control signal. It is preferable to design such systems with control by perturbations (the so-called Poncelet principle).

Active suspensions of the second type have passive resilient-damping elements, which cause a time-variable effort in the suspension. The active elements installed in the suspension generate forces to damp the body's oscillations under the adopted algorithm. At present, this type of active suspensions is arousing the greatest practical interest for the following reasons:

1. Less power is needed to drive active elements since it is possible to restrict the execution speed of the control signal without a performance loss.
2. The control system may not respond quickly.
3. Higher reliability, first, because of a simpler control system, and, second, because the vehicle's performance is not lost at a failure of the control system.

Active suspensions ensure—apart from improved soft driving—less excessive "nosedives" at braking, take-off, and acceleration; smaller rolls at turns; and improved tire operation. All of these features can be decisive for using in SWCs; however, as to designing an active SpS, the following technical problems have not been solved so far:

- For an active suspension it is very important to correctly choose the observation time (averaging). At a short observation time, the evaluation will have a large final sampling error; therefore, the estimates obtained for a stationary process will essentially differ from each other. It is impossible to define whether these distinctions are random, or the static characteristics of the road have changed. At a longer period of observations, the estimate is rather precise, but the automobile may move for a long time on a cobble road with its suspension adjusted to asphalt, before the suspension gets readjusted;

- The decrease in the vibration load on the crew by using an active SpS is not obvious. The known designs of active suspensions operate at maximum frequencies of 2–3 Hz because of their high inertia, while the increased vibration load on the crew is caused by oscillations at frequencies of 4–8 Hz and more;
- Under certain conditions, the power consumption in an active springing system at vertical oscillations is rather high and with the growth in the SWC's travel speed, this consumption is monotonously growing (by its absolute value). The smallest ratio of the power consumed by an active SpS to the total consumed power at motion is registered on uniform roads with an asphalt-concrete pavement and is 3–7% in this case. On the same roads, but with worn-out pavement, and on ground roads in satisfactory condition, this ratio is 15–20%. On roads with cobble pavement and when the chassis moves off-road, this ratio reaches its maximum at 20–50%.

Thus, a broad use of active springing systems, especially for SWCs, in the near future is problematic because of the absence of the theoretical basis for designing them. That is why, at the initial stage, the solution of the problems related to ensuring the required values of loading on the undercarriage and carrying system, as well as the problems of the vibration loads on the crew, should be sought in studying the parameters and designing controlled suspensions.

A suspension is a controlled one if it has additional devices intended for changing its parameters and characteristics in the course of the HDWV's motion.

Passive regulated suspensions are the simplest by design out of the known controlled SpSs. Such suspensions are able to adjust their characteristics of resilient elements and (or) the characteristics of their shock absorbers depending on the full weight, road conditions, and motion modes. Passive suspensions correct characteristics of their individual subsystems in motion automatically or manually by the driver depending on the road type. These suspensions are applicable in SWCs when they operate on routes with long road sections similar in microprofile or when the route character and alternation in the road types are known in advance.

Based on the results of the authors' studies and analyses of the works on oscillations in HDWVs, we can formulate the following requirements for wheel suspensions:

1. High values of the working strokes (dynamic $h_d \geq 500$ mm and dynamic coefficient $K_d \approx 2.5$–3.0), which ensure the HDWV's motion without the suspension striking its stroke limiters (rests) and without a breakaway of the wheels from the support surface;
    a. The permissible values of partial frequencies in the sprung masses' free oscillations should be as follows: vertical oscillations $\omega_z = 10$–12 rad/s and angular oscillations $\omega_\varphi = 6.30$–9.50 rad/s; a maximum amplitude up to 4–5°; and vertical accelerations no more than 1.0–1.5 g;
    b. The required value of damping the oscillations, which ensures the amplitude decreases by 3–8 times per one swing of the free oscillations that corresponds to the decay coefficient $K_{dec} = 1.5$–2.5 rad/s;

2. The suspension parameters should ensure motion at a preset speed: not below 30–40 km/h on ground roads and at least 15–20 km/h on bumpy and worn-out roads and in off-road conditions;
3. The suspension should reliably transmit horizontal and lateral forces and their moments from the wheels to the frame and have the respective kinematics;
4. The suspension should reduce the impact of non-sprung masses and dynamic loads arising from them on the resilient elements and the vehicle's frame. This should result in decreased high-frequency oscillations (vibration) in the vehicle's body;
5. The suspension should ensure small changes to the wheels' rolling trajectories, as well as high stability and passing ability. Vertical displacements of the wheels relative to the frame should not cause any excessive lateral sliding and oscillations in the lateral plane and should not increase the load on the carrying system;
6. The suspension should have at least a capability to regulate the road clearance up to the vehicle's complete landing onto the ground and a capability to stabilize the sprung mass during the vehicle's motion. In certain cases, the suspension should be able to switch off (get blocked), for example, when working with crane equipment.

13.2 Analysis and calculation process of wheel suspensions

13.2.1 Traditional designs of wheel suspension systems in HDWVs

The design layouts and solutions for HDWV supports are greatly borrowed from aviation. Under aviation terminology, a support (trestle) is a structure comprising of a post (column) and supporting elements (wheels). The post is a design or mechanism that ensures the connection and interaction of the support elements (wheels) and carrying structure (the frame of the aircraft or vehicle).

Given the broad spread in scientific and technical literature on the vehicle theory of the notions of a support and post, they will be preferably used in this chapter.

A HDWV support is understood as the undercarriage of a multi-axle vehicle, which ensures the interaction of the carrying system (frame or body) with the road, as well as ensuring the vehicle's motion, braking, and maneuvering. A support is comprised of a post and a wheel mover. A post is a system of levers, shock absorbers, and the wheel's power drive, which ensures an interaction of the wheel mover and the vehicle's frame, as well as transmission of the power flow from the engine (the power unit) to the wheel. This section considers those supports (posts) which only perform the vehicle's wheel suspension functions.

All of the existing kinematical diagrams of the suspension's guiding mechanism can be conventionally split into three types:

- A post or candle-type suspension guiding mechanism, Fig. 13.1*a*;
- A lever-candle suspension (McPherson) or a post-resting suspension, Fig. 13.1*b*; and
- A suspension on double parallel lateral levers, Fig. 13.1*c*.

13.2.2 Post (candle-type) suspension

This suspension is comprised of a non-lever type guiding mechanism and a hydropneumatic resilient element as shown in Fig. 13.2.

The "aircraft" type guiding mechanism of the candle-type suspension consists of a rod connected with the motor-in-wheel's pivot axle and the guiding pipe connected with the vehicle's frame. The rod can move freely inside the guiding pipe in the axial direction and rotate inside it in case of controlled wheels. In the case of uncontrolled wheels, the rod is blocked

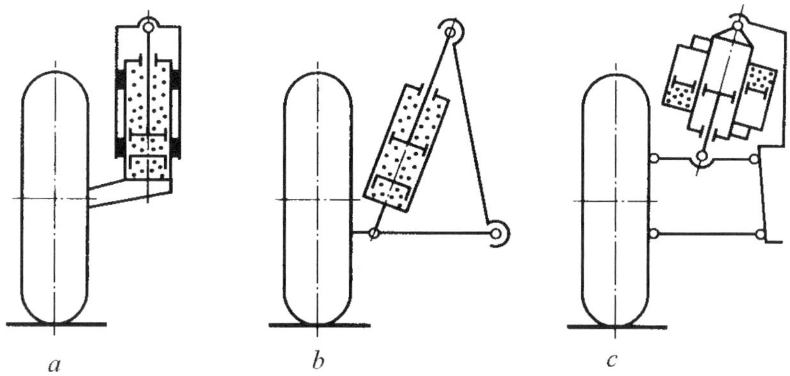

Fig. 13.1. Guiding mechanisms of suspensions in vehicles of high passing ability: a—Post (candle-type suspension); b—Lever-candle suspension; c— Suspension based on two parallel levers

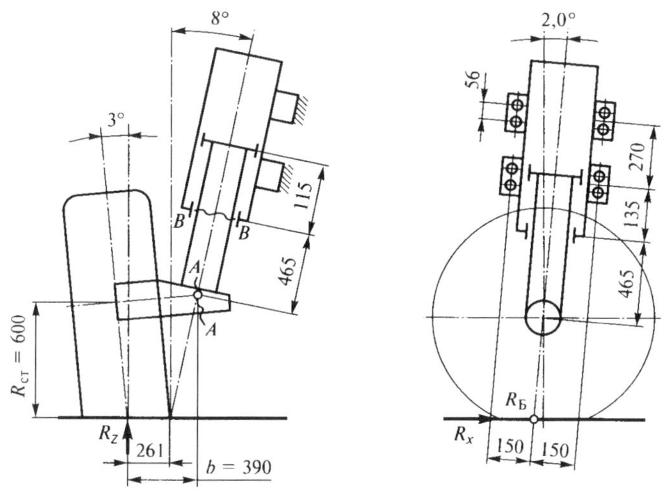

Fig. 13.2. Candle-type (post) suspension

through a two-link mechanism and a draft against a turn in relation to the vehicle's frame.

The basic advantages of this design as compared with the traditional one are as follows:

- A decreased change in the wheelspan at a vertical displacement of the wheel;
- A more favorable kinematics of wheel displacements;
- Compact dimensions and small needed volume;
- A broad range of the controlled wheels' turn angle, up to a full rotation around the guiding pipe;
- Fewer attachment points;
- Shipment to consumers as a single unit; and
- A possibility of using in different vehicles without any essential modernizations.

13.2.3 Suspension with a guiding post (lever-candle type)

In today's technical literature this type of suspension is treated as a suspension on the guiding (spring or shock-absorbing) post. The previous name was the lever-candle suspension or the McPherson suspension. The design and kinematical diagram (as applied to the 36970 chassis model) is presented in Fig. 13.3.

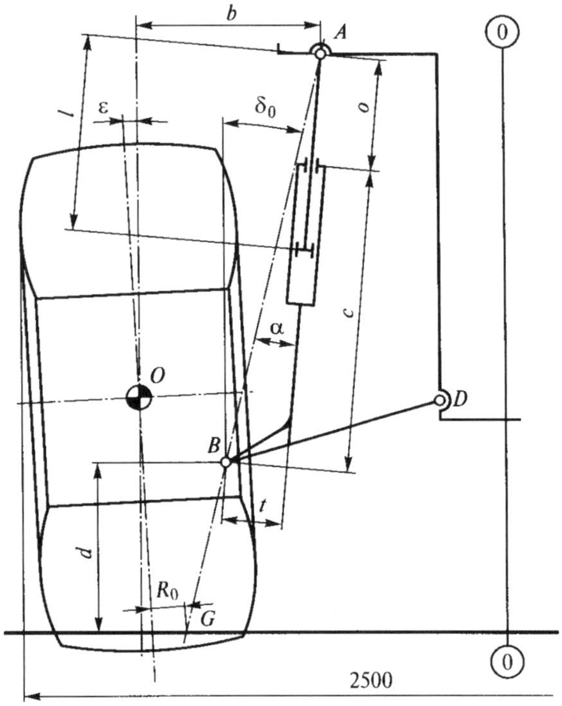

Fig. 13.3. Design-kinematical diagram of the lever-candle suspension

The main advantages of this suspension are as follows:

- The guiding elements are united into one assembly unit;
- Smaller efforts (as compared with a suspension on lateral levers) in attachments points *A* and *D*;
- A small distance between points *B* and *G*;
- A large installation volume in the body in the area of joint *A* (due to the absence of the upper lever);
- A somewhat smaller number of lubrication points; and
- A possibility to execute rather large suspension strokes.

The drawbacks of the suspension are as follows:

- Unfavorable kinematics;
- Impossibility to assemble the whole suspension as a single power element (beam), and, accordingly, an insufficient vibration insulation of the body;
- A weak counteraction to pitching at braking and galloping;
- Essential friction between the rod and the guide and in heavy-duty vehicles an inadmissibly high load on the rod and the need to have a huge guiding mechanism;
- Serious difficulties arising at designing the steering drive with satisfactory kinematics (long and highly located steering rods); and
- A small gap between the tire and post.

In modern suspensions, the roll center is located at about the same height as in suspensions on parallel levers and goes somewhat down when the suspension is loaded, since in order to raise it we need to excessively bend the post inward (see Fig. 13.3).

The suspension's kinematical properties can be improved mainly through elongating the lower levers. On passenger cars, a positive effect is also obtained by displacing point *B* into the wheels and by creating the so-called "roll-off arm."

The need to use a spherical joint in point *B* of heavy-duty vehicles may cause definite technological and performance problems (the experience of finishing GAZ-4905).

13.2.4 Suspension on lateral double levers

The main advantage of this type of suspension (Figs. 13.4 and 13.5) is in its broad kinematical opportunities, which allow for:

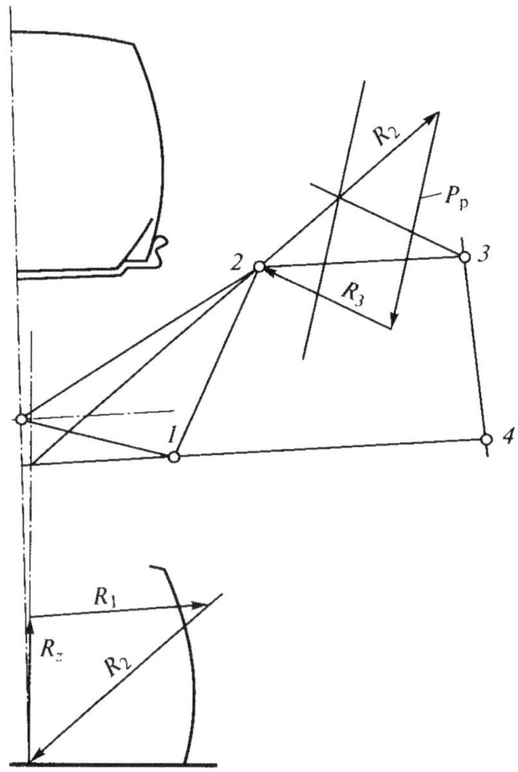

Fig. 13.4. Design and kinematical diagram of suspension on lateral levers

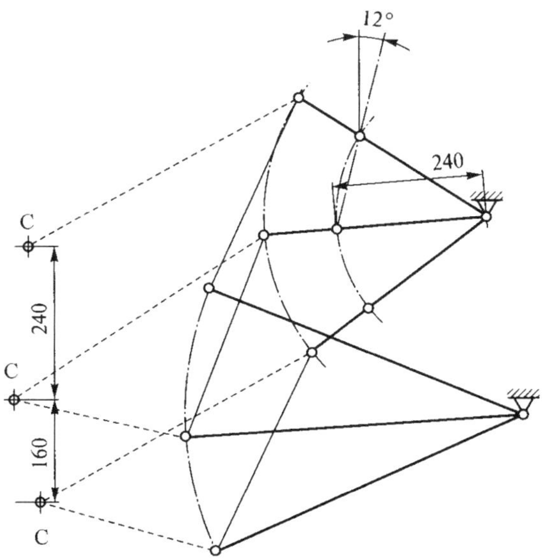

Fig. 13.5. Kinematical diagram of the suspension on double lateral levers

- Affecting the change of the roll and pitching centers' height by changing the mutual positions of the levers;
- Creating different dependences of the camber angle and wheelspan on the wheels' vertical displacements by changing the length of levers; and
- Counteracting the change in the wheel camber at the lateral roll of the body and decreasing the trend of the vehicle's pitching at braking and its tendency to galloping.

It is easy to build a suspension on lateral double levers in the form of an isolated unit, which has its independent noise-resistant attachment to the body.

The traditional steering drive nicely matches this suspension and the rods (drafts) can be placed at any needed height.

If the levers are long enough, we can implement essential suspension strokes as a nice match with the kinematics of the cardan and steering drives.

The main drawbacks of this suspension are as follows:

- The presence of the upper lever, which takes up a huge volume and prevents using a wide frame;
- A great number of lubrication points demanding a lot of time at operation;
- Difficulties in cleaning the guiding mechanism from dust and dirt; and
- A relatively large weight and lots of parts.

Among all of the known independent suspensions, this one is able to accept maximum relative loads at high service life; a known sample, which remained viable at a 15-fold impact overload, is the 019 model.

The most serious circumstance that restricts the progress of suspensions on lateral levers is an impossibility to use long upper levers in vehicles with the traditional girder frame and case-shaped vehicles. The use of the "backbone" frame reduces the effect of this drawback on the general vehicle specifications.

The other drawbacks can be essentially overcome by design and technological measures.

13.2.5 Non-traditional designs of HDWV wheel suspensions

At present, we see a trend toward using aviation chassis, which can take great loads, as long-stroke suspensions. The experience of making such chassis in aviation is huge and their designs are quite diverse. In all of them the power supports can be telescopic or lever-based (Figs. 13.6 and 13.7).

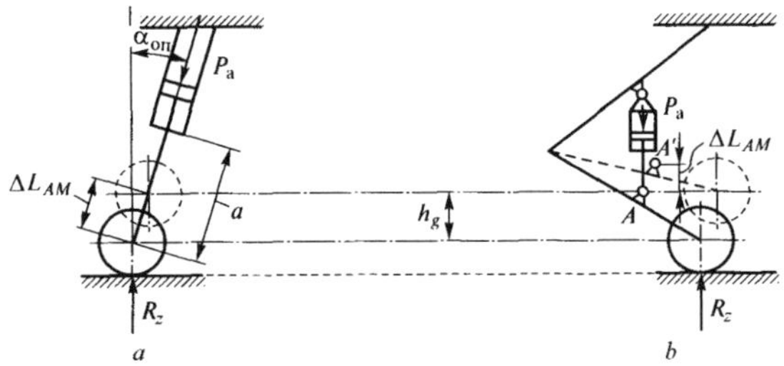

Fig. 13.6. Design of a power post: a—Telescopic; b—Lever-based

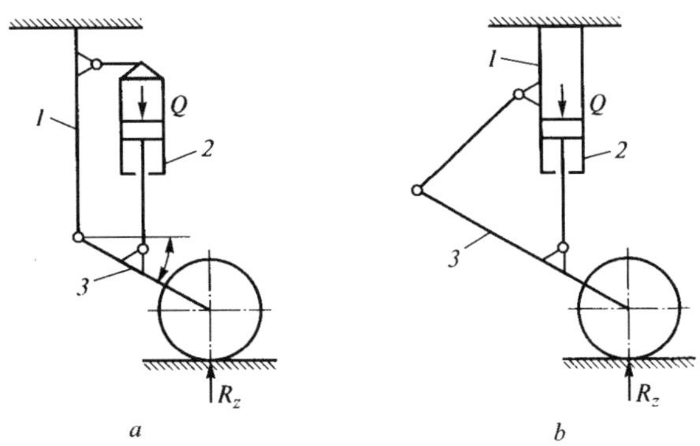

Fig. 13.7. Lever-based post with an external (a) and internal (b) shock absorber: 1—Post; 2—Shock absorber; 3—Lever

In a telescopic power support, a shock absorber is designed as an integral unit with the post. The wheel axle is rigidly linked with the shock absorber's rod and its displacement relative to the chassis' fixing point is equal to the compressions of the shock absorber $\Delta L_{H.s}$; then, the displacement of the wheel axle is $\Delta L_{axis} = \Delta L_{H.s} \cos\alpha_{rel}$. In a lever-type power post, the wheel axle is linked with the shock absorber by a lever. Here, the wheel axle's displacements and the shock absorber's compression may differ essentially: $\Delta L_{axis} = \Delta L_{H.s}\varphi$, where the transmission ratio φ is equal to the ratio of the force compressing the shock absorber to the load on the wheel axle.

Lever-type posts are subdivided into two characteristic groups: with an external shock absorber (made separately from the power post, Fig. 13.7*a*) and with an internal shock absorber (made integrally with the power post, Fig. 13.7*b*). There are also other designs of lever-type posts.

A comparative analysis of designs of automobile suspensions and aviation lever-type posts has shown that the main distinction of lever-type posts is in the fact the system of levers operates in the longitudinal-vertical plane, perpendicular to the wheel axle (Figs. 13.8–13.11); while in automobile suspensions, the levers work in the lateral-vertical plane. This distinction makes lever-type posts more sensitive to the affection of longitudinal forces acting on the wheel than automobile suspensions. That is why when studying HDWVs with supports in the system of differential equations (for example, equations (8.21) and (8.22)), we have to introduce the additional force F_x (see Figs. 13.8–13.10); it can be done by using a smaller value of P_K.

Various designs of posts can be schematically divided into crank-type posts (see Fig. 13.8), semi-lever posts (see Fig. 13.9), lever-type posts (see Fig. 13.10), and lever-balancing posts (see Fig. 13.11). In all figures, point *K* corresponds to the post's fixing point with the frame.

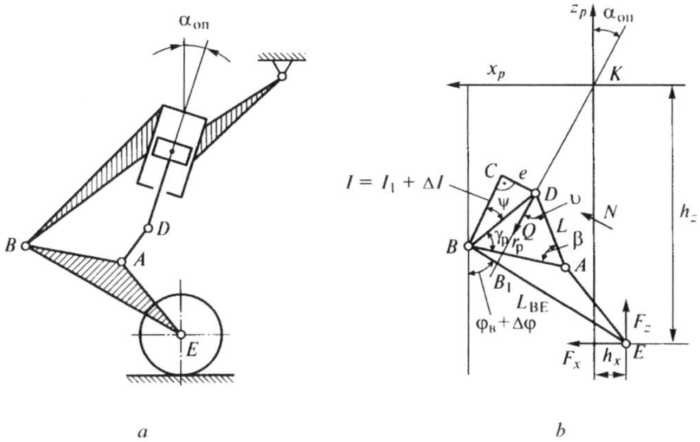

Fig. 13.8. Diagrams of a crank-type wheel post: a—Kinematical; b—Computational

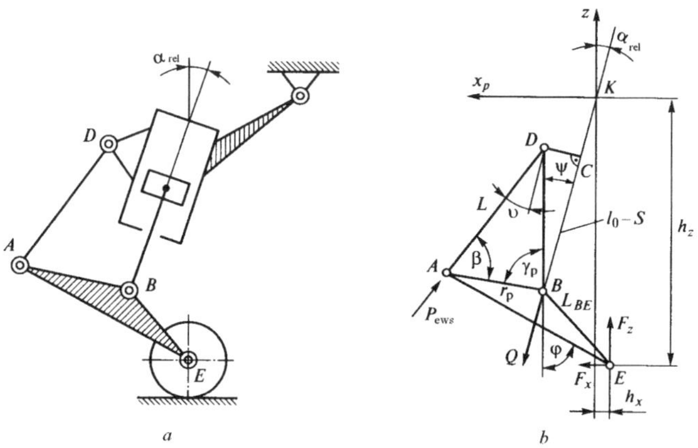

Fig. 13.9. Diagrams of a semi-lever wheel post: a—Kinematical; b—Computational

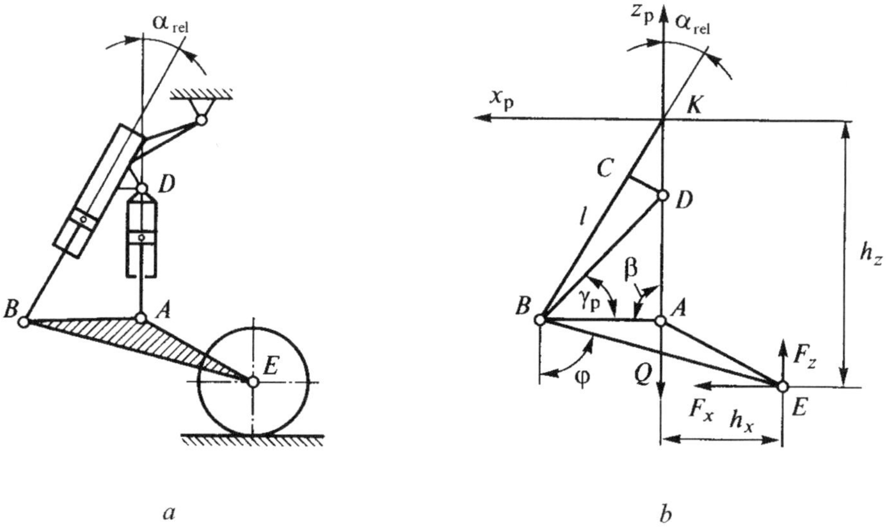

Fig. 13.10. Diagrams of a lever-type wheel post: a—Kinematical; b— Computational

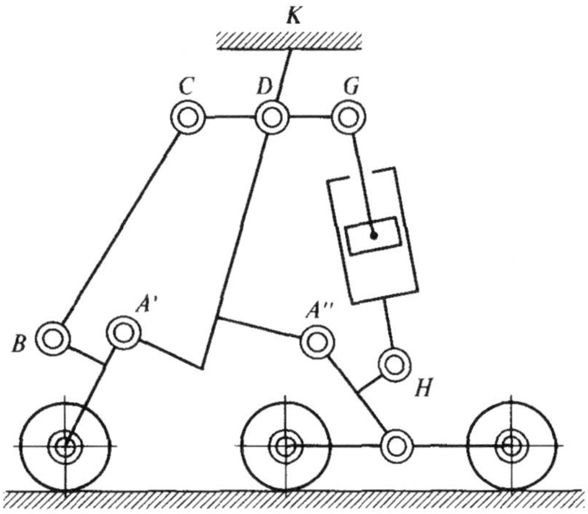

Fig. 13.11. Diagram of a lever-balancing wheel post

13.2.6 Kinematical calculation of support designs

The initial data for calculating and simulating both the lever-type posts and automobile suspensions are the following values: the static load on the wheel; the permissible maximum dynamic factor of vertical loads; the stiffness of the system reduced to the wheel axle; the characteristic of the shock absorber reduced to the wheel axle; the vehicle's road motion conditions; and the permissible dynamic wheel stroke (h_d).

The most important initial data for simulating the HDWV's motion are the kinematical and dynamic correlations of the support post. These correlations are easily defined based on the analysis of the diagram of the post's real design. For example, the derivation of these correlations for a crank-type post (Fig. 13.8) looks as follows.

Fig. 13.8 uses the following notations: points A and D are the connecting rod's fixing points to the lever and the shock absorber's rod respectively; B is the lever's rotation axis; C is the projection of point D onto the straight line passing through point B parallel to the shock absorber's axis; $\varphi = \varphi_0 + \Delta\varphi$ is the lever's turning angle; E is the wheel's rotation axis; h_z is the post's height; h_x is the displacement of axis E relative to the vertical axis passing through the post's fixing point to the frame (point K); φ_0 is the post's setting angle; ψ is the angle CBD; Θ is the angle BDA; γ_p is the angle DBA; β is the angle BAD; and α_{rel} is the adjustable post's setting angle.

The post's geometry dimensions are characterized by the following linear values: $L_{KB} = KB$; $e = CD$; $l = l_0 + \Delta l = CB$; $r_p = BA$; $L_{BE} = BE$; $L = AD$.

The values L_{KB}, l_0, and φ_0 correspond to the post's unloaded condition.

According to the introduced notations, the kinematical correlations of the post can be presented as follows:

$$\gamma_0 = \arccos\frac{r_p^2 + e^2 + L^2 + \ell_0^2}{2r_p\sqrt{e^2 + \ell_0^2}};$$

$$\varphi = \varphi_0 + \Delta\varphi_0 = \varphi_0 + \gamma_0 + \psi_0 - (\gamma_3 - \psi);$$

$$h_z = L_{KB}\cos\alpha_{rel} + L_{BE}\sin\varphi - e\sin\alpha_{rel};$$

$$\gamma_p = \arccos\frac{r_p^2 + e^2 + L^2 + (\ell_0 + \Delta\ell)^2}{2r_p\sqrt{e^2 + (\ell_0 + \Delta\ell)^2}};$$

$$h_x = -L_{KB}\sin\alpha_{rel} - L_{BE}\cos\varphi + e\cos\alpha_{rel};$$

$$\beta = \arccos\frac{L^2 + r_p^2 - e^2 + (\ell_0 + \Delta\ell)^2}{2Lr_p}.$$

In this case, the power correlation will look as follows:

$$Q = \psi_x F_x + \psi_z F_z;$$

$$N = \varphi_x F_x + \varphi_z F_z,$$

where:

$$\psi_x = -L_{BE}\frac{\cos\phi\ \cos\Theta}{r_p\sin\beta};$$

$$\psi_z = L_{BE}\frac{\sin\phi\ \cos\Theta}{r_p\sin\beta};$$

$$\phi_x = L_{BE}\frac{\cos\phi\ \cos\Theta}{r_p \sin\beta};$$

$$\phi_z = -L_{BE}\frac{\sin\phi\ \sin\Theta}{r_p \sin\beta}.$$

In spite of the complex design, lever-type posts most fully respond to the layout conditions and requirements for multi-support vehicles. They have a satisfactory traveling comfort, essential suspension strokes, a possibility to regulate the characteristics of the resilient element and shock absorbers, decrease the dynamic loads in suspensions, etc. That is why supports with lever-type posts have found broad use in multi-support vehicles and can be installed in future perspective HDWV models.

13.3 Active and regulated suspensions and platform stabilization systems

13.3.1 Passive suspensions with regulated characteristics of resilient elements

The aim of regulation in such suspensions is to stabilize the static and dynamic strokes or detuning the body's oscillation frequency from the dominant perturbation frequency.

Stabilizing the static stroke also means stabilizing the oscillation frequency at changing weights—within a broad range—of the sprung parts. The basic suspension diagram, which implements stabilization of the static and dynamic strokes, is presented in Fig. 13.12. Here, the frame (1) rests via pneumatic balloons (2), which are resilient elements of the suspension, on the wheel axle (3), which is connected to the frame with the help of a bracket and levers (4). The upper part of the levers (4) is connected by the resilient connection rod (5) with the cam (6) lever, located in the body (7), a valve, and a retarding mechanism. The cam (6) interacts with the valve's (9) pusher (8) and the retarding mechanism's piston (10).

When the load increases, the volume and height h_0 of the resilient elements decreases; the pusher (8) displaces to the left, opens the valve (9), and connects the balloons to the compressor. A decrease in the load results in the pusher's (8) displacement to the right and forms a gap between the pusher's face and the valve (9) and some air is discharged from the balloons to the atmosphere.

Thus, at any load, the height h_0 and the gas volume remain permanent by replacing the resilient element's mass with its characteristic by another resilient element with a new characteristic.

13.3.2 Suspensions with regulated rigid kinematical link

The purpose of regulation in such suspensions is to stabilize the distance from the body to the road pavement at any static loads at a constant characteristic of the resilient element. A fluid is used as a rigid kinematical link, which transfers the effort from the wheel axle K to the resilient element. That is why such regulation is used exclusively in pneumohydraulic suspensions.

As an example, let us consider the diagram in Fig. 13.13. Here, the frame (1) is connected with the help of the pneumohydraulic balloon (2) and the balancing lever (3) with the wheel axle (4). In its turn, the axle (4) is connected by the resilient connecting rod (5) and the lever (6) with the spool (7) located in the hydraulic regulator's body (8). The face cavities of the spool (7) communicate with the calibrated channel (9), which is damping its displacements. The hydraulic energy comes from the pump (10) and manual control is possible by the handle (11).

When increasing the load on the wheel, the pressure increases and the gas volume in the pneumohydraulic spring (2) goes down respectively, decreasing the distance between the wheel axle and the frame. The regulator's

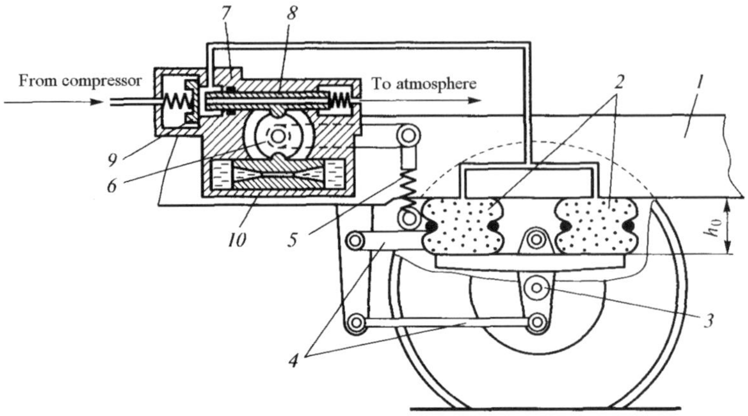

Fig. 13.12. Basic suspension diagram realizing stabilization of static and dynamic strokes

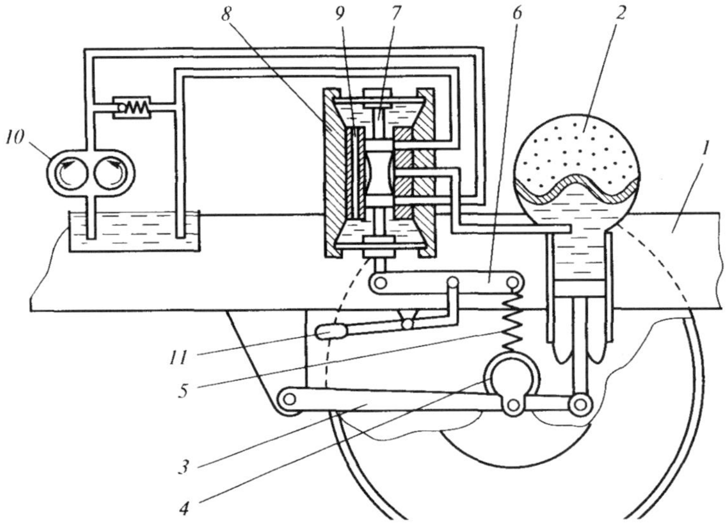

Fig. 13.13. Basic suspension diagram with s regulated rigid kinematical link

spool moves down and oil from the pump (10) is supplied to the pneumohydraulic spring until the earlier position of the body, to which the regulator is tuned, is restored. When decreasing the load, the spool moves up, the oil is drained from the pneumohydraulic spring into the tank, and the body goes down. At a permanent load, the body's position can be regulated manually by the handle (11). For the regulator not to react to dynamic changes in the bend at the vehicle's oscillations on a bumpy road, the spool's face cavities of the spool (7) are filled with oil and covered with rubber diaphragms. Both cavities communicate through the calibrated channel (9) and form a hydraulic damper. The balancing device (3), the connecting rod (5), and the lever (6) comprise a kinematical link, which changes, via the spool (7), the pressure P in the balloon (2).

At the described regulation of the body's height, the air mass remains constant; accordingly, the resilient element's characteristic (stiffness) also remains constant. However, through the kinematical link we can move the constant

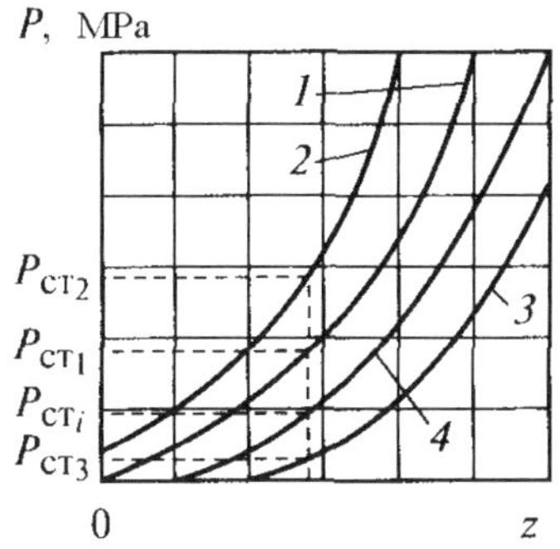

Fig. 13.14. Change of the resilient element's characteristic

characteristic along the axis Z (Fig. 13.14) and the body's position z, depending on the load on it, corresponds to the pressure P and curves 1–4.

13.3.3 Regulation of the shock absorbers' characteristics

The regulation of shock absorbers has the aim of adapting their characteristics to the changeable operation conditions.

The shock absorber's resistance force can be varied by increasing or decreasing the passage area of the valves and the tension of its springs. At a variable load on the vehicle, it is important that the aperiodicity factor remains within the required limits.

The resistance should be also regulated depending on road conditions. When the vehicle moves over small and short irregularities, i.e., in the area of high-frequency resonance, the resistance should reach its maximum.

The basic diagram of the regulated shock absorber is presented in Fig. 13.15. The body (1) with the fluid houses the piston (2) with a hollow rod (3) and the feed washer (4) is installed in the piston (2). The bore (5) connects the rod's (3) internal cavity (6) with the cylinder's rod cavity (7). The feed washer (4) embraces the variable cross-section rod (8), which is attached to the bottom of the body. The jacket (9) of the compensation chamber (10) is attached to the hydraulic cylinder.

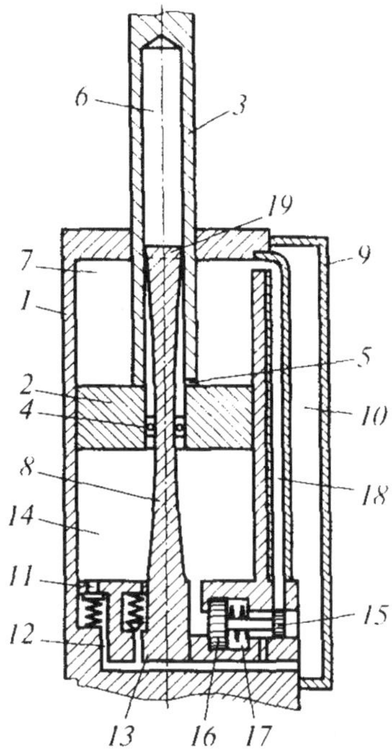

Fig. 13.15. Diagram of a regulated shock absorber

The bottom of the body houses the safety (11) and compensation (12) valves, which connect, via the channel (13), the sub-piston cavity (14) with the compensation chamber (10). The spool valve (15) with the control plunger (16) and return spring (17) is also present. The valve (15) connects, via the connection pipe (18), the rod end (7) with the compensation chamber (10) at the compressions stroke. The rod's (8) guiding belt is provided with cuts (19), which ensure a free flow of the fluid.

At the compressions stroke, the sub-piston cavity's (14) volume decreases, the pressure in it increases, and the control plunger (16) displaces the valve (15) to the right by compressing the spring (17). The rod end (7) is connected, by the pipeline (18), with the compensation chamber. The fluid is throttled from the sub-piston cavity (14) through the circle gap formed by the dispense washer (4) and variable cross section bar (8) to the rod end (7) and further to the compensation chamber (10). Since the rod (8) has a variable lateral cross section along its length, the area of the circular gap, and, accordingly, the resistance, change under a preset law. In case of an excessive pressure increase, the discharge valve (11) opens.

At the retreat stroke, the cavity's (7) volume decreases and its fluid fills in the intra-rod space and is throttled into the sub-piston cavity (14) through the above circular gap. At excessive negative pressure in the sub-piston cavity, the compensation valve (12) opens.

13.3.4 Active suspension with resilient element, slave hydraulic cylinder, and mechanical control

The simplest version of this suspension is presented in Fig. 13.16. The vehicle's body (1) rests in a static position on the suspension's resilient element comprising of two successively positioned parts: the auxiliary resilient element (2) and the main resilient element (3). The auxiliary element (2) acts here as a measuring-converting device, which controls the work of the hydraulic distributor to which the fluid from the pump (4) is supplied from the vessel (23). Any fluid leaks are also brought there through the filter (22). The distributor's spool (5) is attached to the lower support end of the auxiliary resilient element (2) in point **a**, while the distributor's cartridge is attached to the body (1). The distributor controls, via the spool (5) and pump (4) through the channels (22 and 23), the work of the hydraulic cylinder comprising of the body (6) and piston (7). The hydraulic cylinder links are attached to the opposite supporting ends of the main resilient element (3). The hydraulic cylinder's body is attached to the upper end in point **a**, while the piston is connected, through the rod, with the wheel axle (8).

The compensation chamber (C) with the pneumatic compensator (10) is placed separately or is attached to the hydraulic cylinder (6). It is comprised of the compensation valves (11 and 12) and the throttling openings (13 and 14).

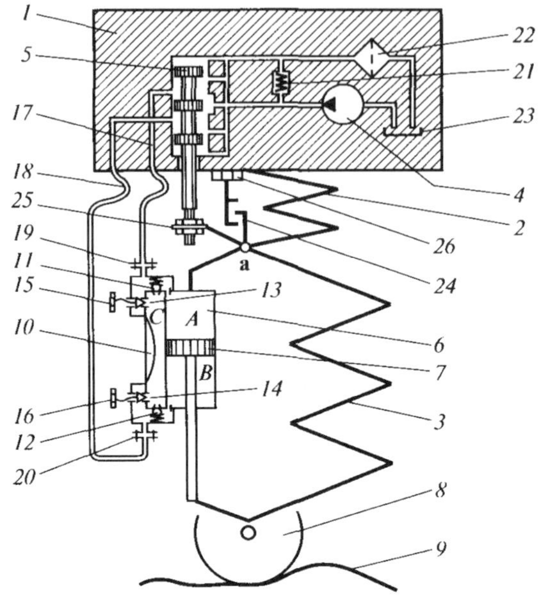

Fig. 13.16. Diagram of an active suspension with resilient element, slave hydraulic cylinder, and mechanical control

When the suspension works in the active mode, the openings (13 and 14) are closed with plugs (15 and 16). The fluid is supplied to the head end (A) and rod end (B) of the hydraulic cylinder from the distributor through the pipelines (17 and 18) and detachable connections (19 and 20), bypassing the compensation valves (11 and 12). The pipelines (17 and 18) have flexible sections, which allow for changing their geometry within the deformation limits of the auxiliary resilient element (2). The suspension's hydraulic drive is also comprised of the safety valve (21) operating within the limits that ensure the maximum displacement of the spool (5), usually measured in millimeters or fractions of millimeters.

The detachable connection (25) allows for regulating the neutral position of the spool (5) when the weight of the body (1) changes. By using a similar connection (26), the mutual position of the stops (24) can also be regulated.

When traveling on a flat, good road, the body (1) is not affected by perturbations. The wheels will affect the body—through the successively located resilient elements (2 and 3)—only with a permanent force which balances the body (1). The deformations of the main and auxiliary resilient elements are equal to static ones. Consequently, the spool (5) is in the neutral position when both hydraulic cylinder (6) cavities are connected with the drain. The pump (4) can be disconnected from the drive engine or run idle. The pump is switched on or engaged before coming onto a bumpy road.

When a wheel runs over a protruding road irregularity, the force acting on the body through the successively located resilient elements goes up and the elements start deforming (compressing). A compression in the resilient element (2) results in decreasing the distance between the body and point **a**; that is why the spool (5) will move up relative to the distributor's cartridge. The middle and the lower ribs of the spool (5) disconnect the rod end (B) from the drain and connect it with the pump's pressure line. Meanwhile, the cavity (A) remains connected with the drain as before. The fluid comes under pressure into the cavity (B) and, by displacing the piston up, compresses the resilient element (3). It should be noted that the resilient element (3) compresses not under the body's inertia force (1), but by the action of the force developed by the hydraulic cylinder.

Thus, in point **a** the auxiliary resilient element is under force from the hydraulic cylinder, which has the opposite direction when compared to the deviation from the main resilient element's (3) surface force static value. As soon as the resultant force from the resilient element (3) and the hydraulic cylinder applied to point **a** becomes equal to the static one or to the preset value, the initial deformation of the auxiliary resilient element (2) is restored and this places the spool (5) into the neutral position and stops the resilient element's (3) compression.

When the wheel runs into a hole, the force acting on the body (1) from the resilient elements goes down and, accordingly, the deformation of the auxiliary resilient element (2) goes down—it gets decompressed. This causes an increase in the distance between the body (1) and point **a**; therefore, the spool (5) goes down relative to the distributor's cartridge. The middle and upper ribs of the spool (5) disconnects the hydraulic cylinder cavity (A) from the drain and connects it with the pump's pressure line. The cavity (B) gets connected with the drain line. The fluid under pressure gets into the cavity (A) and displaces the piston (7) downward, stretching the resilient element (3). Thus, the cylinder will apply a force to point **a** to compensate the decrease in the static force. As soon as the resultant force from the hydraulic cylinder and resilient element (3) becomes equal to the preset value, the auxiliary resilient element's (2) deformation will be restored, the spool (5) takes the neutral position, and the main resilient element's stretching will stop.

Consequently, at motion on a bumpy road with irregularities, which cause intense oscillations in vehicles with passive suspensions, an active suspension (see Fig. 13.16) can ensure a practically stable position of the sprung body along the vertical, while the wheel will make motions relative to the body defined by changes in the road microprofile.

In case the automatic control elements fail, the vehicle will not lose its mobility since this suspension can also operate in the passive mode. In this case, the detachable connections (19 and 20) are used to disconnect the pipelines (17 and 18) from the hydraulic cylinder. The openings in the hydraulic cylinder and the disconnected pipelines are closed with plugs. Then, the plugs (15 and 16) are removed to open the throttling openings (13 and 14). Cut-off valves can also be used instead of detachable connections (19 and 20). In this case, the hydraulic cylinder transforms into an ordinary hydraulic shock absorber, which damps the body's (1) oscillations on the main resilient element (3). The auxiliary resilient element can be blocked by rests (24).

When the piston (7) moves up, the cavity's (A) volume decreases and the fluid is throttled through the calibrated opening (13) to the compensation chamber and then arrives through the compensation valve (12) in the cavity (B), whose volume increases.

At the retreat stroke, the piston goes down, the volume of the cavity (B) goes down, and the fluid is throttled out of it through the calibrated opening (14) and into the compensation chamber (C). The fluid gets into the cavity (A), whose volume goes up, from the compensation chamber through the compensation valve (11). The resistance coefficient is defined by the area of the openings (13 and 14).

Thus, when the suspension works in the passive mode, the main resilient element (3) performs the functions of the spring, while the slave hydraulic cylinder acts as a shock absorber.

13.3.5 Active suspension with resilient element, slave hydraulic cylinder, and electrohydraulic control

The diagram in Fig. 13.16 is presented only for the purpose of demonstrating and explaining the operation principle of an active SpS with a resilient element in the active and passive modes. However, installing two successively connected resilient elements of comparable rigidity in the suspension of a real vehicle can cause design and technological problems.

To avoid them, it is necessary to use an auxiliary resilient element with a much higher rigidity than that of the main one. In particular, it can be the compliance of the vehicle's body (frame) or even the suspension's guiding levers. Naturally, in this case, the sensitivity of the ACS will be decreased and an electrohydraulic control system should be used where strain- or piezo-sensitive transducers are installed as the primary measuring devices of surface forces, which react to insignificant deformations in the auxiliary resilient element or local deformations of the frame in the suspension's attachment points.

In Fig. 13.17 the rectangles represent five possible fixing points for the primary highly-sensitive measuring transducers on the body (1) and active suspension lever with the resilient element (2) and slave hydraulic cylinder (4). The wheel is designated as (3). For simplicity, the automated control elements are not presented. The main condition for correctly placing the measuring transducers (5) is their location outside the power loop comprising of the resilient element and the hydraulic cylinder, i.e., they should increase their readings at the growing deviation in the values of the resilient element's surface force against the static one and decrease their readings when these deviations are compensated for by the slave hydraulic cylinder.

The basic diagram of an active suspension with a resilient element and electrohydraulic system is presented in Fig. 13.18. The vehicle's body (1) is linked by the resilient element (2) to the wheel axle (3) and the slave hydraulic cylinder (4) with the piston (5) is installed in parallel to the resilient element.

The hydraulic cylinder's working cavities are connected to the hydraulic distributor (8) by the hydraulic lines (6 and 7) and have three characteristic positions designated as I, 0, and II. The reverse valves (9 and 10), which open to the side of the hydraulic cylinder's working cavities, connect them with the oil tank (11) and prevent a break in the hydraulic flow when the

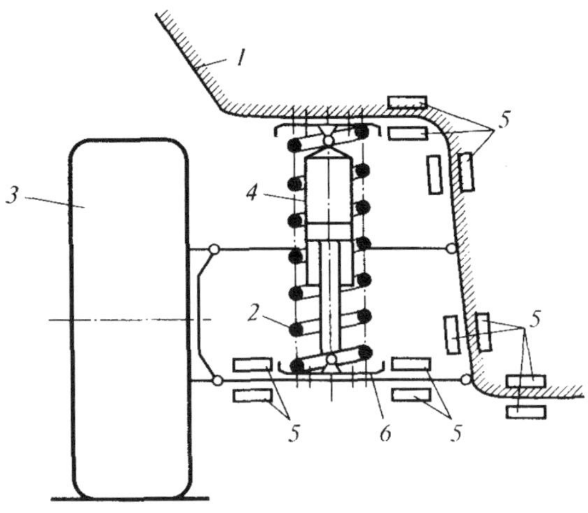

Fig. 13.17. Active suspension with resilient element and slave hydraulic cylinder

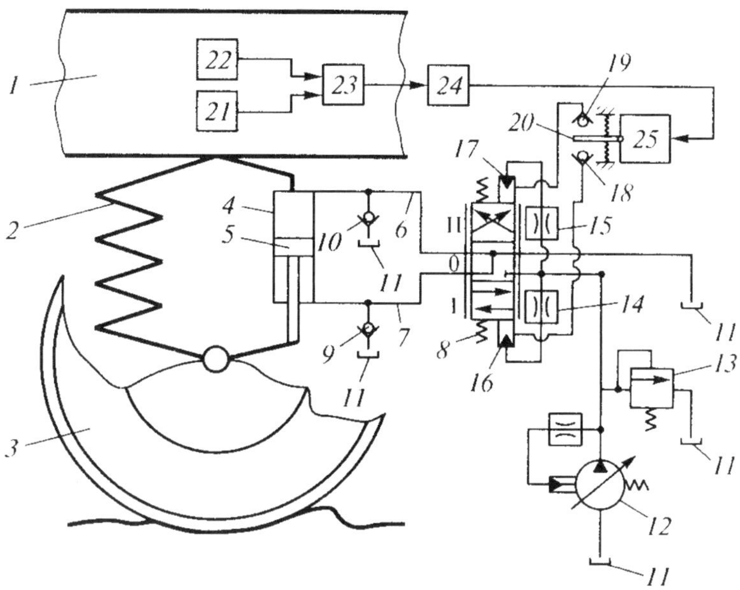

Fig. 13.18. Basic diagram of an active suspension with a resilient element, slave hydraulic cylinder, and an electrohydraulic control system

performance of the pump (12) at motion over short irregularities may be lower than required. The regulated performance pump (12) with a negative pressure feedback has a safety valve (13). The fluid is supplied from the pump's pressure line through the permanent throttles (14 and 15) to the hydraulic distributor's (8) control cavities (16 and 17). The fluid flows out of the cavities (16 and 17) through the nozzles (18 and 19), between which is the damper plate (20) that helps to control the displacements of the distributor's (8) spool.

For the sake of simplicity, Fig. 13.18 shows the supply of the fluid into the control cavities (16 and 17) from the main pressure line. However, with the aim to eliminate the negative effect of pressure fluctuations behind the pump on the control reliability, in servo hydraulic drives, the control lines powering the distributor are often fed from an independent small-power hydraulic energy source.

In series with the resilient element (2) and slave hydraulic cylinder (4), there is a sensitive element or a primary measuring transducer (21) intended to measure the aggregate surface force's value acting on the body (1) from the active suspension elements. A signal from the measuring transducer (21) and setting device (22) is fed to the input of the comparator (23), from which the unbalance signal goes to the amplifier (24). Then, the amplified unbalance signal is fed to the electromechanical converter (25), which is a polarized electromagnetic device controlling the damper plate's (20) displacement.

In the static position, the body's (1) weight is perceived by the resilient element (2). Since both working cavities of the hydraulic cylinder are connected to the drain, in the static position it is unloaded and the piston position is defined by the resilient element's deformation. This initial status of the active suspension excludes any tough impact on the body at a sudden change in the height of irregularities.

The suspension operation presented in Fig. 13.18 is as follows. When the wheel runs over a protruding road irregularity, the resilient element is compressed, the surface force acting on the body (1) increases, and, accordingly, the signal from the measuring converter (21) increases as compared to the signal from the setting device (22). The comparator's output (23) forms a positive unbalance signal, which is intensified by the amplifier (24) and fed to the electromechanical converter's input (25). The latter deflects the damper plate (20) down, blocks one nozzle (18), and opens another (19); thus, the pressure in the distributor's (8) control cavity (16) increases and the pressure in the cavity (17) decreases. The distributor's spool moves to position I and ensures the supply of fluid under pressure into the hydraulic cylinder's rod end. The force from the hydraulic cylinder, when applied to the body (1), has an opposite direction than the force from the resilient element, which ensures a complete or partial compensation for the change in the surface force acting on the body.

If the wheel runs into a hole, the resilient element will stretch up and the force acting on the body decreases against its static value. The signal from the measuring converter will decrease as compared to the signal from the setting device (22). The comparator's (23) output forms a negative unbalance signal, which is intensified by the amplifier (24) and fed to the electromechanical converter's (25) input. The latter deflects the damper plate (20) upward, closing one nozzle (19) and opening another (18). The pressure in one cavity (17) increases and decreases in another (16) and the distributor's (8) spool takes position II. The fluid is supplied under pressure to the above-head end of the hydraulic cylinder (4), which compensates for the decrease in the surface force acting on the body.

At high-frequency oscillations, the pump's performance can be insufficient for the timely filling of the hydraulic cylinder's working cavities. That is why to prevent a break in the hydraulic flow, the remaining amount of fluid needed will be supplied into the hydraulic cylinder through the reverse valves (9 and 10).

Thus, the suspension presented in Fig. 13.18, while ensuring the stabilization of surface forces acting on the vehicle's body, minimizes its oscillations, and, accordingly, the overloads experienced by the body.

13.3.6 Pneumohydraulic active suspension with a slave hydraulic cylinder

The implementation of the above stabilization method of surface forces is also possible on the basis of pneumohydraulic suspensions, which are recently rather popular. In the simplest case, the active suspension's pneumohydraulic resilient element is mounted as presented in Fig. 13.18. However, this may cause certain layout difficulties. That is why we describe here the pneumohydraulic active suspensions that differ from the design presented in Fig. 13.18.

The basic diagram of an active pneumohydraulic suspension with a slave hydraulic cylinder is presented in Fig. 13.19.

The vehicle's body (1) is connected by the pneumohydraulic cylinder (2) and levers (3) to the wheel (4), which rests on the ground (5). The hydraulic cylinder (2) houses a piston with an empty rod (6). The rod's (6) internal cavity (8), partially filled with gas, communicates with the hydraulic cylinder's rod cavity (9) via the connection pipe (7). The piston cavity (10) is connected by the elastic pipeline (11) to the intermediate hydraulic cylinder's (13) piston cavity (12), which houses the piston (14) with the rod (15). The intermediate cylinder's (13) rod cavity (16) is connected by the pipeline (17) with the hydraulic part (18) of the ball-type balloon (19), in which the gas volume (21) is separated from the fluid volume by the elastic diaphragm (20).

The rod (15) is also connected with the slave hydraulic cylinder's (23) piston (22). The piston (22) divides the volume of the hydraulic cylinder (23) into the piston (24) and rod (25) working cavities (the head end and rod end), which are connected by pipelines (26 and 27) to the

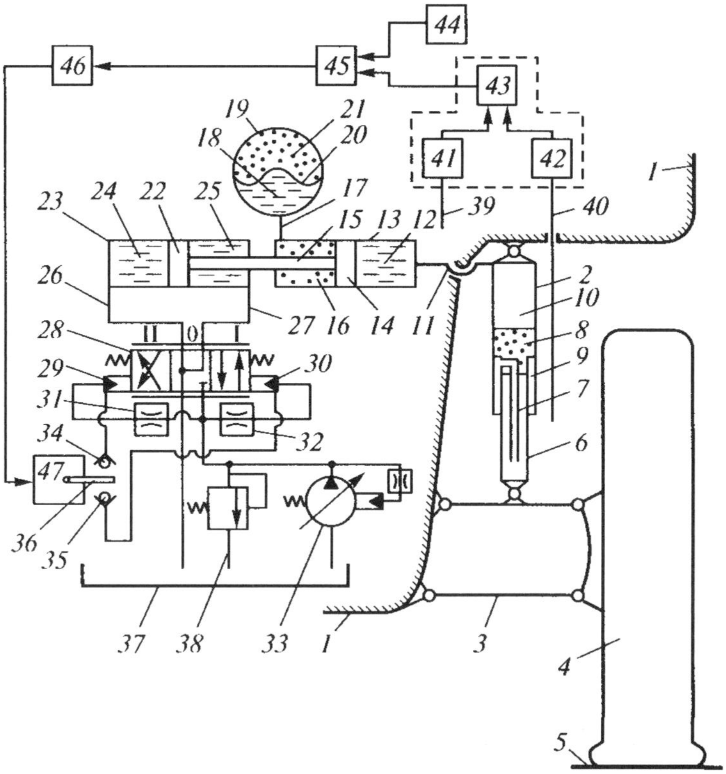

Fig. 13.19. Basic diagram of an active pneumohydraulic suspension with a slave hydraulic cylinder

hydraulic distributor (28). The distributor's (28) hydraulic control cavities (29 and 30) communicate via the permanent throttles (31 and 32) with the regulated performance pump's (33) pressure line with the pressure negative feedback. The fluid flows out of the cavities (29 and 30) through the nozzles (34 and 35), between which the damper plate (36) is located, and is drained into the oil tank (37). The safety valve (38) is placed behind the pump (35).

The pneumohydraulic cylinder's (2) piston cavity (10) is connected by the channel (39) and rod end (9) with the primary measuring pressure converters (41 and 42) via another channel (40). The force exerted by the pneumohydraulic cylinder (2) on the body (1) is a linear function of two variables: the pressures in the piston and rod cavities. That is why the adder (43) forms a signal proportional to the force acting on the body. The signal arrives from the setting device (44) and adder (43) to the comparator (45), from which the unbalance signal is fed through the amplifier (46) to the electromechanical converter (47), which controls the damper plate's (36) displacement.

A characteristic of the pneumohydraulic suspension is the presence of two resilient elements—gas in the direct pressure chamber (21) and in the back-pressure chamber (8). The surface force acting on the body is a linear function of the pressures in these chambers. The permanent value of this force is maintained by regulating the pressure only in the pneumohydraulic cylinder's head end, which allows for simplifying the ACS.

When automation elements are inoperative, the suspension in Fig. 13.19 operates as a passive one. At the compression phase, the rod (6) goes up and the fluid from the pneumohydraulic cylinder's head end (10) goes to the intermediate cylinder's (13) head end (12) via the pipeline (11). The pistons (14 and 22), together with the rod (15), get offset to the left and the volume of the cavities (24 and 16) goes down. The fluid from the intermediary hydraulic cylinder's (13) rod end arrives at the ball vessel (19) and compresses the gas behind the diaphragm (20), while the fluid from the head end of the slave hydraulic cylinder (23) is drained into the tank (37).

The cavity's (9) volume increases and it receives the fluid squeezed through the connecting pipe (7) by the compressed gas in the chamber (8).

During the retreat phase, the fluid from the cavity (9) arrives via the connection pipe (7) into the empty rod (6) and compresses the gas present there. The volume of the head end (10) goes up, the pressure in it goes down, the gas in the direct-pressure chamber (21) squeezes the fluid out of the ball vessel (19) into the intermediate cylinder's (13) rod cavity, and the piston (14) offsets to the right and squeezes the fluid into the pneumohydraulic cylinder. The slave hydraulic cylinder (25) sees a free flow of the fluid from one working cavity to the other through the distributor (28), while the rod's (15) volume is compensated for by adding or draining the fluid from the tank (37).

In its active mode, the suspension in Fig. 13.19 works as follows. When running over a protruding irregularity, the pressure in the pneumohydraulic cylinder's (2) head end (10) increases. The measuring transducers (41 and 42) and the adder (43) supply a signal to the comparator's (45) input that is proportional to the force acting on the body (1). This signal is compared to the one from the setting device (44) and the positive mismatch signal is supplied, through the amplifier (46), to the electromechanical converter (47), which deflects the damper plate (36) downward. One nozzle (35) is closed, while another nozzle (34) is opened, the pressure in the control cavity (30) goes up, and the distributor's (28) spool takes position I. The fluid from the pump (33) arrives into the slave hydraulic cylinder's (23) rod end (25), which moves the pistons (22 and 14) to the left. The piston (14) squeezes the fluid from the intermediate hydraulic cylinder's (13) rod end into the ball vessel and compresses the gas until the surface force acting on the body takes the preset value.

When running into a hole, the suspension with an ACS works in a similar way. The force acting on the body decreases, the amplified negative mismatch signal gets into the electromechanical converter (47), and the damper plate (36) goes up and closes the nozzle (34). The pressure in the control cavity (29) increases and the distributor's (28) spool takes position II, which ensures the supply of the fluid from the pump (33) to the slave hydraulic cylinder's (23) head end (24). With the help of the rod (15), one piston (22) displaces another piston (14) to the right and the latter squeezes the fluid out of the cavity (12) into the above-head end (10) of the pneumohydraulic cylinder (2) until it compensates for the decrease in the surface force acting on the vehicle's body (1).

It should be noted that the automatic compensation system for deviations from the static value of surface forces presented in Fig. 13.19 can be used for several pneumohydraulic cylinders (2), which have a common direct-pressure chamber (21).

If the active suspension's hydraulic system (see Fig. 13.19) is equipped with additional devices, the vehicle's road clearance can be changed within a broad range.

A drawback of the described active pneumohydraulic suspension is the presence of the intermediate and slave hydraulic cylinders, which are comparable—by their dimensions, weight, and cost—with the main pneumohydraulic cylinder (2). This deficiency is absent in active pneumohydraulic suspensions with an execution pump.

The usual composition of a suspension's modern automated control system includes the following sequence of functional elements: sensors, conversion devices, and actuating devices. The rapid progress of microelectronics opens new ways for developing automation devices in the vehicle. The role of the commanding device is performed by a microprocessor.

The coordination of the microprocessor's inputs and outputs with the sensors and actuating devices, respectively, is achieved through input/output converters. A single onboard microcomputer can control many units and systems of the vehicle. In this case, the control programs for different systems are launched in the onboard computer successively, and, depending on the conditions, the priority of executing the programs can vary. In each particular case, the complete structure of the automated control system is unique. As a rule, a combined structure is more reasonable, which is comprised of mechanical regulators, autonomous and interconnected analog and digital control blocks, and the core and peripheral processors.

For execution in one processor, it is desirable to select control programs with approximately the same restart frequency requirements, which should be kept to a minimum to simplify the system. Thus, for most control systems of the engine (motor), brakes, and suspension, a 0.1 second restart pitch of control programs is acceptable. The control programs executed by the processor are downloaded into the read-only memory (ROM).

Most mass-manufactured vehicles equipped with microprocessor-based control systems of the suspension have systems of static regulation. In most cases, the regulated parameters are the body's position, the suspension's rigidity, and the shock absorbers' resistance.

It can be noted here that the response of the regulated shock absorbers' actuating devices used by Toyota is 0.1 s, which is sufficient for the systems of dynamic regulation, while Way-Assauto, which is designing shock absorbers with electronic control, plans to bring the response in the near future down to 0.05 s.

The design of a control program is a rather complicated process involving debugging at all stages. The majority of debugging is done prior to installing it on a vehicle prototype by computer simulation, which makes the process of creating the ACS cheaper.

13.4 Choice of damping and rigidity parameters for long-stroke suspensions

13.4.1 Presuppositions

As shown by the results of the studies presented in Chapter 12, the springing systems' ability to ensure the satisfactory passing ability, soft driving, resistance to roll-over, and loading of the undercarriage and carrying system profile is of paramount importance because of the specific design solutions of special wheeled chassis (essential length and height and substantial axial loads) and their usage conditions.

On the other hand, as demonstrated by the design analysis made in Chapters 1–4, there is a stable trend to improve the load capacity of these wheeled vehicles, and, accordingly, to increase the base, number of axles, and the travel speed, including in off-road conditions. All of this in combination results not only in increased vibration loads, but also in higher dynamic loads, which affect the elements of the vehicle's undercarriage and carrying system. A radical improvement in the soft driving and a decrease in the loads are only possible through applying long-stroke suspensions with dynamic regulation (active and semi-active ones). The progress of studies and designs in the sphere of controlled suspensions of this type is currently promoted by the advance of microprocessor equipment and technologies.

Let us consider an example of justifying the parameters of a long-stroke suspension made on the basis of hydropneumatic and pneumatic springs (HPSs and PSs, accordingly), which have become very popular recently.

As initial data, let us take that using HPSs of wheels results—because of the presence of dry friction—in an increased delivery of high-frequency oscillations (with the resonance at frequencies of 5–6 Hz) to the carrying system (frame). With the increase in the weight and, especially, the increase in the number of axles, frames of HDWVs become less resistant and transfer the oscillations of the vehicle's front section to the cabins. Since the resonances of the frame's elastic oscillations are close in frequencies to those of the cabins mounted on their suspensions, the cabins get swung at 4–6 Hz. The sprung seats of known designs ensure a decrease in the crew's root-mean-square vertical accelerations in the octave bands of 4 and 8 Hz only by 40–50% and leave absolutely unaffected the horizontal vibrations, which in the existing HDWV designs reach the level of vertical accelerations.

Today, suitable systems can be built only on the basis of HPSs, PSs, and hydropneumatic shock absorbers (HPSAs). The structural and functional models of these devices are presented in Figs. 13.20–13.22 and 13.24.

The data of Figs. 13.20–13.22 gives a vivid presentation about the complexity of the processes; however, without understanding these processes it is impossible to correctly select the springing system's parameters. Typical elements of hydropneumatic devices (HPDs) are presented in Fig. 13.23.

In accordance with the modern ideas, the support-travel module should ensure the following:

- Full travel of the wheels of 500–700 mm;
- Maintenance of the mid-axis's vertically static position with the accuracy of 10%;
- Smooth (shock-free) regulation of the road clearance at a fixed running dummy within the limits of the full wheel travel;
- Possibility of automated (automatic) and manual adjustment of the resilient and damping element's characteristic;
- Full power looping inside the module;
- Static loads on the wheel from 50–75 kN and the nominal static load on the wheel up to 73 kN;

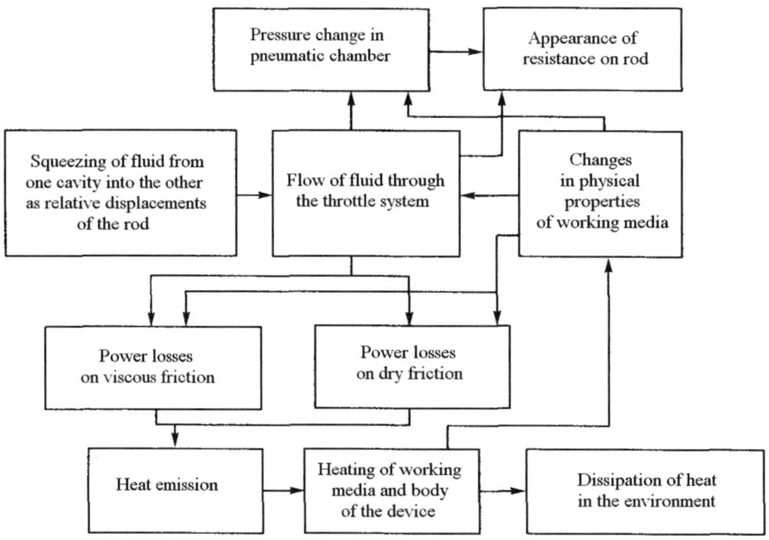

Fig. 13.20. Interconnected physical processes and phenomena typical for operating pneumohydraulic devices

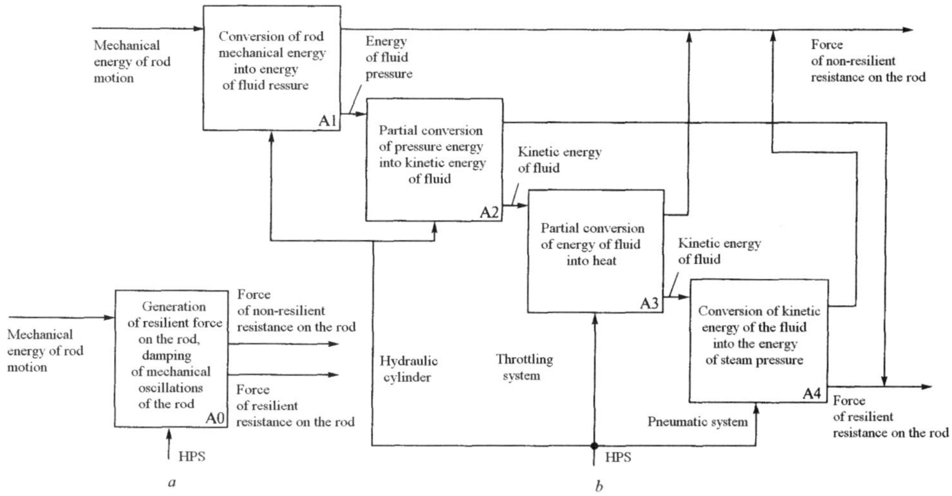

Fig. 13.21. Functional model of HPS: a—HPS diagram as a "black box"; b—Diagram of HPS internal functional structure; A0–A4—Conventional functioning stages of HPS

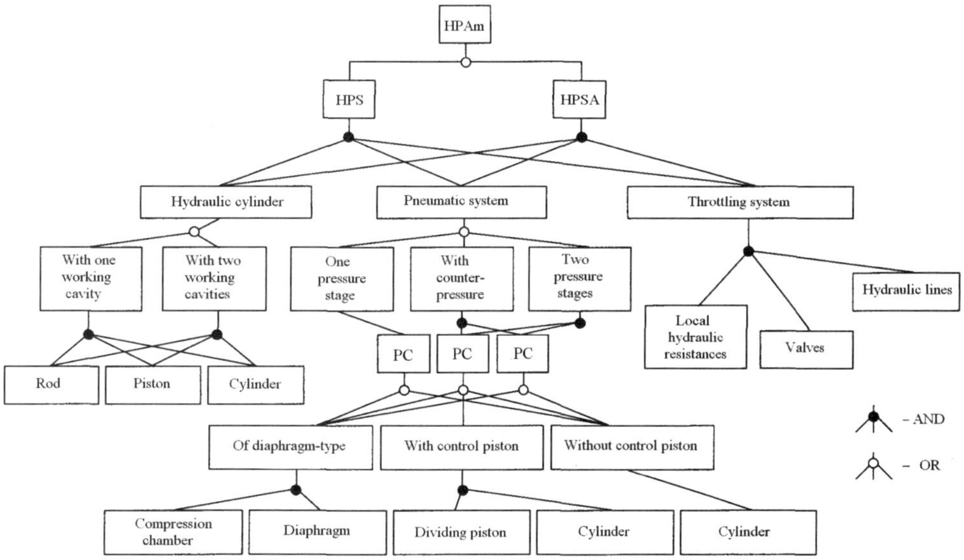

Fig. 13.22. AND-OR graph of structural solutions of HPD: PC—Pneumatic cylinder

- The possibility for the suspension to work when a HDWV model travels on special testing routes and on general-purpose roads;
- A maximum travel speed of up to 80 km/h;
- The possibility to change the resilient-damping parameters; and
- The maximum use of standard units and parts of the motor industry and other branches of machine-building (hydraulic drive, wheel reduction gears, etc.) and the expected service life of up to 10,000 km.

The support and running module (SRM) should include a lever-type guiding device, a wheel gearbox, a wheel hub, a drum braking

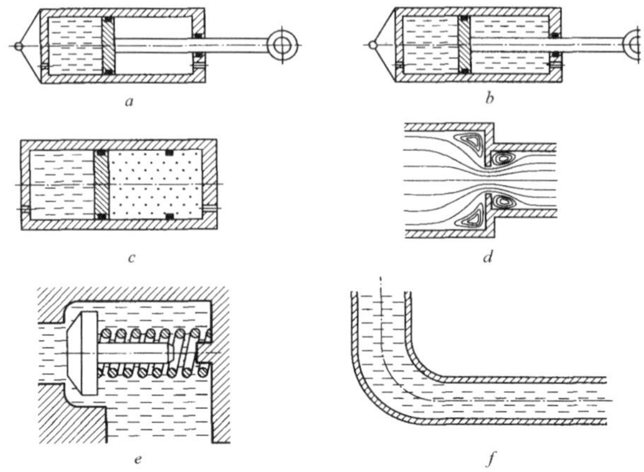

Fig. 13.23. Typical HPD design elements: a—Hydraulic cylinder with one working cavity; b—Hydraulic cylinder with two working cavities; c—Pneumatic cylinder; d— Local hydraulic resistance; e—Valve; f—Hydraulic line

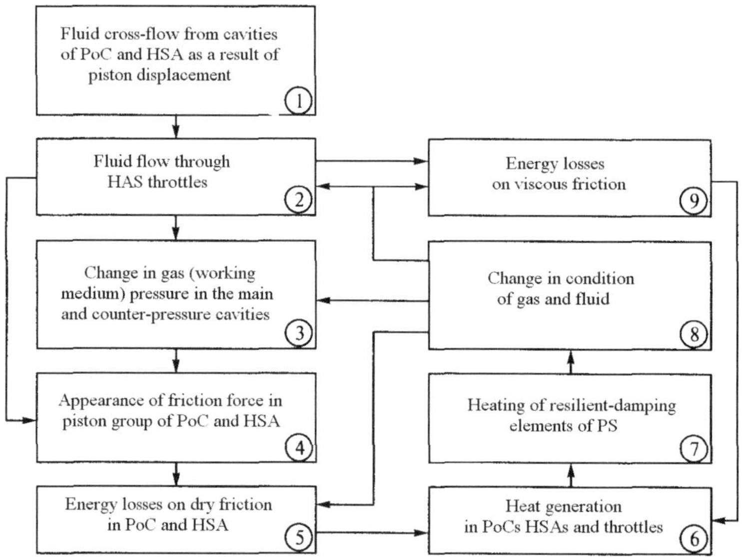

Fig. 13.24. Functional diagram of a PS with a shock absorber

mechanism, tires, regulated resilient elements, shock absorbers, control system elements, hoses, pipelines, and fitting devices.

The lever-type device ensures the operation of resilient elements and shock absorbers at displacement of the wheels in the longitudinal plane.

These requirements can be implemented by a long-stroke regulated pneumatic suspension. These suspensions ensure the operation of resilient elements and shock absorbers at the working ambient pressures of 10 MPa or more. In addition, the high weight of the suspension's unsprung parts requires using devices, which ensure an impact-free retaining of the SRM's wheels at the full retreat travel.

At present, the overwhelming majority of regulated pneumatic springs in heavy-duty vehicles are made as piston resilient elements with separating devices of two types: floating pistons and elastic diaphragms.

It should be noted here that elastic diaphragms have a number of drawbacks that fail to satisfy the designers of high-pressure pneumatic springs. One should also bear in mind that the operation of long-stroke suspensions is associated with a high fluid consumption in the hydraulic accumulators (HAs). Such consumption cannot be ensured under small dimensions of HAs with elastic separators. It is hardly deemed reasonable to install several such hydraulic devices. It is better to use the available designs of HAs with floating pistons since they have a service life sufficient for SRMs. The presence of the friction force in the HA's piston group and its influence on the loading characteristics is evaluated by well-known methods. The suspension under study can use—as a resilient element's power cylinder—cylinders of various devices, which operate at a pressure of 10 MPa or more and have a piston travel of up to 500–700 mm (for example, cylinders of hydraulic hoisting jacks).

In order to ensure an impact-free retaining of the SRM's wheels at the full retreat travel, it is necessary either to install a power-intensive buffer of the retreat on the suspension's guiding device or to use a PS with a counter-pressure. The latter is the most expedient in regulated long-stroke pneumatic springs. It is known that apart from the required PS characteristic at retreat, the counter-pressure allows for changing the load characteristic (stiffness) at decreases in the static load. The counter-pressure chamber can be formed with the help of the available piston hydraulic shock absorbers (HSAs).

In order to successfully damp the experimental suspension's oscillations, it is also necessary to use double-action regulated HSAs. The functional diagram of the suggested regulated PS with counter-pressure is presented in Fig. 13.24 and accounts for the friction forces in the HSA's power cylinders and piston groups, as well as for the fluid cross-flow in the HSA.

13.4.2 Justification for choosing a piston pneumatic spring with counter-pressure

Pneumatic springs with counter-pressure (PSCPs) have the following advantages:

- At changes in the static load, they experience smaller changes in the frequency of the sprung masses' oscillations;
- At the retreat stroke, the suspension's energy intensity increases essentially;
- The suspension can work at retreat under negative forces, which makes installing special limiters unnecessary.

These advantages are quite essential for the suspension of the SRM's prototype. A long-stroke suspension is able to vary the energy intensity at compression and retreat, while changing the power cylinder's (PoC) piston position at the same static load on the resilient element. In case there are regulated shock absorbers available, a broad range of dynamic characteristics of the suspension's resilient and damping elements are obtained. This allows for not only changing the position of the sprung mass's center of gravity, but also making the suspension adaptive to various road surface impacts (motion on roads with hard pavement, passage over irregularities, etc.).

The working pressure of the gas in pneumatic springs reaches—at high load capacities—dozens MPa; therefore, to assess the capabilities of such PSCPs, it is reasonable to use dimensionless characteristics. They are based on the assumption that the atmospheric pressure can be neglected:

$$p^{abs} = p^{exc} + p^{atm} \approx p^{exc},$$

where: p^{abs} is the absolute; p^{exc} is the excessive; and p^{atm} is the atmospheric pressure.

In case of the slow processes of gas compression and expansion, the isotherm equation is used:

$$p = p_0 \varepsilon,$$

where: p_0 is the static pressure in the PSCP; and ε is the compression degree.

At fast processes (impact loading, high travel speed), the adiabatic equation is used:

$$p = p_0 \varepsilon^k,$$

where k is the adiabatic index.

There are methods for calculating load characteristics in which an equivalent polytropic process is considered:

$$p = p_0 \varepsilon^{\Pi},$$

where Π is the polytrope index and $\Pi = 1 \ldots k$.

In this case, in various PSCPs, the polytrope index can vary essentially. This is due to the heat exchange conditions of the compressed gas with the environment. A reliable value of the polytrope index can only be obtained experimentally.

There are plenty of methods for calculating the heat exchange in PSCPs, for which many parameter values are needed: the heat transfer and heat conductivity factors, etc. That is why, when choosing and justifying the choice of the resilient-damping characteristics for a designed PSCP, it is better to use isothermal and adiabatic loading characteristics:

$$p(\chi) = p_0^{exc} \varepsilon_1 F_{pis};$$

$$p(\chi) = p_0^{exc} \varepsilon_2 F_{pis},$$

where: χ is the piston stroke; and F_{pis} is the piston area.

The PSCP's loading characteristic can be built under the well-known formula:

$$p(\chi) = \left|p_{01}^{exc} \varepsilon_1^{\Pi}\right| F_1 - \left|p_{02}^{exc} \varepsilon_2^{\Pi}\right| F_2; \ \Pi = 1 \text{ and } \Pi = k,$$

where: F_1 and F_2 are the piston pressure areas in the first and second chambers, respectively; ε_1 and ε_2 are the respective compression degrees in the first and second chambers; and p_{01}^{exc} and p_{02}^{exc} are the counter-pressures in the first and second chambers, respectively;

$$\varepsilon_1 = \frac{V_{01}}{V_{01} + \chi F_1} = \frac{1}{1 - \chi/h_{01}};$$

$$\varepsilon_2 = \frac{V_{02}}{V_{02} + \chi F_2} = \frac{1}{1 - \chi/h_{02}};$$

$$h_{01} = \frac{V_{01}}{F_1};$$

$$h_{02} = \frac{V_{02}}{F_2},$$

where V_{01}, V_{02} and F_1, F_2 are the volumes of the chambers and piston pressure areas of the PSCP's first and second cavities, respectively; and h_{01} and h_{02} are the chamber lengths of the PS's first and second cavities, respectively.

If the atmospheric pressure's influence is disregarded, then it is convenient to consider the well-known dimensionless characteristics:

$$\frac{p(\chi)}{p_0} = \frac{\left[\dfrac{1}{1 - \chi/h_{01}}\right]^n - \left[\dfrac{1}{1 - \left(\chi/h_{o2}\right)\mu}\right]\xi}{1 - \xi},$$

where: $\mu = \dfrac{V_{01}F_2}{V_{02}F_1}$, $\xi = \dfrac{p_{01}F_2}{p_{02}F_1}$, and n is the polytrope factor.

The coefficient ξ is the forces' ratio acting on the piston of the PS's power cylinder from below and above under the resilient element's static position. The coefficient μ is the ratio of the gas columns' reduced heights in the PS's static position.

The latter equation shows that all of the parameters of the PSCP's loading characteristics (the areas, volumes, and pressures) can be characterized by two coefficients: μ and ξ. At the same values of μ and ξ, the dimensionless characteristics are the same. At the same time, the usual loading characteristics in this case will differ by the scale only. It should be noted here that the dimensionless characteristics have a coordinate origin in point $\chi = 0$, i.e., the ordinate's position corresponds to the static load on the PSCP.

The analysis of the dimensionless characteristics allows for assessing the influence of coefficients μ and ξ on the PSCP's stiffness. When ξ is equal to zero, there is no counter-pressure; with an increase in coefficient ξ, the PSCP's stiffness will grow both at compression $0 < \frac{\chi}{\chi_0} < 1$ and at retreat $-1 < \frac{\chi}{\chi_0} < 0$, where χ_0 is the displacement of the PSCP's piston under the static load.

At high coefficient ξ values, the PSCP's stiffness changes little within a great range of loads, i.e., the characteristic is close to linear.

The coefficient μ is mainly affecting the left-hand part (retreat) of the dimensionless characteristic. The bigger coefficient μ, the steeper the fall of the dimensionless characteristic's curve, the higher the rigidity value, and the higher the energy intensity. The influence of coefficient μ on the characteristic's right-hand part (compression) results mainly in the increased rigidity near the static position $\left(\frac{\chi}{\chi_0} = 0\right)$. At essential compression strokes $\left(\frac{\chi}{\chi_0} > 0\right)$, the influence of coefficient m is insignificant.

By varying the values of coefficients μ and ξ, we can obtain the required loading characteristics in the required direction. The coefficient ξ is proportional to the value of the counter-pressure p_{02}; therefore, there is a possibility to change the value of ξ without changing the dimensions of the PSCP's power cylinder.

It is obvious that at $\frac{\chi}{\chi_0} = 1$, the dimensionless characteristic has no sense and it means that the gas compression degree in the PSCP tends to infinity ($\varepsilon \rightarrow \infty$).

13.4.3 Definition of the main parameters and characteristics of an experimental PS with SRM

The experience of using telescopic PSs in heavy-duty vehicles shows that the maximum ratio of the pressures $\frac{p_{max}}{p_0} = 3$ allows for avoiding a break-through of the suspension. For long-stroke PSs, which have a higher energy intensity, this value is more than sufficient under the resilient element's symmetrical loading characteristic.

The design of the experimental PS with preset parameters was undertaken based on the following initial data and available components:

- The spring's power cylinder has a diameter of 160 mm, a piston stroke of about 700 mm, and the piston rod has a diameter of 140 mm;
- The main resilient element is the ARF 16/320 model hydraulic accumulator (HA1) and the counter-pressure resilient element is the ARF 2.5/320 model hydraulic accumulator (HA2);
- The throttling devices of the built-in double-sided shock absorber are the KVMK 32G1.1 model standard regulated throttles;
- The transmission ratio of the suspension's lever-type guiding device (L) is 1.44.

The PoC's maximum piston travel is restricted by the fluid flow rate in the HA2, which has a volume of 2 liters. The efficient area of the PoC F_1 is 0.02 m^2 and the area under the piston $F_2 = 0.0047$ m^2. The full stroke of the PoC's piston is:

$$x_{max} = \frac{2 \cdot 10^{-3}}{0.0047} = 0.4255 \text{ m} = 425.5 \text{ mm},$$

and, the full travel of the SRM's wheel is:

$$\chi_{max} = 425.5 \cdot 1.44 = 612.7 \text{ mm},$$

which corresponds to modern requirements.

The static load on the PSCP is:

$$p_0 = GL = 73 \cdot 1.44 = 105 \text{ kN},$$

where G is the load on the wheel.

At a symmetrical load characteristic (the piston's middle position), the volume of the compressed gas in HA1 is 9 l and in HA2 is 1 l. In this case, coefficient μ of the PSCP's dimensionless characteristic:

$$\mu = \frac{V_{01}}{V_{02}}\frac{F_2}{F_1} = \frac{9}{1}\frac{0.0047}{0.02} = 2.115.$$

By setting the value of coefficient ξ, we can build the dimensionless characteristic. Fig. 13.25 presents the static (isothermal) dimensionless PS characteristic at the preset coefficient

$$\xi = \frac{p_{02}}{p_{01}}\frac{F_2}{F_1} = 0.2.$$

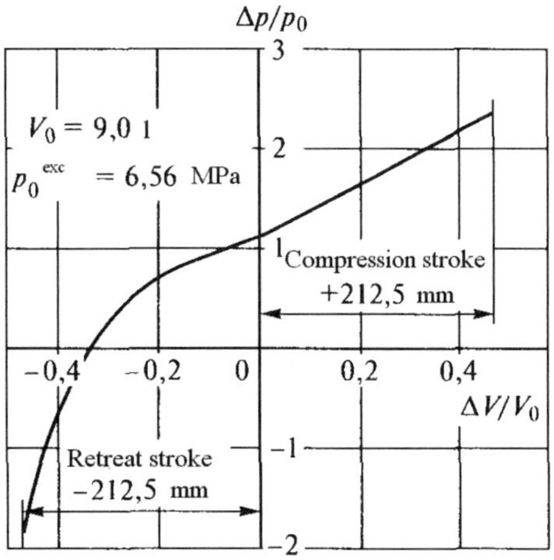

Fig. 13.25. Dimensionless characteristic of a PSCP

The curve of the dimensionless characteristic shows the following:

- The coefficient of the maximum increase in the load exceeds two; and
- At the maximum retreat stroke, the force that holds the unsprung mass exceeds the static load on the PSCP.

The excessive pressure of compressed gas in HA1 in the static position:

$$p_0 = p_{01}^{exc} F_1 - p_{02}^{exc} F_2 = p_{01} F_1 (1 - \xi) \Rightarrow p_{01} = \frac{p_0}{F_1(1-\xi)} =$$

$$= \frac{105 \cdot 10^3}{0.02(1-0.2)} = 6.56 \text{ MPa};$$

the charging pressure in HA1:

$$p_{ch1} = 6.56\frac{9}{16} = 3.69 \text{ MPa} \Rightarrow p_{ch1}^{exc} = 3.59 \text{ MPa}.$$

The excessive pressure in HA2:

$$p_{02} = 0.2\frac{6.56}{0.235} = 5.58 \text{ MPa} \Rightarrow p_{02}^{ch} = 5.48 \text{ MPa},$$

the charging pressure in HA2:

$$p_{ch2} = 5.58\frac{9}{16} = 3.14 \text{ MPa} \Rightarrow p_{ch2}^{exc} = 3.04 \text{ MPa}.$$

Fig. 13.26 presents the curves of changing the charge and work pressures in HA1.

For damping the oscillations of a long-stroke suspension with an SRM, double-sided regulated hydraulic shock absorbers are used. The fluid's flow rate through the throttling devices of these shock absorbers should correspond to the fluid's flow rates at displacement of the PoC's piston. Keeping in mind that the fluid's maximum flow rate is possible at the PSCP's maximum compression stroke or retreat, as well as the fact that the appearance of such a flow rate may happen at low oscillation frequencies of the suspension (running over individual irregularities at a low travel speed), the fluid's maximum flow rate through the throttling devices must be no more than 6 liters per second. This condition is

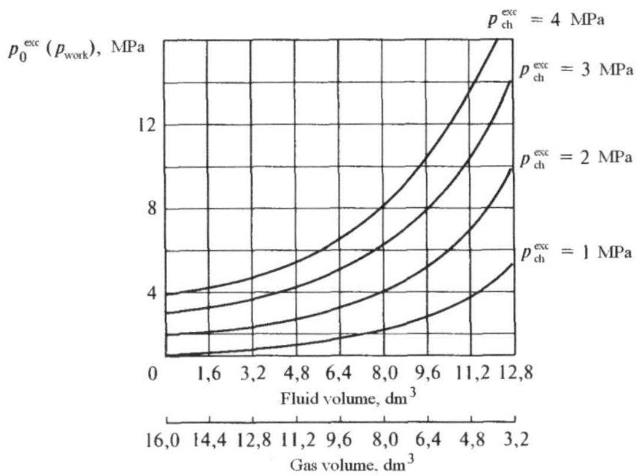

Fig. 13.26. Changes of the charging and working static pressures in HA1

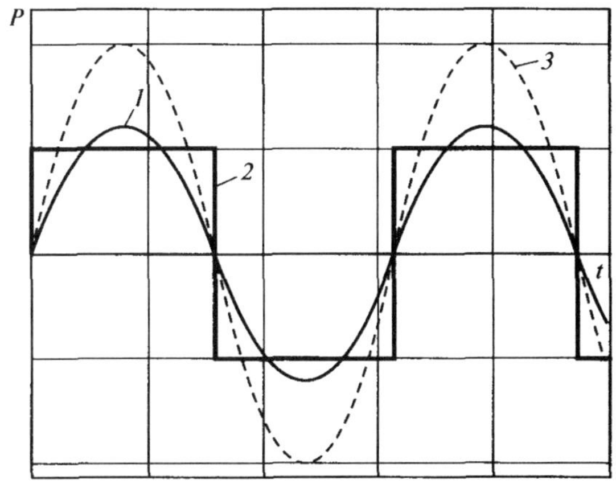

Fig. 13.27. Dependence for replacing the dry friction force by the equivalent hydraulic resistance force: 1—Dry friction force ($P_{fric}(t)$); 2—Equivalent hydraulic resistance force of ($P_{h.res}(t)$); 3—Displacement speeds of the piston ($v(t)$)

met by the KVMK 32G1.1 standard throttles with reverse valves. Such throttles can be mounted directly in the pipelines of the experimental PSCP with the SRM design.

13.4.4 Evaluation of dry friction effects on characteristics of the springing system

When operating a suspension with a hydropneumatic resilient element, one of the power factors acting on the hydraulic cylinder's piston is the dry friction force. The account for dry friction makes the mathematical model non-linear and we need preliminary studies to define whether or not to incorporate dry friction in our mathematical model. For this purpose approximated methods are used. In this case we use the harmonic linearization method based on the equality of the work performed by the dry friction force and by its equivalent force.

Let us define the dry friction force P_{fric} as follows:

$$P_{fric} = P_{fric0}\ \text{sign}\ v,$$

where: $v = \dfrac{dx}{dt}$ is the piston's displacement speed; and P_{fric0} is the force of friction at rest.

Let us now replace the dry friction force P_{fric} with the equivalent hydraulic resistance force $P_{h.res}$, which ensures during the same time interval the same energy absorption as during the action of the dry friction force (Fig. 13.27).

At harmonic oscillations of the hydraulic cylinder's piston:

$$x = a_x \sin \omega t,$$

where: x is the amplitude of oscillations; a_x is the harmonic equation constant; and ω is the oscillation frequency.

Then, the equivalent force will be:

$$P_{h.res} = k_{fric}\frac{dx}{dt} = k_{fric}\, a_x \omega\ \cos \omega t\,.$$

To define k_{fric}, let us put the equation of the hydraulic resistance force and the dry friction force as:

$$A_{h.res} = A_{fric},$$

where

$$A_{h.res} = \int_0^{\frac{2\pi}{\omega}} k_{fric}\omega^2\ a_x^2 \cos^2 \omega t\ dt = \pi k_{fric}\omega a_x^2;$$

$$A_{fric} = 4\int_0^{\frac{\pi}{2\omega}} P_{fric0}\ \omega\ a_x \cos \omega t\ dt = 4a_x\, P_{fric0}\,.$$

Thus, for a harmonically linearized dependence, the equivalent friction coefficient will look as follows:

$$\tilde{k}_{fric} = \frac{4P_{fric0}}{\pi a_x \omega}.$$

To assess the effect of dry friction, let us use a two-mass system equivalent to the springing system (Fig. 13.28), where $c_{\text{pst.w}}$ and c_{tir} are the rigidity coefficients of the suspension and tires; $k_{\text{pst.w}}$ and k_{tir} are the damping coefficients of the suspension and tires; $q(t)$ is the kinematical perturbation from the road surface; and $z_1(t)$ and $z_2(t)$ are the generalized coordinates of the unsprung and sprung masses.

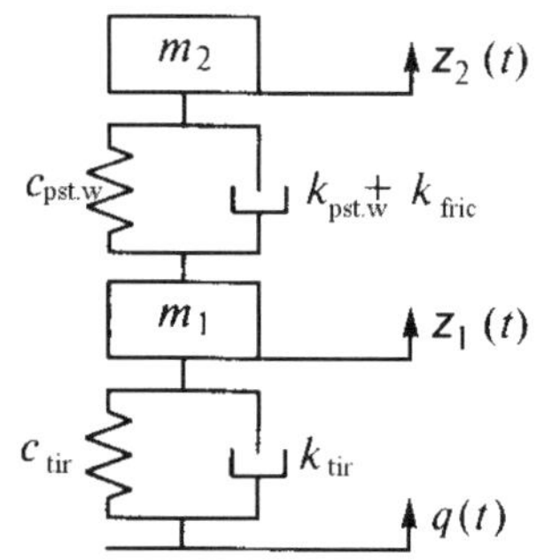

Fig. 13.28. The simplest model of the springing system

The transfer functions for the displacements of the sprung and unsprung masses of this system have the following form:

$$H_{z_{1q}}(p) = \frac{(k_{tir}p + c_{tir})(m_2 p^2 + k_{pst.w}p + c_{pst.w})}{\Delta};$$

$$H_{z_{2q}}(p) = \frac{\left((k_{fric} + k_{tir})p + c_{pst.w}\right)(k_{tir}p + c_{tir})}{\Delta};$$

$$H_{z_2 z_1}(p) = \frac{(k_{pst.w} + k_{fric})\,p + c_{pst.w}}{m_2\,p^2 + k_{pst.w}p + c_{pst.w}},$$

where:

$$\Delta = [m_1\,p^2 + (k_{pst.w} + k_{fric} + k_{tir})\,p + (c_{pst.w} + c_{tir})] \times$$
$$\times (m_2\,p^2 + k_{pst.w}p + c_{pst.w}) - (k_{pst.w}p + c_{pst.w})^2,$$

where p is the fluid pressure corresponding to $k_{\text{pst.w}}$.

Let us set the input affection amplitude a_x in the form of a function that decreases inversely to $\sqrt{\omega}$ within the whole considered frequency range. We take this assumption based on the fact that at low travel speeds the amplitude of the road surface's irregularities can be at maximum; at the same time, travel at a high speed creates tougher requirements for the quality of the road surface. Let us set the function $a_x(\omega)$ in the following form:

$$a_x(\omega) = \frac{2\pi \cdot 0.15}{\sqrt{\omega}},$$

that is, at the frequency of perturbations equal to 1 Hz, the input affection's amplitude will be 0.15 m. This function is just reflecting the general amplitude dependence on the frequency; however, it allows for assessing the influence of dry friction in the springing system without its essential complication.

With account of the fact that $p = j\omega$ and $k_{\text{fric}} = \alpha/\omega$, we obtain the following expressions for the transfer functions:

$$H_{z_{1q}}(\omega) = \frac{(k_{tir}j\omega + c_{tir})(m_2(j\omega)^2 + k_{pst.w}j\omega + c_{pst.w})}{\Delta};$$

$$H_{z_{2q}}(\omega) = \frac{\left((k_{tir} + \frac{\alpha}{\omega})j\omega + c_{pst.w}\right)(k_{tir}j\omega + c_{tir})}{\Delta};$$

$$H_{z_2 z_1}(\omega) = \frac{\left(k_{pst.w} + \frac{\alpha}{\omega}\right)j\omega + c_{pst.w}}{m_2\,(j\omega)^2 + k_{pst.w}j\omega + c_{pst.w}},$$

where:

$$\Delta = \left[m_1\,(j\omega)^2 + \left(k_{pst.w} + \frac{\alpha}{\omega} + k_{tir} \right) j\omega + (c_{pst.w} + c_{tir}) \right]$$

$$\times\,(m_2\,(j\omega)^2 + k_{\text{pst.w}}\,j\omega + c_{\text{pst.w}}) - (k_{\text{pst.w}}\,j\omega + c_{\text{pst.w}})^2,$$

$$\alpha = \frac{4P_{fric0}}{\pi a_x(\omega)}\cdot$$

When assessing the influence of dry friction on the springing system's characteristics, let us assume the following parameters:

1. The load falling on one wheel is equal to the nominal static load, i.e., $M = 73$ kN. We take the unsprung mass equal to 10% of the sprung one, i.e., $m_2 + 0.1m_2 = M$, for example, $m_2 = 6765$ kg, $m_1 = 676.5$ kg;
2. We define the suspension's rigidity coefficient from the condition that the frequency of its own oscillations is 1 Hz:

$$c_{pst.w} = \omega_0^2\; m_2 = (2\pi\cdot 1)^2\cdot 6765 = 267.1\;\;\text{kN/m};$$

3. Then, we find the suspension's damping coefficient by setting the relative decay coefficient ψ equal to 0.25:

$$k_{pst.w} = 2\psi\sqrt{c_{pst.w}m_2} = 2\cdot\, 0.25$$

$$\cdot\,\sqrt{267.1\cdot\, 6765} = 21.25\;\;\text{kN}\times\text{s/m};$$

4. For this example we take the 1600×600–685 tire, which has the following parameters:
   a. The rigidity coefficient $c_{tir} = 1300$ kN/m; and
   b. The damping coefficient $k_{tir} = 0.6$ kN×c/m;
5. Let us set the knowingly high dry friction force $P_{fric0} = 0.2g(m_1 + m_2) = 14.6$ kN.

In the curves of the systems' transfer functions' modules (Figs. 13.29 and 13.30), the solid line presents the dependences obtained without accounting for the dry friction in the hydraulic cylinder, while the dashed line shows the influence of dry friction on the springing system's characteristics. It is clear that dry friction causes an insignificant influence on the suspension's transfer functions in the area of high frequencies. The strongest influence is demonstrated in the resonance sections and is characterized by the change in the values of the transfer functions' module, which makes for individual parameters up to 30%. At high frequencies of perturbation affection, the dry friction has practically no effect on most of the parameters.

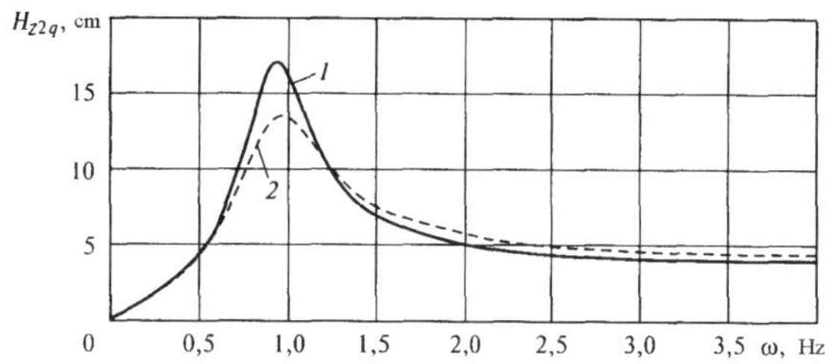

Fig. 13.29. Value of the transfer functions' module H_{z2q} for vibration accelerations of the sprung mass: 1— Without accounting for dry friction; 2—With accounting for dry friction

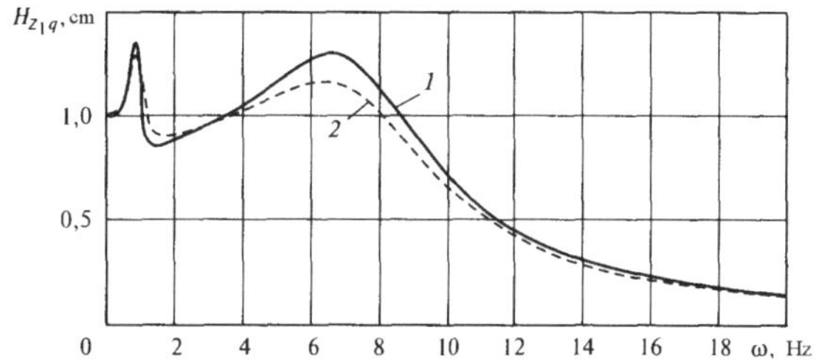

Fig. 13.30. Value of the transfer functions' module H_{z1q} for displacements of the non-sprung mass: 1—Without accounting for dry friction; 2—With accounting for dry friction

13.5 Designing long-stroke wheel suspensions for support-running modules (SRM)

The SRM is intended for transferring all types of forces and moments acting between the wheeled mover and the carrying system (frame), damping the oscillations, and decreasing the dynamic loads arising in the prototype's motion, as well as for controlling the springing system. The SRM is designed as a single assembly unit with a standardized attachment to the carrying system.

The design of the SRM ensures the following:

- The wheel's full travel in the vertical direction is 620 mm;
- The precision of maintaining the static position is ±10%;
- The smooth (impact-free) regulation of road clearance at the standing vehicle within the limits of the full wheel travel;
- The ability to manually regulate the parameters of the resilient and damping elements at the standing vehicle; and
- A complete power looping inside the module.

The SRM (Fig. 13.31) has the following basic elements:

- A suspension guiding device;
- A resilient-damping element with regulated parameters;
- A control system;
- A system of pipelines;
- A wheel gearbox; and
- A tire.

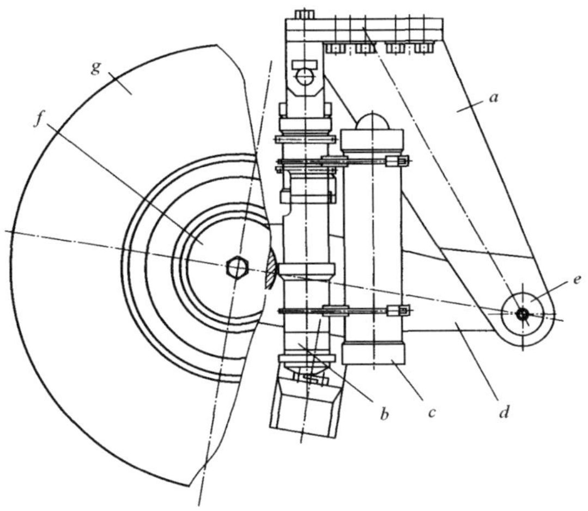

Fig. 13.31. Overall view of the SRM: a—Vertical lever; b—Hydraulic cylinder; c—Hydraulic accumulator; d— Longitudinal lever; e—Cylindrical joint; f—Wheel gearbox; g—Tire

The suspension's guiding device is the springing system's main carrying element. It interconnects the wheeled mover, the frame, and the resilient-damping element. The guiding device consists of the vertical (a) and longitudinal (d) levers. The upper end of the vertical lever is rigidly bolted to the frame. The longitudinal swinging lever is connected with its one end to the fixed lever by means of a cylindrical joint (e). The hub of the wheel gearbox (f) with the tire (g) is rigidly fixed to the other end of the movable lever. The wheel gearbox houses the elements of the braking system of the prototype and the traction electric motor.

The resilient-damping element (Fig. 13.32) consists of the working hydraulic cylinder (a), pneumohydraulic accumulator (or hydraulic accumulator) (b), counter-pressure balloon (c), and throttles with a regulated fluid flow rate (d) ((e) is the opening for the supply and draining of the working fluid).

The working fluid in the hydraulic system is MGE-10A oil according to OST (Branch Standard) 38.01281.

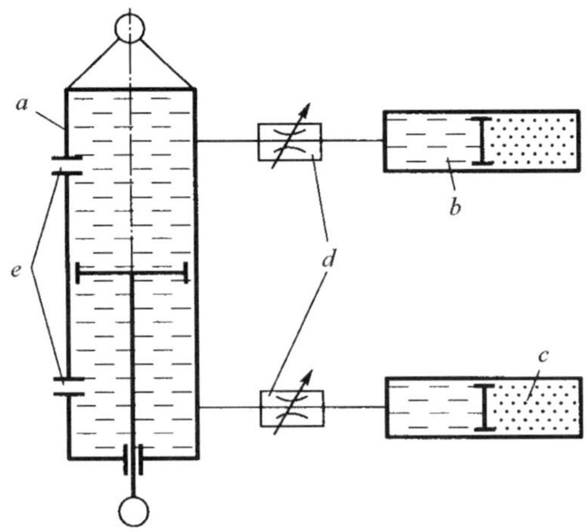

Fig. 13.32. Diagram of the resilient-damping element

The hydraulic accumulator's gas cavities and the counter-pressure balloon are filled with Grade II technical nitrogen (GOST 9293).

The springing system's hydraulic lines are equipped with throttles with KVMK model reverse valves (Fig. 13.33). The throttle is adjusted by rotating its body, which changes the opening's dimensions.

The throttle allows for setting the damping coefficient by regulating the flow rate of the working fluid flowing between the hydraulic cylinder and the hydraulic accumulator. The throttle can be adjusted manually only when the vehicle is at a stop. The throttles are independent from each other and are adjusted separately for each SRM.

The hydraulic cylinder used in the SRM is presented in Fig. 13.34. Its main elements are the cylinder (b) and rod (a), which splits the cylinder's cavity into the chambers of the working pressure (c) and counter-pressure (d).

The hydraulic cylinder is connected with the suspension's longitudinal swinging lever by a spherical joint and with the SRM's stationary part by a cylindrical joint. The working pressure and counter-pressure cavities, connected with the hydraulic accumulator and the counter-pressure balloon, respectively, are filled with MGE-10A grade oil working fluid.

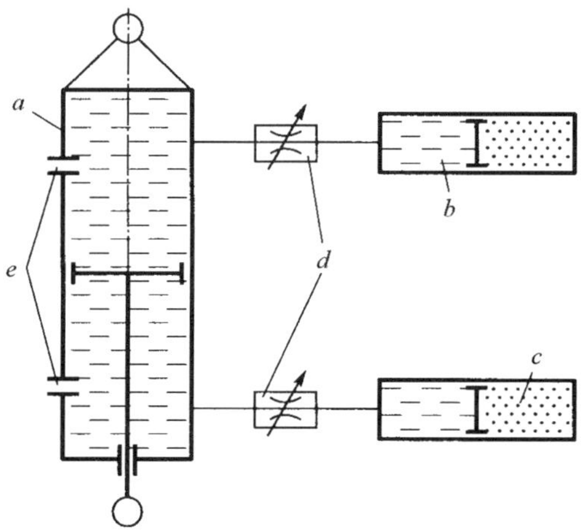

Fig. 13.33. KVMK 32G1.1 TU 2-053-1752 Throttle

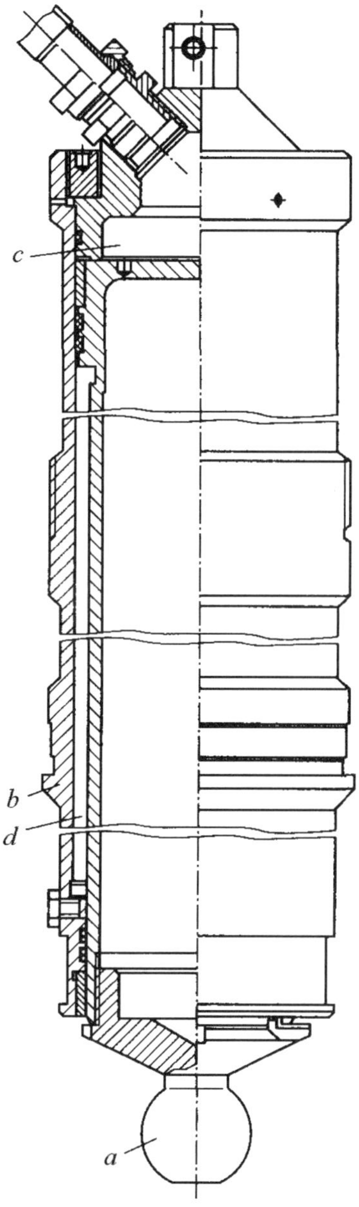

Fig. 13.34. Hydraulic cylinder: a—Rod; b—Body; c— Working pressure cavity; d— Counter-pressure cavity

The SRM's hydraulic system is equipped with an ARKh-16/320 piston pneumohydraulic accumulator (Fig. 13.35), which has a volume of 16 l and operates at the nominal pressure of

32 MPa. The hydraulic accumulator has two cavities: the fluid one (a), connected to the SRM's hydraulic system, and the gas one (b), equipped with a charging unit.

The counter-pressure balloon is an ARKh-2.5/320 pneumohydraulic accumulator, which is similar to the hydraulic accumulator in Fig. 13.35, but has a volume of 2.5 l. The pressure in the hydraulic cylinder's working cavity, which communicates with the hydraulic accumulator, can be regulated by changing the fluid's volume. The fluid is fed into the hydraulic cylinder's working cavity by the pumping station mounted in the rear part of the prototype's frame. A reserve vessel is available for draining the fluid from the hydraulic cylinder's working cavity, which also ensures the fluid reserve necessary to regulate the springing system's parameters within the preset range.

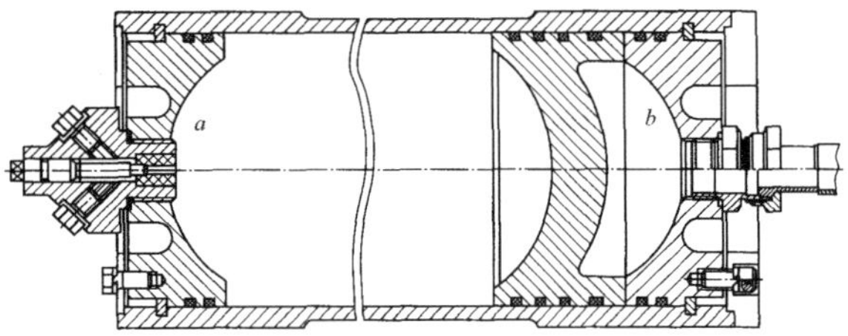

Fig. 13.35. ARKh-16/320 TU 2-053-1410 pneumohydraulic accumulator

By changing the working pressure in the hydraulic system, we can regulate the suspension's stiffness and the cargo platform's height. The working pressure in the hydraulic cylinder can be changed both manually when the HDWV is at stop and automatically for each SRM separately.

We can also change the fluid's volume in the counter-pressure cavity, which allows for regulating the suspension's energy intensity at the retreat stroke.

The control system allows for statically regulating the springing system's parameters: the rigidity and damping coefficients. It is comprised of the KVMK 32G1.1 model regulated throttles (described earlier), the three-position distribution valves, and the pumping station intended to maintain the pressure in the pipelines as needed for the regulation system.

The system of pipelines interconnects the hydraulic system's elements and ensures a reliable supply of the working fluid to the hydraulic cylinder's cavities. It is comprised of cylindrical metal pipes of different diameters and interconnected distribution devices.

The SRM is also equipped with a wheel gearbox and a wheel hub, same as in the MAZ-547 model, and a standard 1600×600–685 tube tire.

In heavy vehicles, the most serious argument in favor of lever-candle suspensions is the possibility to provide sufficient SpS strokes and release of the space in the unit's upper section, which allows for creating efficient modular designs on the basis of the vehicle.

In such applications a drawback of the lever-candle suspension is the guiding mechanism's loading at tire expansion, which restricts the use of large cross-section tires (H/B = 0.7–0.78).

Despite a complicated design, lever-type posts most fully respond to the layout conditions and requirements for multi-support vehicles, namely, satisfactory soft driving, essential suspension strokes, the possibility to regulate the parameters of the resilient element and shock absorbers, and reduced dynamic loads on the undercarriage.

The requirements for the springing systems of high passing ability vehicles prescribed in Standard ISO 2631 can be expanded to embrace vehicles with lever-type posts suspensions.

13.6 Analysis of the results of experimental trials on HDWV hydropneumatic suspensions

The dynamic characteristics of pneumatic springs depend not only on the type of resilient element, but also on the friction forces, the gas compression degree, and the frequency of oscillations. One of the factors affecting the change in the PSCP's dynamic rigidity is the heat exchange between the working medium (gas), the cylinder walls, and the environment.

Let us analyze the results of the bench and laboratory-road trials of the first hydropneumatic suspension, which was built under the diagram in Fig.13.32. The trials were performed by BAZ jointly with MSTU. The test object was the piston telescopic PSCP with an elastic divider intended for high passing ability vehicles made by BAZ.

The PSCP testing sample had the following basic parameters:

- The power cylinder's diameter was 90 mm;
- The maximum piston stroke was 140 mm;
- The static load was 24.7 kN;
- The gas volume in the static position was 0.645 l;
- The charging gas volume was 1.26 l;
- The gas pressure in the static position was 3.9 MPa;
- The gas charging pressure was 2.2 MPa; and
- The PS's stiffness (rigidity) (isothermal) in the static position was 243.8 kN/m.

13.6.1 Bench trials of the pneumatic spring

The pneumatic spring was mounted on a stand, which was able to vary its oscillation frequency from 5.5–13.1 Hz and the piston stroke from 50–100 mm. The stand was designed as a swinging beam and a crank-and-rod mechanism with a gearbox.

Using the results of the trials, we calculated the PS's average dynamic stiffness according to the following formula:

$$c^{ave}_{pst.w} = F^2_{pis}\frac{\Delta p}{\Delta V}.$$

Fig. 13.36 shows the $c^{ave}_{pst.w}$ values versus the frequency of oscillations w and the compression degree ξ. The throttling device mounted between the power cylinder and the PS's gas chamber is essentially affecting the resilient element's dynamic characteristics. This influence can be assessed by the PS's "dynamic toughening" value:

$$K^1_d = \frac{\Delta p}{\Delta y},$$

where: Δp is the maximum change amplitude of the excessive pressure; and Δy is the corresponding maximum piston stroke of the PS.

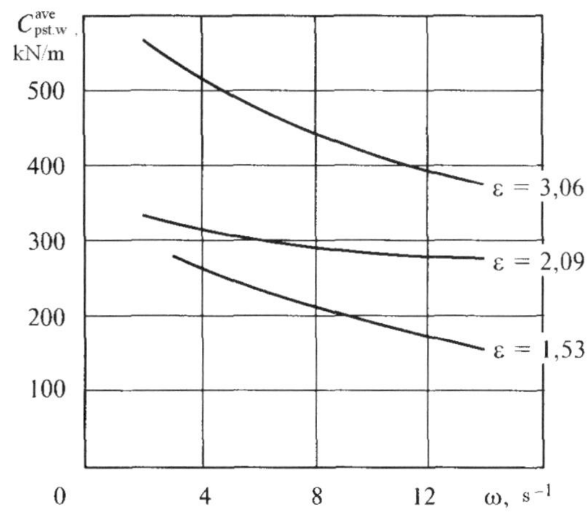

Fig. 13.36. $c^{ave}_{pst.w}$ values versus the frequency of oscillations and gas compression degree in the PSCP

Fig. 13.37 shows the PSCP's dynamic toughening versus the frequency of oscillations, the compression degree ξ, and the diameter of the throttling opening made in the shock absorber.

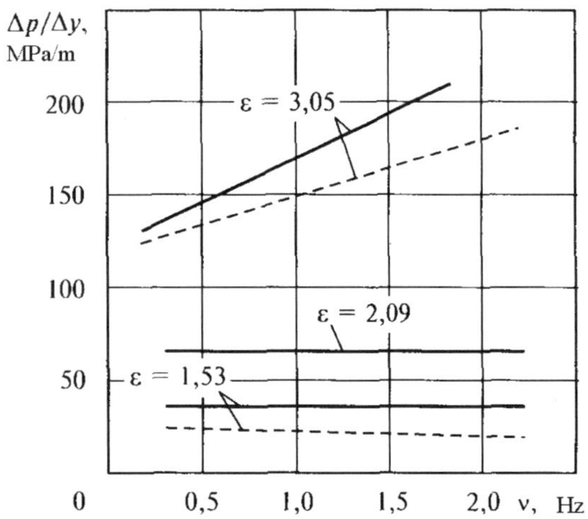

Fig. 13.37. Coefficient $\Delta p/\Delta y$ versus the frequency of oscillations, compression degree, and diameter of the valve built into the shock absorber: -------— valve diameter is 8 mm; - - - - - -—valve diameter is 3.5 mm

A decrease in the diameter of this opening has significantly increased the friction force in the pneumatic spring's power cylinder. The trials showed that the friction force's work was 710–1060 N×m. The friction force value reached 20% of the static load on the PSCP, which was caused by using V-shaped gaskets in the power cylinder.

The results of the bench trials have led to the following conclusions:

1. The PSCP's dynamic stiffness essentially goes up with the increase in the compression degree $\xi > 2$. However, with the increase in the frequency of oscillations under a permanent compression degree, the dynamic stiffness goes down.
2. The PSCP's dynamic toughening also essentially goes up with the increase in the compression degree $\xi > 2$.
3. The telescopic PSCP's dynamic characteristics are affected by friction forces and heat exchange.

Apart from the bench trials, we also held laboratory-road tests on the PSCP integrated into a suspension system of a single-axle trailer prototype.

The prototype's basic characteristics were as follows:

- The trailer's weight at bench trials was 38.5 kN and at road trials was 35.5 kN;
- The weight of the unsprung parts was 0.90 kN;
- The full travel of the trailer's wheels was 300 mm;
- The transmission ratio of the guiding device with the swing of the levers in the trailer's lateral plane was two;
- The tire dimensions were 16.00–20; and
- The throttling openings at the closed compression valve were 3.5 mm; at the open valve there were five openings with diameters of 3.5 mm; at the retreat stroke, there was no valve; and at a disconnected valve, the connection pipeline's diameter was 8 mm.

At the trials, the pressure in the tires varied as follows: 0.1, 0.15, 0.2, 0.225, and 0.25 MPa. At the bench trials, the gas charging pressure in the PSCP was set equal to: 2.42, 3.10, and 3.40 MPa. In the course of the trials, the following charging pressures were used at the testing route: 2.42, 3.02, 3.65, and 4.00 MPa.

Fig. 13.38 presents the 16.00–20 tire characteristics at different air pressures. The tire stiffness under the permanent static load of over 30 kN was 250–580 kN/m. This value is comparable with the PSCP's stiffness, which was an essential drawback in the trailer's suspension.

Another deficiency was the guiding device's design, which caused essential tire displacements in the lateral plane. Their small stiffness and large displacements in the lateral plane decreased the piston stroke in the PSCP and reduced its loading.

Fig. 13.39 presents the dynamic toughening dependence of the PSCP's ratio $\Delta p/\Delta y$ on the

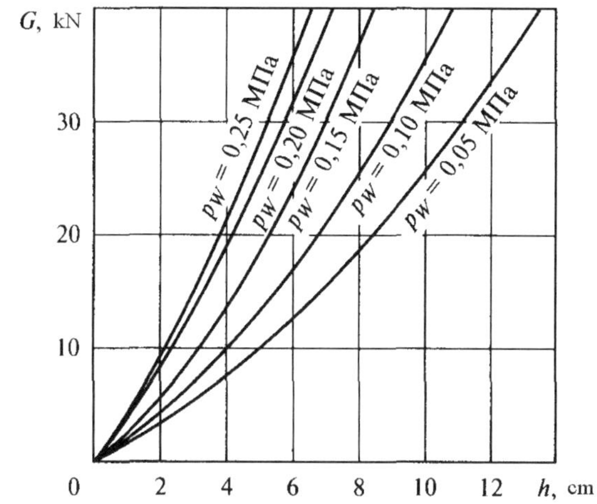

Fig. 13.38. Static characteristics of the 16.00-20 tire

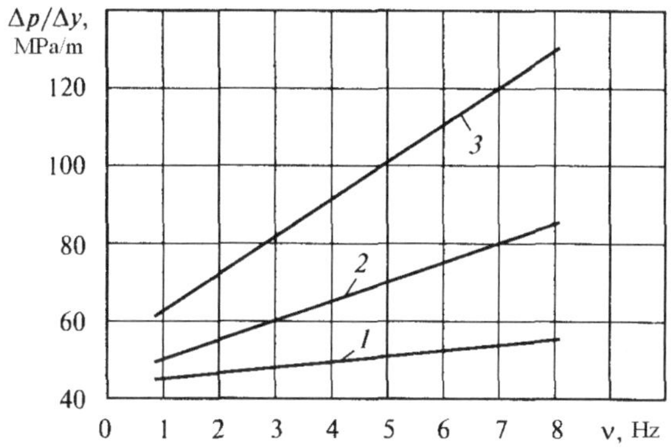

Fig. 13.39. Dynamic toughening of the PSCP's Δp/Δy versus frequency and charging pressure: 1—Charging pressure is equal to 3.4 MPa; 2—3.1 MPa; 3—2.42 MPa. The diameter of the opening in the shock absorber's valve in all the cases was 3.5 mm

frequency and charging pressure. With an increase in the frequency and a decrease in the charging pressure, the PSCP's dynamic toughening will essentially go up.

The road train's motion on the cobbled road was at speeds of 10, 20, 30, and 40 km/h. In the testing travels, we fixed the changes in the compressed gas's pressure and the piston's displacements in the pneumatic spring. Figs. 13.40 and 13.41 present the curves of the gas pressure's dispersion and the relative displacements in the octave frequency bands on the travel speed.

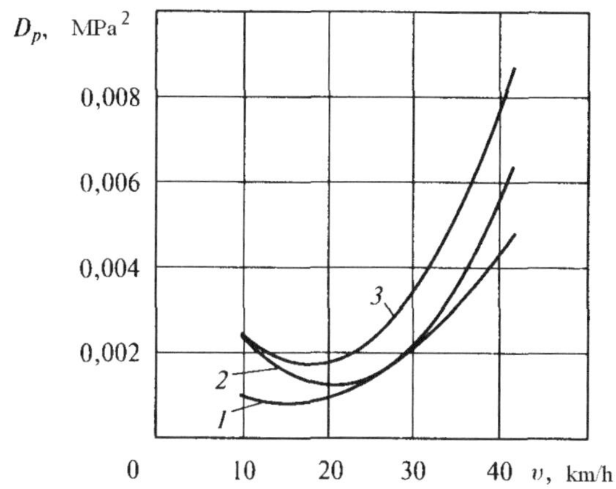

Fig. 13.40. Dispersion of gas pressure in the PSCP in octave frequency bands versus the travel speed: 1 - Octave 0.7-1.4 Hz; 2 - Octave 2.8-5.6 Hz; 3 - Octave 1.4-2.8 Hz

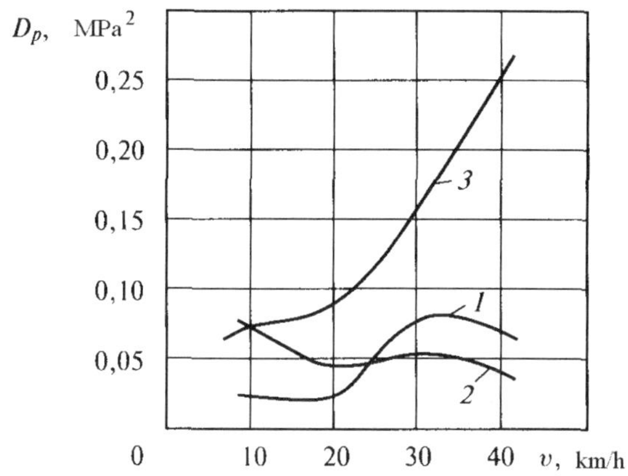

Fig. 13.41. Dispersion of gas pressure in the PSCP in octave frequency bands versus the travel speed: 1 - Octave 0.7-1.4 Hz; 2 - Octave 2.8-5.6 Hz; 3 - Octave 1.4-2.8 Hz

13.6.2 Laboratory and road trials of pneumatic springs

In order to assess the PS's dynamic characteristics, we conducted tests on a trailer on road sections with different sinusoidal irregularities. The results of these trials are presented as curves in Fig. 13.42, where high values for frequencies up to 1 Hz can be explained by the resonance swing of the sprung mass.

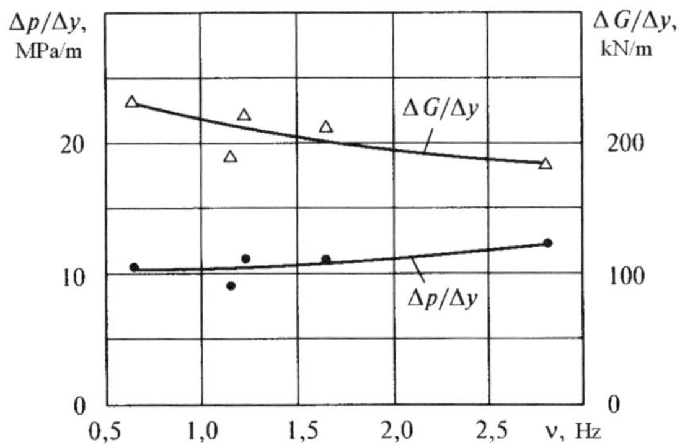

Fig. 13.42. Values of $\Delta p/\Delta y$ and $\Delta G/\Delta y$ versus oscillation frequency: • and Δ are experimental data

It follows from the curves in Fig. 13.42 that with an increase in the frequency of oscillations we saw an increase in the PS's rigidity. The results of the trials evidenced the PS's intensive heat exchange with the environment. The charging pressure was essentially influencing the values of displacements, forces, and pressures in the PS. The maximum amplitudes were obtained at p_{ch} = 2.42 MPa; they corresponded to the piston displacements of 5–6 cm, forces of up to 10,000 N, pressures of up to 0.5 MPa, and to a temperature gradient of up to 20°C.

The analysis of the trial results on the hydropneumatic suspension system intended for a HDWV allows for concluding the following.

1. The design characteristics of telescopic PSs, namely, the presence of friction, the dimensions of the built-in shock absorber's throttling opening, the gas-fluid separator's design, the gas compression level, the charging pressure, the piston stroke, and the oscillation frequency have a strong influence on the spring's dynamic characteristics.
2. Changes in the PSs' dynamic characteristics are related to the presence of the gas's heat exchange with the environment. It is expedient to make comparisons of the working processes in pneumatic springs by using adiabatic (for rapid processes) or isothermal (for slow processes) equations.
3. Running bench and laboratory-road trials on telescopic PSs installed on multi-axle vehicles not only requires plenty of time and money, but also does not allow for conducting certain types of experimental studies. That is why for research purposes it is reasonable to use vehicle testing prototypes, which simulate (imitate) the tested suspension's operation and allows for performing a broad program of tests and trials.

Based on the results of the theoretical and experimental studies, MSTU has designed several models of wheel posts (see point 9.2 and Fig. 13.43) and ACSs for them (Fig. 13.44), which were implemented in HDWV prototypes.

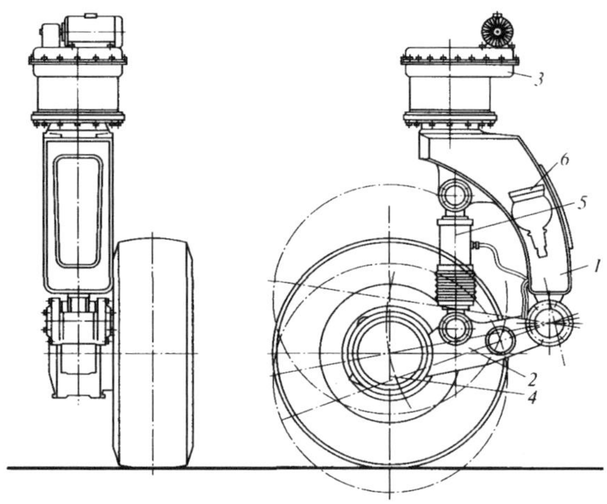

Fig. 13.43. Overall view of the wheel suspension's post: 1—Suspension post; 2—Longitudinal lever; 3—Turning device; 4—Hydropneumatic spring; 5—Hydraulic accumulator; 6—Braking chamber

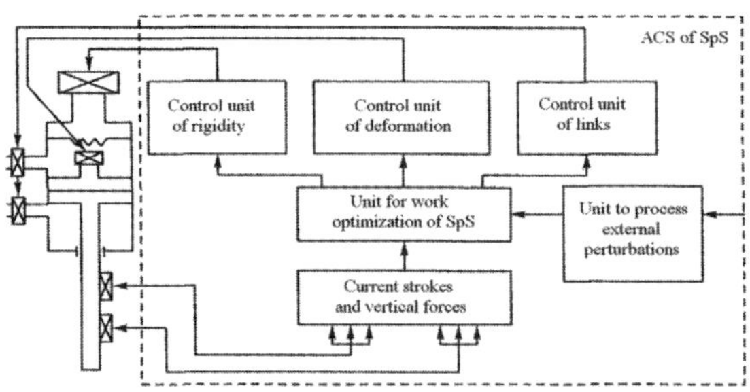

Fig. 13.44. Variant of an ACS for a springing unit

13.7 Designing and calculating suspensions

13.7.1 Calculating the suspension's resilient elements

13.7.1.1 Torsion spring

Designs of torsion springs, which are round metal bars that are twisted in operation, are well-known and broadly used in HDWVs. Traditionally, one end of the spring is rigidly fixed to the frame by a bracket and a spline coupling, while the opposite end is connected to the lever or the guiding device's balancing lever. Calculating a torsion spring is a standard twisting calculation of an individual or bundled rod (bar).

13.7.1.2 Hydropneumatic spring

The calculation diagram of a typical hydropneumatic spring is presented in Fig. 13.45. The spring is comprised of a cylinder (1) with a ball vessel (2) for the fluid and gas (nitrogen) kept there under pressure p and having a volume W. It is separated by a flexible diaphragm and a hollow rod (3) with a piston, which is designed as a vessel with pressure q and volume W_q and has a bypass pipe inside the rod. The fluid's entrance from the cylinder into the ball cavity is equipped with the valve box (4) with throttle openings acting as a hydraulic damper. Various designs of hydropneumatic springs were considered earlier in Chapter 3 and shown in Fig. 13.23.

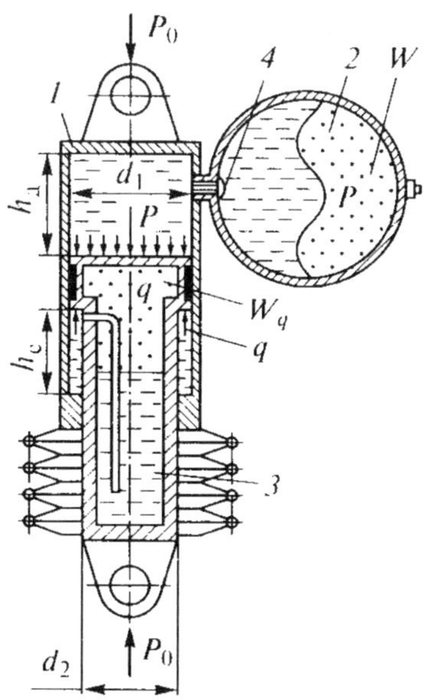

Fig. 13.45. Calculation diagram of a typical hydropneumatic spring: 1—Cylinder; 2—Vessel; 3—Hollow rod; 4—Valve box

The static position of the spring h_c, when it is affected by the force P_0 equal to the vehicle's weight, is regarded to be the initial position. At the spring's static position, its bend is considered to be equal to zero, $h_d = 0$, while the forces, pressures, and volumes correspond to the initial ones: P_0, W_0 and p_0, W_{q0} and q_0 (see Fig. 13.45).

In the course of compression, the rod with the piston goes up and squeezes the fluid out of the head end (piston cavity) through the valve box into the ball vessel. The nitrogen pressure in it increases, while the nitrogen pressure inside the piston decreases because the circular rod end is filled in through the bypass pipe.

In the course of retreat, the process is reversed: the pressure in the piston and ball cavity decreases, while the pressure in the rod end increases, and the retreat force decreases.

The equilibrium equation for the HPS under the action of forces P with account of the fact that the head end's working area is $F_1 = \pi d_1^2/4$ and the rod end's area is $F_2 = \pi/4(d_1^2 - d_2^2)$ can be presented as follows:

$$P = pF_1 - qF_2 - p_{atm}(F_1 - F_2)\ .$$

The last member of the equation is the force acting on the piston rod and caused by the atmospheric pressure p_{atm}; since it is small as compared to other forces, it is usually neglected. Then, the current value of the force P can be written as follows:

$$P = pF_1 - qF_2 \qquad (13.1)$$

In the course of compression, the pressure in the ball cavity changes under a polytrope with index

$\Pi = 1.25$ (at the suspension's oscillation frequency of 0.9–1.0 Hz):

$$p = p_0\left(\frac{W_0}{W_0 - F_1 h_d}\right)^{\Pi} = p_0\left(\frac{1}{1-\frac{h_\partial}{h_0}}\right)^{\Pi}$$

$$= \frac{p_0}{\left(1-\frac{h_d}{h_0}\right)^{\Pi}} \qquad (13.2)$$

where $h_0 = \frac{W_0}{F_0}$ is the column's reduced height in the ball cavity with the area of F_1.

When the rod moves upward, the pressure in the rod end will decrease according to the following dependence:

$$q = q_0\left(\frac{W_{q0}}{W_{q0} - F_2 h_d}\right)^{\Pi} = \frac{q_0}{\left(1+\mu\frac{h_d}{h_0}\right)} \qquad (13.3)$$

where $\mu = \frac{h_0}{h_{q0}}$ is the relative value of the reduced gas column; and $h_{q0} = \frac{W_{q0}}{F_2}$ is the reduced height of the gas column in the rod end with the area of F_2.

By substituting the values of p and q from formulas (13.2) and (13.3) into (13.1), we obtain the following:

$$p = \frac{p_0 F_1}{\left(1-\frac{h}{h_0}\right)^{\Pi}} - \frac{q_0 F_2}{\left(1-\mu\frac{h_d}{h_0}\right)^{\Pi}} \qquad (13.4)$$

The stiffness of spring $C_{pst.w}$ is expressed as follows:

$$C_{pst.w} = \frac{dP}{dh_d} = \frac{\Pi h_0^{\Pi} p_0 F_1}{(h_0 - h_d)^{\Pi+1}} + \frac{\Pi h_q^{\Pi} p_0 F_2}{(h_{q0} - h_d)^{\Pi+1}}.$$

At the spring's static position ($h_d = 0$), we have:

$$P_0 = p_0 F_1 - q_0 F_2.$$

And introducing the force ratio into our calculations gives us:

$$\xi = \frac{q_0 F_2}{p_0 F_1}.$$

Then, we can write:

$$P_0 = p_0 F_1 - \xi p_0 F_1 = p_0 F_1 (1-\xi);$$

$$p_0 F_1 = \frac{p_0}{1-\xi};$$

$$q_0 F_2 = p_0\frac{\xi}{1-\xi}.$$

For the existing designs, $\xi = 0.12$–0.16.

If the spring has no rod end, expression (13.4) will miss the last member. Formula (13.4) is used for building the spring's resilient characteristic. This characteristic will be non-linear and progressive at the presence or absence of the rod end; in the former case it will be convexo-concave and in the latter case it will be concave.

13.7.1.3 Hydraulic spring

One possible design of a hydraulic spring, which works under the principle of volumetric compression in the enclosed volume, is presented in Fig. 13.46. The spring is comprised of the cylinder (1) with two gasket units (3 and 7), as well as the rod (4) with two diameters along its length, and two additional vessels (6 and 8) connected with the cylinder's (1) cavity through valves (5 and 9). The spring's upper attachment unit has pivot axles (2). At the rod's indentation into the cylinder, the fluid's volume in it decreases, the pressure increases, and the reaction of the fluid pressure on the rod also increases. At the presence of one vessel in the cylinder, the spring's resilient characteristic is concave. Then, the spring's static stroke is bigger than the working one. To make the characteristic convexo-concave, two additional vessels (6 and 8) are used.

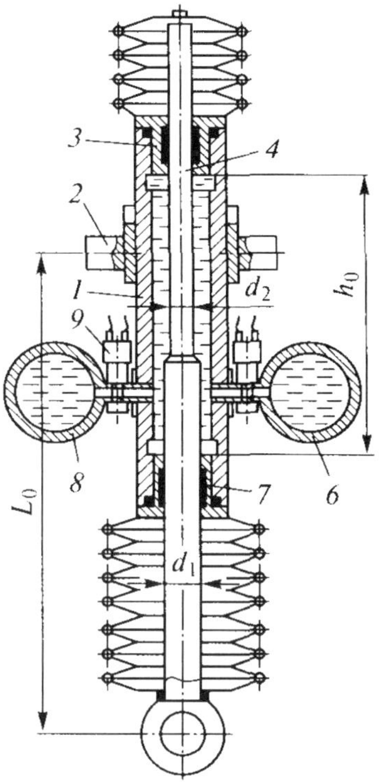

Fig. 13.46. Diagram of the hydraulic spring: 1—Cylinder; 2—Pivot axles; 3 and 7—Gasket units; 4—Rod; 5 and 9— Valves; 6 and 8—Vessels

At a completely unloaded spring, these vessels are blocked by the valves and the spring's stiffness will initially be high; then, both additional vessels are engaged and the stiffness goes down. When the pressure on the spring is above the static load, the vessels are disengaged one by one.

The working area of the rod F for the spring's considered design will be expressed by the formula:

$$F = \frac{\pi}{4}(d_2^1 - d_2^2),$$

where d_1 and d_2 are the diameters of the rod's (4) lower and upper parts.

The change in the fluid's volume at the rod's compressions stroke h_d with the working area F is defined as follows:

$$\Delta W = k_{\text{lig}} W_0 (p - p_0) = \frac{W_0}{E_{\text{lig}}}(p - p_0) = h_d F \quad (13.5)$$

where: W_0 is the initial fluid volume in the cylinder at the initial (atmospheric) pressure; k_{lig} is the volumetric compression coefficient, which is a relative change in the volume $\frac{\Delta W}{W_0}$ falling on the pressure change unit; and E_{lig} is the module of the fluid's volumetric resilience, i.e., the reverse value of the volumetric compression coefficient: $E_{lig} = \frac{1}{k_{lig}}$.

We obtain from expression (13.5) the following:

$$h_d = k_{\text{lig}} h_0 (p - p_0),$$

where $h_0 = \frac{W_0}{F}$ is the reduced height of the fluid column in the cylinder.

The hydraulic spring's force is defined by the well-known formula:

$$P = pF.$$

When calculating the characteristics of hydraulic springs, one should take into account the dependence of the parameters k_{lig} and $\frac{\Delta W}{W_0}$ on the pressure p in the spring and fluid temperature. An example of such dependence is presented in Fig. 13.47.

When calculating the spring's characteristic, one should keep in mind that engagement or disengagement of additional vessels changes the value of h_0.

13.7.2 Calculation of wheel suspensions

13.7.2.1 Kinematics of a two-lever suspension

The most widespread suspension design used in HDWVs, especially in SWCs, is a two-lever suspension, which is a four-link pin-joint

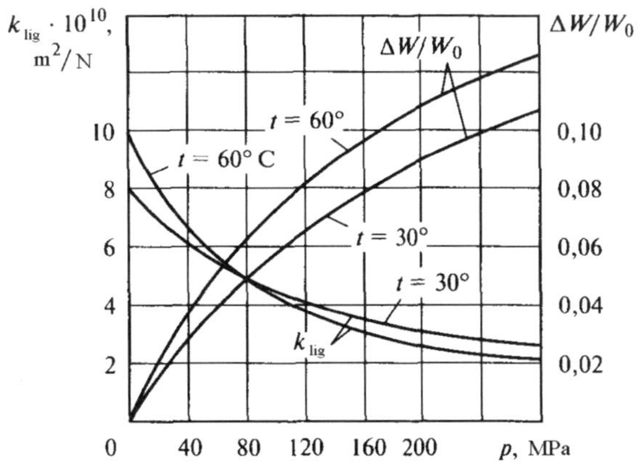

Fig. 13.47. Compressibility curves of the synthetic working fluid

mechanism with two fixed joints that connect the suspension's levers with the frame.

The parameters of independent suspensions in certain SWCs are presented in Table 13.1.

When choosing a suspension design, one should keep in mind the following considerations:

a) Layout aspects:

- The unification degree and design modularity;
- Standardized suspensions used on all wheels;
- The minimum occupation of the body's internal space by the guiding mechanism and coordination (match) with the process equipment's geometry;
- Simplified steering drives; and
- Minimum platform height;

b) Performance aspects:

- The possibility to control parameters of the resilient and damping elements, including the vehicle's loading;
- The possibility to adjust the road clearance;
- A minimum maintenance time;
- A good match of the suspension's service life and maintenance periodicity with those of the chassis;
- A good match of the suspension's storage conditions and periods with the requirements for the chassis;
- Availability of maintenance materials and consumables;
- An ability to withstand impact loads caused by the driver's mistakes without destruction;
- A possibility to be blocked at performance of process operations; and
- An ability to remain operational (without additional maintenance) after crossing a wade or short-term navigation, i.e., in onshore sea areas.

The kinematical diagram of an independent suspension with lateral levers is presented in Fig. 13.48. The letters *A*, *B*, *C*, and *D* designate the pin-joints.

The lengths of levers l_B and l_H are dictated by the vehicle's general layout and the need to obtain kinematics of the wheels' lateral roll at the

Table 13.1 Parameters of independent suspensions in certain SWCs

Parameter	Stolvet	BTR-60PB	BAZ-135LM	MAZ-543
Full travel, mm	254	220	170	281
Static travel, mm	165	110	70	122
Spring rigidity, dyne/cm	84.7	105	208	176
Frequency, Hz	1.32 (axle 1) and 1.28 (axle 3)	1.28	1.38	1.17

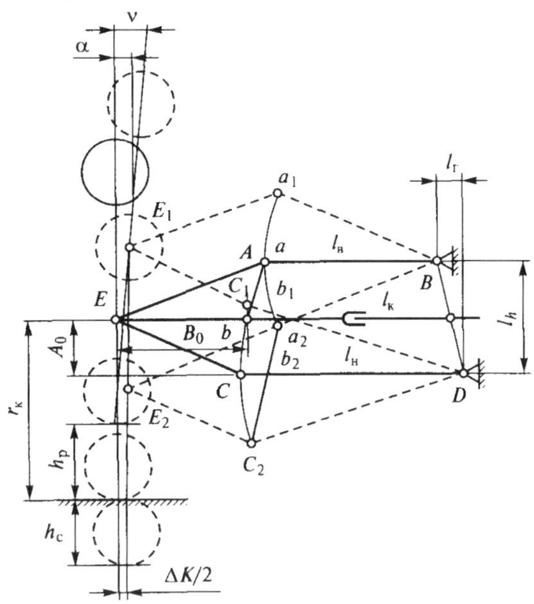

Fig. 13.48. Kinematical diagram of an independent suspension with lateral levers

suspension's up and down travel against its static position, which results in small turning angles of the wheels v and small changes in the wheelspan ΔK at the suspension's required dynamic bends (travels) downward h_c and upward h_p; in this case, $h_d = h_{down} + h_{up}$. Great changes in ΔK cause quick tire wear, while a large swinging angle ($v > 2°$) results in the appearance of a gyroscopic moment on the wheel and the appearance of axial forces, which negatively affect the wheel hubs' bearings.

The cardan transmission's presence in the drive of the driving wheel complicates the selection of the suspension's kinematical diagram and also restricts the possibilities to increase its full travel since cardan transmissions transfer the moment at the tilt angle up to 12–13°. It is also required to find the cardan transmission's position and dimensions that do not change its length l_K at the wheel swinging (only a change in length of Δl_K = 1–2 mm is admitted). Otherwise, we have axial forces in the cardan shaft's spline connection, which translate to the drive's cross-pieces and bearings. At short levers (small l_B and l_H), we see essential changes in the wheelspan and increased dimensions of the resilient element and the cardan shaft's swivel angle, while long levers increase the vehicle's width.

The initial vales of l_B, l_H, and l_K can be defined graphically (Fig. 13.48). Given a certain scale, we can build three positions of the suspension's levers and the wheel: static, upper, and lower. Then, we can find the positions of joints a, a_1, a_2 of the upper (short) lever l_B; of joints c, c_1, c_2 of the lower lever l_H; and positions b, b_1, b_2 of the cardan shaft l_K. By halving the arc a_1a_2, we can find point a, and, by horizontally measuring l_B, we can define the position of joint B. The position of the long lever's joint D is found by measuring the length of l_H from point C under the α angle to the horizon.

Table 13.2 presents the suspension's geometrical parameters obtained in the process of selecting the guiding devices' rational kinematics.

Table 13.2 Kinematical parameters of a two-lever suspension (variants), mm

h_{down}	h_{up}	r_K	A_0	B_0	α, o	l_H
108	112	530	142.3	207	3	390
114	114	630	165	262	1°10'	400
75	100	630	156	230	1	220
l_B	l_K	l_r	l_h	––	v, °	Δp
275	335	82.5	261.5	19.5	2	1.5
280	403	112	325	2	1°30'	2.2
220	220	10	350	18	18	0

After choosing the suspension's kinematical diagram, we can calculate the suspension's resilient (reduced) characteristic as a whole, i.e., the load dependence on the wheel P_K on its travel h_K based on the resilient element's available characteristic $P = f(h_d)$. Thus, it is necessary to establish the change character of the power transmission ratio of the suspension's levers, which connect the power P_K acting on the wheel with the resilient element's force P. This calculation is treated as reducing the suspension to the calculated diagram.

13.7.2.2 Reducing the two-lever wheel suspension with independent longitudinal torsions to the calculated diagram

The wheel suspension's design is presented in Fig. 13.49 and its kinematical diagram is presented in Fig. 13.50. The design comprises lateral levers (12 and 13), which are attached through longitudinal torsions (8 and 10) to the frame by brackets (5 and 6). The wheel (1) bracket (2) is connected to the levers in joints *A* and C. The cardan transmission's (11) presence, which connects the axle gearbox (9) with the wheel gearbox, brings the same deficiencies into the suspension's kinematics as were mentioned for the previous diagram. The hydraulic shock absorber (3), which is attached by a bracket (6) connected to the frame's (7) cross-piece, functions as a damping element. The selection of this suspension's kinematical diagram is similar to the diagram in Fig. 13.48.

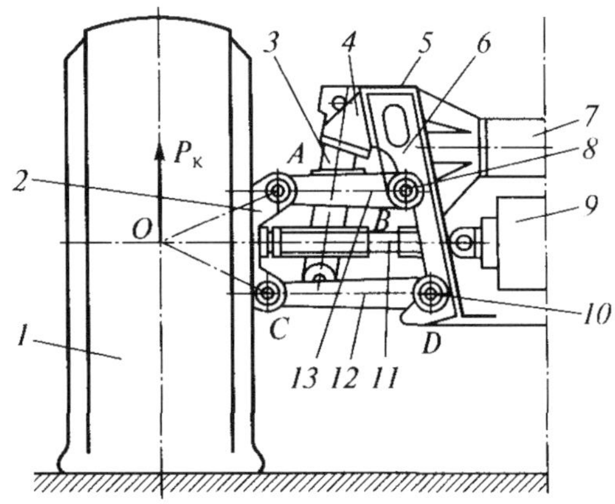

Fig. 13.49. Torsion suspension for HDWVs with a mechanical wheel drive: 1—Wheel; 2—Wheel hub; 3—Shock absorber; 4—Stop; 5—The frame's side member; 6—Bracket; 7—The frame's cross-piece; 8, 10—Resilient elements of the torsions; 9—Driving axle; 11—Cardan shaft; 12, 13—The suspension's guiding levers

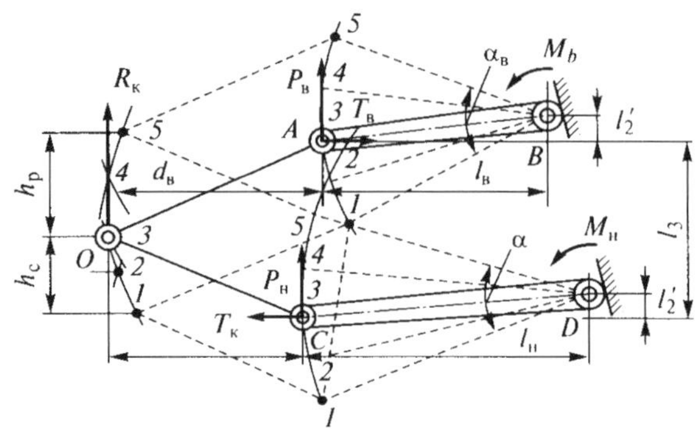

Fig. 13.50. Kinematical diagram of a two-lever wheel suspension with longitudinal independent torsions

Since the longitudinal torsions function as the resilient element, it is important in this case to consider the calculation of torsion parameters.

After selecting the suspension's kinematical diagram, we can define the twisting angles of the upper α_B and lower α_H torsions and the suspension's full travel $h_d = h_{down} + h_{up}$. We also know the geometrical dimensions of the suspension's elements. The unknown values are the horizontal forces T_B and T_n, the vertical forces P_B and P_n applied in joints *A* and *C* of the levers' attachment with the wheel hub, and the reaction of the wheel $R_K = P_K$ (where P_K is the load on the wheel). Thus, we know five parameters. To find the static reaction of the wheel R_K depending on the torsion twisting angles, we can use the system's static equilibrium equations—the equations of the static balance in the levers, hubs, and forces in the vertical and horizontal planes according to Fig. 13.50:

$$M_B + T_B\ell_1^1 = P_B\ell_B;$$

$$M_n + T_n\ell_2^1 = P_n\ell_n;$$

$$R_k d_B + P_n(d_B - d_n) = T_n g;$$

$$R_k d_n + P_2(d_2 - d_n) = T_2 g;$$

$$P_2 + P_n = R_k;$$

$$T_2 = T_n,$$

where: R_K is the wheel's vertical reaction; and d_B and d_H are the wheel touring arms along the suspension's upper and lower levers.

The static equilibrium equations of the moments and forces acting on the hub (link *A0C* in Figs. 13.49 and 13.50) take the directions of the forces P_B, T_B, P_H, and T_H opposite to the similar forces acting on levers *AB* and *CD* in joints *A* and *C*.

By jointly solving the system of static balance equations, we obtain the following:

$$P_k = R_k = \frac{M_2 A + M_n B}{D},$$

where:

$$A = \ell_1^1 d_H - \ell_1^1 d_B + \ell_H \ell_3;$$

$$B = \ell_2^1 d_n - \ell_2^1 d_2 + \ell_2 \ell_3;$$

$$D = \ell_3 \ell_2 \ell_n - \ell_2^1 d_2 \ell_n + \ell_1^1 d_n \ell_2;$$

$$M_B = c_B \alpha_B;$$

$$M_n = c_n \alpha_n,$$

c_B, c_n is the stiffness of the torsions; and α_B, α_n are the twisting angles of the torsions.

The reduced suspension characteristic calculated by using the above dependences is presented in Fig. 13.51. The suspension's stiffness is defined by numerical integration of the reduced characteristic. The basis for the calculation is the static load on the wheel $P_{K.cT}$.

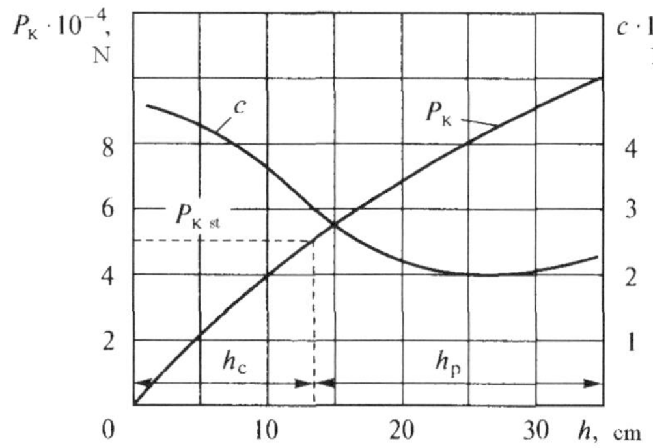

Fig. 13.51. Reduced characteristic of a two-lever torsion suspension

It follows from Fig. 13.51 that at the linear characteristic of resilient elements (torsions), the suspension characteristic is non-linear (convex) because of the non-equal transmission ratios of the kinematical links, which depend on the wheel travel. However, it remains rather close to a linear characteristic with a permanent stiffness.

Many technical solutions are known for obtaining a progressive convexo-concave characteristic at the linear characteristic of resilient elements; one is presented in Fig. 13.52 [3]. This solution differs from the one considered earlier by the fact that

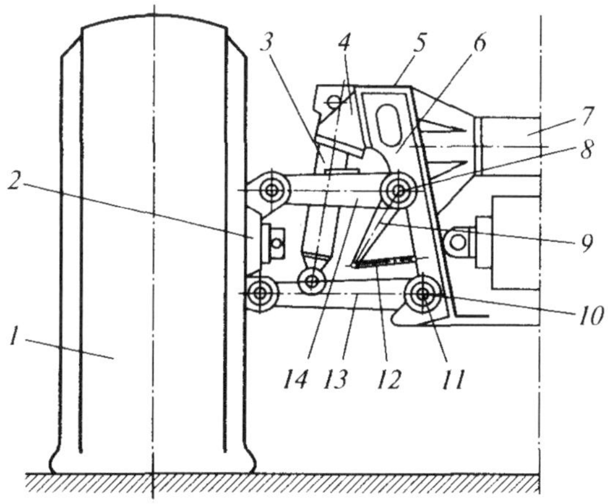

Fig. 13.52. Torsion suspension with hydrostatic (electric) wheel drive: 1—Wheel; 2—Hydraulic (electric) motor; 3—Shock absorber; 4—Stop; 5—The frame's side member; 6—Bracket; 7—Cross-piece; 8, 11—Side members; 9, 10, 12—Levers and rods connecting the torsions; 13, 14—Guiding levers

here the back ends of the upper and lower torsions are interconnected by the levers (9 and 10) and connecting rod (12). In this case, the stiffness of the upper torsion is much higher than the lower torsion, while one lever (9) is much longer than the other (10). When this suspension is loaded, the back ends of the torsions are movable. The lower torsion is the correcting one, while the upper torsion is the carrying one. When the suspension moves up, the lever (9) turns to the same side as the main lever (14), while the other lever (10) moves in the opposite direction of the remaining lever's (13) motion. Thus, the twisting angle of the upper torsion is small and lower one is large. This allows, on one hand, for increasing the suspension's travel under a small length of the bearing upper torsion, and on the other hand, obtaining a progressive convexo-concave characteristic by changing the correlation of the lever arms (9 and 10) at the turn thereof. By choosing the levers' (9 and 10) initial setting, we can obtain any non-linear suspension characteristic.

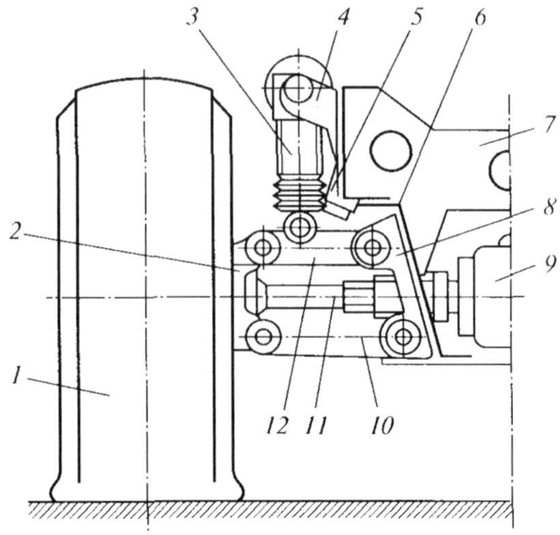

Fig. 13.53. Hydropneumatic suspension with mechanical wheel drive: 1—Wheel; 2—Wheel hub; 3— Hydropneumatic spring; 4, 8—Brackets; 5—Stop; 6— The frame's side member; 7—The frame's cross-piece; 9—Driving axle; 10, 12—The suspension's guiding levers; 11—Wheel drive's cardan shaft

13.7.2.3 Reducing the two-lever wheel suspension with hydropneumatic spring to the calculated diagram

Diagrams of these suspensions are presented in Figs. 13.53 and 13.54. In the first one, the HPS (3) is fixed on the suspension's upper levers (12); in the second one it is fixed on the lower levers (9), which allows for increasing the spring's length and for obtaining a bigger wheel travel in the absence of the cardan shaft in the wheel drive.

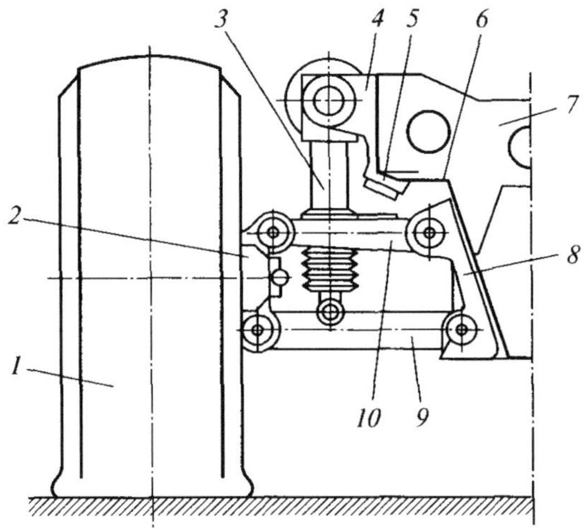

Fig. 13.54. Hydropneumatic suspension with hydrostatic (electric) wheel drive: 1—Wheel; 2—Hydraulic (electric) motor; 3—Hydropneumatic spring; 4, 8— Brackets; 5—Stop; 6—The frame's side members; 7— The frame's cross-piece; 9, 10—The suspension's guiding levers

The suspension's kinematical diagram is presented in Fig. 13.55. The link between the wheels' travel h and the suspension's stroke h_d is established graphically after selecting the kinematical diagram. The transmission ratio of the kinematical links from the load P_K on the wheel to the spring's force P is found with the help of a system of balance equations of the suspension elements, which are under the forces shown in Fig. 13.55. In this figure, all alphanumeric notations are the same as in Fig. 13.48.

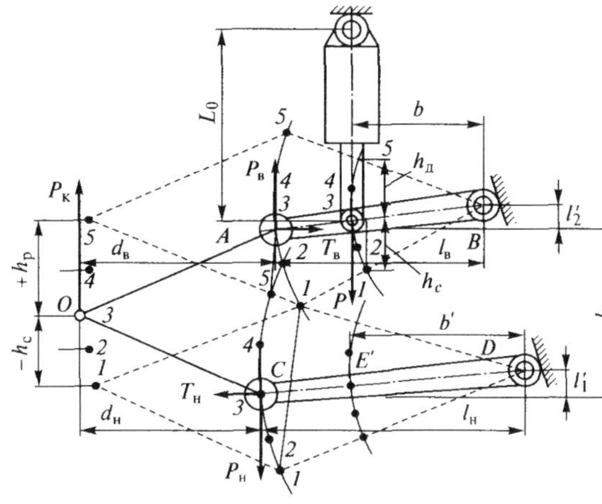

Fig. 13.55. Kinematical diagram of a two-lever suspension with an HPS

When the spring is attached to upper levers:

$$P_B + T_B \ell_2^1 = P_B \ell_B;$$

$$T_B \ell_1^1 = P_H \ell_H;$$

$$R_k + P_n = P_2;$$

$$P_k d_k + P_2(d_2 - d_n) = T_2 \ell_3.$$

After solving this system of equations, we obtain:

$$P_k = R_k = PK,$$

where $K = \dfrac{b'(\ell_n d_2 \ell_1^1 - \ell_n d_n \ell_1^1 - \ell_2 \ell_n \ell_3)}{\ell_n(\ell_n d_2 \ell_1^1 - \ell_2 d_n \ell_2^1 - \ell_2 \ell_n \ell_3)}$ is the transmission ratio of the kinematical links.

When the HPS is attached to the suspension's lower levers (in point E′ of Fig. 13.55), we have:

$$Pb' = T_n \ell_1^1 + P_n \ell_n;$$

$$P_B \ell_B = T_B \ell_2^1;$$

$$R_k = P_2 + P_n;$$

$$R_k d_2 = P_n(d_2 - d_n) + T_2 \ell_3;$$

$$T_2 = T_n,$$

where R_K is the vertical reaction on the wheel.

By jointly solving the equations of the above system, we obtain:

$$P_k = R_k = PK,$$

where:

$$K = \frac{b'(\ell_n d_2 \ell_1^1 - \ell_n d_n \ell_1^1 - \ell_2 \ell_n \ell_3)}{\ell_n(\ell_n d_2 \ell_1^1 - \ell_e d_n \ell_2^1 - \ell_2 \ell_n \ell_3)}.$$

As an example, Fig. 13.56 presents the reduced characteristic of the suspension diagram presented in Fig. 13.55.

13.7.3 Designing a wheel suspension with additional functions

When designing a suspension, the preset initial data are the static load on the wheel, the static stroke, and the full travel of the wheel suspension. That is why the starting calculation and design point is choosing the suspension's kinematical diagram. Once the choice is made, they define the transmission ratio K and find the static load on the spring P_{st}.

The kinematical diagram also gives the spring's strokes h_{st} and h_d. Then, we can calculate the spring's characteristic and the suspension's reduced characteristic.

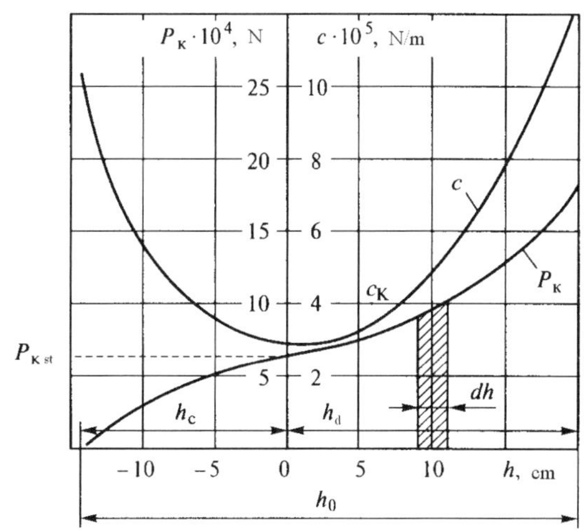

Fig. 13.56. Reduced characteristic of the HPS

As an example of using a wheel suspension for performing additional functions, let us consider the use of a two-lever suspension with a HPSCP in the system intended to improve the HDWV's stability against lateral tipover (tripping).

Fig. 13.57 presents a variant of using a combined suspension for improving the lateral stability of a HDWV [3].

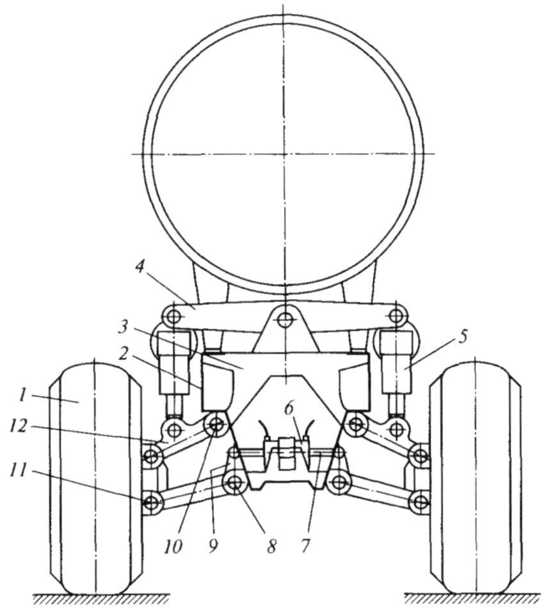

Fig. 13.57. Wheel suspension of a HDWV with a lateral balancing lever: 1—Wheel; 2—Frame; 3—Frame crossbar; 4—Balancing beam; 5—Hydropneumatic spring; 6—Stabilizing hydraulic cylinder; 7—Hydraulic cylinder's rod; 8 and 10—Torsions; 9—Levers; 11, 12—Guiding levers

The suspension is designed with a lateral balancing lever (4). It contains the HPSs (5) fixed above to the balancing lever (4) and below on the guiding device's levers (12). The guiding device's levers (11 and 12) are equipped with longitudinal torsions (8 and 1), the reverse ends of which are linked with the help of levers (9) to the controlling hydraulic cylinder's (6) two-sided rod (7). The vehicle's frame (2) with the cargo is suspended on the balancing lever (4). The HPSs ensure perfect soft driving thanks to their progressive characteristics, while the torsions ensure the lateral stability and the body's stabilization (change in the lateral roll) with the help of the hydraulic cylinder (6), which is controlled from the roll sensors.

A drawback of this suspension's design is the increase in the platforms' loading height because of the lateral balancing levers above the vehicle's frame and because of the complexity in adjusting the road clearance.

In suspensions with HPSs, stabilization diagrams based on the principles of a hydraulic balancing lever are of additional interest. Fig. 13.58 shows a variant of this design [3].

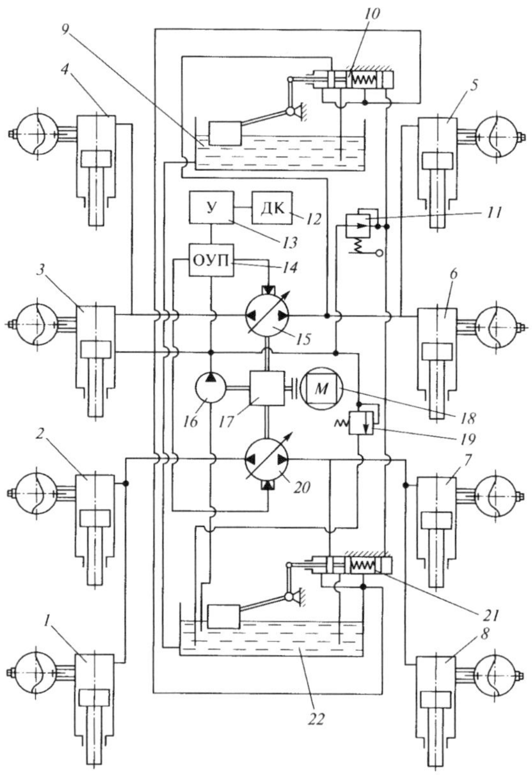

Fig. 13.58. Basic diagram of a HDWV body's lateral stabilization system: 1-8—HPSs; 9 and 22—Beams; 10 and 21—Road clearance regulators; 11—Reducing valve; 12—Roll sensor; 13—Signal amplifier; 14—Pump performance control block (PCB); 15, 16, and 20—Pumps; 17—Reduction gearbox; 18—Engine; 19— Safety valve

The operation of this stabilization design is based on the direct re-pumping of the working fluid from the springs (1–4) of one board of the vehicle to the springs (5–8) of the other board and back in accordance with the lateral oscillations frequency of the vehicle's sprung part by the regulated pumps (15 and 20) driven by the engine (18) through the reduction gear (17). The pumps (15 and 20) are controlled by the PCB (14) under the signal of the roll sensor (12) and signal amplifier (13). To replenish any fluid leaks from the springs that may take place through the glands of the pumps (15 and 20), to feed the PCB, and adjust the road clearance, the system has a control pump (16), topping-up tanks (9 and 22) with road clearance float regulators (10 and 21), a relief valve (11) for adjusting the regulators (10 and 21) to the road clearance's preset height, and the system's safety valve (19).

The system operates as follows: At the body's inclination, for example, to the right, the roll sensor generates a signal proportional to the roll and affects the unit that controls the pumps via the amplifier 0n. The former begins pumping the fluid from the left-hand springs into the right-hand ones to eliminate the roll. At an opposite roll, we see an opposite process: pumping the fluid from the right-hand springs into the left-hand ones.

If the body sets down lower than the preset value (the working fluid's volume in the tanks increases) due to fluid leaks through the pumps (15 and 20), the regulators' (10 and 21) spools will displace to the right and the pump's (16) pressure line will connect to the power lines of the pumps with springs (15 and 20); as a result, the body's sprung mass position will be restored. In order to enforce a change in the road clearance (for example, to increase it), it is necessary to reduce the pressure behind the reducing valve. Then, the fluid will be drained from the springs into the tanks through the regulators (10 and 21).

The considered design differs from other known stabilization structures, on one hand, by its simplicity and small weight, and, on the other hand, by the minimum energy consumption for the sprung mass's lateral stabilization.

We would like to emphasize that there are several suspension types and designs and it is impossible to consider all of them. That is why the book describes only those, which, in the authors' opinions, can be used in designing HDWVs after slight improvements to account for modern achievements. The authors did not touch on the calculations and designs of shock absorbers for suspensions since this topic is nicely covered in scientific and technical literature on vehicles and needs no special consideration as applied to HDWVs.

Chapter 14
Curvilinear motion of multi-axle wheel chassis

14.1 Terms and definitions

In general, the controlled curvilinear motion is understood as the motion of a multi-link wheeled system in the horizontal plane under which the targeted trajectories of the links' center of mass are formed. The orientation of the leading link's longitudinal axis and the mutual orientation of the vehicle's subsequent links are also formed in the horizontal plane.

In particular, we can consider the curvilinear motion of a single-link wheeled chassis, where the driver (operator) or an automatic system forms the targeted trajectory formation of the vehicle's center of mass and ensures in every moment of time the orientation of the carrying system's longitudinal axis. Sometimes this motion is treated as a chassis maneuver in the horizontal plane when a differentiation is made between a speed and turning maneuver. This chapter does not consider the speed control for the chassis' center of mass, which is tangential to its trajectory.

When studying a wheeled chassis' curvilinear motion, two cases are considered: a stationary motion (stationary turning) and a non-stationary motion (non-stationary turning), which is more general. When studying the latter, D.A. Antonov [61] suggested considering two periods:

- Active control (steering) (both the control action intensity and the chassis' controlled motion parameters are changing); and
- Passive waiting (the control action has reached the desired level and the controls are kept in the required position; however, the controlled parameters continue changing).

Sometimes the second case is treated, not quite correctly, as the vehicle's "evolutionary motion" or "evolutionary mode," which assumes a motion at which, in spite of the zero control displacement speeds, the controlled parameters will not be zero. This case is also treated as the after-influence mode or the after-influence condition.

This is undesirable since the driver (operator) is naturally expecting that when he ceases affecting the controls, the controlled parameters are no longer affected. However, this is not the case in reality; therefore, the vehicle's evolutionary characteristics may cause unpredictable and dangerous situations.

In general, the vehicle's maneuverability in the horizontal plane is a complex property. It is manifested in its fullest extent in the vehicle's ability to change its motion direction—within the shortest possible time—to the reverse, while occupying in doing so the support surface's minimum area. It is known that the quality of such a maneuver depends on the vehicle's braking properties, handling quality, maneuverability, turning ability, dynamic properties at acceleration, etc.

In their full scope, these qualities characterize the technical level of both combat and special vehicles. As to multi-axle wheeled chassis, the most important indicator is the vehicle's ability to change the turning parameters when the position of the controls changes to the angle of θ ($\theta = (\theta_1, \theta_2, \ldots, \theta_n)^T$, a parameter characterizing the control device's deviation by turning away from the neutral position).

Following the recommendations of D.A. Antonov [61], let us split the HDWV's turning into three phases:

- The maneuvering phase with an active control ($\dot{\omega} \neq 0$, $\dot{\theta} \neq 0$), which is completely characterized by "the vehicle's handling quality";
- The evolutional phase or a passive aftereffect ($\dot{\omega} \neq 0, \dot{\theta} = 0$); and
- The steady phase ($\dot{\omega} = 0, \dot{\theta} = 0$), which is completely characterized by the "turning ability" (turnability).

The second phase is evaluated separately by holding special trials: "jerk of the steering wheel" and "entrance into the circle."

One should also differentiate between the maneuverability and turning ability of a wheeled vehicle.

It follows from Antonov [61] that the maneuverability of a multi-axle wheeled chassis is understood as the property characterizing the vehicle's ability to make a turn—within the minimum possible time—for moving in the reverse direction. The criterion (evaluation) for the thus defined maneuverability is the minimum possible time of a complete turn executed on a free ground without any external restrictions on performing the maneuver.

The turning ability of a multi-axle wheeled chassis is understood as the property characterizing the vehicle's ability to change the course angle following the control action undertaken by the driver or an automatic control system.

Modern oversized wheeled chassis are sometimes equipped with steering systems (StSs) that allow for creating curvilinear motion trajectories of the center of mass under the vehicle's constant longitudinal axis orientation (an example developed at MSTU is described in Chapter 4). These types of steering systems are also installed on chassis of certain heavy self-propelled cranes. A similar motion character can be found in some modern passenger cars with front and rear steered wheels and compliant rear suspension devices. For clarity, we use a narrower and more widespread understanding of the term "turning ability" as the wheeled vehicle's ability to make an intended turn with a change in the course angle in accordance with the driver's control action.

The sharper the vehicle can turn, the better its turning ability is; that is why the respective characteristics are presented as a dependence of the characteristic points' motion trajectory curvature from the driver's control action. Most often this is the dependence of the center of mass's curvature radius or other characteristic points in the carrying system's trajectory (the turning radius) on the turning angle of the steered wheels. In many cases, for example when making a technical description of a wheeled chassis, they indicate only the border values that characterize the vehicle's turnability, that is, the minimum turning radii of the characteristic points at the controlled wheels' maximum turning angle.

Most of the time the technical requirements and descriptions set the minimum turning radius measured by the trace of the external front wheel since this parameter characterizes the minimum support surface dimensions required for making a complete turn without stopping.

They also use other estimate turnability parameters, including:

- The wheeled chassis' turning radius measured from the center of the instant turn to the vehicle's longitudinal axis (used in most calculation dependences);

- The specific traction force required to turn the chassis in the preset road and ground conditions;
- The chassis' minimum turning radius under which the vehicle or transported cargo do not touch the surrounding articles (the so-called "sweeping radius"); and
- The utilization coefficient of the adhesion strength on the controlled wheels (allows for defining the degree to which the steered wheels' contact loading reaches the adhesion limit with the support surface).

In principle, the wheeled chassis' turnability is better when the value of each of the above parameters is smaller. The rational restrictions imposed on these parameters are defined by the particular chassis' operational conditions and will be considered in point 8.4.

In the further narration of the curvilinear motion theory for HDWVs, we will use the notions of excessive, neutral, and insufficient turnability of multi-axle wheeled chassis (see the simplest stationary turning diagram for a wheeled chassis presented in Fig. 14.1). It is assumed that the reader is aware of the basic theory

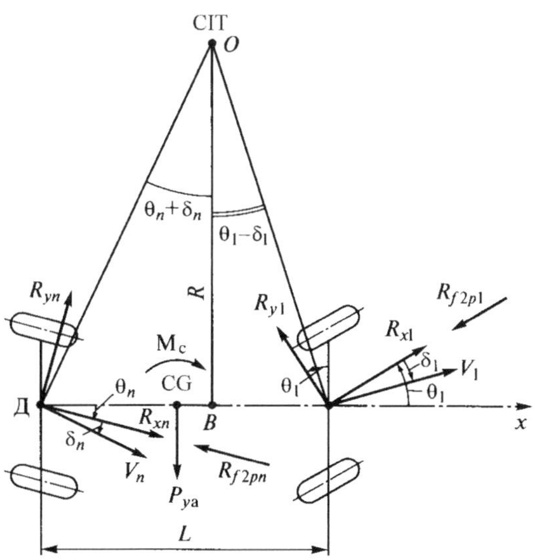

Fig. 14.1. Simplest turning diagram of a wheeled chassis with elastic wheels

of an elastic wheel's lateral skid presented in ordinary training courses on automobile motion theory. The theoretical and empirical models of an elastic wheel's rolling along curvilinear trajectories are presented in more detail in the works of G.A. Smirnov and D.A. Antonov [12 and 61] and are considered in Chapter 8 and the discussions related to designing the StS of multi-axle wheeled chassis.

When designing the diagram in Fig. 14.1, it is assumed that the axle's drift angle is equal to the average drift angle value of the external and internal wheels (the correctness and the limits of using this and similar simplifications, as well as the general rules for the signs and orientation of the coordinate axes are considered later).

It is clear that:

$$\begin{cases} R = \dfrac{L}{tg(\theta_1 - \delta_1) + tg(\theta_n - \delta_n)} \;; \\ x_t = \dfrac{L\, tg(\theta_n + \delta_n)}{tg(\theta_1 - \delta_1) + tg(\theta_n - \delta_n)} \;, \end{cases} \tag{14.1}$$

where: R is the curvature radius of the instant turning center's projection onto the chassis' longitudinal axis (the so-called "turning pole"); x_t is the turning pole's displacement relative to the middle of the chassis' base; θ_1 is the average turning angle of the front axle's wheels; θ_n is the average turning angle of the rear axle's wheels; δ_1 is the drift angle of the front axle's tires; and δ_n is the drift angle of the rear axle's tires.

To define, in principle, the notions of excessive and insufficient turnability, we can restrict ourselves to the description of the chassis at high speeds since in this case we can clearly see the drift phenomenon at relatively small turning angles of the controlled wheels.

At comparable small drifting and turning angles of the controlled wheels, we can linearize the numerator and denominator of the formulas in (14.1):

$$\begin{cases} R = \dfrac{L}{(\theta_1 + \theta_n) + (\delta_n - \delta_1)} \;; \\ x_t = \dfrac{L\,(\theta_n + \delta_n)}{(\theta_1 + \theta_n) + (\delta_n - \delta_1)} \;. \end{cases} \qquad (14.2)$$

- In accordance with (14.2), we introduce the following definitions:
- At the chassis' neutral turnability ($\delta_n - \delta_1 = 0$), the turning radius depends only on the turning angles of the controlled wheels and does not depend on the drift;
- At an insufficient chassis turnability ($\delta_n - \delta_1 < 0$), the turning curvature of the controlled wheels goes down because of drifting; therefore, the turning radius depends not only on the turning angles of the controlled wheels, but also on the drift (in particular, it goes up with an increase in the centrifugal force acting on the turning chassis at an increase in the trajectory curvature);
- At an excessive chassis turnability ($\delta_n - \delta_1 > 0$), the turn made by the controlled wheels is added to the turn made by drifting (in particular, it decreases with the increase in the centrifugal force acting on the turning chassis at an increase in the trajectory curvature).

It was shown in Chapter 4 that in designing single-link multi-axle wheeled chassis, they implement mainly the so-called "steering" methods of making turns—turning the controlled wheels. Certain HDWVs make turns by a forced jackknifing of the links, sometimes in combination with the turn of the controlled wheels. Less frequently, turning the support-running complex's components is used—as a rule, the individual wheel axles or axle groups (bogies). From the viewpoint of the wheels' interaction with the support surface, these designs are described within the steering theory of wheeled vehicles. The so-called "board-type" turning is practically not used in HDWV designs; therefore, it is not considered here.

To characterize the turnability, they usually use the dependence of the relative motion trajectory curvature on the turning angle of the controlled wheels. It follows from formula (14.2) that:

$$\frac{L}{R} = (\theta_1 + \theta_n) + (\delta_n - \delta_1) \qquad (14.3)$$

The chassis' turning component characteristic, which is ensured directly by turning the controlled wheels (the first members in formula (14.3)), is called the steering turnability. The second member of the sum in the formula is treated as the tire turnability.

When a sprung system rolls, the undercarriage elements can change their position relative to the vehicle's longitudinal axis. This effect is typical, for instance, for a spring suspension and a bridge-type design of the chassis' support-running modules; for independent suspensions of guiding mechanisms with a longitudinal-lateral design; and, generally speaking, for most independent suspensions because of the incomplete coordination (match) of the guiding mechanism's kinematics with the steering drive. As a result, additional turning angles of the wheels appear in the chassis' turning because of the centrifugal force, which gives birth to an additional component of the vehicle's turning as a whole. This additional component is called the roll turnability.

Depending on the suspension's design, the roll turnability can increase or reduce the total motion curvature, or not change it whatsoever. Therefore, the roll turnability can be characterized as excessive, insufficient, or neutral. At neutral roll turnability, the chassis can move with a drift, and, although it has no effect on the turning radius, it has an essential effect on the offset of the turning center along the vehicle's longitudinal axis.

If we introduce into the consideration the specific wheel turn, which depends on the lateral force and suspension design, and designate it as θ_{Kp}, the turnability will be assessed by the following dependence:

$$\frac{L}{R} = \frac{L}{R_0} + (\delta_n - \delta_1) + \left[\frac{L}{R}\right]_{ch} = \frac{L}{R_0} + (\delta_n - \delta_1) + \left[\frac{L}{R}\right]_{ch} =$$
$$= (\theta_n + \theta_1) + (\theta_{\text{ch.n}} + \theta_{\text{ch 1}}) + (\delta_n - \delta_1) \qquad (14.4)$$

where: $\frac{L}{R}$ is the relative curvature of the turning pole's trajectory at the absence of the roll turnability effect; $\left[\frac{L}{R}\right]_{ch}$ is the change in the relative curvature of the turning pole's trajectory caused by the roll turnability effect; $\theta_{\text{ch.1}}$ is the additional turn of the front axle's wheels caused by the roll of the chassis' body; $\theta_{\text{ch}.n}$ is the additional turn of the rear axle's wheels caused by the roll of the chassis' body.

Chassis with excessive turnability are the most maneuverable. Nevertheless, as we show below, they are prone to a loss in motion stability; therefore, it is usually recommended to design vehicles with neutral or weakly expressed insufficient turnability (the difference in the drift of the extreme axles is about 2° at the specific lateral force of $P/G_g = 0.4$).

Any specific lateral force acting on the chassis can be assessed according to the formula:

$$\eta_y = \frac{p_y}{G_g} = \frac{M_g\, j_e}{M_g\, g} = \frac{j_y}{g},$$

where: η_y is the specific lateral force; P_y is the lateral force; G_g is the chassis' weight; M_g is the chassis' mass; and j_y is the lateral acceleration.

The results of any lateral force action will be a lateral roll, suspension deformation, and an additional turn of the chassis' wheels. The contact spot of the resilient wheels with the road will see lateral reactions and the corresponding tire breakaway. As a result, the motion trajectory of the center of mass will curve.

The dependence of the specific lateral force acting on the chassis on the turn's relative curvature, caused by a combination of all the factors created by this lateral force, is the chassis' turnability diagram.

If the suspension characteristic is linear, the dependence of the roll turnability on the lateral force will be also linear (usually the distortions introduced into the suspension by the kinematical imperfection or compliance of the guiding device are small). Fig. 14.2 presents the variants of the turnability diagram for two extreme cases:

- Excessive linear roll turnability (line *OA*) and total excessive non-steering turnability (line *OB*); and
- Insufficient linear roll turnability (line *OC*) and the total insufficient non-steering turnability (line *OD*).

Let us suppose that the lateral force is the centrifugal force that appears in the chassis' center of mass motion along a curvilinear trajectory.

In this case:

$$\eta_y = \frac{p_y}{G_g} = \frac{v_g^2}{g\,L} = \frac{v_g^2}{g}\frac{L}{R} \qquad (14.5)$$

where: v_g is the chassis' travel speed.

At v_A = constant, we have a linear dependence of the specific lateral force and the relative motion

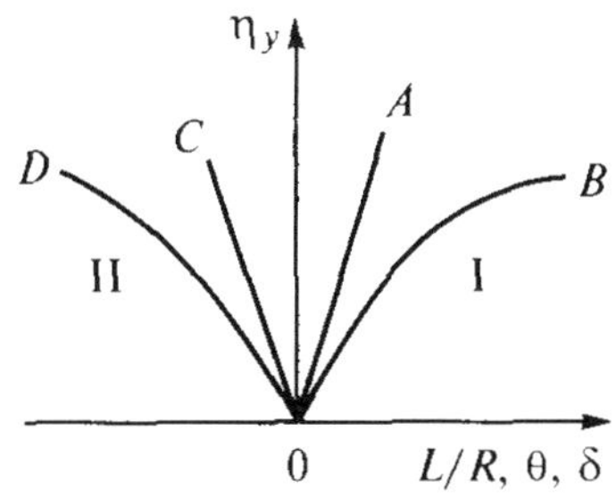

Fig. 14.2. Variants of the chassis' turnability diagram: I—area of excessive turnability; II—area of insufficient turnability

trajectory curvature of the center of mass. The turnability diagram corresponding to this case (a chassis with a clearly expressed excessive total turnability) is presented in Fig. 14.3.

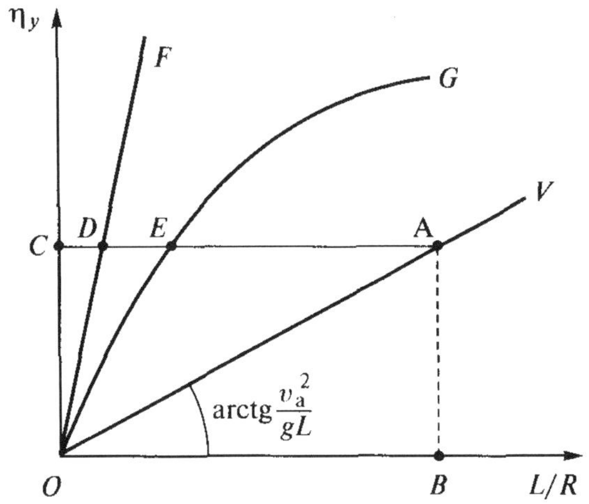

Fig. 14.3. Turnability diagram of a multi-axle wheeled chassis with excessive total turnability

The composition of this diagram allows for defining the turnability components at the chassis' preset motion mode. The chassis with wheelbase *L*, which moves along the trajectory with a relative curvature *OB*, is under the effect of the relative lateral force *BA*. This motion is characterized by the roll turnability *CD* and by the tire turnability *DE*. The chassis' aggregate non-steering turnability, or self-turnability, corresponds to section *CE*. The non-steering turnability is not controlled by the driver and arises spontaneously as a consequence of the curvilinear nature of the chassis' center of mass trajectory.

It is obvious that in order to keep the chassis on a curvilinear trajectory of a preset relative curvature, it is necessary to turn the controlled wheels and create an additional steered turnability, which is characterized by section *EA*. The sections *OF* and *OG* characterize the chassis' turnability executed only for the sake of the roll and tire components accordingly. The line *OV* characterizes the chassis' total turnability.

14.2 Equations of plane curvilinear motion in multi-link wheeled chassis

14.2.1 General approach to mathematical models of plane curvilinear motion in HDWVs

Chapters 1–4 have demonstrated that HDWVs are notable for a great variety of designs. The following are the most frequent:

- Multi-axle vehicles with different numbers and locations of controlled wheels;
- Road trains with several articulated sections;
- Articulated vehicles with active joints.

The tasks addressed at the initial stages of designing HDWVs are also multiple and varying; they demand different levels of detailing in the mathematical model and different depths of describing the occurring phenomena.

The so-called "external problems" are initially addressed, a typical example being the vehicle's navigation. To solve them, we need the minimum level of detailing for the processes that take place in the vehicle's units and systems. The mathematical model, with whose help each of these tasks are solved, is the chassis' plane curvilinear motion equation. To solve it, they use the simplest model of the vehicle's curvilinear motion written down in projections onto the axes of the bound coordinate system:

$$\begin{cases} M_g(\dot{v}_y + v_x\omega_z) = \sum R_{yi} \ ; \\ \mathrm{I}_z\dot{\omega}_z = \sum M_{zi} \ , \end{cases}$$

where: M_g is the vehicle's mass; I_z is the vehicle's inertia moment relative to the vertical axis of the bound coordinate system; v_x is the speed projection of the center of mass onto the longitudinal axis of the bound coordinate system; v_y is the speed projection of the center of mass onto the lateral axis of the bound coordinate system; ω_z is the vehicle's angular velocity projection onto the vertical axis of the bound coordinate system; ΣR_{yi} is the external forces' resultant vector projection onto the lateral axis of the bound coordinate system; and ΣM_{zi} is the external forces' resultant moment vector projection onto the vertical axis of the bound coordinate system.

Although this model disregards a great number of design characteristics for real vehicles (including the centrifugal inertia moments, the members containing the products of the angular velocity's vector components, wheel motion characteristics, etc.), it is quite acceptable for an analysis of elementary cases of a single-link multi-axle wheeled chassis' curvilinear motion on a hard support surface.

At the same time, we can base our general model on the motion equations for the body of a multi-axle wheeled transporter (see section 4.8). All of the remaining variants take intermediate positions between the two extreme cases: the simplest model and the general-type model, which is considered further.

The calculations under the general-type model are extremely cumbersome, which, as a rule, is not reasonable at the initial design phases of multi-axle wheeled vehicles. At the same time, a rather full idea about most of the processes can be obtained from the oscillations analysis of the multi-axle vehicle and from other simple models.

The most general layout of a heavy-duty vehicle is a multi-link road train composed of k n_k-axle links and interconnected by active joints. Every link of the road train can be considered as a multi-axle wheeled section with all driving and all controlled (steered) wheels.

The equations describing the motion of such vehicles are applicable to heavy-duty vehicles of any design and the models for transporters with simple designs will appear as simplified models of the general type. However, this way is not always rational from the viewpoint of the time input required to build a particular model (for example, it is simpler and easier to directly obtain the model describing a plane (two-dimensional) turn of a single-link multi-axle wheeled chassis).

This chapter uses both ways. Initially, we describe the general sequence of building a general-type model [61]. These are quite bulky elaborations, which are of no special interest by themselves; therefore, we present only the main sequence of transformations and the main differences from the methodology presented in *Motion Theory of Military Wheeled Vehicles* [61]. Then, for analyzing the curvilinear motion of a single-link multi-axle wheeled chassis, we present a traditional sequence for building a model.

Since to calculate the control affections we need to know the vehicle's lateral and angular deviations from the preset trajectory, and the derivatives thereof, the solution to the motion equations will be built relative these deviations.

14.2.2 Kinematical relations for general-type models

Let the coordinate system CK_C: (x_C, y_C) be linked with the preset trajectory as shown in Fig. 14.4. The legend is the same as in point 8.4.

Let us assign the serial number i to each of the road train's links starting from the first link and counting the links from the nose to the back. Each ith link is associated with the coordinate system CK_i: (x_i, y_i) with the origin in the link's center of gravity (point C_i). We then orient the coordinate axes of these systems in such a way that axis x_i is parallel to axis x_C.

The longitudinal axis of the ith link makes the angle β_i with the axle x_i.

Given these conditions, the coordinates of any j^{th} point A_{ij} of the ith link's longitudinal axis in the system CK_C will be defined as follows:

$$\begin{cases} x_{ij} = x_{C_i} + a_{ij} \cos\beta_i = x_{C_i} + a_{ij}\ c_i\ ; \\ y_{ij} = y_{C_i} + a_{ij} \sin\beta_i = y_{C_i} + a_{ij}\ s_i\ , \end{cases} \quad (14.6)$$

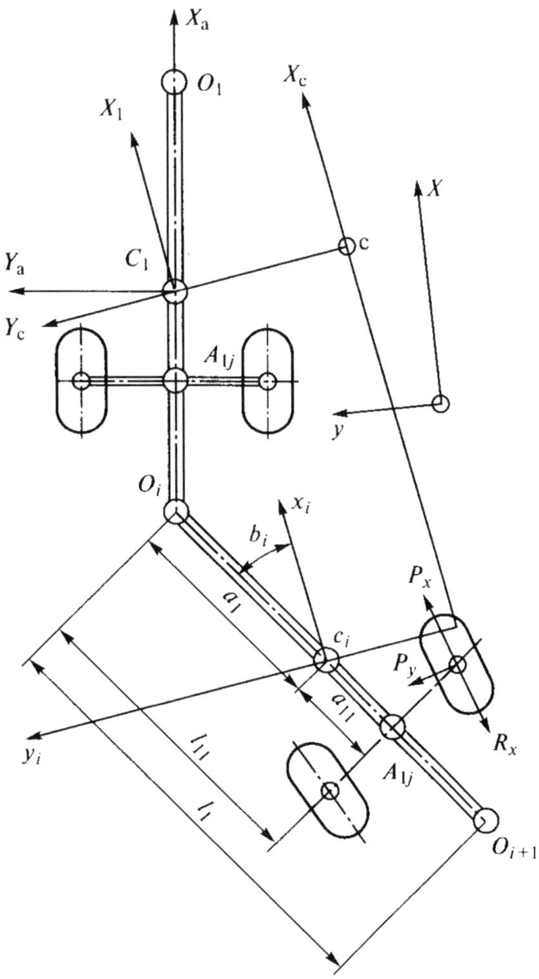

Fig. 14.4. Kinematical diagram of the plane motion of a multi-link wheeled transporter

where: $i = 1, 2, \ldots, k$; $j = 0, 1, 2, \ldots, n$; x_{C_i} is the horizontal longitudinal coordinate of the ith link's center of mass; y_{C_i} is the horizontal lateral coordinate of the ith link's center of mass; C_i is the ith link's center of mass; s_i is the projection of point A_{ij} onto the axis y_i; and a_{ij} is the distance between points C_i and A_{ij} (the positive is the direction from the center of mass to the nose of the link).

Let the l_i be the length of the road train's i^{th} link and a_{C_i} the distance between points C_i (the ith link's center of mass) and O_i (the center of the nose joint of the ith link). By definition, these parameters are always positive.

In this case, in the coordinate system CK_C:

$$x_{C_i} = x_{C_{i-1}} - \left[\left(\ell_{i-1} - a_{C_{i-1}}\right) c_{i-1} + a_{C_i} c_i\right] \quad (14.7)$$

The coordinate system CK_C is located on the first link and its axes are directed forward along the travel of the road train. The coordinates of the following links are negative and increase as the module moves farther from the nose link. However, the geometrical parameters included into formula (14.7) are defined such that they are always positive. To agree with the signs in this formula, we introduce the sign (–) before the square brackets.

Formula (14.7) allows for building the following recurrent diagram:

$$\begin{cases} x_{C_1} = x_{C_1} \ ; \\ x_{C_2} = x_{C_1} - \left[\left(\ell_1 - a_{C_1}\right) c_1 + a_{C_2} c_2\right] ; \\ x_{C_3} = x_{C_2} - \left[\left(\ell_2 - a_{C_2}\right) c_2 + a_{C_3} c_3\right] , \end{cases}$$

or:

$$\begin{cases} x_{C_1} = x_{C_1} \ ; \\ x_{C_2} = x_{C_1} - \left[\left(\ell_1 - a_{C_1}\right) c_1 + a_{C_2} c_2\right] ; \\ x_{C_3} = x_{C_1} - \left[\left(\ell_1 - a_{C_1}\right) c_1 + a_{C_2} c_2\right] - \left[\left(\ell_2 - a_{C_2}\right) c_2 + a_{C_3} c_3\right] . \end{cases}$$

By regrouping the members of these formulas, we get:

$$\begin{cases} x_{C_1} = x_{C_1} ; \\ x_{C_2} = \left[x_{C_1} - a_{C_1} c_1\right] - \left[\ell_1 c_1\right] - \left[a_{C_2} c_2\right] ; \\ x_{C_3} = \left[x_{C_1} + a_{C_1} c_1\right] - \left[\ell_1 c_1 + \ell_2 c_2\right] - \left[a_{C_3} c_3\right] . \end{cases}$$

This correlation can be presented as the following recurrent formula:

$$x_{C_i} = \left[x_{C_1} + a_{C_1} c_1\right] - \sum_{\eta=1}^{i-1}\left[\ell_\eta c_\eta\right] - \left[a_{C_i} c_i\right] \tag{14.8}$$

In the same way, we obtain a recurrent formula for y_{C_i}:

$$y_{C_i} = \left[y_{C_1} + a_{C_1} s_1\right] - \sum_{\eta=1}^{i-1}\left[\ell_\eta s_\eta\right] - \left[a_{C_i} s_i\right] \tag{14.9}$$

When deriving formula (14.9), one should remember that in our adopted situation the road train is moving from left to right and turns to the right. Under this choice of location and motion of the links, all the considerations presented for coordinate x as to the choice of signs are valid also for the coordinate y. This means that in order to obtain formula (14.9), it is enough to replace the function cos (β) for sin (β).

We transform (14.8) as follows:

$$\begin{aligned} x_{C_i} &= \left[x_{C_1} + a_{C_1} c_1\right] - \sum_{\eta=1}^{i-1}\left[\ell_\eta s_\eta\right] - \left[a_{C_i} s_i\right] - \ell_i c_i + \ell_i c_i = \\ &= \left[x_{C_1} + a_{C_1} c_1\right] - \sum_{\eta=1}^{i}\left[\ell_\eta s_\eta\right] + \left[(\ell_i - a_{C_i})\, c_i\right] \end{aligned} \tag{14.10}$$

Let us define $b_i = l_i - a_C$ as the distance from the back joint (hinge) of the ith link (point 0_{i+1}) to its center of mass (point C_i). This allows for presenting formula (14.9) in a more compact form:

$$x_{C_i} = \left[x_{C_1} + a_{C_1} c_1\right] + \left[b_i c_i\right] - \sum_{\eta=1}^{i}\left[\ell_\eta c_\eta\right] \tag{14.11}$$

By analogy, we have:

$$y_{C_i} = \left[y_{C_1} + a_{C_1} s_1\right] + \left[b_i s_i\right] - \sum_{\eta=1}^{i}\left[\ell_\eta s_\eta\right] \tag{14.12}$$

Thus, let us adopt the following agreement about summing up:

$$\begin{cases} \sum_{k=m}^{n} A_k = 0\,, & if \quad m > n\,; \\ \sum_{k=m}^{m} A_k = A_m\,, & if \quad m = n\,, \end{cases}$$

where: n, k, and m are the number of axles in the chassis.

Let us introduce the following auxiliary functions:

$$\begin{cases} S_i = \sum_{\eta=1}^{i} \ell_\eta c_\eta\,; \\ H_i = \sum_{\eta=1}^{i} \ell_\eta s_\eta\,. \end{cases}$$

This allows for presenting the recurrent correlations (14.11) and (14.12) in a more compact form:

$$\begin{cases} x_{C_i} = x_{C_1} + a_{C_1} c_1 + b_i c_i - S_i\,; \\ y_{C_i} = y_{C_1} + a_{C_1} s_1 + b_i s_i - H_i\,. \end{cases} \tag{14.13}$$

From the coordinates of the link's center of mass, it is easy to define the coordinates of any other point on its longitudinal axis. To do so, one should use the equations in (14.6) and exclude from them the coordinates of the link's center of mass with the help of the formulas in (14.13):

$$\begin{cases} x_{ij} = x_{C_1} + a_{C_1} c_1 + b_i c_i - S_i + a_{ij} c_i = x_{C_1} + a_{C_1} c_1 - S_i + (b_i + a_{ij})\, c_i\,; \\ y_{ij} = y_{C_1} + a_{C_1} s_1 + b_i s_i - H_i + a_{ij} s_i = y_{C_1} + a_{C_1} s_1 - H_i + (b_i + a_{ij})\, s_i\,. \end{cases} \tag{14.14}$$

Let us designate $d_{ij} = b_i + a_{ij}$ as the distance between the center of the back joint of the ith link and some random point j on the longitudinal axis of this link.

The coordinate systems CK_C and CK_1 are chosen such that the axes $O_C y_C$ and $O_1 y_1$ coincide. It means that always:

$$x_{C1} = 0.$$

Let us name the distance between the coordinate origins O_C and O_1 as the lateral displacement of the front link relative to the preset trajectory Δ (or briefly, the lateral displacement (offset) of the first link):

$$y_{C1} = \Delta.$$

In this notation, the system in (14.14) takes the form:

$$\begin{cases} x_{ij} = a_{C_1} c_1 + d_{ij} c_i - S_i \\ y_{ij} = \Delta + a_{C_1} S_1 + d_{ij} s_i - H_i \end{cases} \quad (14.15)$$

Motion Theory of Military Wheeled Vehicles [61] offers to simplify the formulas in (14.15) since, in most cases, the vehicle's motion is $\beta 1 < 0.3$ rad. Then (the error is no more than 5%):

$$\begin{cases} x_{ij} = a_{C_1} + d_{ij}\ c_i - S_i\ ; \\ y_{ij} = \Delta + a_{C_1}\ \beta_1 + d_{ij}\ s_i - H_i\ . \end{cases} \quad (14.16)$$

Applying the formulas in (14.16) to derive the motion equations results in a mistake, namely, to the appearance of extra members in the motion equations of the nose link. Physically unexplainable effects appear (for example, the inertia of the nose link becomes dependent on its angular position).

The reason for the error is in the fact that at $i = 1$ in simplified formulas, the equation $x_{c_1} = x_{c_1}$ is violated. This is caused, in its turn, by the fact that in the first (simplified) summand the trigonometry functions of the angle β_1 are absent, while in the second one they are formally preserved.

The best way for further actions is to use the initial formulas at developments, which allows for simplifying the final result.

Let us designate, as it is traditionally done, $2B_{ip}$ as the wheelspan of the pth axle of the road train's i^{th} link and define the wheel by the third index r, taking for the wheels of the left side $r = 1$ and for the wheels of the right side $r = 2$.

For the left-hand board:

$$\begin{cases} x_{ip1} = x_{ip} - B_{ip}\ s_i\ ; \\ y_{ip1} = y_{ip} + B_{ip}\ c_i\ . \end{cases} \quad (14.17)$$

For the right-hand one:

$$\begin{cases} x_{ip2} = x_{ip} + B_{ip}\ s_i ; \\ y_{ip2} = y_{ip} - B_{ip}\ c_i . \end{cases} \quad (14.18)$$

By combining formulas (14.17) and (14.18), we obtain:

$$\begin{cases} x_{ipr} = x_{ip} + B_{ip}\ \mathrm{sgn}(1.5 - r)\ s_i\ ; \\ y_{ipr} = y_{ip} - B_{ip}\ \mathrm{sgn}(1.5 - r)\ A_i\ . \end{cases}$$

Thus, to describe the positions of characteristic points in the space of the multi-link road train, we need to know $k + 1$ parameters (angles β_1–β_k and the lateral displacement of the front link Δ).

4.2.3 Equations of the plane curvilinear motion of multi-link road trains (general model)

The plane motion equations of a multi-link road train looks as follows:

$$\left. \frac{d}{dt}\frac{\partial T}{\partial \dot{q}_s} - \frac{\partial T}{\partial q_s} = Q_s + R_s \right|_{s=1,\ 2,\ \ldots,\ k+1},$$

where: T is the road train's kinetic energy; q_s is the generalized coordinate describing the motion of links; $\dot{q}_s$ is the generalized speed of link s; s is the link number; Q_s is the generalized active forces; R_s is the generalized reactions of nonholonomic links; and q_s is the generalized coordinates.

The road train's kinetic energy T can be presented as a sum of the T_r (the kinetic energy of the motion relative to the coordinate system (x_C,

y_C)) and the T_E (the kinetic energy of the transient motion):

$$T = T_y + T_e.$$

The road train's motion equation relative to the pole *C* is:

$$\frac{d}{dt}\frac{\partial T_r}{\partial \dot{q}_s} - \frac{\partial T_r}{\partial q_s} = L_s + Q_s + R_s \Bigg|_{s=1,\,2,\,\ldots,\,k+1},$$

where: L_s is the transient moments and inertia forces.

The angles β_1–β_k and the front link's lateral displacement relative to the trajectory Δ can be considered as generalized coordinates describing the road train's position and motion.

The road train's kinetic energy in motion relative to the coordinate system (x_C, y_C) in the general case is:

$$T_r = 0.5\sum_{i=1}^{k}\sum_{j=0}^{n}\left[m_{ij}\left(\dot{x}_{ij}^2 + \dot{y}_{ij}^2\right) + I_{zij}\,\dot{\beta}_i^2\right] \quad (14.19)$$

where: m_{ij} is the mass of the jth component of link *i*; and I_{zij} is the inertia moment of the jth component of link *i* relative to the axis z_i.

To open the expression (14.19), we can use either the formulas in (14.13) or their simplified version in (14.16).

In view of the fact that applying simplified formulas may have side effects, let us use here (unlike in *Motion Theory of Military Wheeled Vehicles* [61]) the main variant of (14.15):

$$\begin{cases} \dot{x}_{ij} = -a_{C_1} s_1\,\dot{\beta}_1 - d_{ij}\, s_i\,\dot{\beta}_i - \dot{S}_i\ ; \\ \dot{y}_{ij} = \dot{\Delta} + a_{C_1} c_1\,\dot{\beta}_1 + d_{ij}\, c_i\,\dot{\beta}_i - \dot{H}_i\ . \end{cases}$$

After substitutions and respective transformations, we get the following expression for the kinetic energy in the relative motion:

$$\begin{aligned} 2T_r = {} & \dot{\Delta}^2\sum_{i=1}^{k}\sum_{j=0}^{n} m_{ij} + (a_{C_1}^2)\dot{\beta}_i^2\sum_{i=1}^{k}\sum_{j=0}^{n} m_{ij} + \sum_{i=1}^{k}\dot{\beta}_i^2\sum_{j=0}^{n} m_{ij}(d_{ij}^2) + \\ & + 2a_{C_1}\dot{\beta}_1\sum_{i=1}^{k} c_{(1-i)}\dot{\beta}_1 + \sum_{j=0}^{n} m_{ij}d_{ij} + 2(a_{C_1}c_1)\dot{\Delta}\dot{\beta}_1\sum_{i=1}^{k}\sum_{j=0}^{n} m_{ij} + \\ & + 2\dot{\Delta}\sum_{i=1}^{k} c_i\dot{\beta}_1\sum_{j=0}^{n} m_{ij}d_{ij} + 2(a_{C_1}s_1)\dot{\beta}_1\sum_{i=1}^{k}\dot{S}_i\sum_{j=0}^{n}\left\{m_{ij}\right\} + \\ & + 2\sum_{i=1}^{k} s_i\dot{S}_i\dot{\beta}_1\sum_{j=0}^{n} m_{ij}d_{ij} - 2(a_{C_1}c_1)\dot{\beta}_1\sum_{i=1}^{k}\dot{H}_i\sum_{j=0}^{n} m_{ij} - \\ & - 2\sum_{i=1}^{k} c_i\dot{H}_i\dot{\beta}_1\sum_{j=0}^{n} m_{ij}d_{ij} - 2\dot{\Delta}\sum_{i=1}^{k}\dot{H}_i\sum_{j=0}^{n} m_{ij} + \sum_{i=1}^{k}\dot{S}_i^2\sum_{j=0}^{n} m_{ij} + \\ & + \sum_{i=1}^{k}\dot{H}_i^2\sum_{j=0}^{n} m_{ij} + \sum_{i=1}^{k}\dot{\beta}_i^2\sum_{j=0}^{n} I_{zij} \end{aligned} \quad (14.20)$$

Let us award the symbol *M* to the road train's total mass:

$$M = \sum_{i=1}^{k}\sum_{j=0}^{n} m_{ij}\ ,$$

M_i is the mass of the ith link:

$$M_i = \sum_{j=0}^{n} m_{ij}\ ,$$

M_{ik} is the mass of some of the road train's links, starting from link number *i* and up to link number *k* (without skipping any links):

$$M_{ik} = \sum_{\eta=i}^{k}\sum_{j=0}^{n} m_{\eta j} = \sum_{\eta=i}^{k} M_\eta\ , \text{ and}$$

I_{zi}^* is the inertia moment of the elements in link number *i*:

$$I_{zi}^* = \sum_{j=0}^{n} I_{zij}\ .$$

Now, let us consider the following sum in more detail:

$$\sum_{j=0}^{n}[m_{ij}d_{ij}]=\sum_{j=0}^{n}[m_{ij}(b_i+a_{ij})]=\sum_{j=0}^{n}[m_{ij}b_i]+\sum_{j=0}^{n}[m_{ij}a_{ij}]=$$
$$=b_i\sum_{j=0}^{n}[m_{ij}]+\sum_{j=0}^{n}[m_{ij}a_{ij}]=M_ib_i+\sum_{j=0}^{n}[m_{ij}a_{ij}].$$

The parameter A_{ij} defines the position of mass m_{ij} relative to the center of mass in link i (it is assumed that all of the link's masses are located along its longitudinal axis). This means that:

$$\sum_{j=0}^{n}[m_{ij}a_{ij}]=0$$

and, respectively:

$$\sum_{j=0}^{n}[m_{ij}d_{ij}]=M_ib_i\cdot$$

Then:

$$\sum_{j=0}^{n}m_{ij}d_{ij}^2=\sum_{j=0}^{n}m_{ij}(b_i+a_{ij})^2=\sum_{j=0}^{n}m_{ij}(b_i^2)+2\sum_{j=0}^{n}m_{ij}(b_ia_{ij})+\sum_{j=0}^{n}m_{ij}(a_{ij}^2)=$$
$$=b_i^2\sum_{j=0}^{n}m_{ij}+2b_i\sum_{j=0}^{n}m_{ij}(a_{ij})+\sum_{j=0}^{n}m_{ij}a_{ij}^2=M_ib_i^2+\sum_{j=0}^{n}m_{ij}a_{ij}^2.$$

Let us introduce the following notation:

$$I_{zi}=\sum_{j=0}^{n}m_{ij}a_{ij}^2+\sum_{j=0}^{n}I_{zij}=\sum_{j=0}^{n}m_{ij}a_{ij}^2+I_{zi}^{*}\cdot$$

The obtained results allow for further simplifying the expressions in (14.20):

$$2T_r=M\dot{\Delta}^2+(Ma_{C_1}^2)\dot{\beta}_1^2+2(Ma_{C_1}c_1)\dot{\Delta}\dot{\beta}_1+2a_{C_1}\dot{\beta}_1\sum_{i=1}^{k}M_ib_ic_{(1-i)}\dot{\beta}_i+$$
$$+2\dot{\Delta}\sum_{i=1}^{k}M_ib_ic_i\dot{\beta}_i+\sum_{i=1}^{k}(M_ib_i^2)\dot{\beta}_i^2+2(a_{C_1}s_1)\dot{\beta}_1\sum_{i=1}^{k}M_i\dot{S}_i+$$
$$+2\sum_{i=1}^{k}M_ib_is_i\dot{S}_i\dot{\beta}_i-2(a_{C_1}c_1)\dot{\beta}_1\sum_{i=1}^{k}M_i\dot{H}_i-2\sum_{i=1}^{k}M_ib_ic_i\dot{H}_i\dot{\beta}_i-$$
$$-2\dot{\Delta}\sum_{i=1}^{k}M_i\dot{H}_i+\sum_{i=1}^{k}M_i\dot{S}_i^2+\sum_{i=1}^{k}M_i\dot{S}_i^2+\sum_{i=1}^{k}I_{zi}\dot{\beta}_i^2$$
(14.21)

At the next transformation step, let us calculate the partial derivatives from (14.19):

$$\dot{S}_i=\frac{\partial}{\partial t}\left[\sum_{\eta=1}^{i}\ell_\eta c_\eta\right]=-\sum_{\eta=1}^{i}\ell_\eta s_\eta\dot{\beta}_\eta \qquad 14.22)$$

$$\dot{H}_i=\frac{\partial}{\partial t}\left[\sum_{\eta=1}^{i}\ell_\eta s_\eta\right]=-\sum_{\eta=1}^{i}\ell_\eta c_\eta\dot{\beta}_\eta \qquad (14.23)$$

Then, let us substitute (14.22) and (14.23) into (14.21).

Let us now consider the following structure:

$$\sum_{i=T}^{n}A_i\left[\sum_{\eta=T}^{i}B_\eta\right]=A_T\left(\sum_{\eta=T}^{T}B_\eta\right)+A_{T+1}\left(\sum_{\eta=T}^{T+1}B_\eta\right)+A_{T+1}\left(\sum_{\eta=T}^{T+2}B_\eta\right)+\ldots+$$
$$+A_n\left(\sum_{\eta=T}^{n}B_\eta\right)=A_T(B_T)+A_{T+1}(B_T+B_{T+1})+$$
$$+A_{T+2}(B_T+B_{T+1}+B_{T+2})+\ldots+A_n(B_T+B_{T+1}+\ldots+B_n)=$$
$$=B_T(A_T+A_{T+1}+\ldots+A_n)+B_{T+1}(A_{T+1}+A_{T+2}+\ldots+A_n)+B_nA_n=$$
$$=B_T\sum_{i=T}^{n}A_i+B_{T+1}\sum_{i=T+1}^{n}A_i+B_{T+2}\sum_{i=T+2}^{n}A_i\ldots B_n\sum_{i=n}^{n}A_i=$$
$$=\sum_{\eta=T}^{n}\left[B_\eta\sum_{i=\eta}^{n}A_i\right] \qquad (14.24)$$

It follows from (14.22) that:

$$\sum_{i=T}^{n}A_i\left[\sum_{i=\eta}^{i}B_\eta\right]=\sum_{\eta=T}^{n}\left[B_\eta\sum_{i=\eta}^{n}A_i\right]\cdot$$

In particular:

$$\sum_{i=1}^{k}A_i\left[\sum_{\eta=1}^{i}B_\eta\right]=\sum_{\eta=1}^{k}\left[B_\eta\sum_{i=\eta}^{k}A_i\right]$$

and:

$$\sum_{i=1}^{k} M_i\left[\sum_{\eta=1}^{i}(\ell_\eta c_\eta)\dot\beta_\eta\right]=\sum_{\eta=1}^{i}\left[(\ell_\eta c_\eta)\dot\beta_\eta\sum_{\eta=1}^{k}M_i\right]=\sum_{\eta=1}^{i}(M_{\eta k}\ell_\eta c_\eta)\dot\beta_\eta=$$
$$=\sum_{i=1}^{k}(M_{ik}\ell_i c_i)\dot\beta_i\ .$$

Then, formula (14.21) takes the form:

$$\begin{aligned}2T_r &= M\dot\Delta^2+(Ma_{C_1}^2)\dot\beta_1^2+2(Ma_{C_1}c_1)\dot\Delta\dot\beta_1+2a_{C_1}\dot\beta_1\sum_{i=1}^{k}M_ib_ic_{(1-i)}\dot\beta_i+\\ &+2\dot\Delta\sum_{i=1}^{k}(M_ib_i-M_{ik}\ell_i)\,c_i\dot\beta_i-2(a_{C_1})\dot\beta_i\sum_{i=1}^{k}(M_{ik}\ell_ic_{(1-i)})\dot\beta_1+\\ &+\sum_{i=1}^{k}(M_ib_i^2)\dot\beta_1^2-2\sum_{\eta=1}^{k}\ell_\eta\dot\beta_\eta\sum_{i=\eta}^{k}(M_ib_ic_{(\eta-i)}\dot\beta_i+\\ &+\sum_{i=1}^{k}M_i\left[\sum_{\eta=1}^{i}\ell_\eta s_\eta\dot\beta_\eta\right]^2+\sum_{i=1}^{k}M_i\left[\sum_{\eta=1}^{i}\ell_\eta c_\eta\dot\beta_\eta\right]^2+\sum_{i=1}^{k}I_{zi}\dot\beta_i^2\end{aligned}\qquad(14.25)$$

Let us begin to sum up (14.25) from the second link ($i = 2$ and $\eta = 2$) and combine similar terms, keeping in mind that $M_{1k} = M$, and $C_{(1-1)} = 1$. We can also show that:

$$\sum_{\eta=1}^{k}\ell_\eta\dot\beta_\eta\sum_{i=\eta}^{k}(M_ib_ic_{(\eta-i)}\dot\beta_i=(M_1b_1\ell_1)\dot\beta_1^2+\ell_1\dot\beta_1\sum_{i=\eta}^{k}(M_ib_ic_{(\eta-i)}\dot\beta_i+$$
$$+\sum_{\eta=1}^{k}\ell_\eta\dot\beta_\eta\sum_{i=\eta}^{k}(M_ib_ic_{(\eta-i)}\dot\beta_i\ ,$$

and that:

$$[Ma_{C_1}^2+2M_1b_1a_{C_1}c_{(1-1)}-2M_{ik}\ell_ia_{C_1}c_{(1-1)}+M_1b_1^2-2M_1b_1\ell_1]=$$
$$=M_{2k}b_1^2-M\ell_1^2$$

and:

$$(M_1b_1-M_{1k}\ell_1)\,c_1+(Ma_{C_1}c_1)=$$
$$=(M_1b_1-M\ell_1+Ma_{C_1})\,c_1=-M_{2k}b_1c_1$$

Now, the expression for the relative motion's kinetic energy is transformed as follows:

$$\begin{aligned}2T_r &= M\dot\Delta^2+[M_{2k}b_1^2-M\ell_1^2]\dot\beta_1^2-2[M_{2k}b_1c_1]\dot\Delta\dot\beta_i+\\ &+2a_{C_1}\dot\beta_i\sum_{i=2}^{k}M_ib_ic_{(1-i)}\dot\beta_1+2\dot\Delta\sum_{i=2}^{k}(M_ib_i-M_{ik}\ell_i)\,c_i\dot\beta_i-\\ &-2a_{C_1}\dot\beta_1\sum_{i=2}^{k}(M_{ik}\ell_ic_{(1-i)}\dot\beta_i+\sum_{i=2}^{k}(M_ib_i^2)\dot\beta_1^2-\\ &-2\ell_1\dot\beta_1\sum_{i=2}^{k}(M_ib_ic_{(1-i)})\dot\beta_i-2\sum_{\eta=2}^{k}\ell_\eta\dot\beta_\eta\sum_{i=\eta}^{k}(M_ib_ic_{(\eta-i)}\dot\beta_i+\\ &+\sum_{i=1}^{k}M_i\left[\sum_{\eta=1}^{i}\ell_\eta s_\eta\dot\beta_\eta\right]^2+\sum_{i=1}^{k}M_i\left[\sum_{\eta=1}^{i}\ell_\eta c_\eta\dot\beta_\eta\right]^2+\sum_{i=1}^{k}I_{zi}\dot\beta_i^2\ .\end{aligned}$$

It can be shown that:

$$\sum_{i=1}^{k}M_i\left[\sum_{\eta=1}^{i}\ell_\eta c_\eta\dot\beta_\eta\right]^2=(M\ell_1^2c_1^2)\dot\beta_1^2+2(\ell_1c_1\dot\beta_1)\sum_{\eta=2}^{k}(M_{\eta k}\ell_\eta c_\eta)\dot\beta_\eta+$$
$$+\sum_{i=2}^{k}M_i\left(\sum_{\eta=2}^{i}\ell_\eta c_\eta\dot\beta_\eta\right)^2.$$

After substitutions, isolating the inertial member number $i = 1$, and the respective regrouping of members, the formula for the relative motion's kinetic energy has no components implicitly containing the first link's motion parameters.

To simplify the presentation, it is convenient to make intermediary transformations:

$$\sum_{i=1}^{k}M_i\left[\sum_{\eta=1}^{i}\ell_\eta s_\eta\dot\beta_\eta\right]^2+\sum_{i=1}^{k}M_i\left[\sum_{\eta=1}^{i}\ell_\eta c_\eta\dot\beta_\eta\right]^2=$$
$$=(M\ell_1^2)\dot\beta_1^2+2(\ell_1\dot\beta_1)\sum_{i=2}^{k}(M_{ik}\ell_ic_{(1-i)})\dot\beta_i+\sum_{i=2}^{k}M_i\left(\sum_{\eta=2}^{i}\ell_\eta s_\eta\dot\beta_\eta\right)^2+$$
$$+\sum_{i=2}^{k}M_i\left(\sum_{\eta=2}^{i}\ell_\eta c_\eta\dot\beta_\eta\right)^2$$

and introduce the following notations:

$$M_{2k}b_1c_1=C_0;$$
$$M_{2k}b_1^2+I_{z1}=F_0;$$
$$M_ib_i-M_{ik}\ell_i=MB_{ik}.$$

Finally, we get the following expression for the road train's kinetic energy in the relative motion:

$$2T_r = [M\dot{\Delta}^2 - 2C_0\dot{\Delta}\dot{\beta}_i + F_0\dot{\beta}_1^2] - 2b_1\dot{\beta}_1\sum_{i=2}^{k} MB_{ik}c_{(1-i)})\dot{\beta}_i + 2\dot{\Delta}\sum_{i=2}^{k} MB_{ik}c_i\dot{\beta}_i -$$

$$-2\sum_{\eta=2}^{i}\ell_\eta\dot{\beta}_\eta\sum_{i=\eta}^{k}(M_ib_ic_{(\eta-i)}\dot{\beta}_i + \sum_{i=2}^{k} M_i\left(\sum_{\eta=2}^{i}\ell_\eta s_\eta\dot{\beta}_\eta\right)^2 +$$

$$+\sum_{i=2}^{k} M_i\left[\sum_{\eta=2}^{i}\ell_\eta c_\eta\dot{\beta}_\eta\right]^2 + \sum_{i=2}^{k}(M_ib_1^2 + I_{zi})\dot{\beta}_i^2 \quad (14.26)$$

Following certain recommendations from *Motion Theory of Military Wheeled Vehicles* [61], let us identify in (14.26) three components of the kinetic energy:

a. The one dependent only on the motion parameters of the road train's first (nose) link is:

$$2T^a{}_r(\dot{\Delta}, \dot{\beta}_1) = 2T^a{}_r = M\dot{\Delta}^2 - 2C_0\dot{\Delta}\dot{\beta}_1 + F_0\dot{\beta}_1^2;$$

b. The one dependent both on the motion parameters of the road train's first (nose) link and on the trailed links' motion parameters is:

$$2T^b{}_r(\Delta, \beta_1, \{\beta_i\})\big|_{i=2,k} = 2T^b{}_r =$$

$$= 2\left[\dot{\Delta}\sum_{i=2}^{k} MB_{ik}c_i\dot{\beta}_i - b_1\dot{\beta}_1\sum_{i=2}^{k} MB_{ik}c_{(1-i)}\dot{\beta}_i\right]$$

c. The one dependent only on the trailed links' motion parameters is:

$$2T^c{}_r(\{\beta_i\})\big|_{i=2,k} = 2T^c{}_r = \sum_{i=2}^{k}(M_ib_1^2 + I_{zi})\dot{\beta}_i^2 - 2\sum_{\eta=2}^{i}\ell_\eta\dot{\beta}_\eta\sum_{i=\eta}^{k}(M_ib_ic_{(\eta-i)}\dot{\beta}_i +$$

$$+\sum_{i=2}^{k} M_i\left(\sum_{\eta=2}^{i}\ell_\eta s_\eta\dot{\beta}_\eta\right)^2 + \sum_{i=2}^{k} M_i\left(\sum_{\eta=2}^{i}\ell_\eta c_\eta\dot{\beta}_\eta\right)^2.$$

At the same approach, we have:

$$T_r = T^a{}_0(\dot{\Delta}, \dot{\beta}_1) + 2T^b{}_r(\Delta, \beta_1, \{\beta_i\})\big|_{i=2,k} +$$

$$+ T^c{}_r(\{\beta_i\})\big|_{i=2,k} = T^a{}_r + T^b{}_r + T^c{}_r.$$

While comparing the obtained results with the respective formulas from *Motion Theory of Military Wheeled Vehicles* [61], we can find in the latter additional multipliers of the $c_1 = \cos\beta_1$ type, which appeared because we used a simplified version of the link between the generalized and geometric coordinates of the links. In addition, the formulas from *Motion Theory of Military Wheeled Vehicles* [61] clearly by mistake has the member $a_{11}m_{11}$; in order to eliminate it, we need to change the lower summation limit from $j = 1$ to $j = 0$.

At the next step of our deliberations, we form general expressions for the derivatives of the relative motion's kinetic energy in relation to the generalized speeds and coordinates. In doing so, we postulate the absence of interrelation between the generalized coordinates and their derivatives.

At first, let us perform this operation for the derivative with respect to generalized speeds (we omit the intermediary transformations):

$$\frac{\partial T_r}{\partial\dot{\Delta}} = M\dot{\Delta} - M_{2k}b_1c_1\dot{\beta}_1 + \sum_{i=1}^{k}[MB_{ik}c_i]\dot{\beta}_i\,.$$

In particular, for a single-link vehicle ($i = 1$), we have:

$$\frac{\partial T_r}{\partial\dot{\Delta}} = M\dot{\Delta} - M_{21}b_1c_1\dot{\beta}_1 + \sum_{i=2}^{1}[MB_{ik}c_i]\dot{\beta}_i = M\dot{\Delta} = M_1\dot{\Delta}\,.$$

Then, we find the derivative with respect to the relative motion's kinetic energy by the generalized speed $\dot{\beta}_i$:

$$\frac{\partial T_r}{\partial\dot{\beta}_1} = -\dot{\Delta}\frac{\partial(C_0\dot{\beta}_1)}{\partial\dot{\beta}_1} + 0.5*\frac{\partial(F_0\dot{\beta}_1{}^2)}{\partial\dot{\beta}_1} - \frac{\partial}{\partial\dot{\beta}_1}[b_1\dot{\beta}_1\sum_{i=2}^{k} MB_{ik}c_{(1-i)}\dot{\beta}_i] -$$

$$-\{-\frac{\partial}{\partial\dot{\beta}_1}\sum_{\eta=2}^{k} l_\eta\dot{\beta}_\eta\sum_{i=\eta}^{k}(M_ib_ic_{(\eta-i)})\dot{\beta}_i + 0.5*\frac{\partial}{\partial\dot{\beta}_1}[\sum_{i=2}^{k} M_i(\sum_{\eta=2}^{i} l_\eta s_\eta\dot{\beta}_\eta)^2] +$$

$$+\dot{\Delta}\frac{\partial}{\partial\dot{\beta}_1}[\sum_{i=2}^{k} MB_{ik}c_i\dot{\beta}_i] + 0.5*\frac{\partial}{\partial\dot{\beta}_1}[\sum_{i=2}^{k} M_i(\sum_{\eta=2}^{i} l_\eta c_\eta\dot{\beta}_\eta)^2] +$$

$$+0.5*\frac{\partial}{\partial\dot{\beta}_1}[\sum_{i=2}^{k}(M_ib_i^2 + I_{zi})\dot{\beta}_i{}^2]\}$$

$$(14.27)$$

The members of (14.27) enclosed in curly brackets do not depend on β_1, therefore:

$$\frac{\partial T_r}{\partial \dot\beta_1} = -\dot\Delta \frac{\partial (C_0 \dot\beta_1)}{\partial \dot\beta_1} + 0.5 * \frac{\partial (F_0 \dot\beta_1^{\,2})}{\partial \dot\beta_1} - \frac{\partial}{\partial \dot\beta_1}[b_1 \dot\beta_1 \sum_{i=2}^{k} MB_{ik} c_{(1-i)} \dot\beta_i] =$$

$$= -\dot\Delta \dot\beta_1 \frac{\partial C_0}{\partial \dot\beta_1} - C_0 \dot\Delta \frac{\partial \dot\beta_1}{\partial \dot\beta_1} + 0.5 * \dot\beta_1^{\,2} \frac{\partial F_0}{\partial \dot\beta_1} + F_0 \dot\beta_1 \frac{\partial \dot\beta_1}{\partial \dot\beta_1} - b_1 \frac{\partial}{\partial \dot\beta_1}[\dot\beta_1 \sum_{i=2}^{k} MB_{ik} c_{(1-i)} \dot\beta_i] =$$

$$= \dot\Delta \dot\beta_1 \frac{\partial C_0}{\partial \dot\beta_1} - C_0 \dot\Delta + 0.5 * \dot\beta_1^{\,2} \frac{\partial F_0}{\partial \dot\beta_1} + F_0 \dot\beta_1 - b_1 \frac{\partial}{\partial \dot\beta_1}[\dot\beta_1 \sum_{i=2}^{k} MB_{ik} c_{(1-i)} \dot\beta_i]$$

Let us take into account that C_0 depends on β_1:

$$C_0 = M_{2k} b_1 c_1 = M_{2k} b_1 \cos \beta_1$$

F_0 does not depend on β_1:

$$M_{2k} b_1^{\,2} + I_{z1} = F_0$$

and the fact that we postulate the mutual independence of the generalized coordinate and the respective generalized speed, and, accordingly, that:

$$\frac{\partial (\beta_1 - \beta_i)}{\partial \dot\beta_1} = \frac{\partial \beta_1}{\partial \dot\beta_1} - \frac{\partial \beta_i}{\partial \dot\beta_1} = 0$$

and:

$$\frac{\partial}{\partial \dot\beta_1}[\dot\beta_1 \sum_{i=2}^{k} MB_{ik} c_{(1-i)} \dot\beta_i] = \sum_{i=2}^{k} MB_{ik} c_{(1-i)} \dot\beta_i \, .$$

After all of the above transformations, (14.27) takes a following much simpler form:

$$\frac{\partial T_r}{\partial \dot\beta_1} = -C_0 \dot\Delta + F_0 \dot\beta_1 - b_1 [\sum_{i=2}^{k} MB_{ik} c_{(1-i)} \dot\beta_i] \, .$$

In particular, for a single-link vehicle we have:

$$\frac{\partial T_r}{\partial \dot\beta_1} = I_{z1} \dot\beta_1$$

Finally, for $s > 1$ we have:

$$\frac{\partial T_r}{\partial \dot\beta_s} = -b_1 \dot\beta_1 \frac{\partial}{\partial \dot\beta_s}[\sum_{i=2}^{k} MB_{ik} c_{(1-i)} \dot\beta_i] + \dot\Delta[\frac{\partial}{\partial \dot\beta_s} \sum_{i=2}^{k} MB_{ik} c_i \dot\beta_i] -$$

$$-\frac{\partial}{\partial \dot\beta_s}[\sum_{\eta=2}^{k} l_\eta \dot\beta_\eta \sum_{i=\eta}^{k} (M_i b_i c_{(\eta-i)}) \dot\beta_i] + 0.5 * \frac{\partial}{\partial \dot\beta_s}[\sum_{i=2}^{k} M_i (\sum_{\eta=2}^{i} l_\eta s_\eta \dot\beta_\eta)^2] +$$

$$+0.5 * \frac{\partial}{\partial \dot\beta_s}[\sum_{i=2}^{k} M_i (\sum_{\eta=2}^{i} l_\eta c_\eta \dot\beta_\eta)^2] + 0.5 * \frac{\partial}{\partial \dot\beta_s}[\sum_{i=2}^{k} (M_i b_i^2 + I_{zi}) \dot\beta_i^{\,2}] =$$

$$= -b_1 \dot\beta_1 [\sum_{i=2}^{k} MB_{ik} c_{(1-i)} \frac{\partial \dot\beta_i}{\partial \dot\beta_s}] + \dot\Delta[\sum_{i=2}^{k} MB_{ik} c_i \frac{\partial \dot\beta_i}{\partial \dot\beta_s}] -$$

$$-[\sum_{\eta=2}^{k} (l_\eta \frac{\partial \dot\beta_\eta}{\partial \dot\beta_s}) \sum_{i=\eta}^{k} (M_i b_i c_{(\eta-i)}) \dot\beta_i] - [\sum_{\eta=2}^{k} (l_\eta \dot\beta_\eta) \sum_{i=\eta}^{k} (M_i b_i c_{(\eta-i)}) \frac{\partial \dot\beta_i}{\partial \dot\beta_s}] +$$

$$+[\sum_{i=2}^{k} M_i (\sum_{\eta=2}^{i} l_\eta s_\eta \dot\beta_\eta)(\sum_{\eta=2}^{i} l_\eta s_\eta \frac{\partial \dot\beta_\eta}{\partial \dot\beta_s})] + [\sum_{i=2}^{k} M_i (\sum_{\eta=2}^{i} l_\eta c_\eta \dot\beta_\eta)(\sum_{\eta=2}^{i} l_\eta c_\eta \frac{\partial \dot\beta_\eta}{\partial \dot\beta_s})] +$$

$$+[\sum_{i=2}^{k} (M_i b_i^2 + I_{zi}) \dot\beta_i \frac{\partial \dot\beta_i}{\partial \dot\beta_s}]$$

(14.28)

By using the Kronecker symbol in formula (14.27), we have:

$$\frac{\partial \dot\beta_i}{\partial \dot\beta_s} = \delta_{is}$$

and obtain the following:

$$\frac{\partial T_r}{\partial \dot\beta_s} = -b_1 \dot\beta_1 [\sum_{i=2}^{k} MB_{ik} c_{(1-i)} \delta_{is}] + \dot\Delta[\sum_{i=2}^{k} MB_{ik} c_i \delta_{is}] -$$

$$-[\sum_{\eta=2}^{k} (l_\eta \delta_{\eta s}) \sum_{i=\eta}^{k} (M_i b_i c_{(\eta-i)}) \dot\beta_i] - [\sum_{\eta=2}^{k} (l_\eta \dot\beta_\eta) \sum_{i=\eta}^{k} (M_i b_i c_{(\eta-i)}) \delta_{is}] +$$

$$+[\sum_{i=2}^{k} M_i (\sum_{\eta=2}^{i} l_\eta s_\eta \dot\beta_\eta)(\sum_{\eta=2}^{i} l_\eta s_\eta \delta_{\eta s})] + [\sum_{i=2}^{k} M_i (\sum_{\eta=2}^{i} l_\eta c_\eta \dot\beta_\eta)(\sum_{\eta=2}^{i} l_\eta c_\eta \delta_{\eta s})] +$$

$$+[\sum_{i=2}^{k} (M_i b_i^2 + I_{zi}) \dot\beta_i \delta_{is}]$$

(14.29)

Let us take into account that the delta-function $\delta_{is} = 1$ only at $i = s$. This means that:

$$[\sum_{i=2}^{k} (M_i b_i^2 + I_{zi}) \dot\beta_i \delta_{is}] = (M_s b_s^2 + I_{zs}) \dot\beta_s = F_s \dot\beta_s \quad (14.30)$$

In addition, we transform sums in other elements of formula (14.29), which comprise the delta-function δ_{is}.

We have:

$$\frac{\partial T_r}{\partial \dot{\beta}_s} = F_s\dot{\beta}_s - (MB_{sk}b_1c_{(1-s)})\dot{\beta}_1 + (MB_{sk}c_s)\dot{\Delta} - l_s\sum_{i=s}^{k}(M_ib_ic_{(s-i)})\dot{\beta}_i - $$
$$-[\sum_{\eta=2}^{k}(l_\eta\dot{\beta}_\eta)\sum_{i=\eta}^{k}(M_ib_ic_{(\eta-i)})\delta_{is}] + [\sum_{i=2}^{k}M_i(\sum_{\eta=2}^{i}l_\eta s_\eta\dot{\beta}_\eta)(\sum_{\eta=2}^{i}l_\eta s_\eta\delta_{\eta s})] +$$
$$+[\sum_{i=2}^{k}M_i(\sum_{\eta=2}^{i}l_\eta c_\eta\dot{\beta}_\eta)(\sum_{\eta=2}^{i}l_\eta c_\eta\delta_{\eta s})] \tag{14.31}$$

Let us consider formula (14.31) in more detail with account that:

- The delta-function differs from zero only when the indices coincide;
- The interval *i*∈*[2,k]* surely contains the value of any "*s*";
- The interval *η*∈*[2,i]* surely does not contain the value of any "*s*."

Let us fix a random value of *s* and consider the following structure:

$$B_{si} = \sum_{\eta=2}^{i}\delta_{s\eta}A_\eta \ ; \ i \in [2,k]$$

The following variants (cases) are possible:

a. The value of the upper limit *i* is less than the preset number of equation *s:*

$$i < s : \eta \neq s \rightarrow \delta_{s\eta} \equiv 0 \rightarrow B_{si} \equiv 0 \tag{14.32}$$

b. The values of the upper limit *i* and the number of equation *s* coincide:

$$i = s : \eta = s \rightarrow \delta_{s\eta} = \delta_{ss} = $$
$$= 1 \rightarrow B_{si} = B_{ss} = A_s \tag{14.33}$$

c. The value of the upper limit *i* is bigger than the preset number of equation *s*:

$$i > s : \eta = s \rightarrow \delta_{s\eta} = \delta_{ss} = $$
$$=1 \rightarrow B_{si} = B_{ss} = A_s \tag{14.34}$$

Formulas (14.32)–(14.34) can be interpreted with the help of the unit function:

$$E(i) = \begin{cases} 0; \ i < 0 \\ 1; \ i \geq 0 \end{cases}$$

thus:

$$B_{si} = \sum_{\eta=2}^{i}\delta_{s\eta}A_\eta = E(i-s)A_s$$

In these notations:

$$\frac{\partial T_r}{\partial \dot{\beta}_s} = F_s\dot{\beta}_s - (MB_{sk}b_1c_{(1-s)})\dot{\beta}_1 + (MB_{sk}c_s)\dot{\Delta} - l_s\sum_{i=s}^{k}(M_ib_ic_{(s-i)})\dot{\beta}_i - $$
$$-[\sum_{\eta=2}^{k}(l_\eta\dot{\beta}_\eta)\sum_{i=\eta}^{k}(M_ib_ic_{(\eta-i)})\delta_{is}] + \{\sum_{i=2}^{k}M_i[E(i-s)l_s s_s(\sum_{\eta=2}^{i}l_\eta s_\eta\dot{\beta}_\eta)]\} +$$
$$+\{\sum_{i=2}^{k}M_i[E(i-s)l_s c_s(\sum_{\eta=2}^{i}l_\eta c_\eta\dot{\beta}_\eta)]\} \tag{14.35}$$

Let us consider in more detail the result of the transformations presented by formula (14.35).

The presence of the unit function as a multiplier inside the outer summation operator means that all of its members numbered $i < s$ are equal to zero. In its turn, it means that the summing can start not from $i = 1$, but from $i = s$ (since always $s \leq k$):

$$\frac{\partial T_r}{\partial \dot{\beta}_s} = F_s\dot{\beta}_s - (MB_{sk}b_1c_{(1-s)})\dot{\beta}_1 + (MB_{sk}c_s)\dot{\Delta} - l_s\sum_{i=s}^{k}(M_ib_ic_{(s-i)})\dot{\beta}_i - $$
$$-[\sum_{\eta=2}^{k}(l_\eta\dot{\beta}_\eta)\sum_{i=\eta}^{k}(M_ib_ic_{(\eta-i)})\delta_{is}] + \{\sum_{i=s}^{k}M_i[l_s s_s(\sum_{\eta=2}^{i}l_\eta s_\eta\dot{\beta}_\eta)]\} +$$
$$+\{\sum_{i=s}^{k}M_i[l_s c_s(\sum_{\eta=2}^{i}l_\eta c_\eta\dot{\beta}_\eta)]\} =$$

$$= F_s\dot{\beta}_s - (MB_{sk}b_1c_{(1-s)})\dot{\beta}_1 + (MB_{sk}c_s)\dot{\Delta} - l_s\sum_{i=s}^{k}(M_ib_ic_{(s-i)})\dot{\beta}_i -$$
$$-[\sum_{\eta=2}^{k}(l_\eta\,\dot{\beta}_\eta)\sum_{i=\eta}^{k}(M_ib_ic_{(\eta-i)})\delta_{is}] + \{\sum_{i=s}^{k}M_i[l_s(\sum_{\eta=2}^{i}l_\eta c_{(s-\eta)}\,\dot{\beta}_\eta)]\} \tag{14.36}$$

Let us consider the following structure:

$$\sum_{i=s}^{k}A_i[\sum_{\eta=2}^{i}B_\eta] = \sum_{i=s}^{k}A_i[\sum_{\eta=2}^{s-1}B_\eta\sum_{\eta=s}^{i}B_\eta] = \sum_{i=s}^{k}A_i[\sum_{\eta=2}^{s-1}B_\eta] + \sum_{i=s}^{k}A_i[\sum_{\eta=s}^{i}B_\eta] =$$
$$= \sum_{i=s}^{k}A_i[\sum_{\eta=2}^{s-1}B_\eta] + \sum_{\eta=s}^{k}B_\eta[\sum_{i=\eta}^{k}A_i] \tag{14.37}$$

Let us apply (14.37) to the last member of (14.35):

$$l_s\sum_{i=s}^{k}\{M_i[\sum_{\eta=2}^{i}l_\eta c_{(s-\eta)}\dot{\beta}_\eta]\} = l_s\sum_{i=s}^{k}M_i[\sum_{\eta=2}^{s-1}l_\eta c_{(s-\eta)}\dot{\beta}_\eta] + l_s\sum_{\eta=s}^{k}(l_\eta c_{(s-\eta)}\dot{\beta}_\eta)[\sum_{i=\eta}^{k}M_i] =$$
$$= l_s\sum_{i=s}^{k}M_i[\sum_{\eta=2}^{s-1}l_\eta c_{(s-\eta)}\dot{\beta}_\eta] + l_s\sum_{\eta=s}^{k}M_{\eta k}(l_\eta c_{(s-\eta)}\dot{\beta}_\eta) \tag{14.38}$$

At the fixed s the first member in (14.38) can be converted as follows:

$$l_s\sum_{i=s}^{k}M_i[\sum_{\eta=2}^{s-1}l_\eta c_{(s-\eta)}\dot{\beta}_\eta] = l_s[\sum_{i=s}^{k}M_i][\sum_{\eta=2}^{s-1}l_\eta c_{(s-\eta)}\dot{\beta}_\eta] = M_{sk}l_s[\sum_{\eta=2}^{s-1}l_\eta c_{(s-\eta)}\dot{\beta}_\eta]$$

Since the internal sums become the unified multiplier for all of the members M_i of the outside summation operator, formula (14.38) can be transformed as follows:

$$l_s\sum_{i=s}^{k}\{M_i[\sum_{\eta=2}^{i}l_\eta c_{(s-\eta)}\dot{\beta}_\eta]\} = M_{sk}l_s\sum_{i=2}^{s-1}[l_ic_{(s-i)}]\dot{\beta}_i + l_s\sum_{i=s}^{k}[M_{ik}l_ic_{(s-i)}]\dot{\beta}_i \tag{14.39}$$

After substituting (14.39) into (14.36), we have:

$$\frac{\partial T_r}{\partial\dot{\beta}_s} = F_s\dot{\beta}_s - (MB_{sk}b_1c_{(1-s)})\dot{\beta}_1 + (MB_{sk}c_s)\dot{\Delta} - l_s\sum_{i=s}^{k}MB_{ik}c_{(s-i)}\dot{\beta}_i -$$
$$-[\sum_{\eta=2}^{k}(l_\eta\,\dot{\beta}_\eta)\sum_{i=\eta}^{k}(M_ib_ic_{(\eta-i)})\delta_{is}] + M_{sk}l_s\sum_{i=2}^{s-1}[l_ic_{(s-i)}]\dot{\beta}_i$$

Let us consider the following structure:

$$\sum_{i=\eta}^{k}A_i\delta_{is} = B$$

The following cases are possible:

a. $\eta < s : s \in [i,k] \rightarrow B = A_s$
b. $\eta = s : s = i_{\min} \rightarrow B = A_s$
c. $\eta < s : s \notin [i,k] \rightarrow B \equiv 0$

It means that:

$$B = \sum_{i=\eta}^{k}A_i\delta_{is} = E(s-\eta)A_s \tag{14.40}$$

Then:

$$\frac{\partial T_r}{\partial\dot{\beta}_s} = (F_s - M_sb_sl_s)\dot{\beta}_s - (MB_{sk}b_1c_{(1-s)})\dot{\beta}_1 + (MB_{sk}c_s)\dot{\Delta} -$$
$$-l_s\sum_{i=s+1}^{k}MB_{ik}c_{(s-i)}\dot{\beta}_i - MB_{sk}\sum_{i=2}^{s}(l_i\,c_{(i-s)})\dot{\beta}_i \tag{14.41}$$

To unify the approach, it is expedient to isolate and unite in (14.41) all of the members having the same angular velocity as the link number s:

$$\frac{\partial T_r}{\partial\dot{\beta}_s} = (M_sb_s^2 - 2M_sb_sl_s + M_{sk}l_s^2 + I_{zs})\dot{\beta}_s - (MB_{sk}b_1c_{(1-s)})\dot{\beta}_1 + (MB_{sk}c_s)\dot{\Delta} -$$
$$-l_s\sum_{i=s+1}^{k}MB_{ik}c_{(s-i)}\dot{\beta}_i - MB_{sk}\sum_{i=2}^{s-1}(l_i\,c_{(i-s)})\dot{\beta}_i$$

By introducing an updated designation:

$$F_{0s} = (M_sb_s^2 - 2M_sb_sl_s + M_{sk}l_s^2 + I_{zs}),$$

we finally get ($s > 1$):

$$\frac{\partial T_r}{\partial \dot{\beta}_s} = F_{0s}\dot{\beta}_s - (MB_{sk}b_1c_{(1-s)})\dot{\beta}_1 + (MB_{sk}c_s)\dot{\Delta} -$$
$$-MB_{sk}\sum_{i=2}^{s-1}(l_i\, c_{(i-s)})\dot{\beta}_i - l_s\sum_{i=s+1}^{k} MB_{ik}c_{(s-i)}\dot{\beta}_i$$

Then, we need to calculate the time derivatives from the derivatives of the road train's kinetic energy in the relative motion by generalized speeds.

In accordance with the accepted approach, we have three groups of formulas: for the speed derivative of the nose link's displacement $d\Delta/dt$, for the nose link's angular velocity derivative $d\beta_1/dt$, and the trailing links' angular velocity derivatives $d\beta_s/dt$.

By omitting intermediary transformations, the obtained results are as follows.

For the first link's displacement speed derivative:

$$\frac{d}{dt}\frac{\partial T_r}{\partial \dot{\Delta}} = M\ddot{\Delta} - (M_{2k}b_1c_1)\ddot{\beta}_1 + (M_{2k}b_1s_1)\dot{\beta}_1^{\,2} +$$
$$+\sum_{i=2}^{k}(MB_{ik}c_i)\ddot{\beta}_i - \sum_{i=2}^{k}(MB_{ik}s_i)\dot{\beta}_i^{\,2}$$

For the derivative in relation to the first link's generalized angular velocity ($s = 1$):

$$\frac{d}{dt}\frac{\partial T_r}{\partial \dot{\beta}_1} == F_0\ddot{\beta}_1 - (M_{2k}b_1c_1)\ddot{\Delta} + (M_{2k}b_1s_1)\dot{\Delta}\dot{\beta}_1 + b_1\sum_{i=2}^{k}(MB_{ik}s_{(1-i)})\dot{\beta}_1\dot{\beta}_i -$$
$$-b_1\sum_{i=2}^{k}(MB_{ik}c_{(1-i)})\ddot{\beta}_i - b_1\sum_{i=2}^{k}(MB_{ik}s_{(1-i)})\dot{\beta}_i^{\,2}$$

For the derivatives in relation to the generalized speeds of the second and subsequent links:

$$\frac{d}{dt}\frac{\partial T_r}{\partial \dot{\beta}_s} = F_{0s}\ddot{\beta}_s + (MB_{sk}c_s)\ddot{\Delta} - (MB_{sk}b_1c_{(1-s)})\ddot{\beta}_1 + (MB_{sk}b_1s_{(1-s)})\dot{\beta}_1^{\,2} -$$
$$-(MB_{sk}s_s)\dot{\Delta}\dot{\beta}_s - (MB_{sk}b_1s_{(1-s)})\dot{\beta}_1\dot{\beta}_s - MB_{sk}\sum_{i=2}^{s-1}[(l_i\, c_{(i-s)})\ddot{\beta}_i] -$$
$$-l_s\sum_{i=s+1}^{k}[(MB_{ik}c_{(i-s)})\ddot{\beta}_i] + MB_{sk}\sum_{i=2}^{s-1}[(l_i\, s_{(i-s)})\dot{\beta}_i^{\,2}] + l_s\sum_{i=s+1}^{k}[(MB_{ik}\, s_{(i-s)})\dot{\beta}_i^{\,2}] -$$
$$-MB_{sk}\sum_{i=2}^{s-1}[(l_i\, s_{(i-s)})\dot{\beta}_i\dot{\beta}_s)] - l_s\sum_{i=s+1}^{k}[(MB_{ik}\, s_{(i-s)})\dot{\beta}_i\dot{\beta}_s]$$

Let us deduce formulas for the kinetic energy derivatives in relation to the generalized coordinates.

By differentiating the expressions for the kinetic energy in relation to the motion of the road train's links, it is easy to obtain a system of the following formulas:

$$2\frac{\partial T_r}{\partial \Delta} = \frac{\partial}{\partial \Delta}\{[M\dot{\Delta}^2 - 2C_0\dot{\Delta}\dot{\beta}_1 + F_0\dot{\beta}_1^{\,2}] - 2b_1\dot{\beta}_1\sum_{i=2}^{k}MB_{ik}c_{(1-i)}\dot{\beta}_i + 2\dot{\Delta}\sum_{i=2}^{k}MB_{ik}c_i\dot{\beta}_i -$$
$$-2\sum_{\eta=2}^{k}l_\eta\dot{\beta}_\eta\sum_{i=\eta}^{k}(M_ib_ic_{(\eta-i)})\dot{\beta}_i + \sum_{i=2}^{k}M_i(\sum_{\eta=2}^{i}l_\eta s_\eta\dot{\beta}_\eta)^2 + \sum_{i=2}^{k}M_i(\sum_{\eta=2}^{i}l_\eta c_\eta\dot{\beta}_\eta)^2 +$$
$$+\sum_{i=2}^{k}(M_ib_i^2 + I_{zi})\dot{\beta}_i^{\,2}\} = 0$$

$$2\frac{\partial T_r}{\partial \beta_1} = -\,2M_{2k}b_1\frac{\partial c_1}{\partial \beta_1}\dot{\Delta}\dot{\beta}_1 - 2b_1\dot{\beta}_1\sum_{i=2}^{k}MB_{ik}\frac{\partial c_{(1-i)}}{\partial \beta_1}\dot{\beta}_i =$$
$$= 2(M_{2k}b_1s_1)\dot{\Delta}\dot{\beta}_1 + 2b_1\dot{\beta}_1\sum_{i=2}^{k}(MB_{ik}s_{(1-i)})\dot{\beta}_i$$

$$2\frac{\partial T_r}{\partial \beta_s} = -\,2(MB_{sk}b_1s_{(1-s)})\dot{\beta}_1\dot{\beta}_s - 2(MB_{sk}s_s)\dot{\Delta}\dot{\beta}_s + 2\sum_{\eta=2}^{k}l_\eta\dot{\beta}_\eta\sum_{i=\eta}^{k}[M_ib_is_{(\eta-i)}\delta_{\eta s}]\dot{\beta}_i -$$
$$-2\sum_{\eta=2}^{k}l_\eta\dot{\beta}_\eta\sum_{i=\eta}^{k}[M_ib_is_{(\eta-i)}\delta_{is}]\dot{\beta}_i + 2\sum_{i=2}^{k}M_i[(\sum_{\eta=2}^{i}l_\eta s_\eta\dot{\beta}_\eta)(\sum_{\eta=2}^{i}l_\eta c_\eta\delta_{\eta s}\dot{\beta}_\eta)] -$$
$$-2\sum_{i=2}^{k}M_i[(\sum_{\eta=2}^{i}l_\eta c_\eta\dot{\beta}_\eta)(\sum_{\eta=2}^{i}l_\eta s_\eta\delta_{\eta s}\dot{\beta}_\eta)]$$

Let us consider (at the fixed s) the following structure:

$$B = \sum_{i=\eta}^{k}A_i\delta_{is}$$

The following variants are possible:

a. $s < \eta : \forall_i \rightarrow i \neq s \rightarrow \delta_{is} \equiv 0 \rightarrow \forall_i : \sum_{i=\eta}^{k}A_i\delta_{is} = 0$

b. $s \geq r_i$: exists $s = i \in [\eta, k] \rightarrow$ exists
$\delta_{is} = \delta_{ss} = 1 \rightarrow \sum A_i\delta_{is} = A_s\delta_{ss} = A_s$

This result can be written with the help of the unit function as follows:

$$B = \sum_{i=\eta}^{k} A_i \delta_{is} = E(s-\eta) A_s \qquad (14.42)$$

Then:

$$2\sum_{\eta=2}^{k} l_\eta \dot{\beta}_\eta \sum_{i=\eta}^{k} (M_i b_i s_{(\eta-i)} \dot{\beta}_i \delta_{is}) = 2\sum_{\eta=s}^{k} M_s b_s l_\eta s_{(\eta-s)} \dot{\beta}_s \dot{\beta}_\eta$$

Consequently:

$$2\sum_{\eta=2}^{k} l_\eta \dot{\beta}_\eta \sum_{i=\eta}^{k} [M_i b_i s_{(\eta-i)} \delta_{\eta s}] \dot{\beta}_i - 2\sum_{\eta=2}^{k} l_\eta \dot{\beta}_\eta \sum_{i=\eta}^{k} [M_i b_i s_{(\eta-i)} \delta_{is}] \dot{\beta}_i =$$

$$= 2\sum_{\eta=s}^{k} M_s b_s l_\eta s_{(\eta-s)} \dot{\beta}_s \dot{\beta}_\eta - 2\sum_{\eta=s}^{k} M_s b_s l_\eta s_{(\eta-s)} \dot{\beta}_s \dot{\beta}_\eta = 0$$

and:

$$2\frac{\partial T_r}{\partial \beta_s} = -2(MB_{sk} b_1 s_{(1-s)}) \dot{\beta}_1 \dot{\beta}_s - 2(MB_{sk} s_s) \dot{\Delta} \dot{\beta}_s + 2\sum_{i=2}^{k} M_i [(\sum_{\eta=2}^{i} l_\eta s_\eta \dot{\beta}_\eta)(\sum_{\eta=2}^{i} l_\eta c_\eta \delta_{\eta s} \dot{\beta}_\eta)] -$$

$$-2\sum_{i=2}^{k} M_i [(\sum_{\eta=2}^{i} l_\eta c_\eta \dot{\beta}_\eta)(\sum_{\eta=2}^{i} l_\eta s_\eta \delta_{\eta s} \dot{\beta}_\eta)]$$

If we take into account the structural transformations obtained earlier (formulas (14.37), (14.40), and (14.42)), then, we have:

$$2\sum_{i=2}^{k} M_i [(\sum_{\eta=2}^{i} l_\eta s_\eta \dot{\beta}_\eta)(\sum_{\eta=2}^{i} l_\eta c_\eta \delta_{\eta s} \dot{\beta}_\eta)] - 2\sum_{i=2}^{k} M_i [(\sum_{\eta=2}^{i} l_\eta c_\eta \dot{\beta}_\eta)(\sum_{\eta=2}^{i} l_\eta s_\eta \delta_{\eta s} \dot{\beta}_\eta)] =$$

$$= 2\sum_{i=2}^{k} M_i \{(E(i-s)] l_s \dot{\beta}_s (\sum_{\eta=2}^{i} l_\eta s_{(\eta-s)} \dot{\beta}_\eta)\} = 2\sum_{i=s}^{k} M_i [l_s \dot{\beta}_s (\sum_{\eta=2}^{i} l_\eta s_{(\eta-s)} \dot{\beta}_\eta)]$$

Consequently:

$$2\frac{\partial T_r}{\partial \beta_s} = -2(MB_{sk} b_1 s_{(1-s)}) \dot{\beta}_1 \dot{\beta}_s - 2(MB_{sk} s_s) \dot{\Delta} \dot{\beta}_s +$$

$$+2\sum_{i=s}^{k} M_i [l_s \dot{\beta}_s (\sum_{\eta=2}^{i} l_\eta s_{(\eta-s)} \dot{\beta}_\eta)])]$$

Then (we omit here all of the intermediate transformations as well):

$$2\sum_{i=s}^{k} M_i [l_s \dot{\beta}_s (\sum_{\eta=2}^{i} l_\eta s_{(\eta-s)} \dot{\beta}_\eta)] =$$

$$= 2[M_{sk} l_s \dot{\beta}_s (\sum_{\eta=2}^{s-1} l_\eta s_{(\eta-s)} \dot{\beta}_\eta)] + 2\sum_{\eta=s}^{k} (M_{\eta k} l_\eta l_s s_{(\eta-s)}) \dot{\beta}_s \dot{\beta}_\eta$$

With account of the changing indices in summation operators, we have:

$$2\frac{\partial T_r}{\partial \beta_s} = -2(MB_{sk} b_1 s_{(1-s)}) \dot{\beta}_1 \dot{\beta}_s - 2(MB_{sk} s_s) \dot{\Delta} \dot{\beta}_s +$$

$$+2M_{sk} l_s \dot{\beta}_s (\sum_{i=2}^{s-1} l_i s_{(i-s)} \dot{\beta}_i) + 2\dot{\beta}_s \sum_{i=s}^{k} (M_{ik} l_i l_s s_{(i-s)}) \dot{\beta}_i$$

In accordance with the obtained results, the motion equations of a multi-link road train will look as follows:

a. For describing the motion of the first link relative to the preset trajectory (after intermediary transformations and collection of similar terms):

$$\begin{cases} M\ddot{\Delta} - (M_{2k} b_1 c_1) \ddot{\beta}_1 + (M_{2k} b_1 s_1) \dot{\beta}_1^2 + \sum_{i=2}^{k} (MB_{ik} c_i) \ddot{\beta}_i - \sum_{i=2}^{k} (MB_{ik} s_i) \dot{\beta}_i^2 = \\ = L_\Delta + Q_\Delta + R_\Delta \end{cases}$$

$$\begin{cases} F_0 \ddot{\beta}_1 - (M_{2k} b_1 c_1) \ddot{\Delta} - b_1 \sum_{i=2}^{k} (MB_{ik} c_{(1-i)}) \ddot{\beta}_i - b_1 \sum_{i=2}^{k} (MB_{ik} s_{(1-i)}) \dot{\beta}_i^2 = \\ = L_{\beta_1} + Q_{\beta_1} + R_{\beta_1} \end{cases} \qquad (14.43)$$

b. For describing the motion of the second and subsequent links relative to the preset trajectory (after intermediary transformations, structuring, and collecting similar terms):

$$
\begin{aligned}
&F_{0s}\ddot{\beta}_s + (MB_{sk}c_s)\ddot{\Delta} - (MB_{sk}b_1c_{(1-s)})\ddot{\beta}_1 + (MB_{sk}b_1s_{(1-s)})\dot{\beta}_1^{\,2} - \\
&-MB_{sk}\sum_{i=2}^{s-1}[(l_i\, c_{(i-s)})\ddot{\beta}_i] - l_s\sum_{i=s+1}^{k}[(MB_{ik}c_{(i-s)})\ddot{\beta}_i] + \\
&+MB_{sk}\sum_{i=2}^{s-1}[(l_i\, s_{(i-s)})\dot{\beta}_i^{\,2}] + l_s\sum_{i=s+1}^{k}[(MB_{ik}\, s_{(i-s)})\dot{\beta}_i^{\,2}] - \\
&-MB_{sk}\sum_{i=2}^{s-1}[(l_i\, s_{(i-s)})\dot{\beta}_i\dot{\beta}_s)] - l_s\sum_{i=s+1}^{k}[(MB_{ik}\, s_{(i-s)})\dot{\beta}_i\dot{\beta}_s] - \\
&-M_{sk}l_s\dot{\beta}_s(\sum_{i=2}^{s-1} l_i s_{(i-s)}\dot{\beta}_i) - \dot{\beta}_s\sum_{i=s+1}^{k}(M_{ik}l_il_s s_{(i-s)})\dot{\beta}_i - \dot{\beta}_s(M_{sk}l_sl_s s_{(s-s)})\dot{\beta}_s = \\
&= F_{0s}\ddot{\beta}_s + (MB_{sk}c_s)\ddot{\Delta} - (MB_{sk}b_1c_{(1-s)})\ddot{\beta}_1 + (MB_{sk}b_1s_{(1-s)})\dot{\beta}_1^{\,2} - \\
&-MB_{sk}\sum_{i=2}^{s-1}[(l_i\, c_{(i-s)})\ddot{\beta}_i] + MB_{sk}\sum_{i=2}^{s-1}[(l_i\, s_{(i-s)})\dot{\beta}_i^{\,2}] - MB_{sk}\sum_{i=2}^{s-1}[(l_i\, s_{(i-s)})\dot{\beta}_i\dot{\beta}_s)] - M_{sk}l_s(\sum_{i=2}^{s-1} l_i s_{(i-s)}\dot{\beta}_i\dot{\beta}_s) - \\
&-l_s\sum_{i=s+1}^{k}[(MB_{ik}c_{(i-s)})\ddot{\beta}_i] + l_s\sum_{i=s+1}^{k}[(MB_{ik}\, s_{(i-s)})\dot{\beta}_i^{\,2}] - l_s\sum_{i=s+1}^{k}[(MB_{ik}\, s_{(i-s)})\dot{\beta}_i\dot{\beta}_s] - \sum_{i=s+1}^{k}(M_{ik}l_il_s s_{(i-s)})\dot{\beta}_i\dot{\beta}_s = \\
&= L_{\beta_s} + Q_{\beta_s} + R_{\beta_s} \qquad (14.44)
\end{aligned}
$$

The right-hand parts of equations (14.43) and (14.44) contain generalized active (Q_Δ, $Q_{\beta 1}$, $Q_{\beta s}$) and reactive (R_Δ, $R_{\beta 1}$, $R_{\beta s}$) forces, acting at the virtual displacement related to the respective generalized coordinate and the respective transfer inertia forces (L_Δ, $L_{\beta 1}$, $L_{\beta s}$).

The main vector of the transferred inertia forces corresponding to the generalized coordinate Δ is defined as follows:

$$L_\Delta = -\frac{Mv^2}{\rho_C^2}(\rho_C + \Delta) = -\frac{Mv^2}{\rho_C}\left(1 + \frac{\Delta}{\rho_C}\right)$$

With an accuracy sufficient for practical studies, we can assume:

$$L_\Delta = -\frac{Mv^2}{\rho_C}$$

The main moment of the transfer inertia forces corresponding to the generalized coordinates β_s is described as follows:

$$L_{\beta_s} = -I_{zs}\frac{d^2}{dt^2}\left(\frac{dy_C}{dx_C}\right) \qquad (14.45)$$

The main moment of the Coriolis force is directed along the "x" axis and on the virtual displacement δΔ its work is equal to zero. The Coriolis force's main moment of work on the virtual displacement $\delta\beta_s$ is also equal to zero.

The generalized forces corresponding to the active and reactive forces affecting each of the road train's links are defined in the following succession:

1. The variations of the active and reactive forces' work on their respective virtual displacements are defined;
2. The variations of the active and reactive forces' work are summed up;
3. The obtained sum is referred to as the variation of the respective generalized coordinate.

Let us assume that each link of a multi-link wheeled road train can have *j* axles.

In its turn, each wheel can be turned to the angle $\theta_{\varphi r}$, be under the action of the active motion force $P_{\partial.jr} = (P_k - P_f)_{jr}$, or be under the action of the reactive lateral force $P_{y.jr}$.

The generalized force acting on the road train's i^{th} link and the corresponding generalized coordinate Δ is described as follows:

$$(Q+R)_{i\Delta} = \sum_{j=1}^{n}\sum_{r=1}^{2} \operatorname{sgn}(a_{ij})\{P_{x_{ijr}} \sin[\theta_{ijr} + \operatorname{sgn}(a_{ij})\beta_i] + $$
$$+ P_{y_{jr}} \cos[\theta_{ijr} + \operatorname{sgn}(a_{ij})\beta_i]\}$$

The generalized force acting on the road train's i^{th} link and the corresponding generalized coordinate β_i is described as follows:

$$(Q+R)_{\beta_i} = \sum_{j=1}^{n}\sum_{r=1}^{2} \{P_{x_{ijr}} [a_{ij} \operatorname{sgn}(\theta_{ijr}) + \operatorname{sgn}(a_{ij}) a_{ij} \cos\theta_{ijr}] + $$
$$+ P_{y_{ijr}} [a_{ij} \cos\theta_{ijr} - \operatorname{sgn}(a_{ij}) a_{ij} \sin\theta_{ijr}]\} \tag{14.46}$$

The generalized forces in the right-hand part of the motion equations corresponding to the generalized coordinate Δ are defined by summing up the generalized forces of all of the road train's links:

$$(Q_s + R_s + L_s)_\Delta = \sum_{i=1}^{k} (Q_s + R_s)_{i\Delta} + L_\Delta$$

The generalized coordinates of β_s are independent. This means that the generalized forces in the right-hand parts of the motion equations corresponding to the generalized coordinate β_s can be obtained with the help of formulas (14.45) and (14.46)—in the latter one at $i = s$:

$$(Q_s + R_s + L_s)_{\beta s} = (Q_s + R_s)_{\beta s} + L_{\beta s}$$

4.2.4 Particular cases

4.2.4.1 Multi-axle wheel chassis

The position of a single-link wheeled chassis in the mobile coordinate system is defined by:

- The lateral displacement Δ; and
- The angular deviation β of the longitudinal axis from the direction of the axis Ox_C, tangential to the preset trajectory.

The flat (plane) motion equations of a chassis with a random number of driving and controlled (steered) axles can be obtained from the general equations in (14.41), built for describing the motion of a multi-link road train's nose link at $k = 1$:

$$\begin{cases} M\ddot{\Delta} == L_\Delta + Q_\Delta + R_\Delta \\ I_z\ddot{\beta} = L_\beta + Q_\beta + R_\beta \end{cases}$$

For a single-link vehicle, the description of the generalized forces will be simpler if we compile the motion equations in the projections on the axes (x_a, y_a), rotated relative to the axes (x_c, y_c) such that the axis x_a coincides with the chassis' longitudinal axis.

Taking into account that:

$$L_\Delta = -\frac{Mv^2}{\rho_C}\left(1 + \frac{\Delta}{\rho_C}\right) \approx -\frac{Mv^2}{\rho_C}$$

and:

$$L_\beta = -I_z \frac{d^2}{dt^2}\left(\frac{dy_C}{dx_C}\right) = -I_z\ddot{\psi}$$

we have:

$$\begin{cases} M\left(\ddot{\Delta} + \dfrac{v^2}{\rho_C}\right) = \sum_{j=1}^{n}\sum_{r=1}^{2}\left(P_{xjr}\sin\theta_{jr}\right) + \sum_{j=1}^{n}\sum_{r=1}^{2}\left(P_{yjr}\cos\theta_{jr}\right) \\ I_z(\ddot{\beta} + \ddot{\psi}) = \sum_{j=1}^{n}\left[\left(\left(P_{xj2}\cos\theta_{j2} - P_{yj2}\sin\theta_{j2}\right) - \left(P_{xj1}\cos\theta_{j1} - P_{yj1}\sin\theta_{j1}\right)\right)d_j\right] + \\ \sum_{j=1}^{n}\sum_{r=1}^{2}\left(P_{xjr}\sin\theta_{jr} + P_{yjr}\cos\theta_{jr}\right)a_j \end{cases}$$

It is obvious that to solve these equations one should additionally know the functions that interconnect the lateral and longitudinal forces and wheel turning angles with the motion parameters.

4.2.4.2 *Articulated two-link vehicle of the "single-axle tow car + single-axle semi-trailer" type*

The position of this vehicle on the plane in the mobile coordinate system is defined by the lateral displacement Δ of the nose link's center of mass and the angular deviation $\beta_1 = \beta$ of this link's longitudinal axis relative to the axis x_C in relation to the preset trajectory.

The angle θ between the links is changed forcibly; that is why the function $\theta = \theta(t)$ is assumed known, and, also, $\beta_2 = \theta + \beta$.

For $k = 2$ we have:

$$\begin{cases} M\ddot{\Delta} + (m_1a_1 - m_2b_1)\ddot{\beta} + a_2m_2(\ddot{\beta}_2\cos\beta_2 - \dot{\beta}_2^{\,2}\sin\beta_2) = \\ = L_\Delta + Q_\Delta + R_\Delta \\ (I_z + m_2b_1^{\,2})\ddot{\beta} + (m_1a_1 - m_2b_1)\ddot{\Delta} + a_2b_1m_2(\ddot{\beta}_2\cos\beta_2 - \dot{\beta}_2^{\,2}\sin\beta_2) = \\ = L_{\beta_1} + Q_{\beta_1} + R_{\beta_1} \end{cases}$$

The generalized forces are defined for $k = 2$ with account of the vector and the moment of inertial transfer forces:

$$L_\Delta = -\frac{(m_1 + m_2)(\rho_C + \Delta)v^2}{\rho_C^{\,2}} \approx -\frac{(m_1 + m_2)v^2}{\rho_C}$$

$$L_\beta = -\left\{I_{z1} - I_{z2} + m_2\left[(l_2 - a_2)^2 + (l_1 - a_1)^2\right]\right\}\ddot{\psi}$$

In projections on the front link's axes, we have:

$$\begin{cases} (m_1 + m_2)\left(\ddot{\Delta} + \dfrac{v^2}{\rho_C}\right) + (m_1a_1 - m_2b_2)\ddot{\beta} + (\ddot{\beta} + \ddot{\theta})(m_2a_2 - m_2l_2)\cos\theta - \\ -(\dot{\beta} + \dot{\theta})^2(m_2a_2 - m_2l_2)\sin\theta = \\ = \sum\limits_{r=1}^{2}\left(P_{x1r}\sin\theta_{1r} + P_{y1r}\cos\theta_{1r} + P_{x2r}\sin\theta_{2r}\cos\theta + P_{y2r}\cos\theta_{2r}\cos\theta\right) \\ (m_1a_1 - m_2b_1)\ddot{\Delta} + (I_{z1} + m_2b_1^{\,2})(\ddot{\beta} + \ddot{\psi}) + (I_{z2} + m_2b_2^{\,2})\ddot{\psi} - \\ -(\ddot{\beta} + \ddot{\theta})b_1m_2l_2\cos\theta - (\dot{\beta} + \dot{\theta})^2 b_1m_2l_2\sin\theta = \\ = \sum\limits_{i=1}^{2}\sum\limits_{r=1}^{2}\left(P_{xir}\cos\theta_{ir} - P_{yir}\sin\theta_{ir}\right)\{a_1(2 - i) + [b_1 + l_2\cos\theta - d_2\,\mathrm{sgn}(1.5 - r)\sin\theta](i - 1)\} - \\ -\left(P_{yir}\sin\theta_{ir} + P_{yir}\cos\theta_{ir}\right)\{d_1\,\mathrm{sgn}(1.5 - r)(2 - i) - [d_2 - b_1\,\mathrm{sgn}(1.5 - r)\sin\theta](i - 1)\} \end{cases}$$

14.3 Motion stability of multi-axle wheeled chassis

A turnability diagram of a multi-axle wheeled chassis for different travel speeds is shown in Fig. 14.5.

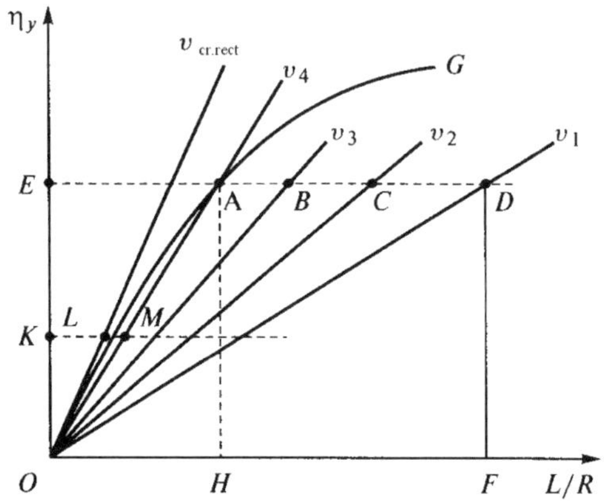

Fig. 14.5. Turnability diagram of a multi-axle wheeled chassis for different travel speeds

Point D corresponds to the motion of the chassis' center of mass at the speed of v_1 along a curvilinear trajectory with the relative curvature OF. To ensure the motion at this curvature, it is necessary to turn the controlled wheels (section DA). Points A, B, and C correspond to the center of mass's motion at speeds v_4, v_3, and v_2 accordingly.

Point A corresponds to the motion at the same specific lateral force as in the point D. However, to make a turn at the relative curvature OH at a speed of $v_4 > v_1$, there is no need to turn the controlled wheels—the vehicle will make its turn only due to the tires' lateral skid and the turn of the wheels due to the carrying system's roll. If we decrease the level of lateral forces and preserve the travel speed $v = v_4$, the arising tire skidding and wheel turn due to the carrying system's roll will be not enough for keeping the vehicle on the trajectory with the relative curvature KM—it will be necessary to additionally turn the controlled wheels (section LM). In this case, the travel speeds ν_2 and ν_3 take the intermediate position between v_1 and ν_4.

By increasing the travel speed, we can reach the situation at which the straight line depicting the dependence of the specific centrifugal lateral force on the trajectory's relative curvature will be tangential to the self-turnability line OG. This speed is treated as the "critical rectilinear motion speed."

Let us assume that the chassis moves straight at a speed exceeding the critical one (Fig. 14.6). In this case, the lateral force is equal to zero, same as the trajectory's relative curvature; therefore, the rectilinear motion corresponds to the coordinates' origin in the turnability diagram. If the chassis is briefly affected by a small lateral force (point 1), then the wheels' contact with the road will experience lateral forces, which, in their turn, will cause tire skidding. In accordance with the

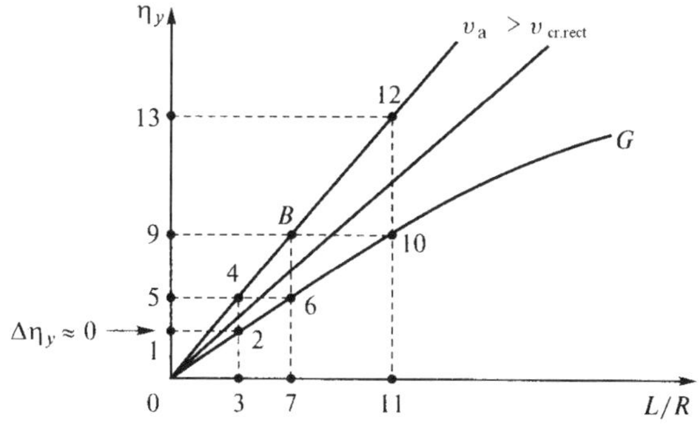

Fig. 14.6. Process of stability loss in rectilinear motion

self-turnability characteristic (line OG), this level of lateral force corresponds to the skid angle difference of the extreme axles, which is equal to the section *1-2*. It also means that the chassis' center of mass begins moving along the trajectory with the relative curvature *0-3* (radius ΔR).

Let us assume that the lateral affection causing this turn has stopped. Despite this fact, the vehicle's center of mass, which moves at a speed of v_a along the arc with the radius ΔR, is under the effect of the lateral centrifugal force corresponding to section *3-4* in the turnability diagram. It is seen from Fig. 14.6 that the value of this force is higher than the value of the lateral force, which keeps the vehicle on the trajectory and is defined by the lateral tire skidding. The balance is restored by increasing the lateral reactions, which act in the tire-road contact area and results in increasing the difference in the skidding angles of the extreme axles up to the level corresponding to section *5-6*, which corresponds to the curvature of trajectory *0-7*. Thus, the process becomes a self-growing one. At a growth in the lateral force up to points *9* and *13*, the lateral centrifugal force will increase up to section *7-8* or *11-2* for ν_A and *7-6* and *11-10* for the self-turnability (*0G*).

This means that after the affection of a random small lateral force, the chassis will start a voluntary and self-developing turn without the driver's participation with an intensively decreasing radius—up to the lateral skidding or even lateral tipover (tripping) (section *0-11*, Fig. 14.6). In other words, a sustainable rectilinear motion at speeds exceeding the critical speed is impossible.

We can show that at a small lateral perturbation in the chassis, which moves at a speed below the critical one, there is a spontaneous restoration of the rectilinear motion—the rectilinear motion is sustainable.

Let us consider now the chassis' rectilinear motion at a speed of $v_{a'}$, which is lower than the critical speed of the rectilinear motion (Fig. 14.7).

To ensure a curvilinear motion at a constant speed $v_{a'}$, it is necessary to additionally turn the controlled

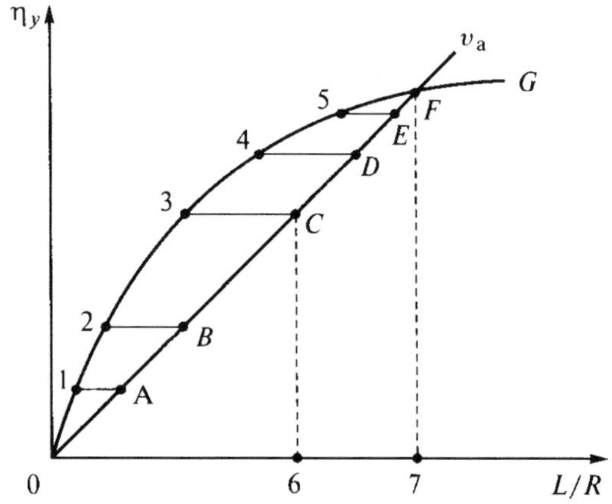

Fig. 14.7. Change in the controlled wheels' turning angle necessary to ensure the chassis' turn at the preset curvature

(steered) wheels. It is seen from Fig. 14.7 that the turn of the controlled wheels depends on the trajectory curvature and has its extreme in point *C*. In particular, to move under the relative curvature *0-7*, there is no need to turn the controlled wheels at all, that is, the driver's interference is not needed.

In point *C* (Fig. 14.8), the chassis is under the effect of the specific lateral force *0-1* and it makes a turn under the relative curvature *0-6*.

Assume that the chassis is now affected by such a small lateral perturbation that its affection is below the driver's sensitivity threshold. In this case, in accordance with the turnability diagram, the motion trajectory's curvature will increase (section *2-4*); however, the driver will not be able to detect it and will keep the controlled wheels in the initial position (section *3-C*, equal to section *4-5*). The position of point *5* shows that the chassis' controlled wheels are excessively turned and the trajectory's curvature is spontaneously increased (section *2-5*). The process is accompanied by an increase in the lateral force acting on the vehicle's center of mass. If this process takes place quickly enough, the driver is not able to react to what happens and turn the controlled wheels; thus, the trajectory's curvature is spontaneously increasing up to the lateral skidding or the chassis' tipover. A further increment in the specific lateral force (sections *0-9* and *0-13*) will cause an increase in the relative curvature (sections *9-10*, *9-8*, *13-14*, and *13-15*) up to *0-7* or *0-16*.

Thus, at the preset travel speed, there is an unstable motion zone to the right of point *C*.

Different travel speeds are associated with different positions of the dependence extreme points in the turning angles of the controlled wheels versus the relative trajectory curvature. It is reasonable to complement the turning diagram with a curve describing the position of extreme point *C* (*0C*) depending on the curvilinear motion speed (line *0-1-2-3-4* in Fig. 14.9). This dependence is drawn by building the tangents *AB*, *CD*, and

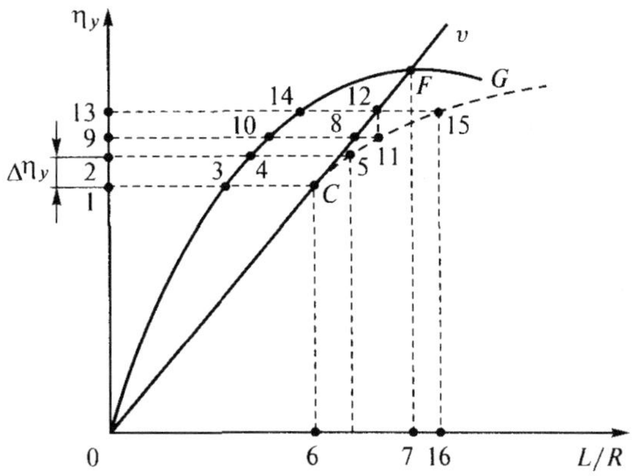

Fig. 14.8. Process of stability loss in the chassis' curvilinear motion

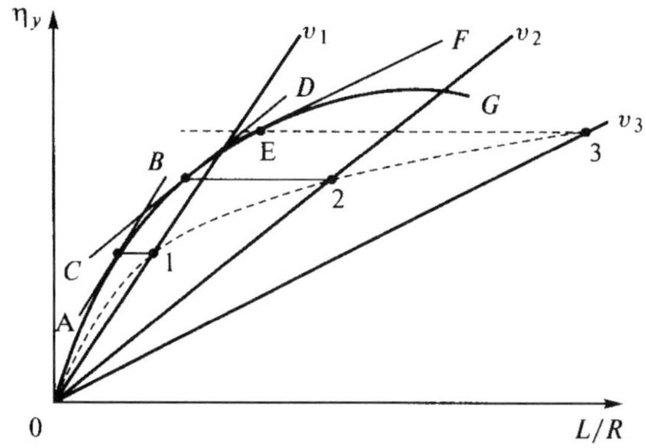

Fig. 14.9. Drawing a line separating stable and unstable motion along the chassis' turning diagram

EB to the curve *0C* and crossing the horizontal straight lines with the respective speed line.

Let us draw the real motion borders over the turning diagram (Fig. 14.10):

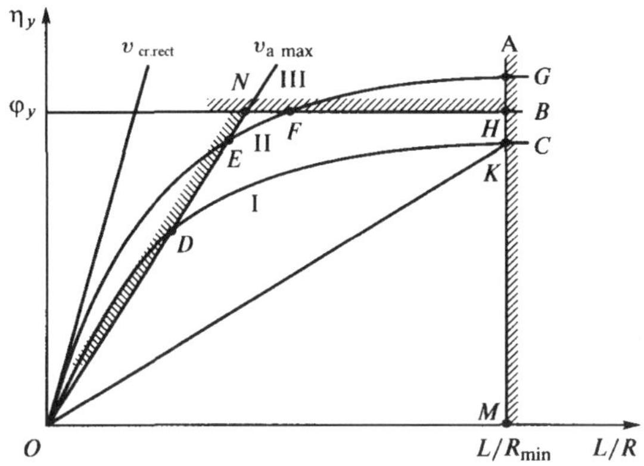

Fig. 14.10. Motion stability diagram of a multi-axle wheeled chassis

Thus, we have added the following:

- The line *ON* corresponding to the chassis' maximum travel speed (found in its specifications);
- The line *OA* corresponding to the chassis' maximum turn curvature (minimum turning radius taken from the specifications); and
- The line *NB* corresponding to the limit by the coupling specific lateral force ϕ_y.

As a result, we have a curve usually treated as a motion stability diagram for a multi-axle wheeled chassis.

The trapezoid *0NBM* is the existence area of the chassis' real motion in the given road and ground conditions. This area can be divided into three zones as follows:

- Stable motion zone-I (*0DKM*—formed by the maximum travel speed's half-line, the stable motion's border, and the turn's maximum possible curvature line);
- Unstable steady turn zone-II (*DEFBM*—formed by the maximum travel speed's half-line, the self-turning characteristic, the adhesion restriction line, the turn's maximum possible curvature line, and the stable motion's border);
- Impossible steady turn zone-III, (*ENF*—located above zone-II).

The chassis' design should ensure that the real motion area and the stable motion area overlap. If this is impossible for whatever reasons, it is necessary to introduce administrative restrictions on the travel speed and narrow the range of motion modes (a similar case is described in Chapters 4 and 5 with the BAZ-135LM chassis as an example).

The rectilinear motion's critical speed is easily defined from expressions (14.4) and (14.6). If we assume that the chassis should move straight at zero turn angles of the controlled wheels, then:

$$\frac{P_{yg}}{G_g} = \frac{v_g^{\,2}}{gL}\left[(\delta_n - \delta_1) + (\theta_{chn} - \theta_{ch1})\right]$$

where: P_{yg} is the lateral force acting on the chassis.

In most cases, multi-axle wheeled chassis with independent suspensions have neutral roll turnability. In such case, it is obvious that:

$$v_{ch\,\lim} = \sqrt{\frac{gL}{\frac{G_g}{P_{eg}}(\delta_n - \delta_1)}}$$

where: $v_{\text{ch.lim}}$ is the critical speed of the chassis' rectilinear motion.

The stability criterion of the chassis' curvilinear motion is the following expression (Fig. 14.11):

$$\frac{v_g^{\,2}}{gL} < \frac{\partial \eta_y}{\partial\left[(\delta_n - \delta_1) + (\theta_{chn} - \theta_{ch1})\right]}$$

The sense of the data presented in Fig. 14.11 is explained in the previous figure.

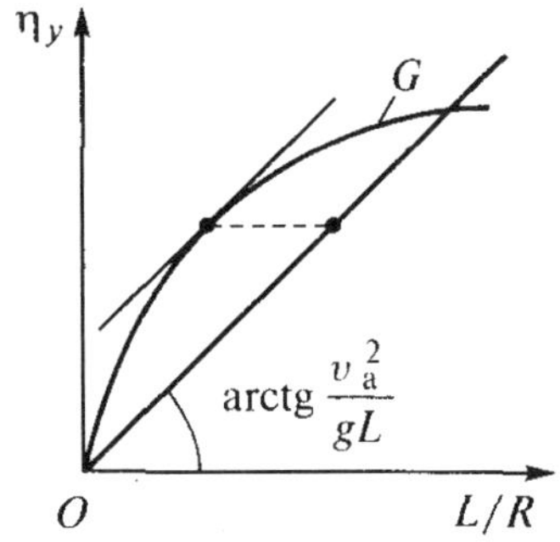

Fig. 14.11. Stability criterion diagram of a multi-axle chassis' curvilinear motion

The sign selection rules are obvious from Fig. 14.12. We assume that the positive direction for all angles is counter-clockwise. An exclusion is the skidding angle, for which positive is the direction against the lateral force acting on the tire in the contact spot (the contact spot is somewhat "squeezed out" by the lateral force). In the diagram in Fig. 14.12, the turning pole is located such that the directions of the forces and angles are equal—it allows for excluding the errors related to the interpretation of the signs for the wheel parameters located on both sides of the turning pole.

To assess the turnability of a multi-axle wheeled chassis versus different factors, it is convenient to use the "bicycle" calculation diagram (Fig. 14.12), which is similar to the one considered in point 8.4.

Let us assume, for the sake of definitiveness, that the lateral forces in the wheels' contact area with the support surface obey the lateral skid hypothesis, applied so far in the most general form, without any special considerations to the structure of the resistance coefficient to lateral skidding. Let us also assume that the resistance to the wheel rolling is proportional to the vertical load on the wheels. The diagram in Fig. 14.12 uses the earlier adopted notations.

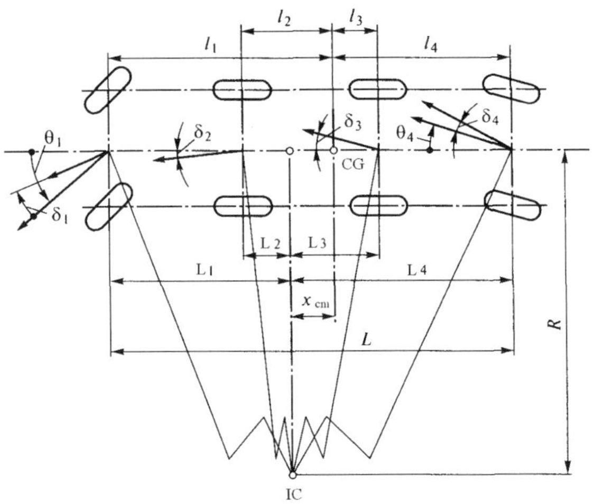

Fig. 14.12. Simplified turning diagram of a multi-axle chassis

$$\begin{cases} \sum_{i=1}^{n}\left[R_{xi}\cos\theta_i - R_{yi}\sin\theta_i - P_{f\,grounti}\cos\theta_i\right] - P_{air} = 0 \\ \sum_{i=1}^{n}\left[R_{xi}\sin\theta_i + R_{yi}\cos\theta_i - P_{f\,grounti}\sin\theta_i\right] - mj_y = 0 \\ \sum_{i=1}^{n}\left[R_{xi}l_i\sin\theta_i + R_{yi}l_i\cos\theta_i - P_{f\,grounti}l_i\sin\theta_i\right] - M_{rtw} - I_z\dfrac{d\omega_z}{dt} = 0 \\ \dfrac{l_i + x_t}{R} = \tan(\theta_i - \delta_i) \end{cases} \tag{14.47}$$

The following notations are used in (14.47): R_{xi} stands for the tangential loads in the contact area of the wheels i^{th} axis (acting on the tire); R_{yi} stands for the lateral loads in the contact area of the wheels i^{th} axis (acting on the tire); M_{rtw} stands for the resistance moment to turning; P_{air} stands for the aerodynamic force resisting the chassis' motion; $P_{f\,groundti}$ stands for the force acting from the ground on the tire and resisting the motion of the ith axle's wheels.

The main equation of the lateral skid hypothesis in the most general form is usually written in the assumption that the lateral force is a reaction in the support surface. The system in (14.47) includes the forces acting on the tire and introduces a specific rule of signs for the skidding angles; that is why the main equation of the lateral skid hypothesis should be used in the following form:

$$R_{yi} = -K_{yi}\delta_i$$

where K_{yi} is the resistance coefficient to the ith axle's lateral tire skidding.

By using the lateral skid hypothesis and the typical presentation about the resistance coefficient to motion, we get:

$$\begin{cases} \sum_{i=1}^{n}\left[R_{xi}\cos\theta_i + K_{yi}\delta_i\sin\theta_i - P_{f\,grount\,i}\cos\theta_i\right] = P_{air} \\ \sum_{i=1}^{n}\left[R_{xi}\sin\theta_i - K_{yi}\delta_i\cos\theta_i - P_{f\,grounti}\sin\theta_i\right] = mj_y \\ \sum_{i=1}^{n}\left[R_{xi}l_i\sin\theta_i - K_{yi}\delta_i l_i\cos\theta_i - P_{f\,grounti}l_i\sin\theta_i\right] = M_{rtw} + I_z\dfrac{d\omega_z}{dt} \\ \dfrac{l_i + x_t}{R} = \tan(\theta_i - \delta_i) \end{cases} \quad (14.48)$$

Let us assume that the motion takes place at small turning angles of the wheels and small skidding angles. This means that the equations in (14.48) can be linearized:

$$\begin{cases} \sum_{i=1}^{n}\left[R_{xi} + K_{yi}\delta_i\theta_i - P_{f\,grount\,i}\right] = P_{air} \\ \sum_{i=1}^{n}\left[R_{xi}\theta_i - K_{yi}\delta_i - P_{f\,grount\,i}\theta_i\right] = mj_y \\ \sum_{i=1}^{n}\left[R_{xi}l_i\theta_i - K_{yi}\delta_i l_i - P_{f\,grount\,i}l_i\theta_i\right] = M_{rtw} + I_z\dfrac{d\omega_z}{dt} \\ \dfrac{l_i + x_t}{R} = (\theta_i - \delta_i) \end{cases} \quad (14.49)$$

In the system in (14.49) it is reasonable to identify the members containing the skidding angles of the front and rear wheels:

$$\begin{cases} K_{y1}\delta_1\theta_1 + K_{yn}\delta_n\theta_n = P_e - \sum_{i=1}^{n}\left[(R_{xi} - f_{3p}R_{zi})\right] - \sum_{i=2}^{n-1}\left[K_{yi}\delta_i\theta_i\right] \\ K_{y1}\delta_1 + K_{yn}\delta_n = -mj_y + \sum_{i=1}^{n}\left[(R_{xi} - f_{grount}R_{zi})\theta_i\right] - \sum_{i=2}^{n-1}\left[K_{yi}\delta_i\right] \\ K_{yi}\delta_i l_i + K_{yi}\delta_i l_i = -M_{rtw} - I_z\dfrac{d\omega_z}{dt} + \sum_{i=1}^{n}\left[(R_{xi} - f_{grount}R_{zi})l_i\theta_i\right] - \sum_{i=2}^{n-1}\left[K_{yi}\delta_i l_i\right] \\ \dfrac{l_i + x_t}{R} = (\theta_i - \delta_i) \end{cases}$$

The second and third equations of this system allow for defining the skidding angles of the chassis' extreme axles and, accordingly, the difference of these angles:

$$\begin{cases} \delta_1 = \dfrac{1}{K_{y1}L}\{l_n mj_y - M_{rtw} - I_z\dfrac{d\omega_z}{dt} + \\ +\sum_{i=1}^{n}\left[(R_{xi} - f_{grount}R_{zi})(l_i - l_n)\theta_i\right] - \sum_{i=2}^{n-1}\left[K_{yi}\delta_i(l_i - l_n)\right]\} \\ \delta_n = \dfrac{1}{K_{yn}L}\{-l_1 mj_y + M_{rtw} + I_z\dfrac{d\omega_z}{dt} + \\ +\sum_{i=1}^{n}\left[(R_{xi} - f_{grount}R_{zi})(l_1 - l_i)\theta_i\right] + \sum_{i=2}^{n-1}\left[K_{yi}\delta_i(l_1 - l_i)\right]\} \end{cases} \quad (14.50)$$

$$\delta_n - \delta_1 = \frac{1}{L}\{-mj_y\left(\frac{l_1}{K_{yn}} + \frac{l_n}{K_{y1}}\right) + M_c\left(\frac{1}{K_{yn}} + \frac{1}{K_{y1}}\right) + I_z\frac{d\omega_z}{dt}\left(\frac{1}{K_{yn}} + \frac{1}{K_{y1}}\right)\} +$$
$$+\frac{1}{L}\{\sum_{i=1}^{n}(R_{xi} - f_{groundt}R_{zi})\theta_i[\left(\frac{l_1}{K_{yn}} + \frac{l_n}{K_{y1}}\right) - l_i\left(\frac{1}{K_{yn}} + \frac{1}{K_{y1}}\right)]\}$$
$$+\frac{1}{L}\{\sum_{i=2}^{n-1}K_{yi}\delta_i[\left(\frac{l_1}{K_{yn}} + \frac{l_n}{K_{y1}}\right) - l_i\left(\frac{1}{K_{yn}} + \frac{1}{K_{y1}}\right)]\} \quad (14.51)$$

Equation (14.51) shows that the tire turnability depends on the traction forces, travel speed, the resistance moment to turning, and the skidding of the chassis' internal axles.

The skidding angles of the internal axles are excluded with the help of the last group of the system equations in (14.46):

$$\delta_i = \frac{(l_1 - l_i)}{R} - (\theta_1 - \theta_i) + \delta_1 \quad (14.52)$$

And we get the following from (14.51):

$$\delta_1 = \frac{1}{K_{y1}L + \sum_{i=2}^{n-1}K_{yi}(l_i - l_n)}\{l_n mj_y - M_{rtw} - I_z\frac{d\omega_z}{dt} +$$
$$+\sum_{i=1}^{n}[(R_{xi} - f_{grount}R_{zi})(l_i - l_n)\theta_i] +$$
$$+\sum_{i=2}^{n-1}[-K_{yi}\frac{(l_i - l_n)(l_1 - l_i)}{R} + K_{yi}(l_i - l_n)(\theta_1 - \theta_i)]\} \quad (14.53)$$

For convenience of further transformations, it is expedient to present (14.53) in the following form:

$$\delta_1 = \frac{l_n m j_y - I_z \dfrac{d\omega_z}{dt}}{\sum_{j=1}^{n} K_{yj}(l_j - l_n)} - \frac{M_{rtw}}{\sum_{j=1}^{n} K_{yj}(l_j - l_n)} + \frac{\sum_{j=1}^{n}(R_{xj} - f_{3p} R_{zj})(l_j - l_n)\theta_j}{\sum_{j=1}^{n} K_{yj}(l_j - l_n)} - $$
$$-\frac{\sum_{j=2}^{n-1} K_{yj}(l_j - l_n)(l_1 - l_j)}{R\sum_{j=1}^{n} K_{yj}(l_j - l_n)} + \frac{\sum_{j=2}^{n-1} K_{yj}(l_j - l_n)(\theta_1 - \theta_j)}{\sum_{j=1}^{n} K_{yj}(l_j - l_n)} \quad (14.54)$$

By using (14.52), let us exclude the skidding angles of the intermediate axles from the expression for the tire turnability:

$$\delta_n - \delta_1 = b + c + \frac{1}{L}\sum_{i=1}^{n} A_i (R_{xi} - f_{grount} R_{zi})\theta_i +$$
$$+\frac{1}{L}\sum_{i=2}^{n-1} A_i K_{yi}\frac{(l_1 - l_i)}{R} -$$
$$-\frac{1}{L}\sum_{i=2}^{n-1} A_i K_{yi}(\theta_1 - \theta_i) + \frac{1}{L}\sum_{i=2}^{n-1} A_i K_{yi}\delta_1 \quad (14.55)$$

where:

$$A_i = \left(\frac{l_1}{K_{yn}} + \frac{l_n}{K_{y1}}\right) - l_i\left(\frac{1}{K_{yn}} + \frac{1}{K_{y1}}\right) = \frac{(l_1 - l_i)}{K_{yn}} - \frac{(l_i - l_n)}{K_{y1}} \quad (14.56)$$

$$b(M_{rtw}) = \left(\frac{1}{K_{yn}} + \frac{1}{K_{y1}}\right)\frac{M_c}{L} \quad (14.57)$$

$$c(j_y, \dot{\omega}_z) = -\left(\frac{l_1}{K_{yn}} + \frac{l_n}{K_{y1}}\right)\frac{m j_y}{L} + \left(\frac{1}{K_{yn}} + \frac{1}{K_{y1}}\right)\frac{I_z}{L}\frac{d\omega_z}{dt} \quad (14.58)$$

Then, with the help of (14.54), we exclude from (14.55) the skidding angle of the first axle. We group the members in such a way to isolate the components that refer to the internal axles only. After substituting and making intermediary transformations, we can obtain the following expression:

$$\delta_n - \delta_1 = a + b + c + d + e + f + g + h \quad (14.59)$$

where:

$$P_{xi} = (R_{xi} - f_{groundt} R_{zi}) \quad (14.60)$$

$$a(P_{x1}, P_{xn}, \theta_1, \theta_n) = \frac{A_1 P_{x1}\theta_1 + A_n P_{xn}\theta_n}{L} \quad (14.61)$$

$$d(j_y, \dot{\omega}) = \frac{1}{L}\left[\sum_{i=2}^{n-1} K_{yi} A_i \frac{l_n m j_y - I_z \dot{\omega}_z}{\sum_{j=1}^{n} K_{yj}(l_j - l_n)}\right] \quad (14.62)$$

$$e(\rho) = \frac{1}{LR}\left\{\sum_{i=2}^{n-1} K_{yi} A_i \left[(l_1 - l_i) - \frac{\sum_{j=2}^{n-1} K_{yj}(l_j - l_n)(l_1 - l_j)}{\sum_{j=1}^{n} K_{yj}(l_j - l_n)}\right]\right\} =$$
$$= \left\{\frac{\sum_{i=2}^{n-1} K_{yi} A_i \left[\sum_{j=1}^{n-1} K_{yj}(l_j - l_n)(l_j - l_i)\right]}{L^2 \sum_{j=1}^{n-1} K_{yj}(l_j - l_n)}\right\}\rho \quad (14.63)$$

$$f(\vec{\theta}) = -\frac{1}{L}\left\{\sum_{i=2}^{n-1} K_{yi} A_i \left[(\theta_1 - \theta_i) - \frac{\sum_{j=2}^{n-1} K_{yj}(l_j - l_n)(\theta_1 - \theta_j)}{\sum_{j=1}^{n} K_{yj}(l_j - l_n)}\right]\right\} \quad (14.64)$$

$$g(\vec{P}_x, \vec{\theta}) = \frac{1}{L}\left\{\sum_{i=2}^{n-1} A_i \left[P_{xi}\theta_i + K_{yi}\frac{\sum_{j=1}^{n} P_{xj}(l_j - l_n)\theta_j}{\sum_{j=1}^{n} K_{yj}(l_j - l_n)}\right]\right\} \quad (14.65)$$

$$h(M_{rtw}) = -\left\{\sum_{i=2}^{n-1}\left[\frac{K_{yi}A_i}{\sum_{j=1}^{n} K_{yj}(l_j - l_n)}\right]\right\}\frac{M_c}{L} \quad (14.66)$$

$$\rho = \frac{L}{R} \quad (14.67)$$

All the summands *d–h* refer to the chassis' internal axles only and in a two-axle vehicle they turn into zero.

In order to use formulas (14.56)–(14.67), we need to know the resistance moment to turning and the distribution of the tangential forces among the chassis' wheels. For the adopted model, the resistance moment to turning will have three components:

- M_{trw1} is the resistance moment to turning arising at the elastic tire's interaction with the support surface;
- M_{rtw2} is the moment of counteracting to turning arising because of the difference between the tangential forces acting on the wheels of the opposite boards; and
- M_{rtw3} is the counteraction moment to turning of the wheels of uncontrolled axles.

When a wheel turns in place (at stop) the first component can reach essential values. Currently, it is known [61] that even at very small motion speeds, this component of the resistance moment to turning is essentially decreasing. Thus, for example, when turning at the minimum radius, the resistance moment to wheel turning goes down 16–20 times as compared to the wheel turning at stop. In motion at high speeds, the trajectory curvature is usually small and this component decreases still further.

The counteraction moment to turning is defined—because of the difference between the tangential forces acting on the wheels of the leaping and the lagging boards—by the design of the transmission and the system to control the power flows' distribution among the chassis' driving wheels. In the simplest case this moment is defined by the degree of blocking of the inter-wheel and inter-axle differentials. As a whole, this moment can be essential; it should not be neglected without making a special analysis, the methodology of which is presented in Chapter 11 dealing with SWC's transmissions.

The third component is automatically accounted for by the methodology of compiling the chassis' motion model.

Let us assume that the chassis is making a stationary turn (ω = constant, v_g = const) at a very small speed ($v_g \sim 0$). In this case it is obvious that $c \sim 0$ and $d \sim 0$. The remaining non-zero members (*a, b, e, f, g, h*) define the difference in the skid angles of the extreme axles, which does not depend on the travel speed. It can be treated as the "static" or "initial" part of the tire turnability [61].

If the chassis moves quickly, it is most likely on a hard road and the turning angles of the controlled wheels, set by the driver, are small (it is clear that at the rectilinear motion always $\theta_i = 0$; however, a curvilinear motion is also possible at the expense of tire turnability also under $\theta_i = 0$). Then, $a = 0, f = 0$, and $g = 0$. In addition, at a rectilinear motion there is no counteraction (resistance) moment to turning; that is why $b = 0$ and $h = 0$.

At any motion regime of a multi-axle wheeled chassis, except for the rectilinear one ($\rho = 0$), $e \neq 0$ and only in a two-axle vehicle is a case of $e = 0$ possible.

In general, we can believe that the total tire turnability is comprised of the "static tire turnability" $\Delta\delta_{tir} = (\delta_n - \delta_1)_{tir}$ (which does not depend on the travel speed) and the "dynamic tire turnability" (another frequently used name is "additional turnability") $\Delta\delta_{tir} = (\delta_n - \delta_1)_{oe}$:

$$\rho = \left[(\theta_n + \theta_1) + \Delta\delta_{tir}\right] + \left[\Delta\delta_{oe} + \left(\theta_{chn} + \theta_{ch1}\right)\right] = \\ = \rho_{tir} + \rho_{oe} \quad (14.68)$$

where $\rho = L/R$ is the relative trajectory curvature of the motion of the chassis' center of mass.

In general, the initial tire turnability is defined by formula (14.56) at the actual zero travel speed. The turn is formally stationary and the angular acceleration is also absent:

$$\Delta\delta_{tir}(\vec{\theta}) = a + b + e + f + g + h$$

Knowing the initial tire turnability, we can build a line characterizing the initial total turnability; that is, build a curve of the relative trajectory curvature's function versus the controlled wheels' turning angle in the form:

$$\rho_{tir}(\vec{\theta}) = (\theta_n + \theta_1) + \Delta\delta_{tir}(\vec{\theta}) \qquad (14.69)$$

On the turnability diagram built in the coordinates ($\eta;\rho$), the points characterizing the initial turnability are obviously on the axis (0-ρ) since $\eta(v_a) = 0$ at $v_a = 0$.

If the controlled wheels are put in the middle position, the line characterizing the tire turnability passes through the point:

$$\rho_{tir}(0) = \Delta\delta_{tir}(0) = b + e + h$$

In the overwhelming majority of cases, the wheels set in the middle position are associated with the rectilinear motion at "creeping" speeds (an exclusion is the chassis' motion on a slope and the chassis' so-called "crabbing motion" with the controlled wheels when the wheels are turned to one side). When the rectilinear motion is $\rho = 0$, then, obviously, the counteraction moment to turning is absent, and, accordingly:

$$\rho_{tir}(0) = \Delta\delta_{tir}(0) = 0\,.$$

Thus, the initial tire turnability line passes through the origin of the turnability diagram (Fig. 14.13). The diagram is built in rectangular coordinates of the turnability diagram: the upper coordinate is the lateral force η_y; the lower coordinate is the relative trajectory curvature L/R_{min}; and the abscissa axis is made of the relative (current) trajectory curvature L/R and the turning angle of the front controlled wheel θ_1. Quadrants I and IV contain the travel speed dependence (v_C) and the turning angle of the front controlled wheel θ_1. Quadrant II contains the turning angle dependence of the first controlled wheel and the relative turning curvature of the controlled wheels' skid angles $(\delta_n - \delta_1)_{tir}$. The order of using the diagram is as follows. For example, at travel speed v_3 (point B), we find point A of the possible trajectory curvature and then—through Quadrants II and III—in Quadrant IV we find the wheel's required turning angle.

Now, assume that the chassis makes a turn at the "creeping speed," that is, the turning angles of the controlled wheels differ from zero. It follows from (14.157) that:

$$\rho_{tir} = (\theta_n + \theta_1) + \Delta\delta_{tir}(\vec{\theta}) = (\theta_n + \theta_1) + [a(P_{x1}, P_{xn}, \theta_1, \theta_n) + b(M_c) + e(\rho_{tir}) + f(\vec{\theta}) + g(\vec{P}, \vec{\theta}) + h(M_{rtw})]$$

Thus:

$$\rho_{tir}(\vec{\theta}) = \frac{(\theta_n + \theta_1)}{1 - E_\rho} + \frac{a(P_{x1}, P_{xn}, \theta_1, \theta_n)}{1 - E_\rho} + \frac{b(M_c)}{1 - E_\rho} + \left[\frac{f(\vec{\theta})}{1 - E_\rho} + \frac{g(\vec{P}, \vec{\theta})}{1 - E_\rho} + \frac{h(M_c)}{1 - E_\rho}\right] \qquad (14.70)$$

where:

$$E_\rho = \left\{ \frac{\sum_{i=2}^{n-1} K_{yi} A_i \left[\sum_{j=1}^{n-1} K_{yj}(l_j - l_n)(l_j - l_i) \right]}{L^2 \sum_{j=1}^{n-1} K_{yj}(l_j - l_n)} \right\}$$

It is obvious that a two-axle chassis has $E_\rho = 0$, the same as the formula member in (14.70) enclosed in square brackets. If in this case we neglect the air resistance ("creeping speed"), assume that the chassis moves in the horizontal plane without a trailer ($\forall i$: $P_{xi} \sim 0$), and that the transmission has no inter-board and inter-axle power circulation ($M_{rtw} \sim 0$), then:

$$P_{tir}(\theta) = (\theta_n - \theta_1).$$

In other words, with the turn of the controlled wheels, the point defining the initial turnability moves along the axis (0-ρ) by (θ_n+θ_1). In general, this displacement needs to be specifically defined (or the procedure indicated in Fig. 14.13 undertaken).

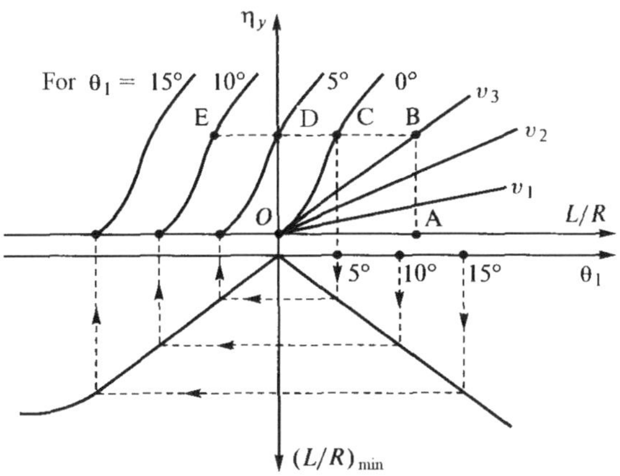

Fig. 14.13. Turnability diagram of a four-axle chassis with insufficient initial and excessive additional turnability

Thus, each value of the extreme controlled wheels' turning angles corresponds to its starting point for the tire turnability curve in the axis ($0 - \rho$) (or in the axis ($0 - L/R$)).

With an increase in the travel speed, we see the speed ("additional") components of the turnability. It follows from (14.59) and (14.68) that:

$$\rho(1-E_\rho) = (\theta_n + \theta_1) + (\theta_{\kappa pn} + \theta_{\kappa p1}) + \\ + a(P_{x1}, P_{xn}, \theta_1, \theta_n) + b(M_c) + c(j_y, \dot{\omega}) + \\ +[d(j_y, \dot{\omega}) + f(\vec{\theta}) + g(\vec{P}, \vec{\theta}) + h(M_{rtw})]$$

from here:

$$\rho = \left\{ \frac{(\theta_n + \theta_1)}{(1-E_\rho)} + \frac{a(P_{x1}, P_{xn}, \theta_1, \theta_n)}{(1-E_\rho)} + \frac{b(M_{rtw})}{(1-E_\rho)} \right\} + \\ + \left\{ \frac{(\theta_{chn} + \theta_{ch1})}{(1-E_\rho)} + \frac{c(j_y, \dot{\omega})}{(1-E_\rho)} + \right. \\ \left. + \left[\frac{f(\vec{\theta})}{(1-E_\rho)} + \frac{d(j_y, \dot{\omega})}{(1-E_\rho)} + \frac{g(\vec{P}, \vec{\theta})}{(1-E_\rho)} + \frac{h(M_{rtw})}{(1-E_\rho)} \right] \right\} \quad (14.71)$$

The members grouped inside the first curly bracket cannot be regarded as the chassis' initial turnability since, in general, the resistance (counteraction) moment to turning and the tangential forces depend on the travel speed. In order to preserve the continuity of narration adopted in this section and to not deepen into the study of the multi-axle chassis' transmission, we can proceed as follows. It is clear that the following presentation is always admissible:

$$\begin{cases} M_{rtw}(v_g) = M_{rtw}(0) - \left[M_{rtw}(0) + M_{rtw}(v_g) \right] = M_{rtw0} - \left[M_{rtw0} + M_{rtw}(v_g) \right] \\ P_{xi}(v_g) = P_{xi}(0) - \left[P_{x0i}(0) + P_{xi}(v_g) \right] = P_{x0i} - \left[P_{x0i} + P_{xi}(v_g) \right] \end{cases}$$

Then:

$$a = \frac{A_1 P_{x01}\theta_1 + A_n P_{x0n}\theta_n}{L} - \frac{A_1(P_{x01} - P_{x1})\theta_1 + A_n(P_{x0n} - P_{xn})\theta_n}{L} = \\ = a_0(P_{x01}, P_{x0n}\theta_1, \theta_n) - a_v[P_{x01}, P_{x0n}, P_{x1}(v), P_{xn}(v), \theta_1, \theta_n]$$

$$b = \left(\frac{1}{K_{yn}} + \frac{1}{K_{y1}} \right) \frac{M_{rtw0}}{L} - \left(\frac{1}{K_{yn}} + \frac{1}{K_{y1}} \right) \frac{(M_{rtw0} - M_{rtw})}{L} = \\ = b_0(M_{rtw0}) - b_v[M_{rtw0}, M_{rtw}(v_{rtw})]$$

$$g = \frac{1}{L} \left\{ \sum_{i=2}^{n-1} A_i \left[P_{x0i}\theta_i + K_{yi} \frac{\sum_{j=1}^{n} P_{x0j}(l_j - l_n)\theta_j}{\sum_{j=1}^{n} K_{yj}(l_j - l_n)} \right] \right\} - \\ - \frac{1}{L} \left\{ \sum_{i=2}^{n-1} A_i \left[(P_{x0i} - P_{xi})\theta_i + K_{yi} \frac{\sum_{j=1}^{n} (P_{x0j} - P_{xj})(l_j - l_n)\theta_j}{\sum_{j=1}^{n} K_{yj}(l_j - l_n)} \right] \right\} = \\ = g_0(\vec{P}_{x0}, \vec{\theta}) - g_v[\vec{P}_{x0}, \vec{P}_x(v_a)\vec{\theta}]$$

$$h = -\left\{\sum_{i=2}^{n-1}\left[\frac{K_{yi}A_i}{\sum_{j=1}^{n}K_{yj}(l_j - l_n)}\right]\right\}\frac{M_{rtw0}}{L} +$$

$$+\left\{\sum_{i=2}^{n-1}\left[\frac{K_{yi}A_i}{\sum_{j=1}^{n}K_{yj}(l_j - l_n)}\right]\right\}\frac{(M_{rtw0} - M_{rtw})}{L} =$$

$$= h_0(M_{rtw0}) - h_v[M_{rtw0}, M_{rtw}(v_g)]$$

Now, formula (14.71) can be presented as a composition of the initial turnability (the chassis is "crabbing" along a curvilinear trajectory set by the controlled wheels) and the additional or speed turnability caused by the influence of the speed factors on the processes that occur in the vehicle's support-running complex and transmission components is:

$$\rho[\vec{P}_x(v_g), M_{rtw}(v_g), \vec{\theta}(v_g), \theta_{ch1}(v_g), \theta_{chn}(v_g), \dot{\omega}, j_y] =$$
$$= \rho_{tir}[P_{x01}, P_{x0n}, M_{rtw0}, \vec{\theta}] +$$
$$+ \rho_{oe}[P_{x01}, P_{x0n}, \vec{P}_x(v_g), M_{rtw0}, M_{rtw}(v_g), \theta_1$$
$$\theta_n, \vec{\theta}, \theta_{ch1}(v_g), \theta_{chn}(v_g), \dot{\omega}, j_y]$$

where:

$$\rho_{oe}[P_{x01}, P_{x0n}, \vec{P}_x(v_g), M_{rtw0}, M_{rtw}(v_g), \theta_1,$$
$$, \theta_n, \vec{\theta}, \theta_{ch1}(v_g), \theta_{chn}(v_g), \dot{\omega}, j_y] =$$

$$= -\left\{\frac{a_v[P_{x01}, P_{x0n}, P_{x1}(v_g), P_{xn}(v_g), \theta_1, \theta_n]}{(1-E_\rho)} + \frac{b_v[M_{rtw0}, M_{rtw}(v_g)]}{(1-E_\rho)} +\right.$$
$$+\left[\frac{g_v[\vec{P}_{x0}, \vec{P}_x(v_g), \vec{\theta}]}{(1-E_\rho)} + \frac{h_v[M_{rtw0}, M_{rtw}(v_g)]}{(1-E_\rho)}\right]\Bigg\}+$$
$$+\left\{\frac{(\theta_{chn} + \theta_{ch1})}{(1-E_\rho)} + \frac{c(j_y, \dot{\omega})}{(1-E_\rho)} + \left[\frac{d(j_y, \dot{\omega})}{(1-E_\rho)}\right]\right\} \tag{14.72}$$

Let us now get back to the formula defining the specific lateral force affecting the chassis' center of mass:

$$\eta_y = \frac{j_y}{g}.$$

The last two members in (14.72) can be transformed so as to explicitly show the specific lateral force:

$$c(j_y, \dot{\omega}_z) = c_{y1}\frac{mg}{L}\eta_y + c_{y2}\frac{I_z}{L}\dot{\omega}$$

$$d(j_y, \dot{\omega}) = d_y\frac{mgl_n}{L}\eta_y - d_y\frac{I_z}{L}\dot{\omega}_z$$

where:

$$c_{y1} = -\left(\frac{l_1}{K_{yn}} + \frac{l_n}{K_{y1}}\right)$$

$$c_{H2} = \left(\frac{1}{K_{yn}} + \frac{1}{K_{y1}}\right)$$

$$d_y = \left[\frac{\sum_{i=2}^{n-1}K_{yi}A_i}{\sum_{j=1}^{n}K_{yj}(l_j - l_n)}\right]$$

Now, by abridging the formulation we have:

$$L(1-E_\rho)\rho = L\rho_{ucx}[P_{x01}, P_{x0n}, M_{c0}, \vec{\theta}] - L\{a_v + b_v +$$
$$+[g_v + h_v]\} + L\left(\theta_{Kpn} + \theta_{Kp1}\right) +$$
$$+I_z(c_{y2} - d_y)\dot{\omega}_z + mg(c_{y1} + d_y l_n)\eta_y$$

From where:

$$\eta_y = \frac{L(1-E_\rho)\rho}{mg(c_{y1} + d_y l_n)} - \frac{L\rho_{ucx}[P_{x01}, P_{x0n}, M_{c0}, \vec{\theta}]}{mg(c_{y1} + d_y l_n)} -$$

$$-\frac{L\{a_v + b_v + [g_v + h_v]\}}{mg(c_{y1} + d_y l_n)} -$$

$$-\frac{L\left(\theta_{Kpn}+\theta_{Kp1}\right)}{mg(c_{y1}+d_y l_n)}-\frac{I_z(c_{y2}-d_y)\dot{\omega}_z}{mg(c_{y1}+d_y l_n)} \quad (14.73)$$

In general, the resistance moment to turning depends on the trajectory curvature and the forces in the tire-road contact zone; therefore, it is a function of the lateral force. The carrying system's roll depends on the lateral force and, accordingly, the roll turnability also depends on the lateral force. The resistance coefficients to the lateral tire skidding depend on at least the vertical loads on the wheels, and, accordingly, on the lateral force.

Thus, formula (14.73) is a nonlinear algebraic equation relative to η_y. Its solution defines the wheel chassis' full turnability diagram and is built by iteration methods. As an initial approximation, we can use a linear approximation at constant resistance coefficients to lateral skidding, the "crabbing travel speed," negligibly small resistance moments to motion, and the free wheel rolling mode (the upper index in the square brackets is the approximation number: [0] means zero or initial approximations):

$$\begin{cases} \eta_y^{[0]} = \dfrac{(1-E_\rho)^2\rho-(\theta_n+\theta_1)-f(\vec{\theta})}{mg(c_{y1}+d_y l_n)(1-E_\rho)}L \\ f(\vec{\theta}) = -\dfrac{1}{L}\left\{\sum_{i=2}^{n-1}K_{yi}A_i\left[(\theta_1-\theta_i)-\dfrac{\sum_{j=2}^{n-1}K_{yj}(l_j-l_n)(\theta_1-\theta_j)}{\sum_{j=1}^{n}K_{yj}(l_j-l_n)}\right]\right\} \end{cases}$$

In many cases, a multi-axle chassis' suspension is designed such that the "roll turnability" is small or absent whatsoever. Also, at a steady turning, ω = constant. Then, formula (14.73) is simplified as:

$$\eta_y=\frac{L(1-E_\rho)\rho}{mg(c_{y1}+d_y l_n)}-\frac{L\rho_{ucx}[P_{x01},P_{x0n},M_{c0},\vec{\theta}]}{mg(c_{y1}+d_y l_n)}-$$

$$-\frac{L[(a_v+b_v)+(g_v+h_v)]}{mg(c_{y1}+d_y l_n)}$$

Figs. 14.13 and 14.14 (quadrant II) show that the sign at the difference in the skidding angles of the extreme axles ($\delta_n-\delta_1$) can be different in different areas of the full turnability diagram. In other words, the chassis, which under some conditions has insufficient turnability, in other conditions can acquire the properties of excessive turnability.

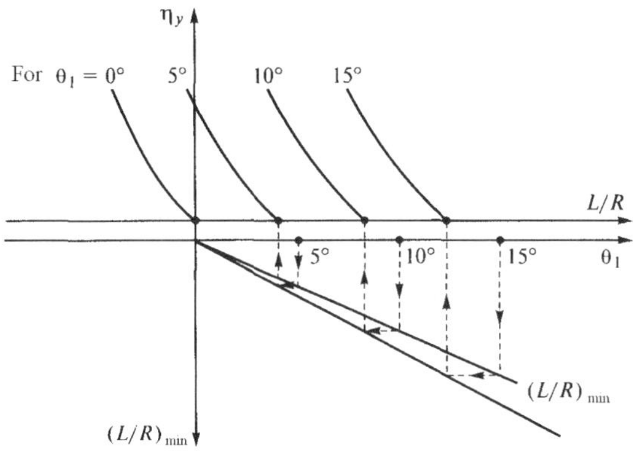

Fig. 14.14. Turnability diagram of a vehicle with excessive initial and insufficient additional turnability

To improve a chassis' turnability, it is necessary to:

- Try to create the maximum possible turning moment by the lateral forces in the contact area by using the maximum number of wheels and applying lateral forces to the maximum possible arm; for which purpose use not only the front but also the rear controlled wheels and, optimally, all of the vehicle's wheels;
- Reduce the resistance moment to turning by bringing the axles with non-controlled wheels closer to the center of the chassis' base and decreasing the base of their bogie; and
- Try to bring the center of instant turn closer to the chassis' longitudinal axis by maximizing the turning angle of the controlled wheels, introducing several axles with controlled wheels, decreasing the wheelbase, making the front and rear wheels controlled, and using

the tire turnability in addition to the steering turnability by using different versions of the so-called "board turning."

Let us consider in more detail a fragment of the full turnability diagram presented in Fig. 14.14 for a multi-axle wheeled chassis with insufficient initial and excessive additional turnability.

In accordance with *Motion Theory of Military Wheeled Vehicles* [61], let us assume that we make a curvilinear motion, which is ensured by some fixed turning angle of the controlled wheels (Fig. 14.15, line *BTN*). Further, let the chassis move at a certain speed v_6, designated as the "speed beam" *OS*. Let us assume also that in point *T* the straight line parallel to this "speed beam" is tangential to the full turnability line. This means that the trajectory curvature of the chassis' center of mass corresponds to point *G* on the speed beam *OS* ($\rho = T^*G$) and that the turning angle of the controlled wheels is defined by section *GT* ($\theta_n + \theta_1 = GT$), while the skidding angle difference in the extreme axles is defined by section TT^* ($\delta_n - \delta_1 = TT^*$).

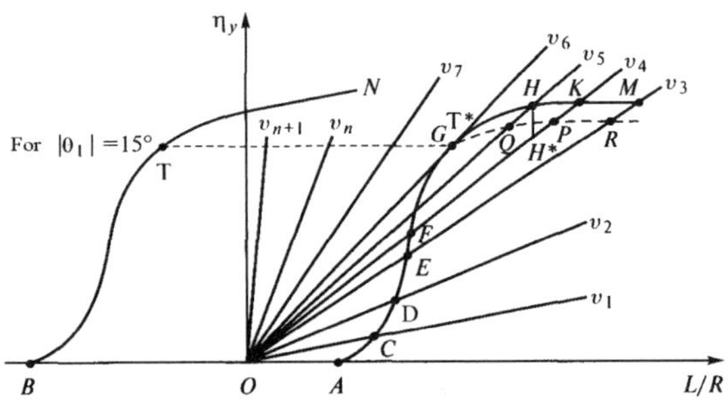

Fig. 14.15. Definition of the stable motion area in the full turnability diagram. It is clear that at lower speeds the chassis' motion is stable (the "speed beams" are below the beam OS).

Let us build in point *G* the curvilinear line *AGM* equidistant from the tire turnability curve *BTN* and suppose that the chassis' center of mass is affected by a small lateral perturbation $\Delta\eta$.

In this case, the lateral force increased up to $OT^* + \Delta\eta$ (at the unchanged position of the controlled wheels), will cause an increase in the turning curvature up to $\rho^* = T_1^*G^*$. Then, we see the already described process of spontaneous growth in the chassis' travel speed. After crossing point *H*, the trajectory curvature's growth process becomes avalanche-type and practically unrestricted (within the considered structure). This means that the trajectory curvature's growth rate may become so large that the driver's physical and physiological capabilities are insufficient to counteract this process by changing the position of the controlled wheels or decreasing the travel speed.

Let us now assume that the driver detected the loss in stability and began decreasing the travel speed. Let us assume further that it happened when the spontaneously growing trajectory curvature reached point *H* and the driver managed to decrease the speed down to v_5 corresponding to this point. The growth in the trajectory curvature will not stop since the new "speed beam" is still much steeper than the tangent to the tire turnability curve in point *H*. If we complement the full turnability diagram with line *GQPR* of the stability loss border, we can see that at the v_5 speed the motion stability condition was broken already at the trajectory curvature corresponding to point *Q*. If the driver manages in point *H* to decrease the speed down to v_4 (point H^*), the situation will be repaired since point H^* is within the stable motion area (below the border line *GQPR*).

This example shows that under certain circumstances a chassis with insufficient turnability can behave as an unstable one.

The described methods for analyzing the motion stability of a multi-axle wheeled chassis are applicable when we know the chassis' full turnability from tests or can build rather precisely the full turnability diagram from calculations. In practice, this way faces essential difficulties; therefore, it is desirable to possess a system of analytical criteria, which allows for making a qualitative characterization of the vehicle in the part that relates to its motion stability at the preset motion conditions.

14.4 Steady motion stability in multi-axle wheeled chassis

Chassis stability is assessed both relative to its position parameters and relative to its motion parameters.

The chassis' stable position is understood as the stability of its balance relative to the bound coordinate system, namely:

- The stability against longitudinal tripping (rotation around axis y_3);
- The stability against lateral tripping (rotation around axis x_3); and
- The stability against skidding and slipping of the front axle (rotation around axis z_3).

The chassis' stable motion is understood as the stability of the parameters characterizing its motion: the advanced travel speed, the carrying system's lateral displacement rate, the angular velocity of the carrying system's turn around axis z_3, the lateral deviation of the center of mass away from the preset trajectory, the angular deviation of the carrying system's longitudinal axis away from the preset course, etc.

At early design stages, the wheeled chassis' motion stability is evaluated only by those parameters which have to do with traffic safety, change in motion direction, or exit from the traffic corridor (namely, the course angle, the course angle's change rate, the center of mass's lateral deviation away from the course, the lateral deviation's change rate, the lateral roll angle, and the change rate of the lateral roll angle).

It is well-known [61] that wheeled chassis with elastic tires are unstable in principle relative to the course angle and lateral displacement; therefore, it is reasonable to assess their motion stability relative to differential parameters that cannot be affected by the driver directly (the change rate of the course angle ω_{z3} and the change rate of the center of mass's lateral deviation away from the motion trajectory v_{y3}).

The problem should be addressed step by step, by having first considered the steady motion stability.

When building the calculation sequence, one usually tries to separately study the motion in the horizontal plane (course maneuver), lateral plane (roll), and longitudinal plane (different or pitch), as it was done in Chapter 8 when building the general curvilinear motion model of a multi-link chassis.

However, when studying the course stability, one should not completely ignore the changes in the vertical reactions on the wheels that arise at the longitudinal-angular and lateral-angular oscillations of the carrying system. As a result, the motions in all planes are interconnected because of the vertical load's influence on the tire breakaway characteristics.

To simplify the solution, D.A. Antonov [61] uses the equal probability hypothesis of the chassis' wheels encountering the same irregularity, which allows for accounting for the influence of the oscillatory process on the motion stability by correcting the breakaway characteristics without departing from the study boundaries of the chassis' isolated plane turn.

In general, the system of motion equations for a single-link multi-axle wheeled chassis (under the described restrictions) looks as follows:

$$\begin{cases} M\left(\dfrac{dv_x}{dt} + \omega_z v_y\right) = \sum_{i=1}^{n}\left[P_{xi}\cos\theta_i - R_{yi}\sin\theta_i\right] - P_{air} - G_g\sin\alpha \\ M\left(\dfrac{dv_y}{dt} - \omega_z v_x\right) = \sum_{i=1}^{n}\left[P_{xi}\sin\theta_i + R_{yi}\cos\theta_i\right] \\ I_z\dfrac{d\omega_z}{dt} = \sum_{i=1}^{n}\left[P_{xi}l_i\sin\theta_i + R_{yi}l_i\cos\theta_i\right] - M_{rtw} \\ \dfrac{l_i + x_t}{R} = \tan\left(\theta_i - \delta_i\right) \end{cases} \tag{14.74}$$

where α is the angle of the conquered slope.

Let us consider the chassis' perturbed motion with the following parameters:

$$\begin{cases} \delta_i = \delta_{0i} + \Delta\delta_i \\ \omega_z = \omega_{0z} + \Delta\omega_z \\ R_{yi} = R_{y0i} + \Delta R_{yi} \\ v_y = v_{y0} + \Delta v_y \\ \rho = \rho_0 + \Delta\rho = \dfrac{1}{R} + \Delta\rho \end{cases}$$

Further, the very statement of the chassis' motion stability problem has sense only for the motion cases at comparatively high speeds; therefore, the equations in (14.73) admit linearization by the turning angles of the controlled wheels:

$$\begin{cases} M\left(\dfrac{dv_x}{dt} + \omega_z v_y\right) = \sum\limits_{i=1}^{n}\left[P_{xi} - R_{yi}\theta_i\right] - P_{air} - G_{ga}\sin\alpha \\ M\left(\dfrac{dv_y}{dt} - \omega_z v_x\right) = \sum\limits_{i=1}^{n}\left[P_{xi}\theta_i + R_{yii}\right] \\ I_z\dfrac{d\omega_z}{dt} = \sum\limits_{i=1}^{n}\left[P_{xi}l_i\theta_i + R_{yi}l_i\right] - M_{rtw} \\ \dfrac{l_i + x_t}{R} = \theta_i - \delta_i \end{cases}$$

Also, since:

$$\tan(\theta_i - \delta_i) = \frac{v_{yi}}{v_x} = \frac{v_y - l_i\omega_z}{v_g\cos\theta_C} \tag{14.75}$$

then:

$$(\theta_i - \delta_i) = \frac{v_{yi}}{v_g} = \frac{v_y - l_i\omega_z}{v_g}$$

The system of equations in (14.74) can be transformed into the system of equations for the perturbed motion:

$$\begin{cases} m\left(\dfrac{d\Delta v_y}{dt} - \Delta\omega_z v_x\right) = \sum\limits_{i=1}^{n}\left[\Delta R_{yi}\cos\theta_i\right] \\ I_z\dfrac{d\Delta\omega_z}{dt} = \sum\limits_{i=1}^{n}\left[\Delta R_{yi}l_i\cos\theta_i\right] \\ (l_i + x_{cm})\Delta\rho = -\Delta\delta_i \end{cases}$$

When defining the perturbed component of the lateral force, the algebraic theory of lateral breakaway, developed by D.A. Antonov, is typically used, and one proceeds from the assumption that the loading of the tire's contact spot with the road is small, and, respectively, the resistance coefficients to lateral breakaway are weakly dependent on the lateral breakaway angle. In these conditions:

$$R_{yi} = R_{y0i} + \left(\frac{\partial R_{yi}}{\partial\delta_i}\right)_{\delta_i=0}\Delta\delta_i$$

It follows from (14.74) that:

$$\Delta\delta_i = \frac{\Delta v_y - l_i\Delta\omega_z}{v_x}$$

Therefore, the equations of perturbed motion are converted to the form:

$$\begin{cases} mv_x\dfrac{d\Delta v_y}{dt} - \Delta v_y\sum\limits_{i=1}^{n}\left[\left(\dfrac{\partial R_{yi}}{\partial\delta_i}\right)_{\delta_i=0}\cos\theta_i\right] - \\ -\Delta\omega_z\left\{mv_x^2 - \sum\limits_{i=1}^{n}\left[\left(\dfrac{\partial R_{yi}}{\partial\delta_i}\right)_{\delta_i=0} l_i\cos\theta_i\right]\right\} = 0 \\ I_z v_x\dfrac{d\Delta\omega_z}{dt} - \Delta v_y\sum\limits_{i=1}^{n}\left[\left(\dfrac{\partial R_{yi}}{\partial\delta_i}\right)_{\delta_i=0} l_i\cos\theta_i\right] + \\ +\Delta\omega_z\sum\limits_{i=1}^{n}\left[\left(\dfrac{\partial R_{yi}}{\partial\delta_i}\right)_{\delta_i=0} l_i^2\cos\theta_i\right] = 0 \end{cases}$$

By designating:

$$F = -\frac{1}{m}\sum_{i=1}^{n}\left[\left(\frac{\partial R_{yi}}{\partial \delta_i}\right)_{\delta_i = 0}\cos\theta_i\right] \tag{14.76}$$

$$H = -\frac{1}{I_z}\sum_{i=1}^{n}\left[\left(\frac{\partial R_{yi}}{\partial \delta_i}\right)_{\delta_i = 0} l_i \cos\theta_i\right] \tag{14.77}$$

$$G = \frac{1}{m}\sum_{i=1}^{n}\left[\left(\frac{\partial R_{yi}}{\partial \delta_i}\right)_{\delta_i = 0} l_i \cos\theta_i\right] \tag{14.78}$$

$$E = \frac{1}{I_z}\sum_{i=1}^{n}\left[\left(\frac{\partial R_{yi}}{\partial \delta_i}\right)_{\delta_i = 0} l_i^2 \cos\theta_i\right] \tag{14.79}$$

we have:

$$\begin{cases} v_x \dfrac{d\Delta v_y}{dt} + F\Delta v_y + (G - v_x^2)\Delta\omega_z = 0 \\ v_x \dfrac{d\Delta\omega_z}{dt} + H\Delta v_y + E\Delta\omega_z = 0 \end{cases} \tag{14.80}$$

By differentiating both equations in (14.80) by time, we have:

$$\begin{cases} \dfrac{dv_x}{dt}\dfrac{d\Delta v_y}{dt} + v_x\dfrac{d^2\Delta v_y}{dt^2} + \dfrac{dF}{dt}\Delta v_y + F\dfrac{d\Delta v_y}{dt} + \\ + \dfrac{d(G - v_x^2)}{dt}\Delta\omega_z + (G - v_x^2)\dfrac{d\Delta\omega_z}{dt} = 0 \\ \dfrac{dv_x}{dt}\dfrac{d\Delta\omega_z}{dt} + v_x\dfrac{d^2\Delta\omega_z}{dt^2} + \dfrac{dH}{dt}\Delta v_y + H\dfrac{d\Delta v_y}{dt} + \\ + \dfrac{dE}{dt}\Delta\omega_z + E\dfrac{d\Delta\omega_z}{dt}\Delta\omega_z = 0 \end{cases} \tag{14.81}$$

To solve (14.81) relative to Δv_y, $\Delta\omega_z$, and their derivatives, let us transform the equations in (14.80) as follows:

$$\begin{cases} \Delta\omega_z = -\dfrac{v_x}{(G - v_x^2)}\dfrac{d\Delta v_y}{dt} - \dfrac{F}{(G - v_x^2)}\Delta v_y \\ \dfrac{d\Delta\omega_z}{dt} = -\dfrac{H}{v_x}\Delta v_y + \dfrac{E}{(G - v_x^2)}\dfrac{d\Delta v_y}{dt} + \dfrac{EF}{(G - v_x^2)v_x}\Delta v_y \end{cases}$$

$$\begin{cases} \dfrac{d\Delta v_y}{dt} = -\dfrac{(G - v_x^2)}{v_x}\Delta\omega_z + \dfrac{EF}{Hv_x}\Delta\omega_z + \dfrac{F}{H}\dfrac{d\Delta\omega_z}{dt} \\ \Delta v_y = -\dfrac{E}{H}\Delta\omega_z - \dfrac{v_x}{H}\dfrac{d\Delta\omega_z}{dt} \end{cases}$$

We introduce $\dfrac{dv_x}{dt} = j_x$, and then:

$$\begin{cases} \Delta\ddot{v}_y + \left[\dfrac{j_x + F + E}{v_x} - \dfrac{\dot{G}}{G - v_x^2} + \dfrac{2v_x j_x}{G - v_x^2}\right]\Delta\dot{v}_y + \\ + \left[\dfrac{\dot{F}}{v_x} - \dfrac{HG + EF}{v_x^2} + H - \dfrac{\dot{G}F}{(G - v_x^2)v_x} + \dfrac{2j_x F}{(G - v_x^2)}\right]\Delta v_y \\ = 0 \\ \Delta\ddot{\omega}_z + \left[\dfrac{(j_x + E + F)}{v_x} - \dfrac{\dot{H}}{H}\right]\Delta\dot{\omega}_z + \\ + \left[\dfrac{\dot{E}}{v_x} - \dfrac{HG - EF}{v_x^2} + H - \dfrac{E\dot{H}}{Hv_x}\right]\Delta\omega_z = 0 \end{cases} \tag{14.82}$$

If, at the evolutional motion, the breakaway angles do not depend on time (for example, at a stationary turn of the chassis) or, in a particular case, the breakaway is insignificant, then:

$$\dot{G} = \dot{H} = \dot{E} = \dot{F} = 0 \tag{14.83}$$

and the equations in (14.82) take a simpler form:

$$\begin{cases} \Delta\ddot{v}_y + \left[\dfrac{j_x + F + E}{v_x} + \dfrac{2v_x j_x}{G - v_x^2}\right]\Delta\dot{v}_y + \left[H - \dfrac{HG + EF}{v_x^2} + \dfrac{2j_x F}{(G - v_x^2)}\right]\Delta v_y \\ = 0 \\ \Delta\ddot{\omega}_z + \left[\dfrac{(j_x + E + F)}{v_x}\right]\Delta\dot{\omega}_z + \left[H - \dfrac{HG - EF}{v_x^2}\right]\Delta\omega_z = 0 \end{cases} \tag{14.84}$$

Let us consider a steady curvilinear motion (a uniform turn), at which $j_x = 0$. The system in (14.84) is simplified still more:

$$\begin{cases} \Delta\ddot{v}_y + \left[\dfrac{F+E}{v_x}\right]\Delta\dot{v}_y + \left[H - \dfrac{HG+EF}{v_x^2}\right]\Delta v_y = 0 \\ \Delta\ddot{\omega}_z + \left[\dfrac{E+F}{v_x}\right]\Delta\dot{\omega}_z + \left[H - \dfrac{HG-EF}{v_x^2}\right]\Delta\omega_z = 0 \end{cases}$$

Here, both equations have the same coefficients, and, respectively, the same forms of general solution. The stability condition of this solution comes from the requirement for the own values of the characteristic equation:

$$-\frac{E+F}{2v_x} \pm \sqrt{\left(\frac{E+F}{2v_x}\right)^2 - H - \frac{EF-GH}{v_x^2}} < 0 \tag{14.85}$$

The values E and F are always positive (with account of the lateral force signs and angle of lateral breakaway), which is the same as the chassis' travel speed. Consequently, to satisfy the condition of (14.85), it is necessary that:

$$H - \frac{EF-GH}{v_x^2} > 0 \tag{14.86}$$

It follows from (14.76)–(14.79) and (14.86) that:

$$\begin{aligned} &\left[-\sum_{i=1}^{n}\left[\left(\frac{\partial R_{yi}}{\partial \delta_i}\right)_{\delta_i=0} l_i^2\cos\theta_i\right]\sum_{i=1}^{n}\left[\left(\frac{\partial R_{yi}}{\partial \delta_i}\right)_{\delta_i=0}\cos\theta_i\right]+\right. \\ &\left.+\sum_{i=1}^{n}\left[\left(\frac{\partial R_{yi}}{\partial \delta_i}\right)_{\delta_i=0} l_i\cos\theta_i\right]^2\right] > \\ &> m v_x^2 \sum_{i=1}^{n}\left[\left(\frac{\partial R_{yi}}{\partial \delta_i}\right)_{\delta_i=0} l_i\cos\theta_i\right] \end{aligned} \tag{14.87}$$

It also follows from (14.87) that the stable motion criterion of a multi-axle wheeled chassis that is performing a stationary turn has the following requirement for its travel speed:

$$v_x < \sqrt{\frac{-\sum_{i=1}^{n}\left[\left(\frac{\partial R_{yi}}{\partial \delta_i}\right)_{\delta_i=0} l_i^2\cos\theta_i\right]\sum_{i=1}^{n}\left[\left(\frac{\partial R_{yi}}{\partial \delta_i}\right)_{\delta_i=0}\cos\theta_i\right]+\sum_{i=1}^{n}\left[\left(\frac{\partial R_{yi}}{\partial \delta_i}\right)_{\delta_i=0} l_i\cos\theta_i\right]^2}{m\sum_{i=1}^{n}\left[\left(\frac{\partial R_{yi}}{\partial \delta_i}\right)_{\delta_i=0} l_i\cos\theta_i\right]}} \tag{14.88}$$

In particular, if the lateral breakaway hypothesis is valid and formalized within the "algebraic theory of lateral breakaway" in the form of:

$$R_{yi} = -q_i K_{y0e}\delta_i,$$

where K_{y0e} is the extreme value of the resistance coefficient to lateral breakaway defined at close-to-zero breakaway angles and q_i is the general correction coefficient for the tires of the ith axle, then we get:

$$v_x < \sqrt{\frac{K_{y0e}}{m}\,\frac{\sum_{i=1}^{n}\left[q_i l_i^2\cos\theta_i\right]\sum_{i=1}^{n}\left[q_i\cos\theta_i\right]-\sum_{i=1}^{n}\left[q_i l_i\cos\theta_i\right]^2}{m\sum_{i=1}^{n}\left[q_i l_i\cos\theta_i\right]}} \tag{14.89}$$

If we suppose that all of the correction coefficients are approximately equal (it is convenient, for example, when making a general analysis of the influence of the chassis' design on its motion stability), then the formula in (14.89) becomes still simpler:

$$v_x < \sqrt{\frac{qK_{y0e}}{m}\,\frac{\sum_{i=1}^{n}\left[l_i^2\cos\theta_i\right]\sum_{i=1}^{n}\left[\cos\theta_i\right]-\sum_{i=1}^{n}\left[l_i\cos\theta_i\right]^2}{m\sum_{i=1}^{n}\left[l_i\cos\theta_i\right]}}$$

The second summand in the numerator of the subradical expression is essentially smaller than the first one since the parameters l_i have opposite signs (the axes are located almost equally on both sides of the center of mass). The first summand is the defining one for the numerator.

We can obtain the qualitative evaluation of the phenomena if we consider a rectilinear motion only and in formula (14.89) neglect the quadratic member in the numerator and assume that $\forall i{:}q_i = q$:

$$v_x < \sqrt{\frac{K_{y0e}}{m}\frac{\sum_{i=1}^{n}[q_i l_i^2]\sum_{i=1}^{n}[q_i]-\sum_{i=1}^{n}[q_i l_i]^2}{\sum_{i=1}^{n}[q_i l_i]}} \approx$$

$$\approx \sqrt{\frac{K_{y0e}qg}{G_a}\frac{\sum_{i=1}^{n}[l_i^2]}{\sum_{i=1}^{n}[l_i]}}$$

The relative resistance coefficient to lateral skidding of most tires used in automobiles of high and very high passage ability is within 4–6 [61]; therefore, the following estimate is valid:

$$v_x < (6.3...7.7)\sqrt{\frac{\sum_{i=1}^{n}[l_i^2]}{\sum_{i=1}^{n}[l_i]}}$$

A change in the position of the center of mass along the longitudinal axis slightly affects the numerator of the subradical expression, but essentially changes the value of the denominator. For example, when the center of mass moves forward, the negative summands dominates, the denominator goes down, and the critical speed goes up. At the same time, the normal load on the wheels located behind the center of mass goes down, and, accordingly, we see a decrease in their resistance coefficients to lateral skidding. We can expect that, starting from some moment, this process will become dominant, and the critical speed will go down again.

We can now state the following general rules:

- The bigger the chassis' base, the higher the critical travel speed (because of the growing quadratic member $\sum l_i^2$);
- The more wheels there are in the chassis, the higher its critical speed is (because of the increase in the number of summands in the quadratic member and its respective increase);
- Under other equal conditions, the chassis whose axles are grouped into the nose and back bogies has a higher critical speed than the same chassis whose wheels are grouped into a central bogie;
- The use of a dual-slope (double-wheel) system improves the motion stability (this effect is clearly manifested when the dual-slope is used on the rear wheels since in this case we can obtain imaginary values of the critical speed).

In general, the smaller the sum $\left|\sum q_i l_i\right|$ is, the higher the chassis' critical travel speed is. The following design-layout solutions allow for decreasing this value:

1. Displacing the chassis' center of mass forward;
2. Creating conditions for increasing the skidding angles of the front axles (decreasing the q_i of the front axles);
3. Creating conditions for decreasing the skidding axles of the rear axles (increasing the q_i of the rear axles);
4. Decreasing the lateral roll moment and redistributing the normal loads between the left- and right-hand wheels (increasing the correction coefficient for the normal load of the wheels on all axles);
5. Overloading the axles located in front of the center of mass ("redistribution" of correction coefficients under the normal load among the wheels of the front and rear axles);
6. Introducing lateral stability stabilizers, which ensure an overload of the external (relative to the turn center) wheels and unloading of the

internal wheels for the axles located in front of the instant turning center;

7. Installing wheels with the maximum possible wheelspan (redistributing the correction coefficients on the normal load among the left- and right-hand wheels);
8. Redistributing the roll moment, which acts on the sprung masses in such a way that its greater part falls on the front axles (increasing the wheelspan of the front axles, increasing the angular rigidity of the front wheels' suspensions, etc.);
9. Equipping the front wheels with shock absorbers with an increased resistance both at the compression and retreat strokes;
10. Using design-layout solutions that ensure a decrease in the correction coefficients for the tangential load on the front wheels and, accordingly, an increase in them for the rear wheels (for example, by using transmissions with front driving and rear stabilizing wheels);
11. Taking measures aimed at decreasing the tangential loading of the contact area, especially of the rear wheels (using transmissions, prohibiting the appearance of power circulation, etc.);
12. Forming the kinematical design of the suspension guiding mechanism in such a manner that it ensures the roll of the wheels in the same direction as the roll of the body on the front axles and against the roll of the body on the rear axles (which ensures smaller correction coefficients on the front wheels than on the rear wheels);
13. Intensive damping of oscillations on the rear wheels through applying shock absorbers with an increased resistance at the retreat stroke (increasing the correction coefficient, which takes into account the wheel oscillations, on the rear wheels);
14. Using devices allowing for slightly braking the rear wheels (for example, braking by the engine, electrodynamic braking, etc.) should there be a threat of losing stability;
15. Using more efficient braking mechanisms on the front wheels than on the rear wheels;
16. Rationally distributing the braking forces among the chassis' wheels; it is advisable to use a special diagram and design of the braking system capable of decreasing the braking load both in case of poor wheel-road adhesion and a decrease in the normal load on the wheel;
17. Using "anti-blocking devices" both on the rear and front wheels capable of forming increased stabilizing properties on the rear wheels;
18. Using a special design of the drive controlling the turn of the rear wheels.

14.5 Steady motion stability in a multi-axle wheeled chassis

Initially, we considered the motion at a varying speed along a trajectory with a permanent curvature. In this case, the condition in (14.83) is observed and we can consider the system of motion equations in the form of (14.84).

Let us write these equations in a more compact form:

$$\begin{cases} \Delta\ddot{v}_y + p_2(t)\Delta\dot{v}_y + q_2(t)\Delta v_y = 0 \\ \Delta\ddot{\omega}_z + p_1(t)\Delta\dot{\omega}_z + q_1(t)\Delta\omega_z = 0 \\ p_1(t) = \left[\dfrac{(j_x + E + F)}{v_x}\right] \\ q_1(t) = \left[H - \dfrac{HG - EF}{v_x^2}\right] \\ p_2(t) = \left[\dfrac{j_x + F + E}{v_x} + \dfrac{2v_x j_x}{G - v_x^2}\right] \\ q_2(t) = \left[H - \dfrac{HG + EF}{v_x^2} + \dfrac{2j_x F}{(G - v_x^2)}\right] \end{cases} \tag{14.90}$$

If we consider the case of j_x = constant, the coefficients' numerators in the equations in (14.90) will be constant values.

The specific power in the majority of wheel chassis is small. In the lightest vehicles it does not exceed 20 hp/t (for example, the BAZ-69501 chassis has a specific power of 14.9 hp/t and the BAZ-135LM chassis has a specific power of 18.0 hp/t). As a result, when moving at a speed of 20–30 km/h, the acceleration intensity does not exceed 0.1–0.2 g.

Reference [61] presents a study of multi-axle wheeled vehicles with different dimensions, general layouts, and design solutions that have shown that $|H| \leq 20$.

In accordance with the study's plan, the solution to equations similar to (14.90) is searched in the following form:

$$y(t) = \exp[\lambda(t)t] \qquad (14.91)$$

After substituting (14.91), the system equations in (14.90) are transformed into the following summarized structure:

$$e^{\lambda t}\left[2\dot{\lambda} + t\ddot{\lambda} + \lambda^2 + 2\lambda\dot{\lambda}t + \left(t\dot{\lambda}\right)^2 + \lambda p(t) + t\dot{\lambda}p(t) + q(t)\right] = 0 \qquad (14.92)$$

According to D.A Antonov [61], at $v_x > 10$ m/s the parameter *λ(t)* slightly varies with the change in the travel speed. It is also shown that:

$$2\dot{\lambda}(t) + t\ddot{\lambda}(t) \approx 0 \qquad (14.93)$$

This means that we can consider further the approximated variant of equation (14.92):

$$\lambda^2 + 2\lambda\dot{\lambda}t + \left(t\dot{\lambda}\right)^2 + \lambda p(t) + t\dot{\lambda}p(t) + q(t) = 0 \qquad (14.94)$$

Yu.A Stepanov [63] does not recommend using equation (14.94) for defining the perturbed motion parameters; however, it is shown that when assessing the conditions of motion stability loss, the error in defining the critical speed lies within 4–6% and the speed values are somewhat understated.

D.A. Antonov has designed the following method for solving equation (14.94).

It follows from this equation that:

$$\lambda(t) + t\dot{\lambda}(t) = -\frac{p(t)}{2} \pm \sqrt{\left(\frac{p(t)}{2}\right)^2 - q(t)} = f(t) \qquad (14.95)$$

By substituting *λ(t)* = *u(t)z(t)*, we bring equation (14.95) to the form:

$$u(t)\dot{z}(t) + z(t)[\dot{u}(t) + u(t)/t] = f(t)/t \qquad (14.96)$$

The function *u(t)* is selected so that the member in the square bracket turns into zero. This equation has a general solution of the form:

$$u(t) = \frac{1}{C_1 t} \qquad (14.97)$$

In this case we have:

$$z(t) = C_1 \int f(t)dt + C_2 \qquad (14.98)$$

from where:

$$\lambda(t) = \frac{1}{C_1 t}\left[C_1 \int f(t)dt + C_2\right] = \frac{1}{t}\left[\int f(t)dt + C_3\right] \qquad (14.99)$$

In order to build the chassis' motion stability criterion relative to the parameter $\Delta\omega_z$, it is necessary to calculate integral (14.99) for the function *f(t)*, which corresponds to the second system equation in (14.90). By introducing the notation $\alpha = j_x + E + H$, $\beta^2 = \alpha^2 - 4(EF - GH)$ and defining the integration constants from the condition $v_x(t = 0) = v_{x0}$, we obtain the following (the routine intermediary transformations are omitted, see details in *Model Planetary Gearbox for Heavy-Duty Vehicles* [63]):

$$\lambda_{1,2}(t) = \frac{1}{2(v_x - v_{x0})}\left[(\alpha \mp \beta)\ln\frac{v_{x0}}{v_x} \pm\right.$$

$$\pm\left(\sqrt{\beta^2 - 4Hv_x^{\ 2}} - \sqrt{\beta^2 - 4Hv_{x0}^{\ 2}} - \right.$$

$$\left. - \beta \ln \frac{\beta + \sqrt{\beta^2 - 4Hv_x^{\ 2}}}{\beta + \sqrt{\beta^2 - 4Hv_x^{\ 2}}}\right)\Bigg] \tag{14.100}$$

Now, the private solutions can be built in the form of (14.91).

It is easy to be convinced that:

$$\begin{vmatrix} y_1(t) & y_2(t) \\ \dot{y}_1(t) & \dot{y}_2(t) \end{vmatrix} = e^{\lambda_1 t} e^{\lambda_2 t} \begin{vmatrix} 1 & 1 \\ \lambda_1(t) + t\dot{\lambda}_1(t) & \lambda_2(t) + t\dot{\lambda}_2(t) \end{vmatrix} =$$
$$= \left[\lambda_2(t) - \lambda_1(t)\right] + t\left[\dot{\lambda}_2(t) - \dot{\lambda}_1(t)\right] \tag{14.101}$$

and:

$$\left[\lambda_2(t) - \lambda_1(t)\right] + t\left[\dot{\lambda}_2(t) - \dot{\lambda}_1(t)\right] = -\frac{\sqrt{\beta^2 - 4Hv_x^{\ 2}}}{v_x}$$

The subradical expression may turn into zero only at positive *H*. However, in this case, the chassis has a negative tire turnability and stability; therefore, by definition there is no need to specifically study this case.

At $H = 0$, the chassis has a neutral tire turnability. The subradical expression turns into zero at $\beta^2 = 0$, which is only possible at the steady motion (this case has already been considered).

Thus, only at a positive tire turnability dynamic, when the chassis is prone to lateral skidding and it makes sense to study its motion stability, is the condition $H < 0$ observed. The determinant in (14.101) differs from zero and the solutions of (14.91)–(14.101) are linearly independent.

The general solution can be built in the following form:

$$\Delta\omega_w = C_1 \exp[\lambda_1(t)] + C_2 \exp[\lambda_2(t)] \tag{14.102}$$

The stability condition for (14.102) looks as follows:

$$\lambda_{1,2}(t) < 0 \tag{14.103}$$

If we assume that acceleration starts from the critical speed, it can be defined from the condition in (14.103) as follows:

$$\lim_{v_x \to v_{x0}} \left(\lambda_{1,2}(t)\right) < 0$$

This limit can be easily defined by using formula (14.100). By omitting technical details, the final result of transformations only is given below:

$$-\frac{\alpha \mp \beta}{v_x} \mp \frac{4Hv_x}{\beta + \sqrt{\beta^2 - 4Hv_x^{\ 2}}} < 0 \tag{14.104}$$

It is easy to see that the first element in (14.104) is always positive. The sign and value of the second summand are defined by the following parameter:

$$H = -\frac{\sum_{i=1}^{n} l_i K_{yi}}{I_z} \tag{14.105}$$

Assume the chassis has an insufficient additional turnability in the steady motion. In this case, the numerator of formula (14.105) is negative and, accordingly, $H > 0$. If the chassis has a neutral turnability in the steady motion, then $H = 0$. At an excessive turnability, $H < 0$.

At a neutral turnability in the steady motion, the condition in (14.104) is always observed.

At insufficient turnability, (14.104) is converted as follows:

$$\pm \frac{4Hv_x^{\ 2}}{\beta + \sqrt{\beta^2 - 4Hv_x^{\ 2}}} < \alpha \pm \beta \tag{14.106}$$

with fulfillment of the mandatory condition:

$$\beta^2 - 4Hv_x^{\ 2} \geq 0 \tag{14.107}$$

The observance of conditions (14.106) and (14.107) at the lower signs is obvious, as $H < 0$. At the upper signs we have:

$$\frac{4Hv_x^2}{\beta+\sqrt{\beta^2-4Hv_x^2}} < \frac{\beta^2}{\beta+\sqrt{\beta^2-4Hv_x^2}} =$$

$$= \frac{\beta}{1+\sqrt{1-\frac{4Hv_x^2}{\beta^2}}} < \beta < \alpha+\beta$$

and the condition (14.104) is also observed.

In case of additional excessive turnability, the condition (14.104) for lower signs is also always observed since $H < 0$. At the upper signs, the condition (14.106) can be written as follows:

$$-\frac{4Hv_x^2}{\alpha-\beta} - \beta < \sqrt{\beta^2-4Hv_x^2} \qquad (14.108)$$

It is shown by Yu.A Stepanov [63] that along with (14.108) the following inequality is valid:

$$\left(-\frac{4Hv_x^2}{\alpha-\beta}-\beta\right)^2 < \left(\sqrt{\beta^2-4Hv_x^2}\right)$$

From here we can obtain an expression for defining the critical travel speed for $\Delta\omega_z$:

$$v_x < \sqrt{\frac{GH-EF}{H}} \qquad (14.109)$$

Thus, for a chassis with a neutral and insufficient additional turnability, the stability criterion is always observed and there is no critical speed.

For a chassis with additional excessive turnability, the criterion for defining the critical speed is not structurally different from the same criterion for the steady motion. Further, the critical speed in this case does not formally depend on the acceleration. Still, because of the difference in the tangential loads acting on the chassis' wheels at a uniform motion and at acceleration, the results of calculating the critical speed by formulas (14.88) and (14.109) will be different, and, accordingly, there is an implicit dependence of the non-stationary motion's critical speed on the chassis' longitudinal acceleration.

In particular, for example, a vehicle with front wheel drive can improve its motion stability characteristics at acceleration by decreasing the correction coefficient on the front wheels' tangential loads and, accordingly, by decreasing the resistance coefficient to lateral skidding. On the contrary, for the same reasons, a vehicle with driven rear wheels will have a decreased critical speed at acceleration.

The structure of the perturbed motion equations relative to parameters $\Delta\omega_z$ and Δv_y is the same; therefore, building the stability criterion of the non-stationary motion for them is the same up to formula (14.99). As applied to parameter $\Delta v_{y,}$ we have:

$$\int f(t)dt = -\int\left(\frac{j_x+E+F}{2v_x}-\frac{j_x v_x}{v_x^2-G}\right)\pm$$
$$\pm\sqrt{\left(\frac{j_x+E+F}{2v_x}-\frac{j_x v_x}{v_x^2-G}\right)^2 - H\frac{v_x^2-G}{v_x^2}-\frac{EF}{v_x^2}+\frac{2j_xF}{v_x^2-G}}dt$$

As earlier, we assume that $v_x^2 - G \approx v_x^2$. In this case:

$$\int f(t)dt = -\int\left(\frac{E+F-j_x}{2v_x}\right)\pm$$
$$\pm\frac{1}{2v_x}\sqrt{(E+F-j_x)^2-4Hv_x^2-4EF+8j_xF}\frac{dv_x}{j_x}=$$
$$= -\int\left(\frac{\alpha-2j_x}{2j_x}\right)\frac{dv_x}{v_x}\pm$$
$$\pm\int\frac{1}{2j_x}\sqrt{(E+F-j_x)^2-4Hv_x^2-4EF+8j_xF}\frac{dv_x}{v_x}=$$
$$= -\frac{\alpha-2j_x}{2j_x}\int\frac{dv_x}{v_x}\pm\int\frac{1}{2j_x}\sqrt{\gamma^2-4Hv_x^2}\frac{dv_x}{v_x}$$

By using integration and defining the integration constants for the initial conditions $v_x(t = 0) = v_{x0}$, we get:

$$\lambda_{1,2}(e) = \frac{1}{2(v_x - v_{x0})}\left\{(\alpha - 2j_x \pm \gamma)\ln\frac{v_{x0}}{v_x} \pm \right.$$

$$\pm\left[\sqrt{\gamma^2 - 4Hv_x^{\ 2}} - \sqrt{\gamma^2 - 4Hv_{x0}^{\ 2}} - \right.$$

$$\left.\left. -\gamma\ln\frac{\gamma + \sqrt{\gamma^2 - 4Hv_x^{\ 2}}}{\gamma + \sqrt{\gamma^2 - 4Hv_{x0}^{\ 2}}}\right]\right\}$$

Following the above methodology, we define the stability criterion from the condition:

$$\lim_{v_{x0}\to v_x}\lambda_{1,2}(t) = -\frac{\alpha - 2j_x \mp \gamma}{2v_x} \pm$$

$$\frac{4Hv_x}{2(-\gamma - \sqrt{\gamma^2 - 4Hv_x^{\ 2}})} < 0$$

It follows from here that:

$$-(\alpha - 2j_x \mp \gamma) \pm \frac{4Hv_x^{\ 2}}{-\gamma - \sqrt{\gamma^2 - 4Hv_x^{\ 2}}} < 0 \tag{14.110}$$

At $H = 0$, we have:

$$\alpha - 2j_x \mp \gamma > 0$$

that is why condition (14.110) is always observed.

At $H > 0$, it is necessary to additionally ensure observance of the condition $\gamma^2 - 4Hv_x^{\ 2} \geq 0$.

It is obvious that:

$$\frac{4Hv_x^{\ 2}}{\gamma + \sqrt{\gamma^2 - 4Hv_x^{\ 2}}} \leq \frac{\gamma^2}{\gamma + \sqrt{\gamma^2 - 4Hv_x^{\ 2}}} \leq \gamma < \alpha - 2j_x \mp \gamma$$

that is why the inequality (14.110) is always valid.

Finally, at $H < 0$, formula (14.110) with lower signs is also always valid.

For the case with upper signs, we have:

$$-\frac{4Hv_x^{\ 2}}{(\alpha - 2j_x - \gamma)} - \gamma < \sqrt{\gamma^2 - 4Hv_x^{\ 2}}$$

It is clear that at negative H:

$$\begin{cases} -\dfrac{4Hv_x^{\ 2}}{(\alpha - 2j_x - \gamma)} > 0 \\ \gamma < \sqrt{\gamma^2 - 4Hv_x^{\ 2}} \end{cases} \tag{14.111}$$

That is why:

$$\left|-\frac{4Hv_x^{\ 2}}{(\alpha - 2j_x - \gamma)} - \gamma\right| < \sqrt{\gamma^2 - 4Hv_x^{\ 2}}$$

and, accordingly, after squaring the right and left parts of (14.111) and simplifying, we can have:

$$-\frac{4Hv_x^{\ 2}}{(\alpha - 2j_x - \gamma)^2} - \frac{2\gamma}{\alpha - 2j_x - \gamma} < 1$$

From here, it is easy to define the critical speed:

$$v_x < \sqrt{\frac{EF - 2j_xF}{-H}} \tag{14.112}$$

Yu.A Stepanov [63] shows that, in case we reject the assumption that $G = 0$, formula (14.112) can be presented as follows:

$$v_x < \sqrt{\frac{EF - GH - 2j_xF}{-H}}$$

This formula can be put down in the following form, given below in 14.113, which can be used at technical calculations and evaluations.

If the acceleration is caused by the effect of tangential forces in the wheels' contact zone with the road, then the dependence of the critical speed from the acceleration can be quite essential because of a decrease in the resistance coefficient to the skidding of the wheels loaded by tangential forces. If the acceleration is caused by other reasons, its effect on the critical speed is small.

It is interesting that at acceleration the critical speed for Δv_y is smaller than the steady motion's critical speed and the critical speed for $\Delta\omega_z$. This is manifested in the fact that at acceleration and stability loss, the chassis initially starts "floating" across the carriageway and only then gets into lateral skidding. At deceleration, on the contrary, we see first the chassis' lateral skidding.

14.6 Non-stationary motion stability in a multi-axle wheeled chassis

Great difficulties arise when studying this problem. To simplify the process, D.A. Antonov has offered the following system of restrictions and assumptions:

- The motion is considered at the moment when stability has not been lost yet ($\omega v_x \approx 0$; $d(\omega v_x)/dt \approx 0$);
- There is no power circulation in the transmission, and, moreover, the distribution of traction forces among the wheels is close to uniform, that is, the transmission diagram is close to the differential one ($\forall i,j : dR_{xi}/dt \approx dR_{xj}/dt$);
- The lateral force is defined in accordance with the lateral skid hypothesis and can be calculated under the "algebraic theory of lateral skid";
- The distribution of normal loads among the wheels is close to uniform, while the tires are practically the same ($\forall i,j : dR_{yi}/dt \approx dR_{yj}/dt$);
- The acceleration is made from the stop ($v_x = 0$); and
- The acceleration is made under a permanent acceleration increment $j_x = (dj_x/dt)_{t=0}t$.

Let us now present only the main idea of the steady (stationary) motion criteria [61].

The time derivatives of the longitudinal and lateral forces in the tire-road contact point are obtained by differentiating the equation describing the forces' projection onto the chassis' longitudinal axis and the equation formalizing the algebraic variant of the lateral breakaway hypothesis (restricting the problem to the cases under which it is admissible to believe that $(nl)/d \approx 1.5$). This allows for defining the derivatives from the correction coefficients of the resistance to lateral breakaway and evaluating the derivative $d(\partial R_{yi}/\partial\delta_i)/dt$. By differentiating (14.80) with respect to time, we obtain (14.81), provided that $\mathring{H} = 0, \mathring{E} \neq 0$. Further, the obtained equations are derived in the same way as before.

For the parameter $\Delta\omega_z$, the stability criterion is obtained in the following form:

$$v_x^2 < \frac{FE - GH}{-H}$$

It is not different in its form from the motion stability criterion at a steady turn. The difference in the speed values is possible because of differences in the wheel's contact loading at the steady and non-steady acceleration and because of the respective difference in their resistance coefficients to lateral breakaway.

For the parameter Δv_y, the stability criterion looks as follows:

$$v_x^2 < \frac{FE - GH - 2j_xF}{-H} - 2\frac{\dot{G}F}{Hj_x^2}\left(\frac{dj_x}{dt}\right)_{t=0} - \frac{\dot{F}j_x^2}{2H\left(\frac{dj_x}{dt}\right)_{t=0}}$$

where the second member is very small and can be neglected.

Yu.A Stepanov [63] shows that the critical speed values at a non-steady acceleration are lower than at a steady acceleration. Further, multi-axle chassis in general accelerate rather slowly (except certain special vehicles like heavy fire-fighting machines intended to extinguish emergency aircraft fires in the airport responsibility zones). Keeping these considerations in mind at the initial design stages of multi-axle wheeled chassis, designers usually limit themselves to the steady motion stability assessment.

$$v_x < \sqrt{\frac{\sum_{i=1}^{n}\left[q_i K_{y0эi} l_i^2 \cos\theta_i\right]\sum_{i=1}^{n}\left[q_i K_{y0эi}\cos\theta_i\right] - \sum_{i=1}^{n}\left[q_i K_{y0эi} l_i \cos\theta_i\right]^2 - 2 j_x I_z \sum_{i=1}^{n}\left[q_i K_{y0эi}\cos\theta_i\right]}{m\sum_{i=1}^{n}\left[q_i K_{y0эi} l_i \cos\theta_i\right]}} \tag{14.113}$$

Chapter 15
Fundamental theory and design of HDWV steering systems

15.1 General provisions

The steering system (StS) is intended to ensure the vehicle's curvilinear motion and maintain the prescribed motion direction. The progress of StSs is associated with their use of automated control systems (ACSs). That is why the main attention in this chapter is given to the design dependences that reflect the control system's operation in a curvilinear motion from the viewpoint of vehicle theory and the justification and development of operational algorithms for ACSs.

The increase in the overall dimensions and the number of axles in HDWVs results in a decrease in their maneuvering properties. The main reason is in the standard approach to selecting the StS's design, irrespective of the number of axles and the vehicle's overall dimensions.

For example, modern chassis with four to seven axles mainly use StSs where the front axles' wheels are controlled (steered). The trend is towards increasing—along with the total number of axles—the number of front axles with controlled (steered) wheels.

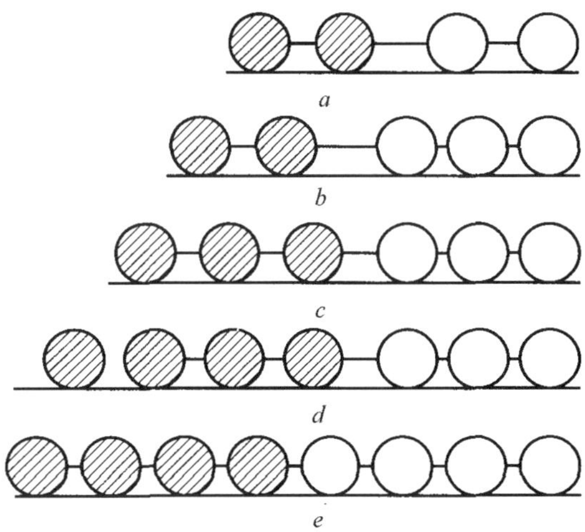

Fig. 15.1. Variants of controlled wheels used in known multi-axle SWCs: a—8×8 chassis, R_{min} = 13.5 (MAZ-543M); b—10×8 chassis, R_{min} = 18.5 (MAZ-7929); c—12×12 chassis, R_{min} = 22.5 (MAZ-547V); d—14×12 chassis, R_{min} = 26.5 m; e—16×16 chassis, R_{min} = 33.6 (MAZ-7906) (controlled wheels are shaded)

Fig. 15.1 presents variants of using controlled wheels in known multi-axle chassis with four to eight axles. The analysis suggests that an increase in the number of front controlled wheels (with an increase in their total number) does not improve the maneuvering properties of this type of SWCs. The minimum turning radius goes from 13.5 m in a four-axle chassis up to 33.6 m in an eight-axle one. The designs with front steered wheels are attractive only for their simple design of the StS drive.

An essential drawback of these designs is in the increased power expense when moving along curvilinear trajectories, especially where the curvature is comparable to the SWC's minimum turning radius. This has to do with the large turning resistance created by the non-controlled wheels of the rear bogie, which in the considered SWCs has a wheelbase of 2.5–3.5 m, i.e., 27–32% of the overall wheelbase.

The topicality of this problem is also confirmed by the fact that along with the growth in the SWC's full weight we see a drop in the specific power—a major indicator that defines the SWC's mobility. Table 10.7 presents the comparison of these indicators for the considered SWCs (indicators of the 8×8 chassis are taken as 100%). The analysis shows that with a 1.5–3.0-fold increase in the SWC's full weight, the power of its power unit goes up 1.24–1.52 times, while the specific power goes down 1.6–2.0 times.

In the SWC's motion, up to 10% of the engine's nominal power is spent on the installed equipment's needs. In this context a problem arises of a rational use of the power unit for maintaining the SWC's mobility in various conditions, including when moving along curvilinear trajectories.

One way to improve the SWC's maneuvering properties and the rational use of the engine's power when moving along curvilinear trajectories is through improved designs of the StS and ways of turning.

The methods to control turning can be dynamic or kinematical.

The dynamic control method implies regulating the angular velocities of the wheels of various boards (sides) without changing their position relative to the body (the so-called "tank-type turning"). In this case, different (in value and direction) longitudinal reactions are formed in the wheels of different boards, which create the turning moment.

The dynamic method simplifies the drive system of the driving wheels, increases the useful volume of the body (platform), and creates conditions for using large-diameter tires with the aim of improving the passing ability. Its application is expedient in vehicles with a short wheelbase. For the considered SWCs where the wheelbase is 7.7 m or more, using the dynamic control method as the main method is not acceptable because of an essential increase in the turning resistance with the increase in the SWC's wheelbase, which requires essential additional power for curvilinear travels.

In multi-axle vehicles with non-turning wheels, in order to reduce the turning resistance when moving along the minimum radii, lifting or lifted extreme axles as compared to the middle axles are sometimes used. However, the use of this concept in SWCs is problematic for the following reasons.

First, lifting the extreme axles will increase the load on the already heavily loaded middle axles. In SWCs with a load capacity of 20 and 50 t, the load on one axle is 100–150 kN; thus, any further increase in this parameter will result in an essential decrease in the passing ability when moving on deformable grounds.

Second, we need to install mechanisms on the chassis which will lift up and bring down the axles depending on the motion conditions. This design solution creates additional problems for SWCs, which have limited free area for installing special equipment and other support and functional systems.

The dynamic control method can be applied complimentarily to the kinematical method. In this case, for example, in vehicles with front controlled wheels and one-side braking of non-controlled wheels, it is possible to reduce the minimum turning radius by 13–15%. The same effect is achieved in short-wheelbase vehicles with no more than three axles. With an increase in the chassis' wheelbase, the efficiency of the additional power control method goes down, while the power needed for turning goes up. Also, to implement this method, we need a control system that ensures regulation of the braking intensity depending on the vehicle's motion mode and conditions.

The kinematical control method can include jackknifing the links of articulated vehicles in the horizontal plane. Its advantage is in an essential increase in the chassis' maneuverability since the turning radius is much smaller as compared to one-link constructions. However, using these vehicles for transporting long monocargoes is hardly possible.

The most widespread kinematical control method is turning the controlled wheels. This process can involve only the front wheels, the front and some of the rear wheels, or all of the vehicle's wheels. Using this control method in SWCs is desirable because of the possibility of increasing the travel speed and improving the maneuvering properties, as well as a simple design of the StS drive.

Turning the controlled wheels of two of the front axles is used in 8×8 SWCs (MAZ-543 and -7911 and their modifications and BAZ-6944, -6950, -69501, and -6953); the controlled wheels of three of the front axles are turned in 12×12 SWCs (MAZ-547V, -7905, and -7916); and the controlled wheels of four of the front axles are turned in 14×12 and 16×16 SWCs (MAZ-7912,-7917, and -7906).

The steering systems of certain SWC models control a number of rear axles along with the front controlled wheels. An example is the 8×8 ZIL-135LM chassis and its modifications, in which the turn is ensured by the controlled wheels of the first and last axles. A similar StS design was used in the 7904 oversized vehicle model. The perspective models of the 16×16 7922 and 7923 chassis have the wheels of three front and three rear axles controlled (steered). Their minimum turning radius is about 17.0 m. The minimum turning radius for the 7912 (17) chassis model, which has fewer axles (14×12) but controlled wheels of four front axles, is 26.5 m. The same minimum turning radius can be found in the 12-axle 7907 chassis with controlled wheels of four front and four rear axles.

The advantages in the maneuvering properties of StS diagrams with controlled wheels of the front and rear axles, as compared to those with only front controlled wheels, are well-known. However, they have an essential drawback: decreased motion stability on transient trajectories.

The reason is in different angular turning speeds—by the value and direction—of the chassis and the wheels of the rear axles, which results in the appearance of an excessive lateral acceleration with a respective short-term sign change in the lateral reactions of the rear axles' wheels at the beginning and an increase in these reactions at the end of this maneuver. These short-term pulses of lateral reactions are perceived by the tires as an impact load and are capable of resulting in the wheels losing contact with the support surface in the lateral direction and skidding.

One way to counteract this phenomenon is to use a turning delay in the rear axles' wheels relative to the front one by an angle of about 0.1 radians (5–6°). However, this design solution leads to a loss in the kinematics coordination during the turn since the wheels of the front and rear controlled axles form their instant turning center O^m_{fa} and O^m_{la} (Fig. 15.2) practically within the turning angle's entire range, while the united turning center of the front axles' wheels O^m is formed because of an additional tire breakaway.

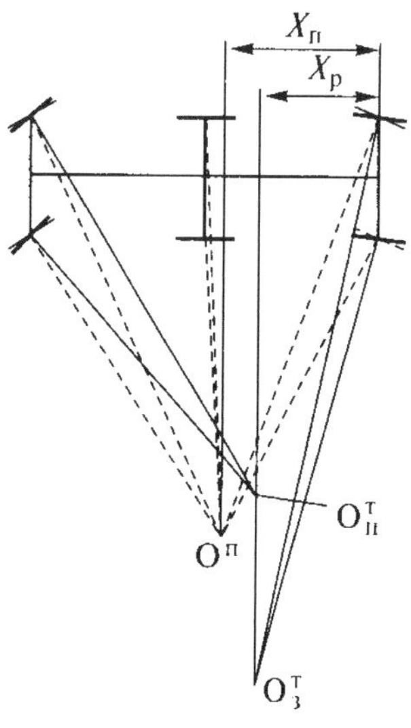

Fig. 15.2. Turning diagram of a SWC with a delay in turning the rear wheels: X_n—turning pole; X_p—the StS pole

As to the design implementation of the StS drive with a turning delay in the rear wheels, it should be noted that it becomes more complicated since it is necessary to add a mechanism for ensuring the delay and a stabilization device for rectilinear motion. If for two- and three-axle vehicles the steered rear wheels' drive does not look very complicated with these design additions, with an increase in the number of axles, the drive's complexity essentially increases. Even without the delay mechanism and stabilization device, the StS drive in a multi-axle chassis (for example, the 12×12 7904 chassis model with controlled wheels of two front and two rear axles) is a rather cumbersome construction (see Fig. 4.15), causing doubts about its ease and maintenance simplicity.

Another way to improve the motion stability of a vehicle with front and rear steered wheels is to regulate the ratio of the controlled wheels' turning angles depending on the motion conditions. The kinematics of this StS diagram is based on using special laws to manage the offset of the StS pole.

The StS pole is understood as the projection of the vehicle's turning pole onto its longitudinal axis.

Here, two main principles to regulate the ratio of the front and rear controlled wheels' turning angles are used, which were justified for the first time by V.G. Kornilov. According to the first, these ratios are regulated depending on the front wheels' turning angle and the travel speed. At a high speed and small turning angles of the front wheels, the rear axles' wheels turn to the same side as the front ones (in the in-phase manner). With an increase in the trajectories curvature and a decrease in the travel speed, the rear axles' wheels turn in the opposite direction relative to the front ones (in the anti-phase manner). The second principle differs from the first one in that it has no speed regulation of the ratios of the front and rear wheels' turning angles.

A StS with a regulated ratio of the controlled wheels' turning angles is of interest from the viewpoint of its possible use in multi-axle SWCs, which in accordance with their function have to move in various road conditions, off-road, in restricted areas (in parks, on construction sites, etc.), and within a broad range of travel speeds.

Using a flexible StS diagram, for example, with an electrohydraulic drive, in SWCs will allow for solving a number of problems: first, improving the maneuvering properties, improving the motion stability, and decreasing the power needed to travel along curvilinear trajectories.

Second, we can do without the bulky system of intermediate levers and rods.

Both considered principles of regulating the ratio of the controlled wheels' turning angles demand using an ACS system. In order to completely utilize the electronics' capabilities, we need a deep justification of its functionality algorithms in the flexible StS system.

That is why the development of functional algorithms of onboard complexes is one of the major problems in designing flexible StSs and their ACSs.

The presence of an onboard computer in the steering system allows for setting various control programs (software). Such StSs are used, for example, in multi-support HDWVs.

15.2 Basic requirements for steering systems

The StS should ensure the preset motion trajectory of its HDWV with observance of the required initial parameters of maneuverability, turning ability, and curvilinear and rectilinear stability travels.

The basic requirements for steering systems are as follows.

1. StSs should meet the requirements for the UNECE Rules, ISO standards, Russian GOSTs and OSTs, and other normative documents. According to all of them, the StS should ensure the following:

   a. An efficient long-term motion in preset conditions with minimum driver fatigability;
   b. A complete realization of the vehicle's performance properties, preset in its design;
   c. Safety motion;
   d. The preset level of quality indicators: handling quality, course motion stability, and maneuverability.
2. The general requirements for an all-wheel StS with a flexible turning geometry should assume the possibility of the vehicle acquiring a number of new properties:
   a. Performance of different turning diagrams with a discrete or continuous transition from one diagram to another;
   b. Preserving the driver's traditional feelings and skills in the process of steering.
3. The special requirements for flexible steering systems apart from those listed in point two are as follows:
   a. The vehicle's motion stability on the course;
   b. The possibility of automated driving under certain programs or in interaction with the motion automated control system (MACS) (in particular, automatic execution of standard performance maneuvers, use of external radio control of the vehicle, support of automatic motion in columns, etc.);
   c. The presence of back-up (reserve) systems;
   d. The presence of built-in control and failure indication systems.

The experience accumulated by MSTU in the course of scientific studies and researches aimed at improving the maneuverability of long-base multi-axle chassis allows us to assert that the task to make such StSs is resolvable at the present level. In this case, the vehicle is equipped with an automated onboard StS, which in general is comprised of the following subsystems:

- A steering control unit (steering column) with an automatic loading device of the steering wheel;
- An onboard computer with a system of analog-digital conversion (ADC) of signals (if necessary);
- A block of executive mechanisms (steering devices or units);
- A work indication system; and
- A power supply system.

The steering control unit is a device that helps to convert the driver's control actions into electric signals that are sent to the StS's electric part.

The loading steering wheel mechanism is needed to imitate the turning resistance of the steering wheel and form in the driver the so-called "feeling of the road." Apart from the mechanical non-controlled loader, this system may include additional controlled devices that change the effort applied to the steering wheel under some preset algorithm.

The onboard computer uses the data arriving from the steering block sensors and forms control signals for the steering aggregates. Depending on the planned maneuver and particular road situation, the driver chooses the required program by using the control algorithm's continuous or discrete setter. This algorithm is intended for supporting the work of the onboard computer when calculating the turning signal for the turn of each controlled wheel depending on the turn of the steering wheel at the varying position of the turning poles.

15.3 Initial data for StS calculation

To calculate a StS and design a functionality algorithm of its ACS, one should know the possible external usage conditions of the respective HDWV (see Chapter 8) and the data from which the design decisions were adopted aimed at meeting the requirements for an all-wheel StS with a flexible turning geometry (see subpoints 15.2.2 and 15.2.3 above). It especially refers to selecting the kinematical transmission ratio and reaching the vehicle's required sensitivity to steering.

These calculations require a thorough comparison with the statistical data results from experimental observations.

Historically, we have a certain sensitivity range for the StSs used in automobile designs. For a long time it was not specially calculated and the needed values of the kinematical transmission ratio were selected based on the required effort on the steering wheel.

15.3.1 Kinematical transmission ratio

The statistical studies conducted at MSTU have defined the dispersion range of the steering wheel's full travel for various vehicles types and different road conditions; this range allows for being more precise in selecting the kinematical transmission ratio (Figs. 15.3–15.6).

15.3.2 Sensitivity of vehicle to steering

The vehicle's sensitivity to steering is the property that reacts to the driver's control action at stationary motion. This is one of the main characteristics

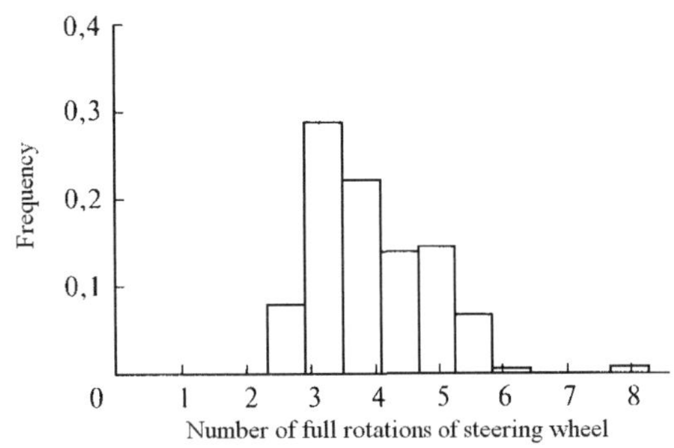

Fig. 15.3. Dispersion of the steering wheel's full travel (number of full rotations) for different vehicle categories

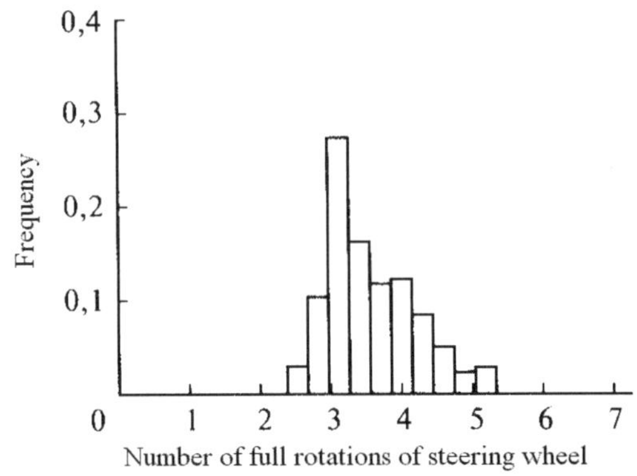

Fig. 15.4. Dispersion of the steering wheel's full travel (number of full rotations) for passenger cars

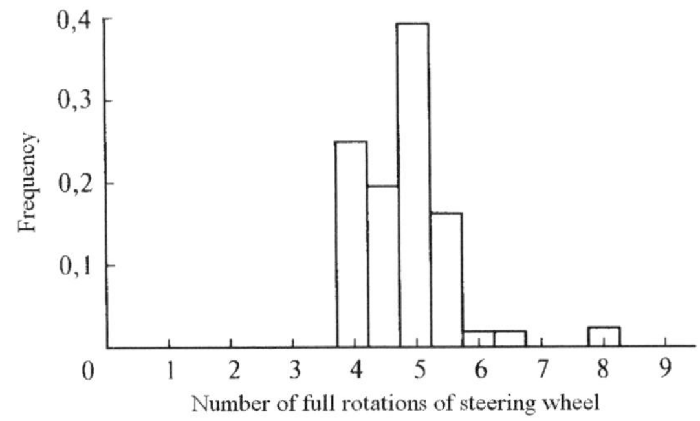

Fig. 15.5. Dispersion of the steering wheel's full travel (number of full rotations) for cargo vehicles

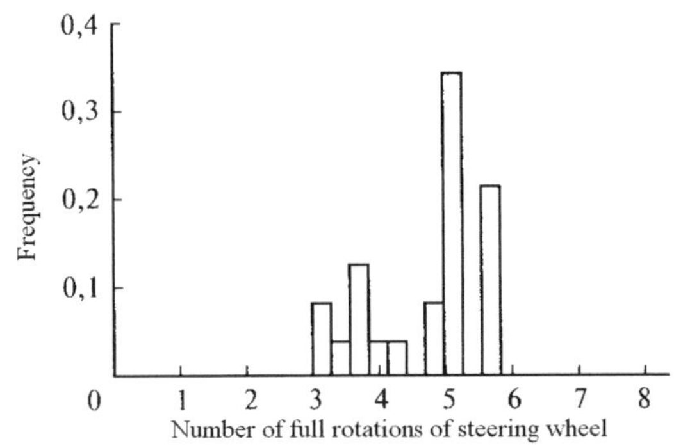

Fig. 15.6. Dispersion of the steering wheel's full travel (number of full rotations) for buses

defining the vehicle's handling quality. The sensitivity links a change in the curvilinear trajectory (or the turning radius) with the steering wheel's rotation; it is defined as follows:

$$C_h = dK_c / d\beta \, , \qquad (15.1)$$

where: K_c is the trajectory's curvature; and β is the steering wheel's turning angle.

The higher the sensitivity, the more actively the vehicle reacts to the driver's control action.

The trajectory curvature's change rate depends not only on the StS's kinematical transmission ratio, but also on other design factors such as the wheelbase, the wheelspan, possible turning angles of the controlled wheels, etc. Thus, the sensitivity allows for assigning particular properties to the vehicle under design.

For instance, when moving at a high speed, a low sensitivity is desirable so that an occasional turn of the steering wheel, for example, caused by a road irregularity, will not result in the vehicle's sharp turn and in creating an accidental situation. When maneuvering, on the contrary, we need a high sensitivity, which ensures the vehicle's quick reaction to the steering wheel's turn. These properties are especially important for long-base multi-axle or multi-support vehicles.

One of the most efficient ways of obtaining the required sensitivity characteristic is to design a StS with a flexible turning geometry, which provides the vehicle with the required properties.

The statistical studies conducted at MSTU have defined a generalized dispersion range of the steering sensitivity values for various vehicles. The results of these studies are presented in Figs. 15.7–15.10, where: M_o is the mathematical expectation, D is the dispersion, and S is the mean root square deviation of sensitivity. The sensitivity of the designed HDWV should fall within the real range available in the existing automobile fleet.

The sensitivity defined according to expression (15.1) above is not a constant value—it slightly varies depending on the steering wheel's turning angle. Fig. 15.11 presents the respective data for

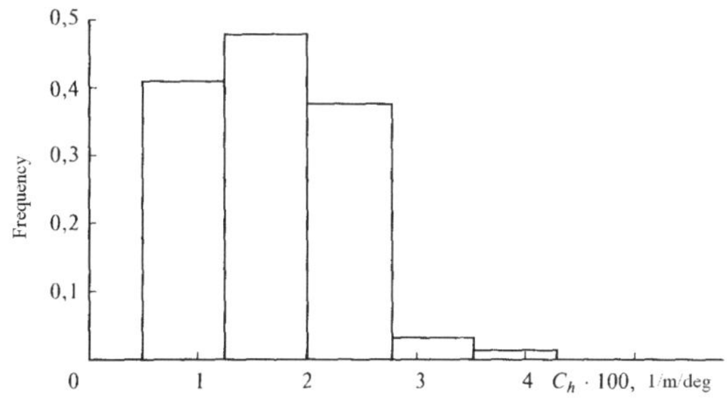

Fig. 15.7. Distribution of sensitivity to steering generalized for all categories of vehicles: M_o = 1.64; D = 0.44; and S = 0.66

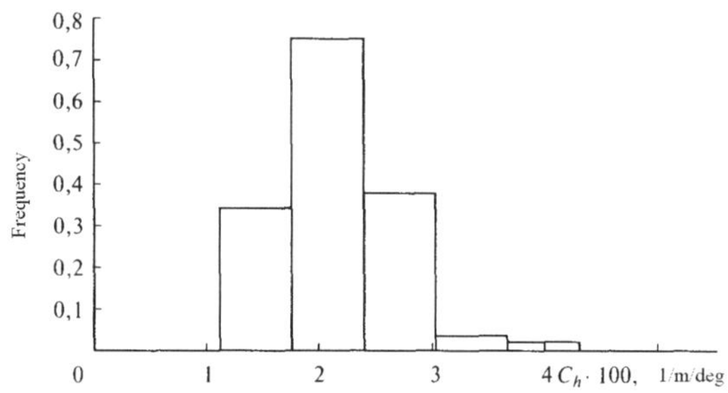

Fig. 15.8. Distribution of sensitivity to steering for passenger cars: M_o = 2.64; D = 0.33; and S = 0.68

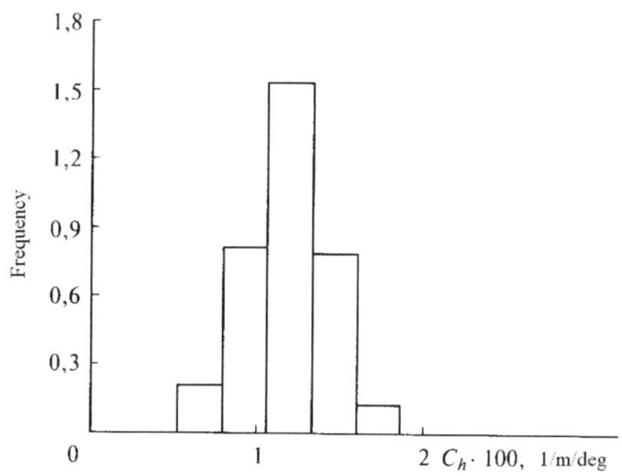

Fig. 15.9. Distribution of sensitivity to steering for cargo vehicles: M_o = 1.17; D = 0.09; and S = 0.30

certain vehicles with increased passing ability, as well as for the 12×12 prototype designed at MSTU for the two most typical motion regimes:

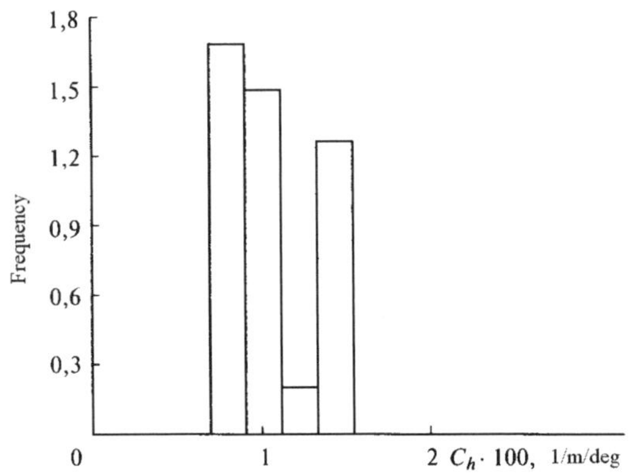

Fig. 15.10. Distribution of sensitivity to steering for buses: M_o = 1.05; D = 0.11; and S = 0.34

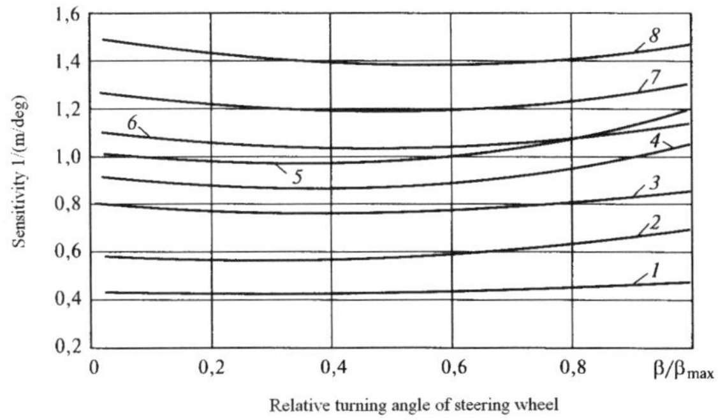

Fig. 15.11. Sensitivity to steering versus the steering wheel's relative turning angle: 1—MAZ-547; 2—12×12 prototype (pole in the middle of the rear bogie on the 5th axle); 3—KrAZ-260; 4—12×12 prototype (pole in the middle of the base); 5—Ural-375; 6—Tatra-148; 7—ZIL-131; 8—GAZ-66

a turn with the minimum radius (the turning pole is in the middle of the base) and the transportation regime (the pole is in the middle of the rear bogie). For comparison, the data for the MAZ-547 prototype are presented.

It follows from Fig. 15.11 that there is the potential to decrease the kinematical transmission ratio of the designed StS, while keeping the sensitivity within reasonable limits.

The issue of choosing the StS's sensitivity is not so simple. One should bear in mind that an increase in the steering sensitivity in large-size vehicles inevitably results in an increased energy demand for the StS. The adequacy of the driver's actions under increased steering sensitivity in an oversized vehicle is in need of additional studies. The rational dependence of the StS's transmission ratio on the steering wheel's turning angle, which forms the sensitivity, is presented in Fig. 15.12.

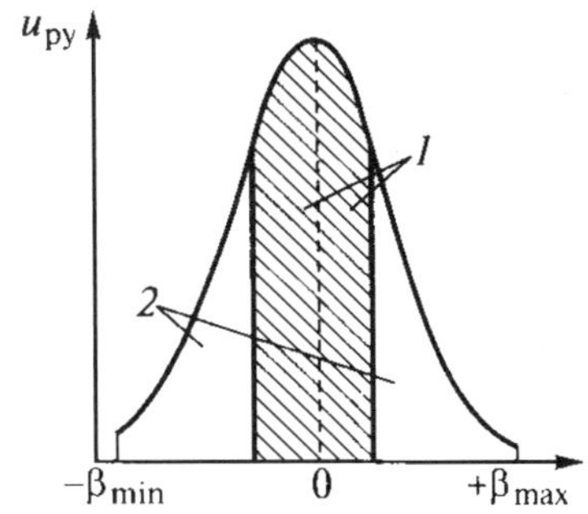

Fig. 15.12. Rational form of the StS's transmission ratio versus the steering wheel's turning angle: 1—low sensitivity area; 2—high sensitivity area

15.3.3 Effort on the steering wheel

The effort needed to turn the steering wheel indirectly informs the driver about the controlled wheels' turning angle via the power feedback. It is regulated by respective standards and should not exceed 245 N. The loading mechanism for the steering wheel should be designed based on the requirements of these documents and the particular steering mechanism (reduction gearbox) used.

The 12×12 running prototype designed at MSTU is equipped with a non-controlled loading device with a linear characteristic and can be regulated. The effort on the steering wheel's rim at its maximum turn is 100–150 N. The required damping is ensured by the worm steering mechanism.

15.3.4 Energy parameters

The issues of the energy demand on the controlled wheels' individual turn drive include the following aspects:

- Evaluating the maximum real wheel resistance-to-turn moment at stop to define the hydraulic drive's required working pressure and the power cylinders' parameters;
- Evaluating the controlled wheels' maximum real possible angular turning rate to define the working fluid's required flow rate and the power cylinders' parameters;
- Evaluating the power needed to turn the controlled wheels.

15.3.5 Evaluation of the maximum resistance moment to turn the wheel

Technical literature offers different approaches to defining the maximum resistance moment to turning the vehicle's controlled (steered) wheels. For example, the moment is evaluated when we turn the wheels of a fixed vehicle standing on a dry asphalt-concrete pavement or we can use the calculated and calculated-experimental dependences at various assumptions, such as disregarding the influence of the tire tread, the wheel roll around the pivot axle, etc. Thus, we can conclude that today there is no unified theoretical justification for calculating the maximum resistance moment to turning the controlled wheels.

To confirm this conclusion, let us consider several expressions presented in various sources.

The following initial data were taken for calculating the evaluation of the 12×12 MAZ-547V chassis prototype:

- The vertical load on the wheel $P_w = 70$ kN;
- The adhesion coefficient $\varphi = 0.8$;
- The contact spot's length $l = 0.5$ m;
- The contact spot's width $s = 0.3$ m;
- The wheel's free radius $r_{free} = 0.8$ m; and
- The roll radius in the free mode $r_k = 0.77$ m.

P.V. Aksyonov [1] presents the expression for the wheel turn's resistance moment borrowed from the motion theory of caterpillar vehicles:

$$M_{rtw} = 0.375\,\varphi\,P_w 3.14 l\,s/4.$$

By substituting our initial data, we obtain the following: $M_{rtw} = 6.6$ kN×m.

By using the following expression from L.A. Ginsburg:

$$M_{rtw} = 0.5\,P_w \varphi\,(r_{free} - r_k),$$

we obtain $M_{rtw} = 6.2$ kN×m.

A.S. Litvinov has offered to calculate the expression with admission of the rectangular epure (curve) of normal reactions in the contact spot:

$$M_{rtw} = \varphi\,l\,P_w/4.$$

From here: $M_{rtw|} = 7$ kN×m.

And, finally, G.A. Smirnov [12] defines the resistance moment to the wheel turn by the following expression:

$$M_{rtw} = 0.14\,\varphi\,P_w r_k.$$

After substituting our initial data, we have: $M_{rtw} = 6.2$ kN×m.

As we can see, the results obtained according to different sources for the value of M_{rtw} are quite close and fall within the limit of 6.2–7.0 kN×m. It should be noted here that the assumption adopted above increases the value of the moment. The specific moment (the moment referring to the vertical load on the wheel) or the specific arm is about 0.09–0.1 m.

A R&D study conducted by MSTU experimentally defined some energy parameters of the controlled wheels' turn. The study objects were the heavy-duty KamAZ-43101 (standard version) with a 43 kN load on the front axle (the energy parameters of the wheels' turn were defined) and the 8×8 79085 running prototype with an individual turn drive for each wheel and a 70 kN load on the wheel.

The experimental data for the KamAZ give the value of M_{rtw} = 3.5–4.2 kN×m, which corresponds to a specific arm of 0.08–0.1 m. This is a good match with the above calculations with account of the above note about assumptions.

As to the 79085 prototype, the resistance moment to the wheel turn was not directly defined. However, we calculated the required and measured angular velocity of the wheels' turn, the drive power needed to turn the wheels, and also the linear speed and displacement of the drive's exit link rod and the effort thereon at turning the steering wheel. As a result of the calculation-experimental studies, we obtained that the StS of the 12×12 prototype with a 75 kN load on the wheel the M_{rtw} = 8 kN×m.

15.3.6 Evaluation of the controlled wheel's angular turning rate

The standards demand the maximum possible rate of displacing the steering wheel without changing the grip of 10 1/s. The transmission ratios of the steering mechanisms in most vehicles are within the limit of 18–22. It is easy to define that the changing range of the wheels' angular turning rate is within the limit of 0.45–0.5 1/s.

The values of the wheels' angular turning rate in various motion regimes, obtained in the course of experimental studies, are as follows: for KamAZ-43101 the maximum registered value was 0.52 1/s and for the 8×8 79085 prototype at turning on the place was 0.43 1/s.

15.4 Selection and mathematical justification of functionality of flexible StS

Designs of steering systems are treated as flexible (non-traditional) when the StS pole can move (unlike traditional systems) along the HDWV's longitudinal axis following a certain law, depending on one or several parameters (Fig. 15.13). The choice of the control law is based on the results of a large number of theoretical studies, the practice of domestic and foreign motor companies, and calculations for various types of laws with the help of the analog computing complex AVK-2 (3) by N.M. Nazarov and V.G. Kornilov. For comparing different StS designs with regulated pole displacement (RPD), they also used some traditional designs. Preliminary studies resulted in the choice of two versions of the law intended to control steering systems.

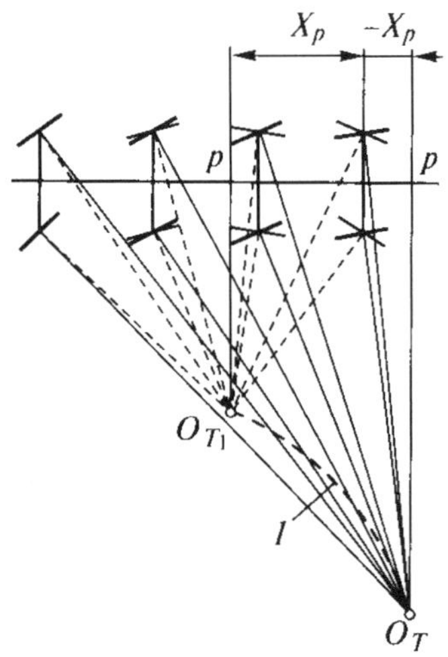

Fig. 15.13. Kinematical diagrams of StS with a mobile pole (software-based change in the turning pole's position): O_T and O_{T1}—extreme positions of the turning center; p and p'—extreme positions of the turning pole; 1—displacement of the turning pole p' depending on the position of the StS pole on the vehicle axis

15.4.1 Mathematical justification of the RPD-AS functioning laws

In the first of these laws, the position of the StS pole on the HDWV's longitudinal axis depends on two parameters: the vehicle's travel speed v_x and the setting turning angle of the setting axle's conventional middle wheel α:

$$X_p = K_1 |\alpha|^K.$$

The respective design was conventionally called RPD-AS (regulation of pole displacement by angle and speed) in the process of the studies. A stable motion of the HDWV along transient trajectories at high speeds can be improved by turning the rear wheels to a certain angle in the same direction as the front wheels. The kinematics of the StS is ensured by displacing the StS pole backwards outside the chassis' base (Fig. 15.13).

With an increase in the turning angle of the setting (first) wheel α_1 or (and) a decrease in the travel speed v_x, the StS pole should move towards the center of the base in order to ensure the turn of the front and rear wheels in counter-phase, thus improving the vehicle's maneuverability and turning ability.

Based on the above deliberations, the mathematical models presented in point 8.4 and Chapter 14, and the data presented in Fig. 15.14, we can formulate the law for displacing the StS pole depending on the setting angle α and the travel speed v_x.

At small travel speeds, the requirements for the turning ability and maneuverability are dominant; therefore, no in-phase turn of the front and rear wheels is required. At travel modes with an increased trajectory curvature, it is desirable to move the StS pole to the base's center. Then, we set the initial displacement of the StS pole for the infinitely small travel speed as equal to zero $X_p = 0$ (point *A*, see Fig. 15.14). With the increase in the setting angle α up to the maximum value a_{max}, the StS pole should move into the middle of the base (point *B*, see Fig. 15.14). Also, for the sake of investigation it was assumed that the StS pole's displacement from the initial position (point *A*) to the base's center (point *B*) occurs along different trajectories (curves *1*, *2*, and *3* in Fig. 15.14). Then, the current

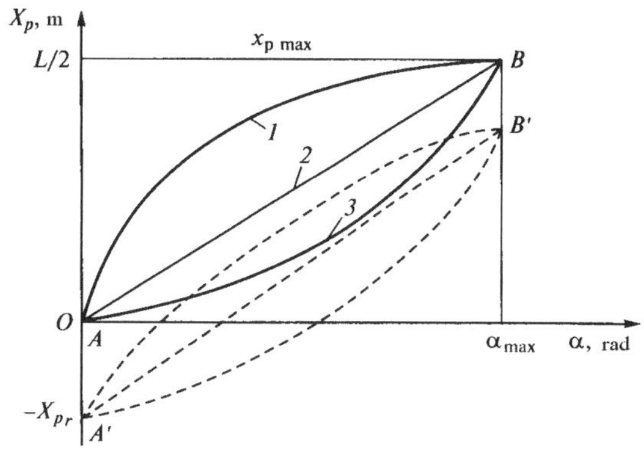

Fig. 15.14. Variants of control by displacing the StS pole: 1, 2, and 3—parabolic, linear, and hyperbolic laws

displacement of the StS pole will be defined as follows:

$$X_p = K_1 |\alpha|^K,$$

where: *K* is the power indicator defining the curve's form; and K_1 is the proportionality coefficient of pole displacement.

With an increase in the travel speed along small curvature trajectories, the requirements for stability are moving more and more to the forefront. In such conditions, we need an in-phase turn of the front and rear wheels at the start of the maneuver and a smooth transition to the turn with these wheels in counter-phase, if it is demanded by the motion conditions. This effect can be reached if we offset the family of curves *AB* into position *A'B'* (see Fig. 15.14) and make the value of the obtained displacement dependent on the travel speed v_x.

With account of this displacement, the current position of the StS pole, depending on the angle and speed, will be written as follows:

$$x_p = K_1 \left|\alpha_{1ave}\right|^K + K_2 v_x^m,$$

where: α_{1ave} is the average turning angle of the wheels of the first axle's right and left boards; K_2 is the proportionality coefficient of the displacement curve for the vehicle's travel speed; v_x is the vehicle's current travel speed; and m is the parameter introduced for research purposes in order to obtain different modifications of the law.

Let us transform this expression:

$$x_p = \left(\left|\alpha_{1ave}\right|^K + K_{v_x} v_x^m\right) K_m \qquad (15.2)$$

where: $K_{v_x} = \frac{K_2}{K_1}$ is the proportionality coefficient of the curve's displacement X_p for the speed; and $K_m = K_1$ is the proportionality coefficient for the displacement.

Thus, with the help of coefficients K_m and K_{V_X} and by selecting the values of the power indicators K and m at parameters α and v_x, we can form different laws for displacing the StS pole. Coefficients K_m and K_{V_X} are defined in equation (15.2) on the basis of the initial conditions:

$$X_{p\ max} = 0.5\ L;\ x_v = 0;\ \alpha = \alpha_{max},$$

where: $X_{p\ max}$ is the maximum possible displacement of the StS pole adopted for calculations; L is the vehicle's base; α is the turning angle's current value for the conventional average wheel of the (setting) first axle; and α_{max} is the maximum value of angle α adopted for calculations. Then:

$$X_{p\ max} = \left|\alpha\right|^K K_m \qquad (15.3)$$

from where:

$$K_m = \frac{X_{p\ max}}{\left|\alpha_{max}\right|^K} \qquad (15.4)$$

Then, by setting the maximum possible speed $v_x = v_{x\ max}$, the zero value of the setting angle $\alpha = 0$, and the expression for K_m obtained from (15.4), let us put down the StS pole's initial displacement as:

$$X_{p.h} = -\frac{v_{x\ max}^m K_{v_x} X_{p\ max}}{\left|\alpha_{max}\right|^K},$$

accordingly:

$$K_{v_x} = -\frac{X_{p.h} \left|\alpha_{max}\right|^K}{v_x^m\ X_{p\ max}}.$$

With account of (15.2) and (15.3), the law for displacing the StS pole takes the form:

$$X_p = \frac{\left|\alpha\right|^K X_{p\ max}}{\left|\alpha_{max}\right|^K} + \frac{v_x^m X_{p.h}}{v_{x\ max}^m}. \qquad (15.5)$$

The graphical form of the function $X_p = f(\alpha, v_x)$ for 8×8 HDWVs with a wheelbase L of about 8.0 m at various parameter values is presented in Fig. 15.14. Its analysis shows that with the growth in angle α from 0 to 0.6 rad and a decrease in the speed v_x from 16 to 0 m/s, the value of the StS pole's displacement tends to the area of positive values and reaches its maximum (X_p – 4.0 m) at the values of $\alpha = 0.6$ rad and $v_x = 0$. Under $\alpha = 0$ and $v_x = v_{x\ max}$, the StS pole gets out of the chassis' base by the value corresponding to the preset value of $X_{p.h}$.

An example of the StS functioning is shown in Fig. 15.15 (at the entrance of a 8×8 vehicle into a turn and at a steady motion along a circle), if its StS pole moves under one of the laws presented in Fig. 15.14. It follows from Fig. 15.16 that the entrance of a four-axle HDWV into a turn and its stable motion along a circle are accompanied—at the initial moment—by an in-phase turn of the controlled wheels (zone I); while the StS pole starts to move from the negative area into the positive area. Further, at the StS pole's transition from the negative area into the positive, the wheels of the rear axles are, starting with the extreme rear axle, successively turning into the counter-phase condition (zones II and III).

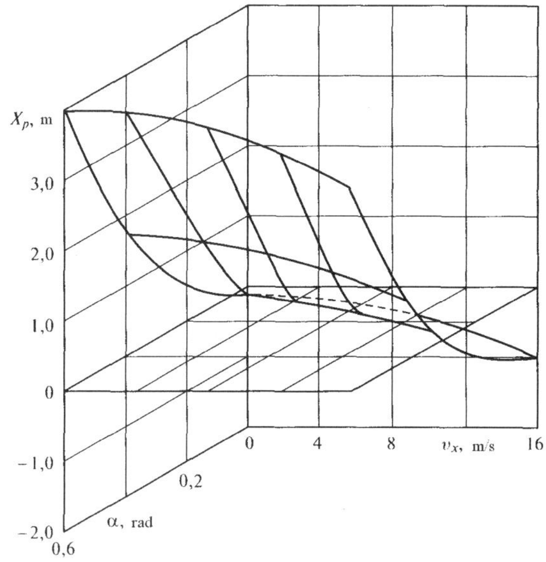

Fig. 15.15. Displacement law of the StS pole in the RPD-AS StS diagram as applied to an 8×8 chassis

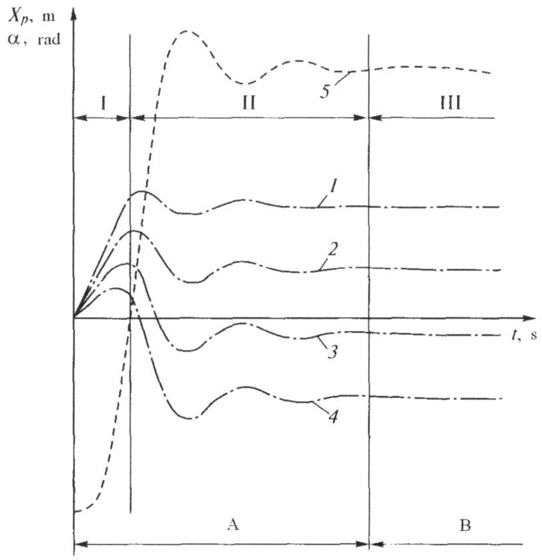

Fig. 15.16. Change character of the displacements in the StS pole X_p and values of the wheels' turning angle in an 8×8 chassis at entering a turn and stable motion along a circle: A—period of entering the circle; B—motion along the circle; I—zone of the wheels' in-phase turn; II— transition zone of the wheels' turn into the counter-phase; III—zone of the wheels' steady turn in the counter-phase; 1, 2, 3, and 4—turn of the wheels of the 1st, 2nd, 3rd, and 4th axles respectively

15.4.2 Mathematical justification of the RPD-A functioning law

The second variant of the law to control the StS pole's displacement is the dependence of its value only on the turning angle of the setting wheel α:

$$X_p = f(\alpha).$$

The law in accordance with Fig. 15.17 can be presented as the following equation:

$$X_p = (|\alpha|^K - \alpha^*)\, K_n, \tag{15.6}$$

where: α^* is the lag angle of the rear controlled wheels' turn into the counter-phase condition in relation to the front wheels; and K_n is the proportionality coefficient of the StS pole's displacement.

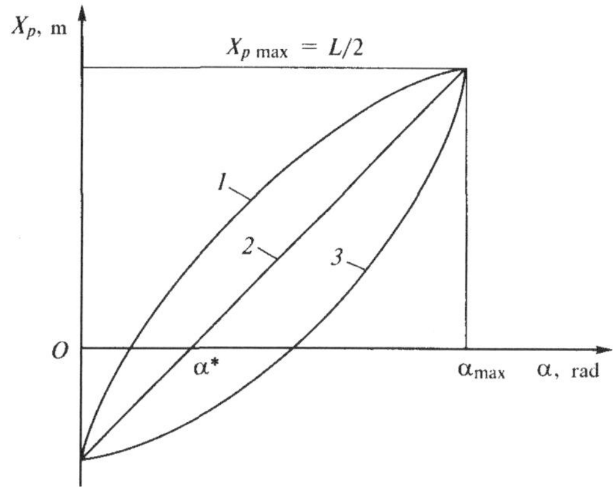

Fig. 15.17. Control variants of the StS pole's displacement in the RPD-A design: 1, 2, 3—parabolic, linear, and hyperbolic laws

This StS diagram is conditionally called RPD-A (regulation of the pole displacement by the angle).

By using the parameters K_n and α^*, as well as by selecting the values of parameter K, we can form different versions of the law to control the StS pole's displacement. The calculation of coefficients K_n and α^* is made based on the initial conditions when $X_{p\,max} = 0.5L$ and $\alpha = \alpha_{max}$. Then, expression (15.6) gets transformed as follows:

$$X_p = (|\alpha|^K - \alpha^*)\, K_n,$$

from where:

$$K_n = \frac{X_{p\ max}}{\left(|\alpha_{max}|^K - \alpha^*\right)} \quad . \qquad (15.7)$$

Then, by setting the StS pole's initial displacement $X_{p.h}$ outside the base at $\alpha = 0$ and the obtained expression for K_n, we find α^* from (15.5)

$$X_{p.h} = \frac{\alpha^* X_{p\ max}}{|\alpha_{max}|^K - \alpha^*} \quad ,$$

from where:

$$\alpha^* = \frac{X_{p.h}\,|\alpha_{max}|^K}{X_{p.h} - X_{p\ max}} \quad . \qquad (15.8)$$

After substituting (15.7) and (15.8) into (15.6), we have:

$$X_p = \left(|\alpha|^K - \frac{X_{p.h}\,|\alpha_{max}|^K}{X_{p.h} - X_{p\ max}}\right) \cdot \frac{X_{p\ max}\, X_{p.h}}{|\alpha_{max}|^K \dfrac{X_{p.h}\,|\alpha_{max}|^K}{X_{p.h} - X_{p\ max}}} \quad .$$

After making some transformations, we obtain the following:

$$X_p = \frac{|\alpha|^K X_{p\ max}}{|\alpha_{max}|^K \left(1 - \dfrac{X_{p.h}}{X_{p.h} - X_{p\ max}}\right)} + X_{p.h} \qquad (15.9)$$

The displacement character of the StS pole X_p depending on the turning angle of the setting wheel α at various values of parameters K, $X_{p\,max}$, and $X_{p.h}$ is presented in Fig. 15.18.

In (15.5) the maximum value of α and $v_{x\,max}$ for various types of HDWVs with four to eight axles are almost same, while the base's length goes up

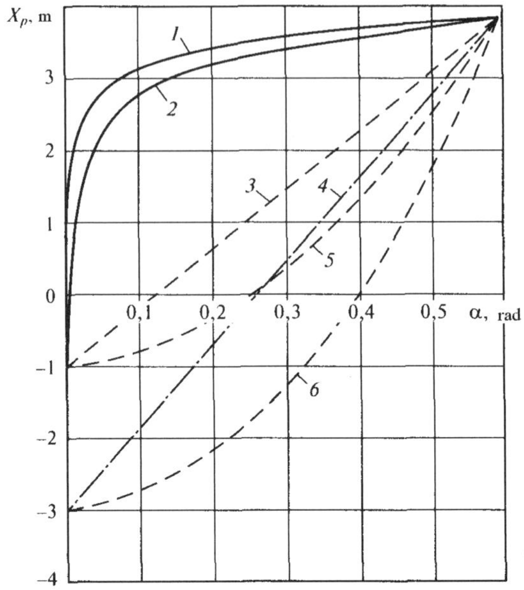

Fig. 15.18. Change character of the StS pole's position depending on the RPD-A control law: 1—parabolic law of X_p displacement inside the vehicle's base; 2—parabolic law of X_p displacement with the StS pole's displacement outside the vehicle's base by one meter; 3—linear law of X_p displacement with the StS pole's displacement outside the vehicle's base by one meter; 4—linear law of X_p displacement with the StS pole's displacement outside the vehicle's base by three meters; 5—hyperbolic law of X_p displacement with the StS pole's displacement outside the vehicle's base by three meters; 6—hyperbolic law of X_p displacement with the StS pole's displacement outside the vehicle's base by three meters

with the increase in the number of axles. That is why, to standardize the control laws for the StS pole's displacements and reduce the volume of studies, it is reasonable to express the StS pole's displacements $X_{p.h}$ and $X_{p\,max}$ in both directions relative to the rear axle in relative values by using the chassis' base:

$$\begin{cases} X_{p\ max} = K_{x\ max}\ L\ ; \\ \%_{p.h} = K_{xu}\ L\ , \end{cases}$$

where $K_{x\,max}$ and K_{xu} are the proportionality coefficients defined for each StS diagram by the kinematical indicators' ratios. The calculation of these coefficients is presented below.

Then, expression (15.5) can be written as follows:

$$X_p = L\left[\frac{\alpha^K \mathrm{K}_{x\ \max}}{|\alpha_{\max}|^K} + \frac{v_x^m \mathrm{K}_{xu}}{v_{x\ \max}^m}\right].$$

Similarly, for (15.9):

$$X_p = L\left[\frac{\alpha^K \mathrm{K}_{x\ \max}}{|\alpha|^K\left(1-\dfrac{K_{xu}}{K_{xu}-K_{x\ \max}}\right)} + K_{xu}\right] \quad (15.10)$$

Thus, the obtained functioning algorithms of StS diagrams with a mobile pole, as well as those that make it possible in studies to pass over StS diagrams with traditional kinematics, allow for describing practically the whole range of currently known designs.

15.5 Study of turning ability, maneuverability, and motion of HDWVs with different StSs

Theoretical studies were conducted on the analog computing complex using expressions (15.2)–(15.10) and the system of differential equations in (8.85). Experimental studies were conducted on the 79085 HDWV model prototype with a flexible StS and the MAZ-543M, -7930, -7906, and -7907 serial and experimental models.

15.5.1 Design with regulated displacement of the StS pole by the turn angle of the setting wheels [RPD-A]

It was already noted that insufficient motion stability in a vehicle with controlled front and rear axle wheels is associated first of all with the phenomenon of excessive lateral acceleration in the rear axle and a change in the sign of the side reaction on the wheels thereof. Studies on the reasons for this phenomenon resulted in several proposed methods for counteracting it. One idea is to use a delay in the turn of the rear axle's wheels relative to the front axle's wheels by an angle equal to about 0.1 radian (rad).

When considering the operation kinematics of designs with a mobile StS pole from this viewpoint, we can note that they implement the same delay principle.

If, at the start of the turn, the StS pole is outside the base and the rear axle's wheels are turning in-phase with the front ones, then, with an increase in the trajectory curvature, the StS pole while moving along the HDWV's longitudinal axis will cross the middle of its last axle causing a certain delay in the turn of its wheels in the opposite direction. Here, the difference with the known delay method is in the fact that the considered design has an in-phase turn of the rear wheels to a certain angle relative to the front ones and that there is a complete kinematical coordination of the controlled wheels in the course of the turn, which is especially important.

For the RPD-A StS design under study, we can find a combination of parameters under which we can ensure the required delay angle of the rear wheels' turn α_3 in counter-phase to the front ones. With this aim in mind, by equating expression (15.10) to zero, we can obtain the following dependence of coefficients from the StS diagram's kinematical indicators:

$$K_{xu} = \frac{|\alpha|^K \mathrm{K}_{x\ \max}}{|\alpha|^K - |\alpha_{\max}|^K}.$$

Taking in this expression $K_{x\max} = 0.53$ and $\alpha_{\max} = 0.6$, let us put K within the limits of 0.1–2.0 and instead of α use the required delay angle α_3. In doing so, we can obtain the value of parameter K_{xu} for various delay angles of the rear wheels' turn, for example, 0.1, 0.2, and 0.3 rad (Fig. 15.19).

The StS pole's displacement from the initial position to the middle of the base for the delay

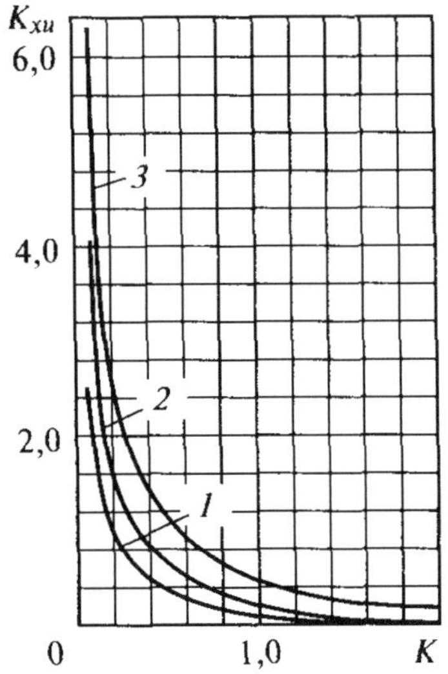

Fig. 15.19. Parameters K_{xu} and K versus different delay angles α_3 of the turn of the rear axle's wheels: 1, 2, 3—α_3 = 0.1; 0.2; 0.3 rad

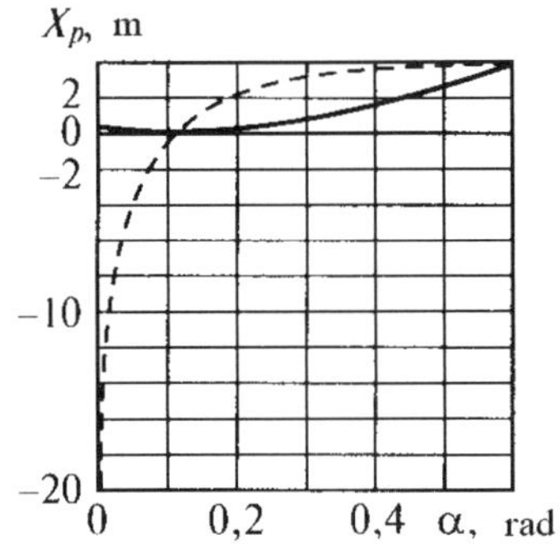

Fig. 15.20. Parameters K_{xu} and K for different turning angles α of the 8×8 chassis' setting axle (L = 0.1) at the delay angle α_1 = 0.1 rad: - - - - - parabolic control law of the StS pole's displacement (K = 2.0; K_{xu} = 0.02); --- - hyperbolic control law of the StS pole's displacement (K = 0.1; K_{xu} = 2.43)

angle α_3 = 0.1 rad is presented in Fig. 15.20. At the minimum K = 0.1 and the maximum K_{xu} = –2.43, the pole's displacement follows a hyperbola, while in the other extreme case when K = 2.0 and K_{xu} = 0.017, the parabolic displacement law is observed.

The change character of the rear axle's lateral acceleration value at the motion of a four-axle HDWV at a left-side passing with a return back to the same lane (the so-called "resetting") at a speed of 9 m/s under the limit values of parameters K and $K_{x\,min}$ for the delay angles of 0.1, 0.2, and 0.3 rad, is presented in Figs. 15.21–15.23. The analysis shows that using the hyperbolic control law of the StS pole's displacement (K = 0.1) within the whole range of parameter K_{xu} = 2.43–6.97 and at all accepted delay angle values of the turn of the rear axle's wheels $\alpha_{3=}$ 0.1–0.3 rad, the curve reflecting the change in the lateral acceleration over the rear axle shows the excess of this value both at the start of "resetting" and at its transition to the second phase. The short time duration of these excesses (0.4–0.5 s) gives grounds to assume that the lateral force pulses appearing in this case and acting on the rear axles' wheels will have an impact load character and reduce the motion stability on transient trajectories.

The reason an excesses of lateral acceleration appears has to do with different values and directions of the angular speeds of the HDWV's turn ω_z and the rear axle's wheels $\dot{\alpha}_l$. An intense displacement of the StS pole X_p under the hyperbolic law (K = 0.1) causes a rapid change in the direction and a sharp increase in the turning speed of the rear axle's wheels $\dot{\alpha}_l$ (Fig. 15.22). The change in the turning speed's direction in the rear axle's wheels $\dot{\alpha}_l$ at the start and in the middle of the maneuver in counter-phase with the angular velocity of the chassis' turn ω_z results in the lateral acceleration's change in direction (an exceed).

A switchover to controlling the StS pole's displacement under the parabolic law with an increase in the value of parameter K to the maximum (2.0) and the respective reduction in the value of parameter K_{xu} results in the complete disappearance of the lateral acceleration's exceeds at all "resetting" phases.

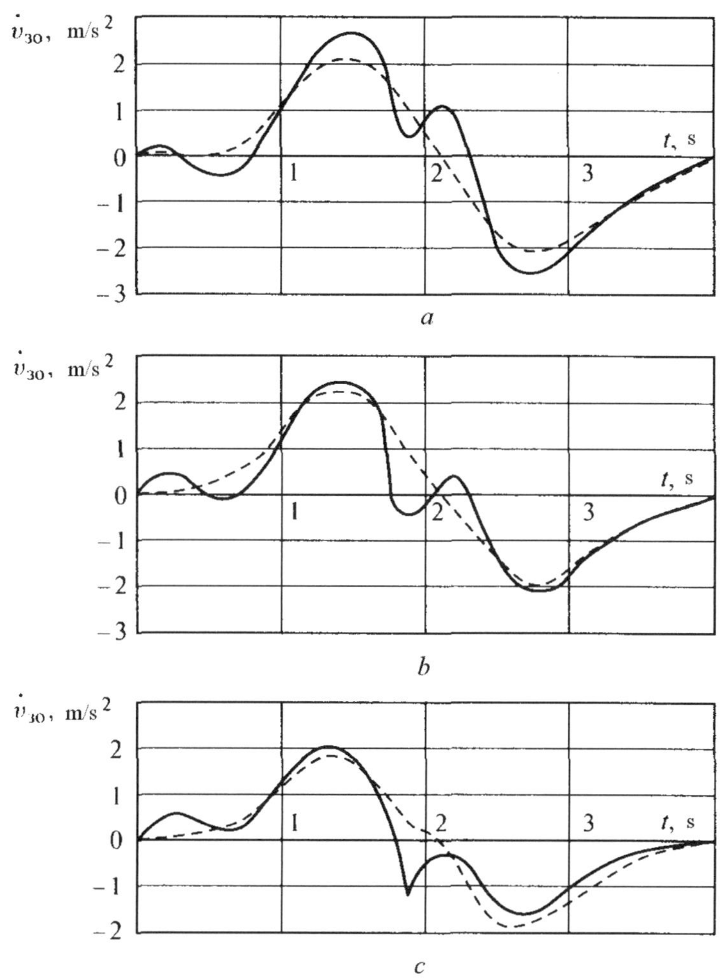

ig. 15.21. Change character of the ıteral acceleration value on the rear axle ;₇₍ of an 8×8 chassis at the "resetting" ıotion: a—K = 0.1; K_{xu} = -2.43; α_l = 0.1; - - - - K = 2.0; K_{xu} = -0.02; α_l = 0.1; —K = 0.1; K_{xu} = -4.31; α_l = 0.2; - - - - - ' = 2.0; K_{xu} = -0.06; α_l = 0.2; c—K = 0.1; $_{xu}$ = -6.97; α_l = 0.3; - - - - - K = 2.0; $_{xu}$ = -0.22; α_l = 0.3

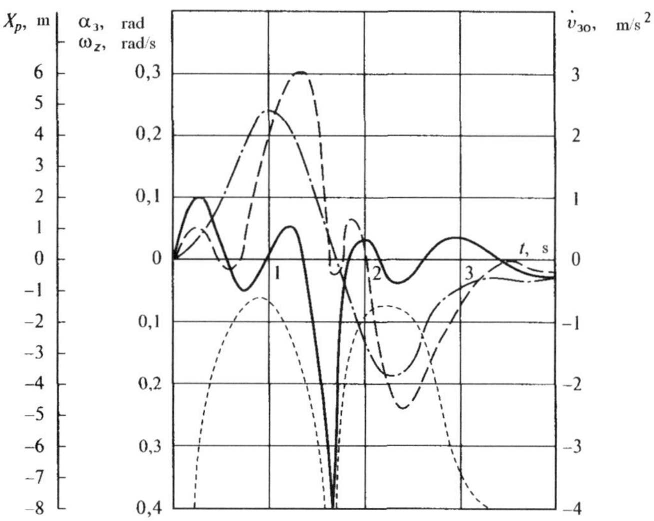

Fig. 15.22. Character of the transient process at performing the "resetting" maneuver and the hyperbolic law of the StS pole's displacement (K = 0.1): ---- the turning speed of the rear axle's wheels, $\dot{\alpha}_l$; - - - - - lateral acceleration over the rear axle, v_{lo}; - - - - angular velocity of the chassis' turn ω_z; -- · -- · -- displacement of the StS pole, X_p

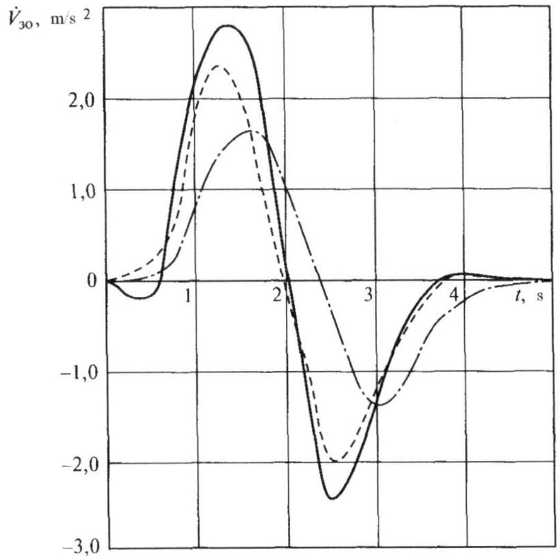

Fig. 15.23. Change character of the lateral acceleration value on the rear axle v_{30} of an 8×8 chassis at "resetting": ---- v_x = 10 m/s; α_l = 0.1 rad; - · - · - v_x = 10 m/s; α_l = 0.3 rad; - - - - - v_x = 8 m/s; α_l = 0.1 rad

The lateral acceleration's smooth change character when "resetting" and a smaller amplitude range of this value than when the hyperbolic control law of the StS pole's displacement is used allows for asserting that using the parabolic law will allow for increasing the motion stability of HDWVs under transient trajectories.

When "resetting" at a higher speed (10 m/s and up) and a delay angle of α_l = 0.1 rad, the lateral acceleration's excess over the lateral axle $\dot{v}_{lo}$ in the start of the maneuver appears again and grows with the increase in speed. An increase in the delay angle to 0.3 rad results in the disappearance of the excess and reduces the lateral acceleration's amplitude (Fig. 15.23). For defining the rational value α_λ, we need to study StS diagrams under other criteria.

Fig. 15.24 presents the results of the maneuverability studies on an 8×8 chassis with the RPD-A StS diagram. The analysis shows that the worst maneuverability was with the HDWV with the rear wheels' turn delay angle of α_3 = 0.3 rad, especially within the turning radii

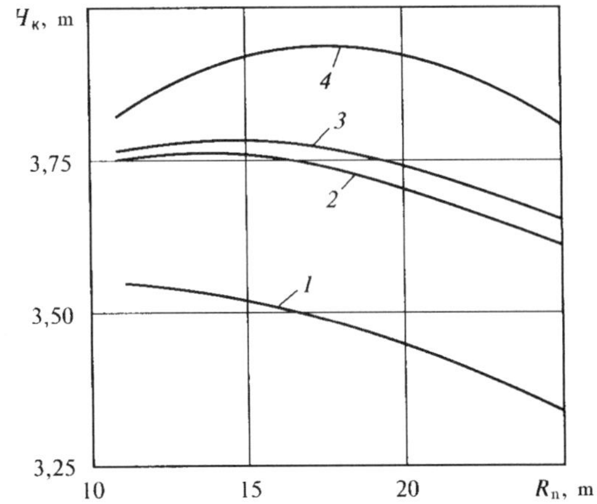

Fig. 15.24. The width of trackway H_K versus the turn radius R_n of the 8×8 chassis with different control laws of the StS pole's displacement: 1—α_l = 0.1; K = 0.1; 2—α_l = 0.1; K = 2.0; 3—α_l = 0.2; K = 2.0; 4—α_l = 0.3; K = 2.0

interval R_n = 12–22 m. This is explained by the fact that the StS pole is outside the base at a turn angle of 0.3 rad for the front wheels; that is why the chassis moves along a circle, while at the same time making a plane-parallel motion and taking a rather broad lane.

With the decrease in the delay angle, the width of trackway H_K goes down within the whole range of the turning radii. The trackway's width at delay angles 0.1 and 0.2 rad differ insignificantly. The best maneuverability on this indicator belongs to the chassis with the StS diagram whose control of the StS pole's displacement is made under the hyperbolic law (K = 0.1) with the delay angle α_λ = 0.1 rad, but this vehicle and StS, as has been demonstrated above, has poorer motion stability.

A study of the RPD-A StS diagram as applied to an 8×8 HDWV under the criteria of maneuverability and motion stability has shown that at the parabolic control law of the StS pole's displacement, which ensures the rear axles' wheels turn relative to the front ones with a delay angle α_λ = 0.1 rad at a travel speed of 10 m/s or more, the lateral acceleration's exceeds $\mathring{v}_{7.>}$ appear over the rear axle at the start of the "resetting" maneuver. At the same control law, but with a higher delay angle equal to 0.3 rad, the stability goes down with an additional degradation of the HDWV's maneuverability. This is manifested through an increase in the trackway's width within a broad range of turning radii (12–22 m).

With account of the aforesaid, we can make the following conclusions.

To improve the maneuverability, turning ability, and motion stability of a four-axle HDWV with a base length of no more than 8.0 m and a RPD-A StS diagram, it is reasonable to use the parabolic control law of the StS pole's displacement with parameter $K = 2.0$ and the delay angle $\alpha_\lambda = 0.2$ rad. This chassis meets the requirements of the normative-technical documentation (NTD) on the criteria for motion stability on "resetting" type transient trajectories and the properties of maneuverability and turning ability.

For a HDWV with a longer base and a large number of axles, using the RPD-A StS diagram without correcting for parameters K and α_λ gives no positive effect in decreasing the trackway's width.

It follows from the data presented in Fig. 15.25 that in a 16×16 HDWV ($L = 13.2$ m) with a RPD-A StS diagram ($K = -2.0$; $\alpha_\lambda = 0.2$ rad), with the increase in the turn radius from 17 m (R_{min}) up to 30.0 m, the H_K does not go down, but, on the contrary, goes up. This is characteristic for the parabolic control law of the StS pole's displacement and high delay angles ($\alpha_l \geq 0.2$ rad). With an increase in the base's length, this trend grows stronger.

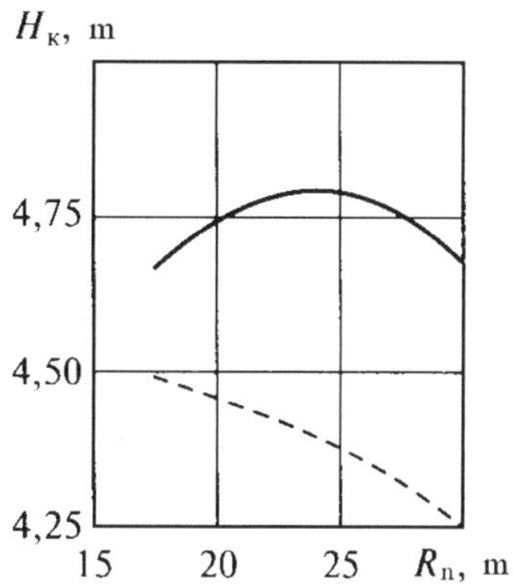

Fig. 15.25. Width of trackway H_K versus the turn radius R_n of a 16×16 chassis with different control laws of the StS pole's displacement: --- $K = 2.0$; $\alpha_l = 0.2$; - - - $K = 0.8$; $\alpha_l = 0.1$

While for an 8×8 HDWV ($L = 7.8$ m) with an RPD-A StS, a rapid growth in the dependence $H_K = f(R)$ was obtained at $K = 2.0$ and $\alpha_\lambda = 0.3$ rad (see Fig. 15.25); in a 16×16 HDWV ($L = 13.2$ m), a similar growth of $H_K = f(R)$ is observed at a smaller delay angle $\alpha_l = 0.2$ rad.

It is obvious that to achieve satisfactory indicators of maneuverability for chassis with longer bases we need a more intense StS pole displacement in the direction of the base's middle and a decrease in the delay angle. A decrease in parameters K and α_λ can tell negatively on the motion stability.

It was established in the course of the studies that a decrease in the width of trackway H_K for a 16×16 HDWV is achieved by decreasing the control law parameters for the StS pole's displacement down to the values of $K = 0.8$ and $\alpha_з = 0.1$ rad (see Fig. 15.24). This does not result in decreased stability when performing the "resetting" maneuver. In this case, the lateral acceleration's amplitude value (Fig. 15.26) on the rear axle goes up by 0.2 m/s in both "resetting" phases as compared to the StS diagram variant where $K = 2.0$ and $\alpha_l = 0.2$ rad, while the lateral acceleration's excess at the start of the maneuver is practically absent.

Accounting for the obtained results, as well as the results of similar studies conducted with five- and six-axle HDWVs where the base dimensions take an intermediate position relative to the dimensions of the considered four- and eight-axle HDWVs, we can state that with an increase in the base L and the number of axles we need to decrease the delay angle α_λ and the control

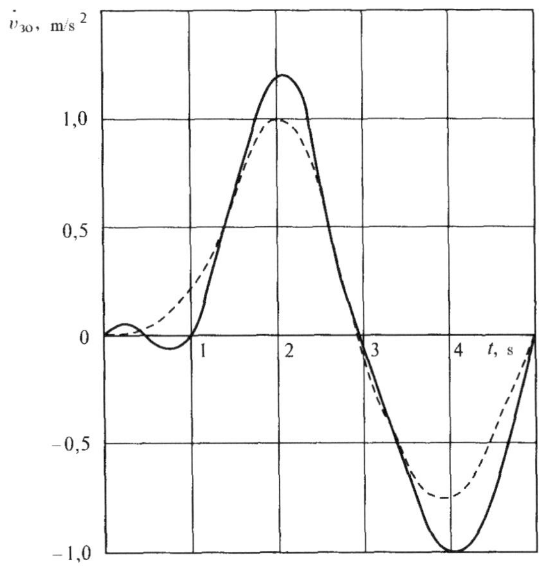

Fig. 15.26. Change in the lateral acceleration values $\dot{v}_{lo}$ on the rear axle of a 16×16 chassis at "resetting": --- K = 0.8; α_l = 0.1; - - - - K = 2.0; α_l = 0.2 (RPD-A StS diagram)

law parameter of the StS pole's displacement K with the aim to improve the maneuverability and turning ability within a broad range of trajectory radii. Here, satisfactory indicators of motion stability are preserved due to the multi-support feature. The rational values of α_λ and K of the RPD-A StS diagram depending on the HDWV's base length are presented in Fig. 15.27.

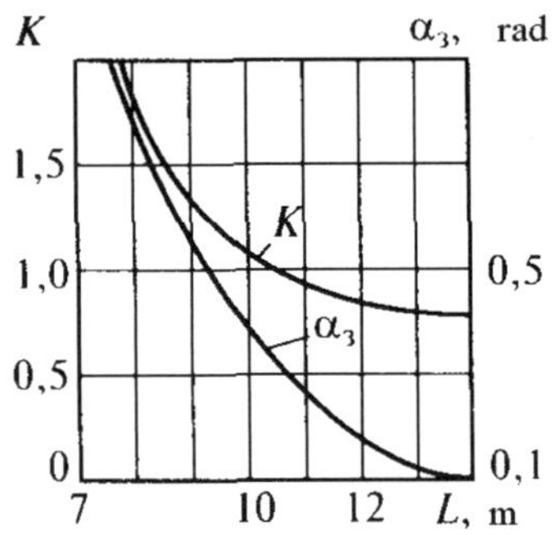

Fig. 15.27. Rational values of the control law parameters K and α_{ll} for the StS pole's displacement in the RPD-A design

15.5.2 Design with regulated displacement of the StS pole on the turn angle and travel speed [RPD-AS]

In the RPD-AS StS diagram, the StS pole's displacement is regulated depending on the setting wheel's turning angle α_λ and the travel speed v_x.

The control law of the StS pole's displacement in designs of this type is presented by expression (15.10). Let us replace the current angle by the delay angle and the maximum possible turning angle by the maximum delay angle. Let us find from the obtained dependence the delay angle from the travel speed, keeping in mind that the StS pole's displacement in this case is equal to zero:

$$\alpha_l = \left(\frac{v_x}{v_{x\ max}}\right)^{\frac{m}{K}} \alpha_{\ell\ max}\,. \qquad (15.11)$$

Accounting for the results of studies on the RPD-A StS diagram, let us assume that for the RPD-AS StS diagram as applied to an 8×8 HDWV that the parabolic control law of the StS pole's displacement depends on the turning angle, i.e., the parameter K = 2.0. Assume that for a maximum travel speed of 60 km/h (16.67 m/s), the maximum delay angle α_λ is 0.3 rad.

The dependence of the delay angle α_λ on the travel speed v_x and the value of parameter m at the preset values of v_{max} and $\alpha_{3\,max}$ is presented in Fig. 15.28. Its analysis shows that with the growth in the speed v_x, the delay angle goes up. The intensity of the growth depends on the value of m. The smaller this parameter is, the stronger the increase of α_l is, especially within the range of small travel speeds—from 0 to 5 m/s.

Based on the outcomes of studies on RPD-A StS diagrams, we can assert that high delay angles are negatively affecting the chassis' turning ability and maneuverability. At the same time, at small motion speeds there is no need to apply

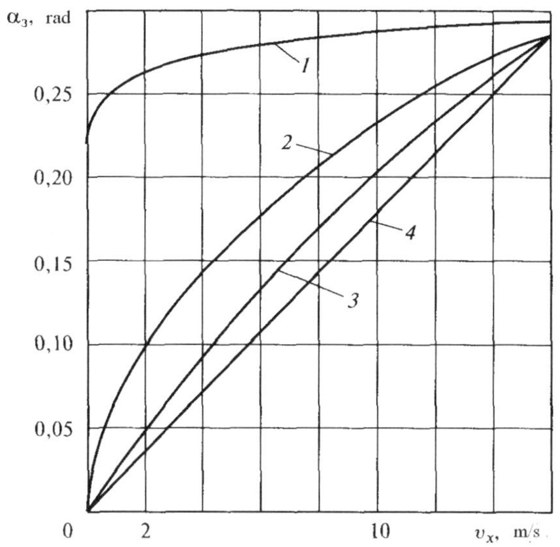

Fig. 15.28. Delay angle α_l versus the travel speed v_x and parameter m in the RPD-AS StS diagram: 1, 2, 3, 4—m = 0.1, 1.0, 1.58, 2.0

large delay angles from the viewpoint of motion stability.

A study on the motion stability of an 8×8 HDWV with a RPD-A StS have shown that at motion at a speed of 10 m/s the delay angle should be 0.2 rad.

Then, $1.0 < m < 2.0$ (see Fig. 15.28); a more precise value of this parameter is calculated from the formula:

$$m = \frac{K \ln \dfrac{\alpha_l}{\alpha_{l\,\max}}}{\ln \dfrac{v_x}{v_{x\,\max}}}.$$

After putting the preset values of $\alpha_\lambda = 0.2$ rad, $\alpha_{l\,\max} = 0.3$ rad, $v_x = 10$ m/s, $v_{x\,\max} = 16.67$ m/s, and $K = 2.0$, we obtain $m = 1.58$. The parameter defining the StS pole's initial displacement at the maximum travel speed $v_{x\,\max}$ and the maximum delay angle $\alpha_{l\,\max}$ can be found from the following expression:

$$K_{xu} = \frac{\alpha_{l\,\max}^K \, K_{x\max}}{\left|\alpha_{\max}\right|^K}.$$

The results of the studies on the motion stability of the 8×8 MAZ-543 chassis with a RPD-AS StS diagram with different values of parameter m are presented in Fig. 15.29, while the indicators of the turning ability and maneuverability are presented in Fig. 15.30.

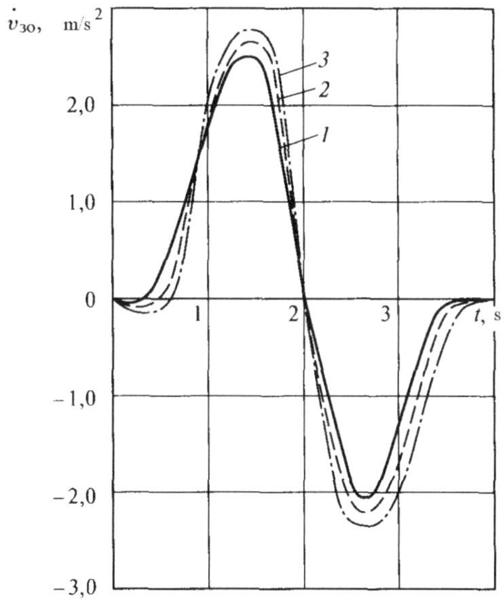

Fig. 15.29. Change in the lateral acceleration value $\dot{v}_{lo}$ on the rear axle of the 8×8 chassis with a RPD-AS StS diagram at the "resetting" maneuver: 1, 2, 3—m = 0.1, 1.58, 2.0

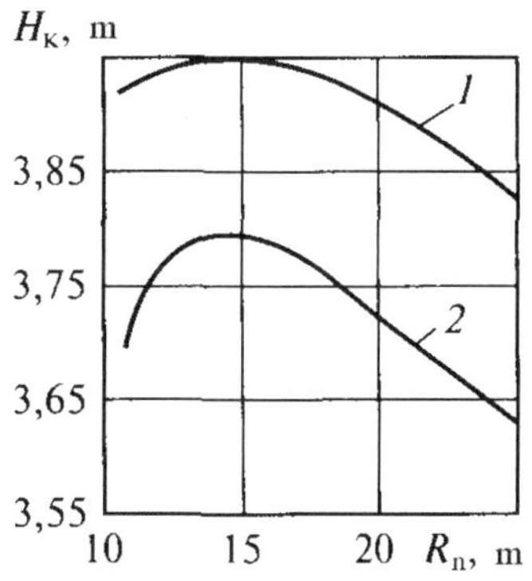

Fig. 15.30. Width of trackway H_K versus the turn radius R_n of the 8×8 chassis with the RPD-AS StS diagram at various values of parameter m: 1, 2—m = 0.1, 1.58

The analysis of these figures shows that the lateral acceleration's amplitudes in all "resetting" phases at different values of parameter m differ insignificantly. The highest lateral acceleration amplitude $\dot{v}_{lo}$ was obtained at $m = 2.0$ and lateral acceleration excess was noticed at the start of "resetting." The smallest lateral acceleration amplitude was obtained at $m = 0.1$ and no lateral acceleration excess was observed; however, the trackway's width in the range of turning radii 10–25 m at $m = 0.1$ (see Fig. 15.30) was bigger than at $m = 1.58$ by 0.1–0.2 m.

Thus, to improve the turning ability, maneuverability, and motion stability of an 8×8 HDWV with a base length of no more than 8.0 m and a RPD-AS StS diagram, it is expedient to use the parabolic control law of the StS pole's displacement with parameters $K = 2.0$ and $m = 1.58$, which ensure a rational change in the turn delay angle of the rear axle's wheels in counter-phase to the front ones. The StS diagram with these control law parameters for the StS pole's displacement meet the established requirements for the motion stability criterion on "resetting" transient trajectories and the properties of maneuverability and turning ability.

Let us consider the RPD-AS StS diagram for a 16×16 7922 (7923) model HDWV with a 13.2 m long base. When studying the RPD-A StS diagram as applied to this chassis, it was established that the parameter K is equal to 0.8 and the delay angle is equal to 0.1 rad. By making calculations under the dependence in (15.11) for a travel speed of 9 m/s, we find that $m = 1.43$.

Thus, as a result of theoretical studies and analysis, we have revealed the influence of the control laws of the StS pole's displacement and the values of their parameters on the properties of turning ability, maneuverability, and motion stability in HDWVs. We have defined the rational regularities of changing the parameters used to control the StS pole's displacement and deduced the analytical formulas for calculating them depending on the chassis' basic dimensions.

The obtained results were implemented in the StS design for the 79085 HDWV prototype and their adequacy to the real system was confirmed experimentally.

15.5.3 Comparison of HDWV stability at different StS diagrams at motion on "resetting" trajectories

Figs. 15.31, 15.32, and 15.33 illustrate the results of a comparative assessment of the motion stability on "resetting" trajectories (length $l_n = 24$ m; width $h_n = 3.5$ m) of the 8×8 7908 MAZ-543M HDWV with flexible StS diagrams (RPD-A and RPD-AS) and with traditional StS diagrams (controlled front axles' wheels only and controlled wheels of the front and rear axles). In the course of the studies, the travel speed was deliberately restricted since the chassis using the StS diagram with controlled wheels of the front and rear axles and the StS pole's fixed position in the base's center is not able to "reset" with the above parameters at speeds above 9 m/s because of the lateral sliding in the rear axles' wheels.

The largest lateral acceleration amplitude over the rear axle $\dot{v}_{lo}$ was seen in the HDWV with

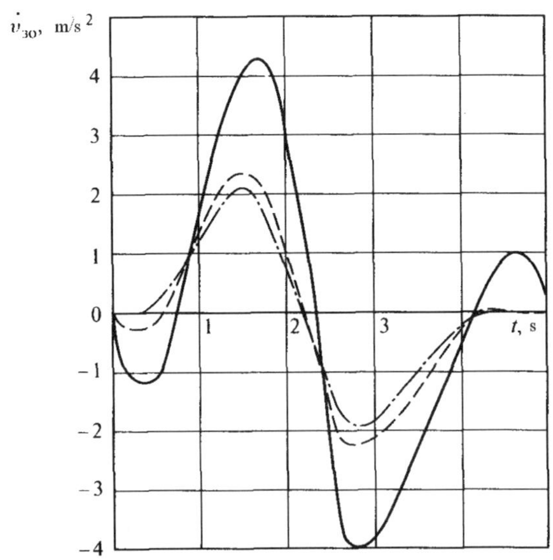

Fig. 15.31. Change in the lateral acceleration value $\dot{v}_{lo}$ on the 8×8 chassis' rear axle with different StS diagrams when "resetting": ---- StS diagram with all-wheel control, X_p=L/2; - - - - - StS diagram with controlled wheels; -- · -- · RPD-A (AS) StS diagram

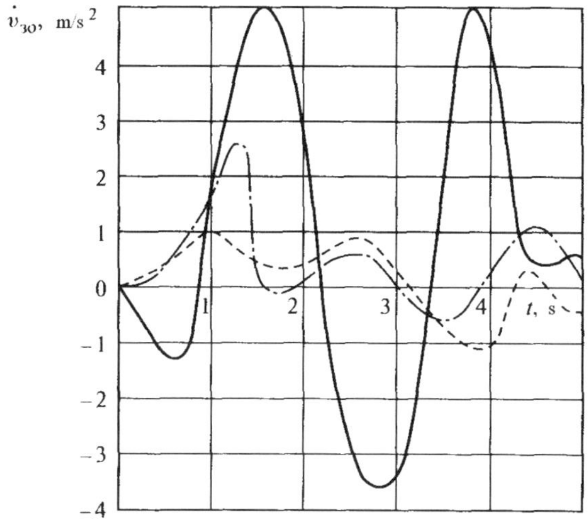

Fig. 15.32. Change character of the lateral acceleration value $\dot{v}_{lo}$ on the 79085 chassis' rear axle when "resetting": ---- pole in the base's center; -- · -- · pole on the fourth axle; - - - - - pole behind the fourth axle

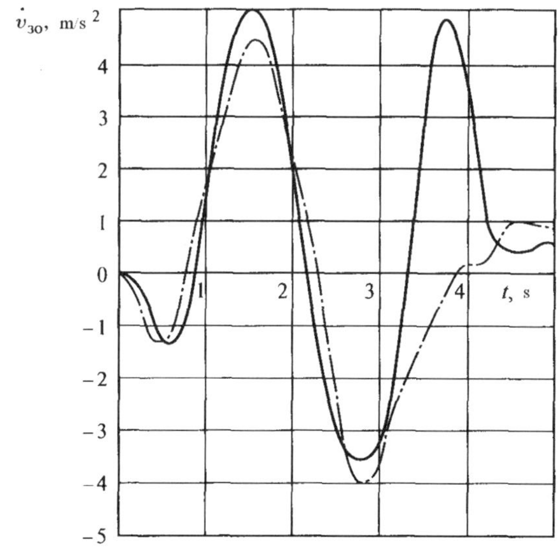

Fig. 15.33. Change character in the lateral acceleration value $\dot{v}_{lo}$ on the 79085 chassis' rear axle when "resetting": ---- under experimental studies; -- · -- · under calculations

controlled wheels of the front and rear axles and the StS pole's fixed position in the middle of the base (see Figs. 15.31 and 15.32). The amplitude value in the first phase reaches 4.0–5.0 m/s² and in the second reaches 3.5–3.9 m/s². This is twice as that in the diagrams with the mobile StS pole.

In the StS diagram with controlled wheels of two front axles, the lateral acceleration amplitude over the rear axle in the first phase of "resetting" is 2.25 m/s² and in the second one is 2.15 m/s². This is 1.13 times higher than in the diagrams with the regulated StS pole position. It is obvious that the probability of stability loss in the HDWV with traditional steering diagrams is higher than the chassis using diagrams with regulated StS pole positions.

Fig. 15.32 presents experimental curves characterizing the change in the lateral acceleration's amplitude over the 79085 chassis' rear axle. The motion was according to the "resetting" maneuver at a speed of 32 km/h using the steering design with all controlled wheels and the StS pole's fixed position on the chassis' longitudinal axis in three points: in the base's center, on the fourth axle, and outside the base.

The analysis of experimental data shows that the worst stability was noticed in HDWVs with all controlled wheels and the StS pole's fixed position in the base's center (solid line in Fig. 15.32). At the entrance into "resetting," we see an essential lateral acceleration excess about 1 second long and an amplitude of 1.3 m/s². In subsequent phases, its value reaches 5 m/s². The obtained results reconfirm the conclusion of several earlier studies about the chassis' inclination with such StS diagrams to lose its stability.

The StS pole's transfer to the fourth axle has an essential influence on the decrease (by the factor of five) in the maximum lateral acceleration amplitude (interrupted curve in Fig. 15.32) as compared with the previous version. It should be also noted that there is an absence of lateral acceleration excess. This diagram has obvious advantages since smaller acceleration values correspond to smaller lateral forces acting on the rear axles' wheels; therefore, the probability of the HDWV's rear part skidding (as the most dangerous form of skidding) goes down.

The StS pole's transfer backward outside the chassis' base also decreases the lateral acceleration amplitude. However, in the first "resetting" phase we see an increase in this value (up to 2.5 m/s^2) (dash-and-dot line in Fig. 15.32), which gives grounds to assume that the StS pole's removal to an essential distance outside the vehicle's base is not reasonable at high speeds for the motion stability conditions.

When the StS pole is located in close vicinity to the chassis' fourth axle, the lateral acceleration over the rear axle when "resetting" is smoothly changing during the course of the whole maneuver and the turn angles of the front axle's wheels are about 0.2 rad.

Fig. 15.33 presents the lateral acceleration curves $\dot{v}_{lo}$ on the rear axle obtained experimentally and by calculations for the 79085 prototype with all controlled wheels and the StS pole's fixed position in the base's center when "resetting" at a speed of 8.4 m/s (30 km/h). The analysis of the curves shows that the calculation data nicely match the experimental ones both for the change in the lateral acceleration values and the maximum acceleration values in the extreme points. The difference in the character of the experimental and theoretical curves at the final "resetting" phase is explained by the driver's additional affections on the controls caused by an increased feeling of danger.

Fig. 15.34 presents the calculation data for the lateral acceleration values on the rear axle $\mathring{v}_{7>}$ of the 16×16 7922 (23) HDWV with different StS diagrams when "resetting" with the parameters of l_n = 30 m and h_n = 3.5 m at a speed of 9 m/s.

The analysis of this data shows that the lateral acceleration amplitude on the rear axle in all motion phases goes down when the pair of axles with uncontrolled wheels closer to the chassis' back section are removed. Thus, for the 1-2-3-0-0-6-7-8 StS diagram the maximum amplitude

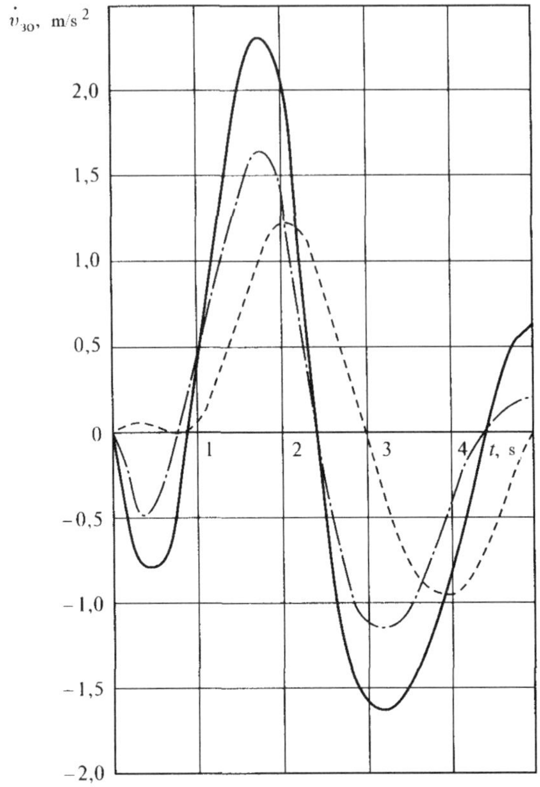

Fig. 15.34. Change in the lateral acceleration value $\dot{v}_{lo}$ on the 16×16 chassis' rear axle with different StS diagrams when "resetting": ----- 1-2-3-0-0-6-7-8 StS diagram; -- · -- · 1-2-3-4-0-0-7-8 StS diagram; - - - - - RPD-AS StS diagram; K = 0.8; m = 1.43

in the first and second motion phases are equal to 2.35 m/s^2 and 1.7 m/s^2, respectively, and the excess value at the start of the first phase is 0.85 m/s^2. In the HDWV with the 1-2-3-4-0-0-78 StS diagram, the amplitude in the first phase is 1.65 m/s^2 and is 1.2 m/s^2 in the second phase. This is 1.42 times smaller than in the chassis with the 1-2-3-0-0-6-7-8 diagram. The excess at the start of the maneuver goes down by the same value.

These variants of StS diagrams were implemented in experimental prototypes of the 7922 (23) SWC model. Their comparison with different StS diagrams with regulated StS pole positions (RPD-A, RPD-AS) shows that using such diagrams allows for eliminating the lateral acceleration excess in the initial phase and a decrease in its amplitude as compared with the considered diagrams in the first phase by 1.32–1.88 times and by 1.25–1.7 times in the second phase.

The analysis gives grounds to assert that StS diagrams with regulated StS pole positions also preserve their advantages in chassis with more than four axles.

15.5.4 Comparison of turning ability and maneuverability in HDWVs with different StS diagrams at steady circle motion

Fig. 15.35 and Fig. 15.36 present the data of theoretical and experimental studies on the turning ability and maneuverability of the 8×8 79085 HDWV using StS diagrams with regulated StS pole positions and with controlled wheels of the two front axles.

The analysis of the data shows that at the same maximum turn angle of the front axle's wheels (0.6 rad), the minimum turning radius using StS diagrams with a regulated StS pole position is 10.7 m, while the HDWV with controlled wheels of the two front axles is 14.5 m, i.e., 1.36 times higher.

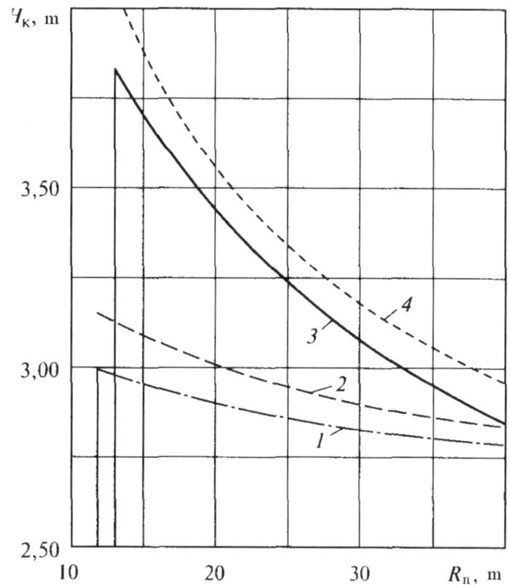

Fig. 15.36. Theoretical and experimental studies on the dependence of the width of trackway H_K on the turn radius R_n of the 79085 chassis with an all-wheel StS: 1—experiment; 2—calculation for the MAZ-543M chassis (1-2-0-0 StS diagram); 3—experiment; 4—calculation

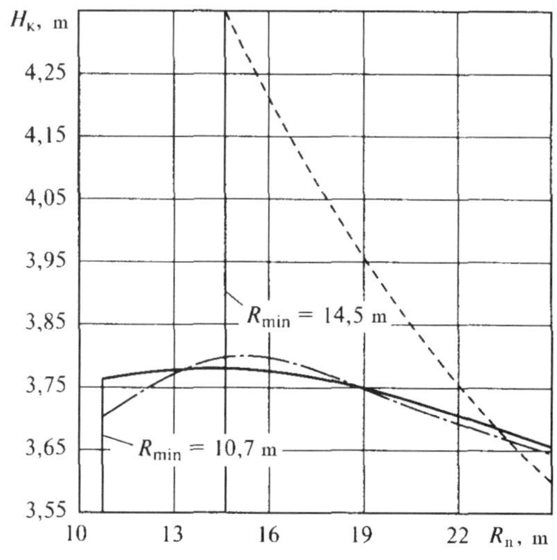

Fig. 15.35. Width of trackway H_K versus the turn radius R_n of the 8×8 chassis at different StS diagrams: ---- 1-2-0-0 StS diagram; - · - · - RPD-A StS diagram; - - - - - RPD-AS StS diagram

For a chassis with the StS diagram as in Fig. 15.36, with the decrease in the turn radius, the width of trackway H_K is increasing rapidly and at the minimum radius R_{min} = 14.5 m reaches the value of 4.35 m. The chassis with the RPD-A StS, when moving along a circle with the same radius, has a trackway width of 3.77 m, i.e., 1.15 times smaller, while the chassis with the RPD-AS diagram has a trackway width of 3.81 m, i.e., by 1.14 times smaller. Practically within the whole range of studied radii, the width of the trackway for the chassis with the regulated StS pole position is smaller than in the chassis with controlled wheels of the two front axles. The results of theoretical and experimental studies give a nice qualitative and quantitative match.

Fig. 15.37 presents the results of studies on the turning ability and maneuverability of a 16×16 7922 (23) HDWV with different StS diagrams.

According to the obtained data, chassis of this class with the 1-2-3-4-0-0-7-8 diagram have—within the whole range of radii (18–25 m)—a larger trackway width than the same chassis with the RPD-A (AS) StS diagrams. When the radius is equal to 18.0 m, the chassis with the 1-2-3-4-0-0-7-8 diagram has a width of 0.58 m, which is 1.13 times larger than the chassis with the RPD-A (AS) diagrams. It should be noted that a more intense decrease in the trackway's width with the increase in the turn radius is observed in the chassis with the RPD-AS diagram than with the RPD-A diagram. This circumstance shows the advantage of the diagrams with additional regulation of the pole's displacement in relation to the speed (RPD-AS), in which the rear wheels' turn delay angle is essentially decreased, improving the HDWV's turning ability, since the StS pole in this case displaces more in the direction of the base's

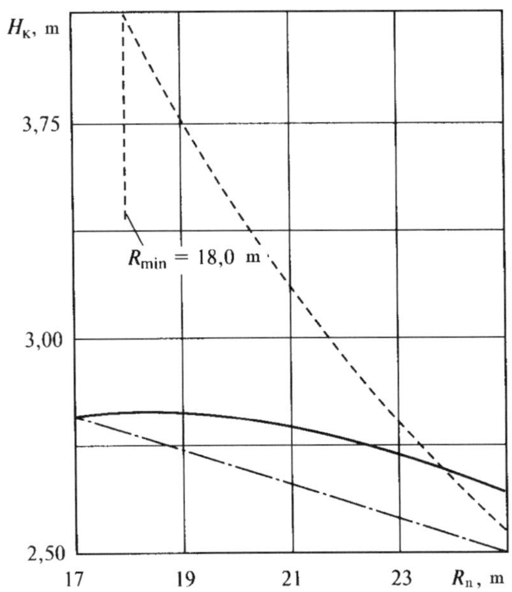

Fig. 15.37. Width of trackway H_K versus the turn radius R_n of the 16×16 chassis with different StS diagrams: - -·- RPD-AS StS diagram; --- RPD-A StS diagram; - - - - 1-2-3-4-0-0-7-8 StS diagram

middle (center). However, as demonstrated by the studies, these advantages begin manifesting themselves and become feasible only with an increase in the base.

In HDWVs with a base of no more than 8.0 m, the H_K value at using the RPD-A and RPD-AS diagrams (see Figs. 15.35 and 15.36) is practically the same. Thus, applying StS diagrams with a regulated pole position in 8×8 multi-axle HDWVs allows for essentially improving the turning ability on radii close to maximum (R_{min} is 1.15 times smaller). For this class of HDWVs (with a base of no more than 8.0 m), it is quite sufficient to have an StS diagram with the pole's regulated displacement only by the angle of the setting wheel's turn (RPD-A). With an increase in the number of axles and the chassis' base length, it is more reasonable to use the StS diagram with additional regulation of the pole's displacement by speed (RPD-AS).

15.5.5 Comparison of power input in HDWVs with different StS diagrams at stable circle motion

Figs. 15.38–15.40 present the results of theoretical and experimental studies on the power input of the 8×8 79805 MAZ-543M HDWV with an StS diagram with the regulated StS pole position and controlled wheels of the two front axles at stable circle motion with various radii in a range of speeds from 1.0–6.0 m/s.

The analysis of Fig. 15.38 shows that the power input of the HDWV with controlled wheels of the two front axles at a circle motion with a radius of 13.7 m at various speeds is 1.3–1.6 times higher than the chassis with all controlled wheels. At a circle motion with a radius of 20.0 m, this difference is 1.15–1.33 times.

The presented data indicates that using all-wheel control allows for obtaining an essential benefit not only through improved stability, turning ability, and maneuverability, but also in the important power input parameter for making a curvilinear motion since it is directly associated

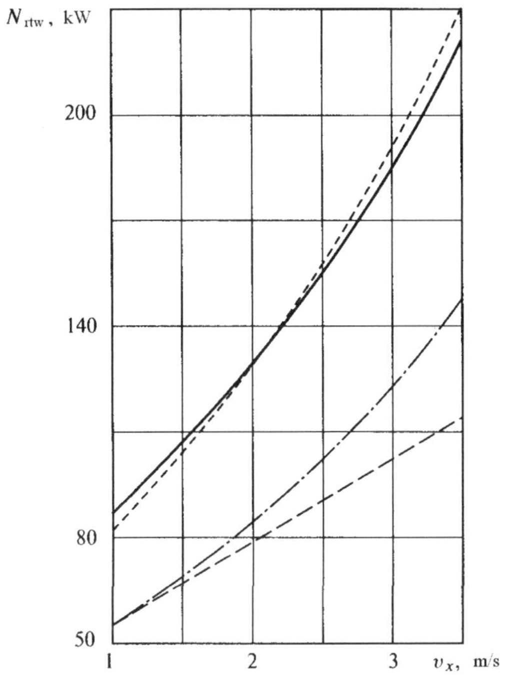

Fig. 15.38. Power input N_{rtw} versus the steady circle motion speed v_x of the 8×8 chassis with different StS diagrams: ---- with controlled wheels of the first and second axles; - - - - - with all controlled wheels; -·-·- rectilinear motion

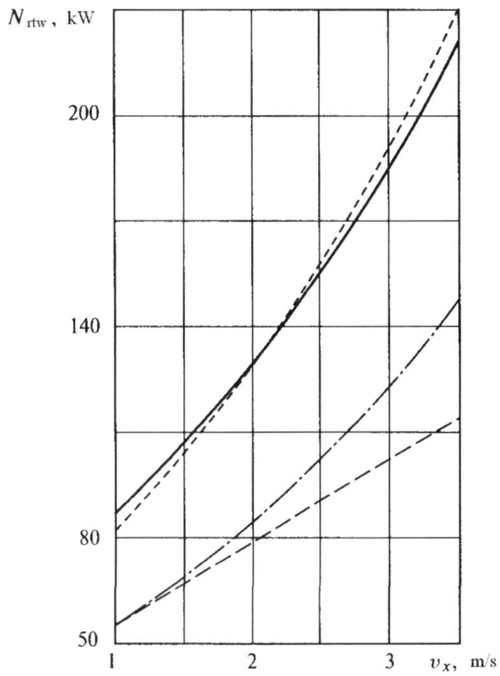

Fig. 15.39. Power input N_{rtw} versus the travel speed v_x of the 7929 chassis along the circle R_n = 18–19 m: ---- experimental data, 1-2-000 StS diagram; - - - - data of calculations, 1-2-000 StS diagram; -·-·- rectilinear motion, experiment; - - - data of calculations, 1-2-345 StS diagram

with the vehicle's efficiency factor and, in particular, with fuel consumption.

Fig. 15.39 presents the results of theoretical and experimental studies on the power input for the steady circle motion of the 10×8 7929 HDWV model, which has a 1-2-000 StS diagram. According to the data, the power required to move this chassis along a circle with a radius of 19.0 m at a speed of 3 m/s is about 137 kW (185 hp). The hourly fuel consumption for such motion is 46 l/h, if we base our reasoning on the load characteristic of the YaSh-8424 engine installed on the chassis. Using an all-wheel StS in the 7929 HDWV decreases the needed power under the same motion conditions down to 89 kW (120 hp) and the hourly fuel flow rate is 34 l/h, i.e., 1.35 times less than the StS diagram with controlled wheels of the two front axles.

Similar experimental studies conducted on the MAZ-543M HDWV (Fig. 15.40) gave similar results.

Thus, the all-wheel control ensures a decrease in the lateral wheel skidding and, due to this, a decrease in the power input required for curvilinear motion and improved economic parameters.

15.6 General design of automated control systems for curvilinear motion

15.6.1 General provisions

The most important structural element in the curvilinear motion control system (that is, in the

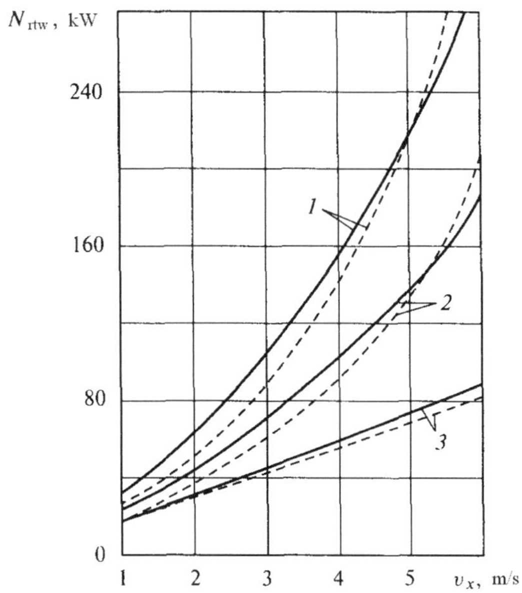

Fig. 15.40. Power input N_{rtw} versus the speed v_x of the MAZ-543M chassis at curvilinear motion: 1—z_n = 13.7 m; 2—R_n = 20.0 m; 3—rectilinear motion; ---- experimental data; - - - - data of calculations

steering system), as a subsystem of the "driver-automobile-road" system (see Fig. 11.1), is the curvilinear motion ACS. Let us consider its diagram, which is based on the above materials and Chapter 11.

In vehicles equipped with an ACS, the StS drive should meet the following requirements:

- Ensure different turning diagrams;
- Maximize the use of the driver's psycho-physiological and power capabilities;
- Use stabilization systems within the vehicle's control loop, which simplify driving and decrease the driver's fatigue; and
- Use systems of automated driving under the software and navigation signals of automated control when moving in columns (in particular, automatic execution of standard maneuvers, external radio and TV control of the vehicle, etc.).

In general, the onboard ACS of the vehicle's StS should include the following systems:

- A steering control unit (steering column) with automatic mechanisms for loading the steering wheel;
- A navigation complex;
- An onboard computer;
- A system of analog-digital-analog control;
- A block of executive mechanisms (steering mechanisms);
- Units and control systems in charge of turning the vehicle's wheels;
- Control and measuring equipment;
- A diagnostic system; and
- A power supply system (sources and power ducts).

The ACS should function in the command mode (under the driver's control) or in the automatic mode (under a program or TV and radio signals).

The basic diagram of the StS's onboard ACS is presented in Fig. 15.41.

The information about the steering wheel's (1) turn angle (in a multi-axle vehicle's variable turning diagram, it is possible for several signals to arrive at the same time) arrives in

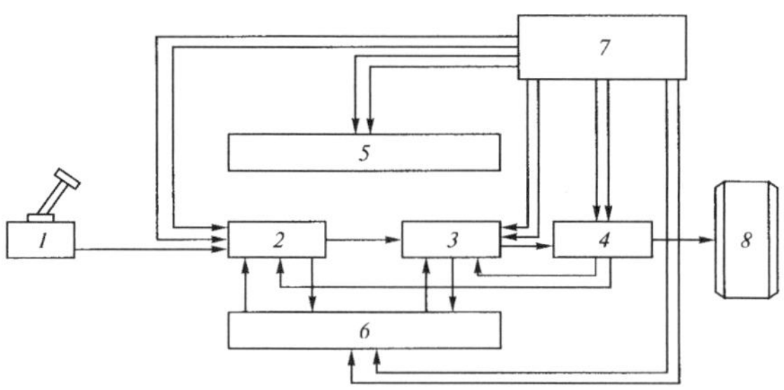

Fig. 15.41. Basic diagram of the StS's onboard ACS: 1—steering column with a loading mechanism (steering unit); 2—navigation complex; 3—onboard computer; 4—block of executive mechanisms; 5—block of instruments and controls; 6—diagnostic system; 7—power supply system; 8—vehicle's wheel

the navigation complex (2), which estimates the vehicle's external parameters (speed, acceleration, etc.). It then goes further to the onboard computer (3), which generates control signals to all executive mechanisms (4).

The computer's main program is set by the driver in accordance with the road situation. Additional information comes from the navigation complex (2).

The input information can be in both digital and analog form. In the latter case, it is initially subject to analog-digital conversion. Communication channels can be electric or optical (fiber-optical).

Control signals arrive at the executive mechanisms (4) through the autopilot units (though not shown in Fig. 15.41, they make up parts of the onboard computer (3) and navigation complex (2)), which makes a comparison with the feedback signal that brings information about the executive mechanisms' actual condition.

The diagnostic system (6) checks the correct operation of the ACS's individual units and aggregates. In case a failure in the system or a non-standard reaction to the control signals is detected, the diagnostic system disconnects the failed unit (element), makes the necessary emergency changes in the system, and issues a signal to the navigation system located in the driver's cabin.

If it is impossible to restore the channel's work by using the internal back-ups, the diagnostic system puts the wheels of the damaged channels into the middle position and switches on the external and internal emergency illumination or (if necessary) the vehicle's emergency stop system.

When preparing for motion, the diagnostic system checks the ACS's operation and sends a signal to the indexing system and the united onboard diagnostics to enable motion. When detecting a failure, the system holds its classification and sends the respective information to the onboard diagnostic network.

In general, the vehicle's automatic navigation with an onboard ACS is possible (automated driving en route, on special platforms, in columns, etc.).

An important ACS feature is the joint use of the steering control unit (1) (steering column) with the so-called loading automatic mechanism.

As in modern aircrafts, where onboard ACSs are widespread, a special loop is provided for stabilizing the loading of the control levers to equip the operator with the "feeling of control" and loop the kinematical force feedback in the pilot's organism, which prevents a loss of stability in the control loop (Fig. 15.42).

Due to huge mistakes in defining the range of the control levers' displacements, this loop, which was not present in StSs with mechanical drives, is crucial in modern high-speed aircrafts since it helps not only to ensure the optimum (by control comfort) handling quality characteristics, but also an increased piloting safety.

Conventionally, we can regard the driver as a part of the control loop—as a system of automatic regulation formed by the following interconnected elements: senses (as sensors), central nervous system (system of data

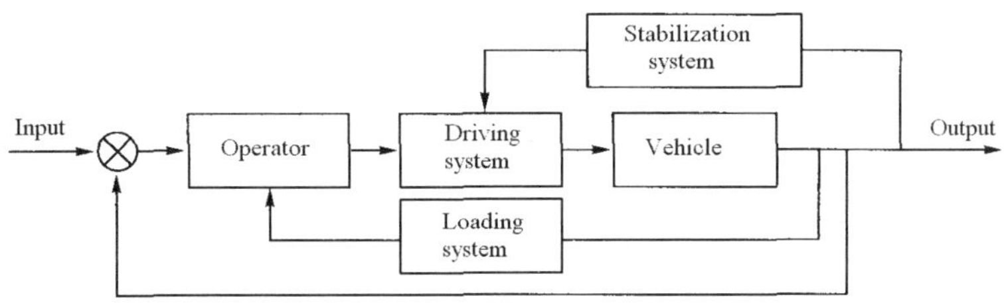

Fig. 15.42. Organization of additional stabilization loops

processing and decision making), and executive organs (arms, legs, back, etc.) (Fig. 15.43).

Information about the executive organs' operation is sent to the central nervous system through kinesthetic receptors, which are responsible for feedbacks in the human body and the human-operator system is looped up. In this context, it is very important to properly define the ranges of the executive organs' operation both by the displacement values and by the forces generated in the process; a change in force is perceived by the driver better than a change in displacement.

As a control link in the loop, the driver has the following properties which affect the control process: a response reaction delay to external signals (by 0.2–0.3 s); an ability to filter off external signals; an ability to change his or her own transfer functions within broad limits, including integration, differentiation, and some others.

A loss of stability in the control loop can happen both at very small and very large changes in the forces on the control lever since the driver is most precisely defining and executing the lever displacements and applying forces to them within the definite range of their values. Fig. 15.44 presents a change in the relative error $\bar{\Delta}$ in displacements and efforts on the control levers under their different values (*X*, *P*).

The simplest loading system (automatic mechanism) is a spring with a linear characteristic that is attached with one end to the control lever. Its other end is usually attached to the output rod of the so-called trimmer effect mechanism.

This mechanism is mounted in those cases when the long-term work with a deflected control lever is possible. In this case, it moves the loading mechanism's fixing point in the direction that decreases the spring effort.

Dry friction in the loading mechanism with springs degrades the handling quality. That is why it is more preferable to use torsion mechanisms, which are notable for a simpler design.

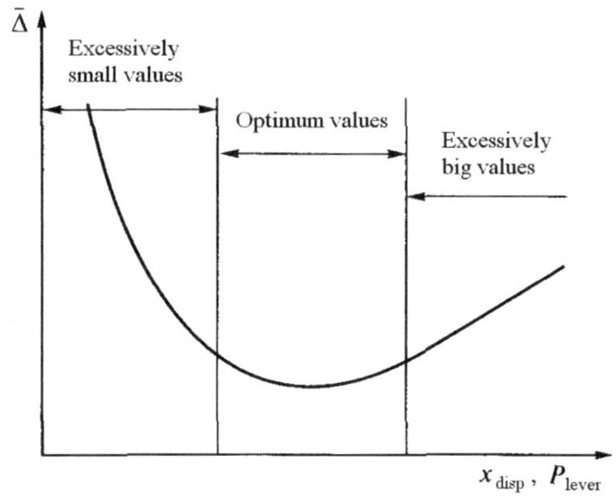

Fig. 15.44. Character of dependence of relative error $\bar{\Delta}$ on the values of necessary displacements X and forces P

Fig. 15.43. Flow-chart of human-operator

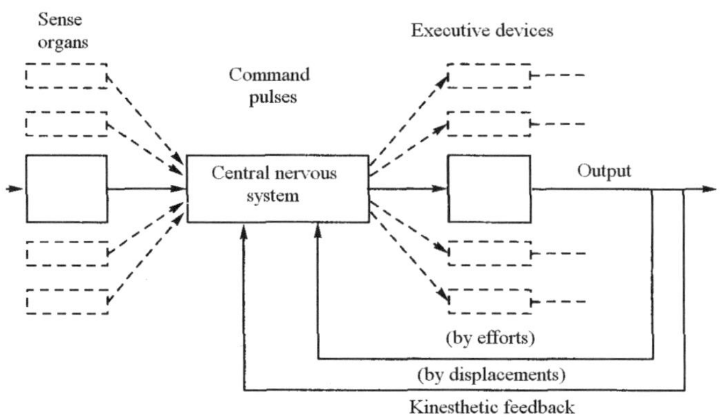

At an increase in speed, for example, in a flight, the forces on the steering surfaces grow quickly because of the spring-based loading mechanisms' increased efficiency, which causes the pilot to have difficulties with a precise defining the range of efforts, which may result in stability loss. Tests of vehicles with flexible StSs have shown that we can expect a similar effect when controlling a vehicle.

One way to improve the handling quality characteristics is to connect additional springs to the control levers.

One of the components of the moment that counteracts the controlled wheels' turn is the stabilizing moment M_s. With the growth in the turning angle, the M_s goes up. It also goes up somewhat with the growth in speed. The loading mechanism should imitate these processes.

A multi-axle chassis with a remote StS drive (RD-StS), created for the first time in our country at MSTU, has applied in the steering control unit a loading mechanism with linear springs for controlling two coordinates. The experiments showed that additional scientific studies are needed for selecting the optimal parameters of the onboard ACS's loading mechanisms.

If the drive of the StS's executive mechanism is equipped with a hydraulic steering booster, then at a sharp increase in the wheel turning's resistance moment it is desirable to have additional steering wheel loading.

An important element of the onboard ACS is the block of executive mechanisms; however, so far, the motor industry has little experience in designing it. In the practice of other branches, electromechanical and electrohydraulic blocks of executive mechanisms are of most interest; each can be designed differently and with different variants of technical solutions. Depending on the type used, we define the type of automated onboard system of the StS (AOS-StS).

The book gives below for the first time the general principles for designing the AOS-StS elements for electromechanical and electrohydraulic vehicles.

15.6.2 Electromechanical steering system

An electromechanical StS is comprised of electromechanical (executive) aggregates; their type almost completely defines the diagram and design of the AOS-StS.

15.6.2.1 Electromechanical steering aggregates

Vehicles with electromechanical AOS-StSs use several types of steering aggregates.

The first group includes steering aggregates with executive electric motors, whose shafts are directly linked with turning wheels by mechanical uncontrolled links.

Both DC and AC motors are used here as executive aggregates.

The former are most frequently motors of independent electromagnetic excitation and magneto-electric excitation (with permanent magnets), and, as a rule, have a tooth armature (series D, MM, and DPM). There are also motors with a smooth and toothless armature (series MIR, PGT); with a smaller moment of rotor inertia and a printed armature (series PYa, MRM); and with a hollow armature (series DPR).

The main advantage of DC executive electric motors (EEMs) is in their linear mechanical and regulation characteristics. The main drawback is in their essential dimensions, weight, and the presence of the commutator-and-brush assembly.

For steering aggregates in vehicles of high passing ability, we need powerful EEMs (3–5 kW per wheel). The motors of all the above series intended for automation systems have much smaller powers (Table 15.1).

Thus, vehicle steering drives demand a special power DC EEM and using a complex multi-stage gearbox with a transmission ratio of 90–450.

Table 15.1 Basic parameters of electric motors for automation systems			
Series	Power, kW	Rotor, rpm	Efficiency factor
DPM-N1, N2, N3	0.46–13.87	1800–9000	0.07–0.33
DPR	0.92–18.50	2500–9000	0.20–0.53
DM	1.60–40.00	6000–8000	0.20–0.67
SL	5.00–230	2300–5500	0.15–0.63
SD, DI	10.00–250	6000–7500	0.26–0.70

Most often, powerful EEMs are generator-controlled (change in the output voltage of the power amplifier). For example, the drives for turning tank towers (power of 3.5–4 kW) use an electric-machine amplifier. These drives have power amplification ratios of 5000–10,000 and quite short time constants. They are nicely automated, rather cheap, and simple in manufacture and operation. Still, the designing practice of such systems has shown that in their response, regulation precision, dimensions, and weight they are inferior to electrohydraulic drives, including those with radial-piston executive motors.

The AC EEMs are represented by specially designed two-phase asynchronous electric motors. They completely meet the main requirements for the steering EEMs (operation reliability and stability, response, uniqueness in regulation, simple control, and impossibility of self-operation). They have no commutator; therefore, they are much more reliable than DC EEMs. These motors are widespread in aviation steering machines and were adopted as the main variant when designing the electromechanical AOS-StS of a new project at MSTU. The executive asynchronous motor's main control method is distorting the stator rotating field.

High-power steering machines (hundreds watts) almost always use short-circuited rotors (squirrel cage), while hollow rotors in the form of an aluminum cup are very seldom used.

The parameters of aviation electric steering mechanisms, as a rule, are insufficient for vehicles. Thus, for example, the RD-7A-7 model steering machine of the AP-40 autopilot has a small limit moment on the driving pinion. This makes it impossible for any direct borrowing or slight modernization of aviation steering mechanisms and demands designing a special automobile steering aggregate.

15.6.2.2 Executive electromagnetic clutches

A different principle is used in drives with electromagnetic clutches (Fig. 15.45). Clutches are not active elements; they just deliver the moment generated by the drive motor. In most cases, such drive motors are not controlled and reversed; their angular velocity is kept nearly constant. The drive's operation principle is clear from Fig. 15.45. Its reversing is executed by using two electromagnetic clutches, whose driving

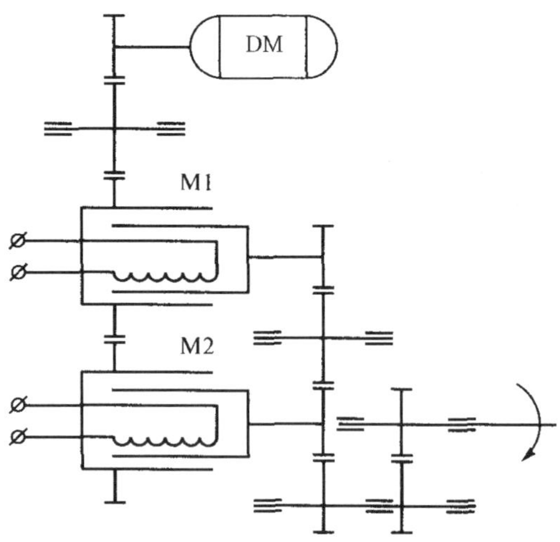

Fig. 15.45. Steering mechanism with electromagnetic clutches: DM—driving motor; M1 and M2—electromagnetic clutches

parts rotate in opposite directions. Thus, we have a rather complicated aggregate comprising of a gearbox with a high transmission ratio (90–450) and at least three electric machines.

The main advantages of this drive are as follows:

- A high power amplification factor;
- High permitted accelerations of the driving clutch parts because of their small inertia;
- Insignificant restrictions on the drive motor's type and characteristics; and
- A comparative simplicity of automating the control process.

Usually, powder and hysteresis electromagnetic clutches are used, which ensure smooth moment regulation on the drive's exit shaft.

Powder clutches admit accelerations of the driven shaft of over 100,000 rad/s^2 and have a power amplification coefficient of 120. The angular velocity of their driving parts should not exceed 200–210 rad/s because of high centrifugal forces acting on the magnetic powder. The powder's quick ageing and wear is a great disadvantage of powder clutches.

The angular velocity of hysteresis clutches is limited only by the bearings performance and reaches 1256 rad/s. Due to the driven parts' high inertia, the permitted accelerations do not exceed 3000 rad/s. The hysteresis clutches' power amplification factor at the driving parts' angular velocity of 620–630 rad/s is within the limits of 5–18.

At present, we are not aware of any designs of automobile and powerful aviation steering machines with electromagnetic clutches; therefore, we cannot reasonable judge on their AOS-StSs applicability to be used in HDWVs.

We deliberately do not consider here steering machines with electromechanical speed variators since the experience of using them in aviation practice as a whole is negative because of their complex design, poor reliability, and problems in reversing the drive.

15.6.2.3 Electromechanical AOS-StS

The Moscow Institute for Electromechanics and Automation (MIEiA) has developed, jointly with the Scientific and Production Center "Special Machine-Building" (S&PC SMB) of the Research and Design Institute "Special Machine-Building" (R&DI SMB) at MSTU, technical solutions for the electromechanical automated onboard system for the steering system (AOS-StS) with a two-phase asynchronous motor within the executive mechanism block.

The basic diagram of the AOS-StS for three controlled axles is presented in Fig. 15.46. The system was designed for n controlled axles and allows for expansion by adding identical modules of built-in control elements.

Fig. 15.47 presents the AOS-StS's general diagram, which is comprised of the following elements.

The steering control block (designated as SCB – (1, 2, 3) in the diagram of Fig. 15.47) is comprised of: the steering wheel mechanism (1) with the loading automatic mechanism (2), gearbox (3), and the block of DPS-4-2 model sensors for the steering wheel's turning angle. The sensors of the steering wheel's turning angle (3) is comprised of three 45D-32-2 model (powered by 36 V, 400 Hz) induction analog sensors of the turning angle or FEP 6G5.315.003 model (powered by a permanent voltage of ±15 V) digital photoelectric sensors.

The block to control the chassis' wheel (BCCW) consists of nine design-functional modules (DFM).

The modules KFM1, KFM2, and KFM3 (4, 5, and 6) are intended to generate control signals, which are sent to the executive mechanisms in the form of time modulation of the output thyristor keys' work depending on the DC input analog signals formed in the modules KFM5, KFM6, and KFM7 (7). Each boosting KFM1, 2, 3 is comprised of elements needed to control a pair of wheels: KFM1 for the first axle (the first and second

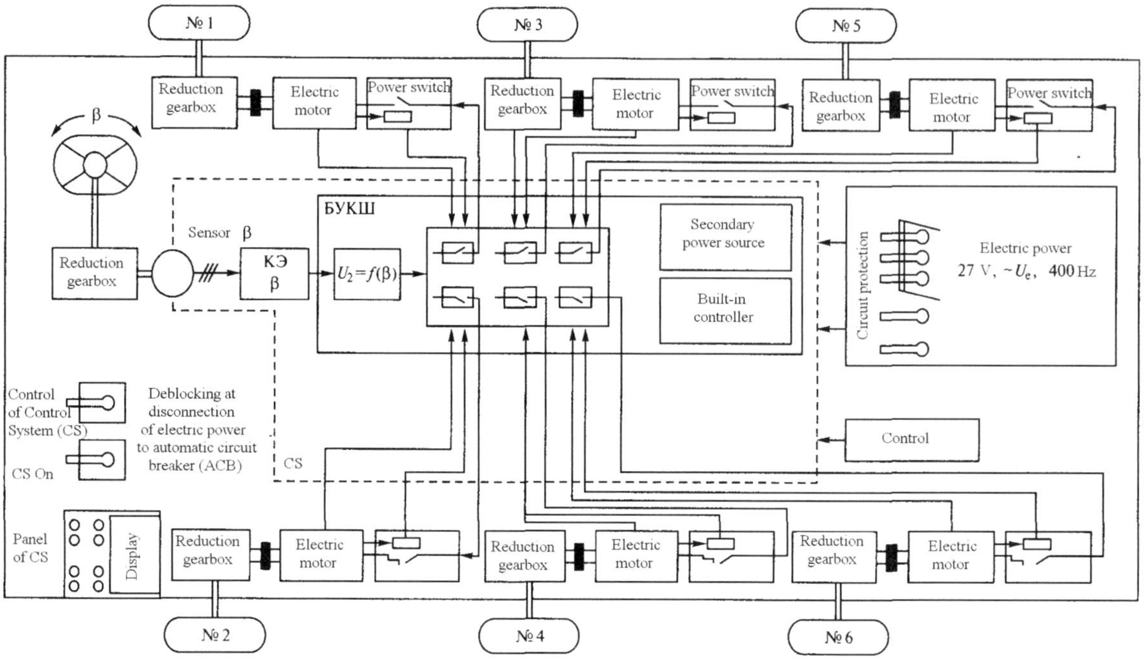

Fig. 15.46 Basic diagram of AOS-StS

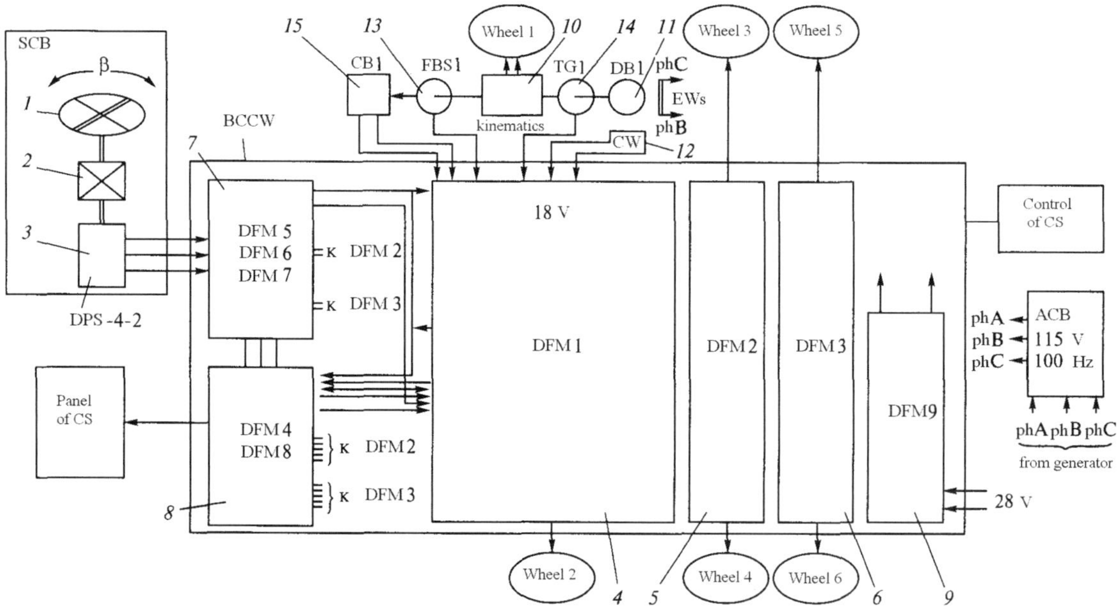

Fig. 15.47. General AOS-StS diagram of a full-drive vehicle: 1—steering wheel mechanism; 2—loading automation; 3—gearbox with a block of sensors for the steering wheel's turning angle; 4, 5, 6—boosting design-functional module; 7—digital computers; 8— internal control module for the computer and control blocks; 9—block of secondary current sources; 10—gearbox; 11—asynchronous electric motor; 12— electric motor's control winding; 13—feedback sensor; 14—tacho-generator; 15—secondary control block; β—steering wheel's turning angle

wheels), KFM2 for the second axle (the third and fourth wheels), and KFM3 for the third axle (the fifth and sixth wheels).

The modules KFM5, KFM6, and KFM7 are digital computers intended for processing and generating DC analog signals.

The modules KFM4 and KFM8 (8) are intended for the internal control of the computers' system (7) and the control blocks (4, 5, and 6).

Module KFM9 (9) is the secondary power sources block (±15 V) and their back-up.

The executive mechanisms' block is comprised of:

- the steering mechanisms system (connecting rods);
- the gearbox (10);
- a DB1 model two-phase asynchronous electric motor (11), the excitation windings (EWs) are connected to phases B and C of the generator and the control winding (CW) (12) is fed by phase A, i.e., a phase shift of 90° is ensured;
- the feedback sensor (FBS 1) (13), the signal from which shows the preset value of the wheels' turning angle depending on the steering wheel's turn (computer signals); and
- the tacho-generator (TG 1) (14).

The power supply system consists of a +27 V DC source (battery), a power generator with the voltage in phase of 115 V, and a frequency $f = 400$ Hz.

The power sources have protection and switch-on automatic devices.

The electric powering of the KFM circuit's elements in the BCCW block is provided from the 6VZ.233.088 secondary source. The source's output is ±15 V with the zero terminal and the input has a voltage of 27 VF from the DC network.

The secondary sources of electric power (the main and reserve ones) with the logic diagram ensure the failed source's switchover into the back-up, which is mounted in KFM9.

The control block is comprised of elements of the secondary control (75) and the control panel.

The design envisions a remote panel of the built-in control elements (quorum-element, voltage signaling device, comparator, OR elements), a container for the AOS-StS, and a cable circuit for connecting the BCCW .

15.6.2.4 Precision evaluation of electromechanical StSs

When designing StSs, we need to evaluate the precision of the preset correspondence of the steering wheel's turning angle β to the turn angle α of the vehicle's controlled wheels.

The precision is defined in static by errors K_β (of the angle sensor β) and $K_{\Phi B\Sigma}$ (of the feedback sensor for α), and in dynamics by the time constant T, which is inversely proportional to the Q-factor of the control loop Q, $T = 1/Q$.

When assessing the precision, as a rule, the following general dependences are used:

$$\alpha/\beta = K\beta/(K_{\text{FBS}}(T_p + 1));$$
$$T_p = 1/Q;$$
$$Q = K\,K_{FBS},$$

where: T_p is the resistors' time constant; and K is the proportionality coefficient defining the change rate of angle β.

We know from the experience of working with a similar control system that the static error is 10% when using the 45D-32-2 model sensors of precision class II (6%) in DPS-4-2, while the dynamic error is defined by the permitted Q-factor Q = 1–5 1/s.

The dynamic error is manifested as the wheel's turn delay α against the steering wheel's turn β.

A signal from the DPS-4-2 sensor is converted into voltage at the computer's output (KFM 5, 6, 7) with a discrete pitch of 0.15° β, if $\beta_{max} = \pm 90°$.

Assuming that at $\beta_{max} = \pm 90°$ and $\alpha_{max} = \pm 48°$ at the operation amplifier's input in KFM 1, (2, 3), the voltage $U_{in} = \pm 10$ V, and $U_{fb} = \pm 10$ V, we obtain the following: $K = 10/90 = 0.11$ V/deg; $K_{fb} = 10/48 = 0.21$ V/deg.

The dynamic error:

$$\Delta(t \to \infty) = K_{\beta}\, \mathrm{K}\, \left(1 - \exp(t/T)\right)/Q\,,$$

where $K = \beta/t$; t is the time of the steering wheel's turn to the angle of β.

Table 15.2 presents the calculated values of the dynamic error $\Delta(t \to \infty)$ and the respective values of the dynamic error reduced to the wheel's turn δ_{ψ} $(t \to \infty)$ at varying the value of the proportionality coefficient (K = 10–50 deg/s) and Q-factor of the loop $Q = 1/T$= 1–5 1/s:

$$\delta_a(t \to \infty) = \Delta\,(\mathrm{t} \to \infty)/K_{\mathrm{FBS}}$$

15.6.3 Electrohydraulic steering system

15.6.3.1 General diagram of electrohydraulic AOS-StS

The electrohydraulic AOS-StS uses the electrohydraulic drive as the executive mechanism. This drive converts the electric control signal into the displacement of the executive mechanism's output link and strengthens the control signal by using the hydraulic power station's energy.

Fig. 15.48 shows the diagram of the electrohydraulic servo drive.

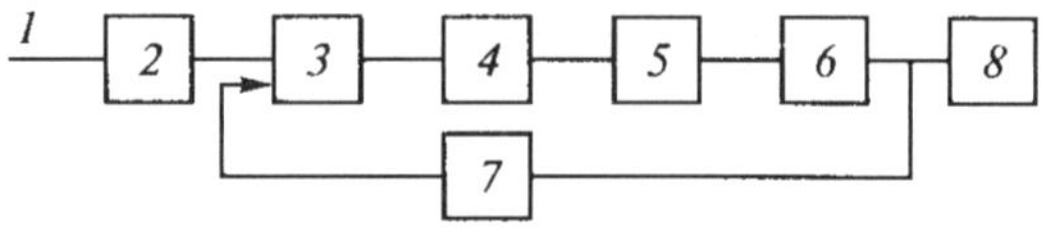

Fig. 15.48. Structural diagram of electrohydraulic servo drive: 1—control signal; 2—input link; 3—comparing link; 4—regulating device; 5—oil pump; 6—output link; 7—feedback; 8—executive link

There are drives with throttle regulation (regulates the pressure in the executive mechanism) and volumetric regulation (regulates the volume of fluid arriving in the executive mechanism).

The main advantage of the StS with volumetric regulation is a higher efficiency factor, especially for drives manufactured autonomously. In this case, we get an essentially simplified maintenance and operation of the steering aggregate, as well as vehicle manufacture since we exclude cumbersome operations of mounting hydraulic systems, which demands the assembly personnel's high qualification and exclusive care.

The dependence of the change in the output link drive's displacement rate on the external load value acting thereon (in particular, of the joint moment M_j required to turn the controlled object) at a permanent mismatch value between the drive's input and output links is called the drive's mechanical (load) characteristic.

Table 15.2 Dynamic errors of the StS $\Delta(t \to \infty)/\delta_a\,(t \to \infty)$ depending on K_{FBS} and Q

K_{FBS}, deg/s	Q, 1/s				
	1	2	3	4	5
10	1.1/5.2	0.55/2.6	0.37/1.73	0.275/1.3	0.22/1.04
20	3.3/15.6	1.1/5.2	1.1/5.2	0.825/1.4	0.66/3.12
30	6.6/31.2	2.2/10.4	2.2/10.4	1.65/7.8	1.32/6.24

Fig. 15.49 presents the dependences of the required power N and the displacement rate of the output link $\dot{\delta}$ for drives with throttle and volumetric regulation depending on joint moment M_j at the output.

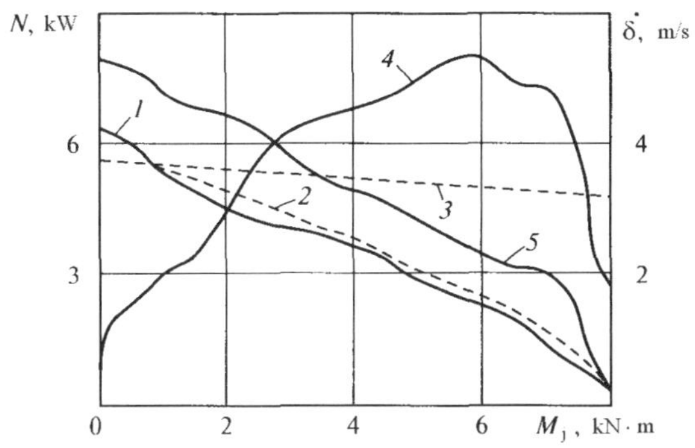

Fig. 15.49. Displacement rate of the output link $\dot{\delta}$ and the required power N versus the joint moment M_l at the output of drives with throttle and volumetric regulation: 1—the drive's required mechanical characteristic;
2—mechanical characteristic of the drive with throttle regulation; 3—mechanical characteristic of the drive with volumetric regulation; 4—required power for the drive with volumetric regulation; 5—required power for the drive with throttle regulation

It is seen from the figure that when ensuring the same mechanical characteristic, the required power of the autonomous drive with volumetric regulation is somewhat lower than the drive with throttle regulation.

However, applying autonomous drives requires solving a number of technical problems, in particular, cooling the working fluid (in centralized hydraulic systems cooling devices are simpler and more reliable).

Another problem is to ensure that the steering aggregate operates in emergency conditions, such as when the engines stop or electric power fails. It is easier to eliminate this problem with the help of the autonomous drive of the StS with volumetric regulation.

The electrohydraulic drive of the StS with throttle regulation was implemented in the remote StS (RD-StS) drive of the E79085 multi-axle chassis prototype designed and built by S&PC SMB at MSTU.

The practice of motor building has no technical solutions for using electrohydraulic steering drives; therefore, it is impossible to give preference to any particular type of steering aggregates. It is necessary to develop all of the known directions and accumulate the experience of designing and operating the automobile type aggregates and undertake scientific studies to accompany experimental design works.

At the same time, the practice accumulated by adjacent engineering areas (aviation) of using steering aggregates with throttle regulation and a centralized system is quite rich.

S&PC SMB at MSTU has conducted a series of works aimed at creating a steering drive with throttle regulation and a centralized system for multi-axle vehicles; thus, this issue is rather deeply studied.

In this context, new HDWV projects should consider a less studied issue—creating an automobile electrohydraulic autonomous steering aggregate with volumetric regulation.

15.6.4 StS with autonomous drive

15.6.4.1 General StS diagram with autonomous drive

The basic diagram of the StS's hydraulic system is presented in Fig. 15.50. It is comprised of the following systems and elements:

- The main system: 1—main axial-piston pump with regulated capacity; 2—pipeline filters; 3, 4, 6, and 16—diverter valves; 5—stop valves; 7—executive mechanism;
- The main control and refilling system: 9—gear-type pump (with protection system); 8— electrohydraulic booster's spool; 13— turning mechanism's executive cylinders of the main pump's support washer

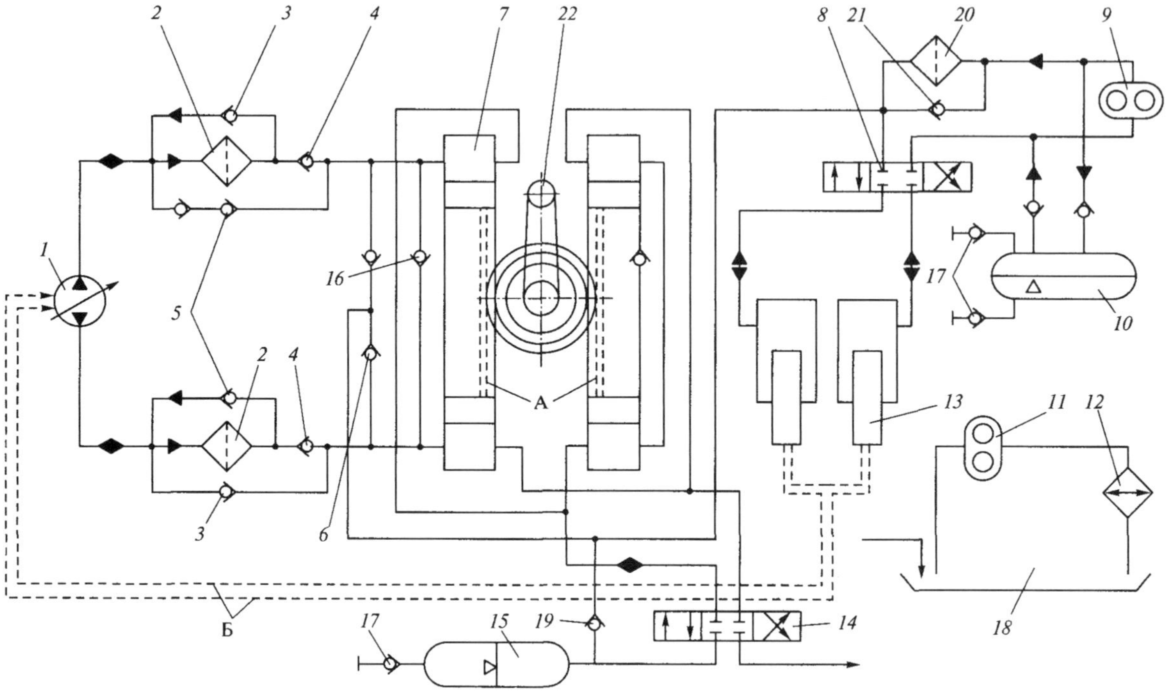

Fig. 15.50. Basic diagram of the StS's hydraulic systems: A—mechanical link; B—drainage

(bolster); 10— refilling system's refilling tank; 17—charging valve; 20 —filter; 21—bypass valve;

- The emergency (backup) control system: 15—hydraulic accumulator; 14—emergency electrohydraulic booster's spool; 19— hydraulic accumulator's charging valve; and
- The cooling and lubrication system: 11—gear-type circulation pump (with protection system); 12—heat exchanger; and 18—draining tank (drive electric motor's casing). (The electrohydraulic converters are not shown in the diagram.)

15.6.4.2 Work of the autonomous steering drive

The main axial-piston pump (1), the refilling system's gear-type pumps, the supply control of the main pump (9), and the cooling system's pump (11) are all driven from one asynchronous electric motor.

The main pump's swivel washer is controlled by the electrohydraulic drive, whose first cascade uses an VS201 electrohydromechanical converter with a "swinging nozzle" (it is possible to use a nozzle-damper plate converter if the drive's manufacturing and operating conditions ensure Class 8–9 of the working fluid's purity (according to GOST 17216) and a filtering fineness of 6 μm). The hydraulic booster's spool (8) moves away from its neutral position under the pressure gradient on the faces and directs the working fluid's flow into the cavities of executive cylinders (13) of the mechanism in charge of controlling the tilt of the main pump's washer.

The refilling tank (10) is connected to the control system's drain pipeline; refilling is done from the refilling pump's pressure union. This tank is used as a compensator for the fluid temperature's expansion. The initial charging of the hydraulic system and the tank's filling with nitrogen are done through the charging valves (17). The fluid in the control system is filtered by a filter (20) with the fineness of 6 μm (10), which is equipped with the bypass valve (21).

The main hydraulic system is looped. The refilling system is connected to it with the help of bypass (refilling) valves (6).

The high pressure and drain lines are interconnected by channels and bypass (protective) valves (16), which operate if the pressure in the manifold is over 32 MPa. Each filter (2) of the main system is provided with a cut-off and lock valve (5), which allows the fluid to flow through the filter in only one direction, as well as with a bypass valve (3), through which the system works when the filters are clogged. The executive mechanism's (7) hydraulic cylinders are connected in parallel, ensuring back up of their functions and improved drive reliability in case one the cylinders loses its air-tightness.

The hydraulic cylinders' (7) pistons are integrated with toothed racks geared with the cylindrical pinion for turning the tie-rod arm's (22) shaft.

The emergency system is designed for 30 seconds of operation and has a hydraulic accumulator (15) and an automatic electrohydraulic booster, which are engaged upon a signal about an emergency pressure drop in the pipelines of the control system and refilling tank or upon a signal about a voltage failure in the driving electric motor's power circuit.

The information about angular displacements of the steering machine's elements is generated by the analog sensors installed:

a. On the tie-rod arm's shaft;
b. On the support shaft of the main pump's swinging washer.

15.6.4.3 Design features of autonomous steering drive

The executive mechanism of the hydraulic system and the electric drive are united into the autonomous steering aggregate.

The asynchronous electric motor with cooling of the hydraulic system's working fluid rotates the main axial-piston pump and both gear-type pumps of the hydraulic system. The engine shaft's nominal rotation rate is rather high (12,000 rpm); that is why the aggregates are connected with the help of the torsion shaft with toroidal splines.

The auxiliary pumps (9 and 11 in Fig. 15.50) have internal meshing, while the hydraulic accumulator and refilling tank are piston type with nitrogen supercharge.

The casing of the electric motor's cooling system is used as a draining tank; the oil is pumped off by the cooling system's pump. There is no indication of the casing overflow; therefore, when modernizing the unit it is expedient to envision pumping off the working fluid from the casing into the supplementary tank through a respective filter and connecting the cooling system to the general hydraulic system.

Maintaining the steering unit is comprised of mainly replacing filters, which is done from outside without the need to disassemble the vehicle. Filter lids are provided with cut-off valves, which minimize losses of the working fluid (WF) at maintenance and the risk of contamination.

The supplementary tank is filled through a special device, which has a reliable WF cleaning system (the filter fineness is at least 6 μm). This design has no check of the working fluid. When finalizing the design, it is necessary to envision a position indicator for the supplementary tank's pistons.

The drive electric motor matches the parameters of a traction electric network by its electric parameters. In case the propulsion engine stops, it is switched over to powering the motors-in-wheel, which work in the generation mode and, at motion at low speed, from the onboard or external DC grid through a frequency converter.

The working fluid is the crude oil-based AMG-10 grade (according to GOST 6734) or the polysiloxane-based 7-5OS-3.

15.6.4.4 Calculation of parts and units of autonomous steering drive

The elements are calculated with account of the maximum moment acting at the turn on the wheel. It is assumed that the maximum load values on the steering system are achieved when we turn the immobilized wheels standing on a concrete foundation. For these calculations, we can use, for example, the Hooke formula:

$$M_w = 10\, P_w^{3/2}\, \mu / A\, p_{tir}^{1/2}\ ,$$

where: μ = 0.7–0.9 for a hard dry ground; P_w is the vertical load on the wheel; A = 2.1 is an empirical coefficient; p_{tir} is the air pressure in the tire; and p_{tir} = 0.05–4.0 MPa for tires sized 1300×530–533. In general, the above calculation, like all other calculations of the loads on the StS presented in this chapter, is rather approximate. The steering drive's elements defined by the maximum moment, as a rule, are oversized. However, currently there is no more precise methodology to calculate the steering drive.

15.7 All-wheel steering system for a 12×12 HDWV

The StS of a new 12×12 HDWV designed and made by S&PC SMB at MSTU sets the vehicle's motion in the required direction.

The prototype is equipped with an all-wheel StS (AWStS), which functions under the algorithm presented in points 15.3 and 15.4 above. It is comprised the following (see Fig. 15.51):

- The steering mechanisms (1) with the turning angle setting units;
- The turning pole setting units (2);
- The onboard computing complex (OBCC) (3);
- The hydraulic system (oil tank (4), pump (5), hydro-pneumo-accumulator (6), and booster (7), while the safety valve, filter, and control-and-measuring instrumentation are not shown in the diagram);
- The feedback sensors (FBS) (8) with the drive;
- The brackets with levers (9);
- The lateral connection rods (10); and
- The tie-rod arms (11).

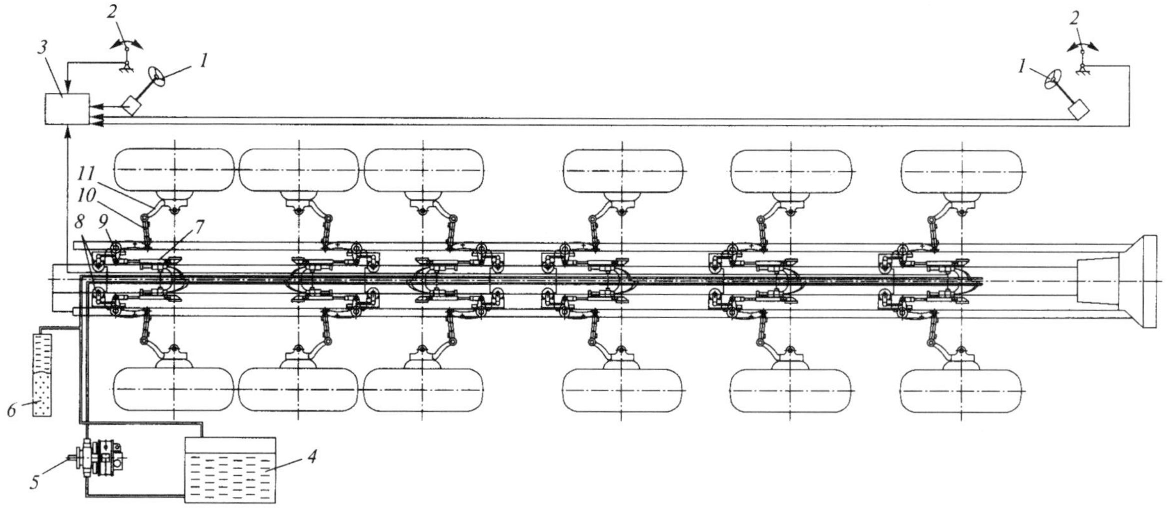

Fig. 15.51. Diagram of the all-wheel StS for a 12×12 HDWV

The OBCC forms, together with the *FBS* and the hydraulic system, a complex of steering servo systems of the wheel mechanism (CSSS-WM), whose functional diagram is presented in Fig. 15.52.

The steering mechanism (Fig. 15.53) is intended for driving the StS into action. It is mounted on the right-hand side in the front and rear cabins. The steering mechanism is attached to the sub-frame (7). The motion from the steering wheel (1) to the turning angle (5) setter is transferred through the steering column (2), cardan transmission (3), and steering gearbox (4), which ensures the required transmission ratio between the steering wheel and the sensor. For stabilizing the motion and creating the necessary effort on the steering wheel, there is the loading automatic device (6), which has one side fixed to the prototype with the help of the bracket (8) and the other side connected to the steering gearbox via the tie-rod arm.

The composition of the steering servo system (SSS), whose functional diagram is presented in Fig. 15.54, is as follows:

- Four adders (Σ1–Σ4) intended for summing up electric signals and the block of electrohydraulic control (EHC);

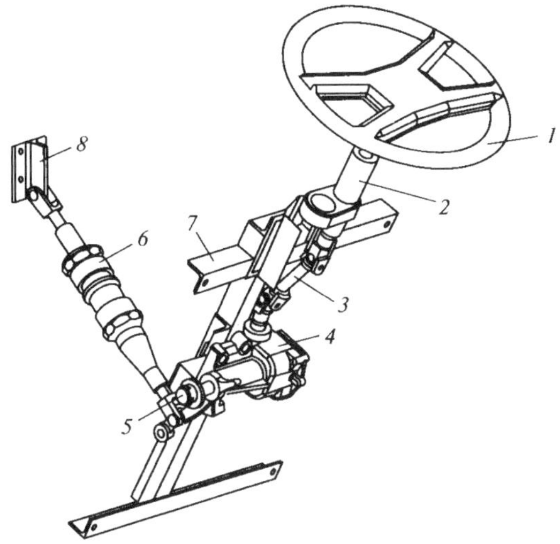

Fig. 15.53. Steering mechanism: 1—steering wheel; 2—tie-rod arm; 3—cardan transmissions; 4—steering gearbox; 5—turning angle setter; 6—loading automation; 7—sub-frame; 8—bracket

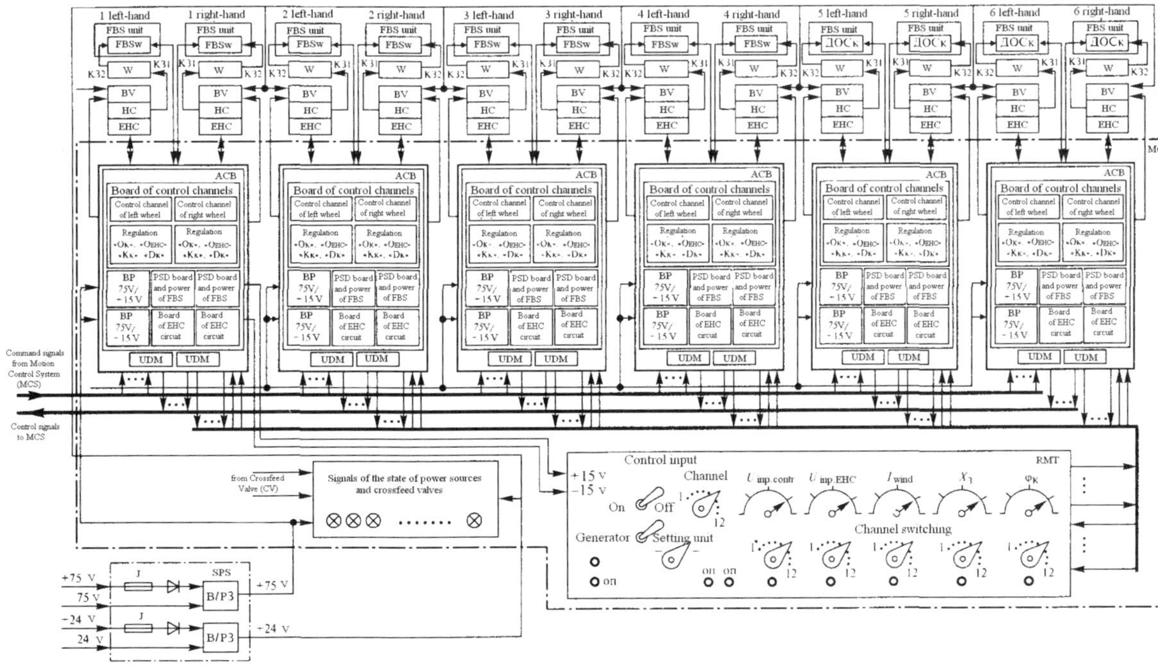

Fig. 15.52. Functional diagram of the CSSS-WM

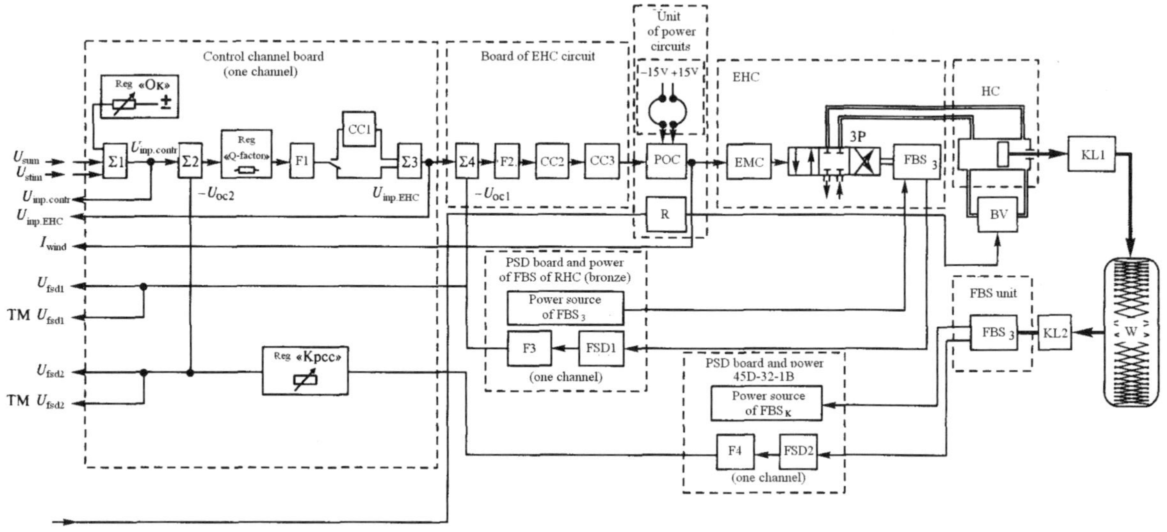

Fig. 15.54. Functional diagram of the SSS

- Four electric filters (F1–F4) for filtering off the noise;
- Three correcting cells (CorrC1–CorrC3) for correcting the SSS's dynamic parameters and increasing its precision;
- A power output cascade (MBK) for amplifying the control signal;
- An electromechanical converter (EMC) for converting the control signal into the mechanical displacement of the output link;
- A spool-type distributor (SDB) for regulating the direction and flow rate of the working fluid arriving in the hydraulic cylinder;
- A hydraulic cylinder (HC) intended for converting the fluid's flow into the rod's linear displacement;
- Feedback sensors of the spool position (FBS_S) and the turn angle of the wheels (FBS_w) for converting mechanical values into electric ones;
- Two phase-sensitive detectors (FD1, FD2) intended for converting AC signals into DC signals;
- A banding (cross-feed) valve (BV) intended for connecting and disconnecting the HC's cavities depending on the presence or absence of an electric signal in its windings;
- A relay (R) for connecting the BV windings to the electric power source;
- Jumpers (J) for disconnecting the MBK from power sources;
- A regulator of the wheels' zero position (Reg. "0K");
- A regulator of the SSS's static transmission factor (Reg. "Kstr");
- A regulator of the SSS circuit's Q-factor (Reg. "Dstr");
- Two power units: +15 V and –15 V (BPP+15 V and VP–15 V), which are common for the SSS of the right- and left-hand wheels of one axle;
- A DC power source for the FBS_s;
- An AC power source for the FBS_w; and
- Kinematical links that connect the HC's rod with the wheel (CL1) and the wheel with the FBC_w (CL2).

The SSS works as follows (see Fig. 15.54). The input of the SSS receives from the motion automated control system (MACS) a control signal (U_{cs}), which passes through Σ1 and is summed up on Σ2 with the wheel turn's angle feedback signal (U_{fb2}). The signal $U_{cye} - U_{fb2}$ passes through F1 and Σ3 and is added at Σ4 with the SDB spool's position feedback signal (U_{fb1}).

The signal $U_{cye} - U_{fb2} - U_{oc1}$ passes through F2, CorrC2, CorrC3 and, after filtering and correction, arrives at the input of the power output cascade (MBK) and from its output goes to the windings of the electromechanical converter (EMC). Under the effect of this signal, the EMC's armature deflects from its neutral position; the output link of the EMC displaces the SDB spool and the FBC_s's armature, which is integral with the spool, away from the zero position. One cavity of the HC gets connected to the pressure line, the second one with the drain line, and the HC's rod starts moving.

At the same time, the wheel (connected through CL1 with the HC's rod) and the FBC_w's armature (connected with the wheel through CL2) start turning. At the armatures' displacement, AC signals appear at the outputs of the FBC_s and FBC_w; these signals are proportional, respectively, to the deviations of the SDB spool and the wheel from the zero position. These signals are converted by FD1 and FD2 into DC signals, filtered by F3 and F4, and arrive at Σ4 and Σ2 (U_{oc1} and U_{fb2}). As a result, the values of $U_{co} - U_{fb2}$ and $U_{co} - U_{fb2} - U_{oc1}$ begin decreasing until the following correlation is reached:

$$U_{co} - U_{oc2} = 0;$$

$$U_{co} - U_{oc2} - U_{oc1} = 0.$$

Here, the SDB spool takes its zero position, the working fluid does not arrive in HC's cavity, HC's rod and the wheel do not move, and the wheel's turning angle corresponds to the control signal value at the input of the SSS:

$$U_{fw} = U_{co} \, K_{pcc},$$

where K_{str} is the system's transmission factor.

If it is necessary to interconnect the cavities of HC, a one-piece command PK_{cs} is delivered to the SSS's input, which triggers the P; the BV's windings are connected to the secondary power source (SPS); and the BV opens and interconnects the HC's cavities. The condition of the BV is indicated by LEDs: green means the BV is disabled and red means the BV is enabled.

A possibility is assumed to regulate the wheel's zero position, the static transition factor, and the Q-factor systems. The two latter parameters are regulated by selecting the constant resistors' values, which are soldered into the board. The wheel's zero position is regulated with the help of a variable resistor and can be adjusted in the course of operation.

The powering of the electric circuits (except for the BV's windings) is ensured by the VP+15 V and BP–15 V; a possibility of simultaneously using the sources for two SSSs of one axle is defined by the MBK's different polarity of these SSSs when the axle's wheels turn in one direction. The power to the VP+15 V and VP–15 V is supplied from the SPS and is indicated by LEDs.

The AC power sources for FBS_s and FBS_w are converting the permanent voltage of ±15 V into the alternating two-pole voltage of rectangular shape.

In the control or adjustment mode, stimulating signals (U_{stims}) are fed to Σ1 from the control and test pendant (C&TP); the SSS's reaction and its compliance with the expected one are evaluated by the values and character of the signals proportional to $U_{in\,contr} = -U_{stims}$ (the signal at the input of the wheel loop), $U_{in\,ehb}$ (the signal at the input of the electrohydraulic booster loop), I_{wind} (the current in the EMC's windings), U_{FD1} (the SDB spool's displacement), and U_{FD2} (the wheel's turning angle), which arrive from the SSS's output to the C&TP.

To make it possible to analyze the status of the SSS, the MACS receive telemetry signals proportionate to the spool's displacements and the wheel's turning angle (TM U_{FD1} and TM U_{FD2}).

The system envisions a possibility to increase the loop's Q-factor (improve the precision) at low-frequency changes of U_{cs}, which is implemented by connecting the CorrC1 and selecting its parameters.

The C&TP is intended for supplying the stimulating U_{stims} signal to the input of one of the SSSs and measuring the values of $U_{in\ contr}$, $U_{in\ ehb}$, I_{wind}, U_{FD1}, and U_{FD2} (U_{FC}) in the control points of that SSS.

The stimulating signal can be fed from the potentiometer "setter" or generator connected to the sockets "generator" and "common wire" (CW). The measuring of the above parameters is made with the help of five pointer instruments included into the C&TP or with the help of external devices connected to the respective sockets and the CW socket.

For connecting the controlled SSS, six rotary wafer switches of the C&TP are set into the position corresponding to the number of the selected SSS.

All of the above electronic cells, regulators, and power units, as well as the C&TP, are integrated into the management and control block (MCB).

Fig. 15.55 presents an example of the assembly drawing for the axle control block (ACB). The ACB is comprised of a two-channel board (1) of the control channels; two boards (2) of the EHC loop; a two-channel board (3), FD, and feedback power of the EHC; a two-channel board (4) of the FD and power FBS45D-32-1B; and two different-polar BP (5). The electric connection of the ACB with the MCB's (6) units and circuits is achieved through blocks X1–X7. The mechanical link of the ACB elements into a united block is ensured by pins (8) and clamps (7). The total number of ACB blocks in the MCB corresponds to the number of axles (six, see Fig. 15.52).

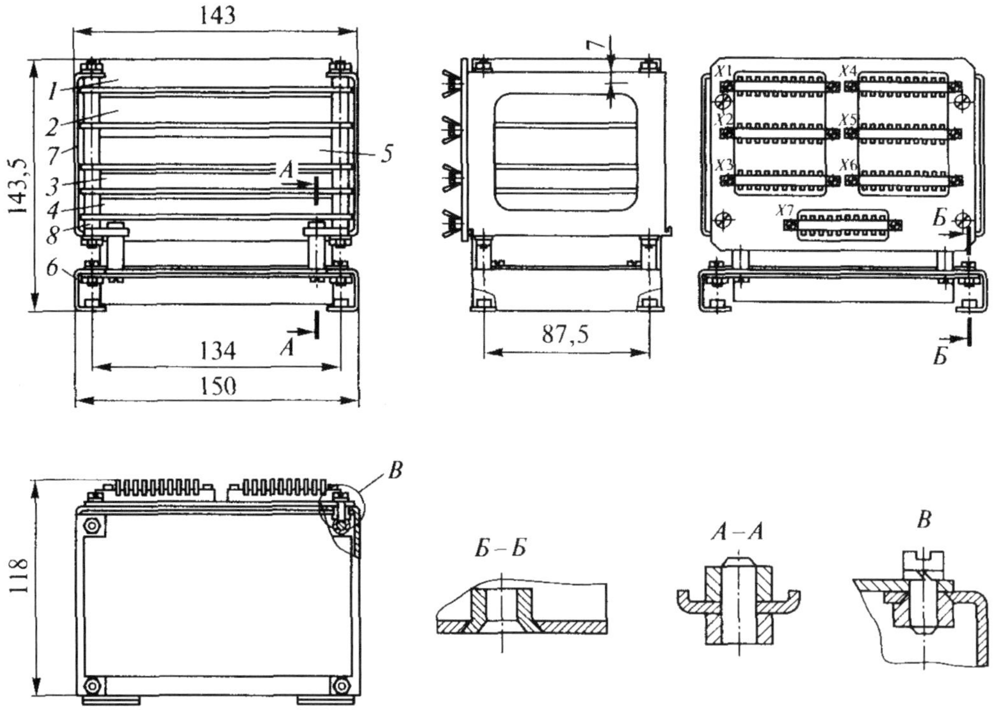

Fig. 15.55. Block of axle control (ACB)

The radiators attached on the lateral sides of the MCB carry 12 (by the number of wheels) MBK (power output cascade) blocks.

The side panel of the MCB (Fig. 15.56) carries the following:

- Two connectors for receiving control signals from the MACS to the SSS's inputs and one-time commands for switching on the KK (INP. URR and INP. KK);
- Two connectors for delivering the telemetry signals to the MACS proportionate to the displacements of the EHC's spool and the wheel turn's angle (TM. EGU and TM. KOL) and two connectors for electric power (POWER, axles 1–3, axles 4–6);
- Twelve connectors for the control signals' transmission to the windings of the EMC and KK, bringing power to the FBS_s and FBS_w, and receiving signals from them (block SSS, 1L-6L, 1P-6P, where the numeral stands for the axle number and L for the left-hand and P for the right-hand wheel);
- Six indicators of the 75 V power supply the ACB's BP (block AXLE, 1–6); and
- Twenty-four indicators of the BV's condition (block LOOPING VALVES, 1L-6L, 1P-6P, green means that the BV is disabled and the red means that the BV is enabled).

The C&TP is fixed on the MCB's lid, which together with the C&TP, is unfolded and fixed by a special clamp in case of adjusting or controlling the CSSS-WM.

The MCB's fixing points are provided with shock absorbers.

The secondary power source (SPS) is intended to protect the electric power circuits, units, and blocks, which make up part of the CSSS-WM, against voltage and current surges and drops, transient processes, pulsations, and noise, which may arise in operation of the onboard power source.

To exclude any mutual influence of power circuits on the channels of the CSSS-WM, the +75 V circuit of the onboard power source is split in the SPS into six two-wire circuits. Each has a safety fuse, a protecting diode, and a capacitive filter, which protect the VP+15 V and VP–15 V SSS power circuit of the right- and left-hand wheels of one axle.

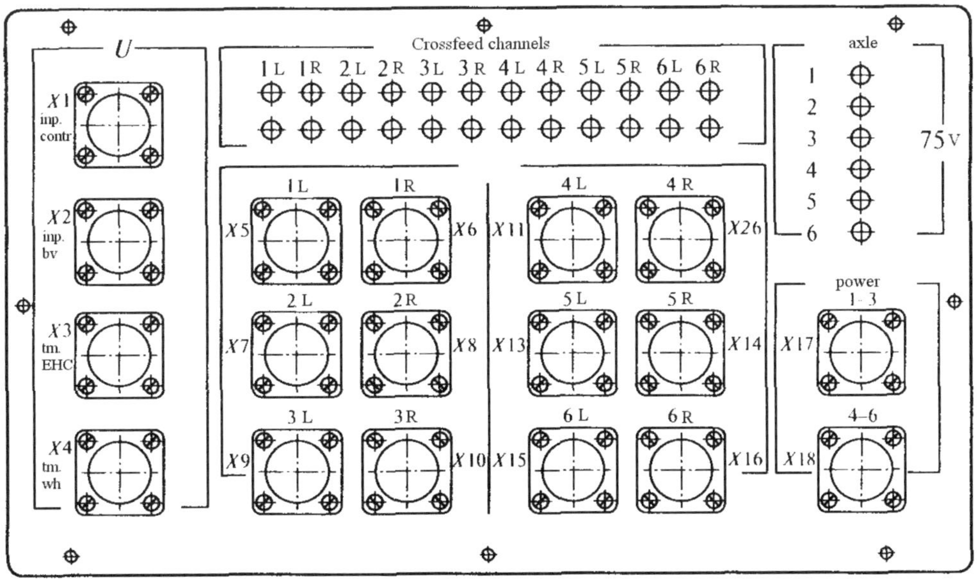

Fig. 15.56. Side panel of the MCB

The +24 V circuits are used only at switching the BV on and consist of a general section and six independent other sections, each is intended to supply power to the SSS's BV of the right- and left-hand wheels of one axle. The general part has a diode and a powerful capacitance filter and each independent section has fuses.

The capacitance filters in the +75 V and +24 V circuits are united into the protection filters block.

The SPS's front panel carries the following:

- Two powerful terminals for connecting the +75 V voltage from the onboard source protected by the lid and fixing screw;
- A connector for bringing the +24 V voltage from the onboard source;
- A switch simultaneously switching on the +75 V and +24 V voltages;
- An indicator of the +75 V power supply;
- Two connectors for supplying electric power to the MCB (POWER, KRSS KM, axles 1–3 and 4–6);
- Fuses in the +24 V and +75 V power circuits in the electric units of two SSSs of each axle (axles 1–6); and
- Fuses in the secondary powering circuits (75 V/15 V and 75 V/24 V).

The SPS's fixing points are provided with shock absorbers.

The oil tank is mounted in the engine section in the vehicle's upper part above the pump. Its volume is 250 liters.

The oil tank has two windows to control the minimum and maximum oil levels. The tank's body has a mesh suction bell.

The oil tank is connected to the drain and pressure pipelines.

The StS pump (Fig. 15.57) is an axial-piston regulated pump with a 313.3.107.3704 model pressure regulator intended for supplying the WF and maintaining the preset pressure in the system.

Functionally, the pump consists of two parts: a swinging unit and a regulator block.

The swinging unit is comprised of the shaft (1) mounted in the body (27) on bearings (5 and 6). From the side, the pump's shaft face is covered by the lid (3), which is sealed by a rubber ring (4) to the gasket (2).

Fig. 15.57. StS pump

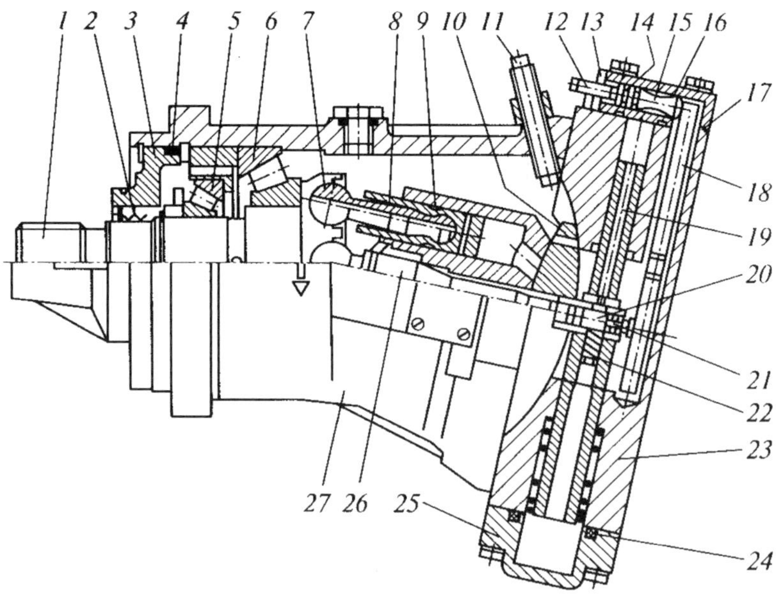

The flange is connected through the connecting rods' (7) spherical heads with the pistons (8) and the pin (26). The pistons move inside the cylinder block (9).

The piston stroke's value is defined by the angle formed by the block cylinders' rotation angle and the shaft (1). The block (9) contacts, by its spherical surface, the distributor (10), which is adjoined by its opposite side with the support surface of the regulator's (23) body.

The regulator block consists of the finger (20), the fixing screw (22), a two-edge spool (21) with a shoe and a thrust block, a two-arm lever (18), and a lid (15) with rubber rings (17), which houses the parts (12, 13, 14 and 16) with different functions mounted inside the stepwise piston's (19) body (23). From the bottom, the regulator block is covered by the cup (25) with the ring (24).

The pneumatic-hydraulic accumulator (PHA) (Fig. 15.58) is intended for maintaining the working pressure in the hydraulic system at a non-working pump.

The PHA consists of a pneumatic valve (1), steel sleeve (2), lids (3 and 4), and piston (5). The lids are fixed by nuts (6), which take the working medium's pressure.

The piston is sealed by rubber rings (10, 11, 12, and 13). The lids are sealed by rubber (7) and protective rings (8). The pneumatic valve (1) is sealed with the help of the rubber ring (9).

The nuts (6) are locked by lock screws (14).

The booster (Fig. 15.59) is the wheel's control element and consists of three parts: the hydraulic cylinder (1), which is the power element; the electrohydraulic distributor (EHD) (2); and the cross-feed valve (3). The EHD is a single-cascade electrohydraulic booster of direct action and consists of the following main units (Fig. 15.60): the moment motor (MM) (1) and the spool distributor (2) with the spool position's induction sensor (3).

The EHD has its own electric connector (4) (the block part).

The MM (Fig. 15.61) is the spool's drive, which is converting a weak current signal into a proportional mechanical displacement. By its design, the MM is an autonomous device; it is mounted and checked separately and is replaceable. This makes the maintenance and repairs easier. The MM is a "dry" motor and is air-tightly separated from the hydraulic part of the EHD by a reinforced gasket (1) mounted on the shaft (2). The armature (3) is rigidly fixed on the shaft and consists of a soft magnetic core and four rare-earth permanent magnets (4) rigidly fixed to the core's faces.

The armature is mounted in the bore formed in the poles (5) with a diameter gap of ~0.2 mm.

The shaft in the bearings (6 and 7) is fixed in the lids (8 and 9), which are rigidly fixed in the poles

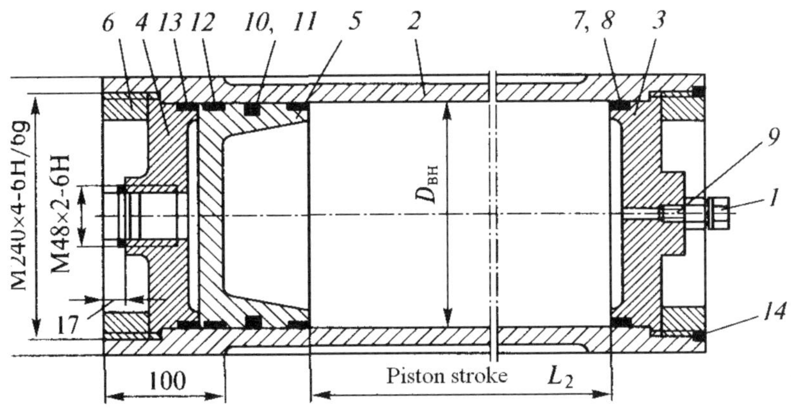

Fig. 15.58. Pneumatic-hydraulic accumulator (PHA)

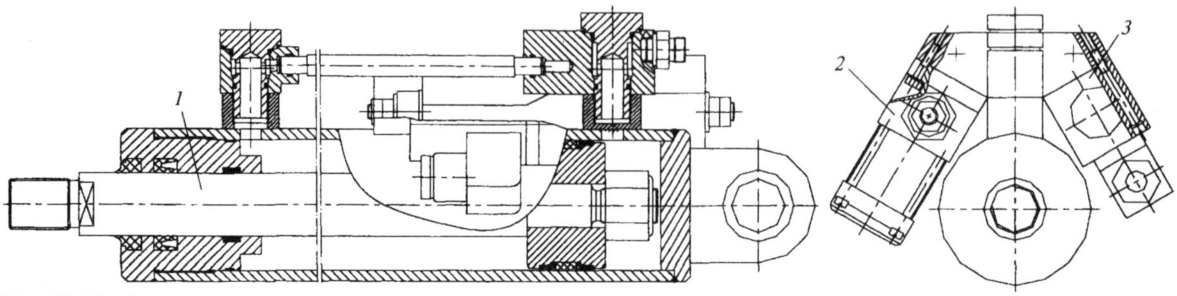

Fig. 15.59. Booster

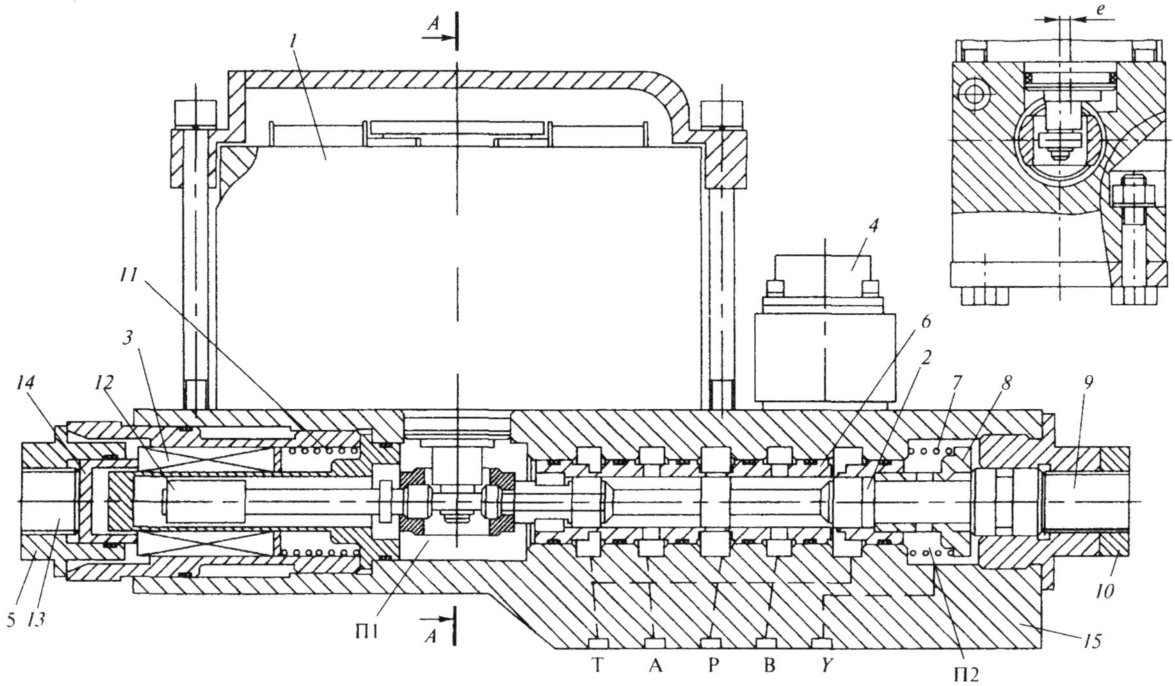

Fig. 15.60. Electrohydraulic distributor

(5) made of a soft magnetic material. The poles, in their turn, are rigidly fixed on the stator (10), also made from a soft magnetic material. The part of the MM shaft (the driving shaft) located below the gasket is made with an eccentricity relative to the shaft's upper part (eccentricity $e = 2.5$ mm, see also the cross-section A-A in Fig. 15.61).

The two coils (11) mounted in the poles (5) form a control magnetic flow in the working gaps with a length of l_B.

Four permanent magnets create the excitation (biasing) magnetic flow in the working air gaps. The power interaction of the control and biasing flows generates the electromagnetic moment acting on the armature, the value of which linearly depends on the current's value in the MM's windings.

The force transmitted to the spool (2) (see Fig. 15.61) corresponds to the ratio of the electromagnetic moment to the arm of length e (see the cross-section A-A in Fig. 15.60), established by the eccentricity of the armature

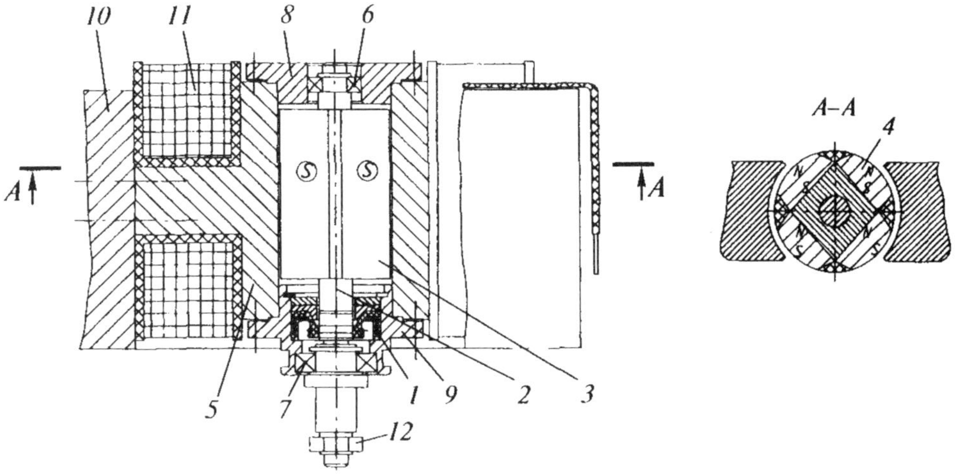

Fig. 15.61. Moment motor

and the driving shaft's axles. The spool's (2) linear displacement is defined by the product of the armature's angular displacement by arm *e* (the cross-section A-A).

The transfer of the force from the armature to the spool (see Fig. 15.61) is made through the bearing (12) installed on the driving shaft. The bearing interacts with the spool by its external ring. The bearing-spool gap is 1–3 μm.

The spool's (2) displacement at the nominal control current is (i_{nom} = ±1.6 A) $x_{s.\ nom}$ = +1.0 mm.

The spool distributor (Fig. 15.60) is a four-slot spool with practically zero overlaps and is rigidly linked with the induction sensor's (4) rotor, which ensures the position feedback. The sleeve (6) has four pairs of rectangular windows, the cut-off edges of which match the overlaps of the spool's the cut-off edges. This ensures a linear dependence of the throttling slot area along the spool travel, i.e., a permanent flow-rate amplification factor within the whole range of spool displacements.

The sleeve (6) (see Fig. 15.60) is kept in the fixed position in the body (75) by the coupling (8) screwed into its face. This position is defined by the spring (7) and regulating screw (9), which interacts with the bushing and the body (15).

The screw is locked by the nut (10). The screw's (9) rotation ensures the zero displacement's hydraulic regulation, which is done on the EHD without a matching block.

The regulation screw's (13) rotation (see Fig. 15.60) ensures regulating the EHD's zero displacement with a connected matching block. At the control signal (U_{contr} = 0), the screw's (13) rotation moves the sprung-up stator of the spool position's feedback sensor (3) by affecting the bushing (12), which allows for achieving the minimum value of the zero displacement. The regulation screw (13) is locked by a nut.

The zero displacement's regulation is achieved by using the technological hydraulic cylinder. Lines A and B of the EHD are connected to the hydraulic cylinder's cavities.

The body's (15) butting plane (see Fig. 15.60) has five circular bores intended for rubber rings, which seal the hydraulic lines' channels.

The draining line *Y* unites the spool's face cavities Cav1 and Cav2 and is not directly connected with the drain line *T*. The presence of *Y* is caused by the pressure restriction in cavity Cav1 by the maximum value of 0.3 MPa. If the pressure in line *T*, defined by the operation conditions, does not exceed this value, this line can be left unused.

At delivery to the matching block's input of the input signal $U_{contr} = 0$, it is compared with the voltage taken from the feedback sensor. The difference in these values $E = U_{contr} - U_{fb}$ goes to the booster's input, is realized in the control block, and is further transformed into the DC signal I at the power booster's output. The power booster's load is made by the two parallel connected windings of the MM. The windings' complex resistance is defined by the combination of the inductive and active resistances. The active resistance of the parallel windings should be 1.85 ± 0.4 Ohm; the maximum value of the control current corresponding to the spool's in the EHD's preliminary amplification cascade is $I_{max} < \pm 1.6$ A.

The spools move respective to the current I; as a result, the control flow rate of the working fluid Q reaches the executive mechanism.

This flow rate at the EHD's output defines the displacement rate of the executive mechanism (hydraulic cylinder, hydraulic motor) and the load associated therewith.

The flow rate's value at a permanent load is practically linearly dependent on the control signal U_{contr}.

The feedback sensor (Fig. 15.62) is intended for issuing a signal to the OBCC about the wheel's position at every time moment in time. The drive's (2) body houses the shaft (6), which rotates in the bearings (11). The lid (1) keeps the shaft (6) from the axial displacement. The angle corresponding to the wheel turn is transferred through an intermediary link, which is comprised of a flywheel (3), from the shaft (6) to the potentiometer (9). The intermediary link is required to decrease the measuring error. The potentiometer is fixed by the pivot axle (8) and the nut (7) in the body (4), which is covered by the lid (5). The socket (10) is intended to connect the cable.

Thus, the design of an all-wheel StS for 12×12 HDWVs implements the latest achievements of MSTU presented in Chapters 14 and 15 of this book. The AWStS is presented here in its real design.

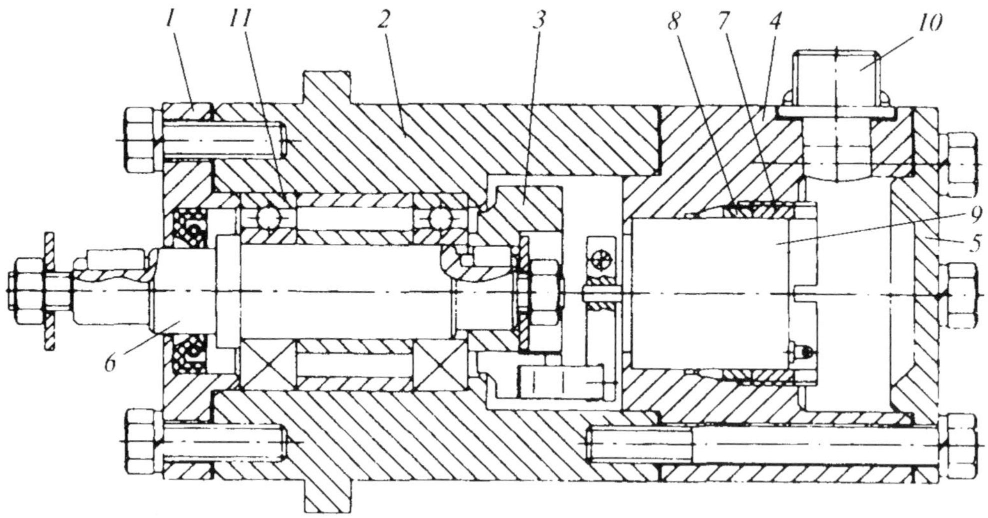

Fig. 15.62. Feedback sensor (FBS)

References

1. Aksyonov, P.V. Multi-Axle Automobiles: Theory of General Design Solutions. M: Mashinostroyenie, 1989. 280 p. (Аксенов П. В. Многоосные автомобили: Теория общих конструктивных решений. – М.: Машиностроение, 1989. 280 с.)

2. Antonov, A.S. Army Vehicles. Design and Calculations. Moscow: USSR Ministry of Defense, 1970. 542 p. (Антонов А. С. Армейские автомобили. Конструкция и расчет. М.: МО СССР, 1970. 542 с.)

3. Stepanchenko, E.P., Falaleev, P.P. Process Equipment: Fundamentals of Design and Calculation of Basic Machines. Moscow: USSR Ministry of Defense, 1986. 364 p. (Степанченко Э. П, Фалалеев П. П. Технологическое оборудование: Основы конструкции и расчета базовых машин. М.: МО СССР, 1986. 364 с.)

4. Vysotsky, M.S., Gileles, L.Kh., Khersonsky, S.G. Trucks: Design and Construction Basics. Moscow: Machine-Building, 1995. 256 p. (Высоцкий М. С., Гилелес Л. Х., Херсонский С. Г. Грузовые автомобили: Проектирование и основы конструирования. М.: Машиностроение, 1995. 256 с.)

5. Belousov, B.N. Popov, S.D., Lovtsov, A.N. All-Wheel Steering System for Heavy-Duty Multi-Axle Vehicles. Motor Industry, 2002, No. 3, p. 11–14. (Белоусов Б. Н., Попов С. Д., Ловцов А. Н. Всеколесное рулевое управление для большегрузных многоосных АТС // Автомобильная промышленность. 2002, № 3. С. 11–14.)

6. Eliseev, A.N., Belousov, B.N., Starikov A.F. et al. Development Trends in Theory and Practice of Delivering Power to Movers of Multi-Axle Full-Drive Vehicles. M.: Academy of Quality Problems, Division of Special Machinery and Conversion. 2000. 76 p. (Елисеев А. Н., Белоусов Б. Н., Стариков А. Ф. и др. Тенденции развития теории и практики систем передачи мощности к движителю многоосных полноприводных транспортных средств. М.: Академия проблем качества, отделение спецтехники и конверсии. 2000. 76 с.)

7. Eliseev, A.N., Belousov, B.N., Starikov A.F. et al. Scientific Approaches to Justification of Technical Appearance of Undercarriage of Future Multi-Axle Vehicles. M.: Academy of Quality Problems, Division of Special Machinery and Conversion. 2000. 48 p. (Елисеев А. Н., Белоусов Б. Н., Стариков А. Ф. и др. Научные подходы к обоснованию технического облика ходовой части перспективных многоосных транспортных средств. М.: Академия проблем качества, отделение спецтехники и конверсии. 2000. 48 с.)

8. Chvyalyov, V.E. Automotive Means Made by SKB MAZ and MZKT. Unknown about Known Things. Motor Industry, 2000, No. 8, p. 7–12. (Чвялев В. Е. Автомобильная техника СКБ МАЗа и МЗКТ. Неизвестное об ивестном // Автомобильная промышленность, 2000, № 8, С. 7–12.)

9. Egorov, A.N. BelAZ: Working for Future. Motor Industry 2000, No. 9, p. 6–9. (Егоров А. Н. БелАЗ: Работа на перспективу // Автомобильная промышленность, 2000, № 9, С. 6–9.)

10. Sirotkin, Z.L. Soviet Heavy-Duty Dumpers in Past, Present and Future. Motor Industry, 1984, No. 10, p. 23–25. (Сироткин З. Л. Советские автомомбили-самосвалы большой и особо

большой грузоподъемности в прошлом, настоящем и будущем. // Автомобильная промышленность. 1984, № 10, С. 23-25.)

11. Sirotkin, Z.L., Osokin, V.A. Perspective Series of Quarry Vehicles. Motor Industry, 1991, No. 1, p. 7–10. (Сироткин З. Л., Осокин В. А. Перспективный типаж карьерных АТС. // Автомобильная промышленность. 1991, № 1, С. 7–10.)

12. Smirnov, G.A. Motion Theory of Wheeled Vehicles. M.: Machine-Building, 1990. 352 p. (Смирнов Г. А. Теория движения колесных машин. М.: Машиностроение, 1990. 352 с.)

13. Kokh, P.I. Climate and Machine Reliability. M.: Machine-Building, 1981. 175 p. (Кох П. И. Климат и надежность машин. М.: Машиностроение, 1981. 175 с.)

14. Vysotsky, M.S., Gileles, L.Kh., Khersonsky, S.G. Trucks: Design Basics. – 2nd revised and added ed., M.: Machine-Building, 2000. 286 p. (Высоцкий М. С., Гилелес Л. Х., Херсонский С. Г. Грузовые автомобили: Проектирование и основы конструирования. – 2-е изд., перераб. и доп. –М.: Машиностроение, 2000. 286 с.)

15. Vantsevich, V.V., Vysotsky, M.S., Gileles L.Kh. Mobile Transport Means: Interaction with Functioning Environment. Minsk: Belarusian Science, 1998. 303 p. (Ванцевич В. В., Высоцкий М. С., Гилелес Л. Х. Мобильные транспортные средства: Взаимодействие со средой функционирования. Минск: Беларуская навука, 1998. 303 с.)

16. Tokarev, A.A. Vehicle Fuel Efficiency and Traction-Speed Properties. M.: Machine-Building, 1982. 282 p. (Токарев А. А. Топливная экономичность и тягово-скоростные свойства автомобилей. М.: Машиностроение, 1982. 282 с.)

17. Grishkevich, A.I. Vehicles: Theory: A Textbook for Universities. Minsk: Higher School Publishers, 1986. 208 p. (Гришкевич А. И. Автомобили: Теория: Учебник для вузов. Минск: Выш. Шк., 1986. 208 с.)

18. Shimkov, A.A. Designing Engine-Transmission Units for Specialized Multi-Axle Heavy-Duty Wheeled Chassis: Thesis for Doctor of Technical Sciences as Scientific Paper: 05.05.03. Minsk, 1990, 59 p. (Шимков А. А. Создание моторно-трансмиссионных установок специализированных многоосных колесных шасси высокой проходимости: Дис… докт. техн. наук в форме научного доклада: 05.05.03. Минск, 1990. –59 с.)

19. Sergeev, V.A. Kornilov, P.Yu. Foreign-Made Vehicles for Transporting Heavy Cargos. Overview. Central Research Institute for Motor Industry, 1988. 46 p. (Сергеев В. А., Корнилов П. Ю. Зарубежные транспортные средства для перевозки крупногабаритных тяжеловесных грузов. Обзорная информация. ЦНИИТЭИавтопром, 1988. 46 с.)

20. Kotik, M.A., Emelianov, A.M. Nature of Human-Operator's Errors (Driving Vehicles as Example). Moscow: Transport, 1993. 252 p. (Котик М. А., Емельянов А. М. Природа ошибок человека-оператора (на примере управления транспортными средствами). М.: Транспорт, 1993. 252 с.)

21. Chumakov, N.M., Serebryany, E.I. Evaluating Effectiveness of Complex Technical Devices. M.: Sov. Radio, 1980. 192 p. (Чумаков Н. М., Серебряный Е. И. Оценка эффективности сложных технических устройств. М.: Сов. радио, 1980. 192 с.)

22. Ipatov, M.I. Problems of Improving Technical and Economic Analysis at Stages of Vehicle Design. Abstract of Thesis for Doctor of Technical Sciences: 05.05.03. M. 1975. 30 p. (Ипатов М. И. Проблемы совершенствования технико-экономического анализа на стадиях проектирования автомобиля. Реферат дис… докт. техн. наук: 05. 05. 03. М. 1975. – 30 с.)

23. Ivanov, A.M. Technical Ways to Improve Design Efficiency of Cargo Vehicles. Abstract of Thesis for Doctor of Technical Sciences: 05.05.03. M. 1995. 33 p. (Иванов А. М. Технические пути

повышения конструктивной эффективности грузовых автотранспортных средств. Реферат дис… докт. техн. наук: 05. 05. 03. М. 1995. 33 с.)

24. V.V. Guskov, N.N. Velev, Yu.A. Atamanov et al; Under General Editorship of V.V. Guskov. Tractors. Theory. Moscow: Machine-Building, 1988. 376 p. (Тракторы. Теория / В. В. Гуськов, Н. Н. Велев, Ю. А. Атаманов и др. Под общей ред. В. В. Гуськова. – М.: Машиностроение, 1988. 376 с.)

25. Vantsevich, V.V. Synthesis of Drives to Driving Axles and Wheels of Multi-Drive Transport-Traction Machines. Thesis of Doctor of Technical Sciences: 05.05.03. Minsk, 1992. 412 p. (Ванцевич В. В. Синтез схем приводов к ведущим мостам и колесам многоприводных транспортно-тяговых машин. Дис. … докт. техн. наук: 05. 05. 03. Минск, 1992. 412 с.)

26. Lefarov, A.Kh., Vysotsky, M.S., Vantsevich, V.V., Kabanov, V.I. Energy Loading and Reliability of Differential Mechanisms of Transport-Traction Machines. Minsk: Navuka i Tekhnika, 1991. 157 p. (Лефаров А. Х., Высоцкий М. С., Ванцевич В. В., Кабанов В. И. Энергонагруженность и надежность дифференциальных механизмов транспортно-тяговых машин. Минск: Навука і тэхніка, 1991. 157 с.)

27. Shalyagin, V.N. Heavy-Duty Transport and Transport-Technological Means. M. Agrarian Industry Publishers, 1986. 186 p. (Шалягин В. Н. Транспортные и транспортно-технологические средства повышенной проходимости. М.: Агропромиздат, 1986. 186 с.)

28. Pogosbekov, M.I. More on Automobile Efficiency Factor. Motor Industry. 1996, No. 12, p. 12–15. (Погосбеков М. И. Еще раз о КПД автомобиля // Автомобильная промышленность. 1996, № 12. С. 12–15.)

29. Platonov, V.F. All-Drive Motor Vehicles. 2nd ed., revised and added. M.: Machine-Building, 1989. 312 p. (Платонов В. Ф. Полноприводные автомобили. 2-е изд., перераб. и доп. – М.: Машиностроение, 1989. 312 с.)

30. Tokarev, A.A. Automobile Fuel Efficiency and Traction-Speed Properties. M.: Machine-Building, 1982. 224 p. (Токарев А. А. Топливная экономичность и тягово-скоростные качества автомобиля. М.: Машиностроение, 1982. 224 с.)

31. Litvinov, A.S., Farobin, Ya.E. Automobile: Theory of Operation Properties. Moscow: Machine-Building, 1989. 240 p. (Литвинов А. С., Фаробин Я. Е. Автомобиль: Теория эксплуатационных свойств. М.: Машиностроение, 1989. 240 с.)

32. Vysotsky, M.S., Vygonnykh A.G., Gileles, L.Kh., Khersonsky, S.G. Ed. by Vysotsky, M.S. Automobiles. Specialized Rolling Stock. Minsk: Vysheishaya Shkola, 1987. 152 p. (Автомобили. Специализированный подвижной состав/ М. С. Высоцкий, А. Г. Выгонный, Л. Х. Гилелес, С. Г. Херсонский; Под ред. М. С. Высоцкого. – Минск: Вышэйшая школа, 1987. 152 с.)

33. Vysotsky, M.S. Designing Basics of Heavy-Duty Automobiles and Road Trains. Minsk: Science and Technology, 1980. 200 p. (Высоцкий М. С. Основы проектирования автомобилей и автопоездов большой грузоподъемности. Минск: Наука и техника, 1980. 200 с.)

34. Wong J., Theory of Ground Vehicles. Tr. from English. Moscow: Machine-Building, 1982. 284 p. (Вонг Дж. Теория наземных транспортных средств. Пер. с англ. М.: Машиностроение, 1982. 284 с.)

35. Aksyonov, P.V., Belousov, B.N., Starikov A.F. Basic Principles of Analysis and Synthesis of Transmissions of Multi-Axle Vehicles. Bulletin of N.E. Bauman MSTU. Series of Machine-Building, 1998, No. 4, p. 88–89. (Аксенов П. В., Белоусов Б. Н., Стариков А. Ф. Основные принципы анализа и синтеза схем трансмиссии многоосных транспортных средств // Вестник МГТУ им. Н. Э. Баумана. Серия Машиностроение, 1998, № 4. С. 88–89.)

36. Shoop T. Solution of Engineering Problems on Computers. Tr. from English. M.: Mir, 1982. 238 p. (Шуп Т. Решение инженерных задач на ЭВМ. Пер. с англ. М.: Мир, 1982. 238 с.)

37. Sirazetdinov, T.K. Methods for Solving Multi-Criteria Problems of Technical System Synthesis. Moscow: Machine-Building, 1988. 160 p. (Сиразетдинов Т. К. Методы решения многокритериальных задач синтеза технических систем. М.: Машиностроение, 1988. 160 с.)

38. Pontryagin, L.S., Boltyansky V.G., Gamkrelidze, R.V., Mischenko, E.F. 4th ed. Mathematical Theory of Optimal Processes. Moscow: Nauka, 1983. 392 p. (Математическая теория оптимальных процессов / Л. С. Понтрягин, В. Г. Болтянский, Р. В. Гамкрелидзе, Е. Ф. Мищенко. – 4-е изд. М.: Наука, 1983. 392 с.)

39. Pontryagin L.S. Maximum Principle in Optimal Control. Moscow: Nauka, 1989. 64 p. (Понтрягин Л. С. Принцип максимума в оптимальном управлении. М.: Наука, 1989. 64 с.)

40. Antomonov, Yu.G. Synthesis of Optimal Systems. Kiev: Naukova Dumka, 1972. 320 p. (Антомонов Ю. Г. Синтез оптимальных систем. Киев: Наукова думка, 1972. 320 с.)

41. Boltyansky, V.G. Mathematical Methods of Optimal Control. 2nd ed. updated and added. Moscow: Nauka, 1969. 408 p. (Болтянский В. Г. Математические методы оптимального управления. 2-е изд., переработ. и доп. М.: Наука, 1969. 408 с.)

42. Voronov, A.A., Titov, V.K., Novogranov B.N. Basics of Theory of Automatic Regulation and Control. Moscow, Higher School, 1977. 519 p. (Воронов А. А., Титов В. К., Новогранов Б. Н. Основы теории автоматического регулирования и управления. М.: Высшая школа, 1977. 519 с.)

43. Petrov, Yu.V. Variation Methods for Optimal Control Theory. 2nd ed. updated and added. L.: Energy, 1977. 280 p. (Петров Ю. В. Вариационные методы теории оптимального управления. 2-е изд., переработ. и доп. Л. : Энергия, 1977. 280 с.)

44. Germintern, V.I., Kagan, B.M. Methods of Optimal Designing. Moscow: Energiya, 1980. 160 p. (Герминтерн В. И., Каган Б. М. Методы оптимального проектирования. М.: Энергия, 1980. 160 с.)

45. Germintern, V.I., Shtilman, M.S. Optimization in Designing. M.: Znaniye, 1982. 64 p. (Герминтерн В. И., Штильман М. С. Оптимизация в задачах проектирования. М.: Знание, 1982. 64 с.)

46. Gaschuk, P.N. Optimization of Automobile Fuel-Speed Properties. Lvov: Vischa Shkola, 1987. 168 p. (Гащук П. Н. Оптимизация топливо-скоростных свойств автомобиля. Львов: Вища школа, 1987. 168 с.)

47. Guskov, V.V. Optimal parameters of Farm Tractors. Moscow: Machine-Building, 1966. 196 p. (Гуськов В. В. Оптимальные параметры сельскохозяйственных тракторов. М.: Машиностроение, 1966. 196 с.)

48. Boikov, V.P., Belkovsky, V.N. Tires for Tractors and Agricultural Machinery. M. Agrarian Industry Publishers, 1988. 240 p. (Бойков В. П., Белковский В. Н. Шины для тракторов и сельскохозяйственных машин. М.: Агропромиздат, 1988. 240 с.)

49. Katsygin, V.V., Gorin G.S. Zenkova, A.A. et al. Future Mobile Energy Resources (MER) for Agriculture. Minsk: Science and Technology, 1982. 272 p. (Перспективные мобильные энергетические средства (МЭС) для сельскохозяйственного производства / В. В. Кацыгин, Г. С. Горин, А. А. Зенькова и др. Минск: Наука и техника, 1982. 272 с.)

50. Kutskov, G.M., Chernykh, A.G., Gabay, E.V., Lukerchik, L.M. Traction Performance of Modular Power-Technology Tool. Tractors and Farm Machinery. 1990, No. 6, p. 19–20. (Тяговые показатели модульного энерготехнологического средства / Г. М. Кутьков, А. Г. Черных, Е. В. Габай, Л. М. Лукерчик // Тракторы и сельхозмашины. 1990, № 6. С. 19–20.)

51. Vantsevich, V.V., Kharitonchik, S.V. Improved Traction Properties of Multi-Drive Wheeled Vehicles in Severe Operating Conditions. Improved Performance of Wheeled and

Caterpillar Vehicles in Severe Operating Conditions: Coll. of Intern. Sc.-Tech. Conf., Tyumen, November 20–22, 1996. Tyumen, 1996, p. 21–22. (Ванцевич В. В., Харитончик С. В. Повышение тяговых свойств многоприводных колесных машин в тяжелых условиях эксплуатации // Повышение эффективности работы колесных и гусеничных машин в суровых условиях эксплуатации: Сб. тр. междунар. научн.-техн. конф., Тюмень, 20–22 ноября 1996. Тюмень, 1996. С. 21–22.)

52. Pirkovsky, Yu.V., Shukhman, S.B. Motion Theory of All-Drive Vehicle (Applied Problems of Chassis Design Optimization). M.: UNITY-DANA, 2001. 230 p. (Пирковский Ю. В., Шухман С. Б. Теория движения полноприводного автомобиля (прикладные вопросы оптимизации конструкции шасси). М.: ЮНИТИ-ДАНА, 2001. 230 с.)

53. Belov, P.M. Engines of Army Vehicles. Part 1. Theory. M.: USSR Military Publishers, 1971. 306 p. (Белов П. М. Двигатели армейских машин. Ч. 1. Теория. М.: Воениздат СССР, 1971. 306 с.)

54. Pirkovsky, Yu.V. Effect of Automobile Design Parameters on Dynamic Features and Fuel Efficiency. M.: Qualification Improvement Institute at the USSR Ministry of Agricultural Machine-Building. 1989. 18 p. (Пирковский Ю. В. Влияние конструктивных показателей автомобилей на их динамические качества и топливныую экономичность. М.: ИПК Минавтосельхозмаша СССР. 1989. 18 с.)

55. Podgaevsky, V.M. Development Trends of Engine Families of Foreign Military Caterpillar Vehicles. Collection of NIID. M., 1992, No. 4. p. 127–146. (Подгаевский В. М. Тенденции развития семейств двигателей зарубежных военно-гусеничных машин. Сборник НИИД. М., 1992, № 4. С. 127–146.)

56. Petrov, L.K. Electronically Controlled Diesel Fuel Equipment. Motor Industry 1993, No. 4, p. 37–39. (Петров Л. К. Дизельная топливная аппаратура с электронным управлением// Автомобильная промышленность. 1993, № 4. С. 37–39.)

57. Denisov, A.V. Fundamentals of Theory of Joint Operation of Gas-Turbine Power Units with Transparent and Non-Transparent Transmissions of Multi-Axle Vehicles: Thesis of Doctor of Technical Sciences: 05.05.03. M., 1991. 395 p. (Денисов А. В. Основы теории совместной работы газотурбинных силовых установок с прозрачной и непрозрачной трансмиссиями многоосных автомобилей: Дис.... докт. техн. наук: 05. 05. 03. М., 1991. 395 с.)

58. Galevsky, E.A. Theory Basics of Mechanical Transmissions with Coordinated Dynamic Vehicle Parameters. Autoabstract of Thesis of Doctor of Technical Sciences: 05.05.03. M., 1998. 36 p. (Галевский Е. А. Теоретические основы создания механических трансмиссий с согласованными динамическими параметрами транспортных машин. Автореферат дис. ...докт. техн. наук: 05. 05. 03. М., 1998. 36 с.)

59. Mazanov, N.D. On Selecting Parameters of Standardized Uniform Hydraulic Transformers for BelAZ Family. Motor Industry. 1966, No. 3. p. 25–27. (Мазанов Н. Д. О выборе параметров унифицированных гидротрансформаторов для семейства автомобилей БелАЗ // Автомобильная промышленность. 1966, № 3. С. 25–27.)

60. Narbut, A.N. Fundamentals of Optimization of Output Characteristics of Hydraulic Transformers. MADI (Moscow Automobile and Road Technical University) Proceedings. Cars / MADI, 1972, No. 42, p. 121–130. (Нарбут А. Н. Основы оптимизации выходных характеристик гидротрансформаторов // Труды МАДИ. Автомобили / МАДИ, 1972. Вып. 42. С. 121 - 130.)

61. Antonov, D.A. Motion Theory of Military Wheeled Vehicles. Moscow, Publishing House of the USSR Ministry of Defense, 1993. 383 p. (Антонов Д. А. Теория движения боевых колесных машин. М., Изд-во Минобороны СССР, 1993. 383 с.)

62. Edited by Khachaturov, A.A. Dynamics of System "Road-Tire-Automobile-Driver." Moscow: Machine-Building, 1976. 535 p. (Динамика системы «дорога-шина-автомобиль-водитель». Под редакцией А. А. Хачатурова. М.: Машиностроение, 1976. 535 с.)

63. Stepanov, Yu.A. Model Planetary Gearbox for Heavy-Duty Vehicles. Motor Industry, 1992, No. 5. p. 7–9. (Степанов Ю. А. Типовая планетарная коробка передач для автомобилей большой грузоподъемности // Автомобильная промышленность. 1992, № 5. С. 7–9.)

64. Silaev, A.A. Spectral Theory of Vehicle Springing. Moscow: Machine-Building, 1972. 192 p. (Силаев А. А. Спектральная теория подрессоривания транспортных машин. М.: Машиностроение, 1972. 192 с.)

65. Khachaturov, A.A., Afanasiev, V.L., Vasiliev, V.S. et al. Calculation of Operational Motion Parameters of Automobile and Train. Moscow: Transport, 1982. 264 p. (Расчет эксплуатационных параметров движения автомобиля и автопоезда / Хачатуров А. А., Афанасьев В. Л., Васильев В. С. и др. М.: Транспорт, 1982. 264 с.)

66. Panovko, Ya.G. Internal Friction at Oscillations of Elastic Systems. Moscow: Physics and Mathematics Publishers, 1960. 114 p. (Пановко Я. Г. Внутреннее трение при колебаниях упругих систем. М.: Физматгиз, 1960. 114 с.)

67. Sorokin, E.S. On Theory of Internal Friction at Oscillations of Elastic Systems. Moscow: State Publishers on Construction, Architecture and Building Materials, 1960. 131 p. (Сорокин Е. С. К теории внутреннего трения при колебаниях упругих систем. М.: Гос. изд. по строительству, архитектуре и строительным материалам, 1960. 131 с.)

68. Kolesnikov, K.S., Sukhov, V.N. Elastic Aircraft as Object of Automatic Control. Moscow: Machine-Building, 1974. 268 p. (Колесников К. С., Сухов В. Н. Упругий летательный аппарат как объект автоматического управления. М.: Машиностроение, 1974. 268 с.)

Index

C

D

E

F

G

H

I

J

K

L

T

U

V

W

About the Authors

Dr. Boris N. Belousov is the head of the Scientific-Research Institute for Special Machine-Building at the Scientific-Production Center for Special Machine-Building of N.E. Bauman Moscow State Technical University (MSTU). He also works with the Russian Foundation for Basic Researches at the Russian Academy of Sciences. He earned his MS degree from Kurgan Machine-Building University and his PhD and Doctor of Technical Sciences from MSTU.

Dr. Sergei D. Popov works at N.E. Bauman Moscow State Technical University, where he is the chief designer at the Scientific-Production Center for Special Machine-Building and associate professor at the MSTU Department of Wheeled Vehicles. He works as vice president of the International Centre for Educational Systems and coordinator of the UNESCO Technical and Vocational Education and Training Network. He earned his MS and PhD degrees, specializing in wheeled vehicles, from MSTU.